Results and Problems in Cell Differentiation

Volume 69

Results and Problems in Cell Differentiation is an up-to-date book series that presents and explores selected questions of cell and developmental biology. Each volume focuses on a single, well-defined topic. Reviews address basic questions and phenomena, but also provide concise information on the most recent advances. Together, the volumes provide a valuable overview of this exciting and dynamically expanding field. This series is indexed in PubMed.

More information about this series at http://www.springer.com/series/400

Malgorzata Kloc
Editor

Symbiosis: Cellular, Molecular, Medical and Evolutionary Aspects

Springer

Editor
Malgorzata Kloc
Department of Surgery, The Houston
Methodist Hospital and
The Houston Methodist Research Institute
Houston, TX, USA

ISSN 0080-1844 ISSN 1861-0412 (electronic)
Results and Problems in Cell Differentiation
ISBN 978-3-030-51848-6 ISBN 978-3-030-51849-3 (eBook)
https://doi.org/10.1007/978-3-030-51849-3

This Springer imprint is published by the registered company Springer Nature Switzerland AG.
The registered company address is: Gewerbestrasse 11, 6330 Cham, Switzerland

Preface

Microbial symbiosis played a pivotal role in the evolution of eukaryotic organisms. While the domesticated endosymbiotic bacteria evolved into cellular organelles such as mitochondria and plastids, the viruses and/or nuclear dwelling bacteria were probably the source of some of the eukaryotic genetic material. The existence of today's plants, some unicellular organisms, and animals, including humans, depends on extremely complex and multifaceted interactions between the symbionts and their hosts. From the perspective of human well-being, we now realize that the symbiotic relationship between the microbiota and the host not only modulates metabolism and immune response but also inhibits pathogens and that a disruption of this symbiotic homeostasis may lead to various diseases. This book covers the current knowledge on different aspects of evolution, mechanisms, and molecular signaling in different types of symbiosis and their potential therapeutic applications. The first part of the volume describes the nuclear symbiosis and how the transfer of genetic information shaped symbiont and host genome and how the symbionts influence cell fate and differentiation. The second part discusses how the symbionts perceive and adjust to the host environment and describes the evolutionary aspects of symbiosis, apoptosis, chemiosmosis, and membrane compartments. The third part summarizes the current knowledge on the evolution and mechanisms of photosynthesis and nitrogen fixation in plants. The fourth part describes the diversity of symbionts in nematodes and insects and their emerging importance for human health and disease. The final (fifth) part describes our dependence on the microbiome integrity, how the symbiosis shapes the adaptive and immune responses, and what are the novel avenues for the therapeutic interventions in balancing the symbiont/host homeostasis and fighting the diseases.

Houston, USA — Malgorzata Kloc

Book Abstract

Symbiosis is an interaction between different organisms. The symbiotic origin of cellular organelles and the exchange of the genetic material between hosts and their bacterial and viral symbionts have shaped the biodiversity of life. Recently, symbiosis has gained a new level of recognition through the realization that all organisms function as a holobiome and that any kind of interference with the hosts influences their symbionts and vice versa and reverberates in profound consequences in the sustenance of both. For example, in humans, the microbiome, which is the collection of all the microorganisms living in association with the intestines, oral cavity, urogenital system, and skin, is inherited during pregnancy and influences maturation and functioning of human immune system, protects against pathogens, and regulates metabolism. The symbionts also regulate cancer development, wound healing, tissue regeneration, and stem cell function. The medical applications of this new realization are vast and largely uncharted. The composition and the status of robustness of human symbionts could be used as the diagnostic tool to predict impending diseases, and the manipulation of symbionts could open new strategies for treatment of incurable diseases.

This volume will be unique in covering molecular, organellar, cellular, immunologic, genetic, and evolutionary aspects of symbiotic interactions in humans and other model systems and fostering new ideas for interdisciplinary research and therapeutic applications.

Contents

Part I
Transfer of Genetic Information, Nuclear Symbiosis, and Cellular Differentiation

Chapter 1
Viral Symbiosis in the Origins and Evolution of Life with a Particular Focus on the Placental Mammals

Frank Ryan

Abstract Advances in understanding over the last decade or so highlight the need for a reappraisal of the role of viruses in relation to the origins and evolution of cellular life, as well as in the homeostasis of the biosphere on which all of life depends. The relevant advances have, in particular, revealed an important contribution of viruses to the evolution of the placental mammals, while also contributing key roles to mammalian embryogenesis, genomic evolution, and physiology. Part of this reappraisal will include the origins of viruses, a redefinition of their quintessential nature, and a suggestion as to how we might view viruses in relation to the tree of life.

1.1 The Historical Perspective

The historical approach to viruses focused on their causative role in specific diseases, leading to their categorization as obligate parasites. Parasitism, in this sense, implied that viruses gained what they needed for their biological cycle and replication at the expense of their hosts while contributing nothing to the relationship. But as early as the 1920s, the pioneering Canadian microbiologist, Felix d'Herelle, observed patterns of behavior involving bacteriophage viruses and their prokaryotic hosts that suggested that the virus–host relationship was more complex than obligate parasitism. In his historic book *The Bacteriophage*, d'Herelle explained how phage viruses exerted a balancing effect in their interactions with their host bacteria (d'Herelle 1926). In his words: "A mixed culture results from the establishment of a state of equilibrium between the virulence of the bacteriophage corpuscles and the resistance of the bacterium. In such cultures a *symbiosis* obtains, in the true sense of the word:

F. Ryan (✉)
The Academic Unit of Medical Education, Faculty of Medicine, Dentistry and Health, The University of Sheffield, Sheffield, UK
e-mail: F.Ryan@sheffield.ac.uk

M. Kloc (ed.), *Symbiosis: Cellular, Molecular, Medical and Evolutionary Aspects*, Results and Problems in Cell Differentiation 69,
https://doi.org/10.1007/978-3-030-51849-3_1

parasitism is balanced by the resistance to infection." This astute observation raised a key question: are viruses symbionts in their relationships with their hosts?

In 1878, the German naturalist Professor Anton De Bary defined symbiosis as the "living together of unlike organisms" (De Bary 1879; Sapp 1994). Today we might interpret this as a significant biological relationship between different species. The concept of symbiosis was further developed by Albert Bernhard Frank (1877), who also pioneered the understanding of mycorrhizae (1885); meanwhile, the evolutionary implications of symbiosis, or "symbiogenesis," were first introduced by Merezhkovskii (1910). Today the interacting "partners" are classed as "symbionts," while the partnership of two or more symbionts is the "holobiont." "Mutualism," where two or more of the interacting symbionts benefit, is a form of symbiosis, but symbiosis also embraces "commensalism," where one or more symbionts benefit without causing harm to their symbiotic partners, and parasitism, where a symbiont benefits at the expense of its partner or partners. The dynamics of symbiosis is further complicated by the fact that mutualism often begins as parasitism and the relationship may evolve to intermediate variations within the extremes of outright parasitism, commensalism, and mutualism.

Symbioses are also categorized by the patterns of biological interaction (Douglas 1994; Margulis and Fester 1991). For example, mycorrhizal symbioses, which are found in all land-based plants, involve the sharing of metabolites between plants and root-interactive fungi, or "mycorrhizae," and so are termed "metabolic symbioses." Other symbioses involve mutually supportive "behavior," such as the feeder stations on the floor of the ocean. Symbiosis, with obvious evolutionary implications, can also involve interactions at the genetic level. "Genetic symbioses" may involve the transfer of a pre-evolved gene, or a package of genes, from one evolutionary lineage to another. At its most powerful, genetic symbiosis involves the union of two or more disparate genomes, to create a novel "holobiontic genome," usually through the union of the genomes of a host with that of a microbial symbiont (Ryan 2009a, 2016).

Given that viruses interact with their hosts at both genomic and physiological levels, it is unsurprising that viruses might sometimes influence the genetic and physiological evolution of the same hosts, and thus play what would normally be classed as symbiogenetic roles in the host–virus relationship. But viruses are not usually classed as living organisms and so, to some, it might appear counterintuitive to apply symbiotic theory to viruses. This is an important dilemma, and it comes down to whether or not viruses should be seen as living organisms from the evolutionary perspective.

1.2 Viruses as Living Organisms and their Relationship to the Tree of Life

This very dilemma provoked López-Garcia and Moreira (2009) to rebut the notion of viruses as living organisms with a paper headed: "Ten reasons to exclude viruses from the tree of life." Their arguments are summarized as follows:

- Viruses, being genetic parasites of living organisms, could not have come into existence until the prokaryotes (Eubacteria and Archaea) had evolved.
- As obligate parasites, and thus being incapable of independent cellular metabolism outside their hosts, viruses are not life forms.
- Viruses do not replicate by themselves.
- Viruses do not evolve through their own mechanisms. They can only evolve through mechanisms borrowed from their cellular hosts.
- Deriving from the previous argument, viruses evolve new genes by "pick-pocketing" genes from their hosts.
- Some of the most important families of viruses originated as mere genetic off-shoots of their host genomes.
- From the above, no meaningful evolutionary tree of life (phylogeny) can be depicted for viruses.
- Viruses are not cellular and, since life can only be defined from a cellular perspective, they should be dismissed from any consideration as life forms.

This resulted in heated discussion and disagreement (Claverie and Ogata 2009; Hedge et al. 2009; Koonin et al. 2009; Ludmir and Enquist 2009; Navas-Castillo 2009; Raoult 2009; Villarreal and Witzany 2010). The key points of debate centered on the existing evidence with regard to the precise nature of viruses. No unanimity emerged from this discourse (López-Garcia and Moreira 2009b), suggesting that it is time we reconsidered the basic nature and of viruses and their contribution to life and biodiversity.

1.3 What Is the Basic Nature of Viruses?

One cannot refute several strands of the López-Garcia and Moreira argument. The life forms they place within the tree of life are indeed confined by cell walls, and they possess ribosomes that enable them to manufacture the proteins needed for their metabolism. These are defining properties that enable their independent life cycles. There is, however, an unstated but implicit caveat here. The tree of life they refer to was specifically defined to fit this same cellular perspective. Viruses are not cellular life forms. Thus, it is inevitable, if we use this definition of the tree of life, that viruses will be excluded. At the same time, when we examine the natural history, physiology, genetics, and evolutionary dynamics of viruses, we find that they possess a great many of the properties we associate with living organisms. This goes to the very heart of the dilemma for both microbiologists and evolutionary virologists. At this point, it might be helpful to re-examine the interactive dynamics of virus and host from a broader evolutionary perspective.

The virion, which is commonly seen as the quintessence of a virus, is actually the propagative stage of the viral cycle, relying on external factors, such as host behavior or other more passive agencies, such as movements of air, wind, and water, to spread the virus in such a way as to discover new hosts. In this virion stage, the virus is essentially inert, rather like the propagative seed of a plant. Of course a virion is not the same as a seed. But the essential nature of a virus is no more defined by the virion than a plant would be defined by examining the seed. The essential biological nature of a virus manifests when it enters the host physiology, there to confront and survive the host's defenses, discover its target cell, or cells, penetrate the cell membrane to enter its interior, and there, in its natural habitat, successfully replicate through utilizing the host's genetic and physiological machinations. Moreira and López-Garcia are correct in their assertion that the virus cannot complete its life cycle, notably its self-replication, without the assistance of its host. But this merely confirms that in its very essence, the virus is an "obligate symbiont."

The fact that an organism is obligate on a symbiotic partner for its survival does not exclude it from consideration as living. Some bacteria, such as the *Rickettsia*, are also obligate intracellular symbionts. Indeed if we widen our perspective to embrace the biological world in general, we discover that it is replete with interdependent symbiotic partnerships.

Bees and hummingbirds and their floral partners depend on one another for food and pollination—a mixture of metabolic and behavioral symbioses. We humans depend on photosynthetic organisms to manufacture the oxygen essential for us to breathe, much as we depend on plants, and a medley of other organisms, to make the essential amino acids, vitamins, fats, and other nutrients that enable our essential physiology every day. In nature, all cellular forms of life are dependent on other organisms for life's essentials with the exception of the relatively rare bacteria known as autotrophs (Ryan 2002). Viewed from the broader perspective, the dependency of viruses on host genetics and physiology for their survival does not preclude them from the definition of "living organisms."

A seemingly irrefutable criticism of the virus-first hypothesis has been the belief that viruses could not have evolved before there were cellular hosts for them to parasitize (Nasir et al. 2012). But this is also based on a mistaken premise. Viruses are not exclusively dependent on cellular hosts for their replication. Parvoviruses, for example, can only replicate in the presence of another virus (Collier and Oxford 1993). As to the Garcia and Moreira suggestion that viruses largely evolve by pick-pocketing host genes, this is refuted by the fact that only a minority of the viral genes found in nature have been acquired by genetic transfer from host genomes. For example, the key genes coding for proteins involved in viral replication are shared by RNA- and DNA-based viruses but are not found in cellular life forms (Koonin et al. 2006). The same goes for the genes encoding the capsid proteins of icosahedral DNA- and RNA-based viruses. These viral equivalents of the cell walls of cellular life are not found in cellular organisms.

While an earlier generation of biologists took the view that large groups of viruses, such as retroviruses and bacteriophages, came into being as offshoots of their prokaryotic hosts, today the evidence suggests otherwise (Villarreal 2005). It would appear that retroviruses and bacteriophages do not derive from host genomes: rather they have their own evolutionary lineages, much as any other organismal groups. This is not to deny that some families of viruses acquired host genes in their evolution. But even this pattern of "horizontal gene transfer" is a feature of the evolution of all life forms, and involved genetic exchange from viruses to host as from host to virus. This argument should thus be dismissed from disqualifying viruses from being categorized as living organisms.

1.4 Theories for the Origins of Viruses

The origins of life are uncertain so that it is hardly surprising that different proposals have been put forward for the origins of viruses. These include the "virus-first" theory, which proposes a primal origin in the prebiotic era of the Earth's evolution; various "reduction" hypotheses, which propose that viruses emerged from the stage of unicellular life; an "escape" hypothesis, which proposes viral origins from genetic material that "escaped" from more developed cellular life-forms to become parasitically self-driven; and the polphyletic theory, which proposes that viruses have multiple origins (Bremermann 1983; Fisher 2010; Forterre 2006; Koonin et al. 2006; Villarreal 2005). Given the complexity and diversity of viruses overall, polyphyletic origins would appear at the very least to offer the most plausible explanation. This would allow for the emergence of prototypical RNA-based viral progenitors during the earliest period of prebiotic life, the presumptive RNA world, with subsequent polyphyletic additions, including a continuous genetic exchange with hosts, the expansion to DNA-based viruses, ultimately to give rise to the diversification of viruses we see today.

1.5 A Virus-First Hypothesis Followed by Subsequent Polyphyletic Expansion in the Prebiotic Stages of Evolution

This hypothesis assumes that a key step in the evolution of genomes was the stage of self-replicating polynucleotide chains. When RNA-based models of this prebiotic stage have been constructed in microbiological laboratories, or as "digital organisms" in computer simulations, parasitic elements have evolved to invade and then interact with the self-replicators (Bansho et al. 2012; Bremermann 1983; Colizzi and Hogeweg 2016; Takeuchi and Hogewoeg 2008; Zaman et al. 2014). Such emergence would fit with the hypothesis of a primary RNA-based virus evolution. This hypothesis does not require a priori evolution of cellular organisms, with its extrapolation to "reduction" and "escape" models for viral emergence. A virus-first origin of RNA-based viruses would also fit with the strongly inverse relationship between genome size and mutation rates across all replication systems, suggesting that the earliest genomes are likely to have been very small and highly error prone—a situation typical of RNA viruses (Holmes 2011). Such a prototypical virus-first model of viral origins would also fit with the models of Forterre (2006) and Koonin et al. (2006). It would suggest that such early RNA-based viruses emerged from an environment of aggressively symbiotic protoviral genetic agents, the prebiotic equivalent of the "virosphere" (Koonin and Dolja 2013; Ryan 2009a, 2019; Villarreal 2014).

Current opinion raises the possibility that life may have originated in the seemingly hostile environments of the deep-sea hydrothermal vents, where searches have revealed a morphological diversity of virus-like particles greatly exceeding the number found in aquatic systems at lower temperatures (Prangashvili and Garrett 2004). Viruses are inherently aggressive, a property, when conferred on a symbiotic partnership of virus and host, has powerful evolutionary potential (Ryan 1997). Behaving as aggressive symbionts, viruses are capable of both killing and protecting their hosts (Ryan 1997, 2009a; Villarreal 2005). The fact that self-replicating entities with properties suggestive of prototypical viruses so readily manifest in laboratory and computer simulations of the presumptive RNA world, and the fact they are such powerful players in the killing or survival sense, combined with their genetic creativity, suggests that virus–host interactivity is likely to have been an inherent and fundamental driving mechanism in the origins of life and in its subsequent diversification.

Viruses are also prolific in transferring genetic "information" between themselves and their hosts. Judging from modern metagenomic analysis, the transfer of genetic information is far commoner from virus to host than from host to virus (Koonin 2001). The only organisms with RNA-coded genomes today are RNA-based viruses, suggesting that we might derive helpful insight into the purported prototypical RNA World from the study of RNA-based viruses.

1.6 The Concept of RNA-Based Quasispecies Behavior

In 1977, Eigen and Schuster introduced the novel concept of "quasispecies" as an explanation for the self-organization of clusters of prebiotic macromolecule chemicals, such as polynucleotides, which presented a prototypical situation that was capable of evolutionary adaptation through Darwinian natural selection (Eigen and Schuster 1977, 1978a, b, 1979). This was extrapolated to "viral quasispecies theory" in which swarms of viruses, closely related through shared mutations, would compete for survival in a highly mutagenic environment. Where one might expect such a high mutation rate to give rise to nonviable virions, quasispecies theory predicts that, in the case of RNA viruses, it would actually create a "cloud of potentially beneficial mutations," which would give the viral quasispecies an advantage in adapting to novel environments and survival challenges (Holland et al. 1992; Vignuzzi et al. 2006). Quasispecies theory also predicts that, in such a scenario, quasispecies will compete with one another. Indeed, it has been observed experimentally that, in the quasispecies situation, less fit viral mutants will suppress what would otherwise constitute fitter mutants (De la Torre and Holland 1990). This suggests that the fitness of the evolving RNA virus population derives from its quasispecies components rather than the conventional evolutionary model, meanwhile inherent to the evolutionary behavior of quasispecies is a collective function and group identity (Villarreal 2014). Quasispecies behavior with group identity has been observed in retroviruses, such as HIV-1, where infected patients harbor a highly diverse viral population with many different mutant strains (Nowak 1992). A similar RNA-mediated group identity may have facilitated the origins of RNA-based viral ancestors during the presumptive RNA World stage. Such a virus-first emergence from the prototypical RNA world would require the acquisition of "group identity" in the form of quasispecies consortia. This group identity, combined with what virologists call "an addiction module", would make possible a prototypical equivalent of "self" (Villarreal 2005, 2014). The sum of key evolutionary features such as group identity, addiction module, regulatory complexity, and virus–host ecology is fundamentally linked, and may have provided a key step in the establishment of life on Earth.

1.7 Virus–Prokaryote Symbiosis and the Homeostasis of the Biosphere

If the RNA world hypothesis is correct, evolution among the primal stage of RNA-based nucleotide chains led not only to the origin of RNA-based viruses but also through an uncertain labyrinth of evolutionary steps, including the emergence of DNA as the perfect memory molecule for heredity, to the emergence of the complexity of cellular life. The evolution retained the versatility of RNA for messenger, transfer, ribosomal and epigenetic functionality. Such evolution will have involved more complex virus evolution, which, in turn, is likely to have been accompanied by a complementary complexity of virus–host genetic interaction, part of which would have been a horizontal genetic transfer in both directions. This never-ending universal genetic symbiosis, involving viruses and hosts, continues to be an integral feature of the Virosphere to the present day.

Earth's planetary origin is thought to date back 4.54 billion years, with a putative arrival of cellular life between 3.77 and 4.28 billion years ago (Dodd et al. 2017). Primal origins are not synonymous with the last universal common ancestor, or LUCA, which denotes the most recent population from which all organisms now living have a common descent. The origins of the LUCA have been tentatively dated to approximately 3.5 to 3.8 billion years ago (Doolittle 2000; Glansdorf et al. 2008).

The Margulis theory of the origin of eukaryotic cells through the endosymbiotic merger of a thermoplasma-like Archaean and Spirochaete-like eubacterium (Margulis 1970; Sagan 1967) has been supported by recent authors (López-García et al. 2017). If true, the concomitant holobiontic genomic merger is likely to have involved viral enzymes involved in genetic transfer. A viral origin has also been proposed for the eukaryotic replication proteins (Villarreal and DeFilippis 2000). The eukaryotic replicative DNA polymerases are dissimilar to those of eubacteria but bear similarities to the polymerases of large DNA viruses that infect eukaryotes as well as the polymerases of T4 bacteriophages. It has been suggested that the LUCA may have been a small single-celled organism, with a DNA-based ring genome (Battistuzzi et al. 2004). Woese, who first proposed the three-domain system of classification based on the analysis of ribosomal RNA, suggested that the LUCA was probably a simpler, more rudimentary entity than the prokaryotes of today (Woese et al. 1990). While virologists proposing the "escape theory" of viral origins assume that this must have followed the origins of LUCA, there is growing evidence that viruses are more ancient than LUCA (Holmes 2011). Luria and Darnell, while acknowledging that bacteria and their viruses are likely to have shared a lengthy genetic exchange relationship, also proposed an origin of viruses after the origins of primitive cells and most likely deriving from mobile cellular genetic elements that were capable of transmissibility and self-replication from one cell to another (Luria and Darnell 1967). While such creative evolutionary interactions between viruses and the prokaryotic domains have been ongoing and universal, there are difficulties with the Luria-Darnell theory of viral origins from mobile cellular genetic entities. Modern metagenomic analysis indicates that the direction of genetic transfer is far more commonly from virus to host (Koonin 2011). This fits with the observation of Forterre (2006), who showed that in an analysis of more than 250 cellular genomes from Archaea, Bacteria, and Eukarya, most of the proteins coded by viral genomes have no cellular homologues. This would suggest that, while viruses and prokaryotes certainly exchanged genes, viruses did not originate as offshoots from prokaryotic cellular life.

In the last decade, microbiologists have belatedly come to realize the critical role of the viruses of prokaryotes as "a major component of the biosphere" (Krupovic et al. 2011). This new perspective, which one authority has labeled "the great virus comeback" (Forterre 2013), has come about through the expanding fields of evolutionary biology, genetics, genomics, metagenomics, and population dynamics, emphasizing the fact that viruses are both essential agents within the tree of life (Villarreal and Witzany 2010) and provide a key to understanding the complex dynamics of major ecologies. Through such examination, it would appear likely that viruses and the three domains of cellular life have been entwined in a complex labyrinth of formative evolutionary interactions since the dawn of evolution (Durzyńska and Goździcka-Jósefiak 2015). This would suggest that the prokaryotes we recognize today evolved during a highly aggressive symbiotic partnership between emerging Archaea and Eubacteria and ancestral forms of Archaeal and Eubacterial viruses (Gribaldo and Brochier-Armanet 2006; Hug et al. 2016; Koonin and Dolja 2013; Shapiro 2019; Villarreal 2014).

Today both the Eubacteria and the Archaea continue to share obligate symbiotic relationship with DNA- and RNA-based bacteriophage viruses (Krupovic et al. 2016). Established patterns of symbioses involving bacteriophage and Archaea-phage viruses and their host bacteria and Archaea include two different patterns of interaction, known as lysogenic and lytic cycles. Both these interactions involve viral reproduction, but while one is "temperate" the other is aggressive. When an invading phage virus enters into a lysogenic, or "latent", cycle, the virus integrates its genome into the host genome, or else it sits outside the genome in the form of a circular "replicon" within the bacterial cytoplasm. While behaving temperately, the virus holds its position as a "prophage," meanwhile suspending the lytic viral cycle of replication. In this way, the viral genome is replicated whenever the bacterium reproduces, with the prophage being passed on to the daughter bacteria. The virus retains its lytic potential and, over time, key stimuli can provoke the more aggressive pattern of lytic behavior.

In the lytic cycle, the virus replicates within the bacterial body independent of bacterial reproduction, hijacking the bacterial genetic physiology for its own selfish purposes. Viruses that behave in this way are known as "virulent" phages. They create a swarm of daughter viruses within the host bacterium, with the ultimate death and rupture of the bacterial cell releasing the daughter viruses into the surrounding medium, where they are poised to infect large numbers of ambient host bacteria. This process of virulence, swarm-type replication, and subsequent release with rupture of the bacterial cell, is known as "lysis" and the cycle is known as "the lytic cycle."

1.8 How Viruses Contribute to the Biosphere

The "virosphere" comprises the junctional zones where viruses interact with their myriad hosts, spanning all environments where life is to be found. Viruses are the most abundant biological entities in all the major environments on Earth, exceeding the numbers of cellular life forms, including prokaryotes, by one or two orders of magnitude (Koonin and Dolja 2013). In addition to this numerical predominance, the genetic diversity of viruses is commensurately enormous and might substantially exceed the genetic diversity of cellular organisms (Hambly and Suttle 2005; Koonin and Dolja 2013; Rosario and Breitbart 2011). The oceans teem with prokaryotic viruses (Suttle 2007). Marine virologists have estimated that there are 10^{31} tailed bacteriophage viruses on Earth (Krupovic et al. 2011). Equally surprising is the discovery, through metagenomic analysis of the marine virosphere of four different oceanic regions, that most of the viral sequences found in these ecologies differ from the sequences in current databases. Global genetic diversity was found to be exceedingly high, suggesting several hundred thousand hitherto-unrecognized species of viruses (Angly et al. 2006).

Such colossal numbers of prokaryotic viruses, with high genetic diversity, make it likely that viruses play key roles in the planet's ecosystems, exerting a significant force on the evolution of their bacterial and archaeal hosts. Viruses are also drivers of global geochemical cycles (Suttle 2005; Rosario and Breitbart 2011). Through prokaryotic lytic and lysogenic cycles, phage viruses have a major impact on the availability of nutrients in the marine ecosphere, including the termination of algal blooms (Fuhrman 1999; Wilhelm and Suttle 1999; Wommack and Colwell 2000). Identical bacteriophage sequences have been found in a wide variety of different marine environments, suggesting that there is an extensive circulation of viral genes among distantly related host populations. It is unlikely that the presence and importance of viruses are confined to oceanic ecologies. Up to the present, we simply have not searched sufficiently for them elsewhere. But with the expansion of metagenomics, this situation is changing.

Only recently, with the introduction of metagenomic studies, have we come to realize that this same gargantuan cycling of prokaryotes and their aggressively symbiotic viruses are also playing a fundamental role in a wide variety of terrestrial ecologies, including agricultural soils (Williamson et al. 2005), forest floors, and even the dry soil of the relatively lifeless Antarctic valleys (Williamson et al. 2007), enabling transfer of key elements such as carbon and iron and other micronutrients from the bacterial biomass to smaller prokaryotic life-forms within the same environments. In the words of these pioneering investigators, it would appear that soil virus diversity is very much underestimated and the impact of viruses on soil ecosystems poorly understood, emphasizing the need for more extensive metagenomic viral studies (Roossinck et al. 2015; Rosario and Breitbart 2011; Williamson et al. 2017). But it would appear likely that further study will confirm similarities in behavior of viruses in soil to those in oceanic environments, with major biospheric importance.

This pattern of "stable aggressive symbiosis" between phage viruses and bacteria at every depth of the oceans and every soil tested bears striking testimony to d'Herelle's conclusions a century earlier, with his pioneering introduction of the term "symbiosis" in the "establishment of a state of equilibrium between the virulence of the bacteriophage corpuscles and the resistance of the bacterium."

1.9 A Huge Unknown

The role of viruses in the evolutionary origins and subsequent proliferation of eukaryotic life is largely unexplored. But viral roles have been discovered wherever researchers have looked for them, whether in the origins of photosynthesis (Lindell et al. 2004), the nitrogen cycle (Long 2001; Sullivan and Ronson 1998), the potential origins of the eukaryotic nucleus (Bell 2009; Chaikeeratisak et al. 2017), key mitochondrial and nuclear enzymes (Filée et al. 2003), and numerous evolutionary developments in invertebrates and vertebrates (Villarreal 2005), making it likely that a dynamic and highly interactive virosphere is likely to have played a key role in both the origins of and subsequent diversification of life. One of the most comprehensively, if still incomplete, studies of virus–host symbiogenesis is that of the mammals, which might offer an exemplar of what is likely to be discovered more widely in biodiversity with future study.

1.10 Retroviral Symbiogenesis

Viruses play a vital role in the biosphere in terms of sheer physical abundance and genetic diversity, infecting and interacting with the evolution and life cycles of every cellular form of life. But the relative abundance of different classes of viruses differs with regard to eukaryotes in comparison to prokaryotes. Most viruses of prokaryotes have double-stranded DNA-based genomes, while most eukaryotic viruses have RNA-based genomes (Koonin et al. 2015). One such RNA-based family of viruses, the retroviruses, have entered the life cycles, and genomes, of sea slugs (Pierce et al. 1999; Pierce et al. 2016), fish (Naville and Volff 2016), basal amphibians (Aiewakun and Katzourakis 2017), reptiles and birds (Martin et al. 1997). Aiewsakun and Katzourakis also concluded that ray-finned fish foamy-like endogenous retroviruses, or FLERVs, exhibited an overall cospeciation pattern with their hosts, suggesting that retroviruses as a whole have an ancient marine origin at the time of, if not before, their jawed vertebrate hosts more than 450 million years ago in the Paleozoic Era. A wide range of genetic symbioses between exogenous retroviruses and animals during their pre- and postmammalian evolutionary lineages led to the colonizing of animal genomes with endogenous retroviruses, with considerable implications for mammalian evolution (Ryan 2016).

The genomes of retroviruses are composed of RNA. To insert themselves into the DNA-coded genome of their hosts, they employ a key enzyme, reverse transcriptase, that copies the viral RNA genome to its complementary DNA, which is then integrated into the host chromosomes. As part of its exogenous infectious cycle, the retrovirus inserts its genome into the genome of the target cell, such as a T-lymphocyte, causing the target cell to become a factory for the generation of daughter viruses. But sometimes, during some retroviral epidemics, the exogenous retrovirus uses the same replication mechanisms to insert its genome, perhaps in multiple different sites and chromosomes, into that of the host germline. Such viral genomic inserts are known as "endogenous retroviruses." This process of endogenization has the evolutionary potential of uniting the two disparate genomes, virus and host, to form a novel "holobiontic genome." The endogenous viral inserts, now known as "proviruses," will initially be suppressed by epigenetic mechanisms. But epigenetic silencing is not permanent. In time, many of these proviruses will be permanently silenced by indels, but some proviral genomes, in whole or in part, may be selectively conserved if they come to enhance holobiontic survival. To understand what this means, we need to grasp the concept and implications of holobiontic genomic evolution.

1.11 The Concept of Holobiontic Genomic Evolution

A "holobiont" is the symbiological term for the overall partnership in a symbiotic relationship. A "holobiontic genome" arises through the symbiological merger of two or more genomes arising from different evolutionary lineages. The origins of mitochondria and plastids are examples arising from the symbiological mergers of the genomes of prokaryotic and eukaryotic lineages. The evolutionary mechanisms implicit in the "endogenization" of an exogenous retroviral genome is another example of holobiontic genomic evolution, giving rise to a holobiontic genome incorporating both host and viral evolutionary lineages.

While the neo-Darwinian term "capture" is often applied to the incorporation of the viral genome, this should not be interpreted as the conversion of the viral genome to become host sequences. On the contrary, the usefulness of the viral genome is through its pre-evolved quintessentially viral properties, which are now available for adoption by the holobiontic organism. This implies that natural selection will no longer operate at the level of host lineage but at the level of the holobiontic genome, selecting for viral or vertebrate sequences that enhance holobiontic survival and selecting against viral or vertebrate sequences that reduce the potential for survival. Since viral genomes are minuscule in size when compared to those of their hosts, it is easy to underestimate the contribution retroviral genomes might make to such a union. We should recall that retroviruses have evolved over vast time periods to interact with, and manipulate, specific host physiological, biochemical, genetic, and immunological mechanisms and pathways.

The study of the holobiontic merger of endogenous retroviruses and host genome has been most extensively studied in humans, which might therefore offer an exemplar of what is likely to be discovered throughout the mammals.

1.12 Holobiontic Genetic and Epigenetic Regulation

Over the course of our evolutionary history, the historic human genomic lineage is thought to have been "colonized" by roughly 200 or so exogenous retroviruses, which have been incorporated as endogenous retroviruses (ERVs), assigned to different viral lineages, with the most recently incorporated primate viral lineages known as HERV-Ks. All retroviruses are regulated by their terminal genomic regions, known as "long terminal repeats," or "LTRs." Once incorporated into a host genome, these LTRs have the potential of taking over the regulation of formerly host genetic and other regulatory functions. For example, at least 50% of HERV-K LTRs are active promoters for host DNA transcription and, depending on what fraction of the human genome is surveyed, there may be hundreds of HERV-K solo LTRs that are human-specific (Buzdin et al. 2006). Most of the HERV-K loci found throughout the genome are solo LTRs, but a subset has persisted as full-length proviruses (van de Laagemaat et al. 2003). The so-called retrotransposon elements, which include HERVs in addition to ERV-related elements known as LINEs, SINEs, and alus, appear to work in a complex pattern of coordination in regulation (Villarreal 2005). In a systematic screening of different classes of regulatory regions in the human genome, Jordan and colleagues found transposable elements within 533 human genes, often in intron regions, where they appeared to control function, or splicing (Jordan et al. 2003). This included almost 25% of the analyzed promoter regions. In an instructive encapsulation of the way selection operates at the level of a holobiontic genome, the LTR of ERV-9 has replaced the host controls of the β-globin gene cluster—a cluster of five genes coding for the beta globin part of our hemoglobin (Routledge and Proudfoot 2002). The viral regulator has displaced several former host promoters and has been conserved over 15 million years of primate evolution. Other examples include LTRs acting as alternative promoters of the endothelin B receptor, the apolioprotein C-I genes (Medstrand et al. 2001), and the leptin receptor (Kapitonov and Jurka 1999).

1.13 Holobiontic Roles in Reproduction

Retroviruses have evolved strategies aimed at evading and nullifying host immunological attack. These include fusing host phagocytes into useless clumps of multinucleated giants and blocking humoral attack nullifying the effectiveness of antiviral antibodies. In the year 2000, two groups reported the discovery that the envelope gene of a human endogenous retrovirus, ERVWE1, coded for a protein, called syncytin-1, which plays a critical role in the lining cells of the human placenta (Blond et al. 2000; Mi et al. 2000). In essence, the expression of syncytin-1 in the placental trophoblast cells changes their fate to develop into syncytiotrophoblasts, which results in the fusion of the placental interface into a syncytial monolayer. This affords the placenta a more efficient filter for nutrients crossing from maternal to fetal circulation, and waste crossing from fetus to mother. The syncytin-1 locus is common to the great apes, including gorillas, orangutans, chimpanzees, and humans; hence, the ERV nomenclature rather than HERV. The envelope gene of a second endogenous retrovirus, HERV-FRD, was subsequently found to code for the protein syncytin-2. This protein is expressed deeper in the placenta and contributes both to placental fusion and to immunosuppression of maternal immunity to fetal antigens (Blaise et al. 2003). Another endogenous retrovirus, ERV-3, has a syncytin-1 type of action, and others, including an HERV-K, may be involved with many others playing as yet not fully defined roles in reproduction in humans (Villarreal and Ryan 2011).

Mice possess syncytin-A and syncytin-B, which play similar roles to syncytin-2 and syncytin-2 in great apes. Knockout mice for syncytin-A and syncytin-B show grossly defective placentas with impaired embryo survival, confirming that the syncytins are vital to placental structure and function (Dupressoir et al. 2009). Heidmann and colleagues screened a number of additional mammalian clades for endogenous placental retroviral *syncytin* genes, confirming that syncytin variants are associated with placentation in all of the other mammals tested, including Lagomorphs, which include the rabbits, *(syncytin-Orl)*, Carnivora (*syncytin-Car1*), Perissodactyls, which includes the horses, Chiroptera, which are bats, higher Ruminantia (*syncytin-Rum1*), Cetacea, Suina, which includes the pigs, Insectivora, which includes hedgehogs and shrews, Afrothera (*syncytin-Ten1*), which include elephants, aardvarks, and sea cows and Xenarthra, which include anteaters, sloths, and armadillos (Cornelis et al. 2012, 2013, 2014). The same researchers demonstrated the presence of a syncytin in the South American opossum (*syncytin-Opo1*), a marsupial that undergoes short-lived placentation (Cornelis et al. 2015). They also discovered a second nonfusogenic retroviral envelope gene that had been selectively conserved for more than 80 million years among all marsupials, including the South American opossum and the Australian tammar wallaby, and which possesses an immunosuppressive domain. This latter appeared to function in a similar way to syncytin-2 in great apes, to help suppress maternal rejection of paternal and thus potentially alien antigens in the fetus.

The discovery of two retroviruses in marsupials showing transient placentation is surely a key finding. Up to this point there was uncertainty whether the ERVs had played a key role in the origins of placentation, or whether they arrived after the evolution of a more primitive placenta, merely helping to improve its efficiency. But now, to quote these authors: "The capture of a founding syncytin by an oviparous ancestor was pivotal for the emergence of placentation more than 150 million years ago."

1.14 Holobiontic Roles in Embryogenesis

Syncytin-1 may be involved in the spermatozoa–oocyte fusion that results in the zygote (Soygur and Sati 2016). Methylation is one of the key epigenetic regulators capable of switching off the expression of unwanted genes and other genetic sequences in the genomes of animals and plants. Thus, it was intriguing to discover that a major wave of genome-wide demethylation takes place throughout a substantial portion of the genome during the earliest stage of embryonic development of plants and humans (Eckardt 2006; Guo et al. 2014). In humans, as in mammals generally, this might allow the activation of normally suppressed endogenous retroviral entities and their related retrotransposons. When Spadafora exposed early murine embryos to the antireverse transcriptase drug, Nevirapine, he discovered that it led to an irreversible arrest of development up to the four-cell stage (Spadafora 2008). Other researchers discovered that inhibition of the retroviral enzyme, reverse transcriptase, during early murine embryonic development resulted in substantial reprogramming of gene expression, involving both developmental and translational genes (Pittoggi et al. 2003). These findings suggested that endogenous retroviruses, and/or LINE-type products, are involved in regulation at the earliest stages of mammalian development.

Multiple HERV-K loci with intact open reading frames are scattered throughout the human chromosomes. DNA hypomethylation of the LTRs of the most recently acquired HERV-K integrations, accompanied by transactivation by OCT4 (a key definer of pluripotency during development), facilitates HERV-K reactivation and expression during the early stages of human fetal development (Grow et al. 2015a, b). Such retroviral expression appears to play a key role in embryogenesis from the eight-cell stage and continuing through the emergence of epiblast cells in preimplantation blastocysts, and ceasing at the stage of embryonic stem cell derivation from blastocyst outgrowths.

There are roughly 1000 loci of endogenous retrovirus HERV-H scattered through the human chromosomes. Some 231 of these loci are highly expressed in embryonic stem cells, where their LTRs, working in a coordinated way, act as a long noncoding RNA essential to human embryonic stem cell identity. This endogenous retroviral expression causes the upregulation of *OCT4*, *SOX2*, and *NANOG*, which maintain stem cell pluripotency. If HERV-H expression is blocked, using RNAi, the stem cells lose their pluripotency and differentiate into cells that resemble fibroblasts (Lu et al. 2014). This arena of research is still relatively new. But it seems likely that, given more time and exploration, many more such intimate and important interactions with host embryonic development will be discovered.

1.15 A Wider Role in Mammalian Evolution and Health

Recombination of viral loci on different chromosomes during meiosis has contributed to large-scale genetic deletions and duplications throughout our mammalian, hominid, and hominin evolution (Hughes and Coffin 2001; Medstrand and Mager 1998). Duplications of flanking endogenous retroviral may have contributed to the evolution of the extended major histocompatibility complex (Dawkins et al. 1999). A similar virus-induced genomic plasticity may also have contributed to the surprising level of genetic variation currently being observed between individual humans (Redon et al. 2006). A recent study of regulatory evolution of innate immunity showed that ERVs have shaped the evolution of a transcriptional network underlying the interferon response, which is a key element of innate immunity (Chuong et al. 2016). Meanwhile, endogenous retroviruses have also contributed to many symbiotic interactions at postdevelopmental level, including the physiological expression of viral *env* genes as proteins, in many different primate and human tissues (Fei et al. 2014; Kim et al. 2006). This has led to the proposal of an HERV Transcriptome Project as a separate enterprise from the published Human Proteome Project (Pontén et al. 2009).

1.16 The Debit Potential of Viral Symbiogenesis

The human genomic inheritance is shared with, or shows considerable similarities to, the genomes of other mammals, in containing hundreds of thousands of endogenous retroviral elements. Since these retain the legacy of exogenous retroviral sequences, they can threaten the resultant holobiontic organism through their inherent virus-specific pathological potential (Ryan 2009b). This includes unwanted expression, insertion, and dysregulation of established symbiotic viral gene, regulatory sequence or other genetic pathway, or through the cooption of evolved viral roles in complex, multistep disease progressions. In humans, such unwanted expression of retroviral genes and sequences contributes to placental pathologies, Down syndrome, pre-eclampsia, liver dysfunction, low platelet syndrome/intrauterine growth retardation, and gestational trophoblastic diseases, including hydatidiform mole and choriocarcinoma (Bolze et al. 2017).

HERVs have been associated with a miscellany of diseases, including Sertoli-cell only syndrome, or SCOS, a form of male infertility (Bosch and Jobling 2003), hemophilia (Kazazian et al. 1988), muscular dystrophy, and miscellaneous other diseases (Ryan 2009b). More specifically, HERV expression has been linked to many, if not all, of the autoimmune disorders (Volkman and Stetson 2014), notably multiple sclerosis (Mameli et al. 2009), where the disease may be caused by inappropriate levels of expression of syncytin-1 in astrocytes in the pathological lesions. It is possible that the malexpression of syncytin-1 may be triggered by exogenous viruses, notably Epstein Barr virus or Human Herpesvirus 6A (Mameli et al. 2012; Fierz 2017). Trials of specific T-cell therapy using monoclonal antibody formulations in MS are currently in progress (Pender et al. 2018).

However, we need to be cautious in interpreting the significance of HERV sequences in pathological conditions since, at present, we do not fully understand the role of HERV proteins and other sequences in normal physiology and their recruitment as part of the protective immunological responses to pathological insult. In accommodating the potential of ERVs in mammals as "both creators and destroyers," we also need to consider their potential as protectors against cancer in addition to their potential as perpetrators (Bannert et al. 2018).

1.17 Conclusions

The origins of life on Earth are inevitably obscure, but there are theoretical grounds for proposing a primal RNA-based protoviral origin during a hypothetical RNA World stage. Viruses are defined as noncellular capsid-encoding obligate symbionts. It is further proposed that this obligate symbiosis working both ways with evolving viruses and cellular hosts is key to understanding the further evolution of viruses as well as the influence of viruses on host evolution throughout the tree of life. This same obligate symbiosis has resulted in viruses playing an important in the health and maintenance of the biosphere.

References

Aiewakun P, Katzourakis A (2017) Marine origin of retroviruses in the early Palaeozoic era. Nat Comm 8:13954. https://doi.org/10.1038/ncomms13954|www.nature.com/naturecommunications

Angly FE, Felts B, Breitbart M, Salamon P, Edwards RA et al (2006) The marine Viromes of four oceanic regions. PLoS Biol 4(11):2121–2131

Bannert N, Hofmann H, Block A, Hohn O (2018) HERVs new role in cancer: from accused perpetrators to cheerful protectors. Frontiers in Microbiol 9:178

Bansho Y, Ichihashi N, Kazuta Y, Matsuura T, Suzuki H, Yomo T (2012) Importance of parasite RNA species repression for prolonged translation-coupled RNA self-replication. Chem Biol 19:478–487

Battistuzzi FU, Feijao A, Hedges SB (2004) A genomic timescale of prokaryotic evolution: insights into the origin of methanogenesis, phototrophy, and the colonization of land. BMC Evol Biol 4:44. https://doi.org/10.1186/1471-2148-4-44

Bell PJL (2009) The viral eukaryogenesis hypothesis: a key role for viruses in the emergence of eukaryotes from a prokaryotic world environment. Ann NY Acad Sci 1178:91–105

Blaise S, de Parseval N, Bénit L, Heidmann T (2003) Genomewide screening for fusogenic human endogenous retrovirus envelopes identifies syncytin 2, a gene conserved on primate evolution. Proc Natl Acad Sci U S A 100:13013–13018

Blond J-L, Lavillette D, Cheynet V, Bouton O, Oriol G et al (2000) An envelope glycoprotein of the human endogenous retrovirus HERV-W is expressed in the human placenta and fuses cells expressing the type D mammalian retrovirus receptor. J Virol 74:3321–3329

Bolze P-A, Mommert M, Mallet F (2017) Contributions of syncytins and other endogenous retroviral envelopes to human placenta pathologies. Prog Mol Biol and Trans Sci 1877-1173:111–162. https://doi.org/10.1016/bs.pmbts.2016.005

Bosch E, Jobling MA (2003) Duplications of the *AZFa* region of the human Y chromosome are mediated by homologous recombination between HERVs and are compatible with male fertility. Hum Mol Genet 12:341–347

Bremermann HJ (1983) Parasites at the origin of life. J Math Biol 16:165–180

Buzdin A, Kovalskaya-Alexandrova E, Gogvadze E, Sverdlov E (2006) At least 50% of human-specific HERV-K (HML-2) long terminal repeats serve in vivo as active promoters for host nonrepetitive DNA transcription. J Virol 80:10752–10762

Chaikeeratisak V, Nguyen K, Khanna K, Brilot AF, Erb ML et al (2017) Assembly of a nucleus-like structure during viral replication in bacteria. Science 355:194–197

Chuong EB, Elde NC, Feschotte C (2016) Regulatory evolution of innate immunity through co-option of endogenous retroviruses. Science 351:1083–1087

Claverie JM, Ogata H (2009) Ten good reasons not to exclude viruses from the evolutionary picture. Nat Rev Microbiol 7(8):615. https://doi.org/10.1038/nrmicro2108-c3

Colizzi ES, Hogeweg P (2016) Parasites sustain and enhance RNA-like replicators through spatial self-organisation. PLoS Comput Biol 12(4):e1004902. https://doi.org/10.1371/journal.pcbi.1004902

Collier L, Oxford J (1993) Human Virology. Oxford University Press, Oxford

Cornelis G, Heidmann O, Bernard-Stoecklin S, Reynaud K, Véron G et al (2012) Ancestral capture of syncytin-*Car1*, a fusogenic endogenous retroviral envelope gene involved in placentation and conserved in Carnivora. Proc Natl Acad Sci U S A 109(7):E432–E441. https://doi.org/10.1073/pnas.1115346109

Cornelis G, Heidmann O, Degrelle SA, Vernochet C, Lavialle C et al (2013) Captured retroviral envelope syncytin gene associated with unique placental structure of higher ruminants. Proc Natl Acad Sci U S A 110(9):E828–E837. https://doi.org/10.1073/pnas.1215787110
Cornelis G, Vernochet C, Carradec Q, Souquere S, Mulot B et al (2015) Retroviral envelope gene captures and syncytin exaptation for placentation in marsupials. Proc Natl Acad Sci U S A 111 (41):E4332–E4341. https://doi.org/10.1073/pnas.1417000112
Cornelis G, Vernochet C, Malicorne S, Souquere S, Tzika AC et al (2014) Retroviral envelope syncytin capture in an ancestrally diverged mammalian clade for placentation in the primitive Afrotherian tenrecs. Proc Natl Acad Sci U S A 111(41):E4332–E4341. www.pnas.org/cgi/doi/10.1073/pnas.1412268111
d'Herelle F (1926) The bacteriophage and its behaviour. Ballière, Tindall and Cox, London. Chapter V: 211. See also pp 326, 343 and 354
Dawkins R, Leelayuwat C, Gaudieri S, Tay G, Hui J et al (1999) Genomics of the major histocompatibility complex: haplotypes, duplication, retroviruses and disease. Immunol Rev 167:275–304
De Bary A (1879) Die Erscheinung der Symbiose. In: Vortrag auf der Versammlung der Naturforscher und Ärtze zu Cassel. Verlag von Karl J Trubner, Strasburg, pp 1–30
De La Torre JC, Holland JJ (1990) RNA virus quasispecies populations can suppress vastly superior mutant progeny. J Virol 64(12):6278–6281
Dodd MS, Papineau D, Grenne T, Slack JF, Rittner M et al (2017) Evidence for early life in Earth's oldest hydrothermal vent precipitates. Nature 543:60–64
Doolittle WF (2000) Uprooting the tree of life. Sci Amer 282:90–95
Douglas A (1994) Symbiotic interactions. Oxford University Press, New York
Dupressoir AC, Vernochet O, Bawa Harper F, Pierron G et al (2009) Syncytin-a knockout mice demonstrate the critical role in placentation of fusogenic, endogenous retrovirus-derived, envelope gene. Proc Natl Acad Sci U S A 106(2):12127–12132
Durzyńska J, Goździcka-Jósefiak A (2015) Viruses and cells intertwined since the dawn of evolution. Virol J 12:169. https://doi.org/10.1186/s12985-015-0400-7
Eckardt NA (2006) Genetic and epigenetic regulation of embryogenesis. Plant Cell 18:781–784
Eigen M, Schuster P (1977) The Hypercycle. A principle of natural self-organisation. Part a: emergence of the Hypercycle. Naturwissenschaften 64:541–565
Eigen M, Schuster P (1978a) The Hypercycle. A principle of natural self-organisation. Part B: the abstract Hypercycle. Naturwissenschaften 65:7–41
Eigen M, Schuster P (1978b) The Hypercycle. A principle of natural self-organisation. Part C: the realistic Hypercycle. Naturwissenschaften 65:341–369
Eigen M, Schuster P (1979) The Hypercycle: a principle of natural self-organization. Springer ISBN 0-387-09293-5
Fei C, Atterby C, Edqvist P-H, Ponten F, Zhang WW et al (2014) Detection of the human endogenous retrovirus ERV3-encoded Env-protein in human tissues using antibody-based proteomics. J Roy Soc Med 107(1):22–29. https://doi.org/10.1177/0141076813509981
Fierz W (2017) Multiple sclerosis: an example of pathogenic viral interaction? Virol J 14:42–48. https://doi.org/10.1186/s12985-017-0719-3
Filée J, Forterre P, Laurent J (2003) The role played by viruses in the evolution of their hosts: a view based on informational protein phylogenies. Research in Virol 154:237–243
Fisher S (2010) Are RNA viruses vestiges of an RNA world? J Gen Philos Sci 41:121–141
Forterre P (2006) The origin of viruses and their possible roles in major evolutionary transitions. Virus Res 117:5–16
Forterre P (2013) *The Great Virus Comeback* (translated from the French). Biol Aujourdhui 207 (3):153–168
Frank AB (1877) Über die biologischen Verhältnisse des Thallus eineger Krustenflechten. Beitrage zur Biologie der Pflanzen 2:123–200
Frank AB (1885) Über die auf Wurzelsymbiose beruhende Ernährung gewisser Bäume durch unterirdische Pilze. Berichte der Deutschen Botanischen Gesellschaft 3:128–145

Fuhrman JA (1999) Marine viruses and their biogeochemical and ecological effects. Nature 399:541–548
Glansdorf N, Xu Y, Labedan B (2008) The last universal common ancestor: emergence, constitution and genetic legacy of an elusive forerunner. Biol Direct 3:29. https://doi.org/10.1186/1745-6150-3-29
Gribaldo S, Brochier-Armanet C (2006) The origin and evolution of Archaea: a state of the art. Phil Trans Roy Soc B 361:1007–1022. https://doi.org/10.1098/rstb.2006.1841
Grow EJ, Flynn RA, Chavez SL, Bayless NL, Wossidlo M, Wesche DJ, Martin L, Ware CB, Blish CA, Chang HY, Pera RA, Wysocka J (2015a) Intrinsic retroviral reactivation in human preimplantation embryos and pluripotent cells. Nature 522:221–225
Grow EJ, Flynn RA, Chavez SL, Bayless NL, Wossidlo M et al (2015b) Intrinsic retroviral reactivation in human preimplantation embryos and pluripotent cells. Nature 522:221–225
Guo H, Zhu P, Yan L, Li R, Hu B et al (2014) The DNA methylation landscape of human early embryos. Nature 511:606–610
Hambly E, Suttle CA (2005) The virosphere, diversity, and genetic exchange within phage communities. Curr Opinion Microbiol 8:444–450
Hedge NR, Maddur MS, Kaveri SV, Bayry J (2009) Reasons to include viruses in the tree of life. Nat Rev Microbiol 7(8):615. https://doi.org/10.1038/nrmicro2108-c1
Holland JJ, De La Torre JC, Steinhauer DA (1992) RNA virus populations as quasispecies. Curr Top Microbiol Immunol 176:1–20
Holmes EC (2011) What does virus evolution tell us about virus origins? J Virol 85(11):5427–5251
Hug LA, Baker BJ, Anantharaman K, Brown CT, Probst AJ et al (2016) A new view of the tree of life. Nat Microbiol 1:16048. https://doi.org/10.1038/nmiccrobiol.2016.48
Hughes JF, Coffin JM (2001) Evidence for genomic rearrangements mediated by human endogenous retroviruses during primate evolution. Nat Genet 29:487–489
Jordan K, Rogozin IB, Glazko GV, Koonin EV (2003) Origin of a substantial fraction of human regulatory sequences from transposable elements. Trends Genet 19(2):68–72
Kapitonov VV, Jurka J (1999) The long terminal repeat of an endogenous retrovirus induces alternative splicing and encodes an additional carboxy-terminal sequence in the human leptin receptor. J Mol Evol 48:248–251
Kazazian HH, Wong C, Youssoufian H, Scott AF, Phillips DG, Antonarakis SE (1988) Haemophilia a resulting from de novo insertion of L1 sequences represents a novel mechanism for mutation in man. Nature 332:164–166
Kim H-S, Yi J-M, Hirai H, Huh J-W, Jeong M-S et al (2006) Human endogenous retrovirus (HERV)-R in primates: chromosomal location, gene expression, and evolution. Gene 370:34–42. https://doi.org/10.1016/j.gene.2005.11.008
Koonin EV (2001) The logic of chance: the nature and origin of biological evolution. F.T. Press Science, Upper Saddle River, NJ
Koonin EV (2011) The logic of chance: the nature and origin of biological evolution. F.T. Press Science, New Jersey
Koonin EV, Dolja VV (2013) A virocentric perspective on the evolution of life. Curr Opin Virol 3 (5):546–557
Koonin EV, Dolja VV, Krupovic M (2015) Origins and evolution of viruses of eukaryotes: the ultimate modularity. Virol 479-480:2–25
Koonin EV, Senkevich TG, Dolja VV (2006) The ancient virus world and the evolution of cells. Biol Direct 1:29. https://doi.org/10.1186/1745-6150-1-29
Koonin EV, Senkevich TG, Dolja VV (2009) Compelling reasons why viruses are relevant for the origin of cells. Nat Rev Microbiol 7(8):615. https://doi.org/10.1038/nrmicro2108-c5
Krupovic M, Dutilh BE, Adriaenssens EM, Wittmann J, Vogensen FK et al (2016) Taxonomy of prokaryotic viruses: update from the ICTV bacterial and archaeal viruses subcommittee. Arch Virol 161:1095–1099

Krupovic M, Prangishvil D, Hendrix RW, Bamford DH (2011) Genomics of bacterial and Archaeal viruses: dynamics within the prokaryotic Virosphere. Microbiol and Mol Biol Rev 75 (4):610–635

Lindell D, Sullivan MB, Johnson ZI, Tolonen AC, Rohwer F, Chisholm W (2004) Transfer of photosynthesis genes to and from Prochlorococcus viruses. Proc Natl Acad Sci U S A 101 (30):11013–11018

Long SR (2001) Genes and signals in the rhizobium-legume symbiosis. Plant Physiol 125:69–72

López-García P, Eme L, Moreira D (2017) Symbiosis in eukaryotic evolution. J Theor Biol 434:20–33

López-Garcia P, Moreira D (2009) Ten reasons to exclude viruses from the tree of life. Nat Rev Microbiol 7:306–311

López-Garcia P, Moreira D (2009b) Yet viruses cannot be included in the tree of life. Nature Rev Microbiol 7:306–311. https://doi.org/10.1038/nrmicro2108-c1

Lu X, Sachs F, Ramsay L, Jacques PE, Jacques PÉ et al (2014) The retrovirus HERV-H is a long noncoding RNA required for human embryonic stem cell identity. Nat Struct Mol Biol 21 (4):423–425

Ludmir EB, Enquist LW (2009) Viral genomes are part of the phylogenetic tree of life. Nat Rev Microbiol 7(8):615. https://doi.org/10.1038/nrmicro2108-c4

Luria SE, Darnell JE (1967) General virology. John Wiley & Sons, INC., New York

Mameli G, Poddighe L, Astone V, Delogu G, Arru G et al (2009) Novel reliable real-time PCR for differential detection of MSRV*env* and syncytin-1 in RNA and DNA from patients with multiple sclerosis. J Virol Methods 161:98–106

Mameli G, Poddighe L, Mei A, Uleri E, Sotgiu S et al (2012) Expression and activation by Epstein Barr virus of human endogenous retroviruses-W in blood cells and astrocytes: inference for multiple sclerosis. PLoS One 7(9):e44991. https://doi.org/10.1371/journal.pone.0044991

Margulis L (1970) Origin of eukaryotic cells. Yale University Press, New Haven and London

Margulis L, Fester R (eds) (1991) Symbiosis as a source of evolutionary innovation. MIT Press, Cambridge Massachusetts and London

Martin J, Herniou E, Cook J, O'Neill RW, Tristem M (1997) Human endogenous retrovirus type I-related viruses have an apparently widespread distribution within vertebrates. J Virol 71:437–443

Medstrand P, Landry JR, Mager DL (2001) Long terminal repeats are used as alternative promoters for the endothelin B receptor and apolipoprotein C-I genes in humans. J Biol Chem 276:1896–1903

Medstrand P, Mager DL (1998) Human-specific integrations of the HERV-K endogenous retrovirus family. J Virol 72:9782–9787

Merezhkovskii C (1910) Retrospectively quoted in his paper, La plante considérée comme un complexe symbiotique. Bull de la Société Naturelles 6:17–98

Mi S, Lee X, Li X-P, Veldman GM, Finnerty H et al (2000) Syncytin is a captive retroviral envelope protein involved in human placental morphogenesis. Nature 403:785–789

Nasir AK, Kim M, Caetano-Annoles G (2012) Viral evolution. Primordial cellular origins and late adaptation to parasitism. Mob Genet Elem 2(5):247–252

Navas-Castillo J (2009) Six comments on the ten reasons for the demotion of viruses. Nat Rev Microbiol 7(8):615. https://doi.org/10.1038/nrmicro2108-c2

Naville M, Volff J-N (2016) Endogenous retroviruses in fish genomes: from relics of past infections to evolutionary innovations? Front Microbiol 7:1197. https://doi.org/10.3389/fmicb.2016.01197

Nowak MA (1992) What is a Quasispecies? TREE 7:118–121

Pender MP, Csurhes PA, Smith C, Douglas NL, Neller MA et al (2018) Epstein-Barr virus-specific T cell therapy for progressive multiple sclerosis. JCI Insight 3(22):e124714. https://doi.org/10.1172/jci.insight.124714

Pierce SK, Mahadevan P, Massey SE, Middlebrooks ML (2016) A preliminary molecular and phylogenetic analysis of the genome of a novel endogenous retrovirus in the sea slug *Elysia Chlorotica*. Biol Bull 231:236–244

Pierce SK, Maugel TK, Rumpho ME, Hanten JJ, Mondy WL (1999) Annual viral expression in a sea slug population: life cycle control and symbiotic chloroplast maintenance. Biol Bull 197:1–6

Pittoggi C, Sciamanna I, Mattei E, Beraldi R, Lobascio AM et al (2003) Role of endogenous reverse transcriptase in murine early embryo development. Mol Reprod Dev 66:225–236

Pontén F, Gry M, Fagerberg L, Lundberg E, Asplund A et al (2009) A global view of protein expression in human cells, tissues, and organs. Mol Systems Biol 5:337. https://doi.org/10.1038/msb.2009.93

Prangashvili D, Garrett RA (2004) Exceptionally diverse morphotypes and genomes of crenarcheal hyperthermophilic viruses. Biochem Soc Trans 32(2):204–208

Raoult D (2009) There is no such thing as a tree of life (and of course viruses are out!). Nat Rev Microbiol 7(8):615. https://doi.org/10.1038/nrmicro2108-c6

Redon R, Ishikawa S, Fitch KR, Feuk L, Perry GH et al (2006) Global variation in copy number in the human genome. Nature 444:444–454

Roossinck MJ, Martin DP, Roumagnac P (2015) Plant virus Metagenomics: advances in virus discovery. Phytopath Rev 105:716–727

Rosario K, Breitbart M (2011) Exploring the viral world through metagenomics. Curr Opin Virol 1 (1):289–297

Routledge SJ, Proudfoot NJ (2002) Definition of transcriptional promoters in the human beta globin locus control region. J Mol Biol 323:601–611

Ryan F (1997) Virus X. Little Brown Publishers, Boston, MA

Ryan F (2002) Darwin's blind spot. Houghton Mifflin, Boston and New York

Ryan F (2009a) Virolution. Collins Publishers, London, UK

Ryan F (2009b) An alternative approach to medical genetics based on modern evolutionary biology. Part 3: HERVs in diseases. J R Soc Med 102:415–424

Ryan F (2016) Viral symbiosis and the holobiontic nature of the human genome. APMIS 124:11–19

Ryan F (2019) Virusphere. Collins Publishers, London, UK

Sagan L (1967) On the origin of mitosing cells. J Theor Biol 14:255–274

Sapp J (1994) Evolution by association: a history of Symbiosis. Oxford University Press, New York

Shapiro JA (2019) No genome is an island: towards a 21st century agenda for evolution. Ann N Y Acad Sci 1447(1):21–52. https://doi.org/10.1111/nyas.14044

Soygur B, Sati L (2016) The role of syncytins in human reproduction and reproductive organ cancers. Reproduction 152(5):R167–R178. Soc for Reprod and Fertility ISSN 1470-1626 (paper) 1741-7899 (online). https://doi.org/10.1530/REP-16-0031

Spadafora C (2008) A reverse transcriptase-dependent mechanism plays central roles in fundamental biological processes. Systems Biol in Reprod Medi 54:11–21

Sullivan JT, Ronson CW (1998) Evolution of rhizobia by acquisition of a 500-kb symbiosis island that integrates into a phe-tRNA gene. Proc Natl Acad Sci U S A 95:5145–5149

Suttle CA (2005) Viruses in the sea. Nature 437:356–361

Suttle CA (2007) Marine viruses – major players in the global ecosystem. Nat Rev Microbiol 5:801–812

Takeuchi N, Hogewoeg P (2008) Evolution of complexity in RNA-like replicators systems. Biol Direct 3:11. https://doi.org/10.1186/1745-6150-3-11

van de Laagemaat LN, Landry J-R, Mager DL, Medstrand P (2003) Transposable elements in mammals promote regulatory variation and diversification of genes with specialized functions. Trends Genet 19(10):530–536

Vignuzzi M, Stone JK, Arnold JJ, Cameron CE, Andino R (2006) Quasispecies diversity determines pathogenesis through cooperative interactions in a viral population. Nature 439:344–348

Villarreal LP (2005) Viruses and the evolution of life. ASM Press, Washington, DC

Villarreal LP (2014) Force for ancient and recent life: viral and stem-loop RNA consortia promote life. Ann N. Y. Acad Sci 1341:25–34

Villarreal LP, DeFilippis VR (2000) A hypothesis for DNA viruses as the origin of eukaryotic replication proteins. J Virol 74(15):7079–7084

Villarreal LP, Ryan F (2011) Viruses in host evolution: general principles and future extrapolations. Curr Topics in Virol 9:79–90

Villarreal LP, Witzany G (2010) Viruses are essential agents within the roots and stem of the tree of life. J Theoret Biol 262:698–710

Volkman HE, Stetson DB (2014) The enemy within: endogenous retroelements and autoimmune disease. Nat Immun 15(5):415–422

Wilhelm SW, Suttle CA (1999) Viruses and nutrient cycles in the sea. Bioscience 49:781–788

Williamson KE, Fuhrmann JJ, Wommack KE, Radosevich M (2017) Viruses in soil ecosystems: an unknown quantity within an unexplored territory. Ann Rev Virol 4:201–219

Williamson KE, Radosevich M, Smith DW, Wommack KE (2007) Incidence of lysogeny within temperate and extreme soil environments. Environ Microbiol 9:2563–2574

Williamson KE, Radosevich M, Wommack ME (2005) Abundance and diversity of viruses in six Delaware soils. Appl Environ Microbiol 71(60):3119–31125

Woese CR, Kandler O, Wheelis ML (1990) Towards a natural system of organisms: proposal for the domains, Archaea, bacteria and Eukarya. Proc Nat Acad Sci 87(12):4576–4579

Wommack KE, Colwell RR (2000) Viroplankton: viruses in aquatic ecosystems. Microbiol Mol Biol Rev 64:69–114

Zaman L, Meyer JL, Devangam S, Bryson DM, Lenski RE, Ofria C (2014) Coevolution drives the emergence of complex traits and promotes evolvability. PLoS Biol 12(12):e1002023. https://doi.org/10.1371/journal.pbio.1002023

Chapter 2
Gene Transfer Agents in Symbiotic Microbes

Steen Christensen and Laura R. Serbus

Abstract Prokaryotes commonly undergo genome reduction, particularly in the case of symbiotic bacteria. Genome reductions tend toward the energetically favorable removal of unnecessary, redundant, or nonfunctional genes. However, without mechanisms to compensate for these losses, deleterious mutation and genetic drift might otherwise overwhelm a population. Among the mechanisms employed to counter gene loss and share evolutionary success within a population, gene transfer agents (GTAs) are increasingly becoming recognized as important contributors. Although viral in origin, GTA particles package fragments of their "host" genome for distribution within a population of cells, often in a synchronized manner, rather than selfishly packaging genes necessary for their spread. Microbes as diverse as archaea and alpha-proteobacteria have been known to produce GTA particles, which are capable of transferring selective advantages such as virulence factors and antibiotic resistance. In this review, we discuss the various types of GTAs identified thus far, focusing on a defined set of symbiotic alpha-proteobacteria known to carry them. Drawing attention to the predicted presence of these genes, we discuss their potential within the selective marine and terrestrial environments occupied by mutualistic, parasitic, and endosymbiotic microbes.

2.1 Introduction

As a response to colonizing diverse and changing environments, as well as competing against threats within those environments, microbes employ either generalist or specialist strategies (Futuyma and Moreno 1988; Sriswasdi et al. 2017; van Tienderen 1997). Generalist prokaryotes carry a large and diverse complement of genes, which ensure autonomy and flexibility in the face of changing environmental

S. Christensen · L. R. Serbus (✉)
Department of Biological Sciences, Florida International University, Miami, FL, USA

Biomolecular Sciences Institute, Florida International University, Miami, FL, USA
e-mail: Lserbus@fiu.edu

M. Kloc (ed.), *Symbiosis: Cellular, Molecular, Medical and Evolutionary Aspects*,
Results and Problems in Cell Differentiation 69,
https://doi.org/10.1007/978-3-030-51849-3_2

conditions (Guieysse and Wuertz 2012). These organisms also serve as an important ongoing contributor to speciation, by enabling the introduction of new "specialist" microbes that thrive in niche environments (Sriswasdi et al. 2017). However, maintenance of a large genome comes at considerable energetic cost (Lever et al. 2015), with decreasing benefits of additional genes in response to increasing genome size (Sela et al. 2016). Metagenomic analyses increasingly point to genome reduction as not only common to prokaryotic systems but more the rule than the exception (Ochman 2005; Wolf and Koonin 2013). Affording efficiency in genome replication, coding elements are regularly pseudogenized and removed, while intervening noncoding sequences are minimized (Goodhead and Darby 2015; Kuo et al. 2009; Kuo and Ochman 2010). Genetic drift can also serve as a significant contributor to genome reduction (McCutcheon and Moran 2011; Sabater-Muñoz et al. 2017).

For symbiotic prokaryotes, genome reduction is particularly exaggerated. Association with a host organism as a surface "ectosymbiont," or as a cytoplasmic resident "endosymbiont," inherently provides richer nutrient availability for the prokaryote (Bulgheresi 2016). For both types of symbioses, reduced selection for retention of metabolic pathways results in loss of the corresponding genes (Moran and Bennett 2014; Nicks and Rahn-Lee 2017). For endosymbionts, loss of DNA repair genes, general lack of recombination, and population bottlenecks further accelerate and generalize gene loss in a process referred to as Muller's ratchet (Dale et al. 2003; Moran 1996; Moran et al. 2008). This genome decay converts endosymbionts into an organelle-like state, such as for *candidatus Hodgkinia cicadicola, candidatus Tremblaya princeps*, and *candidatus Nasuia deltocephalinicola* (McCutcheon and Keeling 2014; McCutcheon and Moran 2011; Moran and Bennett 2014). Although there are success stories, such as for mitochondria, chloroplasts, and certain other plastids (O'Malley 2015), the default outcome of genome reduction is endosymbiont replacement or extinction (Brown 2018; Hackstein and de Graaf 2018; Husnik and Keeling 2019; McCutcheon et al. 2019).

While a genome reduction model is broadly supported for evolution of symbiotic prokaryotes, several recent reports have challenged its general premise. Examples of genome expansion, representing an increase in the number of functional genes, have been reported for a range of symbiotic bacteria, including *Acinetobacter baumannii* (Yakkala et al. 2019), *Chlamydia trachomatis* (Bohlin 2015), *Spiroplasma clarkii* (Tsai et al. 2018), and *Wolbachia pipientis w*Fol strain (Kampfraath et al. 2019), ancient *Mycoplasma*-related endosymbionts of fungi (Naito and Pawlowska 2016) and a number of others (López-Madrigal and Gil 2017). Considering that prokaryotic symbioses span a range of parasitic to beneficial interactions, obligate as well as facultative, and cross all scales of organism size and complexity (Bulgheresi 2016; Moran et al. 2008; Wernegreen 2012), the similarity of these reports converges on a core principle. Horizontally transferred genetic content is necessary to offset genetic losses that are otherwise intrinsic to symbiosis (Sela et al. 2016; Takeuchi et al. 2014). In relation to natural selection, genetic drift, and gene mutation, horizontal gene transfer (HGT) is generally recognized as a direct and extremely effective way of maintaining and distributing genetic ability (Hosseini and Wagner 2018; Koonin

2016; Zhaxybayeva and Doolittle 2011). Often used interchangeably with the term "lateral gene transfer," the term HGT is most often used to describe the acquisition of genes by similar or closely related species, yet can also occur between phylogenetically distant microbes (Koonin 2015).

HGT in bacteria is known to occur through the following three major mechanisms: (1) transformation: the uptake of free DNA from the surrounding environment by competent bacteria, (2) conjugation: the transfer of mobile genetic elements by pili structures that span two adjacent bacteria, usually of similar mating type, and (3) transduction: the transfer of bacterial DNA between a bacteriophage-infected cell and an otherwise susceptible bacterium (García-Aljaro et al. 2017; Johnston et al. 2014; Mell and Redfield 2014; Takeuchi et al. 2014). These mechanisms have been described for aquatic and terrestrially abundant microbes, as well as important pathogens of higher eukaryotes and mammals (Christie and Vogel 2000; Davison 1999; Fillol-Salom et al. 2019; Husnik and McCutcheon 2018; Mathur and Singh 2005; Parkinson 2016; Sun 2018; Ye et al. 2019). However, as bacterial endosymbionts commonly reside within vesicle-derived membrane compartments inside eukaryotic host cells, this largely precludes the uptake of exogenous microbial DNA (López-Madrigal and Gil 2017). Phage-like elements have considerable potential to facilitate HGT for ecto- and endosymbiotic bacteria. Although the amount of nucleic acid deliverable by a phage-like particle is physically limited, such particles can make significant, collective contributions to genome size (Gao et al. 2019).

HGT is also facilitated by phage-like elements referred to as gene transfer agents (GTAs), which have received relatively little attention to date. GTAs are distinct from typical viruses in that GTA particles distribute host-specific traits via transfer of host genomic DNA, generally with little preference for the sequences encoding their phage-specific gene clusters (Berglund et al. 2009; Bertani and Baresi 1987; Humphrey et al. 1997; Marrs 1974; Rapp and Wall 1987). GTAs appear to be the result of bacteriophage genes having been "domesticated" by their microbial hosts, resulting in the production of phage-like particles under the regulatory control of endogenous host mechanisms (Bobay et al. 2014; Harrison et al. 2017; Olszak et al. 2017). Typically, the size range of DNA that is transferred is far less than that encoding the GTA particle components, thus precluding the type of selection pressure associated with most bacteriophages and maintenance of prophage gene clusters. A variety of GTA types have now been identified in association with bacteria as well as archaea, combining the key aspects of transducing phage and recipient cell transformation machinery (Lang and Beatty 2007; Québatte and Dehio 2019).

The abundance of GTAs in nature is expected to be vast. Sampling data suggest that marine bacterial genomes contain a significant number of GTA genes (Biers et al. 2008; Roux et al. 2013; Zhang et al. 2018), and a recent in silico analysis suggests that more than half the "prophage" predictions may instead be GTAs (Kogay et al. 2019). Testing GTAs from one type of alpha-proteobacteria of the *Roseobacter* clade showed a wide range of host targeting and interspecific gene transfer frequencies, in the range of 10^{-3} to 10^{-1} (McDaniel et al. 2010). This frequency represents ~6×10^5 to 3×10^7 times that of transduction, or from 2000 to

5×10^8 times the frequency of transformation, previously measured in marine environments (Jiang and Paul 1998; McDaniel et al. 2010; Williams et al. 1997). While these likely represent overestimates of naturally occurring GTA-based exchange frequencies, their maintenance in various branches of microbial life over large evolutionary time scales, as well as the presence of similar homologs in viral genomes, attest to a functional role (Redfield and Soucy 2018) (Biers et al. 2008; Shakya et al. 2017).

Here, we provide an overview of what is understood about GTAs and outline the range of symbiotic prokaryotes where active GTA particles and GTA gene homologs have been experimentally or bioinformatically detected. As prokaryotic symbionts form complex, environmentally relevant relationships, and many are considered pathogenic to higher eukaryotes, we highlight organisms known to form well-defined symbiotic associations. To date, the overwhelming majority of molecular and genetic information, as well as completely annotated genomic sequence data, is available for symbionts of the alpha-proteobacteria class, which range from marine microbes and plant-associated nitrogen-fixers to endosymbiotic mutualists and blood parasites. As such, this review highlights what is known about GTAs and GTA genes and speculates on other potential functions outside of HGT, for this large set of symbiotic bacteria. The established link between the earliest diverging alpha-proteobacteria and the formation of a proto-mitochondrial lineage is considered, in the context of estimated introductions of GTA-related homologs, and we discuss what is understood for GTAs as far removed as archaea.

2.2 Prototypical Gene Transfer Agent: *Rhodobacter capsulatus* GTA

The most well-understood GTA system, first discovered in 1974, is that of the purple-nonsulfur marine bacterium *Rhodobacter capsulatus* (Marrs 1974). This model has been intensively studied due to its highly branched electron transport chain and the ability to grow under a wide variety of conditions. The cell cultures of *R. capsulatus* were found to transfer genes between bacterial strains without the detectable presence of bacteriophage or cell–cell contact (Solioz et al. 1975). Detailed electron microscopy images revealed that the cultures of *R. capsulatus* in vitro formed phage-like particles that structurally resembled *Siphoviridae* bacteriophage (Yen et al. 1979). Siphoviruses, such as Lambda and HK97, are typically characterized by an un-enveloped icosahedral capsid, on the order of 60 nm in diameter, that carries approximately 50 kb of a double-stranded DNA (dsDNA)-based viral genome (Helgstrand et al. 2003; Hendrix and Johnson 2012). The capsid is attached to a flexible, noncontractile tail structure, typically in the 135 to 260 nm length range (Duda et al. 1995; Plisson et al. 2003; Vegge et al. 2005). The phage-like particles produced by *R. capsulatus* represent a smaller variation on *Siphoviridae*, with a head size of ~30 nm, carrying <10% the amount of DNA

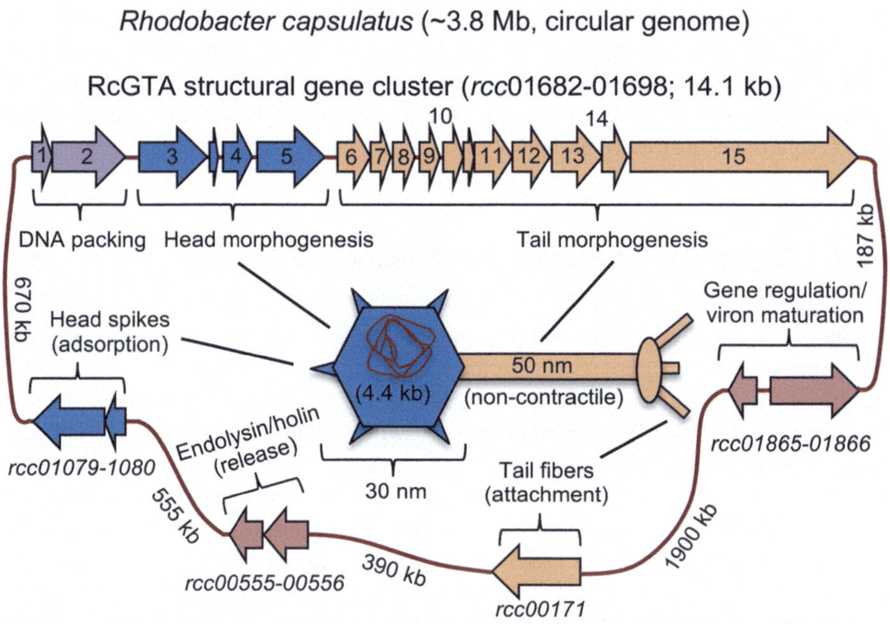

Fig. 2.1 Genetic elements involved in GTA production in *Rhodobacter capsulatus*. The red line represents the genome of *R. capsulatus*, while shaded arrows indicate predicted ORFs associated with RcGTA production. Locus information and the lengths of predicted ORFs, gene clusters, and intervening genomic regions are indicated in kilobases (kb). ORF colors indicate an association with DNA packing (purple), head morphogenesis (blue), tail morphogenesis (orange), and maturation/release of active particles (red). A model of a representative RcGTA particle is shown in the center, with capsid and tail sizes indicated in nanometers (nm) and the length of DNA packaged within the capsid is indicated (kb). Genome and particle representations are not to scale

(~4.5 kb), with an ~50 nm tail length (Fig. 6.1) (Yen et al. 1979). As these phage-like particles are responsible for carrying out HGT activity in vivo, these particles are now directly referred to as "gene transfer agents" or GTAs in the literature.

The *R. capsulatus* GTA (RcGTA) is primarily encoded by a cluster of 15 genes spanning a ~ 14 kb genomic region, conserved in the order *Rhodobacterales* (Biers et al. 2008; Lang and Beatty 2000, 2007). Akin to siphoviruses and tailed bacteriophages of the order *Caudovirales*, this structural "head-tail" gene cluster is organized discretely in units that involve open-reading frames (ORFs) predicted to encode DNA head-packing enzymes, head morphogenesis, and tail morphogenesis proteins (Fig. 2.1) (Lang and Beatty 2007; Lang et al. 2012). Even ORFs annotated as "conserved hypothetical protein" at this time show clear homology to other phage proteins. While the majority of the structural cluster gene products are involved in the formation of GTA particles, several products are not found in association with purified particles, two of which were not originally ascribed gene numbers (Chen et al. 2009; Lang et al. 2017). In addition to these structural cluster RcGTA genes, other essential structural genes have been identified elsewhere in the *R. capsulatus* genome. These include two ORFs associated with the formation of head spikes and one ORF encoding a tail protein required for the attachment of RcGTA particles to other *R. capsulatus* cells (Hynes et al. 2016; Westbye et al. 2017). Several regulatory genes are thought to have an integral role in the assembly and release of active RcGTA particles, as described below.

2.2.1 Regulation of Particle Production and Release

Production of *R. capsulatus* GTA particles shares important features with the replication cycle of *Siphoviridae* bacteriophage, though substantial distinguishing features exist. Lysogenic prophages and GTAs are similar in that particles are not produced until a specific trigger is met. However, while the lytic bacteriophage phase can be induced at any time during the cell cycle, production of RcGTA particles is tied to intracellular *R. capsulatus* signaling processes and regulation of the cell cycle (Fig. 6.2) (Lang and Beatty 2007; Lang et al. 2017). It was recently reported that a helix-turn-helix DNA binding protein, now called GafA, is the central regulator of RcGTA structural gene cluster expression (Fogg 2019). Mutational analyses of *gafA* (*rcc01865*), along with an adjacent gene involved in virion maturation (*rcc01866*), revealed a requirement for these protein products in the production of functional RcGTAs (Hynes et al. 2016; Lang et al. 2017). A number of reports indicate that RcGTA production is governed by an integrated signaling network that regulates *gafA* expression, starting with the sensing of environmental signals through the quorum sensing system, as described below.

Quorum sensing (QS) and, specifically, the proteins GtaI and GtaR play an important role in triggering RcGTA production (Leung et al. 2010; Schaefer et al. 2002; Solioz et al. 1975). Homologous to the well-studied LuxI and LuxR proteins of *Escherichia coli*, these systems regulate multiple group-oriented behaviors, including pathogenicity in humans, symbiosis in plants, HGT, extracellular polysaccharide production, and biofilm formation (Bottomley et al. 2007; Koppenhöfer et al. 2019; Leung et al. 2012; Mellbye et al. 2017; Nadell et al. 2008; Tang et al. 2020; Tun-Garrido et al. 2003). GtaI promotes the synthesis of two long-chain N-acyl-homoserine lactone (HSL) signaling molecules. These acyl-HSLs are diffusible, membrane-permeable compounds that act as cues for intra- and inter-cellular signaling. In the absence of acyl-HSL produced by GtaI, GtaR acts as a negative repressor of the *gtaIR* operon and indirectly represses transcription of the RcGTA structural cluster (Leung et al. 2012; Leung et al. 2010). When GtaR binds to HSL ligands in the cytoplasm, this leads to a range of gene expression changes including upregulation of *gafA* (Fig. 2.2) (Lazdunski et al. 2004). In addition, RcGTA particle production can also be stimulated by several acyl-HSLs not produced by *R. capsulatus* itself (Leung et al. 2012).

Cell cycle signaling processes also have a major impact on *gafA* expression. In the classic cell cycle paradigm, defined in the asymmetrically dividing, dimorphic marine *Caulobacter crescentus* system, the sensor histidine-kinase CckA activates

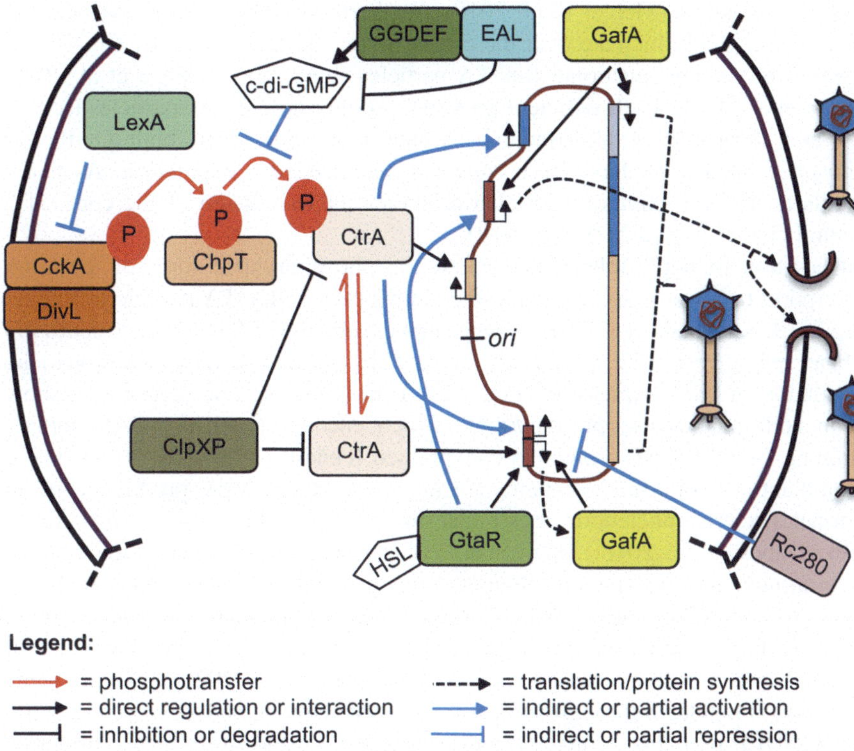

Fig. 2.2 Summary model for the integration of *R. capsulatus* GTA expression, two-component system phosphorelay regulation and the QS/DNA damage repair pathway. A donor cell is depicted with inner and outer membranes shown in purple and black. A signaling cascade, starting from the left, leads to activation of RcGTA gene transcription, the production/assembly of functional GTA particles, and, ultimately, membrane lysis depicted to the right. The *R. capsulatus* genome is indicated in maroon, with gene/ORF start positions indicated with hooked arrows, using the same color conventions as in Fig. 2.1. Lines that represent different types of mechanistic interactions are color-coded as indicated by the legend

the response-regulator CtrA via a Chp-T-mediated phosphotransfer relay (Fig. 2.2) (Iniesta et al. 2006; Jacobs et al. 2003; Laub et al. 2002). This relay, conserved for a majority of alpha-proteobacteria, is additionally dependent on the membrane-associated DivL pseudo-kinase (Childers and Shapiro 2014; Westbye et al. 2018). Depending on a variety of intrinsic factors, the "master regulator," a transcription factor called CtrA, is responsible for upregulating and downregulating specific sets of genes, as well as modulating access of the chromosomal replication initiation protein DnaA to the origin of replication (*ori*) (Gorbatyuk and Marczynski 2005; Laub et al. 2002; Ozaki 2019). Despite sharing 71% sequence identity with *Caulobacter* CtrA, *R. capsulatus* CtrA function may have diverged, as it is no longer essential for cell cycle progression in these alpha-proteobacteria (Lang and Beatty 2000). Nonetheless, CtrA is involved in the pleiotropic regulation of ~6% of

R. capsulatus genes (Mercer et al. 2010). Furthermore, levels of the cytoplasmic CtrA protein, which fluctuate over the cell cycle due to degradation by ClpXP, are integral to producing functional RcGTA particles (Fig. 2.2) (Westbye et al. 2018; Westbye et al. 2013). Cells deficient for ClpX produce and release particles lacking DNA and incapable of transduction. At the molecular level, both GtaR and unphosphorylated CtrA have been shown to activate *gafA* expression. As GafA upregulates its own expression, it also promotes the expression of the core GTA structural cluster (Fig. 2.2) (Fogg 2019).

In addition to unphosphorylated CtrA, the phosphorylated form of the CtrA transcription factor is also essential to the completion of RcGTA particle formation (Fogg 2019; Kuchinski et al. 2016). Genes that encode RcGTA tail fibers (*rcc00171*) and head spikes (*rcc01079–01080*) are regulated by phospho-CtrA, in addition to the upregulation of *gafA* expression (Fig. 2.2). Transcriptomic microarray data show that an additional transcript, *rcc00645*, which encodes a sensory PAS domain containing GGDEF/EAL protein, is coexpressed (Peña-Castillo et al. 2014). These predominantly cytoplasmic enzymes are known to regulate the production of an important bacterial nucleotide messenger bis-(3′–5′) cyclic dimeric GMP (c-*di*-GMP) (Dubey et al. 2016; Jenal et al. 2017; Lori et al. 2015). In the case of *R. capsulatus*, four such GGEDF-containing proteins, encoded additionally by *rcc00620, rcc02629,* and *rcc02857*, have been demonstrated to inhibit both RcGTA particle production and flagellar motility, whereas a lower abundance of c-*di*-GMP favors both processes (Fig. 2.2) (Pallegar et al. 2020). This dynamic mechanism is regulated through competition between the diguanylate cyclase activity of GGDEF domains, which synthesize c-*di*-GMP, and the phosphodiesterase activity of EAL domains, which inactivate c-*di*-GMP (Simm et al. 2004).

EAL domain proteins linearize the c-*di*-GMP molecule to generate 5′-phosphoguanylyl-(3′–5′)-guanosine, or pGpG. Subsequent cleavage of pGpG into two molecules of GMP, by an oligoribonuclease, helps to maintain homeostasis of c-*di*-GMP within the cytoplasm (Orr et al. 2015; Ross et al. 1990). As levels of c-*di*-GMP are well known in *C. crescentus* for their impact on cell cycle progression, via feedback onto CckA signaling, regulation of GGDEF/EAL protein activity is central to CtrA phosphorylation status (Fig. 2.2) (Jenal et al. 2017; Römling and Galperin 2017). Mutation of CtrA results in decreased expression of GGDEF/EAL domain-containing proteins (Mercer et al. 2010), and c-*di*-GMP levels are inversely correlated with gene transfer activity (Pallegar et al. 2020). Thus, intriguingly, this class of enzymes ties together both cell cycle and GTA fields of investigation (Fig. 2.2) (Dubey et al. 2016; Lori et al. 2015; Mann et al. 2016).

The mechanism by which CtrA serves as both transcriptional activator and repressor of RcGTA gene expression has also recently been elucidated. Opposing developmental phenotypes and transcriptional activities induced by CtrA occur through a conserved determinant in the DNA-binding domain, sensitive to an intrinsic signal produced during the stationary phase (Delaby et al. 2019). This domain serves to reprogram CtrA-promoter preference during growth transitions under the "stringent response"-based signal quanosine tetra/penta-phosphate, (p)ppGpp. The fact that RcGTA expression requires a confluence of both phosphor-CtrA forms, as well as GafA, speaks to the extent of integration of this HGT-mechanism with the stationary phase transition. Consistent with this view, it was shown that HSLs promote RcGTA particle production by inactivating the transcriptional repressor LexA, which normally suppresses *cckA* gene expression (Fig. 2.2) (Kuchinski et al. 2016; Schaefer et al. 2002). LexA is considered central to the so-called SOS DNA damage response; however, in this case, LexA autoproteolytic cleavage occurs under oxidative stress in the absence of DNA damage (Kuchinski et al. 2016). While the specific trigger for the membrane-associated histidine kinase CckA is not yet resolved, GTA particle production is responsive to phosphate concentration (Leung et al. 2010; Westbye et al. 2013), salinity (McDaniel et al. 2012), and nutrient depletion (Westbye et al. 2017). The role of a "partner switching" pathway, involving the conserved alpha-proteobacterial rho-regulation proteins RbaV and RbaW, which are believed to recruit RNA polymerase, has also been implicated (Mercer and Lang 2014). Ultimately, however, the release of mature GTA particles is only observed upon entry of bacterial cultures into the stationary phase (Fogg et al. 2012; Mercer and Lang 2014).

After RcGTA particles are formed, their release occurs upon accompanying cell lysis. This process requires a cluster of RcGTA holin and endolysin genes, *rcc00555–00556* (Chen et al. 2009; Hynes et al. 2012; Lang et al. 2012; Westbye et al. 2013). Holins are multipass transmembrane proteins that can cluster together to form a pore in the bacterial inner membrane (Saier and Reddy 2015; Wang et al. 2000; Young 2002). They are thought to allow endolysin enzymes to pass from the cytosol into the periplasm, where they access the peptidoglycan-based cell wall. Destabilization of the cell wall structure by endolysin leads to the rupture of the bacterial plasma membrane, which expels RcGTA particles into the extracellular environment. In addition to driving expression of the core RcGTA structural cluster genes, GafA, in concert with GtaR, directs expression of the holin/endolysin gene cluster (Fig. 2.2) (Sherlock et al. 2019). Cell culture experiments have indicated that between 0.1 and 3% of cells within a given *R. capsulatus* population express the RcGTA gene cluster and undergo lysis, while over 30% of stationary-phase cells lyse in mutant overproducer strains (Fogg et al. 2012; Hynes et al. 2012). Recently, a mutation in the *rcc00280* gene was identified as responsible for this difference (Ding et al. 2019). The Rc280 protein is believed to encode an extremely negatively charged, Ca^{2+}-binding, extracellular protein secreted through a Type I secretion system-based process. While the mechanism is unknown, Rc280 appears to signal a pathway repressing expression of the *gafA* gene, dampening GafA feedback, and thereby inhibiting the switch to particle production (Fig. 2.2) (Ding et al. 2019). Thus, in summary, the literature indicates that the initial production of RcGTA particles, assembly of mature particles, and release are highly coordinated in a step-wise and stochastic fashion (Lang and Beatty 2000; Mercer et al. 2012; Westbye et al. 2017).

The primary RcGTA structural gene cluster is ~14 kb, well in excess of the ~4.5 kb carrying capacity of RcGTA particles (Lang and Beatty 2000; Yen et al. 1979). Empirical observation further shows significant underrepresentation of the RcGTA-encoded structural gene cluster in the particles that are produced. This is attributed to the hyperexpression of the structural cluster genes in the subset of cells, which blocks access by DNA-processing "terminase" enzymes (Hynes et al. 2012). The RcGTA terminase has homology to nonsequence-specific T4-like phages, such as P22, Mu, and T4 (Hynes et al. 2012; Yen et al. 1979). In such cases, the terminase makes an endonucleolytic cut in host DNA, joining it to the assembling capsid through interaction with the portal protein and packaging linear dsDNA genome into the capsid to form a highly ordered, condensed structure. In a process termed "headful packaging," a second cut terminates DNA packaging and allows dissociation from the packaged structure, then processively initiates linkage to another prohead (Alam et al. 2008; Black and Rao 2012; Rao and Black 2010). While most siphoviruses produce blunt end phage dsDNA (Casjens and Gilcrease 2009), it is interesting to note that RcGTA particles contain 3′ overhangs, thus DNA inside particle-producing cells may form "concatamers" for packaging inside particles (Hynes et al. 2012). Data are consistent with a model in which the RcGTA-*g2* terminase cleaves host nucleic acid with little-to-no sequence specificity and proceeds in a phage-like "headful packaging" mechanism (Casjens and Weigele 2005; Feiss and Rao 2012; Hynes et al. 2012).

2.2.2 Recipient Cell Regulation and Natural Homologous Recombination

Similar to active bacteriophages, which target a specific range of recipient cells, RcGTA particles appear to also have limited species specificity (Wall et al. 1975). Many surface-exposed structures, such as lipopolysaccharides, outer membrane proteins, and extracellular polysaccharide (EPS), have been shown to function as phage receptors (Choy et al. 1975; Rakhuba et al. 2010). As EPS surrounds most *R. capsulatus* strains and phage-resistant strains lack a polysaccharide capsule, it is thought that EPS functions in RcGTA binding and adsorption (Brimacombe et al. 2013; Flammann and Weckesser 1984; Weaver et al. 1975). At the level of the RcGTA particle, head spikes (*rcc01079–rcc01080*) are required for polysaccharide binding, and tail fibers (*rcc00171*) are expected to mediate direct attachment to the cell surface, as is typical for siphoviruses (Hynes et al. 2016; Westbye et al. 2017).

Transfer of DNA into recipient cells by RcGTA particles is thought to occur in a manner similar to siphovirus transduction (Fig. 2.3) (Brimacombe et al. 2015; Westbye et al. 2017). Infection by noncontractile, tailed phages typically involves a tape measure protein (TMP) structure that initially nests within the tail fiber. After a target cell is bound, the TMP acts as a telescoping extension of the tail into the recipient cell, traversing the outer membrane, periplasm, and cytoplasmic membrane (Davidson et al. 2012). In RcGTA particles, however, the TMP is much shorter than typical noncontractile tailed phage (only 219 amino acids vs. 600 amino acids), suggesting that it does not access the cytosol. Thus, donor DNA passage through the periplasm is thought to be facilitated by the peptidoglycan layer degrading enzyme

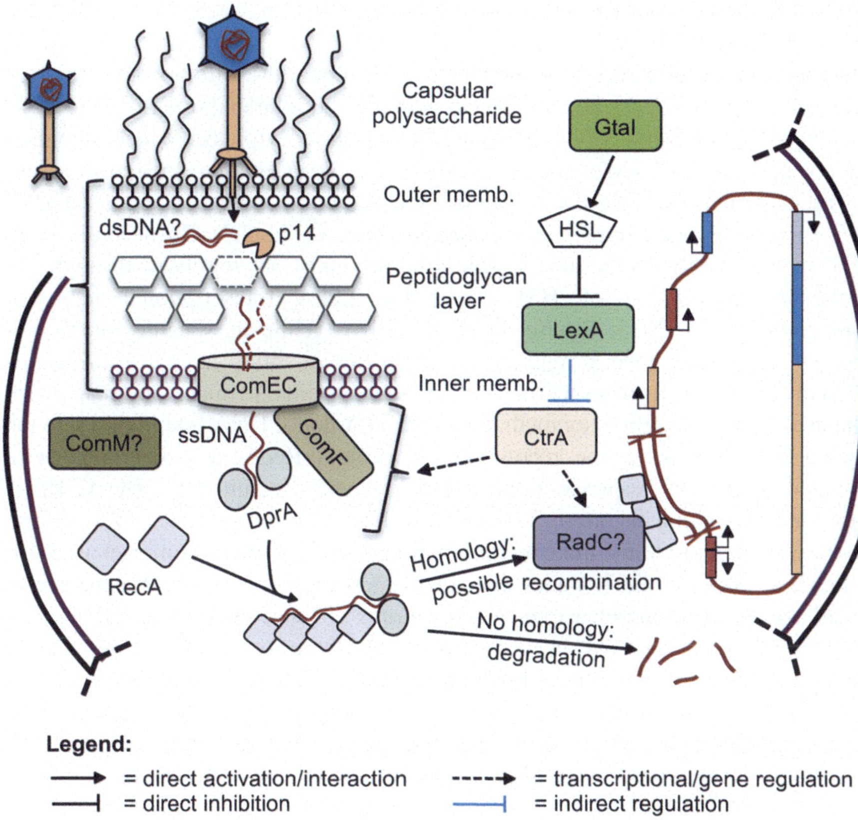

Fig. 2.3 Summary model for the integration of *R. capsulatus* recipient cell GTA adsorption, two-component phosphorelay, and the QS/natural transformation pathway. Recipient cell depicts uptake of functional RcGTA particles, magnified at upper left, through the inner/outer membranes and periplasm, and expressing natural transformation competence proteins associated with efficient GTA uptake and potential recombination into a new host. The *R. capsulatus* genome is shown as in Fig. 2.1, with gene/ORF start positions indicated by hooked arrows. Interactions are listed in the legend, color-coded to indicate the interaction type supported by experimental data. A question mark indicates an unknown role for that protein. A portion of this figure was adapted from Brimacombe et al. 2015, Fig. 2.5, with permission of the authors

p14, encoded by ORF *g14* of the RcGTA structural cluster (Figs. 2.1 and 2.3) (Brimacombe et al. 2015; Fogg et al. 2012). RcGTA-delivered dsDNA further requires the inner membrane transporters ComEC and ComF for recipient cell infectivity (Draskovic and Dubnau 2005; Johnston et al. 2014; Mell and Redfield 2014). Passage via these transport proteins results in single-stranded donor DNA (ssDNA) transfer into the cytoplasm (Fig. 2.3). Therefore, although the precise mechanism is unclear, import of donor DNA from RcGTA particles is thought to involve natural transformation mechanisms, distinct from temperate phages that inject dsDNA directly into the cytoplasm (Brimacombe et al. 2015; Fogg et al. 2014; Groth and Calos 2004).

The integration of RcGTA donor DNA into a new host chromosome involves proteins that are also classically associated with the natural transformation (Fig. 2.3). Incoming ssDNA is initially bound by the DNA-specific, cytoplasmic recombination mediator protein DprA (Johnston et al. 2014). The *dprA* gene, expressed in essentially all cells in a population, is coregulated along with competence genes *comEC, comF,* and *comM* by the transcriptional regulator CtrA (Fig. 2.3) (Brimacombe et al. 2015). In turn, CtrA levels and phosphorylation status are regulated through QS signals, the production of HSLs, and the LexA transcriptional repressor (Brimacombe et al. 2013; Kuchinski et al. 2016). While its specific function is unknown, the *comM* gene is required for recipient capability in *R. capsulatus*, as well as maximal transformation efficiency in other systems (Brimacombe et al. 2015; Sinha et al. 2012). Another key mediator of RcGTA donor DNA integration into the recipient cell genome is the recombination protein RecA, a protein otherwise associated with SOS response-induced DNA repair. Recruited by DprA, RecA forms filaments along incoming ssDNA and, based on the presence of sufficient homology, is thought to facilitate recombination (Fig. 2.3) (Brimacombe et al. 2014; Mortier-Barrière et al. 2007; Yadav et al. 2014). Expression of the *radC* gene is also upregulated during competence in *R. capsulatus* (Brimacombe et al. 2014), and homologs of this factor have been shown to contribute to recombination in other systems (Attaiech et al. 2008; Lloyd and Rudolph 2016). The observation that efficiency of transduction by RcGTAs is much higher than that of generalized transducing phage is attributed to the synchrony of donor and recipient cell populations, through precise coordination of particle release and uptake.

2.3 Alpha-Proteobacterial Homologs of the *R. capsulatus* GTA System

Awareness that RcGTAs facilitate HGT has opened questions about the prevalence of this mechanism across organismal systems. Two recent, comprehensive bioinformatics studies have revealed that close homologs of RcGTA genes are carried by many alpha-proteobacterial species, likely as a result of multiple integration events (Kogay et al. 2019; Shakya et al. 2017). The primary structural cluster of RcGTA genes appears to have been introduced to the alpha-proteobacteria after the divergence of *Rhodospirillales* (Fig. 2.4), estimated to have occurred between 777 and 1710 million years ago (Kogay et al. 2019; Luo et al. 2013; Shakya et al. 2017). This cluster of RcGTA genes, referred to as the “Large Cluster” (LC), has markedly higher GC content than host genomes on average. The LC is also predominantly flanked by the same genes, not only across alpha-proteobacteria, but also for other systems that have horizontally acquired these LCs from alpha-proteobacteria. This

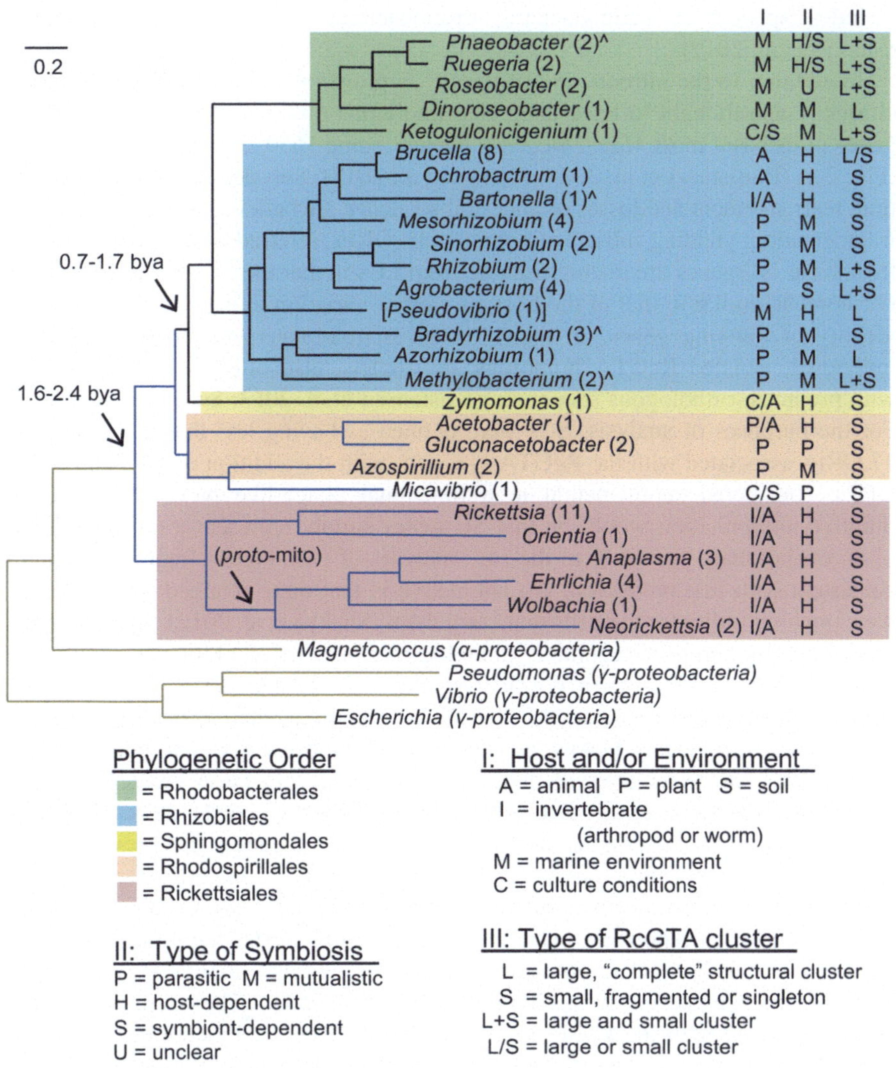

Fig. 2.4 Distribution of RcGTA gene clusters within alpha-proteobacterial endosymbiont species. Only species reported to form symbiotic associations in the published literature are presented (see text for description), along with associated outgroups (unshaded). Genus designations are listed, with the number of species containing either RcGTA large clusters, small clusters, or both, given in parentheses. ^ indicates a genus that also contains BaGTA homologs (see Fig. 2.5 and text for further information). Host, environment, and symbiosis type are indicated in the columns on the right side. Arrows indicate the predicted evolutionary introductions of progenitor RcGTA-like phage, in billions of years ago (bya), as well as an approximation for the origin of the protomitochondrial ancestor (*proto*-mito). Brackets indicate cases in which the NCBI classification is not reflected by this display. Scale bar indicates substitutions per site. Phylogenetic relationships and distances are based upon the maximum likelihood phylogeny of 255 α-proteobacteria species shown in Shakya et al. 2017, Supplemental Fig. S3

includes species of actinobacteria, cyanobacteria, and gamma-proteobacteria (Shakya et al. 2017).

In addition to the introduction of the LC, a progenitor to the RcGTA structural cluster is also thought to have integrated into alpha-proteobacteria after the divergence of its basal-most *Magnetococcus* lineage, some 1650–2390 million years ago (Fig. 2.4) (Battistuzzi et al. 2004; Shakya et al. 2017). Subsequent to this introduction, rearrangement and losses of ancestral sequence are believed to have resulted in fragmentation, yielding subsets of homologous ORFs, referred to as "small clusters" (SCs). The SC genes are distinguishable from LCs in that the SCs generally exhibit GC content similar to that of the host, as well as variation in genomic position and identity of flanking genes. In many instances, rearrangements have resulted in individual, isolated RcGTA homologs, or singlets, also often closely associated with prophage-orthologous sequences (Kristensen et al. 2013; Shakya et al. 2017). For the purposes of analysis, SCs were defined as having less than 9 of the total 17 ORFs associated with the RcGTA (*g1–g15*, with the addition of *g3.5* and *g10.1* ORFs, considered supplemental to a functional phage-like particle). While an intuitive interpretation would be that SC genes simply represent a decaying form of an evolutionary precursor to the LC, analyses of pseudogene frequency are not consistent with that prediction. An alternative is that these retained SC homologs have ongoing utility for the cells that carry them (Shakya et al. 2017). An increasing loss of the structural and regulatory homologs has been reported for orders removed from *Rhodobacterales,* when considering the GTA-related genes outside of the "head-tail" structural LC, namely, the ORFs encoding putative head spikes, tail fibers, holin/lysin release factors, as well as the GafA transcriptional regulator. It is possible that the faster rates of evolution associated with these genes may have limited their detection across diverse bacterial systems thus far (Hynes et al. 2016).

As alpha-proteobacteria include many symbiotic lineages, the prevalence of RcGTA homologs within this class implies that many symbiotic bacteria must necessarily also retain a subset of these homologs. In carrying out this review, all named strains that reportedly carry RcGTA structural cluster homologs were collated (Biers et al. 2008; García-Aljaro et al. 2017; Hynes et al. 2016; Lang and Beatty 2007; Paul 2008; Shakya et al. 2017; Tamarit et al. 2018). The strains were then cross-referenced with the Kyoto Encyclopedia of Genes and Genomes (Kanehisa et al. 2017; Kanehisa and Goto 2000; Ogata et al. 1999) to identify those confirmed as completely sequenced, with an assembled and annotated genome. Following this, hundreds of research papers and field databases were mined for evidence of the symbiotic properties associated with each strain. We considered that symbionts may associate with either the exterior or interior of host cells, within a variety of environmental, industrial, and organismal settings. As symbiosis represents a spectrum of interactions, we further considered that these organisms potentially act as mutualists, commensals, or parasites, with these outcomes often directed by the symbiont/host combination (Fig. 2.4) (Ferri et al. 2011; Gillespie et al. 2014; Moran et al. 2008; Wernegreen 2012). Results were consolidated at the species level, with strain examples identified for each, to fit within the scope of this review.

In summary, this analysis indicated that 66 out of 117 alpha-proteobacterial species reported to carry RcGTA homologs ultimately fall under the classification of "symbiont" (Fig. 2.4). All *Rickettsiales* species reported to date as carrying RcGTA homologs ($n = 22$), and 28 of 47 *Rhizobiales* species, were identified as symbionts. This was also the case for 1 of 8 *Sphingomonadales* species, 9 of 25 *Rhodobacterales* species, 5 of 12 *Rhodospirillales* species, and 1 of 3 unclassified species. The full range of mutualism to parasitism was generally represented across lineages (Fig. 2.4). No symbiotic species were identified for lineages of *Parvularculales* ($n = 1$) or *Caulobacterales* ($n = 6$) predicted to retain RcGTA homologs. Here, we highlight selected organisms from each order of symbiotic alpha-proteobacteria from this analysis and provide greater detail on the RcGTA homologs associated with each taxonomic class.

2.3.1 Order Rhodobacterales

In parallel to the model system *R. capsulatus*, the majority of *Rhodobacterales* are thought to carry RcGTA structural genes (Shakya et al. 2017). Pelagic genera of *Rhodobacterales* have been commonly discussed in the literature (Elifantz et al. 2013; Ghai et al. 2012), with many species expected to act as symbionts, particularly with respect to the *Roseobacter* clade (Crenn et al. 2018; Geng and Belas 2010; Riclea et al. 2012). Of the RcGTA-carrying *Rhodobacterales* that have been specifically identified as symbionts to date, many have been found in association with algal species (Crenn et al. 2018). These include *Ruegeria pomeroyi* DSS-3, *Ruegeria* sp. TM1040, *Phaeobacter gallaeciensis* DSM 26640, *Phaeobacter inhibens* DSM 17395, *Roseobacter litoralis* Och 149, and *Roseobacter denitrificans* Och 114. On a case-by-case basis, these species have been found to act as algal mutualists (Buchan et al. 2005; Durham et al. 2017; Durham et al. 2015; Geng and Belas 2010; Kalhoefer et al. 2011; Miller and Belas 2004; Seyedsayamdost et al. 2011; Swingley et al. 2007) or switch from mutualist to algicidal in response to algal cues (Bramucci et al. 2018; Riclea et al. 2012; Seyedsayamdost et al. 2011; Wang and Seyedsayamdost 2017). *Pseudovibrio* sp. FO-BEG1 has been found in association with black-band disease samples from coral (Bondarev et al. 2013) and *Ketogulonicigenium vulgare* WSH-001 takes part in metabolic symbioses with *Bacillus megaterium* in the context of industrial fermenters (Jia et al. 2015; Zhou et al. 2013). As both algal blooms and biofilms are ideal environments for communication via diffusible molecules, QS signals are believed to play an intricate role in the coordination of *Rhodobacterales* populations (Cude and Buchan 2013).

The majority of *Rhodobacterales* are predicted to carry sets of RcGTA LC genes, accompanied by SCs in some cases (Fig. 2.4) (Tables 2.1 and 2.2). This is exemplified by *Dinoroseobacter shibae*, the most well-studied alpha-proteobacterial GTA system outside of *R. capsulatus*, which carries a homologous cluster termed the DsGTA (Tomasch et al. 2018; Wang et al. 2014a). Believed to play a substantial role in global biogeochemical sulfur cycling, *D. shibae* was originally isolated in association with the dinoflagellate alga *Prorocentrum lima* (Biebl et al. 2005; Wagner-Döbler et al. 2010). It also acts as a mutualist of toxic, red-tide forming *P. minimum* (Wang et al. 2014a). The DsGTA structural operon carried by this species is a well-

Table 2.1 RcGTA "Large Cluster" homologs in symbiotic alpha-proteobacteria

Organism ORFs	*g1*	*g2*	*g3*	*g3'*	*g4*	*g5*	*g6*	*g7*	*g8*	*g9*	*g10*	*g10'*	*g11*	*g12*	*g13*	*g14*	*g15*
Order *Rhodobacterales*																	
Phaeobacter gallaeciensis DSM-26640	+	+	+	+	+	+	+	+	+	+	+	+	+	+	+	+	+
Phaeobacter inhibens DSM-17395	+	+	+	+	+	+	+	+	+	+	+	+	+	+	+	+	+
Ruegeria sp. TM1040	+	+	+	+	+	+	+	+	+	+	+	+	+	+	+	+	+
Ruegeria pomeroyi DSS3	-	+	+	+	+	+	+	+	+	+	+	+	(+)	+	+	+	+
Roseobacter denitrificans OCh 114	(+)	+	+	+	+	+	+	+	+	+	+	+	+	+	+	+	+
Roseobacter litoralis Och 149	+	(+)	+	+	+	+	+	+	+	+	+	+	+	+	+	+	+
Dinoroseobacter shibae DFL12	(+)	+	+	+	+	+	+	+	+	+	+	+	+	+	+	+	+
Ketogulonicigenium vulgare WSH001^	+	(+)	+	+	+	+	+	+	+	+	+	+	+	+	+	+	+
Pseudovibrio sp. FO-BEG1	-	+	+	-	+	+	+	+	+	+	+	+	+	+	+	+	+
Order *Rhizobiales*																	
Brucella suis 1330^	-	+	+	-	+	+	+	+	+	+	+	-	+	+	+	+	(+)
Brucella canis HSK A52141^	-	+	+	-	+	+	+	+	+	+	+	-	+	+	+	+	+
Brucella ovis ATCC 25840	-	+	+	-	+	(+)	+	+	+	(+)	+	(+)	+	(+)	(+)	+	+
Brucella microti CCM 4915	-	+	+	-	+	+	+	+	+	+	+	+	+	+	+	+	+
Brucella pinnipedialis B294	-	+	+	-	+	+	+	+	+	+	+	(+)	+	+	+	+	+
Brucella melitensis bv. 1 str 16M^	-	+	+	-	-	-	-	-	-	+	+	-	+	+	+	+	+
Brucella abortus 2308^	-	+	+	-	+	+	+	+	+	+	+	(+)	+	+	(+)	+	(+)
Rhizobium sp. IRBG74	-	+	+	-	-	-	-	-	-	+	+	-	+	+	+	+	+
Agrobacterium fabrum str. C58	-	+	+	-	+	+	-	-	+	(+)	(+)	(+)	+	+	+	+	+
Agrobacterium vitis S4 & sp. H13-1	-	+	+	-	+	+	+	+	+	+	+	+	+	+	+	+	+
Azorhizobium caulinodans ORS-571	-	+	+	-	(+)	+	+	+	+	+	+	+	+	+	+	+	+
Methylobacterium nodulans ORS-2060*	-	+	4+	-	-	4+	+	4+	5+	5+	5+	-	+	5+	5+	5+	5+

+ indicates presence of homolog; - indicates absence; (+) indicates a putatively pseudogenized ORF

* *M nodulans* RS_2060 contains a total of 5 LCs; 4+ and 5+ indicate the total number of homologs of each ORF within 5 LCs total

^ some variation between strains exists, see Shakya et al., 2017; Supplemental Figure S6

Table 2.2 Alpha-proteobacterial symbiont RcGTA "small cluster" homologs, part 1

Organism type ORFs	*g1*	*g2*	*g3*	*g4*	*g5*	*g6*	*g7*	*g8*	*g9*	*g10*	*g10'*	*g11*	*g12*	*g13*	*g14*	*g15*
Order *Rhodobacterales*																
Phaeobacter gallaeciensis DSM-26640	-	-	A,B*	-	A,B*	A	A	-	-	-	-	-	-	-	-	-
Phaeobacter inhibens DSM-17395	-	-	A,B	A,B	A,C	A	A	-	-	-	-	-	-	-	-	-
Ruegeria sp TM1040	-	-	A,C	A,B	A,B,C	A,C	A	-	-	-	-	-	-	-	-	-
Ruegeria pomeroyi DSS3	-	-	-	-	-	-	-	-	-	-	-	-	-	-	A	-
Roseobacter litoralis Och-149	-	-	A	-	-	-	-	-	-	-	-	-	-	-	-	-
Ketogulonicigenium vulgare WSH001^	-	-	A,B	(+)	-	A	A	A	A,B	A	-	-	-	-	-	-
Order *Rhizobiales*																
Brucella ceti TE28753-12^	-	-	A	-	-	-	-	-	A	A	(+)	-	(+)	A	A	A
Ochrobacterium anthropi ATCC 49188	-	-	A,B	A	A	-	A	A	A	A	-	-	-	-	-	-
Bartonella grahamii as4aup^	-	-	A,B	-	-	-	-	-	-	-	-	-	-	-	-	-
Mesorhizobium cireri bv. biserrulae str. WSM1271	-	-	-	-	-	-	-	A	-	-	-	-	-	-	-	-
Mesorhizobium loti str. NZP2037	-	-	A	-	A	-	-	-	-	-	-	-	-	-	-	-
Mesorhizobium australicum WSM2073	-	-	-	-	-	-	-	A	-	-	-	-	-	-	-	-
Mesorhizobium opportunistum str. WSM2075	-	-	A	-	-	-	-	B	-	-	-	-	-	-	-	-
Sinorhizobium meliloti AK83^	-	A	B,C,D	B	B	B	B	-	B,C,D	B,C,D	-	-	-	-	-	-
Sinorhizobium medicae WSM419	-	A	B,C	B	B	B	B	B	B,C	B,C	-	-	-	-	-	-
Rhizobium tropici str. CIAT 899	-	A	B	B	B	B	B	B	B	B	-	-	-	-	-	-
Rhizobium sp. IRBG74	-	-	A	-	A	-	A	A	A	-	-	-	-	-	-	-
Agrobacterium sp. H13-3	-	-	A,B	B	A,B	-	A,B	A,B	A	A,B	-	-	-	-	-	-
Agrobacterium radiobacter K84	-	A	B,C	B	B,C	B	B	B	B	-	-	-	-	-	-	-
Agrobacterium vitis S4	-	-	A,B,C*	A,B	A,B,C*	B,C*	A,C*	B	B,C*	C*	-	-	-	-	-	-
Bradyrhizobium sp. BTAi1	-	-	A	-	-	-	-	-	-	-	-	-	-	-	-	-
Bradyrhizobium sp. ORS 278	-	-	A	-	A	A	-	-	-	-	-	-	-	-	-	-
Bradyrhizobium japonicum USDA-6	-	-	-	A	-	-	-	-	-	-	-	-	-	-	-	-
Methylobacterium nodulans ORS-2060	-	A,B,C	D,E,F	E,F	E,F	E,F	-	-	-	-	-	-	-	-	-	-
Methylobacterium sp. 446	-	A	-	B	-	-	-	A	A	A	-	-	A	A	A	A

Letters A thru F indicate predicted SC ORFs or singlets of RcGTA *g1-g15* , where ORFs with the same letter are in the same cluster
Linear order of ORFs in the same cluster are not necessarily maintianed in accord with that of LC arrays
(+) indicates a putatively pseudogenized ORF
* indicates ORF(s) are located on an extrachromosomal plasmid
^ indicates some variation exists between strains, see Shakya et al., 2017; Supplementary Figure S5

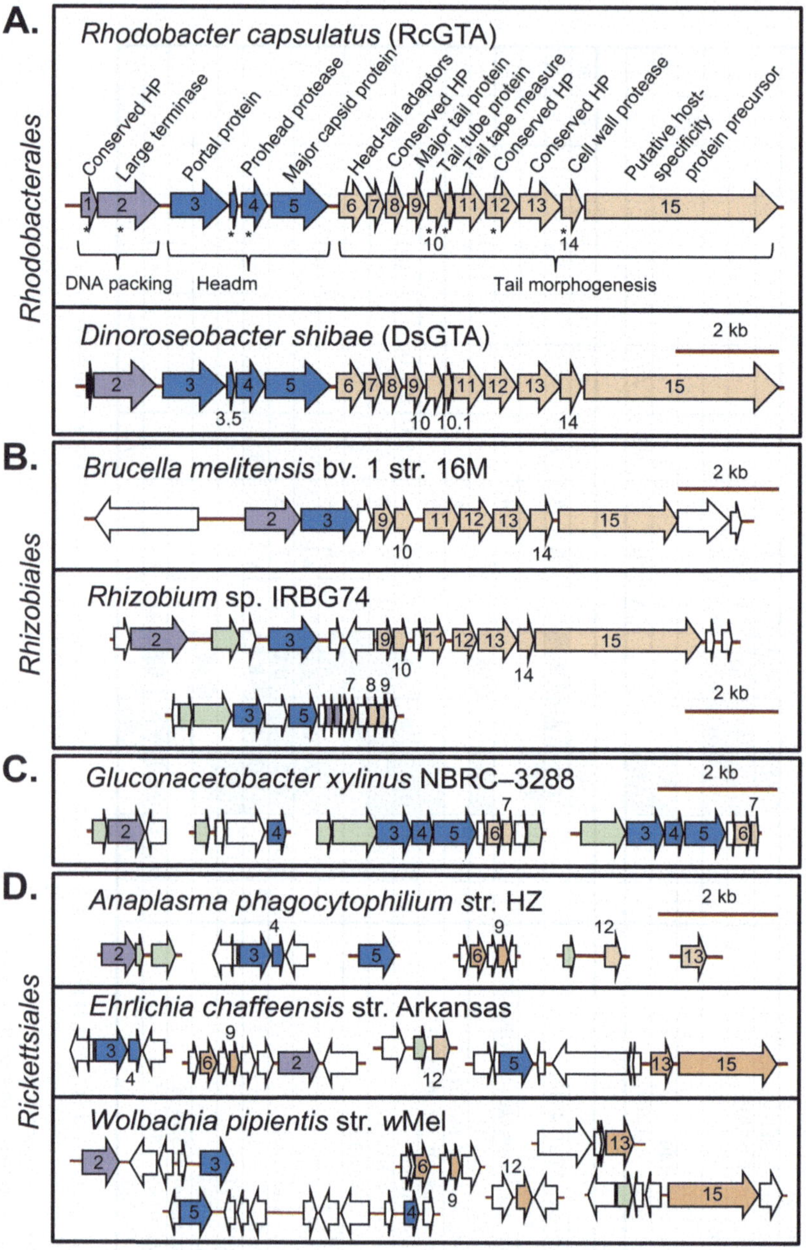

Fig. 2.5 RcGTA structural gene cluster variations within symbiotic alpha-proteobacterial orders. Arrows represent the predicted RcGTA ORFs, with coloration used to indicate similar predicted particle functions. Green coloration indicates ORF with homology to POG database, of presumed viral origin. Numbers within or below arrows indicate RcGTA homologs where, for simplicity, "g" is left out of annotations (i.e., *g1*, *g2* etc.). (**a**) Large cluster gene representations of two

preserved genetic module within the *Roseobacter* clade (Fig. 2.5a) (Table 2.1) (Luo and Moran 2014; Newton et al. 2010; Shakya et al. 2017).

Although the production of HGT-competent particles has not been described for many alpha-proteobacteria in general, *D. shibae* has been shown to produce particles capable of horizontal transfer. Analogous to RcGTAs, DsGTA particles contain ~4.2 kb DNA fragments with a head diameter of 33 nm and a tail length of 48 nm (Tomasch et al. 2018). DsGTA particle production is regulated by the QS system through Lux-type autoinducer synthesis, which coordinates expression of other highly adaptive traits such as flagella, expression of the Type IV secretion system (T4SS), and morphological heterogeneity (Patzelt et al. 2013; Wang et al. 2015; Wang et al. 2014b). The central CtrA transcription factor and c-*di*-GMP levels play a role in differentiation into particle producing and nonproducing cells (Koppenhöfer et al. 2019). DsGTA particles are also thought to proceed via "headful-type" packaging (Patzelt et al. 2013; Tomasch et al. 2018). In contrast to *R. capsulatus*, the packaging of the *D. shibae* genome into DsGTA particles does not appear random. Multiple chromosomal regions are over-represented in packaged particles, with sequencing indicating that peak coverage is initiated at seven sites in the 4.4 Mb genome (Patzelt et al. 2013; Tomasch et al. 2018). In addition, it is thought that GC content, DNA modification, and chromatin structure influence GTA packing at initiation sites (Tomasch et al. 2018; Wagner-Döbler et al. 2010). As *D. shibae* also harbors *rcc00555–rcc00556* homologs, it has been suggested that the release of DsGTA particles proceeds through the activity of endolysin/holin enzymes (Lang et al. 2017). Parallel to *R. capsulatus*, induction of the SOS response is implicated in recipient cell capability and chromosomal integration (Koppenhöfer et al. 2019).

Eight other symbiotic *Rhodobacterales* species have also been predicted to carry LC homologs (Fig. 2.4). Considered to be vertically inherited in *Rhodobacterales* (Hynes et al. 2016) and predominantly capable of species-specific HGT (Weaver et al. 1975), this group of organisms carries the most complete RcGTA sets for all symbionts examined to date (Table 2.1). DNA head packing, head morphogenesis, and tail morphogenesis genes are all represented, in some cases entirely, such as for *P. gallaeciensis* DSM-26640, *P. inhibens* DSM-17395, and *Ruegeria* sp. TM1040. Empirical data help to further frame the significance of these LC ORF predictions, as GTA gene expression and particle release have also been demonstrated for *Ruegeria pomeroyi* DSS-3 (Biers et al. 2008). As neither *R. pomeroyi* DSS-3 nor *D. shibae* DFL-12 carry detectable homologs of the conserved hypothetical protein *g1*

Fig. 2.5 (continued) Rhodobacterales species. *Rhodobacter capsulatus* GTA structural cluster ORFs are shown alongside homologous loci of *Dinoroseobacter shibae*. Asterisks indicate products not detected in association with purified particles. HP indicates coding for a "hypothetical protein" (Chen et al. 2009; Hynes et al. 2016). (**b**) Large and small cluster gene representations of *Rhizobiales* species *Brucella melitensis* bv. 1 str. 16 M and *Rhizobium* sp. IRBG74. (**c**) Small/fragmented cluster gene representations of the *Rhodospirillales* species *Gluconacetobacter xylinus* NBRC 3288. (**d**) Small/fragmented cluster gene representatives of the *Rickettsiales* species *Anaplasma phagocytophilum* str. HZ, *Ehrlichia chaffeensis* str. Arkansas and the *w*Mel strain of *Wolbachia pipientis*. All homologous ORF cluster representations are to scale by length, indicated in kilobases (kb), based upon Shakya et al. 2017, Supplemental Fig. S5 and S6

(Table 2.1), otherwise essential for RcGTA production in *R. capsulatus* (Hynes et al. 2016), this finding opens the possibility that "incomplete" LCs may encode functional GTAs for other systems.

In addition to the already extensive complement of LC genes, *Rhodobacterales* symbionts have been predicted to carry SC genes as well (Fig. 2.4) (Table 2.2). Some of the *Phaeobacter* and *Ruegeria* symbionts are predicted to carry capsid and tail-related genes ranging from *g3 to g7*, at times in multiple copies. In the case of *Phaeobacter gallaeciensis* DSM – 26,640, one such cluster that contains both the predicted portal protein and the major capsid protein was found residing on an extrachromosomal plasmid. Minimal complements of SC genes are predicted in other *Ruegeria* and *Roseobacter* species, such as a single copy of the cell wall protease *g14* in *R. pomeroyi* DSS-3. The SC set for *K. vulgare* WSH-001 differs from the marine symbionts described above, with predicted inclusion of more tail-related ORFs (Table 2.2). Given the robust presence of LCs in these organisms, the data suggest that SC homologs are not being retained as strategic compensation for LC gene loss. *Pseudovibrio* sp. FO-BEG1, which in this analysis happens to cluster phylogenetically with the *Rhizobiales*, contains a single LC locus and no detectable SC genes at all (Fig. 2.4) (Table 2.1).

2.3.2 *Order* Rhizobiales

Rhizobiales, an order representing an array of microbes that act as pathogens of mammals and plants (*Brucella*, *Ochrobactrum,* and *Bartonella* species), as well as a range of soil and plant rhizome-associated nitrogen-fixing mutualists, are predicted to contain a broad range of RcGTA homologs. *Brucella* species infect erythrocytes in a wide range of mammals, including pinnipeds, rodents, ruminants, canines, swine, and horses, with humans regarded mainly as incidental hosts (de Figueiredo et al. 2015). Communicated predominantly by contact with infected animal tissues and/or fluids, including unpasteurized milk, *Brucella* species cause various fevers and diseases, but are perhaps most known for inducing contagious/spontaneous abortions in animal herds (Olsen and Palmer 2014; Xavier et al. 2010). Species of *Bartonella* are primarily regarded as vector-borne pathogens of mammals (Chomel et al. 2009b), transmitted by sandflies, biting flies, lice, and possibly ticks. Ultimately, vector-based transmission leads to infection of mammalian erythrocytes (Cheslock and Embers 2019; Chomel et al. 2009a). For both *Brucella* and *Bartonella,* some extent of commensalism is suggested by reports of asymptomatic canines, felines, rats, and other small mammals carrying these infections (Cheslock and Embers 2019; Olsen and Palmer 2014). By contrast, *Ochrobactrum anthropi* str. ATCC 49188 is an emerging pathogen of immunocompromised humans, and the subtype of a species that is otherwise found in association with soil, plants, invertebrates, and vertebrate animals (Romano et al. 2009).

Despite the shared pathogenic properties of these symbionts, considerable variation is evident in terms of their predicted RcGTA homologs. Of the RcGTA-carrying *Brucella* species identified to date, seven of eight carry LC genes only, and in most cases, the cluster appears largely intact (Fig. 2.4) (Table 2.1) (Shakya et al. 2017). Currently available data suggest that losses of *g1* and *g3.5* have already occurred and losses of *g10.1* are ongoing (Table 2.1). The significance of these losses is not fully understood, yet *g3.5* and *g10.1* are considered dispensable for particle production (Hynes et al. 2012). The LC of *B. ovis* ATCC 25840 is noteworthy in that five predicted ORFs appear to have become pseudogenized (Shakya et al. 2017). *B. melitensis* bv. 1 str. 16 M also bears mentioning, as the loss of predicted head and tail ORFs *g4–g8* differs from other strains that predominantly retain intact LCs (Fig. 2.5b) (Table 2.1). The predicted RcGTA homologs reported for *B. ceti* TE28753–12, otherwise given an "SC" designation, also lack *g4–g8* as well as *g11–g12*, implying derivation from a similarly degraded LC homolog (Table 2.2). While this may also be the case for the largest SC predicted in *O. anthropi* ATCC 49188, more accurate phylogenetic resolution is required. Interestingly, both *O. anthropi* ATCC 49188 and *Bartonella grahamii* as4aup contain predicted singlet ORFs of the portal protein *g3* (Table 2.2). All other strains of *B. grahamii* are notable for the absence of any RcGTA homologs, yet they contain a *Bartonella*-specific GTA, which will be described in greater detail below in Sect. 2.4.1 (Québatte and Dehio 2019; Tamarit et al. 2018).

Predicted RcGTA homologs are also noted for *Rhizobiales* species predominantly found as nitrogen-fixating symbionts of leguminous plants. Bacteria of the genera *Mesorhizobium, Sinorhizobium, Rhizobium, Bradyrhizobium, Azorhizobium, and Methylobacterium* colonize plant roots, driving the formation of nodule structures. For many *Rhizobiales* species, this involves inducing the root to create cellulosic tube structures, facilitating entry into the root by endosymbionts (Huisman et al. 2012; Oldroyd et al. 2011). For some *Bradyrhizobium, Azorhizobium and Methylobacterium* species, an alternate "crack entry" mechanism is implicated, in which the bacteria invade small, natural cracks near root junctions, and then enter root cells via endocytosis (Coba de la Peña et al. 2017; Reddy et al. 1997; Senthilkumar et al. 2009). Through more invasive mechanisms, *Agrobacterium* species *A. vitis* str. S4 *and A. fabrum* str. C58, formerly named *Agrobacterium tumefaciens* strain C58 (Huo et al. 2019), induce tumorous growths referred to as "crown gall disease" in major crop plants (Escobar and Dandekar 2003; Slater et al. 2009). Notably, some *Agrobacterium* species are avirulent, such as *Agrobacterium* sp. H13–3 (Wibberg et al. 2011), and others antagonize pathogenic *Agrobacteria,* as is done by *A. radiobacter* str. K84 (Slater et al. 2009).

With respect to predicted RcGTAs homologs, the *Rhizobiales* are the most diverse order under consideration (Fig. 2.4). While phylogenetically most proximal to the *Rhodobacterales*, the two representative *Methylobacterium* species with completed genome sequences available for comparison, *Methylobacterium* sp. 446 and *M. nodulans* ORS 2060, appear to have diverged substantially with respect to RcGTA ORF content. While *Methylobacterium* sp. 446 retains a collection of predicted ORFs more closely resembling *Brucella ceti* strains, also given the "SC" designation, *M. nodulans* ORS 2060 contains not only five LCs with similar composition but also multiple SCs (Tables 2.1 and 2.2). Four of the five LCs contain the putative head components *g3* and *g5*, while all five contain not only the tail-associated *g8–g10* and *g12–g15* ORFs but also copies of the *rcc00171* tail fiber and *rcc00555* endolysin homologs immediately downstream at the 3′-end of the *g15* ORF (Shakya et al. 2017). While only one of the *M. nodulans* ORS 2060 LCs has retained a copy of the putative *g2* terminase, three singlet ORFs for *g2* exist elsewhere in the genome, along with two *g3–g6* capsid-containing SCs, and the only *g4* putative prohead-protease homologs. Among the symbionts discussed in this review, *M. nodulans* ORS 2060 thus displays the greatest potential for complementation with regard to RcGTA LC and SC ORFs, allowing for functional speculation in this regard. The anecdotal evidence of LC linkage to other predicted tail-fiber and endolysin homologs suggests that a progenitor RcGTA-like element may have included these additionally relevant ORFs, in what is now considered the main structural "head-tail" cluster (Shakya et al. 2017).

Of the Rhizobial lineages branching from *Methylobacterium*, *Azorhizobium caulinodans* ORS-571 has retained a predicted LC, curiously with a pseudogenized *g4* prohead protease ORF. By contrast, *Bradyrhizobium* species carry either singlets or, at most, one SC consisting of predicted *g3*, *g5*, and *g6* ORFs. For two other symbiotic *Bradyrhizobium* species from this analysis, *B. oligotrophicum* S58 and *B. diazoefficiens* USDA 110, there is a complete absence of predicted RcGTA homologs (Shakya et al. 2017). This pattern of RcGTA LC-complement retention in earlier diverging lineages, followed by the loss in latter branching lineages, is essentially recapitulated for the remaining *Rhizobiales* symbionts. While *Agrobacterium* species predominantly retain both LCs and SCs, the single "LC" of *Rhizobium* sp. IRBG74 lacks central ORFs (Fig. 2.5b), and other *Rhizobium* species retain only SC versions of *g3–g10* ORFs or are entirely devoid of predicted RcGTA homologs (Tables 2.1 and 2.2). It is worth mentioning that while some of the LC ORFs of *Agrobacterium fabrum* str C58 are predicted to have been pseudogenized, one of the SCs in *Agrobacterium vitis* S4 can be found on an extrachromosomal plasmid, indicative of the vast genetic/allelic variation and rearrangement occurring throughout this genus. Finally, while the mutualistic *Sinorhizobium* species contain SCs with a complete set of predicted head-morphogenesis ORFs, i.e., *g3–g5*, as well as the tail-associated *g6–g10*, *Mesorhizobium* species have retained very few homologs, predominantly only singlets. In the case of two other *Sinorhizobium* species identified as symbionts in this study, *S. meliloti* and *S. fredii,* the loss of detectable RcGTA homologs is conspicuous (Shakya et al. 2017).

2.3.3 *Orders* Sphingomonadales *and* Rhodospirillales

Symbionts from an array of other alpha-proteobacterial orders have predicted homologs of RcGTA genes. From the order *Sphingomonadales*, *Zymomonas mobilis* shows a capacity for mutualism, by protecting mammals against infections by yeast, schistosomal worms, and bacteria in experimental settings (Campos et al. 2013; Santos et al. 2004). From the order *Rhodospirillales, Gluconobacter oxydans* and *Acetobacter pasteurianus* 386B are opportunistic parasites of plant matter that lead to fruit rot (Gupta et al. 2001), wine spoilage (Campaniello and Sinigaglia 2017), and cocoa bean fermentation (Illeghems et al. 2013). *Gluconacetobacter xylinus* NBRC-3288 is a constituent of microbial mat communities in kombucha and supports those associations by contributing cellulosic structural content (Jayabalan and Waisundara 2019). The *Azospirillum* species *A. brasilense* sp. 245 and *A. lipoferum* sp. 4B are nitrogen-provisioning ectosymbionts that use pili to attach to the surface of wheat and rice plant roots (Steenhoudt and Vanderleyden 2000; Wisniewski-Dyé et al. 2011; Wisniewski-Dyé et al. 2012). Both species are thought to provide phytohormones to host plants (Spaepen et al. 2007), and *A. brasilense* sp. 245 has also been found to promote symbiosis between *Rhizobium* and host plants such as *Vicia sativa* (Sarig et al. 1986; Star et al. 2012). By contrast, *Micavibrio aeruginosavorus* is an ectoparasite and predator of other microbes, including *Pseudomonas aeruginosa, Burkholderia cepacia*, and *Escherichia coli* (Dashiff et al. 2011; Kadouri et al. 2007). Although technically unclassified, *Micavibrio* is grouped in with *Rhodospirillales* for this analysis (Shakya et al. 2017).

The paucity of complete sequence data informing symbionts of these collective orders somewhat constrains the extent of analysis at this time. However, while the divergence of *Sphingomonadales* occurs after the predicted introduction of the RcGTA LC, and thus includes genera with both LC and SC homologs, *Rhodospirillales* genera contain only SCs (Fig. 2.4) (Shakya et al. 2017; Viswanathan et al. 2017). For the purposes of this review, while genera such as *Sphingomonas*, *Novosphingobium*, and *Erythrobacter* are predicted to retain LCs, they are considered nonsymbiotic. For the only symbiotic *Sphingomonadales,* and mirroring the absence of detectable homologs for sister species or strains discussed thus far, one strain of *Z. mobilis* subsp. *mobilis* str. CP4 has retained a singlet copy of *g3*, while other documented strains have retained no predicted RcGTA ORFs (Table 2.3). For the *Rhodospirillales* genera that contain only SCs, most lack predicted tail-related ORFs. *G. xylinus* NBRC-3288 stands in contrast to other species, as it is predicted to contain two SCs with the *g3–g7* ORFs, in addition to two singlet homologs of the predicted terminase *g2* and prohead protease *g4* (Fig. 2.5c). Oddly for this group of organisms, *A. brasilense* Sp245 carries only two single ORFs, *g2* and *g11*, while *A. lipoferum* 4B contains two copies of the predicted portal protein *g3* ORFs on individual plasmids. Similarly, the ectoparasite *M. aeruginosavorus* and *G. oxydans 621H* carry only singlet copies of the *g2* terminase (Table 2.3). In summary, particularly for the soil-associated organisms in these groups, interpretations will become increasingly well informed in the future as additional symbiont lineages are identified, sequenced, and compared to the existing datasets.

2.3.4 *Order* Rickettsiales

Rickettsiales, the most basal order in which RcGTA homologs have been predicted, are all considered endosymbionts and are regarded as the divergence point of a mitochondrial-progenitor species (Archibald and Richards 2010). *Rickettsiales* are by definition "obligate intracellular bacteria" because they can only replicate within the cytoplasm of eukaryotic host cells (Hackstadt 1996; Thomas et al. 2017). *Rickettsiales* infections are facultative with respect to their invertebrate hosts, though there are exceptions. *Rickettsia felis* URRWXCal2, carried by booklice, and *Wolbachia pipientis*, in certain insects and filarial nematodes, are required for host oogenesis, as well as to support host viability in nematodes (Dedeine et al. 2001; Gillespie et al. 2014; Hoerauf et al. 2000; Landmann et al. 2011; Rao et al. 2002). *Wolbachia, Rickettsia, Orientia,* and *Neorickettsia* are propagated by endosymbiont loading into eggs of the host organism (Azad and Beard 1998; Brumin et al. 2012; Elliott et al. 2019; Greiman et al. 2016; Landmann 2019; Serbus et al. 2008). *Ehrlichia* and *Anaplasma* may use horizontal or vertical transmission, depending on the species (Baldridge et al. 2009; Long et al. 2003; Moore et al. 2018; Perlman et al. 2006; Walker 2017). In many cases, *Rickettsiales* species are also transmitted to mammals by the saliva or feces of ticks, fleas, lice, and chigger mites (Day and Newton 2017; Raoult 2015; Walker 2017). *Neorickettsia* are acquired by animals via ingestion of parasitized material (Greiman 2015; Paris and Day 2014). Extended releases of *Wolbachia* bacteria by dying filarial nematodes into animal hosts induce inflammation that underlies the neglected tropical diseases African river blindness and lymphatic Filariasis (discussed in the review by Slatko et al. within this same volume) (Gillette-Ferguson et al. 2004; Saint André et al. 2002; Taylor 2003). *Wolbachia pipientis* strains that are endogenous to insect hosts are not transmitted to mammals (Kamtchum-Tatuene et al. 2017; Popovici et al. 2010).

The *Rickettsiales* order, like the *Rhodospirillales*, retain no LCs (Fig. 2.4) (Shakya et al. 2017). However, in addition to the predicted terminase and head morphogenesis homologs *g2–g5*, a large proportion of the species carrying SC ORFs have specifically retained the tail-associated homologs *g6*, *g9*, *g12*, *g13,* and *g15* (Table 2.3). For the *Anaplasmataceae* genera *Anaplasma, Ehrlichia*, and *Wolbachia*, this includes SCs that contain the predicted head-tail adaptor *g6* and major-tail protein *g9* ORFs on the same fragment. By contrast, *g12*, *g13,* and *g15*, whose role in RcGTA biology other than potentially contributing to host specificity is less certain, are all independently located (Fig. 2.5d). *Neorickettsia risticii* str. Illinois and *Neorickettsia sennetsu* Miyayama have retained the ORFs *g2–g5* as scattered, independent singlets (Table 2.3). As the branch leading to the divergence of the *Anaplasmataceae* most closely approximates the phylogenetic origin of a proto-mitochondrial species (Fig. 2.4), the retention of any ancestrally introduced bacteriophage-related homologs is of particular interest.

For the most recently diverged members of the family *Rickettsiaceae*, specifically *Rickettsia* and *Orientia* species, a similar pattern of retention followed by divergence and loss can be seen, as described for *Rhizobiales* and *Sphingomonadales* species

Table 2.3 Alpha-proteobacterial symbiont RcGTA "small cluster" homologs, part 2

Organism type ORFs	*g1*	*g2*	*g3*	*g4*	*g5*	*g6*	*g7*	*g8*	*g9*	*g10*	*g11*	*g12*	*g13*	*g14*	*g15*
Order *Sphingomondales*															
Zymomonas mobilis subsp. *mobilis* str. CP4^"	-	-	A	-	-	-	-	-	-	-	-	-	-	-	-
Order *Rhodospirillales*															
*Acteobacter pasteurianus 386B^*	-	-	A	B	-	-	-	-	-	-	-	-	-	-	-
Gluconobacter oxydans 621H^	-	A,B	-	-	-	-	-	-	-	-	-	-	-	-	-
Gluconacetobacter xylinus NBRC-3288	-	A	B,C	B,C,D	B,C	B,C	B,C	-	-	-	-	-	-	-	-
Azospirillum brasilense Sp245	-	A	-	-	-	-	-	-	-	-	B	-	-	-	-
Azospirillum lipoferum 4B	-	A	A,B*,C*	-	-	-	-	-	-	-	-	-	-	-	-
Micavibrio aeuruginosavorus (2 strains)	-	A	-	-	-	-	-	-	-	-	-	-	-	-	-
Order *Rickettsiales*															
Rickettsia rickettsii (+8 other species)"	-	-	-	-	-	-	-	-	-	-	-	A	-	-	-
Rickettsia felis URRWXCal2	-	A	B	-	-	-	-	-	-	-	-	C	-	-	-
Rickettsia bellii OSU-85389^	-	A	B	(+)	C	C	-	-	-	-	-	C	C	-	-
Orientia tsutsugamushi (2 strains)	-	A	B	C	C	C	-	-	-	-	-	C	C	-	D
Anaplasma phagocytophilum (4 strains)^	-	A	B	B	C	D	-	-	D	-	-	E	D	-	-
Anaplasma centrale str. Israel	-	A	B	B	C	A	-	-	A	-	-	D	C	-	C
Anaplasma marginale (4 strains)^	-	A	B	B	C	A	-	-	A	-	-	D	C	-	C
Ehrlichia ruminantium (2 strains)	-	A	B	B	C	A	-	-	A	-	-	D	C	-	C
Ehrlichia canis str. Jake	-	A	B	B	C	A	-	-	A	-	-	D	C	-	C
Ehrlichia muris AS145	-	A	B	B	C	A	-	-	A	-	-	D	C	-	C
Ehrlichia chaffeensis str. Arkansas	-	A	B	B	C	A	-	-	A	-	-	D	C	-	C
Wolbachia pipientis (8 strains)^	-	A	A	B	B	C	-	-	C	-	-	D	E	-	F
Neorickettsia risticii str. Illinois	-	A	B	C	D	-	-	-	-	-	-	-	-	-	-
Neorickettsia sennetsu Miyayama	-	A	B	C	D	-	-	-	-	-	-	-	-	-	-

Letters A thru F indicate predicted SC ORFs or singlets of RcGTA *g1-g15* , where ORFs with the same letter are in the same cluster
Linear order of ORFs in the same cluster are not necessarily maintianed in accord with that of LC arrays
(+) indicates a putatively pseudogenized ORF
* indicates ORF(s) are located on an extrachromosomal plasmid
^ indicates some variation exists between strains, see Shakya et al., 2017; Supplementary Figure S5
" = *Rickettsia typhi, R. canadensis, R. akari, R. heilongjiangensis, R. japonica, R. australis, R. slovaca* 13B*
^" = *Zymomonas mobilis* strain CP4=NRRL B-14023; strains ATCC 10988, NCIMB 11163 and ZM4 ATCC 31821 have no RcGTA homologs
[Note: 9 strains have no RcGTA homologs: *Rickettsia prowazekii* , *R. rickettsii* (8 strains), *R. africae, R. parkeri, R. conorii, R. slovaca R. rhipicephali, R. montanensis, R. philipii*

above (Table 2.3). *Orientia tsutsugamushi* and *Rickettsia belli,* which more closely approximate the majority of *Anaplasmataceae* in overall homolog content, show a curious coupling of the putative major capsid protein *g5* with the head-tail connector *g6* and tail component *g12* and *g13* on a single SC. However, *Rickettsia felis* has retained multiple singlet ORFs, *Rickettsia rickettsia,* and eight other species retain only a single *g12* ORF, and the remaining of the most recently diverged species, including *Rickettsia prowazekii* and nine other species, are all presumed to have lost singlet or SC ORFs entirely (Table 2.3). For *Wolbachia pipientis*, the nematode-associated variants *w*Oo and *w*Bm completely lack detectable RcGTA homologs (Shakya et al. 2017), however for the purposes of this analysis, the multitude of *Wolbachia* variants extends beyond the scope of this review.

2.3.5 Possible Functions for Incomplete Sets of RcGTA Homologs

As discussed for the various symbiotic orders above, divergence and loss of *detectable* homologous RcGTA ORFs have occurred both after the introduction of the progenitor LC, estimated to be ~1 billion years ago, and after the otherwise deeply branching progenitor "SC" introduction, around ~2 billion years ago (Battistuzzi et al. 2004; Kogay et al. 2019; Luo et al. 2013; Shakya et al. 2017). As the trend toward genome reduction in symbionts should hold true within individual bacterial groupings, these general trends would be expected. Thus in numerous cases, we may be viewing snapshots of LC loss and consequently, loss of functional GTA particles. While this analysis by no means suggests that RcGTA homologs are indeed expressed, the ubiquitous nature of the ppGpp "stringent response," SOS DNA damage response, two-component environmental sensors, CtrA "master" transcriptional regulator, and the second messenger c-*di*-GMP within these symbiotic alpha-proteobacterial orders lends some support to the possibility that regulatory similarities may exist. This, of course, awaits further confirmations for each symbiont discussed in this regard.

The implication of carrying SC genes, or an incomplete set of LC genes, as reflected within the diverse alpha-proteobacterial symbiont lineages, remains unclear at this time. Emerging reports that bacterial symbionts are recipients of HGT open an array of speculative possibilities (García-Aljaro et al. 2017; Husnik and McCutcheon 2016, 2018; Koonin 2016). One potential scenario is for products of endogenous RcGTA genes to assemble into alternate particle conformations, with or without inclusion of additional prophage-like gene products substituting for missing RcGTA components. Another possibility is that lysogenic phages supplement regulatory and/or structural genes that facilitate de novo RcGTA assembly. For example, roseophages, RDJLΦ1, and RDJLΦ2 have been shown to encode a CtrA-like transcription factor, as well homologs of *rcc01865* and *rcc01866*, which encode the GafA transcription factor and virion maturation protein otherwise necessary for the production of functional RcGTA particles (Hynes et al. 2016; Lang et al. 2017). Interestingly, almost all marine *Siphoviridae* family roseophages discovered thus far encode the RcGTA homologs *g12–g15* predicted to be involved in tail morphogenesis and host recognition (Huang et al. 2011; Liang et al. 2016; Yang et al. 2017; Zhan et al. 2016). The roseophages RDJLΦ1 and vB_DshS-R5C have also been shown to harbor a gene with homology to the GTA "tail-fiber/attachment" *rcc00171* (Hynes et al. 2016; Zhan and Chen 2019).

It also remains possible that retention of RcGTA homologs by symbionts is uncoupled from HGT to some extent, with gene products potentially repurposed for other uses. One possibility is the use of capsid-like proteins to create compartmentalized structures. The largest of such compartments reported so far are carboxysomes, which are important for carbon fixation (Rae et al. 2013). A smaller-sized set of microcompartments has been identified that drive certain B_{12}-dependent catabolic reactions and otherwise toxic metabolic processes (Bobik et al. 2015). Nanocompartments called "encapsulins" that carry oligomerized enzymes may be vastly more common, with bioinformatic predictions for encapsulin proteins numbering in the thousands for bacteria and archaea (Nichols et al. 2017). Another possibility is that RcGTA tail genes are repurposed as tailocin-like structures. Tailocins have been widely regarded as bactericidal due to their activity as uncoupling agents or by transferring toxins into recipient cells (Ghequire and De Mot 2015). Analogous to RcGTA particles, tailocins are only produced by a small proportion of prokaryotic cells within a population, and tailocin release ultimately requires cell lysis of a cell subpopulation (Scholl 2017). Tailocins are also becoming recognized in terms of new roles, such as ectosymbiont release of tailocin-like structures that are critical for metamorphosis of settling tubeworm larvae (Shikuma et al. 2014). Given that the range of RcGTA ORF predictions across symbiotic lineages reflects an emphasis on capsid-related genes, tail-related genes, or both (Tables 2.1, 2.2, and 2.3), this leaves open many potential avenues for use of RcGTA-related gene products in symbiotic interactions.

2.4 Additional GTA and Phage-like Particles Identified across Diverse Microbial Systems

Horizontal gene transfer has been documented for many symbionts not reported to carry RcGTA genes. At this time, large-scale analyses seeking to identify GTA-related genes have only just begun and remain constrained by the current limits of predictive analyses (Kogay et al. 2019; Shakya et al. 2017). Since the initial discovery of RcGTAs, researchers have identified a number of other putative GTA systems, as well as additional types of phage-like particles that are not believed to participate in HGT. These non-*R. capsulatus* GTA-like systems are distinctive at the level of sequence homology, gene cluster size, DNA packing capacity within the particle capsid, and overall particle dimensions (Lang et al. 2017; Lang et al. 2012).

As they have been identified across a sweeping taxonomic range, we will consider each in turn, beginning with a GTA type found in *Bartonella* and a few other alpha-proteobacterial symbionts. Then, we briefly discuss information available for GTA-like particles identified in delta- and gamma-proteobacteria, spirochetes, firmicutes, and even archaea. To further frame the capabilities of GTA-like particles, we conclude with examples from microbial systems not strictly considered under a "symbiotic" designation.

2.4.1 BaGTA from **Bartonella** *Spp., Class Alpha-Proteobacteria*

"Bacteriophage-like particles," originally dubbed "BLPs," were discovered in 1994 from the facultative rhizobial symbiont *Bartonella henselae*, previously *Rochalimaea henselae* (Anderson et al. 1994). After this phage-like particle was demonstrated to be functional for gene transfer, the particles were renamed BaGTAs (Guy et al. 2013; Québatte et al. 2017). The highly conserved, ~32 kb region including the BaGTA structural gene cluster (Fig. 2.6b) is located within an ~80 kb region that contains multiple copies of a putatively phage-derived origin of replication, known as a run-off replication (ROR) gene cassette (Berglund et al. 2009). Flanked by various Type IV and V secretion system gene clusters, the BaGTA and ROR cassettes are strictly maintained in all modern *Bartonella* species (Alsmark et al. 2004; Guy et al. 2013; Tamarit et al. 2018). Unlike RcGTAs, it is evident that DNA packaging into BaGTA capsids is nonrandom. BaGTA particles show bias toward packing content located near the *Bartonella* runoff origin of replication (BaROR), which includes disproportionate representation of the BaGTA structural cluster genes (Berglund et al. 2009; Guy et al. 2013; Lindroos et al. 2006; Québatte et al. 2017). Bidirectional amplification of DNA surrounding the BaROR ensures that the frequency of inclusion of host genes into BaGTA particles is directly related to the distance from the ROR (Québatte et al. 2017; Québatte and Dehio 2019).

While the *Bartonella* GTA particles isolated to date have uniformly been demonstrated to carry ~14 kb fragments of host DNA, their capsid sizes have ranged from 40 to 80 nm (Fig. 2.6e). However, some of this variation may be due in part to the isolation of early proheads or structural intermediates formed during capsid maturation (Hernando-Pérez et al. 2014; Roos et al. 2012). While associated tail structures have not yet been strictly demonstrated for *B. henselae* (Anderson et al. 1994; Barbian and Minnick 2000), 16 nm tails were found to be associated with phage particles from *B. bacilliformis* that were capable of forming plaques on blood agar plates (Umemori et al. 1992). As contractile tailed particles were also isolated from *Bartonella vinsonii* subsp. *berkhoffii*, in that case 60–80 nm in length, it is uncertain whether all BaGTA-related particles require tail structures or the lack of associated tails is tied to isolation methodologies (Carvalho et al. 2010; Maggi and

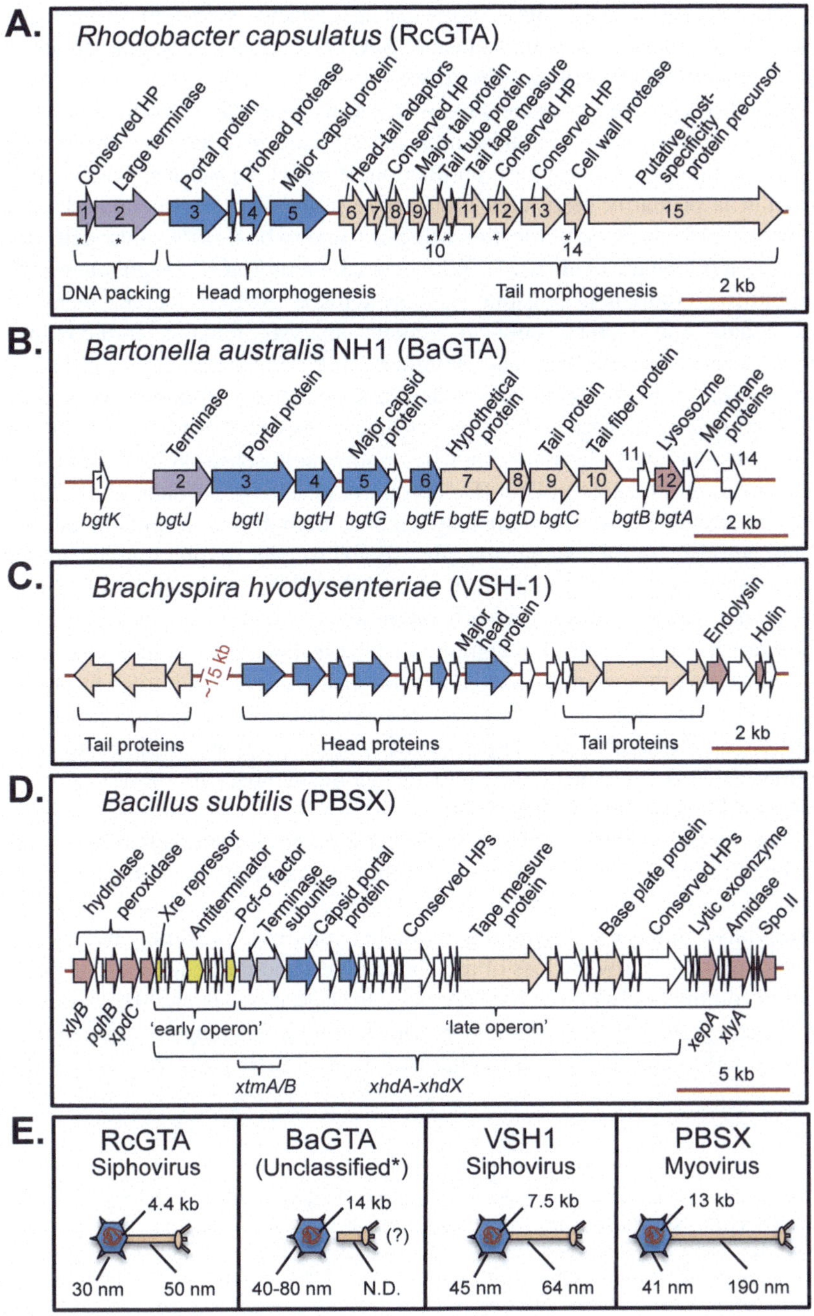

Fig. 2.6 Gene transfer agents in symbiotic bacteria. Arrows represent annotated ORF predictions, using the same color scheme as for Fig. 2.5. (**a**) RcGTA structural gene cluster of *Rhodobacter capsulatus* included for reference as in previous Figs. (**b**) BaGTA structural gene cluster from

Breitschwerdt 2005). Regardless, with capsid capacity sufficient to package the entire ~14 kb BaGTA structural cluster, BaGTA particles may yet remain more virus-like than the GTAs of *R. capsulatus*.

A subset of *Bartonella* species, for which BaGTA particle production has been demonstrated, have provided substantial mechanistic insights into the biology of BaGTAs. Unlike stationary-phase production of RcGTA particles, BaGTA particle expression is primarily associated with actively dividing, exponential-phase cells (Guy et al. 2013). In *B. henselae*, this was estimated to be 6–17% of the cells in such a population (Québatte et al. 2017). While coordinating BaGTA regulation with cell cycle and two-component phospho-relay systems is likely to include many of the same signals and factors, detailed genetic analysis remains forthcoming for *Bartonella* systems (Barbian and Minnick 2000; Québatte et al. 2017). Both feline-associated *B. henselae* and rodent-associated *B. grahamii* production of BaGTA particles are regulated by the "stringent response" (Québatte et al. 2017). This response involves the BatR/BatS two-component sensor system and the ubiquitous signaling molecule ppGpp (Dalebroux et al. 2010; Québatte et al. 2013). In contrast to RcGTAs, it was demonstrated that high levels of ppGpp inhibit the BaGTA activity, further suggesting that BaGTA gene expression is limited to actively replicating cells (Québatte et al. 2017). While the mechanism of particle docking onto recipient cells is currently unknown, BaGTA uptake into recipient cells appears restricted by the Tol/Pal complex, which is thought to maintain the integrity of the outer membrane in *E. coli* (Québatte et al. 2017; Walburger et al. 2002). Similar to *R. capsulatus*, host homologous recombination machinery, including the ComEC, ComF, ComM, and DrpA proteins (Fig. 2.3), are believed to regulate incoming dsDNA and integration into the recipient host chromosome; reviewed recently in (Québatte and Dehio 2019).

Modern parasitic *Bartonella* species infect a range of mammalian hosts and are transmitted primarily via insect vectors, though also directly, as in the case of "cat-scratch" fever (Eicher and Dehio 2012; Engel et al. 2011; Nelson et al. 2016). Comparisons with earlier diverging mutualist relatives, including honeybee and ant gut-symbionts, show the loss of metabolic genes and the acquisition of virulence factors in conjunction with the BaGTA cassette (Kešnerová et al. 2016; Kosoy et al. 2008; Segers et al. 2017). *Bartonella* genomes themselves range in size from 1.4 to 2.6 Mb, small in comparison with other soil or plant associated *Rhizobia*, and even

Fig. 2.6 (continued) Bartonella australis NH1 (NCBI IDs BAnh113370–113,220) (Tamarit et al. 2018). (**c**) VSH1 structural gene cluster from *Brachyspira hyodysenteriae* (tail protein IDs: BHWA1_RS08855-RS08865 and VSH1 IDs BHWA1_BHWA1_RS08920-RS09010, which replace discontinued locus annotations) (Matson et al. 2005; Stanton et al. 2009). (**d**) PBSX structural gene cluster from *Bacillus subtilis* subsp. *subtilis* str. 168 (NCBI genome NC_000964.3; IDs BSU_12460–12,830). (**e**) Depictions of GTAs or phage-like particles purified from previously listed organisms. Packaged DNA size, head diameter, and tail length estimates are given; depictions not to scale. N.D. indicates "not determined." * while formally unclassified, these particles bear features otherwise associated with podovirus-type particles

with the ~3.3 Mb genomes of their *Brucella* relatives (Boussau et al. 2004; Ettema and Andersson 2009). It is thought that the maintenance of BaGTAs and the influence of the BaROR were driven by selection to increase HGT and counter gene loss (Batut et al. 2004; Segers et al. 2017; Tamarit et al. 2018). The introduction of accumulated mutation, insertion, deletion, and rearrangement would then have affected the integrity of progenitor prophage regulation, effectively "domesticating" the ancestral BaGTA (Bobay et al. 2014; Québatte et al. 2017; Tamarit et al. 2018). Coupled with the acquisition of T4SSs and pathogenesis factors, adaptive evolution would then have driven discrete lineages to match a divergent set of host cells (Chomel et al. 2009a; Engel et al. 2011; Guy et al. 2013; Harms et al. 2017; Québatte et al. 2017; Tamarit et al. 2018).

Through an extensive analysis of BaGTA homologs within the alpha-proteobacterial orders *Rhodobacterales, Rhizobiales,* and *Caulobacterales* (Tamarit et al. 2018), BaGTA ORFs have been predicted in association with 13 symbionts of the *Rhizobiales* order (Table 2.4). Nine of these are pathogenic *Bartonella* that infect mammalian hosts (Eicher and Dehio 2012; Engel et al. 2011) and are predicted to carry a nearly or completely intact cluster of BaGTA genes (Tamarit et al. 2018). Other BaGTA-carrying *Rhizobiales* are mutualists that form root symbioses, specifically *A. caulinodans* ORS 571, *B. sp.* BTAi1, *Rhizobium etli*, and *R. leguminosarum* bv. *viciae* 3841 (Tamarit et al. 2018). BaGTAs were also detected in one mutualistic *Rhodobacterales*, specifically in the scallop symbiont *Phaeobacter gallaeciensis* DSM 26640 (Genard et al. 2014). Unlike pathogenic *Bartonella,* these five mutualistic species were predicted to carry a minimal number of BaGTA homologs (Table 2.4) (Tamarit et al. 2018). It is further notable that three of these have been identified as carriers of both BaGTA and RcGTA homologs. *P. gallaeciensis* DSM 26640 is predicted to carry a complete RcGTA LC, and *Azorhizobium caulinodans* ORS 571 also has most of the LC genes (Tables 2.1 and 2.4), whereas *Bradyrhizobium sp.* BTAi1 carries only the *g3* SC gene (Tables 2.2 and 2.4). While the selective advantage of the BaGTA/BaROR region has been discussed for modern, phylogenetically divergent *Bartonella* ssp. (Québatte and Dehio 2019), any specific fitness advantage of BaGTA homologs, or their dual occupancy with RcGTA homologs, has yet to be experimentally addressed in these mutualistic symbionts.

2.4.2 *VSH-1 from* Brachyspira hyodysenteriae, *Phylum Spirochaetes*

Alternate GTA forms have been found outside of the phylum proteobacteria. An important example comes from *Brachyspira hyodysenteriae* (Humphrey et al. 1997), an anaerobic spirochete known to infect the large intestine of swine, causing dysentery and substantial herd morbidity/mortality (Mirajkar and Gebhart 2014; Taylor and Alexander 1971; Whiting et al. 1921). *B. hyodysenteriae* produces

Table 2.4 *Bartonella* BaGTA homologs in symbiotic alpha-proteobacteria

	Terminase		Capsid-associated				Tail-associated				Lysozyme/Memb. prot.			
Organism type / **ORFs**	1	2	3	4	5	6	7	8	9	10	11	12	13	14
Order *Rhodobacterales*														
Phaeobacter gallaeciensis DSM 26640	+	+	+	-	-	-	-	-	-	-	-	-	-	-
Order *Rhizobiales*														
Bartonella australis Aust/NH1	+	+	+	+	+	+	+	+	+	+	+	+	+	+
Bartonella bacilliformis (2 strains)	+	+	+	+	+	+	+	+	+	+	+	+	+	+
Bartonella clarridgeiae strain 73	+	+	+	+	+	+	+	+	+	+	+	+	+	+
Bartonella grahamii as4au	+	+	+	+	+	+	+	+	+	+	+	+	-	-
Bartonella henselae (3 strains)	+	+	+	+	+	+	+	+	+	+	+	+	-	+
Bartonella quintana RM-11^	+	+	+	+	+	+	+	+	+	+	+	-	-	+
Bartonella tribocorum^	+	+	+	+	+	+	+	+	+	+	+	-	-	+
Bartonella vinsonii subsp. *berkhoffii* str. Winnie	+	+	+	+	+	+	+	+	+	+	+	-	-	+
Candidatus Bartonella ancashi strain 20.00	+	+	+	+	+	+	+	+	+	+	+	+	+	+
Rhizobium etli CFN 4	-	+	-	-	-	-	-	-	+	-	-	+	-	-
Rhizobium leguminosarum bv. viciae 3841	-	+	-	-	-	-	-	-	+	-	-	+	-	-
Bradyrhizobium sp. BTAi	-	+	-	-	-	-	+	-	-	+	-	-	-	-
Azorhizobium caulinodans ORS 571	-	-	+	+	+	+	+	-	+	+	-	-	-	-

\+ indicates presence of homolog; - indicates absence
^ indicates some variation exists between strains

GTA-like particles, classified as the siphovirus-type, now referred to as VSH-1 (for *virus of Serpulina hyodysenteriae*). These VSH-1 particles are known to mediate the transfer of a variety of markers between cells, including virulence genes and antibiotic resistance (Humphrey et al. 1997; Stanton et al. 2008; Stanton et al. 2001; Trott et al. 1997). Particles similar to VSH-1 have been observed in association with other species of *Brachyspira* including *B. intermedia* and *B. pilosicoli* (Calderaro et al. 1998a; Calderaro et al. 1998b; Motro et al. 2009). As an indication of host range, VSH-1-like GTAs also appear to be produced by spirochetes that infect humans (Calderaro et al. 1998a; Calderaro et al. 1998b; Humphrey et al. 1997).

The VSH-1 GTA particle has a 45 nm head diameter and 64 nm noncontractile tail and packs ~7.5 kb of DNA (Fig. 2.6e) (Humphrey et al. 1995; Humphrey et al. 1997). The major gene cluster of VSH-1 is a 16.3 kb fragment which contains ORFs for head, tail, and lysis (Fig. 2.6c). A smaller 3.6 kb cluster of three tail proteins has been identified apart from the main cluster, as well as putative phage endolysin and holin genes, which appear to be coregulated (Matson et al. 2005; Stanton et al. 2009). Quantitative measurements indicate that although VSH-1-specific transcription increases over 200-fold after induction, there is no gene copy number disparity in the packing of VSH-1 genes relative to non-VSH-1 genes (Stanton et al. 2009). Unlike RcGTA production, constrained to few cells within a population, VSH-1 particle production and release by cell lysis may be more generalized (Humphrey et al. 1995; Lang et al. 2017). Particle production is inducible by DNA-damaging agents such as mitomycin C, hydrogen peroxide, and antibiotics widely used in the swine industry, such as carbadox and metronidazole (Humphrey et al. 1995; Matson et al. 2007; Stanton et al. 2008). As studies have shown transfer of antibiotic resistance between these spirochete strains, this suggests that VSH-1 GTAs affect the pathogenic properties and population structure of *Brachyspira*, while also impacting the host intestinal microbiome (Stanton et al. 2008; Trott et al. 1997; Zuerner et al. 2004).

2.4.3 PBSX from Bacillus *Spp., Phylum Firmicutes*

As *Bacillus* species have been shown to form metabolic symbioses with other microbes (Jia et al. 2015; Zhou et al. 2013), a further unconventional phage-like particle reported from *Bacillus* also warrants discussion. These spore-forming firmicutes include pathogenic species such as *B. anthracis* and *B. cereus*, as well as the model system and industrially relevant species, *B. subtilis*. The phage-like particles produced by *Bacillus*, termed PBSX, are 41 nm in diameter with an associated 190 nm contractile tail, fitting the structural classification of a myovirus (Fig. 2.6e) (Lang et al. 2012; Wood et al. 1990a). Similar phage-like elements have been identified for other *Bacillus ssp.* and termed PBSW, PBSY, and PBSZ (Glaser et al. 1966; Karamata et al. 1987; Young et al. 1989). The 28 kb PBSX gene cluster is completely unrelated to any of the GTA clusters considered thus far (Fig. 2.6d). It contains phage head, tail, lysis, and lysogeny genes and lacks replication-related functions (McDonnell et al. 1994; Seaman et al. 1964; Wood et al. 1990a). The induction of PBSX is controlled by the helix-turn-helix transcriptional repressor Xre, encoded in the "early operon" of the PBSX cluster (Buxton 1976; Wood et al. 1990a; Wood et al. 1990b). Xre binds to multiple promoters within the *Bacillus* genome and controls expression of a positive control factor (Pcf), needed for the expression of genes from a "late promoter" (McDonnell et al. 1994). Structural and lytic proteins, including autolysin and holin-like protein ORFs, encoded in the "late operon" (Fig. 2.6d), are thought to facilitate host cell lysis (Foster 1993; Krogh et al. 1996; Longchamp et al. 1994).

PBSX particles have also been shown to damage the peptidoglycan layer of neighboring cells, presumably in a manner typical of myoviruses (Toyofuku et al. 2017). Myovirus-type phages, such as T4 and P″, are generally known as lytic, rather than temperate phages, whose contractile sheath acts like a syringe, piercing the cell wall with a central tube and injecting the genetic material into the host. PBSX particles are thought to contain DNA, and their production is induced by mitomycin C and the SOS response, analogous to other GTAs. Counterintuitively, PBSX particles do not actively participate in HGT, but instead kill other *B. subtilis* cells that are nonlysogenic for PBSX (Okamoto et al. 1968a; Okamoto et al. 1968b). Previously termed "phage-like bacteriocins" (McDonnell et al. 1994), PBSX particles may act primarily to limit competing cells. Interestingly, *B. subtilis* encodes a second unusual genetic element termed "sKin" (sigK intervening sequence), which shares high homology to PBSX and other potentially cryptic ancestral phage and plasmid genes (Krogh et al. 1996; Takemaru et al. 1995). As many *B. subtilis* strains have been maintained under laboratory conditions, even subject to mutagenesis, tying laboratory studies to environmentally isolated variants will remain an important aspect of PBSX investigation.

2.4.4 *Dd1 from* Desulfovibrio desulfuricans, *Class Delta-Proteobacteria*

Historically, the second GTA to be identified by electron microscopy was from the strictly anaerobic, sulfur-reducing delta-proteobacterium *Desulfovibrio desulfuricans* (Rapp and Wall 1987). To date, *D. desulfuricans* is generally regarded as a nonsymbiotic bacterium. It is primarily found in environmental samples and only rarely causes infection in humans (Goldstein et al. 2003). However, the GTA found in this system, termed Dd1, is distinctive in that it resembles a tailed podovirus, rather than a siphovirus-like structure. Dd1 particles exhibit a 43 nm head diameter and only a short, 7 nm tail (Rapp and Wall 1987). The American Type Culture Collection (ATCC) strain 27,774 contains a ~ 17.8 kb ORF region, *Ddes_0706* through *Ddes_0726* that is presumed to encode the Dd1 particle components (Lang et al. 2012). Phage-mediated, intra-species transfer of ~13.6 kb linear fragments of DNA occurs with a frequency of 10^{-5} to 10^{-6} per recipient cell, capable of carrying multiple antibiotic resistance markers (Rapp and Wall 1987). Akin to RcGTAs, Dd1 particle production is not induced by DNA-damaging mitomycin C. Regulation of particle production for this GTA type remains unclear at this time (Krupovic et al. 2010; Rapp and Wall 1987).

2.4.5 *VTA from* Methanococcus voltae, *Phylum Euryarchaeota*

While the distantly related archaeon *Methanococcus voltae* is not reported to form symbiotic associations, in order to underscore the large evolutionary distance in which GTAs have been reported, we include what is currently known regarding the VTA (for *v*oltae-*t*ransfer *a*gent). *M. voltae* is a heterotrophic, H_2-oxidizing methanogenic bacterium, which was initially collected from estuary sediment (Whitman et al. 1982). Genetic exchange occurs through phage-like particles that contain ~4.4 kb of host DNA in a 40 nm head capsid with an associated 61 nm tail (Eiserling et al. 1999). A major gene cluster region of ~14 kb, which spans ~12 ORFs (*Mvol_0401–Mvol_0414*), with homologs of siphovirus components, is thought to encode the VTA (Krupovic et al. 2010). While DNA packing appears predominantly random, there may be partial enrichment of one 0.9 kb genomic region (Bertani 1999). Similar to PBSX and VHS1 particles, VTA particle production is induced by DNA damage, and transfer is measured between ~10^{-5} and 10^{-2} events per donor cell (Bertani 1999). While *M. voltae* undergoes low-frequency natural transformation (Bertani and Baresi 1987), the use of this GTA structure is somewhat unconventional. Archaea are commonly infected by a variety of morphologically diverse viruses, but usually not tailed phages (Krupovic et al. 2018; Pina et al. 2011; Prangishvili et al. 2017).

2.5 Conclusions and Considerations

Whether by fault or design, the intrinsic nature of genetic exchange renders the life histories of cells, viruses, bacteriophages, prophages, and, in the case of this review, GTAs, innately linked. As our awareness of the mechanisms underlying "phage domestication" matures, the role that phage-related sequences and phage-like particles play in the evolution of microbes becomes increasingly evident. The presence of GTAs in marine environments, highlighted by *R. capsulatus* and *D. shibae*, has informed an appreciation of their role in HGT within the context of the ectosymbiotic association. Yet, numerous other symbiotic relationships and environments have received very little attention in this regard. Paralleling our insufficient understanding of viromes (Pratama and van Elsas 2018; Trubl et al. 2018), perhaps none is more evident than for the broad range of soil-associated and N_2-fixing symbiotic *Rhizobiales*. In the case of the facultative parasitic relatives of the genus *Bartonella*, the novel integration of a GTA and a phage-derived origin of replication is believed to have facilitated adaptive radiation and dramatically diversified host range. While it is also believed that the initial integration of the RcGTA-like progenitor, estimated to be ~1 bya, fueled radiation of the *Rhodobacterales*, it remains an open question just what advantage, if any, there is for earlier diverging lineages to have retained similar homologs. Albeit from an integration event, estimated some ~1 billion years prior, their predicted presence as far back as basal *Rickettsiales* species bears far greater scrutiny.

Bacterial and archaeal members of the dsDNA virus order *Caudovirales*, i.e., *Myoviridae*, *Siphoviridae*, and *Podoviridae*, contain the characteristic major capsid protein HK97-fold (Suhanovsky and Teschke 2015). As the HK97-like capsid proteins are among the oldest, most abundant, and widespread viral proteins on the planet (Chow and Suttle 2015; Cobián Güemes et al. 2016), it stands to reason that not only proto-bacterial cells but also eukaryotic-progenitor cells coevolved alongside HK97-fold proteins. Outside of the virosphere, the HK97-like fold is only found in a class of bacterial and archaeal nanocompartments called encapsulins (Giessen 2016). That these nanocompartments contain cargo proteins related to oxidative stress makes for several lines of interesting speculation on the potential benefit of such proteins, particularly in the emergence of endosymbiotic or even proto-eukaryotic cell lineages. Overall, while there are several well-defined systems with which to understand the mechanisms of GTA regulation, and their benefit within certain symbiotic associations, our knowledge of their function and potential is only just being uncovered.

References

Alam TI, Draper B, Kondabagil K, Rentas FJ, Ghosh-Kumar M, Sun S, Rossmann MG, Rao VB (2008) The headful packaging nuclease of bacteriophage T4. Mol Microbiol 69(5):1180–1190

Alsmark CM, Frank AC, Karlberg EO, Legault BA, Ardell DH, Canbäck B, Eriksson AS, Näslund AK, Handley SA, Huvet M, La Scola B, Holmberg M, Andersson SG (2004) The louse-borne human pathogen Bartonella quintana is a genomic derivative of the zoonotic agent Bartonella henselae. Proc Natl Acad Sci U S A 101(26):9716–9721

Anderson B, Goldsmith C, Johnson A, Padmalayam I, Baumstark B (1994) Bacteriophage-like particle of Rochalimaea henselae. Mol Microbiol 13(1):67–73

Archibald JM, Richards TA (2010) Gene transfer: anything goes in plant mitochondria. BMC Biol 8:147

Attaiech L, Granadel C, Claverys JP, Martin B (2008) RadC, a misleading name? J Bacteriol 190 (16):5729–5732

Azad AF, Beard CB (1998) Rickettsial pathogens and their arthropod vectors. Emerg Infect Dis 4 (2):179–186

Baldridge GD, Scoles GA, Burkhardt NY, Schloeder B, Kurtti TJ, Munderloh UG (2009) Transovarial transmission of Francisella-like endosymbionts and Anaplasma phagocytophilum variants in Dermacentor albipictus (Acari: Ixodidae). J Med Entomol 46(3):625–632

Barbian KD, Minnick MF (2000) A bacteriophage-like particle from Bartonella bacilliformis. Microbiology 146(Pt 3):599–609

Battistuzzi FU, Feijao A, Hedges SB (2004) A genomic timescale of prokaryote evolution: insights into the origin of methanogenesis, phototrophy, and the colonization of land. BMC Evol Biol 4:44

Batut J, Andersson SG, O'Callaghan D (2004) The evolution of chronic infection strategies in the alpha-proteobacteria. Nat Rev Microbiol 2(12):933–945

Berglund EC, Frank AC, Calteau A, Vinnere Pettersson O, Granberg F, Eriksson AS, Näslund K, Holmberg M, Lindroos H, Andersson SG (2009) Run-off replication of host-adaptability genes is associated with gene transfer agents in the genome of mouse-infecting Bartonella grahamii. PLoS Genet 5(7):e1000546

Bertani G (1999) Transduction-like gene transfer in the methanogen Methanococcus voltae. J Bacteriol 181(10):2992–3002

Bertani G, Baresi L (1987) Genetic transformation in the methanogen Methanococcus voltae PS. J Bacteriol 169(6):2730–2738

Biebl H, Allgaier M, Tindall BJ, Koblizek M, Lünsdorf H, Pukall R, Wagner-Döbler I (2005) Dinoroseobacter shibae gen. Nov., sp. nov., a new aerobic phototrophic bacterium isolated from dinoflagellates. Int J Syst Evol Microbiol 55(Pt 3):1089–1096

Biers EJ, Wang K, Pennington C, Belas R, Chen F, Moran MA (2008) Occurrence and expression of gene transfer agent genes in marine bacterioplankton. Appl Environ Microbiol 74 (10):2933–2939

Black LW, Rao VB (2012) Structure, assembly, and DNA packaging of the bacteriophage T4 head. Adv Virus Res 82:119–153

Bobay LM, Touchon M, Rocha EP (2014) Pervasive domestication of defective prophages by bacteria. Proc Natl Acad Sci U S A 111(33):12127–12132

Bobik TA, Lehman BP, Yeates TO (2015) Bacterial microcompartments: widespread prokaryotic organelles for isolation and optimization of metabolic pathways. Mol Microbiol 98(2):193–207

Bohlin J (2015) Genome expansion in bacteria: the curios case of chlamydia trachomatis. BMC Res Notes 8:512

Bondarev V, Richter M, Romano S, Piel J, Schwedt A, Schulz-Vogt HN (2013) The genus Pseudovibrio contains metabolically versatile bacteria adapted for symbiosis. Environ Microbiol 15(7):2095–2113

Bottomley MJ, Muraglia E, Bazzo R, Carfì A (2007) Molecular insights into quorum sensing in the human pathogen Pseudomonas aeruginosa from the structure of the virulence regulator LasR bound to its autoinducer. J Biol Chem 282(18):13592–13600

Boussau B, Karlberg EO, Frank AC, Legault BA, Andersson SG (2004) Computational inference of scenarios for alpha-proteobacterial genome evolution. Proc Natl Acad Sci U S A 101 (26):9722–9727

Bramucci AR, Labeeuw L, Orata FD, Ryan EM, Malmstrom RR, Case RJ (2018) The bacterial Symbiont *Phaeobacter inhibens* shapes the life history of its algal host *Emiliania huxleyi*. Front Mar Sci 5(188):1–12

Brimacombe CA, Ding H, Beatty JT (2014) Rhodobacter capsulatus DprA is essential for RecA-mediated gene transfer agent (RcGTA) recipient capability regulated by quorum-sensing and the CtrA response regulator. Mol Microbiol 92(6):1260–1278

Brimacombe CA, Ding H, Johnson JA, Beatty JT (2015) Homologues of genetic transformation DNA import genes are required for Rhodobacter capsulatus gene transfer agent recipient capability regulated by the response regulator CtrA. J Bacteriol 197(16):2653–2663

Brimacombe CA, Stevens A, Jun D, Mercer R, Lang AS, Beatty JT (2013) Quorum-sensing regulation of a capsular polysaccharide receptor for the Rhodobacter capsulatus gene transfer agent (RcGTA). Mol Microbiol 87(4):802–817

Brown AMV (2018) Endosymbionts of plant-parasitic nematodes. Annu Rev Phytopathol 56:225–242

Brumin M, Levy M, Ghanim M (2012) Transovarial transmission of rickettsia spp. and organ-specific infection of the whitefly Bemisia tabaci. Appl Environ Microbiol 78(16):5565–5574
Buchan A, González JM, Moran MA (2005) Overview of the marine roseobacter lineage. Appl Environ Microbiol 71(10):5665–5677
Bulgheresi S (2016) Bacterial cell biology outside the streetlight. Environ Microbiol 18 (8):2305–2318
Buxton RS (1976) Prophage mutation causing heat inducibility of defective Bacillus subtilis bacteriophage PBSX. J Virol 20(1):22–28
Calderaro A, Dettori G, Collini L, Ragni P, Grillo R, Cattani P, Fadda G, Chezzi C (1998a) Bacteriophages induced from weakly beta-haemolytic human intestinal spirochaetes by mitomycin C. J Basic Microbiol 38(5–6):323–335
Calderaro A, Dettori G, Grillo R, Plaisant P, Amalfitano G, Chezzi C (1998b) Search for bacteriophages spontaneously occurring in cultures of haemolytic intestinal spirochaetes of human and animal origin. J Basic Microbiol 38(5–6):313–322
Campaniello D, Sinigaglia M (2017) Wine spoiling phenomena, The microbiological quality of food: foodborne spoilers. Woodhead Publishing, Cambridge, pp 237–255
Campos IA, Ximenes EA, Carvalho Júnior CH, de Mesquita AR, Silva JB, Maia MB, Franco ES, Medeiros PL, Peixoto CA, da Silva TG (2013) Zymomonas mobilis culture protects against sepsis by modulating the inflammatory response, alleviating bacterial burden and suppressing splenocyte apoptosis. Eur J Pharm Sci 48(1–2):1–8
Carvalho C, Susano M, Fernandes E, Santos S, Gannon B, Nicolau A, Gibbs P, Teixeira P, Azeredo J (2010) Method for bacteriophage isolation against target campylobacter strains. Lett Appl Microbiol 50(2):192–197
Casjens S, Weigele P (2005) DNA packaging by bacteriophage P22. Viral genome packaging machines: genetics, structure, and mechanism. Springer, Boston, MA
Casjens SR, Gilcrease EB (2009) Determining DNA packaging strategy by analysis of the termini of the chromosomes in tailed-bacteriophage virions. Methods Mol Biol 502:91–111
Chen F, Spano A, Goodman BE, Blasier KR, Sabat A, Jeffery E, Norris A, Shabanowitz J, Hunt DF, Lebedev N (2009) Proteomic analysis and identification of the structural and regulatory proteins of the Rhodobacter capsulatus gene transfer agent. J Proteome Res 8(2):967–973
Cheslock MA, Embers ME (2019) Human Bartonellosis: an underappreciated public health problem? Trop Med Infect Dis 4(2):69
Childers WS, Shapiro L (2014) A pseudokinase couples signaling pathways to enable asymmetric cell division in a bacterium. Microb Cell 2(1):29–32
Chomel BB, Boulouis HJ, Breitschwerdt EB, Kasten RW, Vayssier-Taussat M, Birtles RJ, Koehler JE, Dehio C (2009a) Ecological fitness and strategies of adaptation of Bartonella species to their hosts and vectors. Vet Res 40(2):29
Chomel BB, Kasten RW, Williams C, Wey AC, Henn JB, Maggi R, Carrasco S, Mazet J, Boulouis HJ, Maillard R, Breitschwerdt EB (2009b) Bartonella endocarditis: a pathology shared by animal reservoirsand patients. Ann N Y Acad Sci 1166:120–126
Chow CE, Suttle CA (2015) Biogeography of viruses in the sea. Annu Rev Virol 2(1):41–66
Choy YM, Fehmel F, Frank N, Stirm S (1975) Escherichia coli capsule bacteriophages. IV. Primary structure of the bacteriophage 29 receptor, the E. coli serotype 29 capsular polysaccharide. J Virol 16(3):581–590
Christie PJ, Vogel JP (2000) Bacterial type IV secretion: conjugation systems adapted to deliver effector molecules to host cells. Trends Microbiol 8(8):354–360
Coba de la Peña T, Fedorova E, Pueyo JJ, Lucas MM (2017) The Symbiosome: legume and rhizobia co-evolution toward a nitrogen-fixing organelle? Front Plant Sci 8:2229
Cobián Güemes AG, Youle M, Cantú VA, Felts B, Nulton J, Rohwer F (2016) Viruses as winners in the game of life. Annu Rev Virol 3(1):197–214
Crenn K, Duffieux D, Jeanthon C (2018) Bacterial Epibiotic communities of ubiquitous and abundant marine diatoms are distinct in short- and Long-term associations. Front Microbiol 9:2879

Cude WN, Buchan A (2013) Acyl-homoserine lactone-based quorum sensing in the Roseobacter clade: complex cell-to-cell communication controls multiple physiologies. Front Microbiol 4:336

Dale C, Wang B, Moran N, Ochman H (2003) Loss of DNA recombinational repair enzymes in the initial stages of genome degeneration. Mol Biol Evol 20(8):1188–1194

Dalebroux ZD, Svensson SL, Gaynor EC, Swanson MS (2010) ppGpp conjures bacterial virulence. Microbiol Mol Biol Rev 74(2):171–199

Dashiff A, Junka RA, Libera M, Kadouri DE (2011) Predation of human pathogens by the predatory bacteria Micavibrio aeruginosavorus and Bdellovibrio bacteriovorus. J Appl Microbiol 110(2):431–444

Davidson AR, Cardarelli L, Pell LG, Radford DR, Maxwell KL (2012) Long noncontractile tail machines of bacteriophages. Adv Exp Med Biol 726:115–142

Davison J (1999) Genetic exchange between bacteria in the environment. Plasmid 42(2):73–91

Day NPJ, Newton PN (2017) Scrub typhus and other tropical Rickettsioses. In: Cohen J, Powderly WG, Opal SP (eds) Infectious diseases, 4th edn. Elsevier, Amsterdam, pp 1091–1097

de Figueiredo P, Ficht TA, Rice-Ficht A, Rossetti CA, Adams LG (2015) Pathogenesis and immunobiology of brucellosis: review of Brucella-host interactions. Am J Pathol 185 (6):1505–1517

Dedeine F, Vavre F, Fleury F, Loppin B, Hochberg ME, Bouletreau M (2001) Removing symbiotic Wolbachia bacteria specifically inhibits oogenesis in a parasitic wasp. Proc Natl Acad Sci U S A 98(11):6247–6252

Delaby M, Panis G, Viollier PH (2019) Bacterial cell cycle and growth phase switch by the essential transcriptional regulator CtrA. Nucleic Acids Res 47(20):10628–10644

Ding H, Grüll MP, Mulligan ME, Lang AS, Beatty JT (2019) Rhodobacter capsulatus gene transfer agent gene expression is a Bistable stochastic process repressed by an extracellular calcium-binding RTX protein homologue. J Bacteriol 201(23):e00430–e00419

Draskovic I, Dubnau D (2005) Biogenesis of a putative channel protein, ComEC, required for DNA uptake: membrane topology, oligomerization and formation of disulphide bonds. Mol Microbiol 55(3):881–896

Dubey BN, Lori C, Ozaki S, Fucile G, Plaza-Menacho I, Jenal U, Schirmer T (2016) Cyclic di-GMP mediates a histidine kinase/phosphatase switch by noncovalent domain cross-linking. Sci Adv 2 (9):e1600823

Duda RL, Martincic K, Xie Z, Hendrix RW (1995) Bacteriophage HK97 head assembly. FEMS Microbiol Rev 17(1–2):41–46

Durham BP, Dearth SP, Sharma S, Amin SA, Smith CB, Campagna SR, Armbrust EV, Moran MA (2017) Recognition cascade and metabolite transfer in a marine bacteria-phytoplankton model system. Environ Microbiol 19(9):3500–3513

Durham BP, Sharma S, Luo H, Smith CB, Amin SA, Bender SJ, Dearth SP, Van Mooy BA, Campagna SR, Kujawinski EB, Armbrust EV, Moran MA (2015) Cryptic carbon and sulfur cycling between surface ocean plankton. Proc Natl Acad Sci U S A 112(2):453–457

Eicher SC, Dehio C (2012) Bartonella entry mechanisms into mammalian host cells. Cell Microbiol 14(8):1166–1173

Eiserling F, Pushkin A, Gingery M, Bertani G (1999) Bacteriophage-like particles associated with the gene transfer agent of methanococcus voltae PS. J Gen Virol 80(Pt 12):3305–3308

Elifantz H, Horn G, Ayon M, Cohen Y, Minz D (2013) Rhodobacteraceae are the key members of the microbial community of the initial biofilm formed in eastern Mediterranean coastal seawater. FEMS Microbiol Ecol 85(2):348–357

Elliott I, Pearson I, Dahal P, Thomas NV, Roberts T, Newton PN (2019) Scrub typhus ecology: a systematic review of Orientia in vectors and hosts. Parasit Vectors 12(1):513

Engel P, Salzburger W, Liesch M, Chang CC, Maruyama S, Lanz C, Calteau A, Lajus A, Médigue C, Schuster SC, Dehio C (2011) Parallel evolution of a type IV secretion system in radiating lineages of the host-restricted bacterial pathogen Bartonella. PLoS Genet 7(2): e1001296

Escobar MA, Dandekar AM (2003) Agrobacterium tumefaciens as an agent of disease. Trends Plant Sci 8(8):380–386
Ettema TJ, Andersson SG (2009) The alpha-proteobacteria: the Darwin finches of the bacterial world. Biol Lett 5(3):429–432
Feiss M, Rao VB (2012) The bacteriophage DNA packaging machine. Adv Exp Med Biol 726:489–509
Ferri E, Bain O, Barbuto M, Martin C, Lo N, Uni S, Landmann F, Baccei SG, Guerrero R, de Souza Lima S, Bandi C, Wanji S, Diagne M, Casiraghi M (2011) New insights into the evolution of Wolbachia infections in filarial nematodes inferred from a large range of screened species. PLoS One 6(6):e20843
Fillol-Salom A, Alsaadi A, Sousa JAM, Zhong L, Foster KR, Rocha EPC, Penadés JR, Ingmer H, Haaber J (2019) Bacteriophages benefit from generalized transduction. PLoS Pathog 15(7): e1007888
Flammann HT, Weckesser J (1984) Characterization of the cell wall and outer membrane of Rhodopseudomonas capsulata. J Bacteriol 159(1):191–198
Fogg PC, Colloms S, Rosser S, Stark M, Smith MC (2014) New applications for phage integrases. J Mol Biol 426(15):2703–2716
Fogg PC, Westbye AB, Beatty JT (2012) One for all or all for one: heterogeneous expression and host cell lysis are key to gene transfer agent activity in Rhodobacter capsulatus. PLoS One 7(8): e43772
Fogg PCM (2019) Identification and characterization of a direct activator of a gene transfer agent. Nat Commun 10(1):595
Foster SJ (1993) Analysis of Bacillus subtilis 168 prophage-associated lytic enzymes; identification and characterization of CWLA-related prophage proteins. J Gen Microbiol 139(12):3177–3184
Futuyma DJ, Moreno GT (1988) The evolution of ecological specialization. Annu Rev Ecol Syst 19:207–233
Gao NL, Chen J, Wang T, Lercher MJ, Chen WH (2019) Prokaryotic genome expansion is facilitated by phages and plasmids but impaired by CRISPR. Front Microbiol 10:2254
García-Aljaro C, Ballesté E, Muniesa M (2017) Beyond the canonical strategies of horizontal gene transfer in prokaryotes. Curr Opin Microbiol 38:95–105
Genard B, Larouche O, Nicolas JL, Miner P, Beaudin ML, Tremblay R (2014) Effect of the probiotic strain P*haeobacter gallaeciensis* after bacterial challenge on the complete larval development of P*ecten maximus*. Aquatic Living Resources 27(1):27–34
Geng H, Belas R (2010) Molecular mechanisms underlying roseobacter-phytoplankton symbioses. Curr Opin Biotechnol 21(3):332–338
Ghai R, Hernandez CM, Picazo A, Mizuno CM, Ininbergs K, Díez B, Valas R, DuPont CL, McMahon KD, Camacho A, Rodriguez-Valera F (2012) Metagenomes of Mediterranean coastal lagoons. Sci Rep 2:490
Ghequire MGK, De Mot R (2015) The Tailocin tale: peeling off phage tails. Trends Microbiol 23 (10):587–590
Giessen TW (2016) Encapsulins: microbial nanocompartments with applications in biomedicine, nanobiotechnology and materials science. Curr Opin Chem Biol 34:1–10
Gillespie JJ, Driscoll TP, Verhoeve VI, Utsuki T, Husseneder C, Chouljenko VN, Azad AF, Macaluso KR (2014) Genomic diversification in strains of rickettsia felis isolated from different arthropods. Genome Biol Evol 7(1):35–56
Gillette-Ferguson I, Hise AG, McGarry HF, Turner J, Esposito A, Sun Y, Diaconu E, Taylor MJ, Pearlman E (2004) Wolbachia-induced neutrophil activation in a mouse model of ocular onchocerciasis (river blindness). Infect Immun 72(10):5687–5692
Glaser L, Ionesco H, Schaeffer P (1966) Teichoic acids as components of a specific phage receptor in Bacillus subtilis. Biochim Biophys Acta 124(2):415–417
Goldstein EJ, Citron DM, Peraino VA, Cross SA (2003) Desulfovibrio desulfuricans bacteremia and review of human Desulfovibrio infections. J Clin Microbiol 41(6):2752–2754

Goodhead I, Darby AC (2015) Taking the pseudo out of pseudogenes. Curr Opin Microbiol 23:102–109

Gorbatyuk B, Marczynski GT (2005) Regulated degradation of chromosome replication proteins DnaA and CtrA in Caulobacter crescentus. Mol Microbiol 55(4):1233–1245

Greiman, S. E. (2015) Parasites within parasites: transmission and evolution of Neorickettsia in Digeneans. PhD University of North Dakota

Greiman SE, Rikihisa Y, Cain J, Vaughan JA, Tkach VV (2016) Germs within Worms: localization of Neorickettsia sp. within life cycle stages of the Digenean Plagiorchis elegans. Appl Environ Microbiol 82(8):2356–2362

Groth AC, Calos MP (2004) Phage integrases: biology and applications. J Mol Biol 335 (3):667–678

Guieysse B, Wuertz S (2012) Metabolically versatile large-genome prokaryotes. Curr Opin Biotechnol 23(3):467–473

Gupta A, Singh VK, Qazi GN, Kumar A (2001) Gluconobacter oxydans: its biotechnological applications. J Mol Microbiol Biotechnol 3(3):445–456

Guy L, Nystedt B, Toft C, Zaremba-Niedzwiedzka K, Berglund EC, Granberg F, Näslund K, Eriksson AS, Andersson SG (2013) A gene transfer agent and a dynamic repertoire of secretion systems hold the keys to the explosive radiation of the emerging pathogen Bartonella. PLoS Genet 9(3):e1003393

Hackstadt T (1996) The biology of rickettsiae. Infect Agents Dis 5(3):127–143

Hackstein JHP, de Graaf R (2018) Anaerobic ciliates and their Methanogenic Endosymbionts. In: Hackstein JHP (ed) (Endo)symbiotic Methanogenic Archaea, 2nd edn. Springer Nature Switzerland, Cham, pp 19–21

Harms A, Segers FH, Quebatte M, Mistl C, Manfredi P, Körner J, Chomel BB, Kosoy M, Maruyama S, Engel P, Dehio C (2017) Evolutionary dynamics of Pathoadaptation revealed by three independent acquisitions of the VirB/D4 type IV secretion system in Bartonella. Genome Biol Evol 9(3):761–776

Harrison E, Hall JPJ, Paterson S, Spiers AJ, Brockhurst MA (2017) Conflicting selection alters the trajectory of molecular evolution in a tripartite bacteria-plasmid-phage interaction. Mol Ecol 26 (10):2757–2764

Helgstrand C, Wikoff WR, Duda RL, Hendrix RW, Johnson JE, Liljas L (2003) The refined structure of a protein catenane: the HK97 bacteriophage capsid at 3.44 a resolution. J Mol Biol 334(5):885–899

Hendrix RW, Johnson JE (2012) Bacteriophage HK97 capsid assembly and maturation. Adv Exp Med Biol 726:351–363

Hernando-Pérez M, Pascual E, Aznar M, Ionel A, Castón JR, Luque A, Carrascosa JL, Reguera D, de Pablo PJ (2014) The interplay between mechanics and stability of viral cages. Nanoscale 6 (5):2702–2709

Hoerauf A, Volkmann L, Hamelmann C, Adjei O, Autenrieth IB, Fleischer B, Büttner DW (2000) Endosymbiotic bacteria in worms as targets for a novel chemotherapy in filariasis. Lancet 355 (9211):1242–1243

Hosseini SR, Wagner A (2018) Genomic organization underlying deletional robustness in bacterial metabolic systems. Proc Natl Acad Sci U S A 115(27):7075–7080

Huang S, Zhang Y, Chen F, Jiao N (2011) Complete genome sequence of a marine roseophage provides evidence into the evolution of gene transfer agents in alphaproteobacteria. Virol J 8:124

Huisman R, Ovchinnikova E, Bisseling T, Limpens EHM (2012) Endocytic accommodation of microbes in plants. In: Šamaj J (ed) Endocytosis in plants. Springer-Verlag, Heidelberg, pp 271–295

Humphrey SB, Stanton TB, Jensen NS (1995) Mitomycin C induction of bacteriophages from Serpulina hyodysenteriae and Serpulina innocens. FEMS Microbiol Lett 134(1):97–101

Humphrey SB, Stanton TB, Jensen NS, Zuerner RL (1997) Purification and characterization of VSH-1, a generalized transducing bacteriophage of Serpulina hyodysenteriae. J Bacteriol 179 (2):323–329

Huo N, Gu Y, McCue KF, Alabed D, Thomson JG (2019) Complete genome sequence of *Agrobacterium fabrum* strain 1D159. Microbiol Resour Announc 8(19):e00207–e00219

Husnik F, Keeling PJ (2019) The fate of obligate endosymbionts: reduction, integration, or extinction. Curr Opin Genet Dev 58-59:1–8

Husnik F, McCutcheon JP (2016) Repeated replacement of an intrabacterial symbiont in the tripartite nested mealybug symbiosis. Proc Natl Acad Sci U S A 113(37):E5416–E5424

Husnik F, McCutcheon JP (2018) Functional horizontal gene transfer from bacteria to eukaryotes. Nat Rev Microbiol 16(2):67–79

Hynes AP, Mercer RG, Watton DE, Buckley CB, Lang AS (2012) DNA packaging bias and differential expression of gene transfer agent genes within a population during production and release of the Rhodobacter capsulatus gene transfer agent, RcGTA. Mol Microbiol 85 (2):314–325

Hynes AP, Shakya M, Mercer RG, Grüll MP, Bown L, Davidson F, Steffen E, Matchem H, Peach ME, Berger T, Grebe K, Zhaxybayeva O, Lang AS (2016) Functional and evolutionary characterization of a gene transfer Agent's multilocus "genome". Mol Biol Evol 33 (10):2530–2543

Illeghems K, De Vuyst L, Weckx S (2013) Complete genome sequence and comparative analysis of Acetobacter pasteurianus 386B, a strain well-adapted to the cocoa bean fermentation ecosystem. BMC Genomics 14:526

Iniesta AA, McGrath PT, Reisenauer A, McAdams HH, Shapiro L (2006) A phospho-signaling pathway controls the localization and activity of a protease complex critical for bacterial cell cycle progression. Proc Natl Acad Sci U S A 103(29):10935–10940

Jacobs C, Ausmees N, Cordwell SJ, Shapiro L, Laub MT (2003) Functions of the CckA histidine kinase in Caulobacter cell cycle control. Mol Microbiol 47(5):1279–1290

Jayabalan R, Waisundara VY (2019) Kombucha as a functional beverage. In: Grumezescu A, Holban AM (eds) Functional and medicinal beverages. Woodhead Publishing, Cambridge, pp 413–446

Jenal U, Reinders A, Lori C (2017) Cyclic di-GMP: second messenger extraordinaire. Nat Rev Microbiol 15(5):271–284

Jia N, Du J, Ding MZ, Gao F, Yuan YJ (2015) Genome sequence of bacillus endophyticus and analysis of its companion mechanism in the Ketogulonigenium vulgare-bacillus strain consortium. PLoS One 10(8):e0135104

Jiang SC, Paul JH (1998) Gene transfer by transduction in the marine environment. Appl Environ Microbiol 64(8):2780–2787

Johnston C, Martin B, Fichant G, Polard P, Claverys JP (2014) Bacterial transformation: distribution, shared mechanisms and divergent control. Nat Rev Microbiol 12(3):181–196

Kadouri D, Venzon NC, O'Toole GA (2007) Vulnerability of pathogenic biofilms to Micavibrio aeruginosavorus. Appl Environ Microbiol 73(2):605–614

Kalhoefer D, Thole S, Voget S, Lehmann R, Liesegang H, Wollher A, Daniel R, Simon M, Brinkhoff T (2011) Comparative genome analysis and genome-guided physiological analysis of Roseobacter litoralis. BMC Genomics 12:324

Kampfraath AA, Klasson L, Anvar SY, Vossen RHAM, Roelofs D, Kraaijeveld K, Ellers J (2019) Genome expansion of an obligate parthenogenesis-associated Wolbachia poses an exception to the symbiont reduction model. BMC Genomics 20(1):106

Kamtchum-Tatuene J, Makepeace BL, Benjamin L, Baylis M, Solomon T (2017) The potential role of Wolbachia in controlling the transmission of emerging human arboviral infections. Curr Opin Infect Dis 30(1):108–116

Kanehisa M, Furumichi M, Tanabe M, Sato Y, Morishima K (2017) KEGG: new perspectives on genomes, pathways, diseases and drugs. Nucleic Acids Res 45(D1):D353–D361

Kanehisa M, Goto S (2000) KEGG: kyoto encyclopedia of genes and genomes. Nucleic Acids Res 28(1):27–30

Karamata D, Pooley HM, Monod M (1987) Expression of heterologous genes for wall teichoic acid in Bacillus subtilis 168. Mol Gen Genet 207(1):73–81

Kešnerová L, Moritz R, Engel P (2016) Bartonella apis sp. nov., a honey bee gut symbiont of the class Alphaproteobacteria. Int J Syst Evol Microbiol 66(1):414–421

Kogay R, Neely TB, Birnbaum DP, Hankel CR, Shakya M, Zhaxybayeva O (2019) Machine-learning classification suggests that many Alphaproteobacterial Prophages May instead be gene transfer agents. Genome Biol Evol 11(10):2941–2953

Koonin EV (2015) The turbulent network dynamics of microbial evolution and the statistical tree of life. J Mol Evol 80(5–6):244–250

Koonin, E. V. (2016) Horizontal gene transfer: essentiality and evolvability in prokaryotes, and roles in evolutionary transitions. *F1000 Res*, 5

Koppenhöfer S, Wang H, Scharfe M, Kaever V, Wagner-Döbler I, Tomasch J (2019) Integrated transcriptional regulatory network of quorum sensing, replication control, and SOS response in. Front Microbiol 10:803

Kosoy M, Morway C, Sheff KW, Bai Y, Colborn J, Chalcraft L, Dowell SF, Peruski LF, Maloney SA, Baggett H, Sutthirattana S, Sidhirat A, Maruyama S, Kabeya H, Chomel BB, Kasten R, Popov V, Robinson J, Kruglov A, Petersen LR (2008) Bartonella tamiae sp. nov., a newly recognized pathogen isolated from three human patients from Thailand. J Clin Microbiol 46 (2):772–775

Kristensen DM, Waller AS, Yamada T, Bork P, Mushegian AR, Koonin EV (2013) Orthologous gene clusters and taxon signature genes for viruses of prokaryotes. J Bacteriol 195(5):941–950

Krogh S, O'Reilly M, Nolan N, Devine KM (1996) The phage-like element PBSX and part of the skin element, which are resident at different locations on the Bacillus subtilis chromosome, are highly homologous. Microbiology 142(Pt 8):2031–2040

Krupovic M, Cvirkaite-Krupovic V, Iranzo J, Prangishvili D, Koonin EV (2018) Viruses of archaea: structural, functional, environmental and evolutionary genomics. Virus Res 244:181–193

Krupovic M, Forterre P, Bamford DH (2010) Comparative analysis of the mosaic genomes of tailed archaeal viruses and proviruses suggests common themes for virion architecture and assembly with tailed viruses of bacteria. J Mol Biol 397(1):144–160

Kuchinski KS, Brimacombe CA, Westbye AB, Ding H, Beatty JT (2016) The SOS response master regulator LexA regulates the gene transfer agent of Rhodobacter capsulatus and represses transcription of the signal transduction protein CckA. J Bacteriol 198(7):1137–1148

Kuo CH, Moran NA, Ochman H (2009) The consequences of genetic drift for bacterial genome complexity. Genome Res 19(8):1450–1454

Kuo CH, Ochman H (2010) The extinction dynamics of bacterial pseudogenes. PLoS Genet 6(8): e1001050

Landmann F (2019) The Wolbachia Endosymbionts. Microbiol Spectr 7(2):BAI-0018-2019

Landmann F, Voronin D, Sullivan W, Taylor MJ (2011) Anti-filarial activity of antibiotic therapy is due to extensive apoptosis after Wolbachia depletion from filarial nematodes. PLoS Pathog 7 (11):e1002351

Lang AS, Beatty JT (2000) Genetic analysis of a bacterial genetic exchange element: the gene transfer agent of Rhodobacter capsulatus. Proc Natl Acad Sci U S A 97(2):859–864

Lang AS, Beatty JT (2007) Importance of widespread gene transfer agent genes in alpha-proteobacteria. Trends Microbiol 15(2):54–62

Lang AS, Westbye AB, Beatty JT (2017) The distribution, evolution, and roles of gene transfer agents in prokaryotic genetic exchange. Annu Rev Virol 4(1):87–104

Lang AS, Zhaxybayeva O, Beatty JT (2012) Gene transfer agents: phage-like elements of genetic exchange. Nat Rev Microbiol 10(7):472–482

Laub MT, Chen SL, Shapiro L, McAdams HH (2002) Genes directly controlled by CtrA, a master regulator of the Caulobacter cell cycle. Proc Natl Acad Sci U S A 99(7):4632–4637

Lazdunski AM, Ventre I, Sturgis JN (2004) Regulatory circuits and communication in gram-negative bacteria. Nat Rev Microbiol 2(7):581–592

Leung MM, Brimacombe CA, Spiegelman GB, Beatty JT (2012) The GtaR protein negatively regulates transcription of the gtaRI operon and modulates gene transfer agent (RcGTA) expression in Rhodobacter capsulatus. Mol Microbiol 83(4):759–774

Leung MM, Florizone SM, Taylor TA, Lang AS, Beatty JT (2010) The gene transfer agent of Rhodobacter capsulatus. Adv Exp Med Biol 675:253–264

Lever MA, Rogers KL, Lloyd KG, Overmann J, Schink B, Thauer RK, Hoehler TM, Jørgensen BB (2015) Life under extreme energy limitation: a synthesis of laboratory- and field-based investigations. FEMS Microbiol Rev 39(5):688–728

Liang Y, Zhang Y, Zhou C, Chen Z, Yang S, Yan C, Jiao N (2016) Complete genome sequence of the siphovirus Roseophage RDJLΦ 2 infecting Roseobacter denitrificans OCh114. Mar Genomics 25:17–19

Lindroos H, Vinnere O, Mira A, Repsilber D, Näslund K, Andersson SG (2006) Genome rearrangements, deletions, and amplifications in the natural population of Bartonella henselae. J Bacteriol 188(21):7426–7439

Lloyd RG, Rudolph CJ (2016) 25 years on and no end in sight: a perspective on the role of RecG protein. Curr Genet 62(4):827–840

Long SW, Zhang X, Zhang J, Ruble RP, Teel P, Yu XJ (2003) Evaluation of transovarial transmission and transmissibility of Ehrlichia chaffeensis (Rickettsiales: anaplasmataceae) in Amblyomma americanum (Acari: Ixodidae). J Med Entomol 40(6):1000–1004

Longchamp PF, Mauël C, Karamata D (1994) Lytic enzymes associated with defective prophages of Bacillus subtilis: sequencing and characterization of the region comprising the N-acetylmuramoyl-L-alanine amidase gene of prophage PBSX. Microbiology 140 (Pt 8):1855–1867

Lori C, Ozaki S, Steiner S, Böhm R, Abel S, Dubey BN, Schirmer T, Hiller S, Jenal U (2015) Cyclic di-GMP acts as a cell cycle oscillator to drive chromosome replication. Nature 523 (7559):236–239

Luo H, Csuros M, Hughes AL, Moran MA (2013) Evolution of divergent life history strategies in marine alphaproteobacteria. MBio 4(4):e00373–e00313

Luo H, Moran MA (2014) Evolutionary ecology of the marine Roseobacter clade. Microbiol Mol Biol Rev 78(4):573–587

López-Madrigal S, Gil R (2017) Et tu, brute? Not even intracellular mutualistic Symbionts escape horizontal gene transfer. Genes (Basel) 8(10):E247

Maggi RG, Breitschwerdt EB (2005) Isolation of bacteriophages from Bartonella vinsonii subsp. berkhoffii and the characterization of Pap31 gene sequences from bacterial and phage DNA. J Mol Microbiol Biotechnol 9(1):44–51

Mann TH, Seth Childers W, Blair JA, Eckart MR, Shapiro L (2016) A cell cycle kinase with tandem sensory PAS domains integrates cell fate cues. Nat Commun 7:11454

Marrs B (1974) Genetic recombination in Rhodopseudomonas capsulata. Proc Natl Acad Sci U S A 71(3):971–973

Mathur S, Singh R (2005) Antibiotic resistance in food lactic acid bacteria--a review. Int J Food Microbiol 105(3):281–295

Matson EG, Thompson MG, Humphrey SB, Zuerner RL, Stanton TB (2005) Identification of genes of VSH-1, a prophage-like gene transfer agent of Brachyspira hyodysenteriae. J Bacteriol 187 (17):5885–5892

Matson EG, Zuerner RL, Stanton TB (2007) Induction and transcription of VSH-1, a prophage-like gene transfer agent of Brachyspira hyodysenteriae. Anaerobe 13(2):89–97

McCutcheon JP, Boyd BM, Dale C (2019) The life of an insect Endosymbiont from the cradle to the grave. Curr Biol 29(11):R485–R495

McCutcheon JP, Keeling PJ (2014) Endosymbiosis: protein targeting further erodes the organelle/symbiont distinction. Curr Biol 24(14):R654–R655

McCutcheon JP, Moran NA (2011) Extreme genome reduction in symbiotic bacteria. Nat Rev Microbiol 10(1):13–26

McDaniel LD, Young E, Delaney J, Ruhnau F, Ritchie KB, Paul JH (2010) High frequency of horizontal gene transfer in the oceans. Science 330(6000):50

McDaniel LD, Young EC, Ritchie KB, Paul JH (2012) Environmental factors influencing gene transfer agent (GTA) mediated transduction in the subtropical ocean. PLoS One 7(8):e43506

McDonnell GE, Wood H, Devine KM, McConnell DJ (1994) Genetic control of bacterial suicide: regulation of the induction of PBSX in Bacillus subtilis. J Bacteriol 176(18):5820–5830

Mell JC, Redfield RJ (2014) Natural competence and the evolution of DNA uptake specificity. J Bacteriol 196(8):1471–1483

Mellbye BL, Spieck E, Bottomley PJ, Sayavedra-Soto LA (2017) Acyl-Homoserine lactone production in nitrifying bacteria of the genera Nitrosospira, Nitrobacter, and Nitrospira identified via a survey of putative quorum-sensing genes. Appl Environ Microbiol 83(22):e01540–e01517

Mercer RG, Callister SJ, Lipton MS, Pasa-Tolic L, Strnad H, Paces V, Beatty JT, Lang AS (2010) Loss of the response regulator CtrA causes pleiotropic effects on gene expression but does not affect growth phase regulation in Rhodobacter capsulatus. J Bacteriol 192(11):2701–2710

Mercer RG, Lang AS (2014) Identification of a predicted partner-switching system that affects production of the gene transfer agent RcGTA and stationary phase viability in Rhodobacter capsulatus. BMC Microbiol 14:71

Mercer RG, Quinlan M, Rose AR, Noll S, Beatty JT, Lang AS (2012) Regulatory systems controlling motility and gene transfer agent production and release in Rhodobacter capsulatus. FEMS Microbiol Lett 331(1):53–62

Miller TR, Belas R (2004) Dimethylsulfoniopropionate metabolism by Pfiesteria-associated Roseobacter spp. Appl Environ Microbiol 70(6):3383–3391

Mirajkar NS, Gebhart CJ (2014) Understanding the molecular epidemiology and global relationships of Brachyspira hyodysenteriae from swine herds in the United States: a multi-locus sequence typing approach. PLoS One 9(9):e107176

Moore TC, Pulscher LA, Caddell L, von Fricken ME, Anderson BD, Gonchigoo B, Gray GC (2018) Evidence for transovarial transmission of tick-borne rickettsiae circulating in northern Mongolia. PLoS Negl Trop Dis 12(8):e0006696

Moran NA (1996) Accelerated evolution and Muller's rachet in endosymbiotic bacteria. Proc Natl Acad Sci U S A 93(7):2873–2878

Moran NA, Bennett GM (2014) The tiniest tiny genomes. Annu Rev Microbiol 68:195–215

Moran NA, McCutcheon JP, Nakabachi A (2008) Genomics and evolution of heritable bacterial symbionts. Annu Rev Genet 42:165–190

Mortier-Barrière I, Velten M, Dupaigne P, Mirouze N, Piétrement O, McGovern S, Fichant G, Martin B, Noirot P, Le Cam E, Polard P, Claverys JP (2007) A key presynaptic role in transformation for a widespread bacterial protein: DprA conveys incoming ssDNA to RecA. Cell 130(5):824–836

Motro Y, La T, Bellgard MI, Dunn DS, Phillips ND, Hampson DJ (2009) Identification of genes associated with prophage-like gene transfer agents in the pathogenic intestinal spirochaetes Brachyspira hyodysenteriae, Brachyspira pilosicoli and Brachyspira intermedia. Vet Microbiol 134(3–4):340–345

Nadell CD, Xavier JB, Levin SA, Foster KR (2008) The evolution of quorum sensing in bacterial biofilms. PLoS Biol 6(1):e14

Naito M, Pawlowska TE (2016) Defying Muller's ratchet: ancient heritable Endobacteria escape extinction through retention of recombination and genome plasticity. MBio 7(3):e02057–e02015

Nelson CA, Saha S, Mead PS (2016) Cat-scratch disease in the United States, 2005-2013. Emerg Infect Dis 22(10):1741–1746

Newton RJ, Griffin LE, Bowles KM, Meile C, Gifford S, Givens CE, Howard EC, King E, Oakley CA, Reisch CR, Rinta-Kanto JM, Sharma S, Sun S, Varaljay V, Vila-Costa M, Westrich JR,

Moran MA (2010) Genome characteristics of a generalist marine bacterial lineage. ISME J 4 (6):784–798
Nichols RJ, Cassidy-Amstutz C, Chaijarasphong T, Savage DF (2017) Encapsulins: molecular biology of the shell. Crit Rev Biochem Mol Biol 52(5):583–594
Nicks T, Rahn-Lee L (2017) Inside out: archaeal Ectosymbionts suggest a second model of reduced-genome evolution. Front Microbiol 8:384
O'Malley MA (2015) Endosymbiosis and its implications for evolutionary theory. Proc Natl Acad Sci U S A 112(33):10270–10277
Ochman H (2005) Genomes on the shrink. Proc Natl Acad Sci U S A 102(34):11959–11960
Ogata H, Goto S, Sato K, Fujibuchi W, Bono H, Kanehisa M (1999) KEGG: kyoto encyclopedia of genes and genomes. Nucleic Acids Res 27(1):29–34
Okamoto K, Mudd JA, Mangan J, Huang WM, Subbaiah TV, Marmur J (1968a) Properties of the defective phage of Bacillus subtilis. J Mol Biol 34(3):413–428
Okamoto K, Mudd JA, Marmur J (1968b) Conversion of Bacillus subtilis DNA to phage DNA following mitomycin C induction. J Mol Biol 34(3):429–437
Oldroyd GE, Murray JD, Poole PS, Downie JA (2011) The rules of engagement in the legume-rhizobial symbiosis. Annu Rev Genet 45:119–144
Olsen SC, Palmer MV (2014) Advancement of knowledge of Brucella over the past 50 years. Vet Pathol 51(6):1076–1089
Olszak T, Latka A, Roszniowski B, Valvano MA, Drulis-Kawa Z (2017) Phage life cycles behind bacterial biodiversity. Curr Med Chem 24(36):3987–4001
Orr MW, Donaldson GP, Severin GB, Wang J, Sintim HO, Waters CM, Lee VT (2015) Oligoribonuclease is the primary degradative enzyme for pGpG in Pseudomonas aeruginosa that is required for cyclic-di-GMP turnover. Proc Natl Acad Sci U S A 112(36):E5048–E5057
Ozaki S (2019) Regulation of replication initiation: lessons from Caulobacter crescentus. Genes Genet Syst 94(5):183–196
Pallegar P, Peña-Castillo L, Langille E, Gomelsky M, Lang AS (2020) Cyclic di-GMP-mediated regulation of gene transfer and motility in Rhodobacter capsulatus. J Bacteriol 202(2):e00554–e00519
Paris DH, Day NPJ (2014) Tropical Rickettsial infections. In: Farrar J, Hotez PJ, Junghanss T, Kang G, Lalloo D, White NJ (eds) Manson's tropical infectious diseases, 23rd edn. Elsevier, Amsterdam, pp 273–291
Parkinson JS (2016) Classic spotlight: the discovery of bacterial transduction. J Bacteriol 198 (21):2899–2900
Patzelt D, Wang H, Buchholz I, Rohde M, Gröbe L, Pradella S, Neumann A, Schulz S, Heyber S, Münch K, Münch R, Jahn D, Wagner-Döbler I, Tomasch J (2013) You are what you talk: quorum sensing induces individual morphologies and cell division modes in Dinoroseobacter shibae. ISME J 7(12):2274–2286
Paul JH (2008) Prophages in marine bacteria: dangerous molecular time bombs or the key to survival in the seas? ISME J 2(6):579–589
Perlman SJ, Hunter MS, Zchori-Fein E (2006) The emerging diversity of rickettsia. Proc Biol Sci 273(1598):2097–2106
Peña-Castillo L, Mercer RG, Gurinovich A, Callister SJ, Wright AT, Westbye AB, Beatty JT, Lang AS (2014) Gene co-expression network analysis in Rhodobacter capsulatus and application to comparative expression analysis of Rhodobacter sphaeroides. BMC Genomics 15:730
Pina M, Bize A, Forterre P, Prangishvili D (2011) The archeoviruses. FEMS Microbiol Rev 35 (6):1035–1054
Plisson C, Drucker M, Blanc S, German-Retana S, Le Gall O, Thomas D, Bron P (2003) Structural characterization of HC-pro, a plant virus multifunctional protein. J Biol Chem 278 (26):23753–23761
Popovici J, Moreira LA, Poinsignon A, Iturbe-Ormaetxe I, McNaughton D, O'Neill SL (2010) Assessing key safety concerns of a Wolbachia-based strategy to control dengue transmission by Aedes mosquitoes. Mem Inst Oswaldo Cruz 105(8):957–964

Prangishvili D, Bamford DH, Forterre P, Iranzo J, Koonin EV, Krupovic M (2017) The enigmatic archaeal virosphere. Nat Rev Microbiol 15(12):724–739
Pratama AA, van Elsas JD (2018) The 'Neglected' soil Virome - potential role and impact. Trends Microbiol 26(8):649–662
Québatte M, Christen M, Harms A, Körner J, Christen B, Dehio C (2017) Gene transfer agent promotes Evolvability within the fittest subpopulation of a bacterial pathogen. Cell Syst 4 (6):611–621.e6
Québatte M, Dehio C (2019) Bartonella gene transfer agent: evolution, function, and proposed role in host adaptation. Cell Microbiol 21(11):e13068
Québatte M, Dick MS, Kaever V, Schmidt A, Dehio C (2013) Dual input control: activation of the Bartonella henselae VirB/D4 type IV secretion system by the stringent sigma factor RpoH1 and the BatR/BatS two-component system. Mol Microbiol 90(4):756–775
Rae BD, Long BM, Badger MR, Price GD (2013) Functions, compositions, and evolution of the two types of carboxysomes: polyhedral microcompartments that facilitate CO2 fixation in cyanobacteria and some proteobacteria. Microbiol Mol Biol Rev 77(3):357–379
Rakhuba DV, Kolomiets EI, Dey ES, Novik GI (2010) Bacteriophage receptors, mechanisms of phage adsorption and penetration into host cell. Pol J Microbiol 59:145–155
Rao RU, Moussa H, Weil GJ (2002) Brugia malayi: effects of antibacterial agents on larval viability and development in vitro. Exp Parasitol 101(1):77–81
Rao VB, Black LW (2010) Structure and assembly of bacteriophage T4 head. Virol J 7:356
Raoult D (2015) Introduction to Rickettsioses, Ehrlichioses, and Anaplasmosis. In: Bennett JE, Dolin R, Blaser MJ (eds) Mandell, Douglas, and Bennett's principles and practice of infectious diseases, 8th edn. Elsevier Saunders, Philadelphia, pp 2194–2197
Rapp BJ, Wall JD (1987) Genetic transfer in Desulfovibrio desulfuricans. Proc Natl Acad Sci U S A 84(24):9128–9130
Reddy PM, Ladha JK, So RB, Hernandez RJ, Ramos MC, Angeles OR, Dazzo FB, de Bruijn FJ (1997) Rhizobial communication with rice roots: induction of phenotyphic changes, mode of invasion and extent of colonization. Plant Soil 194:81–98
Redfield RJ, Soucy SM (2018) Evolution of bacterial gene transfer agents. Front Microbiol 9:2527
Riclea R, Gleitzmann J, Bruns H, Junker C, Schulz B, Dickschat JS (2012) Algicidal lactones from the marine Roseobacter clade bacterium Ruegeria pomeroyi. Beilstein J Org Chem 8:941–950
Romano S, Aujoulat F, Jumas-Bilak E, Masnou A, Jeannot JL, Falsen E, Marchandin H, Teyssier C (2009) Multilocus sequence typing supports the hypothesis that Ochrobactrum anthropi displays a human-associated subpopulation. BMC Microbiol 9:267
Roos WH, Gertsman I, May ER, Brooks CL, Johnson JE, Wuite GJ (2012) Mechanics of bacteriophage maturation. Proc Natl Acad Sci U S A 109(7):2342–2347
Ross P, Mayer R, Weinhouse H, Amikam D, Huggirat Y, Benziman M, de Vroom E, Fidder A, de Paus P, Sliedregt LA (1990) The cyclic diguanylic acid regulatory system of cellulose synthesis in Acetobacter xylinum. Chemical synthesis and biological activity of cyclic nucleotide dimer, trimer, and phosphothioate derivatives. J Biol Chem 265(31):18933–18943
Roux S, Krupovic M, Debroas D, Forterre P, Enault F (2013) Assessment of viral community functional potential from viral metagenomes may be hampered by contamination with cellular sequences. Open Biol 3(12):130160
Römling U, Galperin MY (2017) Discovery of the second messenger cyclic di-GMP. Methods Mol Biol 1657:1–8
Sabater-Muñoz B, Toft C, Alvarez-Ponce D, Fares MA (2017) Chance and necessity in the genome evolution of endosymbiotic bacteria of insects. ISME J 11(6):1291–1304
Saier MH, Reddy BL (2015) Holins in bacteria, eukaryotes, and archaea: multifunctional xenologues with potential biotechnological and biomedical applications. J Bacteriol 197 (1):7–17
Saint André A, Blackwell NM, Hall LR, Hoerauf A, Brattig NW, Volkmann L, Taylor MJ, Ford L, Hise AG, Lass JH, Diaconu E, Pearlman E (2002) The role of endosymbiotic Wolbachia bacteria in the pathogenesis of river blindness. Science 295(5561):1892–1895

Santos JFM, Vasconcelos J, de Souza JR, Coutinho EM, Montenegro SML, Azevedo-Ximenes E (2004) The effect of Zymomonas mobilis culture on experimental Schistosoma mansoni infection. Rev Soc Bras Med Trop 37(6):502–504

Sarig S, Kapulnik Y, Okon Y (1986) Effect of Azospirillum inoculation on nitrogen fixation and growth of several winter legumes. Plant Soil 90:335–342

Schaefer AL, Taylor TA, Beatty JT, Greenberg EP (2002) Long-chain acyl-homoserine lactone quorum-sensing regulation of Rhodobacter capsulatus gene transfer agent production. J Bacteriol 184(23):6515–6521

Scholl D (2017) Phage tail-like Bacteriocins. Annu Rev Virol 4(1):453–467

Seaman E, Tarmy E, Marmur J (1964) Inducible phages of Bacillus subtilis. Biochemistry 3:607–613

Segers FH, Kešnerová L, Kosoy M, Engel P (2017) Genomic changes associated with the evolutionary transition of an insect gut symbiont into a blood-borne pathogen. ISME J 11 (5):1232–1244

Sela I, Wolf YI, Koonin EV (2016) Theory of prokaryotic genome evolution. Proc Natl Acad Sci U S A 113(41):11399–11407

Senthilkumar M, Madhaiyan M, Sundaram S, Kannaiyan S (2009) Intercellular colonization and growth promoting effects of Methylobacterium sp. with plant-growth regulators on rice (Oryza sativa L. cv CO-43). Microbiol Res 164(1):92–104

Serbus LR, Casper-Lindley C, Landmann F, Sullivan W (2008) The genetics and cell biology of Wolbachia-host interactions. Annu Rev Genet 42:683–707

Seyedsayamdost MR, Case RJ, Kolter R, Clardy J (2011) The Jekyll-and-Hyde chemistry of Phaeobacter gallaeciensis. Nat Chem 3(4):331–335

Shakya M, Soucy SM, Zhaxybayeva O (2017) Insights into origin and evolution of α-proteobacterial gene transfer agents. Virus Evol 3(2):vex036

Sherlock D, Leong JX, Fogg PCM (2019) Identification of the first gene transfer agent (GTA) small Terminase in Rh*odobacter capsulatus a*nd its role in GTA production and packaging of DNA. J Virol 93(23):e01328–e01319

Shikuma NJ, Pilhofer M, Weiss GL, Hadfield MG, Jensen GJ, Newman DK (2014) Marine tubeworm metamorphosis induced by arrays of bacterial phage tail-like structures. Science 343(6170):529–533

Simm R, Morr M, Kader A, Nimtz M, Römling U (2004) GGDEF and EAL domains inversely regulate cyclic di-GMP levels and transition from sessility to motility. Mol Microbiol 53 (4):1123–1134

Sinha S, Mell JC, Redfield RJ (2012) Seventeen Sxy-dependent cyclic AMP receptor protein site-regulated genes are needed for natural transformation in Haemophilus influenzae. J Bacteriol 194(19):5245–5254

Slater SC, Goldman BS, Goodner B, Setubal JC, Farrand SK, Nester EW, Burr TJ, Banta L, Dickerman AW, Paulsen I, Otten L, Suen G, Welch R, Almeida NF, Arnold F, Burton OT, Du Z, Ewing A, Godsy E, Heisel S, Houmiel KL, Jhaveri J, Lu J, Miller NM, Norton S, Chen Q, Phoolcharoen W, Ohlin V, Ondrusek D, Pride N, Stricklin SL, Sun J, Wheeler C, Wilson L, Zhu H, Wood DW (2009) Genome sequences of three agrobacterium biovars help elucidate the evolution of multichromosome genomes in bacteria. J Bacteriol 191(8):2501–2511

Solioz M, Yen HC, Marris B (1975) Release and uptake of gene transfer agent by Rhodopseudomonas capsulata. J Bacteriol 123(2):651–657

Spaepen S, Vanderleyden J, Remans R (2007) Indole-3-acetic acid in microbial and microorganism-plant signaling. FEMS Microbiol Rev 31(4):425–448

Sriswasdi S, Yang CC, Iwasaki W (2017) Generalist species drive microbial dispersion and evolution. Nat Commun 8(1):1162

Stanton TB, Humphrey SB, Bayles DO, Zuerner RL (2009) Identification of a divided genome for VSH-1, the prophage-like gene transfer agent of Brachyspira hyodysenteriae. J Bacteriol 191 (5):1719–1721

Stanton TB, Humphrey SB, Sharma VK, Zuerner RL (2008) Collateral effects of antibiotics: carbadox and metronidazole induce VSH-1 and facilitate gene transfer among Brachyspira hyodysenteriae strains. Appl Environ Microbiol 74(10):2950–2956

Stanton TB, Matson EG, Humphrey SB (2001) Brachyspira (Serpulina) hyodysenteriae gyrB mutants and interstrain transfer of coumermycin a(1) resistance. Appl Environ Microbiol 67 (5):2037–2043

Star L, Matan O, Dardanelli MS, Kapulnik Y, Burdman S, Okon Y (2012) The Vicia sativa spp. nigra - rhizobium leguminosarum bv. Viciae symbiotic interaction is improved by Azospirillum brasilense. Plant Soil 356(1–2):165–174

Steenhoudt O, Vanderleyden J (2000) Azospirillum, a free-living nitrogen-fixing bacterium closely associated with grasses genetic, biochemical and ecological aspects. FEMS Microbiol Rev 24:487–506

Suhanovsky MM, Teschke CM (2015) Nature's favorite building block: deciphering folding and capsid assembly of proteins with the HK97-fold. Virology 479-480:487–497

Sun D (2018) Pull in and push out: mechanisms of horizontal gene transfer in bacteria. Front Microbiol 9:2154

Swingley WD, Sadekar S, Mastrian SD, Matthies HJ, Hao J, Ramos H, Acharya CR, Conrad AL, Taylor HL, Dejesa LC, Shah MK, O'huallachain ME, Lince MT, Blankenship RE, Beatty JT, Touchman JW (2007) The complete genome sequence of Roseobacter denitrificans reveals a mixotrophic rather than photosynthetic metabolism. J Bacteriol 189(3):683–690

Takemaru K, Mizuno M, Sato T, Takeuchi M, Kobayashi Y (1995) Complete nucleotide sequence of a skin element excised by DNA rearrangement during sporulation in Bacillus subtilis. Microbiology 141(Pt 2):323–327

Takeuchi N, Kaneko K, Koonin EV (2014) Horizontal gene transfer can rescue prokaryotes from Muller's ratchet: benefit of DNA from dead cells and population subdivision. G3 (Bethesda) 4 (2):325–339

Tamarit D, Neuvonen MM, Engel P, Guy L, Andersson SGE (2018) Origin and evolution of the Bartonella gene transfer agent. Mol Biol Evol 35(2):451–464

Tang M, Bouchez O, Cruveiller S, Masson-Boivin C, Capela D (2020) Modulation of quorum sensing as an adaptation to nodule cell infection during experimental evolution of legume Symbionts. MBio 11(1):e03129–e03119

Taylor DJ, Alexander TJ (1971) The production of dysentery in swine by feeding cultures containing a spirochaete. Br Vet J 127(11):58–61

Taylor MJ (2003) Wolbachia in the inflammatory pathogenesis of human filariasis. Ann N Y Acad Sci 990:444–449

Thomas S, Alexander W, Gilligan J, Rikihisa Y (2017) The importance of Rickettsiales infections. In: Thomas S (ed) Rickettsiales: biology, molecular biology, epidemiology, and vaccine development. Springer International Publishing, AG Switzerland, Cham, pp 3–22

Tomasch J, Wang H, Hall ATK, Patzelt D, Preusse M, Petersen J, Brinkmann H, Bunk B, Bhuju S, Jarek M, Geffers R, Lang AS, Wagner-Döbler I (2018) Packaging of Dinoroseobacter shibae DNA into gene transfer agent particles is not random. Genome Biol Evol 10(1):359–369

Toyofuku M, Cárcamo-Oyarce G, Yamamoto T, Eisenstein F, Hsiao CC, Kurosawa M, Gademann K, Pilhofer M, Nomura N, Eberl L (2017) Prophage-triggered membrane vesicle formation through peptidoglycan damage in Bacillus subtilis. Nat Commun 8(1):481

Trott DJ, Oxberry SL, Hampson DJ (1997) Evidence for Serpulina hyodysenteriae being recombinant, with an epidemic population structure. Microbiology 143(Pt 10):3357–3365

Trubl G, Jang HB, Roux S, Emerson JB, Solonenko N, Vik DR, Solden L, Ellenbogen J, Runyon AT, Bolduc B, Woodcroft BJ, Saleska SR, Tyson GW, Wrighton KC, Sullivan MB, Rich VI (2018) Soil viruses are underexplored players in ecosystem carbon processing. mSystems 3(5): e00076–e00018

Tsai YM, Chang A, Kuo CH (2018) Horizontal gene acquisitions contributed to genome expansion in insect-symbiotic Spiroplasma clarkii. Genome Biol Evol 10(6):1526–1532

Tun-Garrido C, Bustos P, González V, Brom S (2003) Conjugative transfer of p42a from rhizobium etli CFN42, which is required for mobilization of the symbiotic plasmid, is regulated by quorum sensing. J Bacteriol 185(5):1681–1692

Umemori E, Sasaki Y, Amano K, Amano Y (1992) A phage in Bartonella bacilliformis. Microbiol Immunol 36(7):731–736

van Tienderen PH (1997) Generalists, specialists, and the evolution of phenotypic plasticity in sympatric populations of distinct species. Evolution 51(5):1372–1380

Vegge CS, Brøndsted L, Neve H, Mc Grath S, van Sinderen D, Vogensen FK (2005) Structural characterization and assembly of the distal tail structure of the temperate lactococcal bacteriophage TP901-1. J Bacteriol 187(12):4187–4197

Viswanathan V, Narjala A, Ravichandran A, Jayaprasad S, Siddaramappa S (2017) Evolutionary genomics of an ancient Prophage of the order Sphingomonadales. Genome Biol Evol 9 (3):646–658

Wagner-Döbler I, Ballhausen B, Berger M, Brinkhoff T, Buchholz I, Bunk B, Cypionka H, Daniel R, Drepper T, Gerdts G, Hahnke S, Han C, Jahn D, Kalhoefer D, Kiss H, Klenk HP, Kyrpides N, Liebl W, Liesegang H, Meincke L, Pati A, Petersen J, Piekarski T, Pommerenke C, Pradella S, Pukall R, Rabus R, Stackebrandt E, Thole S, Thompson L, Tielen P, Tomasch J, von Jan M, Wanphrut N, Wichels A, Zech H, Simon M (2010) The complete genome sequence of the algal symbiont Dinoroseobacter shibae: a hitchhiker's guide to life in the sea. ISME J 4 (1):61–77

Walburger A, Lazdunski C, Corda Y (2002) The Tol/pal system function requires an interaction between the C-terminal domain of TolA and the N-terminal domain of TolB. Mol Microbiol 44 (3):695–708

Walker DH (2017) Rickettsia. In: Quah SR (ed) International encyclopedia of public health, 2nd edn. Academic Press, Cambridge, pp 370–377

Wall JD, Weaver PF, Gest H (1975) Gene transfer agents, bacteriophages, and bacteriocins of Rhodopseudomonas capsulata. Arch Microbiol 105(3):217–224

Wang H, Tomasch J, Jarek M, Wagner-Döbler I (2014a) A dual-species co-cultivation system to study the interactions between Roseobacters and dinoflagellates. Front Microbiol 5:311

Wang H, Tomasch J, Michael V, Bhuju S, Jarek M, Petersen J, Wagner-Döbler I (2015) Identification of genetic modules mediating the Jekyll and Hyde interaction of Dinoroseobacter shibae with the Dinoflagellate Prorocentrum minimum. Front Microbiol 6:1262

Wang H, Ziesche L, Frank O, Michael V, Martin M, Petersen J, Schulz S, Wagner-Döbler I, Tomasch J (2014b) The CtrA phosphorelay integrates differentiation and communication in the marine alphaproteobacterium Dinoroseobacter shibae. BMC Genomics 15:130

Wang IN, Smith DL, Young R (2000) Holins: the protein clocks of bacteriophage infections. Annu Rev Microbiol 54:799–825

Wang R, Seyedsayamdost MR (2017) Roseochelin B, an Algaecidal natural product synthesized by the Roseobacter Phaeobacter inhibens in response to algal Sinapic acid. Org Lett 19 (19):5138–5141

Weaver PF, Wall JD, Gest H (1975) Characterization of Rhodopseudomonas capsulata. Arch Microbiol 105(3):207–216

Wernegreen JJ (2012) Endosymbiosis. Curr Biol 22(14):R555–R561

Westbye AB, Beatty JT, Lang AS (2017) Guaranteeing a captive audience: coordinated regulation of gene transfer agent (GTA) production and recipient capability by cellular regulators. Curr Opin Microbiol 38:122–129

Westbye AB, Kater L, Wiesmann C, Ding H, Yip CK, Beatty JT (2018) The protease ClpXP and the PAS domain protein DivL regulate CtrA and gene transfer agent production in Rhodobacter capsulatus. Appl Environ Microbiol 84(11):e00275–e00218

Westbye AB, Leung MM, Florizone SM, Taylor TA, Johnson JA, Fogg PC, Beatty JT (2013) Phosphate concentration and the putative sensor kinase protein CckA modulate cell lysis and release of the Rhodobacter capsulatus gene transfer agent. J Bacteriol 195(22):5025–5040

Whiting RA, Doyle LP, Spray RS (1921) Swine dysentery. Purdue Univ Agr Exp Sta Bull 257:1–15

Whitman WB, Ankwanda E, Wolfe RS (1982) Nutrition and carbon metabolism of Methanococcus voltae. J Bacteriol 149(3):852–863

Wibberg D, Blom J, Jaenicke S, Kollin F, Rupp O, Scharf B, Schneiker-Bekel S, Sczcepanowski R, Goesmann A, Setubal JC, Schmitt R, Pühler A, Schlüter A (2011) Complete genome sequencing of agrobacterium sp. H13-3, the former rhizobium lupini H13-3, reveals a tripartite genome consisting of a circular and a linear chromosome and an accessory plasmid but lacking a tumor-inducing Ti-plasmid. J Biotechnol 155(1):50–62

Williams HG, Benstead J, Frischer ME, Paul JH (1997) Alteration in plasmid DNA following natural transformation to populations of marine bacteria. Mol Mar Biol Biotechnol 6 (3):238–247

Wisniewski-Dyé F, Borziak K, Khalsa-Moyers G, Alexandre G, Sukharnikov LO, Wuichet K, Hurst GB, McDonald WH, Robertson JS, Barbe V, Calteau A, Rouy Z, Mangenot S, Prigent-Combaret C, Normand P, Boyer M, Siguier P, Dessaux Y, Elmerich C, Condemine G, Krishnen G, Kennedy I, Paterson AH, González V, Mavingui P, Zhulin IB (2011) Azospirillum genomes reveal transition of bacteria from aquatic to terrestrial environments. PLoS Genet 7 (12):e1002430

Wisniewski-Dyé F, Lozano L, Acosta-Cruz E, Borland S, Drogue B, Prigent-Combaret C, Rouy Z, Barbe V, Herrera AM, González V, Mavingui P (2012) Genome sequence of Azospirillum brasilense CBG497 and comparative analyses of Azospirillum Core and accessory genomes provide insight into niche adaptation. Genes (Basel) 3(4):576–602

Wolf YI, Koonin EV (2013) Genome reduction as the dominant mode of evolution. BioEssays 35 (9):829–837

Wood HE, Dawson MT, Devine KM, McConnell DJ (1990a) Characterization of PBSX, a defective prophage of Bacillus subtilis. J Bacteriol 172(5):2667–2674

Wood HE, Devine KM, McConnell DJ (1990b) Characterisation of a repressor gene (xre) and a temperature-sensitive allele from the Bacillus subtilis prophage, PBSX. Gene 96(1):83–88

Xavier MN, Paixao TA, den Hartigh AB, Tsolis RM, Santos RL (2010) Pathogenesis of B*rucella* spp. Open Vet Sci J 4(1):109–118

Yadav T, Carrasco B, Serrano E, Alonso JC (2014) Roles of Bacillus subtilis DprA and SsbA in RecA-mediated genetic recombination. J Biol Chem 289(40):27640–27652

Yakkala H, Samantarrai D, Gribskov M, Siddavattam D (2019) Comparative genome analysis reveals niche-specific genome expansion in Acinetobacter baumannii strains. PLoS One 14(6): e0218204

Yang Y, Cai L, Ma R, Xu Y, Tong Y, Huang Y, Jiao N, Zhang R (2017) A novel Roseosiphophage isolated from the oligotrophic South China Sea. Viruses 9(5):E109

Ye M, Sun M, Huang D, Zhang Z, Zhang H, Zhang S, Hu F, Jiang X, Jiao W (2019) A review of bacteriophage therapy for pathogenic bacteria inactivation in the soil environment. Environ Int 129:488–496

Yen HC, Hu NT, Marrs BL (1979) Characterization of the gene transfer agent made by an overproducer mutant of Rhodopseudomonas capsulata. J Mol Biol 131(2):157–168

Young M, Mauël C, Margot P, Karamata D (1989) Pseudo-allelic relationship between non-homologous genes concerned with biosynthesis of polyglycerol phosphate and polyribitol phosphate teichoic acids in Bacillus subtilis strains 168 and W23. Mol Microbiol 3 (12):1805–1812

Young R (2002) Bacteriophage holins: deadly diversity. J Mol Microbiol Biotechnol 4(1):21–36

Zhan Y, Chen F (2019) Bacteriophages that infect marine roseobacters: genomics and ecology. Environ Microbiol 21(6):1885–1895

Zhan Y, Huang S, Voget S, Simon M, Chen F (2016) A novel roseobacter phage possesses features of podoviruses, siphoviruses, prophages and gene transfer agents. Sci Rep 6:30372

Zhang X, Xu W, Liu Y, Cai M, Luo Z, Li M (2018) Metagenomics reveals microbial diversity and metabolic potentials of seawater and surface sediment from a Hadal biosphere at the yap trench. Front Microbiol 9:2402

Zhaxybayeva O, Doolittle WF (2011) Lateral gene transfer. Curr Biol 21(7):R242–R246

Zhou J, Zheng Q, Liu J, Du G, Chen J (2013) Indigenous plasmids of Bacillus megaterium WSH-002 involved in mutualism with Ketogulonicigenium vulgare WSH-001. Plasmid 70 (2):240–246

Zuerner RL, Stanton TB, Minion FC, Li C, Charon NW, Trott DJ, Hampson DJ (2004) Genetic variation in B*rachyspira:* chromosomal rearrangements and sequence drift distinguish B. *pilosicoli* from B. *hyodysenteriae*. Anaerobe 10(4):229–237

Chapter 3
Evolution from Free-Living Bacteria to Endosymbionts of Insects: Genomic Changes and the Importance of the Chaperonin GroEL

Beatriz Sabater-Muñoz and Christina Toft

Abstract Major insect lineages have independently acquired bacterial species, mainly from Gamma-proteobacteria and Bacteroidetes class, which could be nutritional mutualistic factories, facultative mutualists that protect against biotic and abiotic stresses, or reproductive manipulators (which alter the fertility of the host species in its benefit). Some of them are enclosed in bacteriocytes to assure their maternal transmission over generations. All of them show an increased level of genetic drift due to the small population size and the continuous population bottlenecking at each generation, processes that have shaped their genome, proteome, and morphology. Depending on the nature of the relationship, the degree of genome plasticity varies, i.e., obligate nutritional mutualistic symbionts have extremely small genomes lacking mobile elements, bacteriophages, or recombination machinery. Under these conditions, endosymbionts face high mutational pressures that may drive to extinction or symbiont replacement. How do then they survive for such long evolutionary time, and why do they show a genome stasis? In this chapter, after a brief introduction to the problem, we will focus on the genome changes suffered by these endosymbionts, and on the mutational robustness mechanisms, including the moonlighting chaperone GroEL that could explain their long prevalence from an evolutionary perspective by comparing them with free-living bacteria.

B. Sabater-Muñoz (✉)
Institute for Molecular and Cellular Biology of Plants (IBMCP), Spanish National Research Council (CSIC)–Polytechnic University of Valencia (UPV), Valencia, Spain

Smurfit Institute of Genetics, Trinity College of Dublin (TCD), Dublin, Ireland
e-mail: sabaterb@tcd.ie; beasamuo@upvnet.upv.es

C. Toft
Smurfit Institute of Genetics, Trinity College of Dublin (TCD), Dublin, Ireland

Institute for Integrative and Systems Biology (I2SysBio), Spanish National Research Council (CSIC)–University of Valencia (UV), Paterna, Spain
e-mail: toftc@tcd.ie; christina.toft@uv.es

M. Kloc (ed.), *Symbiosis: Cellular, Molecular, Medical and Evolutionary Aspects*, Results and Problems in Cell Differentiation 69,
https://doi.org/10.1007/978-3-030-51849-3_3

3.1 Introduction: A Brief History about Symbiosis and its Importance in Insects' Biology and Control

Symbiosis—from the Greek συμβίωσις (συμ = sym, within; βίωσις = biosis, living)—refers to any type of *close* and *long-term* interaction between two different organisms. The use of the term "symbiosis" became a controversy to describe the phenomenon of "living together" as reviewed in (Martin and Schwab 2012, 2013; Gontier 2016; Oborník 2019). Despite this controversy, symbiosis has been redefined in terms of the relationship between and physical location of partners. Depending on the type of relationships, symbiosis has been further classified as mutualism, commensalism, parasitism, parasitoidism, predation, amensalism, antagonism, or neutralism, whereas the partners are considered ectosymbionts or endosymbionts, depending on the relative physical location to one another. Symbiosis has been a process that from the evolutionary perspective has shaped the Life on Earth, being at the origin of the eukaryotic cell (in several bursts of endosymbiosis), facilitating one of the largest evolutionary leaps along with gene duplication, or at the birth of Plant kingdom. Some examples (eukaryotic cell and plants) are not being treated in this chapter as each of them deserves their own chapter or even own book (Deschamps et al. 2008; Zimmer 2009; Gontier 2016; Eme et al. 2017; Melnikov et al. 2019; Bowles et al. 2020; Fernández and Gabaldón 2020). As indicated, symbiosis refers to the close and long-term interaction between two different organisms, usually one microbe (mainly bacteria) and one eukaryote. Microbes are everywhere, around us, on us, and within us; thus, they have a high propensity to establish symbiotic relationships with any other organism on earth. The endosymbiotic process had enthralled scientists from several fields of Biology, but also from Chemistry, Physics, and Mathematics; here, we will focus on one of the most diverse groups on Earth, insects (arthropods), and their symbiotic bacteria and yeasts (Buchner 1965; Schwemmler and Gassner 1989; Sapp 2002; Bourtzis and Miller 2006, 2009; Zchori-Fein and Bourtzis 2012).

Within the Molecular-Genomic Era, the advances in molecular and 'omic technologies have been useful to determine "who is who" in the microbe–insect symbiotic relationship. They have also become an indispensable tool to identify and name the myriad of unculturable or as-yet-uncultured bacteria, yeasts, fungi, and viruses that were previously identified as symbionts by microscopy techniques early in the first three decades of the XX century (Buchner 1965; Sapp 2002; Engel and Moran 2013; De Cock et al. 2019). Likewise, they are at the center for determining the *long-term relationships* and their coevolutionary patterns (Fig. 3.1), along with their effects on the genomes of the bacterial endosymbiont, as we will explain later (Moran 2001; Russell et al. 2003; Gil et al. 2004; Moran et al. 2008; Chevignon et al. 2018). Demonstrating that unculturability is mainly due to gene loss and/or transfer of genes to the insect host, among other factors, metabolic interactions with the host limit both host and endosymbiont reproduction and success as independent partners, and what is more important, indicate the missing metabolites that would make these microbes being culturable outside of insect host–cell cultures (Darby

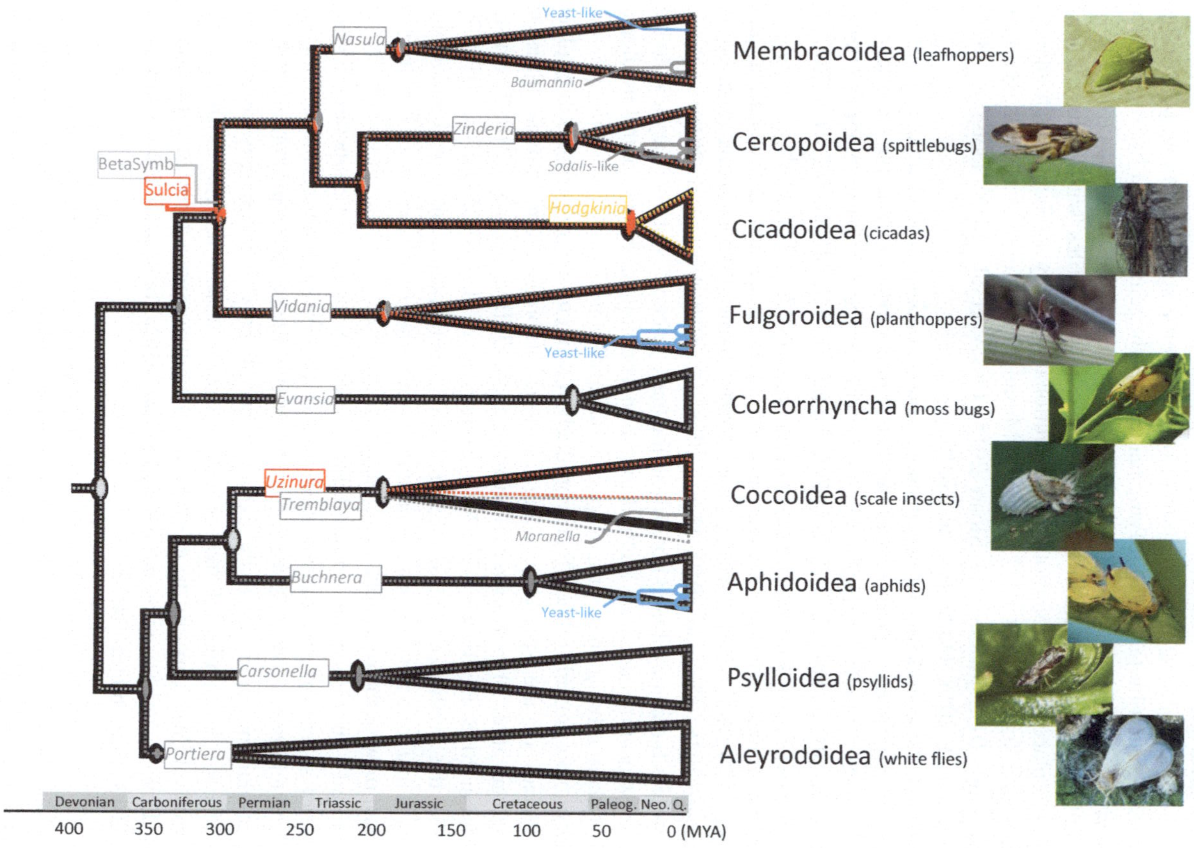

Fig. 3.1 Dated phylogeny of hemipteroid insects (Insecta: Hemiptera) based on genome or transcriptome sequences of 193 samples, along with the timescale in millions of years estimated using 23 fossil calibration points on a reduced dataset. Insect phylogeny is extracted from (Glenner et al. 2006; Johnson et al. 2018),

et al. 2005; Pontes and Dale 2006; Stewart 2012; Xu et al. 2016; Chevrette and Handelsman 2020). These new techniques (NGS, microbiome, and metabolome studies) have helped in separating between primary and secondary endosymbionts, identifying facultative symbionts, or giving importance to the viruses and yeasts. They also improved the identification of other bacterial species forming part of the insect stable microbiota, by identifying bacterial pathogens vectored by insects, and by describing the long-term nature of the coevolution with the host. Overall they shed light on the complexity of the endosymbiotic profiles with endosymbiont lineages replacement throughout the evolutionary history of hemipteran insects (Liesack and Stackebrandt 1992; Dale and Moran 2006; Moran et al. 2008; Bennett and Moran 2013; Moran and Sloan 2015; Ordax et al. 2015; Douglas 2015; Matsuura et al. 2018; Darboux et al. 2019).

Some of these endosymbionts, primary and secondary endosymbionts mainly, have been identified as the drivers of biological innovation (phenotypic complexity) of their hosts through adaptation, allowing many of them to survive abiotic (temperature mainly, as nowadays driving factor under the climate change scenario) and biotic stresses (parasitoid egg encapsulation, resistance to entomopathogenic bacteria and/or fungi), leading to diversification of host species (Schwemmler and Gassner 1989; Bourtzis and Miller 2006, 2009; Moran 2007; Desneux et al. 2018; Vorburger 2018; McLean 2019; Monticelli et al. 2019; Volf et al. 2019). Indeed, these studies have contributed not only to the basic sciences, but also they lead to the rise of new pest management methods, the biotechnological-based control methods, many of them relying on the use of specific bacterial symbionts (*Wolbachia* spp. and *Rickettsia* spp., mainly) that affect reproduction of their host to control either plant pest species or the species causing zoonotic outbreaks, and human illnesses (Berasategui et al. 2016; Lopez-Fernandez et al. 2017; Boucias et al. 2018; Raymann and Moran 2018; Deutscher et al. 2019; Somerville et al. 2019; Xie et al. 2019). In addition, the prediction of essential nutrients, metabolites, or shared metabolic pathways that indicate consortia requirement with other (endo-, ecto-, or free-living) symbiont or with the own host, is an ongoing outcome of these 'omics technologies that will give answers to the "culturability" processes for the as-yet-uncultured ones that are of special interest in bioremediation or pest/disease management (Douglas 2018). Obviously, by the aid of these 'omic techniques, the contribution of these reduced-genome symbionts to the extraordinary phenotypic complexity observed in their host species can be inferred. And, as we will show at the end of the chapter, their evolvability (capacity to evolve) and their persistence through evolutionary times (Liesack and Stackebrandt 1992; Stewart 2012; Hays et al. 2015; Chevignon et al. 2018; Sarhan et al. 2019).

Fig. 3.1 (continued) primary endosymbionts estimated symbiosis establishment is indicated under the names of the main genera (in red from Firmicutes/Bacteroidetes class, and in gray from Proteobacteria), along with some more recent symbiotic events (yeast-like symbionts, in blue) (adapted from Sudakaran et al. 2017). Insect pictures at the right column are representative from the Iberian Fauna collection, with special focus on pest species

In this chapter, we will focus on the molecular and evolutionary processes that have shaped the genomes of the bacterial endosymbionts of several insect species, comparing these processes, with the aid of experimental evolution, with free-living microorganisms.

3.2 The Process of (Endo)Symbiosis: Interaction between Bacteria and Host Cells

As said previously, insects are the home of a complex community of microorganisms, mainly belonging to, but not restricted to, bacteria. But how did some of these bacteria become (endo)symbionts? To achieve this fate, bacteria should establish a strict linkage with its host (mainly metabolic linkage), being perpetuated in time through efficient transgenerational transmission (mainly by vertical transfer with occasional horizontal transfer events) (Fig. 3.1) (Antonovics et al. 2017). Comparative studies have illuminated general patterns of insect–microbiota associations, finding that insect endosymbionts were derived from insect gut communities or phytobiomes in some cases, being shaped by the host diet, to keep the perfect metabolic functioning of the linked community and their host (Bright and Bulgheresi 2010; Colman et al. 2012; Augustinos et al. 2019; Itoh et al. 2019). Indeed, many of these symbiotic bacteria improve host's metabolism, by supplying the nutritionally imbalanced food source with the missing key components as aromatic amino acids, vitamins, or cofactors, as occurring in the aphid-*Buchnera aphidicola* system (Moran 2001; Colman et al. 2012; Gil and Latorre 2019; Bell-Roberts et al. 2019). The insect gut is generally divided into three regions, the foregut, the midgut, and the hindgut, each with a different pH and other physicochemical characteristics that constrict the microbiota diversity, and hence shaped the actual endosymbiont portfolio of each insect/arthropod species (Engel and Moran 2013; Lanan et al. 2016; Blow and Douglas 2019).

3.2.1 Transmission Model Affects Classification of Symbionts

To assure vertical transmission to the next generation, hosts have developed "organelles" (not true bacteria-derived organelles as defined in Theissen and Martin 2006 and Oborník 2019) that harbor the bacteria. From the evolutionary perspective, the starting point should be the infection of insect gut lumen, and from there, the development of bacteriocytes from gut cells, especially from those of the midgut, as many of these bacterial species show "free-living" relatives within the insect gut microbiome. The shape, distribution, and location of these bacteriocytes depend greatly on the insect host, finding them as intercalated cells within midgut cells, to being grouped into specific organs, the bacteriome. One of the simplest and youngest forms of these "organelles" is simply formed by the engulfment of the bacteria by the host cell membrane (called in some works, symbiosomal membrane), an of this example is the spheroid bodies of the diatom *Rhopalodia gibba* (not an insect species, but to date, we have not found any other early-stage symbiotic example); these are cell inclusions separated from the cytoplasm by an additional membrane that harbor the diazotrophic cyanobacteria (Adler et al. 2014). In this case, the "nitrosome" (nitrogen-fixing "organelle") has been described as an engulfment derived from the diatom membrane, produced ~25 MYA, and could be considered (the bacteria engulfment) the starting point of the more complex endosymbiosis of insects, from the host cell—symbiont interaction point of view (Adler et al. 2014). As many of the insect bacterial symbionts are derived from the gut microbiota or the phytobiome, the insect midgut has become the host organ of bacteriocytes (cells containing the bacterial symbionts) or the developmental source for the bacteriome or for bacterial pockets or crypts (Wallin 1927; Buchner 1965; Colman et al. 2012). One of the simplest forms is present in carpenter ants; their bacteriocytes are intercalated between midgut cells, hosting *Candidatus* Blochmannia floridanus cells free within its cytoplasm, without any symbiosomal membrane (Stoll et al. 2010). At an advanced stage of the endosymbiotic process, we can find an organ derived from the midgut of insects, the bacteriome or mycetome. In the case of aphids (as in other hemipteran insects), the bacteriome is a bilobular organ located between the foregut and ovaries, formed by 6 to 8 bacteriocytes (mycetocytes) hosting *Buchnera aphidicola* within host-derived symbiosomal membrane, and covered by flat-type cells that usually harbor secondary symbionts. Its proximity to the ovaries allows the endosymbiont to evade insect innate immunity system, at the same time allows a quick transfer to the next generation. In aphids, it has been observed that some of the bacteriocyte cells closest to the ovaries break, releasing *Buchnera* cells that enter the newly developing embryo or egg (Buchner 1965; Braendle et al. 2003; Shigenobu and Stern 2013; Simonet et al. 2018).

Other forms of vertical transmission imply the transovarial infection by bacteria from the surrounding infected tissues or from the insect hemolymph, as occurred with the infamous *Wolbachia* or other secondary endosymbionts of aphids; by coprophagy from their next of kin, or by feeding on their own infected-eggshell (see Bourtzis and Miller 2006, 2009). Vertical transmission through coprophagy or feeding of infested eggshell should also be done before the proventriculus valve is closed, as this system filters ingested bacteria and avoids its establishment in the gut lumen, a system that could protect insects from plant-borne bacteria in the past, limiting nowadays the vectoring of plant bacterial pathogens by insects (Lanan et al. 2016; Lopez-Fernandez et al. 2017). Vertical transmission of whole bacteriocytes instead of isolated bacteria through female ovary to oocytes has been described in the whitefly *Bemisia tabaci* (Hemiptera: Aleyrodoidea), to assure the transmission of the primary endosymbiont *Candidatus* Portiera aleyrodidarum and the secondary endosymbiont *Halmintonella defensa* hosted in the same bacteriocyte by protrusion, elongation, and separation of the mother bacteriocyte (Luan et al. 2016; Santos-Garcia et al. 2020).

Wolbachia, another intracellular symbiont, has enthralled scientists for decades, is a ubiquitous bacterial endosymbiont belonging to the alpha-proteobacteria phylum, with more than 60% of insect species being infected, mainly due to reproductive manipulation of its hosts (see Zchori-Fein and Bourtzis 2012 for a revision on this genus, and also chapters by Szklarzewicz et al.; and Lefoulon et al., in this book). Recently, it has been stated that under the name *Wolbachia* there are several clades (lettered named as A, B, C, D, K, L, M, ...; up to 12 clades whose differentiation is based on multiple-locus sequence alignment analyses). Each of these *Wolbachia* clades differs in the phenotype induced, tissue localization (extra- and intracellularly), genome composition, and metabolic capabilities. *Wolbachia* is usually transmitted vertically (through egg infection at ovaries). However, *Wolbachia* can be also transmitted horizontally (between insect species or among insect orders) via the insect host plant or via insect parasitoid species, as occurs with some facultative symbionts (Lopez-Fernandez et al. 2017; Hafer and Vorburger 2019; Hannula et al. 2019). Altogether, these facts have raised the concern about the symbiotic fate of *Wolbachia*, from reproductive manipulator to nutritional mutualist (revised in Newton and Rice 2020).

Another exceptional case of alteration of symbiont classification (as intra- or extracellular symbiont) is the bacterial symbiont of olive fruit fly, *Candidatus* Erwinia dacicola. This symbiont is located within midgut cells (by definition endosymbiont) when the host is at its larval stage, whereas *Ca.* E. dacicola is located extracellularly in the foregut of adults at molting from larvae to pupae. This change in lifestyle affects symbiont transmission mode and genome structure. But this is not a common case; it has also been observed in another holometabolous insect, the carpenter ants, affecting their primary endosymbiont *Candidatus* Blochmania floridanus, highlighting the relationship between host developmental type (holometabolous or hemimetabolous) and symbiont lifestyle (endo- or ectosymbiont) with the corresponding transmission mode and genome structure as explained earlier (Stoll et al. 2010; Augustinos et al. 2019).

Opposite to intracellular symbionts, extracellular symbionts as *Burkholderia* spp. or *Pantoea* spp. (Pentatomid insect symbionts located in midgut crypts) face issues, as mode of transmission, ecological, and evolutionary relationships with their hosts (horizontal transfer between host groups). These insects' extracellular symbionts are the perfect example of the initial stages of symbiont colonization, interaction with the insect host, ex-host symbiont culturability and of symbiont genome evolution (as reviewed in Salem et al. 2015 and Otero-Bravo et al. 2018). However, many of these extracellular symbionts are considered pathogens and will not be discussed further in this chapter.

3.2.2 Symbionts and the Insect Immune System

Irrespective of the symbiont location, during transmission to next insect generation, the bacterial symbiont should face the insect immunity system and melanization response, whereas, on the other side, the host should deal with the effects of inherent virulence of the bacteria (Login et al. 2011; Eleftherianos et al. 2013; Shigenobu and Stern 2013; Freitak et al. 2014; Wong et al. 2016; Itoh et al. 2019; Liberti and Engel 2020). The equilibria of both processes (insect immunity system and bacteria virulence) are the result of the evolutionary race faced by host and bacteria to adapt to the relationship (Red Queen effect) (Zchori-Fein and Bourtzis 2012; Bennett and Moran 2015; Wernegreen 2017). On the other hand, once established, the relationship could lead to mutual dependence over evolutionary time. As we will see later on, the degeneration of symbiont genome may drive the host to extinction due to the accumulation of deleterious mutations that limit their beneficial contributions and environmental tolerance provided by the symbiont (Dale and Moran 2006; Bourtzis and Miller 2009; Bennett and Moran 2015). Endosymbionts interact directly with the secreted molecules of the humoral insect host immune response, especially at their "extracellular stage" when moving from mother to offspring bacteriocytes, but how they do it is something still debatable (Login et al. 2011; Eleftherianos et al. 2013; Pérez-Brocal et al. 2013; Leonard et al. 2020). Previous and recent studies have determined that the interaction of endosymbionts with its insect host is similar to those of microbial pathogens. Some endosymbionts still have the molecular machinery involved in cell-to-cell communication or pathogenicity induction as toxin, pathogenicity islands, type III secretion systems (T3SS), ureases among others (Degnan et al. 2009; Pérez-Brocal et al. 2013; Liberti and Engel 2020; Leonard et al. 2020). Genome-wide transcriptome analyses in several endosymbiotic models (like *Buchnera aphidicola*/aphids or primary endosymbiont of *Sitophilus zeamais* (SPE)/stored-food weevils) have indicated that peptidoglycan recognition protein (PGRP) is upregulated in the bacteriome, at the same time as some virulence genes are also upregulated in the endosymbiont, and the host overproduces the AMP gene coleoptericin A. However, this is not the rule, as aphids carrying *Regiella insecticola, Serratia symbiotica,* or *Hamiltonella defensa*, as secondary symbionts in addition to *B. aphidicola*, respond differentially to host defenses, including the modification of their intracellular location (as depending on the combination of *B. aphidicola* with the other secondary symbionts, *B. aphidicola* were found in phagolysosomes of adherent bacteriocytes in their way from mother to offspring), due to the differential genomic content (T3SS, ureases, etc.) of the secondary symbiont. *Wolbachia* and *Spiroplasma*, two widely distributed secondary endosymbionts that go through this extracellular stage, also evaded a broad range of immune defense mechanisms not even involving upregulation or downregulation of diptericin, cecropin, defensins, and other AMP genes as is usually done by the primary endosymbiont of *Sitophilus zeamais* (SPE). Maybe due to this immune response avoidance, these two bacterial endosymbionts are so prevalent and are able to infect and manipulate several insect orders, while at the same time render the host susceptible to entomopathogenic bacteria. This immune response avoidance had led to a biotechnological application of endosymbionts, especially referring to *Wolbachia* and *Spiroplasma*, as biological control agent of pest insects (Eleftherianos et al. 2013; Liberti and Engel 2020; Leonard et al. 2020).

As the NGS technologies advances, the greater number of genomes from unculturable endosymbiotic microorganism is being released, contributing to the knowledge of how these microbes interact with the host, highlighting the similarity between pathogenic and symbiotic processes at a global evolutionary scale, such as plants and their rhizobial symbionts. However, still, some questions remain open, like how pathogens and symbionts communicate within the same host? Or, how primary and secondary symbionts communicate? Questions of special importance for the cicadas group, where two coprimary endosymbionts, one within the other, belong to different bacterial orders raising the importance of interbacteria communication.

3.3 Genome Evolution: Changes in Bacterial Endosymbionts

As presented, insects harbor different microorganisms, some becoming examples of the different stages of the process of endosymbiosis, each reflecting evolutionary time since the first successful encounter between a free-living bacterium and an insect species, some dating back to the Pangea geological time (Fig. 3.1). As we will see, each stage of host adaptation is affected by the transmission mode and will affect the final fate of the bacterial species involved, i.e., facultative and horizontally acquired from the environment to obligate and vertically transmitted bacterial species (Toft and Andersson 2010; Gorovits and Czosnek 2013).

3.3.1 The Stages of Host Adaptation

The transition from extracellular free-living bacteria to intracellular symbiont represents a dramatic environmental change, which was even postulated as beneficial to the symbiont due to the stability of the new environment (inside insect eukaryotic cells), had imposed constrictions to their evolvability since the establishment of the relationship (reviewed in (Toft and Andersson 2010)). As indicated by Toft and Andersson (2010), the symbiont genome went through stages of host adaptation. The early stage of symbiont–host adaptation involves the reduction of mobile elements (phages, plasmids, genomic islands), recombination machinery, reduction of gene duplication, pseudogenization and gene loss, as showed by many facultative intracellular symbionts (as some *Wolbachia* spp., or *Cardinium* spp). In a second stage, bacteria become an obligate intracellular species on which gene loss rate is increased involving gene transfer to host genome and pseudogenization, whereas the reduction of recombination and rearrangement of gene clusters start to shape the final structure of the genome, as seen in *Hamiltonella defensa*, *Regiella insecticola*, *Rickettsiella viridis,* or *Serratia symbiotica* (Table.3.1). As the relationship continues

Table 3.1 Genome characteristics of different insect endosymbionts and some free-living relatives mainly used in comparative analyses

Bacterial Class	Bacterial species (strain)	Endosymbiont type (P, S)	Host	Host diet	Genome size (Mb)	% G + C	Genes/ CDSs (n)	ψ-genes (n)	rRNAs (n)	tRNAs (n)	Ref.
Bacteroidetes	*Blattabacterium* (MADAR)	P	Basal termites	Wood	0.590	27.5	544	9	3	34	Sabree et al. (2009)
Bacteroidetes	*Blattabacterium* (BBGE)	P	Cockroaches	Omnivorous	0.637	27.1	590	1	3	34	Sabree et al. (2009)
Bacteroidetes	*Blattabacterium* (BPLA)	P	Cockroaches	Omnivorous	0.637	28.2	581	4	3	33	Sabree et al. (2009)
Bacteroidetes	Ca. *Skilesia alterna* (Endosymbiont of Geopemphigus S2)	P	Aphids (Geopemphigus sp.)	Phloem sap	1.320	37.0	1611	0	3	35	Chong and Moran, (2018)
Bacteroidetes	Ca. *Sulcia muelleri*	Co-P	Spittlebugs (Cercopoidea)	Xylem sap	0.276	21	246	0	3	29	McCutcheon et al. (2010)
Bacteroidetes	Ca. *Sulcia muelleri* (ALF)	Co-P	Leafhopper (Cicadellidae: Deltocephalinae)	Phloem sap	0.190	24.0	190	2	3	30	Bennett and Moran (2013)
Bacteroidetes	Ca. *Uzinura diaspidicola* (ASNER)	P	Scale insects (Coccoidea)	Parenchyma	0.263	30.2	227	12	3	31	Sabree et al. (2013)
Bacteroidetes	Ca. *Walczuchella monophlebidarum*	P	Scale insects	Phloem sap	0.309	32.6	271	27	3	33	Rosas-Pérez et al. (2014)
Bacteroidetes	*Bacteroidetes vulgatus ATCC_8482*	Free-living	–	–	5.16	42.2	4077	146	22	83	NC009614
Proteobacteria	*Escherichia coli K12*	Free-living	–	–	4.64	50.8	4140	184	22	89	NC000913
α-Proteobacteria	Ca. *Hodgkinia cicadicola* Dsem	Co-P	Cicadas	Xylem sap	0.143	58.4	169	0	3	15	McCutcheon et al. (2009a, b)
α-Proteobacteria	Ca. *Hodgkinia cicadicola* (MAGSEP)	P	Cicadas	Xylem/ phloem sap	1.640[a]	28.37	179	Nd	30[a]	11	Campbell et al. (2017)
β-Proteobacteria	Ca. *Nasuia deltocephalinicola* (ALF)	Co-P	Leafhopper (Cicadellidae: Deltocephalinae)	Xylem/ phloem sap	0.112	17.1	137	Nd	3	29	Bennett and Moran (2013)

β-Proteobacteria	Ca. *Tremblaya princeps* (PCIT)	P	Mealybugs	Phloem sap	0.139	59	121	2	6	12	McCutcheon and von Dohlen (2011)
β-Proteobacteria	Ca. *Vidania fulgoroideae* (OLIH)	Co-P	Fulgorid planthoppers	Phloem sap	0.140	18.2	154	Nd	3	26	Bennett and Mao (2018)
β-Proteobacteria	Ca. *Zinderia insecticola* CARI	Co-P	Spittlebugs	Xylem sap	0.208	14	202	0	3	25	McCutcheon and Moran (2012)
γ-Proteobacteria	Ca. *Baumannia cicadellinicola* (Hc)		Sharpshooter leafhoppers	Xylem sap	0.69	33.2	598	10	6	39	Wu et al. (2006)
γ-Proteobacteria	Ca. *Baumannia cicadellinicola* (BGSS)		Sharpshooter leafhoppers	Xylem sap	0.76	39.0	668	14	6	39	Bennett et al. (2014)
γ-Proteobacteria	*Buchnera aphidicola* (APS)	P	Aphids	Phloem sap	0.656	26	575	1	3	32	Shigenobu et al. (2000)
γ-Proteobacteria	*Buchnera aphidicola* (bpi)	P	Aphids	Phloem sap	0.618		560	3	3	31	van Ham et al. (2003)
γ-Proteobacteria	*Buchnera aphidicola* (cc)	P	Aphids	Phloem sap	0.422	20.2	362	3	3	31	Pérez-Brocal et al. (2006)
γ-Proteobacteria	Ca. *Carsonella ruddii* (PV)	P	Psyllids	Phloem sap	0.159	16.6	182	0	3	24	Nakabachi et al. (2006)
γ-Proteobacteria	Ca. *Carsonella ruddii* (DC)	P	Psyllids	Phloem sap	0.174	17.6	201	3	3	27	Nakabachi et al. (2013)
γ-Proteobacteria	Ca. *Hamiltonella defensa* (APS 5AT)	S / co-P[b]	Aphids	Phloem sap	2.169	40.1	2100	188	9	42	Degnan et al. (2009)
γ-Proteobacteria	Ca. *Moranella endobia*	P	Mealybugs	Phloem sap	0.538	43.5	406	28	5	41	McCutcheon and von Dohlen (2011)
γ-Proteobacteria	Ca. *Purcelliella pentastirinorum* (OLIH)		Planthopper	Phloem sap	0.479	21.2	430	6	3	31	Bennett and Mao (2018)
γ-Proteobacteria	Ca. *Regiella insecticola* (LSR1)	S / co-P[b]	Aphids	Phloem sap	2.07	42.4	1658	469	12	41	Degnan et al. (2009)

(continued)

Table 3.1 (continued)

Bacterial Class	Bacterial species (strain)	Endosymbiont type (P, S)	Host	Host diet	Genome size (Mb)	% G + C	Genes/ CDSs (n)	ψ-genes (n)	rRNAs (n)	tRNAs (n)	Ref.
γ-Proteobacteria	Ca. *Rickettsiella viridis* (Ap-RA04)	S	Aphids	Phloem sap	1.579		1349	30	6	42	Nikoh et al. (2018)
γ-Proteobacteria	Ca. *Serratia symbiotica* TUC	S	Aphids	Phloem sap	2.789	52.1	2098	550	15	44	Burke and Moran (2011)
γ-Proteobacteria	Ca. *Serratia symbiotica* STs	Co-P	Aphids	Phloem sap	0.650	20.7	492	7	3	33	Manzano-Marín et al. (2016)
γ-Proteobacteria	Ca. *Serratia symbiotica* SCt	Co-P	Aphids	Phloem sap	2.49	52.0	1601	916	13	47	Manzano-Marín et al. (2016)
γ-Proteobacteria	Ca. *Serratia symbiotica* SCc	Co-P	Aphids	Phloem sap	1.762	29.2	677	98	3	36	Lamelas et al. (2011)

[a]Distributed in 39 replicons, as from the description at NCBI, see reference for further details.
[b]coprimary status determined without complete genome sequencing, see Gil and Latorre 2019 for further references

in life history, bacteria become obligate intracellular mutualist (referred also as the extreme stage of endosymbiosis), the rate of gene loss or transfer to host genome and the number of retained pseudogenes are reduced, and their genome is no longer rearranged keeping a gene synteny between sibling endosymbiotic strains, as seen in Ca. *Baumannia cicadellinicola, Blattabacterium spp., Blochmannia spp., Buchnera aphidicola, Carsonella ruddii, Hodgkinia cicadicola, Moranella endobia, Nasuia deltacephalinicola, Sulcia muelleri, Tremblaya princeps, Uzinura diaspidicola, Vidania fulgoroideae, Walczuchella monophlebidarum* and *Zinderia insecticola* (Table 3.1). The final endosymbiotic step is the conversion into true cellular organelles, as occurring with mitochondria and chloroplasts (see Kaczanowski's chapter in this book), by an increased gene transfer to host cell nuclei and increased gene loss rate, which reduces considerably the genome content and size. However, even the tiniest endosymbionts characterized so far still retain most of the genes required for transcription, translation, and replication, except some *Tremblaya princeps* strains, as indicative of their "independent" bacterial nature (McCutcheon and Von Dohlen 2011; McCutcheon and Moran 2012; Moran and Bennett 2014; Campbell et al. 2015, 2017, 2018; Łukasik et al. 2017; Matsuura et al. 2018; Gil and Latorre 2019). Table 3.1 describes the characteristics of bacterial genomes of some of the mentioned endosymbiont species, and some of the free-living bacterial species used for comparisons, showing the difference in genome size between primary and facultative secondary symbionts, along with the number of ribosomal RNAs and tRNAs genes (van Ham et al. 2003; Pérez-Brocal et al. 2006; Burke and Moran 2011; Rosas-Pérez et al. 2014; Campbell et al. 2015, 2017, 2018; Manzano-Marín et al. 2016; Gil and Latorre 2019).

3.3.2 The Impact of Host Adaptation to Symbiont Genome Composition

As explained, the long-term endosymbiosis usually leads to massive gene losses, genome shrinkage (with genomes going from ~4 Mb in size of free-living bacteria to the tiniest *Tremblaya princeps* or *Nasuia deltocephalinicola* (ALF) with ~0.139 and ~ 0.112 Mb; in Table 3.1), and eventually, genome stability (synteny) in related endosymbiont lineages (Gil et al. 2004; Silva et al. 2007; Moran et al. 2008; Toft and Andersson 2010; McCutcheon and Moran 2012; Bennett and Moran 2013; Moran and Bennett 2014; Oakeson et al. 2014; Van Leuven et al. 2014; Campbell et al. 2015, 2017, 2018). What are the molecular mechanisms that shaped the genome architecture and composition? Stated generally, evolution trends to remove redundant functions in the cells to keep their metabolic balance between metabolic capabilities and energy provisioning from the environment. Insect bacterial symbionts have constant provisioning of nutrients, leading to some redundant metabolic capabilities. However, this redundancy relies on the insect host diet. Many of the examples described here feed on unbalanced diets, as plant xylem and phloem, whose quality has driven the bacterial species selection and shaped partially their metabolic capabilities to become nutritional symbionts; i.e., *Buchnera aphidicola* provides the essential amino acids' tryptophan and leucine to their aphid hosts, which are feeding on low-quality phloem (reviewed in Moran et al. 2003). The other part of metabolism has been modified by the mode of transmission of these bacteria, which affected both genome structure and composition (reviewed in Silva et al. 2007 and Gil and Latorre 2019). In addition, the process of intracellularization has limited the bacterial capability to interact with relatives or nonrelatives, due to the loss of interacting proteins (surface proteins, pumps, and other transporters involved in cell-to-cell communication). One of the first molecular patterns observed in these symbionts was the reduction of GC content (both in genes and intergenic regions, despite being at differential reduction rate), with some minor exceptions as *Hodgkinia cicadicola* (DESEM), *Tremblaya princeps,* or the primary endosymbiont of *Sitophilus oryzae* (SOPE). This general tendency was correlated with the loss of DNA repair genes and with an increase of mutational bias from GC to AT and their effect on small effective population size. Indeed, the effective population size of a symbiont species is limited by the number of hosts, the number of infected cells (forming the bacteriome), and the available space to grow within them, which makes selection less efficient to remove slightly deleterious mutations. Adding the continuous populational bottlenecks to the equation, symbionts are subjected to strong evolutionary pressures that accelerate their evolutionary rate leading to divergent populations between insect species. The lack of horizontal gene transfer or recombination events between these divergent populations has also been correlated with the genome content reduction and genome GC content, and by the physical limits imposed by the intracellular lifestyle. It has been also linked to the great gene synteny observed in the genomes of long-standing endosymbionts (Tamas et al. 2002; Dale and Moran 2006; Moran et al. 2008; Toft and Andersson 2010; Bennett and Moran 2015; Lopez-Madrigal and Gil 2017). If this evolutionary race leads to specific and adapted bacterial lineages linked to each insect host species food selection, how can one explain the coexistence of several reduced symbiont lineages not being replaced for new ones? And how can one explained the *long-term* (measured in millions of years) relationships of almost all these bacterial species?

3.4 GroEL and Other Chaperones in the Preservation of Bacterial Endosymbionts

As indicated in the preceding section, many insect bacterial symbionts have been restricted to an intracellular lifestyle for millions of years. This lifestyle has shaped their genome by means of genome shrinkage, gene loss, acceleration of mutational rates, increased AT content, by modification of the translational codes (especially in the tiniest genomes), among others, overall, rendering many of these bacterial symbionts to constrained biofactories of essential nutrients missing from the insect

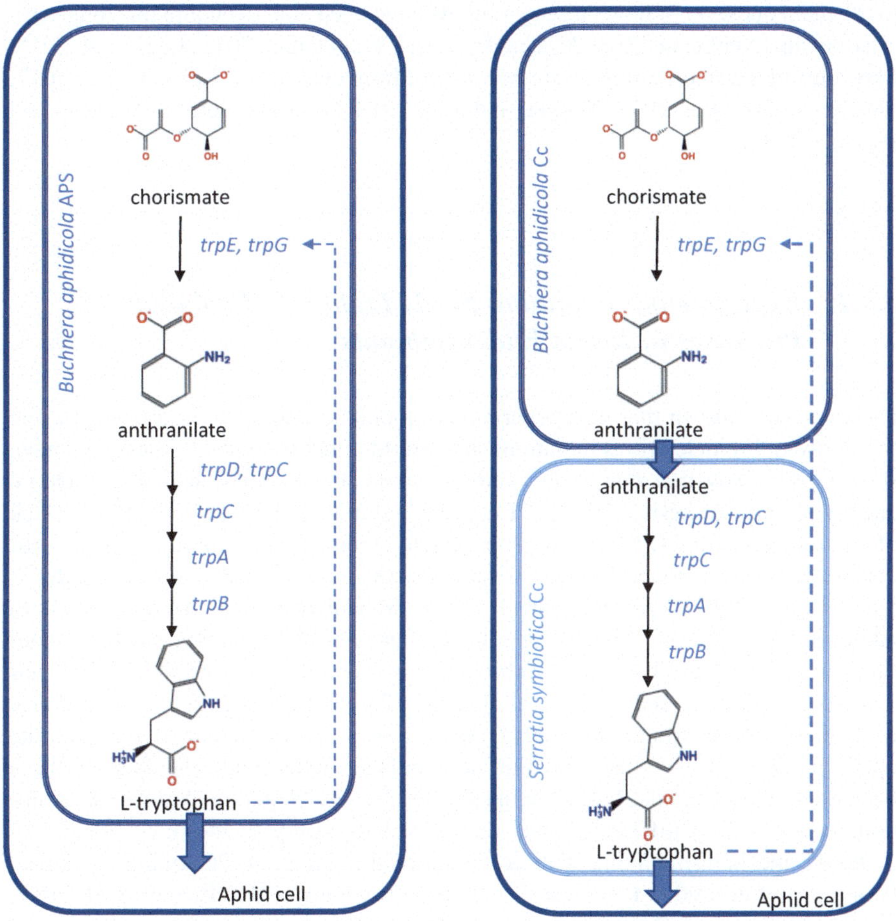

Fig. 3.2 An example of metabolic complementation between insect bacterial endosymbionts. At the left panel, the tryptophan amino acid synthesis pathway coded by the primary endosymbiont *Buchera aphidicola* of the pea aphid (*Acyrthosiphon pisum*). In this case, the first genes *trpE, trpG* are located in multicopy plasmid, whereas the remaining genes of the pathway (*trpDCBA*) are encoded in the main chromosome. In the cedar aphid *Cinara cedri*, this essential amino acid is synthetized by both coprimary endosymbionts, *B. aphidicola* Cc and *Serratia symbiotica* Cc (redrawn from Lamelas et al. 2008; Mori et al. 2016; Gil and Latorre 2019; other examples of metabolic complementation at Zientz et al. 2004; Manzano-Marín et al. 2016; Campbell et al. 2018; Gil et al. 2018; Gil and Latorre 2019 and references herein)

specialized diets. However, due to the population structure of these intracellular symbionts, some of them have lost key enzymes on the routes they were selected for, relying on their interaction with ancient facultative symbionts, nowadays known as coprimary endosymbionts (Fig. 3.2). This metabolic coevolution, by metabolic complementation of essential pathways, has been recurrent, especially affecting those species with the smallest/reduced genomes, but is not the complete answer to the main question of how these symbionts have been able to endure this long-term relationship (Neef et al. 2010; McCutcheon and Von Dohlen 2011; Luan et al. 2015; Manzano-Marín and Latorre 2016; Manzano-Marín et al. 2016; Łukasik et al. 2017; Sabater-Muñoz et al. 2017; Bennett and Mao 2018; Gil and Latorre 2019; Mao and Bennett 2020).

3.4.1 Experimental Evolution Seeds Light into Evolutionary Processes in Insect Endosymbionts

It is generally known that transgenerational population bottlenecks affect selective forces swapping deleterious mutations that would affect individual fitness. We have shown previously that due to the vertical mode of transmission of the bacterial endosymbionts, bacteria increase AT content and mutational rates linked to the lack of recombination and DNA repair genes, driving their genomes through genetic drift instead of being removed by natural selection (Mira et al. 2001; Fares et al. 2002; Kuo et al. 2009; Wernegreen 2011, 2012; McCutcheon and Moran 2012). In addition, the signature of purifying selection and positive selection has been identified in some of the bacterial symbionts, in genes not directly linked with their metabolic biofactory status (Fares et al. 2002; Toft and Fares 2008; Alvarez-Ponce et al. 2016; Sabater-Muñoz et al. 2017). In the past years, we have tried to answer the open question of how the selection-drift balance determines the fate of these bacterial endosymbionts, by following the simple and elegant experimental evolution model of Richard Lenski that has probed its utility to the direct insight of evolutive mechanisms along with computational methods to determine its impact (Commins et al. 2009; Toft et al. 2009; Toft and Fares 2010; Rainey et al. 2017; Sabater-Muñoz et al. 2017).

Working with bacteria is a routine in the lab, unless you try to work directly with yet unculturable insect bacterial endosymbionts or you want to work with plant-fastidious bacteria as *Xylella fastidiosa*, which is not the case. Mimicking two population dynamics, one with single-cell bottlenecks resembling the endosymbiont intergenerational passage, and the other with 1% of population bottleneck resembling the free-living bacteria, we obtained a fossil record of *Escherichia coli* K12 MG1655 ΔmutS lines evolved through 4400 generations (Sabater-Muñoz et al. 2017). We observed a genome reduction in the drifted populations, with differences in the rate of indels between protein-coding and intergenic regions, indicating that genes were shrinking, some being pseudogenized, and genome was being eroded by losing a large chunk of DNA containing 42 genes involved in prophage movement and other IS (insertion sequence) transposases. In addition, genome-wide deregulation was observed, with central metabolism genes upregulated, and genes of cell localization, cellular components, and biogenesis processes downregulated. This gradual trend of genetic drift with punctuated events of big deletions has also been demonstrated in bacterial symbionts, specifically in *Buchnera aphidicola* (as reviewed in Silva et al. 2007). But it opens up new questions, as the implication of drift and genome erosion on genome architecture and complexity in clonal bacteria (as seen in endosymbionts or in digital microorganisms), and how protein evolvability could affect or relax the effects of drift (Mira et al. 2001; Toft and Fares 2010; Whitney and Garland 2010; LaBar and Adami 2016; Bobay and Ochman 2017). Due to the increased number of available genomes, evolutionary studies have questioned the neutral theory of evolution, especially when determining selective constraints. Previous methods ignored the evolvability of amino acids within proteins, as they don't consider the quaternary structure of the same, and the established atomic connections (Toft and Fares 2010). Once rates of evolution are calibrated, identification of selection signatures becomes easily transferred to other drifting systems like the bacterial endosymbionts of insects, being able to distinguish between adaptive evolution from relaxed constraints of biological systems under genetic drift (Toft and Fares 2010).

Comparing all these facts (mutational spectrum, population size, population dynamics) with the selective constraints observed in pairs of bacterial symbionts compared against pairs of free-living bacteria, we observed a convergent host-independent evolution (constricted genes not being involved in host–bacteria interaction) of endosymbiotic bacteria. Finding six genes (*groES, groEL, rplP, rpsJ, rpsM, and rpsS*) highly constrained in each of the five endosymbiotic bacterial groups tested at that time (*Buchnera aphidicola, Blochmania* spp., *Wigglesworthia glossinidia, Baumania cicadellinicola,* and *Blattabacterium* spp.). The number of highly constrained genes was variable between endosymbiotic bacteria, but among them, we would like to highlight the chaperones or buffering proteins coded by genes: *groEL, groES, dnaK, clpB, clpX, cspE, ahpC,* and *ptsH*, especially the first three ones, in the next heading (Sabater-Muñoz et al. 2017).

3.4.2 Chaperones, Moonlighting Proteins, with Mutational Robustness Properties Protect Drifting Genomes

Despite not finding many highly constrained genes conserved between all the symbiotic bacteria here presented, some of the constrained genes identified have been related to maintenance of drifting systems, as the chaperonin GroEL (Fares et al. 2002, 2004; Toft and Fares 2010).

The chaperonin GroEL and its cochaperoning GroES (also known as cpn60 and cpn10, respectively) belong to the heat-shock protein HSP60 family. A ubiquitous and essential protein in bacteria and eukaryotic organelles is implied in the proper folding of other proteins, either naturally or after heat stress, in the proper assembly of the protein complex and protein transport (reviewed in Horwich et al. 2007). These two proteins form a homotetradecamer organized in two heptamer rings back-to-back oriented, while the GroEL monomer is divided into three domains: the apical (binds unfolded proteins), intermediate (allows structural transition between trans and cis conformation), and equatorial (ATPase and folding activities). Despite this protein being highly conserved among bacteria, some amino acid sites not implied in the canonical folding activity have been identified through coevolution analysis, supporting the evolutionary plasticity of GroEL across the entire bacterial phylogeny, and the implication of some regions within the protein implied in other functions, which was a proof of the moonlighting nature of this protein (Fares et al. 2002; Ruiz-González and Fares 2013; Kupper et al. 2014; Fares 2015).

As indicated previously, population dynamics affect endosymbiont genome structure, composition, and by hence symbiont fate, rising clonality as a key microbial trait due to lack of gene exchange mechanisms (Shapiro 2016). These bacterial clones are destined to disappear unless a molecular mechanism helps them evade this fate. Nearly 40 years ago, Ishikawa already identified the protein involved in the maintenance of symbiosis, called *symbionin*, which was further identified as GroEL (cpn60) (Ishikawa and Yamaji 1985). But its moonlighting activity as mutational buffering systems was not unveiled until nearly 20 years later (Fares et al. 2002, 2004; Fares 2015). Again, by using experimental evolution with clonal dynamics, Fares and coworkers demonstrated that the system GroEL/ES buffers against deleterious mutations, recovering fitness by overexpressing the chaperone (Fares et al. 2002). This is precisely the link with the bacterial endosymbionts of insects, the mutational robustness property of GroEL/ES that allows genetic drifting genomes to persist during evolutionary times if overexpressed. The Fares' group have proved that there is a link between mutational robustness provided by GroEL/ES and DnaK (another chaperone with moonlighting activities involved in early protein folding) and evolvability by unveiling the mutational landscapes, transcriptomic profiles, and phenotypic characterization of clonal lines subjected to several populational dynamics (Fares et al. 2002, 2004; Williams et al. 2010; Williams and Fares 2010; Fares 2015; Sabater-Muñoz et al. 2015; Aguilar-Rodríguez et al. 2016; Alvarez-Ponce et al. 2016; Fares 2016). These two proteins, GroEL/ES and DnaK, are naturally overexpressed in many endosymbiont species, not only in the highly studied *Buchnera aphidicola*. GroEL/ES has been found in the hemolymph of the host insects, with not-yet-unveiled function, but is implied in several processes linked to host immunity avoidance to assure appropriate transmission of bacterial endosymbionts from mother to insect offspring (Chaudhary et al. 2014; Freitak et al. 2014; Jeffery 2018). These "environmental services" are somehow exploited by other plant pathogens, the plant viruses vectored by insects (Eleftherianos et al. 2013; Kliot and Ghanim 2013; Jeffery 2015). Naturally, the endosymbionts overexpress GroEL/ES with a survivorship purpose, but this has a strong metabolic cost for the bacteria. As indicated, the active form of GroEL is a homotetradecamer organized in two heptamer rings, plus a heptamer ring of GroES, whose production cost to the cell is 28,364 ATP molecules per active complex (4 ATP molecules per synthetized amino acid, with 7×97 amino acids for GroES ring and 14×458 amino acids for GroEL barrel). Its overproduction (either by overexpression of operon from plasmid or by amplification of chromosome numbers) is energetically costly to the cell, imposing a fitness cost to the organism. We demonstrated that overexpression of either GroEL or DnaK imposes a metabolic cost to the cell, only supported by drifting populations as the ones observed in insect bacterial endosymbionts. *Escherichia coli* lines, subjected to experimental evolution under two populational dynamics, tend to remove and silence the chaperone plasmids in short term. Whereas when applying clonal dynamics, only the lines with overexpression of GroEL or DnaK can survive long term, until plasmid is lost by unpaired distribution between daughter cells (Sabater-Muñoz et al. 2015; Aguilar-Rodríguez et al. 2016; unpublish results). It remains to be explored: What happens then, if we reach the overexpression through chromosome amplification or by promoter modulation?

3.5 Future Directions

As described in this chapter, insects harbor a plethora of bacterial species, many of them since memorial times, when the life on the earth was evolving in the Pangea continent. This long evolutionary relationship has shaped both host and endosymbiont, while the first symbiotic partner, the host, can innovate thanks to the biologic capabilities (not restricted to metabolism) raised by the bacterial partner as abiotic and biotic stress resistance. The bacterial partner is subjected to evolutionary constraints that despite what was thought still allows for biological innovation (through transcriptional and protein divergence). Thanks to these studies, some bacterial endosymbionts or some of their main protein complexes have been identified as putative targets for insect pest control measures (like endosymbiont-drive population control, or virus trapping with GroEL/ES) or for biomedical applications (Kupper et al. 2014). However, as highlighted in this chapter, the innovative potential of these endosymbionts may/would disrupt the devised biotechnological or biomedical applications deserving further research.

References

Adler S, Trapp EM, Dede C et al (2014) *Rhopalodia gibba*: the first steps in the birth of a novel organelle? In: Loffelhardt W (ed) Endosymbiosis. Springer, Berlin Heidelberg, Vienna (Austria), pp 167–179

Aguilar-Rodríguez J, Sabater-Muñoz B, Montagud-Martínez R et al (2016) The molecular chaperone *dnak* is a source of mutational robustness. Genome Biol Evol 8:2979–2991. https://doi.org/10.1093/gbe/evw176

Alvarez-Ponce D, Sabater-Munoz B, Toft C et al (2016) Essentiality is a strong determinant of protein rates of evolution during mutation accumulation experiments in *Escherichia coli*. Genome Biol Evol 8:2914–2927. https://doi.org/10.1093/gbe/evw205

Antonovics J, Wilson AJ, Forbes MR et al (2017) The evolution of transmission mode. Philos Trans R Soc B Biol Sci 372:7–11. https://doi.org/10.1098/rstb.2016.0083

Augustinos AA, Tsiamis G, Cáceres C et al (2019) Taxonomy, diet, and developmental stage contribute to the structuring of gut-associated bacterial communities in tephritid pest species. Front Microbiol 10:1–13. https://doi.org/10.3389/fmicb.2019.02004

Bell-Roberts L, Douglas AE, Werner GDA (2019) Match and mismatch between dietary switches and microbial partners in plant sap-feeding insects. Proc R Soc B Biol Sci 286:20190065. https://doi.org/10.1098/rspb.2019.0065

Bennett GM, Mao M (2018) Comparative genomics of a quadripartite symbiosis in a planthopper host reveals the origins and rearranged nutritional responsibilities of anciently diverged bacterial lineages. Environ Microbiol 20:4461–4472. https://doi.org/10.1111/1462-2920.14367

Bennett GM, Moran NA (2013) Small, smaller, smallest: the origins and evolution of ancient dual symbioses in a phloem-feeding insect. Genome Biol Evol 5:1675–1688. https://doi.org/10.1093/gbe/evt118

Bennett GM, Moran NA (2015) Heritable symbiosis: the advantages and perils of an evolutionary rabbit hole. Proc Natl Acad Sci U S A 112:10169–10176. https://doi.org/10.1073/pnas.1421388112

Berasategui A, Shukla S, Salem H, Kaltenpoth M (2016) Potential applications of insect symbionts in biotechnology. Appl Microbiol Biotechnol 100:1567–1577. https://doi.org/10.1007/s00253-015-7186-9

Blow F, Douglas AE (2019) The hemolymph microbiome of insects. J Insect Physiol 115:33–39. https://doi.org/10.1016/j.jinsphys.2019.04.002

Bobay LM, Ochman H (2017) The evolution of bacterial genome architecture. Front Genet 8:1–6. https://doi.org/10.3389/fgene.2017.00072

Boucias DG, Zhou Y, Huang S, Keyhani NO (2018) Microbiota in insect fungal pathology. Appl Microbiol Biotechnol 102:5873–5888. https://doi.org/10.1007/s00253-018-9089-z

Bourtzis K, Miller TA (2006) Insect symbiosis, vol 2. CRC Press, Boca Raton, FL

Bourtzis K, Miller TA (2009) Insect symbiosis, vol 3. CRC PRess, Taylor & Francis Group, Boca Raton, FL

Bowles AMC, Bechtold U, Paps J (2020) The origin of land plants is rooted in two bursts of genomic novelty. Curr Biol:1–7. https://doi.org/10.1016/j.cub.2019.11.090

Braendle C, Miura T, Bickel R et al (2003) Developmental origin and evolution of bacteriocytes in the aphid–*Buchnera* symbiosis. PLoS Biol 1:e21. https://doi.org/10.1371/journal.pbio.0000021

Bright M, Bulgheresi S (2010) A complex journey: transmission of microbial symbionts. Nat Rev Microbiol 8:218–230. https://doi.org/10.1038/nrmicro2262

Buchner P (1965) Endosymbiosis of animals with plant microorganisms, 4th edn. Wiley, Interscience Publishers, New York, NY

Burke GR, Moran NA (2011) Massive genomic decay in *Serratia symbiotica*, a recently evolved symbiont of aphids. Genome Biol Evol 3:195–208. https://doi.org/10.1093/gbe/evr002

Campbell MA, Łukasik P, Meyer MC et al (2018) Changes in endosymbiont complexity drive host-level compensatory adaptations in cicadas. MBio 9:8–11. https://doi.org/10.1128/mBio.02104-18

Campbell MA, Łukasik P, Simon C, McCutcheon JP (2017) Idiosyncratic genome degradation in a bacterial endosymbiont of periodical cicadas. Curr Biol 27:3568–3575.e3. https://doi.org/10.1016/j.cub.2017.10.008

Campbell MA, Van Leuven JT, Meister RC et al (2015) Genome expansion via lineage splitting and genome reduction in the cicada endosymbiont *Hodgkinia*. Proc Natl Acad Sci U S A 112:10192–10199. https://doi.org/10.1073/pnas.1421386112

Chaudhary R, Atamian HS, Shen Z, Briggs SP, Kaloshian I (2014) GroEL from the endosymbiont *Buchnera aphidicola* betrays the aphid by triggering plant defense. Proc Natl Acad Sci U S A 111:8919–8924. https://doi.org/10.1073/pnas.1407687111

Chevignon G, Boyd BM, Brandt JW et al (2018) Culture-facilitated comparative genomics of the facultative symbiont *Hamiltonella defensa*. Genome Biol Evol 10:786–802. https://doi.org/10.1093/gbe/evy036

Chevrette MG, Handelsman J (2020) From metagenomes to molecules: innovations in functional metagenomics unlock hidden chemistry in the human microbiome. Biochemistry:0–1. https://doi.org/10.1021/acs.biochem.0c00033

Chong RA, Moran NA (2018) Evolutionary loss and replacement of *Buchnera*, the obligate endosymbiont of aphids. The ISME Journal 12(3):898–908. https://doi.org/10.1038/s41396-017-0024-6

Colman DR, Toolson EC, Takacs-Vesbach CD (2012) Do diet and taxonomy influence insect gut bacterial communities? Mol Ecol 21:5124–5137. https://doi.org/10.1111/j.1365-294X.2012.05752.x

Commins J, Toft C, Fares MA (2009) Computational biology methods and their application to the comparative genomics of endocellular symbiotic bacteria of insects. Biol Proced Online 11:52–78. https://doi.org/10.1007/s12575-009-9004-1

Dale C, Moran NA (2006) Molecular interactions between bacterial symbionts and their hosts. Cell 126:453–465. https://doi.org/10.1016/j.cell.2006.07.014

Darboux I, Cusson M, Volkoff AN (2019) The dual life of ichnoviruses. Curr Opin Insect Sci 32:47–53. https://doi.org/10.1016/j.cois.2018.10.007

Darby AC, Chandler SM, Welburn SC, Douglas AE (2005) Aphid-symbiotic bacteria cultured in insect cell lines. Appl Environ Microbiol 71:4833–4839. https://doi.org/10.1128/AEM.71.8.4833-4839.2005

De Cock M, Virgilio M, Vandamme P et al (2019) Impact of sample preservation and manipulation on insect gut microbiome profiling. A test case with fruit flies (Diptera, Tephritidae). Front Microbiol 10:1–13. https://doi.org/10.3389/fmicb.2019.02833

Degnan PH, Yu Y, Sisneros N et al (2009) *Hamiltonella defensa*, genome evolution of protective bacterial endosymbiont from pathogenic ancestors. Proc Natl Acad Sci U S A 106:9063–9068. https://doi.org/10.1073/pnas.0900194106

Deschamps P, Colleoni C, Nakamura Y et al (2008) Metabolic symbiosis and the birth of the plant kingdom. Mol Biol Evol 25:536–548. https://doi.org/10.1093/molbev/msm280

Desneux N, Asplen MK, Brady CM et al (2018) Intraspecific variation in facultative symbiont infection among native and exotic pest populations: potential implications for biological control. Biol Control 116:27–35. https://doi.org/10.1016/j.biocontrol.2017.06.007

Deutscher AT, Chapman TA, Shuttleworth LA et al (2019) Tephritid-microbial interactions to enhance fruit fly performance in sterile insect technique programs. BMC Microbiol 19:1–14. https://doi.org/10.1186/s12866-019-1650-0

Douglas AE (2015) Multiorganismal insects: diversity and function of resident microorganisms. Annu Rev Entomol 60:17–34. https://doi.org/10.1146/annurev-ento-010814-020822

Douglas AE (2018) Omics and the metabolic function of insect–microbial symbioses. Curr Opin Insect Sci 29:1–6. https://doi.org/10.1016/j.cois.2018.05.012

Eleftherianos L, Atri J, Accetta J, Castillo JC (2013) Endosymbiotic bacteria in insects: guardians of the immune system? Front Physiol 4:1–10. https://doi.org/10.3389/fphys.2013.00046

Eme L, Spang A, Lombard J et al (2017) Archaea and the origin of eukaryotes. Nat Rev Microbiol 15:711–723. https://doi.org/10.1038/nrmicro.2017.133

Engel P, Moran NA (2013) The gut microbiota of insects – diversity in structure and function. FEMS Microbiol Rev 37:699–735. https://doi.org/10.1111/1574-6976.12025

Fares MA (2015) The origins of mutational robustness. Trends Genet 31:373–381. https://doi.org/10.1016/j.tig.2015.04.008

Fares MA (2016) Coevolution analysis illuminates the evolutionary plasticity of the chaperonin system GroES/L. Stress Environ Regul Gene Expr Adapt Bact 2:796–811. https://doi.org/10.1002/9781119004813.ch77

Fares MA, Barrio E, Sabater-Muñoz B, Moya A (2002) The evolution of the heat-shock protein GroEL from *Buchnera*, the primary endosymbiont of aphids, is governed by positive selection. Mol Biol Evol 19:1162–1170

Fares MA, Moya A, Barrio E (2004) GroEL and the maintenance of bacterial endosymbiosis. Trends Genet 20:413–416. https://doi.org/10.1016/j.tig.2004.07.001

Fares MA, Ruiz-González MX, Moya A et al (2002) GroEL buffers against deleterious mutations. Nature 417:398–398. https://doi.org/10.1038/417398a

Fernández R, Gabaldón T (2020) Gene gain and loss across the metazoan tree of life. Nat Ecol Evol. https://doi.org/10.1038/s41559-019-1069-x

Freitak D, Schmidtberg H, Dickel F et al (2014) The maternal transfer of bacteria can mediate trans-generational immune priming in insects. Virulence 5:547–554. https://doi.org/10.4161/viru.28367

Gil R, Latorre A (2019) Unity makes strength: a review on mutualistic symbiosis in representative insect clades. Life 9:1–24. https://doi.org/10.3390/life9010021

Gil R, Latorre A, Moya A (2004) Bacterial endosymbionts of insects: insights from comparative genomics. Environ Microbiol 6:1109–1122. https://doi.org/10.1111/j.1462-2920.2004.00691.x

Gil R, Vargas-Chavez C, López-Madrigal S, Santos-García D, Latorre A, Moya A (2018) Tremblaya phenacola PPER: an evolutionary beta-gammaproteobacterium collage. The ISME Journal 12(1):124–135. https://doi.org/10.1038/ismej.2017.144

Glenner H, Thomsen PF, Hebsgaard MB et al (2006) The origin of insects. Science 80 (314):1883–1884. https://doi.org/10.1126/science.1129844

Gontier N (2016) History of symbiogenesis. Encycl Evol Biol 4:261–271. https://doi.org/10.1016/B978-0-12-800049-6.00016-0

Gorovits R, Czosnek H (2013) Insect symbiotic bacterial GroEL (Chaperonin 60) and plant virus transmission. In: Henderson B (ed) Moonlighting cell stress proteins in microbial infections. Springer Netherlands, Dordrecht, pp 173–187

Hafer N, Vorburger C (2019) Diversity begets diversity: do parasites promote variation in protective symbionts? Curr Opin Insect Sci 32:8–14. https://doi.org/10.1016/j.cois.2018.08.008

Hannula SE, Zhu F, Heinen R, Bezemer TM (2019) Foliar-feeding insects acquire microbiomes from the soil rather than the host plant. Nat Commun 10:1–9. https://doi.org/10.1038/s41467-019-09284-w

Hays SG, Patrick WG, Ziesack M et al (2015) Better together: engineering and application of microbial symbioses. Curr Opin Biotechnol 36:40–49. https://doi.org/10.1016/j.copbio.2015.08.008

Horwich LA, Fenton WA, Chapman E, Farr GW (2007) Two families of chaperonin: physiology and mechanism. Annu Rew Cell Dev Biol 23:115–145. https://doi.org/10.1146/annurev.cellbio.23.090506.123555

Ishikawa H, Yamaji M (1985) Symbionin, an aphid endosymbiont-specific protein-I. production of insects deficient in symbiont. Insect Biochem 15:155–163. https://doi.org/10.1016/0020-1790(85)90003-4

Itoh H, Jang S, Takeshita K et al (2019) Host–symbiont specificity determined by microbe–microbe competition in an insect gut. Proc Natl Acad Sci U S A:116. https://doi.org/10.1073/pnas.1912397116

Jeffery CJ (2015) Why study moonlighting proteins? Frontiers Genet 6:211. https://doi.org/10.3389/fgene.2015.00211

Jeffery CJ (2018) Intracellular proteins moonlighting as bacterial adhesion factors. AIMS Microbiol 4(2):362–376. https://doi.org/10.3934/microbiol.2018.2.362

Johnson KP, Dietrich CH, Friedrich F et al (2018) Phylogenomics and the evolution of hemipteroid insects. Proc Natl Acad Sci U S A 115:12775–12780. https://doi.org/10.1073/pnas.1815820115

Kliot A, Ghanim M (2013) The role of bacterial chaperones in the circulative transmission of plant viruses by insect vectors. Viruses 5(6):1516–1535. doi: https://doi.org/10.3390/v5061516

Kuo CH, Moran NA, Ochman H (2009) The consequences of genetic drift for bacterial genome complexity. Genome Res 19:1450–1454. https://doi.org/10.1101/gr.091785.109

Kupper M, Gupta SK, Feldhaar H, Gross R (2014) Versatile roles of the chaperonin GroEL in microorganism-insect interactions. FEMS Microbiol Lett 353:1–10. https://doi.org/10.1111/1574-6968.12390

LaBar T, Adami C (2016) Different evolutionary paths to complexity for small and large populations of digital organisms. PLoS Comput Biol 12:1–19. https://doi.org/10.1371/journal.pcbi.1005066

Lamelas A, Pérez-Brocal V, Gómez-Valero L, Gosalbes MJ, Moya A, Latorre A (2008) Evolution of the secondary symbiont "*Candidatus* Serratia symbiotica" in aphid species of the subfamily *Lachninae*. Appl Environ Microbiol 74(13):4236–4240. doi:https://doi.org/10.1128/AEM.00022-08

Lamelas A, Gosalbes MJ, Manzano-Marin A, Peretó J, Mora A, Latorre A (2011) *Serratia symbiotica* from the aphid *Cinara cedri*: a missing link from facultative to obligate insect endosymbiont. PLoS Genet 7(11):e1002357. https://doi.org/10.1371/journal.pgen.1002357

Lanan MC, Rodrigues PAP, Agellon A et al (2016) A bacterial filter protects and structures the gut microbiome of an insect. ISME J 10:1866–1876. https://doi.org/10.1038/ismej.2015.264

Leonard SP, Powell JE, Perutka J et al (2020) Engineered symbionts activate honey bee immunity and limit pathogens. Science 367:573–576. https://doi.org/10.1126/science.aax9039

Liberti J, Engel P (2020) The gut microbiota — brain axis of insects. Curr Opin Insect Sci 39:6–13. https://doi.org/10.1016/j.cois.2020.01.004

Liesack W, Stackebrandt E (1992) Unculturable microbes detected by molecular sequences and probes. Biodivers Conserv 1:250–262. https://doi.org/10.1007/BF00693763

Login FH, Balmand S, Vallier A et al (2011) Antimicrobial peptides keep insect endosymbionts under control. Science 334:362–365. https://doi.org/10.1126/science.1209728

Lopez-Fernandez S, Mazzoni V, Pedrazzoli F et al (2017) A phloem-feeding insect transfers bacterial endophytic communities between grapevine plants. Front Microbiol 8:1–17. https://doi.org/10.3389/fmicb.2017.00834

Lopez-Madrigal S, Gil R (2017) Et tu, brute? Not even intracellular mutualistic symbionts escape horizontal gene transfer. Genes (Basel) 8:1–16. https://doi.org/10.3390/genes8100247

Luan JB, Chen W, Hasegawa DK et al (2015) Metabolic coevolution in the bacterial symbiosis of whiteflies and related plant sap-feeding insects. Genome Biol Evol 7:2635–2647. https://doi.org/10.1093/gbe/evv170

Luan J-B, Shan H-W, Isermann P et al (2016) Cellular and molecular remodelling of a host cell for vertical transmission of bacterial symbionts. Proc R Soc B Biol Sci 283:20160580. https://doi.org/10.1098/rspb.2016.0580

Łukasik P, Nazario K, Van Leuven JT et al (2017) Multiple origins of interdependent endosymbiotic complexes in a genus of cicadas. Proc Natl Acad Sci U S A 115:E226–E235. https://doi.org/10.1073/pnas.1712321115

Manzano-Marín A, Latorre A (2016) Snapshots of a shrinking partner: genome reduction in *Serratia symbiotica*. Sci Rep 6:1–11. https://doi.org/10.1038/srep32590

Manzano-Marín A, Simon JC, Latorre A (2016) Reinventing the wheel and making it round again: evolutionary convergence in *Buchnera-Serratia* symbiotic consortia between the distantly related *Lachninae* aphids *Tuberolachnus salignus* and *Cinara cedri*. Genome Biol Evol 8:1440–1458. https://doi.org/10.1093/gbe/evw085

Mao M, Bennett GM (2020) Symbiont replacements reset the co-evolutionary relationship between insects and their heritable bacteria. ISME J. https://doi.org/10.1038/s41396-020-0616-4

Martin BD, Schwab E (2012) Symbiosis : " living together " in chaos. Stud Hist Biol 4:7–25

Martin BD, Schwab E (2013) Current usage of symbiosis and associated terminology. Int J Biol 5:32–45. https://doi.org/10.5539/ijb.v5n1p32

Matsuura Y, Moriyama M, Łukasik P et al (2018) Recurrent symbiont recruitment from fungal parasites in cicadas. Proc Natl Acad Sci U S A 115:E5970–E5979. https://doi.org/10.1073/pnas.1803245115

McCutcheon JP, McDonald BR, Moran NA (2009a) Convergent evolution of metabolic roles in bacterial co-symbionts of insects. Proc Natl Acad Sci U S A 106:15394–15399

McCutcheon JP, McDonald BR, Moran NA (2009b) Origin of an alternative genetic code in the extremely small and GC-rich genome of a bacterial symbiont. PLoS Genet 5:e1000565

McCutcheon JP, Moran NA (2010) Functional convergence in reduced genomes of bacterial symbionts spanning 200 My of evolution. Genome Biol Evol 2:708–718
McCutcheon JP, Moran NA (2012) Extreme genome reduction in symbiotic bacteria. Nat Rev Microbiol 10:13–26. https://doi.org/10.1038/nrmicro2670
McCutcheon JP, Von Dohlen CD (2011) An interdependent metabolic patchwork in the nested symbiosis of mealybugs. Curr Biol 21:1366–1372. https://doi.org/10.1016/j.cub.2011.06.051
McLean AH (2019) Cascading effects of defensive endosymbionts. Curr Opin Insect Sci 32:42–46. https://doi.org/10.1016/j.cois.2018.10.003
Melnikov S, Kwok H-S, Manakongtreecheep K et al (2019) Archaeal ribosomal proteins possess nuclear localization signal-type motifs: implications for the origin of the cell nucleus. Mol Biol Evol 37:124–133. https://doi.org/10.1093/molbev/msz207
Mira A, Ochman H, Moran NA (2001) Deletional bias and the evolution of bacterial genomes. Trends Genet 17:589–596. https://doi.org/10.1016/S0168-9525(01)02447-7
Monticelli LS, Outreman Y, Frago E, Desneux N (2019) Impact of host endosymbionts on parasitoid host range — from mechanisms to communities. Curr Opin Insect Sci 32:77–82. https://doi.org/10.1016/j.cois.2018.11.005
Moran NA (2001) The coevolution of bacterial endosymbionts and phloem-feeding insects. Ann Missouri Bot Gard 88:35–44
Moran NA (2007) Symbiosis as an adaptive process and source of phenotypic complexity. Proc Natl Acad Sci 104:8627–8633. https://doi.org/10.1073/pnas.0611659104
Moran NA, Bennett GM (2014) The tiniest tiny genomes. Annu Rev Microbiol 68:195–215. https://doi.org/10.1146/annurev-micro-091213-112901
Moran NA, McCutcheon JP, Nakabachi A (2008) Genomics and evolution of heritable bacterial symbionts. Annu Rev Genet 42:165–190. https://doi.org/10.1146/annurev.genet.41.110306.130119
Moran NA, Plague GR, Sandström JP, Wilcox JL (2003) A genomic perspective on nutrient provisioning by bacterial symbionts of insects. Proc Natl Acad Sci U S A 100:14543–14548. https://doi.org/10.1073/pnas.2135345100
Moran NA, Sloan DB (2015) The hologenome concept: helpful or hollow? PLoS Biol 13:1–10. https://doi.org/10.1371/journal.pbio.1002311
Mori M, Ponce-de-León M, Peretó J, Montero F (2016) Metabolic complementation in bacterial communities: Necessary conditions and optimality. Front Microbiol 7. https://doi.org/10.3389/fmicb.2016.01553
Nakabachi A, Ueoka R, Oshima K, Teta R, Mangoni A, Gurgui M, Oldham NJ, van Echten-Deckert G, Okamura K, Yamamoto K, Inoue H, Ohkuma M, Hongoh Y, Miyagishima SY, Hattori M, Piel J, Fukatsu T (2013) Defensive bacteriome symbiont with a drastically reduced genome. Curr Biol 23(15):1478–1484. https://doi.org/10.1016/j.cub.2013.06.027
Nakabachi A, Yamashita A, Toh H, Ishikawa H, Dunbar HE, Moran NA, Hattori M (2006) The 160-kilobase genome of the bacterial endosymbiont *Carsonella*. Science 314(5797):267. https://doi.org/10.1126/science.1134196
Neef A, Delaye L, López-Sánchez MJ et al (2010) Stages of metabolic adaptation of endosymbiotic bacteria. Nov Acta Leopoldina 378:141–149
Newton ILG, Rice DW (2020) The Jekyll and Hyde symbiont: could *Wolbachia* be a nutritional mutualist? J Bacteriol 202:1–8. https://doi.org/10.1128/JB.00589-19
Nikoh N, Tsuchida T, Maeda T, Yamaguchi K, Shigenobu S, Koga R, Fukatsu T (2018) Genomic insight into symbiosis-induced insect color change by a facultative bacterial endosymbiont, "*Candidatus* Rickettsiella viridis". MBio 9:e00890–e00818. https://doi.org/10.1128/mBio.00890-18
Oakeson KF, Gil R, Clayton AL et al (2014) Genome degeneration and adaptation in a nascent stage of symbiosis. Genome Biol Evol 6:76–93. https://doi.org/10.1093/gbe/evt210
Oborník M (2019) In the beginning was the word: how terminology drives our understanding of endosymbiotic organelles. Microb Cell 6:134–141. https://doi.org/10.15698/mic2019.02.669

Ordax M, Piquer-Salcedo JE, Santander RD et al (2015) Medfly *Ceratitis capitata* as potential vector for fire blight pathogen Erwinia amylovora: survival and transmission. PLoS One 10. https://doi.org/10.1371/journal.pone.0127560

Otero-Bravo A, Goffredi S, Sabree ZL (2018) Cladogenesis and genomic streamlining in extracellular endosymbionts of tropical stink bugs. Genome Biol Evol 10(2):680–693. https://doi.org/10.1093/gbe/evy003

Pérez-Brocal V, Gil R, Ramos S et al (2006) A small microbial genome: the end of a long symbiotic relationship? Science 314:312–313. https://doi.org/10.1126/science.1130441

Pérez-Brocal V, Latorre A, Moya A (2013) Symbionts and pathogens: what is the difference? Curr Top Microbiol Immunol 358:215–243. https://doi.org/10.1007/82_2011_190

Pontes MH, Dale C (2006) Culture and manipulation of insect facultative symbionts. Trends Microbiol 14:406–412. https://doi.org/10.1016/j.tim.2006.07.004

Rainey PB, Remigi P, Farr AD, Lind PA (2017) Darwin was right: where now for experimental evolution? Curr Opin Genet Dev 47:102–109. https://doi.org/10.1016/j.gde.2017.09.003

Raymann K, Moran NA (2018) The role of the gut microbiome in health and disease of adult honey bee workers. Curr Opin Insect Sci 26:97–104

Rosas-Pérez T, Rosenblueth M, Rincón-Rosales R et al (2014) Genome sequence of "*Candidatus* Walczuchella monophlebidarum" the flavobacterial endosymbiont of *Llaveia axin axin* (Hemiptera: Coccoidea: Monophlebidae). Genome Biol Evol 6:714–726. https://doi.org/10.1093/gbe/evu049

Ruiz-González MX, Fares MA (2013) Coevolution analyses illuminate the dependencies between amino acid sites in the chaperonin system GroES-L. BMC Evol Biol 13. https://doi.org/10.1186/1471-2148-13-156

Russell JA, Latorre A, Sabater-Muñoz B et al (2003) Side-stepping secondary symbionts: widespread horizontal transfer across and beyond the Aphidoidea. Mol Ecol 12:1061–1075. https://doi.org/10.1046/j.1365-294X.2003.01780.x

Sabater-Muñoz B, Prats-Escriche M, Montagud-Martínez R et al (2015) Fitness trade-offs determine the role of the molecular chaperonin GroEL in buffering mutations. Mol Biol Evol 32:2681–2693. https://doi.org/10.1093/molbev/msv144

Sabater-Muñoz B, Toft C, Alvarez-Ponce D, Fares MA (2017) Chance and necessity in the genome evolution of endosymbiotic bacteria of insects. ISME J 11:1291–1304. https://doi.org/10.1038/ismej.2017.18

Sabree ZL, Huang CY, Okusu A, Moran NA, Normak BB (2013) The nutrient supplying capabilities of *Uzinura* and endosymbiont of armoured scale insects. Environ Microbiol 15 (7):1988–1999. https://doi.org/10.1111/1462-2920.12058

Sabree ZL, Kambhampati S, Moran NA (2009) Nitrogen recycling and nutritional provisioning by *Blattabacterium*, the cockroach endosymbiont. Proc Natl Acad Sci U S A 106:19521–19526. https://doi.org/10.1073/pnas.0907504106

Salem H, Florez L, Gerardo N, Kaltenpoth M (2015) An out-of-body experience: the extracellular dimension for the transmission of mutualistic bacteria in insects. Proc R Soc B Biol Sci 282. https://doi.org/10.1098/rspb.2014.2957

Santos-Garcia D, Mestre-Rincon N, Zchori-Fein E, Morin S (2020) Inside out: microbiota dynamics during host-plant adaptation of whiteflies. ISME J. https://doi.org/10.1038/s41396-019-0576-8

Sapp J (2002) Paul Buchner (1886-1978) and hereditary symbiosis in insects. Int Microbiol 5:145–150. https://doi.org/10.1007/s10123-002-0079-7

Sarhan MS, Hamza MA, Youssef HH et al (2019) Culturomics of the plant prokaryotic microbiome and the dawn of plant-based culture media – a review. J Adv Res 19:15–27. https://doi.org/10.1016/j.jare.2019.04.002

Schwemmler W, Gassner G (1989) Insect endocytobiosis: morphology, physiology, genetics, evolution. CRC Press, Boca Raton, FL

Shapiro BJ (2016) How clonal are bacteria over time? Curr Opin Microbiol 31:116–123. https://doi.org/10.1016/j.mib.2016.03.013

Shigenobu S, Stern DL (2013) Aphids evolved novel secreted proteins for symbiosis with bacterial endosymbiont. Proc R Soc B Biol Sci 280(1750):20121952. https://doi.org/10.1098/rspb.2012.1952

Shigenobu S, Watanabe H, Hattori M, Sakaki Y, Ishikawa H (2000) Genome sequence of the endocellular bacterial symbiont of aphids *Buchnera* sp. APS Nature 407(6800):81–86. https://doi.org/10.1038/35024074

Silva FJ, Latorre A, Gómez-Valero L, Moya A (2007) Genomic changes in bacteria: from free-living to endosymbiotic life. In: Bastolla U, Porto M, Roman HE, Vendruscolo M (eds) Structural approaches to sequence evolution: molecules, networks, populations. Springer, Berlin, pp 149–165

Simonet P, Gaget K, Balmand S et al (2018) Bacteriocyte cell death in the pea aphid/*Buchnera* symbiotic system. Proc Natl Acad Sci U S A 115:E1819–E1827. https://doi.org/10.1073/pnas.1720237115

Somerville J, Zhou L, Raymond B (2019) Aseptic rearing and infection with gut bacteria improve the fitness of transgenic diamondback moth *Plutella xylostella.* Insects 10(4): 89. https://doi.org/10.3390/insects10040089

Stewart EJ (2012) Growing unculturable bacteria. J Bacteriol 194:4151–4160. https://doi.org/10.1128/JB.00345-12

Stoll S, Feldhaar H, Fraunholz MJ, Gross R (2010) Bacteriocyte dynamics during development of a holometabolous insect, the carpenter ant *Camponotus floridanus*. BMC Microbiol 10. https://doi.org/10.1186/1471-2180-10-308

Sudakaran S, Kost C, Kaltenpoth M (2017) Symbiont acquisition and replacement as a source of ecological innovation. Trends Microbiol 25:375–390. https://doi.org/10.1016/j.tim.2017.02.014

Tamas I, Klasson L, Canbäck B et al (2002) 50 million years of genomic stasis in endosymbiotic bacteria. Science 296:2376–2379. https://doi.org/10.1126/science.1071278

Theissen U, Martin W (2006) The difference between organelles and endosymbionts. Curr Biol 16: R1016–R1017. https://doi.org/10.1016/j.cub.2006.11.020

Toft C, Andersson SGE (2010) Evolutionary microbial genomics: insights into bacterial host adaptation. Nat Rev Genet 11:465–475. https://doi.org/10.1038/nrg2798

Toft C, Fares MA (2008) The evolution of the flagellar assembly pathway in endosymbiotic bacterial genomes. Mol Biol Evol 25:2069–2076. https://doi.org/10.1093/molbev/msn153

Toft C, Fares MA (2010) Structural calibration of the rates of amino acid evolution in a search for Darwin in drifting biological systems. Mol Biol Evol 27:2375–2385. https://doi.org/10.1093/molbev/msq123

Toft C, Williams TA, Fares MA (2009) Genome-wide functional divergence after the symbiosis of Proteobacteria with insects unraveled through a novel computational approach. PLoS Comput Biol 5(4):e1000344. https://doi.org/10.1371/journal.pcbi.1000344

van Ham RCHJ, Kamerbeek J, Palacios C et al (2003) Reductive genome evolution in *Buchnera aphidicola*. Proc Natl Acad Sci 100:581–586. https://doi.org/10.1073/pnas.0235981100

Van Leuven JTT, Meister RCC, Simon C, McCutcheon JPP (2014) Sympatric speciation in a bacterial endosymbiont results in two genomes with the functionality of one. Cell 158:1270–1280. https://doi.org/10.1016/j.cell.2014.07.047

Volf M, Salminen JP, Segar ST (2019) Evolution of defences in large tropical plant genera: perspectives for exploring insect diversity in a tri-trophic context. Curr Opin Insect Sci 32:91–97. https://doi.org/10.1016/j.cois.2018.12.005

Vorburger C (2018) Symbiont-conferred resistance to parasitoids in aphids – challenges for biological control. Biol Control 116:17–26. https://doi.org/10.1016/j.biocontrol.2017.02.004

Wallin IE (1927) Symbionticism and the origin of species. Williams & Wilkins Company, Baltimore, MD

Wernegreen JJ (2011) Reduced selective constraint in endosymbionts: elevation in radical amino acid replacements occurs genome-wide. PLoS One 6:1–15. https://doi.org/10.1371/journal.pone.0028905

Wernegreen JJ (2012) Strategies of genomic integration within insect-bacterial mutualisms. Biol Bull 223:112–122. https://doi.org/10.1086/BBLv223n1p112
Wernegreen JJ (2017) In it for the long haul: evolutionary consequences of persistent endosymbiosis. Curr Opin Genet Dev 47:83–90. https://doi.org/10.1016/j.gde.2017.08.006
Whitney KD, Garland T (2010) Did genetic drift drive increases in genome complexity? PLoS Genet 6:1–6. https://doi.org/10.1371/journal.pgen.1001080
Williams TA, Codoñer FM, Toft C, Fares MA (2010) Two chaperonin systems in bacterial genomes with distinct ecological roles. Trends Genet 26:47–51. https://doi.org/10.1016/j.tig.2009.11.009
Williams TA, Fares MA (2010) The effect of chaperonin buffering on protein evolution. Genome Biol Evol 2:609–619. https://doi.org/10.1093/gbe/evq045
Wong ACN, Vanhove AS, Watnick PI (2016) The interplay between intestinal bacteria and host metabolism in health and disease: lessons from *Drosophila melanogaster*. DMM Dis Model Mech 9:271–281. https://doi.org/10.1242/dmm.023408
Wu D, Daugherty SC, Van Aken SE et al (2006) Metabolic complementarity and genomics of the dual bacterial symbiosis of sharpshooters. PLoS Biol 4:e188. https://doi.org/10.1371/journal.pbio.0040188
Xie S, Lan Y, Sun C, Shao Y (2019) Insect microbial symbionts as a novel source for biotechnology. World J Microbiol Biotechnol 35:1–7. https://doi.org/10.1007/s11274-019-2599-8
Xu Y, Buss EA, Boucias DG (2016) Culturing and characterization of gut symbiont *Burkholderia* spp. from the southern chinch bug, *Blissus insularis* (Hemiptera: Blissidae). Appl Environ Microbiol 82:3319–3330. https://doi.org/10.1128/AEM.00367-16
Zchori-Fein E, Bourtzis K (2012) Manipulative tenants: Bacteria associated with arthropods. CRC PRess, Taylor & Francis Group, Boca Raton, FL.
Zientz E, Dandekar T, Gross R (2004) Metabolic interdependence of obligate intracellular bacteria and their insect hosts. Microbiology and Molecular Biology Reviews 68(4):745–770. https://doi.org/10.1128/MMBR.68.4.745-770.2004
Zimmer C (2009) On the origin of eukaryotes. Science 325:666–668. https://doi.org/10.1126/science.325_666

Chapter 4
Epidemiology of Nucleus-Dwelling *Holospora*: Infection, Transmission, Adaptation, and Interaction with *Paramecium*

Martina Schrallhammer and Alexey Potekhin

Abstract The chapter describes the exceptional symbiotic associations formed between the ciliate *Paramecium* and *Holospora*, highly infectious bacteria residing in the host nuclei. *Holospora* and *Holospora*-like bacteria (Alphaproteobacteria) are characterized by their ability for vertical and horizontal transmission in host populations, a complex biphasic life cycle, and pronounced preference for host species and colonized cell compartment. These bacteria are obligate intracellular parasites; thus, their metabolic repertoire is dramatically reduced. Nevertheless, they perform complex interactions with the host ciliate. We review ongoing efforts to unravel the molecular adaptations of these bacteria to their unusual lifestyle and the host's employment in the symbiosis. Furthermore, we summarize current knowledge on the genetic and genomic background of *Paramecium*–*Holospora* symbiosis and provide insights into the ecological and evolutionary consequences of this interaction. The diversity and occurrence of symbioses between ciliates and *Holospora*-like bacteria in nature is discussed in connection with transmission modes of symbionts, host specificity and compatibility of the partners. We aim to summarize 50 years of research devoted to these symbiotic systems and conclude trying to predict some perspectives for further studies.

M. Schrallhammer (✉)
Microbiology, Biological Institute II, Albert-Ludwigs University of Freiburg, Freiburg, Germany
e-mail: martina.schrallhammer@biologie.uni-freiburg.de

A. Potekhin
Department of Microbiology, Faculty of Biology, Saint Petersburg State University, Saint Petersburg, Russia

M. Kloc (ed.), *Symbiosis: Cellular, Molecular, Medical and Evolutionary Aspects*, Results and Problems in Cell Differentiation 69,
https://doi.org/10.1007/978-3-030-51849-3_4

4.1 Characteristics of *Holospora* and *Holospora*-Like Bacteria

Symbioses and especially intracellular symbioses are drivers of evolution and biological innovation. The interaction between unicellular eukaryotic host and intracellular symbionts is naturally rather intimate as any interruption leading to the decay of the symbiotic system may ultimately cause the death of the host cell itself. In unicellular eukaryotes, we find a tremendous amount of organismal and functional diversity regarding intracellular symbionts. Hence, we can use them as fascinating models to study the mechanisms of bacterial transmission on individual and the host population level, regulation of interactions between eukaryotic and prokaryotic cells, and coevolution of host and symbiont. It is no surprise that symbioses between different groups of protists and prokaryotes continuously attract scientific attention (Gast et al. 2009; Nowack and Melkonian 2010; Dziallas et al. 2012; Edgcomb 2016; Samba-Louaka et al. 2019). Ciliates, one of the most numerous taxa of protists, harbor a plethora of diverse prokaryotes. The first-discovered endosymbionts of ciliates were *Holospora*, and now these and related *Holospora*-like bacteria (HLB) are, probably, among the best-studied prokaryotic symbionts of protists.

Holospora and HLB exhibit several fascinating features. The most conspicuous are the occupation of the host's nucleus (Fig. 4.1), their cellular dimorphism with the eye-catching long infectious form, and their complex life cycle with an infectious stage. Thanks to the prominent localization and the atypical cell shape, these bacteria were reported already in very early microscopic studies devoted to *Paramecium* (Bütschli 1887). 130 years ago, bacteria in *Paramecium* nuclei were observed and described as *Holospora* ("whole spore") by Wladimir Hafkine (Hafkine 1890), a well-known bacteriologist from the laboratory of Louis Pasteur. His descriptions were confirmed and formalized according to taxonomic rules a century later (Gromov and Ossipov 1981).

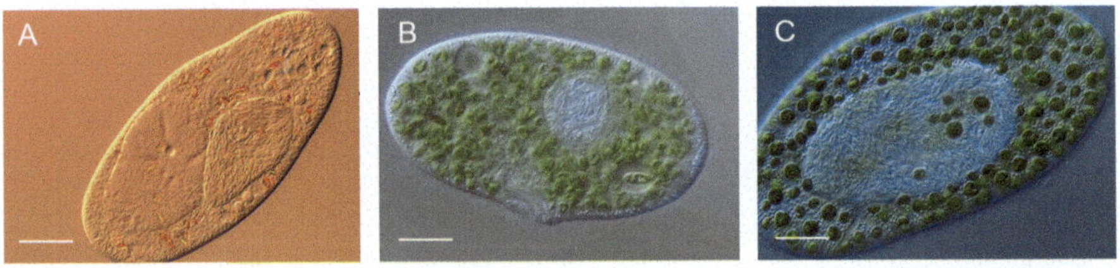

Fig. 4.1 Different paramecia infected with various *Holospora* species. *Holospora undulata* (**a**) in the *Paramecium caudatum* micronucleus and *Holospora parva* (**b**) and *Holospora curviuscula* (**c**) in the macronuclei of their *Paramecium* hosts. The nuclei are heavily infected and appear swollen in size. Note the algal symbionts additionally harbored by *Paramecium chlorelligerum* (**b**) and *Paramecium busaria* (**c**). Living cells observed by differential interference contrast microscopy. Scale bars: 25 μm (**a**), 10 μm (**b**, **c**)

Holospora and HLB form a monophyletic group within the *Holosporales* (see Sect. 4.2.1). All of them are obligate intranuclear bacteria as they live "inside the control center" of their eukaryotic hosts (Schulz and Horn 2015). They are capable of vertical and horizontal transmission; thus, they are distributed from mother to daughter cells and can also infect new hosts after uptake from the environment (see Sect. 4.1.2).

The only known hosts for *Holospora* and HLB are ciliates. Relevant features of ciliates for the symbiotic interaction with HLB are their filter-feeding followed by phagocytosis, which provides to bacteria a possibility to enter the cell, and their nuclear dimorphism, as HLB reside in either somatic polyploid macronuclei or in the germline micronuclei and are restricted to one type or the other.

In this review, we will discuss new insights into adaptation, evolution, and host interactions of the following *Holospora* species: *H. undulata* (type species), *H. obtusa, H. elegans, H. acuminata, H. curviuscula,* and "*Candidatus*[1] H. parva." As there has been no update since the most recent reviews on the diversity of *Holospora* (Fokin and Görtz 2009; Fujishima and Kodama 2012), we will skip *H. recta* (Fokin 1991), *H. curvata* (Fokin and Sabaneyeva 1993), and *H. bacillata* (Fokin 1989). The group here termed HLB includes the following bacteria: "*Ca.* Preeria caryophila" (Potekhin et al. 2018, basonym: *Holospora caryophila*), "*Ca.* Gortzia infectiva" (Boscaro et al. 2013), "*Ca.* G. shahrazadis" (Serra et al. 2016), "*Ca.* G. yakutica" (Beliavskaia et al. 2020), and "*Ca.* Hafkinia simulans" (Fokin et al. 2019).

"*Ca.* Paraholospora nucleivisitans" (Eschbach et al. 2009) is only distantly related and lacks typical HLB characteristics (see Sect. 4.2.1). Thus, we will not consider it here as HLB.

4.1.1 *Symbioses between* Paramecium *and* Holospora*-like bacteria*

Holospora are obligate intracellular bacteria, i.e., cultivation attempts on artificial media outside the host have not been successful so far (Fokin and Görtz 2009). Inside *Paramecium*, they can elicit dramatic alterations of the host's nuclear structure (Fig. 4.1) and impact host growth and fitness (see Sect. 4.3.2). There are no indications that they are required by their host under any circumstance. Thus, they can be considered parasites. Nevertheless, we use the more general term "symbiont" in this review as *Paramecium* and *Holospora* may form an intimate, long-term

[1] *Candidatus* indicates that these bacteria cannot be cultivated outside their host and thus are not deposited as a pure culture at two culture collections, preventing their full valid description according to the International Code of Nomenclature of Prokaryotes. For brevity's sake, we will omit *Candidatus* further on in the text. Organisms originally described before 1980 were given valid names even when cultivation could not be accomplished.

interaction; hence, the definition of symbiosis according to de Bary, 1878 (de Bary 1879; Oulhen et al. 2016) applies.

All *Holospora* species are strictly associated with a *Paramecium* species. This pattern of host specificity indicates that the intranuclear symbionts coevolved with their paramecia hosts. Very rarely *Holospora* can enter the "wrong" host species and even complete its infection cycle (Fokin et al. 2005), but yet under laboratory conditions such associations are very unstable and quickly disappear.

In the following, we will review publications and ongoing efforts conducted to unravel the molecular adaptations of these bacteria to their unusual lifestyle and the host's employment in the symbiosis. Furthermore, we aim to provide insights into the ecological and evolutionary consequences of this interaction.

4.1.2 Infection, Life Cycle, and Cellular Dimorphism

Holospora and all HLB (see Sect. 4.2.3) display a complex life cycle connected to the infection process and are characterized by two different cell morphologies that serve distinct functions (Figs. 4.2 and 4.3). The reproductive forms (RF) are typical bacterial rod-shaped cells (0.4–1.0 × 2.0–4.0 μm). They can be found multiplying inside the host nucleus. At some point, they differentiate into infectious forms (IF). These cells are much longer than RF and can reach up to 20.0 μm. IF shapes can be straight, spindle-shaped, curved, or sigmoidal (Fig. 4.2; reviewed by Fokin and Görtz 2009; Potekhin et al. 2018). For several decades, these differences served as one major diagnostic character for discrimination of *Holospora* and HLB species (Gromov and Ossipov 1981; Fokin et al. 1996; Görtz and Schmidt 2005; Fokin and Görtz 2009; Schweikert et al. 2013).

On the ultrastructural level (Fig. 4.3), IF are subdivided into recognition tip (also termed infection tip), an enlarged periplasmic lumen, and the remaining condensed cytoplasm (Ossipov 1981; Görtz and Wiemann 1989). The recognition tip plays an important role during the escape of IF from the phagosome (Fig. 4.3a) and in penetration of the nuclear membrane.

IF are the agents of horizontal transmission (Fig. 4.4a). They are released from the nucleus either during cell division (Fig. 4.4b) or at cell death and can persist for a certain time outside a host cell (Fujishima et al. 1991). A new infection cycle starts after phagocytosis of IF by a *Paramecium* cell. Usually, bacteria inside the phagosome are digested, but IF can avoid this fate by escaping the digestive vacuole with the recognition tip spearheading the exit (Fig. 4.4a). This process is triggered by the acidification of the phagosome. Inhibitors of vacuolar-type ATPases, which block acidification, prevent IF from leaving the vacuole (Fujishima et al. 1997). The importance of acidification for the maturation of IF was also shown by experiments where IF of *H. obtusa* were microinjected into the macronucleus of *Paramecium caudatum* bypassing all intermediate stages of the infection cycle. These IF did not form constrictions and failed to differentiate into RF (Skovorodkin et al.

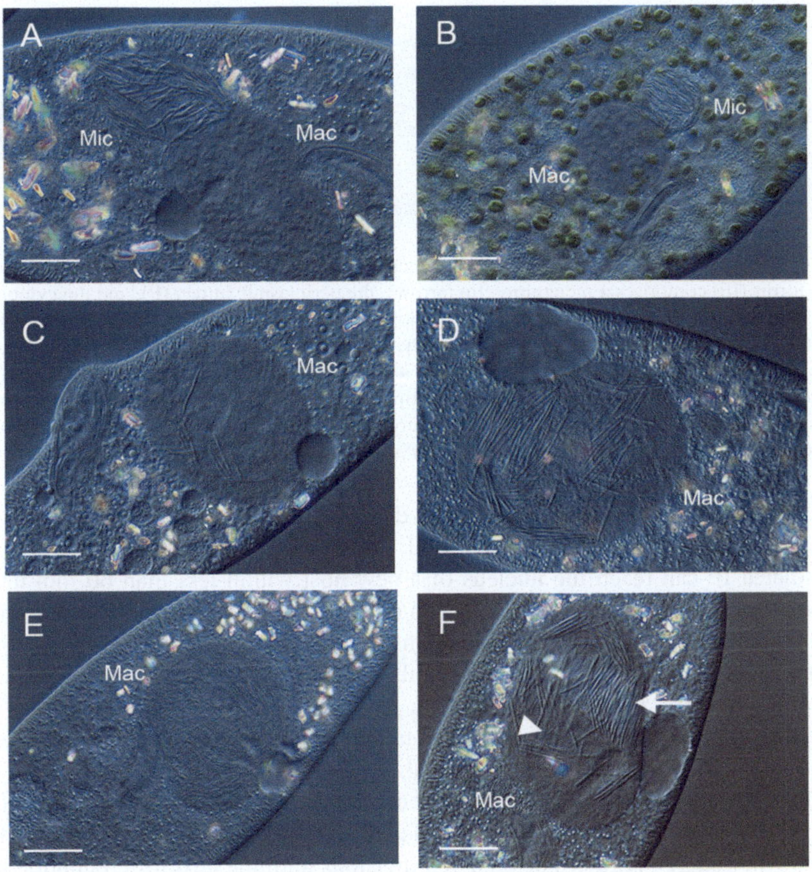

Fig. 4.2 Nuclear infections by *Holospora* and *Holospora*-like bacteria. *Holospora undulata* (**a**) and *Holospora acuminata* (**b**) in the micronucleus (Mic) of their hosts. Infected macronuclei (Mac) harboring *Holospora obtusa* (**c**, **d**), *Preeria caryophila* (**e**), and a double infection (**f**) with *H. obtusa* (arrow head) and *Preeria caryophila* (white arrow). The majority of *H. obtusa* are present as reproductive forms in C, while in D the number of infectious forms has increased. Scale bars: 10 μm

2001). Acidification of isolated IF of *H. obtusa* induces the production of an IF-specific antigen (Kawai and Fujishima 2000).

It is debated and remains unclear if the phagosome membrane is collapsed in the process of IF escape and the bacteria are naked in the *Paramecium* cytoplasm (Fujishima 2009) or if the symbionts remain surrounded by remnants of the vacuole membrane (Ossipov 1981; Görtz and Wiemann 1989). Similar to intracellular pathogens like *Listeria monocytogenes, Shigella flexneri,* and *Rickettsia conorii* (Stevens et al. 2006; de Souza and Orth 2015), *Holospora* repurposes the host's actin cytoskeleton in order to move intracellularly and to reach the target nucleus. The actin tail polymerizes at the side of the IF, occasionally nearly perpendicular to

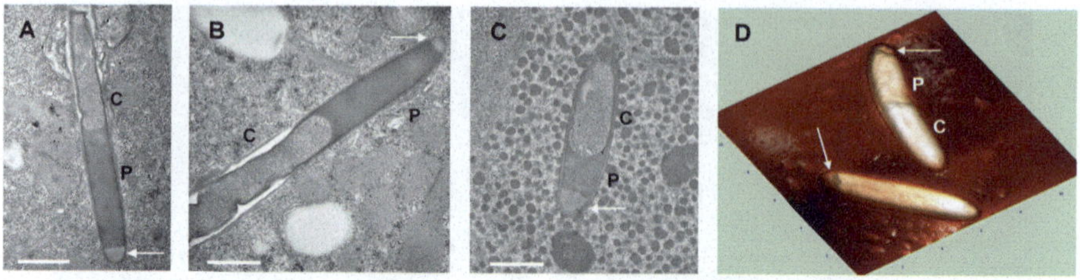

Fig. 4.3 Tripartite compartmentalization of infectious forms. The typical cell structure of infectious forms (IF) comprises a recognition tip (arrow), the periplasmic lumen (P), and the cytoplasm (C). Transmission electron micrographs (**a**-**c**) show IF of *Holospora obtusa* exiting from a food vacuole (**a**), inside the host cytoplasm (**b**) and the macronucleus (**c**). A three-dimensional atomic force microscopy image (**d**) depicts two IF of *Gortzia infectiva*. Note a slight depression of the surface of both bacteria at the polar recognition tip (arrow). Scale bars: 2 μm. Images were kindly provided by Dr. Elena Sabaneyeva, St. Petersburg State University (**d**), and Prof. Sergei Fokin, St. Petersburg State University & University of Pisa (**a**-**c**)

the longitudinal cell axis, and serves as driving force for propulsion through the *Paramecium* cytoplasm (Sabaneyeva et al. 2009).

Isolated IF can reach the nucleus of a new host within less than 60 min under laboratory conditions. The quick invasion of host nuclei by *Holospora* is accompanied by extensive ruffling and perturbations of the nuclear envelope (Ossipov 1981; Görtz and Fokin 2009). It implies highly specific recognition mechanisms and bacterial effectors. Indeed, some specific proteins crucial for the escape from the digestive vacuole and invasion of the target nucleus have been biochemically identified (Dohra et al. 1994; Iwatani et al. 2005; Abamo et al. 2008).

Once the nucleus is reached, the IF penetrates the nuclear envelope recognition tip oriented forward. Inside the organelle, IF constrict and differentiate into RF, which will undergo regular bacterial cell divisions until the next differentiation into IF (Fig. 4.4a).

Species-specific targeting to only one type of the host nuclei is another major feature used for *Holospora* diagnosis (Fokin and Görtz 2009). Rarely, the nucleus recognition is not precise and symbionts may end up in the nontarget compartment, especially in course of massive infections (Borchsenius et al. 1990; Ossipov et al. 1993). Still, with exception of *H. curviuscula,* which was able to colonize simultaneously both nuclei and to stay in the nonspecific micronucleus for 3–5 months (Borchsenius et al. 1990), *Holospora* species stop their proliferation and are very quickly lost from the nontarget nucleus (Ossipov et al. 1993; Lebedeva, Skoblo, Ossipov, pers. comm.).

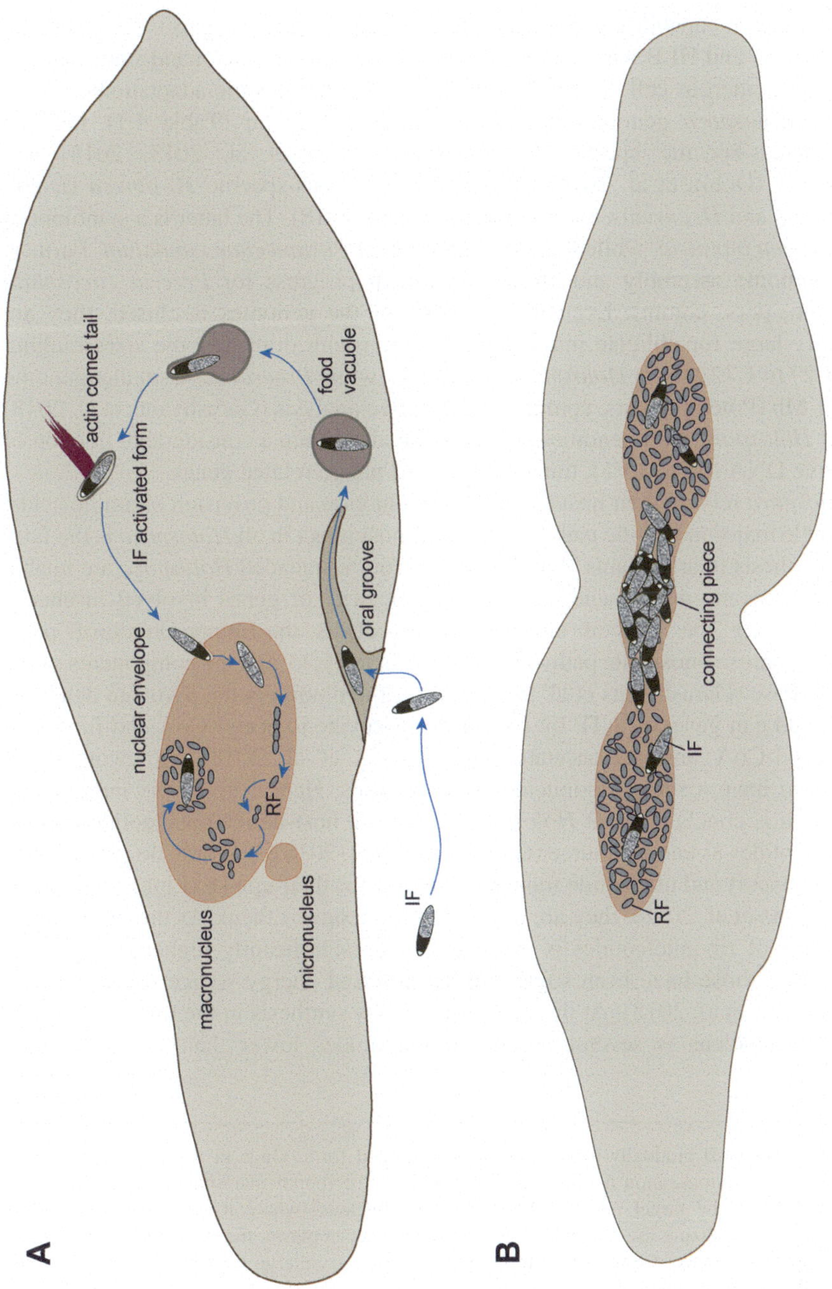

Fig. 4.4 Horizontal and vertical transmission of *Holospora*. (**a**) Major stages of the infection cycle of *Holospora* starting with the uptake of an extracellular infectious form (IF) via phagocytosis followed by incorporation in a food vacuole. The IF escapes from the vacuole with recognition tip oriented forward and

4.1.3 Molecular Adaptations to an Intranuclear Lifestyle

The uncommon capability to live and replicate inside a host nucleus is characteristic to *Holospora* and HLB. Obviously, successful infection of nuclei and stable symbiosis (over numerous cell divisions of the host) requires specific adaptations.

Four *Holospora* genomes have been sequenced so far (Table 4.1), i.e., the micronucleus-specific species *H. undulata* (Dohra et al. 2013, 2014) and *H. elegans* (Dohra et al. 2014) and the macronucleus-specific *H. obtusa* (Dohra et al. 2014) and *H. curviuscula* (Garushyants et al. 2018). The latter is a symbiont of *Paramecium bursaria,* while the other three infect *Paramecium caudatum.* Furthermore, genome assembly and annotation are in progress for *Preeria caryophila* (Potekhin, pers. comm). Even though none of the genomes is closed, they are relatively large for obligate intracellular bacteria with draft genome sizes ranging from 1.27 to 1.72 Mb for *Holospora* (Table 4.1), while *Preeria* has a smaller genome of ca. 1 Mb (Potekhin, pers. comm). Comparative analysis (Garushyants et al. 2018) of four *Holospora* draft genomes revealed that all contain a considerable fraction of repetitive DNA (up to 15%), transposases, and phage-related genes.

Holospora rely on their host for energy production and provision of amino acids. The single major metabolic pathway that is almost intact in all *Holospora* is the fatty acids synthesis (Garushyants et al. 2018). All four sequenced *Holospora* are unable to synthesize any amino acid and lack the majority of genes involved in energy production, e.g., basically all enzymes for glycolysis, the Entner-Doudoroff pathway, the pentose phosphate pathway, the citric acid cycle, and the components of the F_1F_0-ATPase (Garushyants et al. 2018). All of them possess the pyruvate dehydrogenase and can generate ATP by converting pyruvate to acetyl-CoA and further to acetoacetyl-CoA and acetoacetate (Garushyants et al. 2018). Noteworthy, all *Holospora* have a set of ribonucleotide reductases. Hence, given the intranuclear lifestyle, it is conclusive that *Holospora* might use host-derived nucleotides and/or ribonucleotides as energy source (Garushyants et al. 2018). Not only do they encode ribose transport and nucleotide transport proteins for their uptake (Linka et al. 2003; Garushyants et al. 2018), they are capable to interconvert them. As the intracellular abundance of ribonucleotides is estimated to be significantly higher than that of nucleotides, those have been suggested as preferred energy source for *Holospora* (Garushyants et al. 2018). At the same time, RNA synthesis in the transcriptionally inert micronucleus is several orders of magnitudes lower than in the somatic

---

Fig. 4.4 (continued) gradually matures into the activated form. Once in the cytoplasm, the IF recruits the host cytoskeleton for intracellular motility. A perpendicular situated actin comet tail moves the IF to the target nucleus (here the macronucleus) where it penetrates the nuclear membrane again utilizing the recognition tip. Inside the nucleoplasm, the IF constricts and differentiates into reproductive forms (RF). These multiply and can undergo further differentiation from RF into IF. (**b**) Cell division of *Paramecium* and division of *Holospora*-infected nucleus. IF accumulate in the connecting piece, a structure bridging the dividing nuclei, while RF remain attached to the chromatin in the nuclei of daughter cells

Table 4.1 Comparison of genome sizes and GC content of *Holospora* and related bacteria

Organism	Chromosome size (Mb)	GC content	Life style
Holospora			
Holospora undulata HU1 (NZ_ARPM00000000.3)	1.40	36.1	Intranuclear, micronucleus of *Paramecium caudatum*
Holospora elegans E1 (NZ_BAUP00000000.1)	1.27	36.0	Intranuclear, micronucleus of *Paramecium caudatum*
Holospora obutsa F1 (NZ_AWTR00000000.2)	1.33	35.2	Intranuclear, macronucleus of *Paramecium caudatum*
Holospora curviuscula NRB217 (NZ_PHHC00000000.1)	1.72	36.1	Intranuclear, macronucleus of *Paramecium bursaria*
Members of *Holosporaceae*			
Cytomitobacter primus 1604LC (NZ_CP043316.1)	0.62	30.0	Intracellular, host: *Diplonema japonicum*
Cytomitobacter indipagum 1605 (NZ_CP043315.1)	0.63	29.7	Intracellular, host: *Diplonema aggregatum*
Nesciobacter abundans 1604HC (NZ_CP043314.1)	0.62	29.8	Intracellular, host: *Diplonema japonicum*
Members of *Holosporales*			
Caedimonas varicaedens (NZ_BBVC00000000.1)	1.69	42.1	Intracellular, host: *Paramecium* spp. (macronucleus of *P. caudatum*)
Paracaedibacter acanthamoebae PRA3 (NZ_CP008941.1)	2.47	41.0	Intracellular, host: *Acanthamoeba* sp. UWC9
Members of *Rickettsiales*			
Anaplasma marginale Florida (NC_012026.1)	1.2	49.8	Intracellular, life cycle alternates between ticks and mammals
Wolbachia pipientis wMel_N25 (NZ_CP042446.1)	1.27	35.2	Intracellular, host: *Drosophila melanogaster*
Deianiraea vastatrix CyL4–1 (NZ_CP029077.1)	1.21	32.9	Extracellular, attached to host: *Paramecium primaurelia*
Midichloria mitochondrii IricVA (NC_015722.1)	1.18	36.6	Intramitochondrial, host: Ticks
Rickettsia prowazekii Madrid E (NC_000963.1)	1.11	29.0	Intracellular, life cycle alternates between arthropods and mammals

(continued)

Table 4.1 (continued)

Organism	Chromosome size (Mb)	GC content	Life style
Rickettsia typhi Wilmington (NC_006142.1)	1.11	28.9	Intracellular, life cycle alternates between arthropods and mammals
Members of *Rhodospirillales*			
Rhodospirillum rubrum ATCC 11,170 (NC_007643.1)	4.35	65.4	Free-living, autotroph
Other Alphaproteobacteria			
Bartonella henselae Houston-1 (NC_005956.1)	1.93	38.2	Intracellular, life cycle alternates between fleas and mammals
Other bacteria			
Buchnera aphidicola APS (NC_002528.1)	0.64	26.3	Intracellular, host: Aphids
Escherichia coli K-12 MG1655 (NC_000913.3)	4.64	50.8	Free-living, heterotroph

macronucleus (Freiburg 1988). This could represent the crucial difference between the two types of nuclei in terms of nucleus preference by *Holospora* species. Their ability to propagate exclusively in one type is, probably, connected with metabolic peculiarities of the respective species.

Regarding infection-related adaptations, there have been substantial efforts to characterize engaged components both on ultrastructural level and biochemically (for a summary see Fujishima 2009). Probably the best-characterized protein involved in the infection of *Holospora* is the secreted 89-kDa protein of the recognition tip (Iwatani et al. 2005). It interacts with the phagosome membrane during IF escape and forms a fine fibrous structure between bacterial and vacuolar membranes. Comparative genomics revealed that the corresponding gene is conserved in all four *Holospora* genomes (Garushyants et al. 2018). Another interesting, interspecies-conserved gene encodes the 5.4-kDa periplasm-specific peptide, which has been described as a major protein of the IF periplasmic region likely playing a crucial role in the differentiation from RF to IF (Dohra et al. 1997). Genome mining (Garushyants et al. 2018) in *Holospora* genomes revealed about fifty proteins either containing transmembrane helixes or predicted to be secreted. Among them is the outer membrane protein A (OmpA), which is encoded in multiple copies in all analyzed *Holospora* species. OmpA interacts with host glycoproteins and is required for efficient entry into the host cell in some *Rickettsiales* (Ojogun et al. 2012), well-studied obligate intracellular pathogens, which are related to *Holospora* (Fig. 4.5).

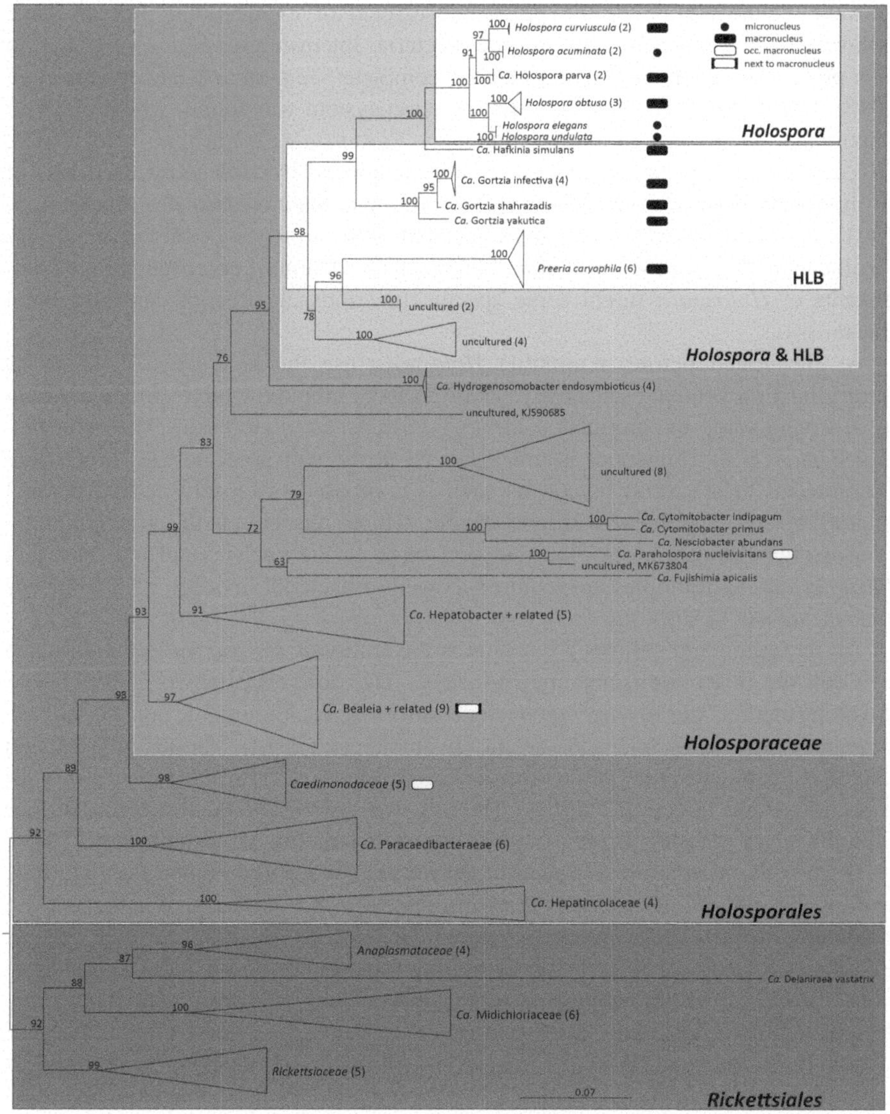

Fig. 4.5 Phylogenetic reconstruction of the order *Holosporales* demonstrating that *Holospora* and *Holospora*-like bacteria (HLB) share a common ancestor. Maximum likelihood tree was calculated with IQ-TREE based on an alignment of 94 16S rRNA gene sequences obtained from GenBank comprising1377 characters. The applied best fit evolutionary model is TVMe+R5. Numbers near nodes indicated Ultrafast Bootstrap support values of IQ-TREE. Numbers in brackets indicate sequences in collapsed groups. Symbols depict occupied intracellular localization (black circle – micronucleus; black ellipse – macronucleus; white ellipse – occasionally macronucleus; white ellipse with bars – close to macronucleus; no symbol – cytoplasm). Scale bar corresponds to 0.07 sequence divergence. *Ca.* stands for *Candidatus*

However, almost none of the predicted secreted or extracellular proteins have orthologs with known function in other bacteria. Interestingly, *Holospora* seem to lack protein secretion systems besides the complete Sec system (Garushyants et al. 2018). In particular, no type IV VirB secretion system was found, which plays an important role in host interaction in many *Rickettsiales* (Gillespie et al. 2015). Glycosylation of surface structures is likely employed by *Holospora,* as a loss of IF infectivity was observed after exposure to alpha-mannosidase (Fujishima et al. 1991). As many glycoproteins are associated with virulence factors of medically significant pathogens (Schmidt et al. 2003), glycosylation of outer membrane components of *Holospora* might serve specific functions in infection and horizontal transmission.

As mentioned before, nonmotile *Holospora* use the host's cytoskeleton for intracellular movement (Sabaneyeva et al. 2009) by employing actin polymerization as a driving force for propulsion through *Paramecium* cytoplasm. The actin tail, which in case of Holospora is not localized at the cell pole as, e.g., in *Listeria* (Lambrechts et al. 2008) but on its side, is composed of closely packed parallel microfilaments. Treatment with nocodazole, which interferes with the polymerization of microtubules, blocks the transport of IF to the nucleus and indicates that *Paramecium* microtubules are required as well in bacterial invasion of the nucleus (Sabaneyeva et al. 2005).

An intriguing and still open question is the ability of *Holospora* to discriminate between the host micro- and macronucleus. The difference between the nuclear envelope markers of *Paramecium* nuclei is still elusive, although the pore complexes have been proposed as nucleus-specific (Iwamoto et al. 2017). Various physical and chemical treatments (e.g., pH, temperature, detergents, etc.) revealed no effect on IF recognition and infection abilities. Thus, it was speculated that bacterial surface proteins might not play a crucial role in organelle targeting (Fujishima et al. 1991).

Once *Holospora* are inside their target nucleus and differentiated into RF, they exhibit a strong affinity to host chromatin that IF lack (Görtz and Wiemann 1989; Fokin et al. 1996). This difference between RF and IF is important for their vertical transmission (see Sect. 4.1.5). It is intriguing to speculate about additional interactions of *Holospora* with components of the host nucleus and their potential outcome, e.g., alteration of host gene expression (see Sect. 4.1.4, 4.3.2) or symbiont distribution at host division. Potential factors involved in intranuclear symbiont-host crosstalk might be IF surface proteins 25 kDa and 50 kDa of size that were identified to specifically bind to *Paramecium* nuclear proteins (Ehrsam and Görtz 1999).

4.1.4 Host-Symbiont Compatibility and Dissection of the Infection Process

The complex series of events leading to the successful establishment of a symbiotic association can be interrupted at different stages. It is well known that the outcome of an infection depends on the combination of both partners' genotypes (Lambrechts et al. 2006), which applies also to symbioses between ciliates and *Holospora*. Some *Paramecium* strains can be considered as universal recipients for all strains of a certain symbiont species, while others are resistant to infection (Barhey and Gibson 1984; Fujishima and Fujita 1985; Rautian et al. 1990, 1993; Skoblo et al. 1996, 2001; Bella et al. 2016). The molecular mechanisms of crosstalk between *Paramecium* and *Holospora* are unknown, though it was shown that *H. obtusa* alters the expression of multiple host genes after establishing endosymbiosis (Nakamura et al. 2004).

The infection cycle of *Holospora* species has been studied in great detail (Borchsenius et al. 1992; Rautian et al. 1993; Skoblo and Lebedeva 1993; Skoblo et al. 1996, 2001; Kawai and Fujishima 2000). It comprises following distinct stages (Fig. 4.4): (1) entrance by phagocytosis, (2) escape from the food vacuole, (3) transport to the nuclei and nucleus penetration, (4) differentiation of IF into RF, (5) propagation of RF in the nucleus, and (6) maturation of RF into next generation of IF. Each of these stages can be blocked in certain combinations of partners. The most controversial is selective feeding of paramecia and thereby avoiding the ingestion of *Holospora* that has not been firmly proven. On the other hand, carbohydrate residues on the IF surface are important for engulfment by *Paramecium* (Sabaneyeva, pers. comm.) and some strains indeed do not engulf *Holospora* (Skoblo et al. 1996).

An interesting outcome in some incompatible host–symbiont combinations is the so-called symbiogenic lysis, in which *Holospora* simultaneously disintegrate in the host nuclei (Ossipov et al. 1993; Skoblo et al. 2001). Symbiont cells swell, their outer and cytoplasmic membranes visibly separate, and ribosomes disappear. Then bacterial outer membranes are disrupted and the protoplasts finally lyse (Ossipov et al. 1993). This phenomenon might be related to an unknown *Paramecium* defense mechanism or it could be operated by the bacteria themselves. Virus-induced lysis cannot be ruled out but seems unlikely insofar as no viruses have been observed in *Holospora* by transmission electron microscopy. However, in the genome of *H. undulata,* a possibly functional prophage is encoded (Garushyants et al. 2018).

Symbiosis establishment between *Holospora* and *Paramecium* is a discrete process that can be interrupted at different stages, confirming that these are independently controlled. The regular arresting in a particular combination of partners is a strong evidence of its genetic determination. Thus, the blockages may be considered as phenotypic markers of genes involved in symbiosis control. Genetic analysis of *Paramecium bursaria* susceptible and resistant to *H. curviuscula* confirmed that some infection stages are controlled by several host genes (Makarov, Skoblo and Ossipov, unpublished). Transplantation of macronuclear karyoplasm from susceptible *Paramecium* strains to resistant ones conferred the latter the ability to be infected by *Holospora* and allowed to deduce at least three *Paramecium* genes involved in susceptibility to infection (Rautian et al. 1996).

4.1.5 Holospora-*Induced Changes of Host Cellular Machineries*

Infections with *Holospora* (and HLB) comprise numerous consequences for affected *Paramecium* cells. First we will discuss morphological and ultrastructural alterations, while ecological and evolutionary consequences will be discussed later (see Sect. 4.3).

During various steps of their infection cycle, *Holospora* interfere with host membranes, the cytoskeleton, and even host chromatin. Most prominent alterations induced by *Holospora* are changes in size and shape of infected nuclei (Fig. 4.1). Besides, *Holospora* infection often leads to a complete loss of micronucleus, micronucleus aberrations, or appearance of additional micronuclei in the host (Ossipov 1981).

The nuclei can be completely filled with bacterial cells and enormously swollen in size and volume. In micronuclear infections by *H. undulata*, the infected organelle can increase its volume up to 80 times (Fig. 4.1a; Ossipov 1981). Astonishingly, paramecia do not necessarily always suffer from such an occupation of their nuclei, but the effects can differ dramatically. Hyperinfections, when the macronucleus occupies the major volume of the cell and is densely packed with IF, almost always end lethally for the host cell in case of *H. obtusa* or *H. curviuscula* (Ossipov 1981; Borchsenius et al. 1983). It was also one cause of failure in the formation of a stable symbiotic system between *Paramecium* strains and *Preeria caryophila*, while in some other cases paramecia could not survive exposure to *P. caryophila* at early stages of infection development for unknown reasons (Potekhin et al. 2018).

The universal consequence of *Holospora* presence in the nuclei is a decrease of DNA content, dispersion of chromatin and nuclear aberrations, even when the infection was cured or disappeared (Ossipov 1981; Rautian et al. 1993). It is unknown how the intranuclear bacteria interact with the genetic material of the ciliate, but they do not cause significant damage to the integrity of the macronuclear genome (Potekhin et al. 1999).

Holospora can impact the regular course of sexual processes in their hosts. These are autogamy, a process of self-fertilization, and conjugation (Mulisch 2003). During conjugation, two ciliate cells adhere to each other and build a temporary cytoplasmic bridge. The micronuclei of each conjugant cell undergo meiosis and then mitosis, and haploid gametic pronuclei are exchanged between the paired cells. In each cell, they fuse to form zygotic nucleus, which divides mitotically. Anlagen of new micronuclei and macronuclei start to develop, while the old macronucleus degrades gradually. Similarly, autogamy involves meiosis and further mitosis of the micronuclei and fusion of haploid pronuclei but in the same cell, followed by development of the new macronucleus and disintegration of the old one. Thus, the intracellular habitat of *Holospora* species is destroyed at each sexual event, which can occur more or less frequently, depending on the biology of the host species. For example, species of the *Paramecium aurelia* complex pass autogamy every 25–30 vegetative divisions (Potekhin et al. 2018), while for *Paramecium caudatum* or

Paramecium bursaria autogamy has never been observed and conjugation can be rare. The intranuclear bacteria have evolved different strategies to cope with sexual processes of their host (see Sect. 4.3.2). For example, *H. undulata* inhibits conjugation as ultimate consequence (Görtz and Fujishima 1983; Fokin and Görtz 2009). *Preeria caryophila*, instead, does not prevent sexual processes in its host (Potekhin et al. 2018), but reinfects the new macronuclear anlagen (see Sect. 4.2.2). Elimination of the infection with *H. undulata* resulted in retrieval of host's ability for sexual processes (Ossipov 1981). Interestingly, *H. elegans* occasionally produce irreversible changes in the micronucleus that when ciliates were cured of the infection, they could not proceed with regular conjugation (Fujishima and Görtz 1983).

Transmission of *Holospora* also involves the modification of typical *Paramecium* cell structures and processes. During its vegetative cell cycle, *Paramecium* reproduces by binary fission and its two types of nuclei undergo mitosis (micronuclei) or amitotic division (macronuclei). Once the cell divides, the bacterial symbionts are transmitted along with their host organelle. Over the course of host nucleus division, *Holospora* induce formation of the so-called connecting piece, resulting of IF concentration in a particular median body of the dividing nucleus (Fig. 4.4b). This process has been intensively studied in *Holospora* and HLB (Fokin et al. 1996; Fokin and Görtz 2009). *Holospora* IF remain in the connecting piece linking the parts of dividing nucleus. While RF are accumulated in the new nuclei due to their high chromatin affinity and are distributed to the clonal offspring, the IF are collected in the connecting piece in order to maximize their exit from the host and further transmission success (Fig. 4.4b). Indeed, after the karyokinesis, the connecting piece gets in cyclosis and is eventually expelled from the cytoproct, so that IF can start a new infection cycle (Wiemann and Görtz 1989; Fokin et al. 1996). The formation of the connecting piece has been used to differentiate between "classic" *Holospora* species and other bacteria, here termed HLB (Fokin et al. 1996). None of the latter are able to provoke connection piece formation in their hosts (see Sect. 4.2.2). However, since *H. parva*, the most recently described *Holospora* species found in the extremely rare *Paramecium chlorelligerum*, also does not induce connecting piece formation (Lanzoni et al. 2016), it cannot be considered as an apomorphic feature for all *Holospora* species.

4.2 Differences and Similarities between Classic *Holospora* and *Holospora*-Like Bacteria

Our understanding of the diversity, occurrence, and phylogeny of symbionts, not only those of ciliates and other protists, is constantly increasing. For *Holosporaceae* (Fig. 4.5), ten new reports were published recently (Boscaro et al. 2013, 2019; Lanzoni et al. 2016; Serra et al. 2016; Tashyreva et al. 2018; Potekhin et al. 2018; Fokin et al. 2019; Konecka and Olszanowski 2019; Takeshita et al. 2019; Beliavskaia et al. 2020). Characteristic features of HLB are the cellular dimorphism connected to the diphasic infectious life cycle, the special ultrastructural organization of IF shared with *Holospora* species, and occupancy of the host nucleus as a major niche in the host cell. The question if all these new symbionts should be considered as HLB or if they are simply a group of related bacteria with different characteristics is discussed (see Sect. 4.2.1).

4.2.1 Evolutionary History and Systematics of Holosporaceae

At the time of its description, the family *Holosporaceae* (Görtz and Schmidt 2005) was included in the order *Rickettsiales* within Alphaproteobacteria. Recently, the order *Holosporales* (Szokoli et al. 2016) was establised as a sister group to *Rickettsiales* and has been confirmed according to several phylogenetic reconstructions (Boscaro et al. 2019; Castelli et al. 2019; Fokin et al. 2019). This interpretation was then called into question (Muñoz-Gómez et al. 2019). Whether phylogenomics based on increasing data sets affiliates *Holosporales* with *Rickettsiales* or *Rhodospirillales* is awaiting future studies.

Still, *Rickettsiales* and *Holosporales* have many features in common. Both contain exclusively intracellular bacteria (with the prominent exception of the epibiotic parasite *Deianiraea*; Castelli et al. 2019) colonizing hosts from various groups of protists. *Holosporales* currently includes four families (Fig. 4.5) and all HLB are members of the family *Holosporaceae*. However, we recommend to avoid using the term HLB synonymously with *Holosporaceae*. The latter additionally comprises several recently detected symbionts, e.g. *Mystax* (Korotaev et al. 2020), *Nesciobacter* (George et al. 2019), *Cytomitobacter* (Tashyreva et al. 2018), *Hydrogenosomobacter* (Takeshita et al. 2019), and *Fujishimia* (Boscaro et al. 2019), which live and replicate within their host's cytoplasm and apparently are not characterized by two morphological stages and do not clearly exhibit a life cycle with horizontal transmission.

On the other hand, two *Holosporaceae* members besides HLB show a certain degree of affinity for the host nucleus (Fig. 4.5): *Bealeia paramacronuclearis* (Szokoli et al. 2016) that generally accumulates in close proximity to the host macronucleus, and, more prominent, *Paraholospora nucleivisitans* (Eschbach et al. 2009). This symbiont of *Paramecium sexaurelia* alternates between the cytoplasm and the nucleus but never occupies both subcellular compartments simultaneously. Thus, this symbiont shares certain features associated to HLB but lacks the HLB-typical infectivity and cellular dimorphism. Furthermore, *Paraholospora nucleivisitans* branches separately in phylogenetic reconstructions (Fig. 4.5). Thus, it should not be considered as HLB.

4.2.2 *Occurence of* Holospora-*Like Bacteria: Host Range and Cellular Compartments*

Potential hosts for *Holospora* are members of the genus *Paramecium*, while HLB may also be harbored by other ciliates (reviewed by Fokin and Görtz 2009; Fujishima 2009). Next to IF morphology and occupied host compartment, the host species was used as a pivotal feature for the discrimination between *Holospora* species. As with the type of host nuclei, each *Holospora* species can infect only a single *Paramecium* species (Fokin and Görtz 2009; Fujishima and Kodama 2012). HLB, as in case of *Preeria caryophila* and potentially *Gortzia infectiva,* are not restricted to a single host species (Boscaro et al. 2013; Potekhin et al. 2018). *Hafkinia simulans* can infect hosts other than *Paramecium* (Fokin et al. 2019).

Interestingly, species infecting the huge polyploid macronucleus are more numerous than those colonizing the much smaller micronucleus (Fig. 4.5) with just *H. elegans, H. undulata*, and *H. acuminata* as micronuclear symbionts. It should be mentioned that there is increasing doubt if *H. elegans* and *H. undulata* truly represent two distinct species (Garushyants et al. 2018), especially as *H. undulata* is known for a high degree of morphological plasticity (Skoblo et al. 1996; Lebedeva, pers. comm). All described HLB infect exclusively macronuclei (Figs. 4.2 and 4.5).

The HLB phylogenetically closest to the genus *Holospora* is *Hafkinia simulans.* It does not infect *Paramecium* but has been found in the brackish water ciliate *Frontonia salmastra* (Fokin et al. 2019). Both *Paramecium* and *Frontonia* belong to the order Peniculida. *Hafkinia* differentiates into RF and IF, the latter showing compartmentalization typical for *Holospora.* Still, *Hafkinia* IF differ from *Holospora* IF as they exhibit ultrastructural variability and present occasionally two recognition tips (Fokin et al. 2019). Furthermore, the IF of *Hafkinia* are the largest described so far (up to 30 μm). They have a very peculiar spindle form, which strongly resembles in shape and dimensions the diatom *Phaeodactylum tricornutum*, a prey organism of *Frontonia salmastra.* This morphology might have evolved to increase the likelihood of phagocytosis and, thus, horizontal transmission success (Fokin et al. 2019).

The only HLB genus yet with more than a single described species is *Gortzia*. It comprises *G. infectiva* from *Paramecium jenningsi* (Boscaro et al. 2013), *G. shahrazadis* from *Paramecium multimicronucleatum* (Serra et al. 2016), and *G. yakutica* from *Paramecium putrinum* (Beliavskaia et al. 2020). *G. infectiva* was isolated from a habitat in which its host organism, *Paramecium jenningsi,* co-occurred with *Paramecium quadecaurelia* cells. The latter carried *G. infectiva* in the macronucleus, but when monoclonal strains were established the infection was lost from *Paramecium quadecaurelia.* Reinfection experiments revealed that the bacteria could enter the nucleus but failed to complete their life cycle (Boscaro et al. 2013). All three *Gortzia* species infect the macronuclei of their hosts and present two distinct morphologies. Their IF have typical appearance of *Holospora* IF, as observed by light (Beliavskaia et al. 2020), transmission electron microscopy (Boscaro et al. 2013; Serra et al. 2016), and atomic force microscopy (Fig. 4.3d).

A special case is the cytoplasmic extrusion in the periplasmic space observed in IF of *G. shahrazadis* (Serra et al. 2016).

Preeria caryophila [basonyms: *Holospora caryophila*; alpha particles] infecting the macronucleus of *Paramecium aurelia* is known since the 1960s (Preer 1969) and has been recently redescribed as type species of the new genus (Potekhin et al. 2018). *Preeria caryophila* also alternates tiny RF and short IF (max. 6 μm, Fig. 4.2e) in its life cycle. It exhibits the broadest host range described for HLB comprising at least eleven *Paramecium* species (Potekhin et al. 2018).

Interestingly, HLB show a higher degree of flexibility not only in regards of ciliate host species but also in the confinement to the nuclear compartment. All HLB have been observed occasionally in the cytoplasm of their hosts (e.g., *Preeria caryophila* in Fig. 4.6). IF in the cytoplasm might occur as a result of inversion of the infection process, which allows the release of IF from the infected nucleus (Fokin et al. 2019). The latter may facilitate the exit from the ciliate cell for intranuclear symbionts unable to induce the connecting piece formation (Fig. 4.3b). Another, nonexclusive explanation is that occasional, potentially temporary, visits to the cytoplasm are a part of the life cycle of these HLB. Evidence therefore has been obtained in *G. shahrazadis* and *P. caryophila*. For *G. shahrazadis,* numerous IF and even multiplying RF were observed in *Paramecium* cytoplasm in long-persisting associations (Serra et al. 2016). In case of *P. caryophila,* singular IF often roam outside the macronucleus as observed during conjugation (Fig. 4.6a) or autogamy (Fig. 4.6b; Potekhin et al. 2018). This ability is likely responsible for the fact that some macronuclear HLB (but not *Holospora*) can infect *Paramecium* species that regularly undergo autogamy. Those are for example members of the *Paramecium*

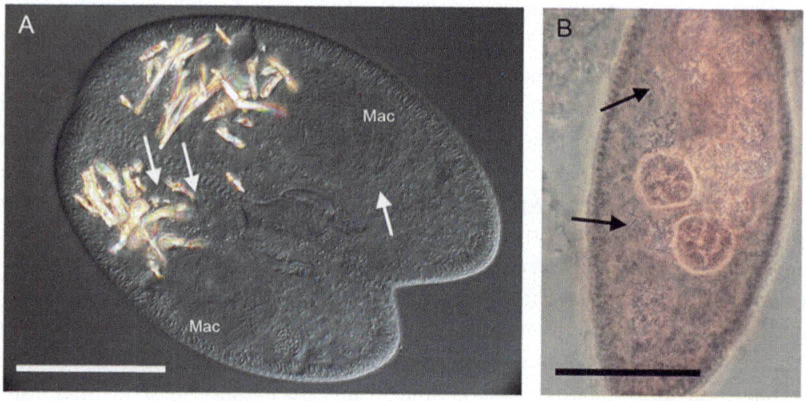

Fig. 4.6 Behavior of *Preeria caryophila* during sexual processes of their host. (**a**) Conjugating couple of *Paramecium novaurelia*, *Preeria caryophila* cells are present in both macronuclei (Mac) and some infectious forms (IF) are visible in cytoplasm (white arrows). (**b**) Postautogamous *Paramecium biaurelia* cell stained with lacto-aceto-orcein, IF are visible in old macronuclear fragments and in the cytoplasm (black arrows). Bright yellow structures are crystals, typical for paramecia Living (**a**) and fixed (**b**; with glutaraldehyde) cells observed by differential interference contrast microscopy. Scale bars: 30 μm (**a**) and 50 μm (**b**)

aurelia species complex and *Paramecium jenningsi,* which can harbor *Preeria caryophila* and *Gortzia infectiva.* Some *Preeria* IF are not enclosed in fragments of the old macronucleus but appear in the cytoplasm. These can immediately reinfect the new macronucleus once it is formed (Potekhin et al. 2018). Thus, with such an apparently effective strategy at hand, it is not surprising that *Preeria* does not prevent autogamy or conjugation (Fig. 4.6) of infected *Paramecium* strains (Potekhin et al. 2018).

4.3 Ecological and Evolutionary Consequences of Symbiosis with *Holospora* and *Holospora*-Like Bacteria for *Paramecium*

4.3.1 Low Frequency of Paramecium–Holospora *Symbioses in Nature*

The low frequency of associations between *Paramecium* and their infectious bacterial symbionts in nature is paradoxical. Indeed, *Paramecium caudatum,* many species of the *Paramecium aurelia* complex, *Paramecium bursaria,* and *Paramecium multimicronucleatum,* the natural hosts of *Holospora, Preeria,* and *Gortzia,* are rather common ciliates, and potentially many strains are capable of harboring symbionts. Extensive infection studies (Potekhin et al. 2018) demonstrated that at least 20–30% of *Paramecium aurelia* strains can host *P. caryophila.* In parallel, all *Holospora* and HLB are highly infectious bacteria able to colonize their hosts quickly and efficiently. However, there are only dozens of *Paramecium-Holospora* and HLB associations known to ever have been isolated from the environment and maintained in laboratory collections. Why are infected ciliates not more prevalent?

One explanation is that symbionts may easily get lost under changing environmental conditions. Indeed, ciliates may face periods of nutrient surplus when they can divide much faster than their symbionts, thus critically diluting the number of intracellular bacteria per host. On the contrary, they may sometimes face starvation. Ciliates carrying parasitic bacteria as a burden would be outcompeted or, possibly, would not be able to supply the symbionts sufficiently with necessary metabolites. At the same time, under constant laboratory conditions *H. obtusa* can persist in *Paramecium caudatum* strains for at least 30 months, which corresponds to more than 1000 ciliate vegetative divisions (Ossipov 1981). Moreover, the associations between *Paramecium* and *P. caryophila* may last for years in the laboratory, for example *Paramecium biaurelia* strain 562 maintains *P. caryophila* already for more than 50 years (Preer 1969; Potekhin et al. 2018).

This discrepancy might be explained by the variation in environmental factors influencing the populations of paramecia carrying *Holospora* as symbionts. *Paramecium caudatum* populations became unstable and declined when exposed to variable temperature conditions. Furthermore, the impact of infection by *H. undulata* was additive and enhanced the overall negative effect of the variable environment on *Paramecium* (Duncan et al. 2011). Environmental fluctuations also caused a decrease in *H. undulata* prevalence in the host population (Duncan et al. 2011). Moreover, patterns of temporal and spatial environmental fluctuations can impact parasite spread and host population abundance in nature and should be considered in prediction of parasite transmission and epidemics (Duncan et al. 2013). Clearly, such fluctuations are avoided in laboratory culture maintenance.

Another possible cause of a low prevalence of symbiont-bearing ciliates in nature is that infected cells might face a higher risk of extinction. Misbalanced symbiosis leading to hyperinfection is one cause of death of a ciliate (see Sect. 4.1.5). It was shown that *H. undulata* infection of *Paramecium caudatum* frequently leads to karyopyknosis (irreversible condensation of chromatin) and further loss of the micronucleus during the first day of infection, which is not always lethal for a ciliate but decreases its fitness (Ossipov et al. 1983). Similarly, up to 54–90% of *Paramecium bursaria* cells were losing micronuclei after experimental infection by *H. acuminata* (Skoblo and Lebedeva 1993). The most likely explanation for this phenomenon is the damage of micronuclear membranes due to multiple events of bacterial penetration in experimental infection conditions. It is important to emphasize that multiple penetrations by *Holospora* under environmental conditions are rather unlikely according to the assumed low frequency of IF. Thus, any infected ciliate isolated from nature presumably contains a monoclonal strain of symbionts, as infection probably mostly starts with single IF entering the host (Ossipov 1981; Skoblo and Lebedeva 1993).

Finally, infected paramecia may get outcompeted in nature by symbiont-free ciliates, which do not have to share resources with bacterial residents. On the other hand, benefits provided by HLB under certain conditions might balance the cost of infection, e.g., the observed increased exponential growth rate in *Paramecium* when infected by *P. caryophila* (Bella et al. 2016) or increased stress tolerance (see Sect. 4.3.2).

Of course, it is also possible that infections with *Holospora* and HLB are not as rare in nature as perceived. The standard approach to search for bacterial infections is sampling and further isolation and cultivation of ciliates. Infected specimens might get quickly lost or simply overlooked during initial picking cells from environmental samples and introducing them into laboratory maintenance. In this regard, it is worth noting that in water samples collected in the last 7 years from ponds, streams, and ditches of Peterhof, a small suburb of Saint Petersburg, eight *Paramecium* species and all seven matching *Holospora* species and *P. caryophila* were retrieved (Lebedeva, pers. comm.). Continued efforts in the assessment of the diversity and occurrence of symbionts of protists will provide a better insight in this puzzling aspect of HLB epidemiology.

4.3.2 Interference of Intranuclear Symbionts with the Host Nuclei and Host Stress Response

Symbionts influence individual hosts as well as their populations. One of the most important criteria defining a population is interbreeding of its members. In ciliates, sexual processes result in the complete renovation of the nuclear apparatus in a very short time period, and there is no continuous line of either micronuclei or macronuclei in sexual generations. The micronucleus passes through a series of meiotic and mitotic divisions, while the old macronucleus gets completely demolished at each sexual process and is formed de novo from micronucleus derivatives. Thus, *Holospora* species either have to prevent conjugation of their host or get lost (Fokin 1998). Still, conjugation in presence of *H. elegans* was reported (Fujishima and Görtz 1983). Some bacteria managed to remain in the pronuclei, but survival of infected exconjugants was severely reduced compared to aposymbiotic cells as they were not able to form new macronuclei and regenerated the old ones (Fujishima and Görtz 1983).

While micronuclear symbionts may mechanically interfere with meiosis, there is no plausible explanation of inhibition of host conjugation by *H. obtusa* residing in the macronucleus than its influence on host gene expression. As discussed, *Preeria caryophila* does not prevent sexual processes in its host and, instead, temporarily escapes from the transforming nuclei into cytoplasm; there are no data concerning conjugation of ciliates infected by *Gortzia* or *Hafkinia*.

Holospora are parasites. A heavy infection of the *Paramecium* macronucleus by different *Holospora* results in a decreased fission rate of the ciliates (Ossipov 1981; Borchsenius et al. 1983). Only in case of the symbiosis between *Paramecium chlorelligerum* and *H. parva* the slow growth of infected cells was consistent with that of uninfected ones (Lanzoni et al. 2016). At the same time, no retarding effects on host divisions rates were reported for other symbionts (Ossipov 1981; Kaltz and Koella 2003; Castelli et al. 2016). Interestingly, if the host culture experiences unfavorable cultivation conditions and thus reduces cell division rate, *H. undulata* apparently becomes more virulent (Magalon et al. 2010; Dusi et al. 2015). Elevated host division rates, on the other hand, increased the levels of parasite vertical transmission and resulted in a near-complete loss of infectivity (Dusi et al. 2015). Insufficient time for the bacteria to mature into IF could explain at least partially these observations, but, obviously, the balance between host division rate as well as prevalence and infectivity of symbionts is rather delicate.

Paramecium have been shown to acquire heat-shock resistance (Hori and Fujishima 2003) and osmotic shock tolerance (Smurov and Fokin 1998) when infected by *Holospora*. This was considered as an advantageous effect of the symbiosis (Hori et al. 2008). An increase of Hsp70 expression is also known from other symbiotic systems (Kodama et al. 2014; Grosser et al. 2018). It might be either specifically induced by the symbionts or represent part of the *Paramecium* stress response to a large-scale infection. Nevertheless, elevated levels of Hsp70 allowed paramecia infected by *H. obtusa* to survive at nonpermissive temperatures (Hori and

Fujishima 2003). Infected cells were able to maintain the ciliary movement and continued active swimming at temperatures above and below the physiological range of *Paramecium* (Fujishima et al. 2005). Heat resistance was not acquired by *Paramecium caudatum* infected by *H. undulata*, but this symbiont conferred osmotic shock tolerance to some strains (Duncan et al. 2010).

4.3.3 Epidemiology of **Paramecium-Holospora** *Symbioses and Impact of Environmental Factors*

Many epidemiological parameters of *Paramecium-Holospora* associations are still unknown. Success of infection is higher if more bacteria simultaneously enter the target nucleus, as each IF generates several multiplying RF. Even such small initial differences can strongly influence the subsequent intensity of infection (Fels et al. 2008). Interestingly, direct transmission from infected cell to recipient as occurring in nature is at any rate not less efficient than experimental infections utilizing homogenates of heavily infected paramecia. The presence of a single infected donor cell was sufficient to infect a population of naïve paramecia with the same rate and prevalence (Potekhin et al. 2018). Optimal parasite strategies may depend on the balance between local transmission and the capacity to reach new habitats through dispersal (Lion and Boots 2010). Surprisingly, *Holospora*-bearing ciliates tended to disperse less in interconnected microcosms (Fellous et al. 2011).

In the *Paramecium–Holospora* interaction, a negative correlation between the growth rate of the host and the parasite's investment in horizontal transmission has been observed. The results suggest a tradeoff between efficient vertical and horizontal transmission. If conditions for *Paramecium* replication decline, the symbionts switch to horizontal transmission (Kaltz and Koella 2003). Addressing the effects of early and late stages of infection, parasite load, and food abundance, it was shown that a reduced availability of food and thus a lower division rate of the host correlates with a higher *Holospora* virulence (Restif and Kaltz 2006).

Paramecium offers sufficient resources to host multiple bacterial infections. Paramecia with double infections by *Caedimonas varicaedens* (Preer 1969; Schrallhammer et al. 2018) or *Megaira polyxenophila* (Schrallhammer et al. 2013) with *P. caryophila, Megaira polyxenophila,* and *H. undulata* (Lanzoni et al. 2019), and even *H. obtusa* and *P. caryophila* (Fig. 4.2f) are rarely but repeatedly detected in environmental samples. Simultaneous infection of both nuclei of *Paramecium bursaria* with *H. curviuscula* and *H. acuminata* was achieved many times (Borchsenius et al. 1983). Similarly, the presence of *Caedimonas varicaedens* in the macronucleus did not prevent an infection with micronuclear-specific *Holospora* (Skoblo et al. 1996). Even experimental double infections of naïve paramecia were obtained (Duncan et al. 2018), albeit at rather low frequencies. The most exceptional case was likely that of *H. undulata* (normally restricted to micronuclei) infecting a macronucleus already inhabited by *H. obtusa* (Lebedeva et al. 1992). *Holospora* can even serve as shuttle transporting free-living bacteria to the macronucleus (Fokin et al. 2004).

However, antagonistic interactions between different bacterial symbionts have also been observed. Resident symbionts might prevent the efficient colonization of the same host by other bacteria, even if both occupy different compartments (Fokin et al. 1987; Görtz 1987). Mixed infection experiments showed that competitive exclusion is more common than coexistence (Duncan et al. 2018). It is tempting to speculate that certain symbionts may provide their host colonization resistance against invasion by other, potentially harmful microorganisms (Plotnikov et al. 2019). Bacterial competition for the host cell, interactions of two different symbiont species in one host, and tradeoffs of multiple bacterial symbioses remain to be further studied.

4.4 Outlook and Perspectives

In this chapter, we aimed to summarize currently available data on the formation and maintenance of very peculiar symbiotic systems, where *Holospora* and HLB reside directly in the nucleus of their host. This field has experienced tremendous progress in the last decade. The expansion of state-of-the-art technologies, first of all Next-Generation Sequencing together with current microscopy and molecular biology techniques, now opens extremely interesting directions for further studies of *Paramecium-Holospora* and HLB symbioses.

The genomes of several *Paramecium* species have been sequenced and are available at ParameciumDB (Arnaiz et al. 2019), and the genomes of several *Holospora* species are either sequenced (Dohra et al. 2013, 2014; Garushyants et al. 2018) or in progress. These are the prerequisites for in-depth interaction analyses by transcriptomics. Comparative transcriptomics together with genetic dissection of the symbiotic systems will allow to detect the genes of host and symbiont differentially expressed at each stage of symbiosis development and maintenance. Further studies of such genes' functions will approach the molecular interaction mechanisms of both partners and potentially may lead to the identification of new bacterial effectors.

Even in mutually beneficial symbiotic associations, excessive number of symbionts may become a heavy burden for the host decreasing its fitness and leading to defeat in local competition (Cunning and Baker 2014; Parkinson et al. 2017). In case of *Holospora* and HLB, which are not mutualistic, this problem of symbiont population control becomes crucial. A possible pathway for the regulation of symbionts could be production of antimicrobial peptides (AMP) by ciliates. AMP are an ancient defensive weapon of the eukaryotic cell (Wollman 2016) and have been reported from *Paramecium caudatum* (Cui et al. 2016). Examples of AMP targeting bacterial symbionts, not eliminating the microbial population but rather keeping it in check, are known from different host organisms (Mergaert 2018). Quorum sensing (QS) may be part of self-regulation mechanisms of the symbiont's population in the host. Possibly, some of the numerous short (<100 amino acids) peptides with unknown function encoded in *Holospora* genomes (Garushyants et al. 2018) might be involved as QS signal.

The intimate localization of HLB in the nuclei of their hosts might offer possibilities for crosskingdom horizontal gene transfer (HGT). As the symbionts live and die in the nucleus, their DNA may occasionally be integrated into the host genome. While symbiont DNA would not be fixed in the somatic macronuclear genome, it can become a part of the generative micronucleus genome, which is a "safe haven" for noncoding DNA (Bétermier and Duharcourt 2014). If then bacterial genes are somehow retained in the developing macronuclear genome, they may get a chance to be expressed. *Holospora* are deficient for nearly all major pathways due to genome reduction. In addition, up to 15% of their genomes is represented by noncoding sequences (Garushyants et al. 2018). At the same time, *Holospora* switch between several stages and environments in their life cycle and perform complex interactions with the host during the infection process. Hence, *Holospora* belong to the same category of obligatory bacterial symbionts whose genomes are irreversibly shrinking (Wernegreen 2017; Husnik and Keeling 2019). Severely limited metabolic capacities put *Holospora* and HLB in absolute dependence of the host making the search for HGT promising. As *Holosporales* are considered as close relatives of free-living ancestors of mitochondria (Wang and Wu 2015), insights into *Paramecium-Holospora* and HLB symbioses might provide clues for initial stages in the transition from symbiont to organelle.

Acknowledgments The research of endosymbiosis is currently supported by Research Innovation Fund of the Albert-Ludwigs University of Freiburg (M.S.) and by Russian Foundation for Basic Research grant 19-04-01256a (A.P.). We are grateful to Dr. E. Sabaneyeva and Prof. S. Fokin for the kind provision of images and to Dr. I. Nekrasova for useful comments and feedback while working on the manuscript. We wish to use an opportunity to thank the founders of *Holospora* studies, our teachers Prof. Hans-Dieter Görtz and Prof. Dmitry Ossipov, and members of their laboratories, M. Schweikert, T. Brigge, M. Rautian, and I. Skoblo, for introducing us to ciliate-bacterial symbiosis biology; our colleagues and collaborators G. Petroni, O. Kaltz, S. Fokin, E. Sabaneyeva, and N. Lebedeva for inspiring ideas and sharing of knowledge and data; last but not least our current and former groups members and students for stimulating discussions and continuous interest in intranuclear symbiosis.

References

Abamo F, Dohra H, Fujishima M (2008) Fate of the 63-kDa periplasmic protein of the infectious form of the endonuclear symbiotic bacterium *Holospora obtusa* during the infection process. FEMS Microbiol Lett 280:21–27. https://doi.org/10.1111/j.1574-6968.2007.01023.x

Arnaiz O, Meyer E, Sperling L (2019) ParameciumDB 2019: integrating genomic data across the genus for functional and evolutionary biology. Nucleic Acids Res 48:D599–D605. https://doi.org/10.1093/nar/gkz948

Barhey K, Gibson I (1984) A study on the conditions for infection of *Holospora caryophila*, a macronuclear symbiont of *Paramecium biaurelia*. Micron Microsc Acta 15:261–268

Beliavskaia AY, Predeus AV, Garushyants SK et al (2020) New intranuclear symbiotic bacteria from macronucleus of *Paramecium putrinum* — "*Candidatus* Gortzia yakutica". Diversity 12 (5):198. https://doi.org/10.3390/d12050198

Bella C, Koehler L, Grosser K et al (2016) Fitness impact of obligate intranuclear bacterial symbionts depends on host growth phase. Front Microbiol 7:2084. https://doi.org/10.3389/fmicb.2016.02084

Bétermier M, Duharcourt S (2014) Programmed rearrangement in ciliates: *Paramecium*. Microbiol Spectr 2:MDNA3-0035–2014

Borchsenius O, Skoblo I, Lebedeva N, Ossipov D (1990) Occasional appearance and short-term persistence of symbiotic bacteria *Holospora acuminata* in macronucleus of ciliates *Paramecium bursaria*. Tsitologiia 32:578–582

Borchsenius O, Skoblo I, Lebedeva NA, Ossipov D (1992) The infectiousness of *Holospora acuminata* the intranuclear symbiont of the ciliate *Paramecium bursaria* 2. The impossibility of a symbiont system formation is a result of disturbances occuring at the stage when symbiotic bacterium leaves the. Tsitologiia 34:97

Borchsenius O, Skoblo I, Ossipov D (1983) *Holospora curviuscula* - a new species of macronuclear symbiotic bacteria of *Paramecium bursaria*. Tsitologiya 25:91–97

Boscaro V, Fokin SI, Schrallhammer M et al (2013) Revised systematics of *Holospora*-like bacteria and characterization of "*Candidatus* Gortzia infectiva", a novel macronuclear symbiont of *Paramecium jenningsi*. Microb Ecol 65:255–267. https://doi.org/10.1007/s00248-012-0110-2

Boscaro V, Husnik F, Vannini C, Keeling PJ (2019) Symbionts of the ciliate *Euplotes*: diversity, patterns and potential as models for bacteria-eukaryote endosymbioses. Proc R Soc B Biol Sci 286:20190693. https://doi.org/10.1098/rspb.2019.0693

Bütschli O (1887) Protozoa. H. G. Bronns Klassen und Ordnungen des Tierreichs. Akadem. Verl.-Ges. Geest & Portig, Leipzig

Castelli M, Lanzoni O, Rossi L et al (2016) Evaluation of enrichment protocols for bacterial endosymbionts of ciliates by real-time PCR. Curr Microbiol 72(6):723–732. https://doi.org/10.1007/s00284-016-1006-z

Castelli M, Sabaneyeva E, Lanzoni O et al (2019) *Deianiraea*, an extracellular bacterium associated with the ciliate *Paramecium*, suggests an alternative scenario for the evolution of *Rickettsiales*. ISME J 13:2280–2294. https://doi.org/10.1038/s41396-019-0433-9

Cui P, Dong Y, Li Z et al (2016) Identification and functional characterization of an uncharacterized antimicrobial peptide from a ciliate *Paramecium caudatum*. Dev Comp Immunol 60:53–65. https://doi.org/10.1016/j.dci.2016.02.016

Cunning R, Baker AC (2014) Not just who, but how many: the importance of partner abundance in reef coral symbioses. Front Microbiol 5:400

de Bary A (1879) Die Erscheinungen der Symbiose. In: Vers. Deut. Naturforscher und Aerzte zu Cassel 1878. Strassburg: Strassburg

de Souza SM, Orth K (2015) Subversion of the cytoskeleton by intracellular bacteria: lessons from *Listeria, Salmonella* and *Vibrio*. Cell Microbiol 17:164–173. https://doi.org/10.1111/cmi.12399

Dohra H, Fujishima M, Hoshide K (1994) Monoclonal antibodies specific for periplasmic materials of the macronuclear specific bacterium *Holospora obtusa* of the ciliate *Paramecium caudatum*. Eur J Protistol 30:288–294. https://doi.org/10.1016/S0932-4739(11)80075-1

Dohra H, Suzuki H, Suzuki T et al (2013) Draft genome sequence of *Holospora undulata* strain HU1, a micronucleus-specific symbiont of the ciliate *Paramecium caudatum*. Genome Announc 1:1–2. https://doi.org/10.1128/genomeA.00664-13

Dohra H, Tanaka K, Suzuki T et al (2014) Draft genome sequences of three *Holospora* species (*Holospora obtusa, Holospora undulata*, and *Holospora elegans*), endonuclear symbiotic bacteria of the ciliate *Paramecium caudatum*. FEMS Microbiol Lett 359:16–18. https://doi.org/10.1111/1574-6968.12577

Dohra H, Yamamoto K, Fujishima M, Ishikawa H (1997) Cloning and sequencing of gene coding for a periplasmic 5.4 kDa peptide of the macronucleus-specific symbiont *Holospora obtusa* of the ciliate *Paramecium caudatum*. Zool Sci 14:69–75

Duncan AB, Dusi E, Schrallhammer M et al (2018) Population-level dynamics in experimental mixed infections: evidence for competitive exclusion among bacterial parasites of *Paramecium caudatum*. Oikos 127:1380–1389. https://doi.org/10.1111/oik.05280

Duncan AB, Fellous S, Accot R et al (2010) Parasite-mediated protection against osmotic stress for *Paramecium caudatum* infected by *Holospora undulata* is host genotype specific. FEMS Microbiol Ecol 74:353–360. https://doi.org/10.1111/j.1574-6941.2010.00952.x

Duncan AB, Fellous S, Kaltz O (2011) Reverse evolution: selection against costly resistance in disease-free microcosm populations of paramecium caudatum. Evolution (N Y) 65:3462–3474. https://doi.org/10.1111/j.1558-5646.2011.01388.x

Duncan AB, Gonzalez A, Kaltz O (2013) Stochastic environmental fluctuations drive epidemiology in experimental host-parasite metapopulations. Proc Biol Sci 280:20131747. https://doi.org/10.1098/rspb.2013.1747

Dusi E, Gougat-Barbera C, Berendonk TU, Kaltz O (2015) Long-term selection experiment produces breakdown of horizontal transmissibility in parasite with mixed transmission mode. Evolution (N Y) 69:1069–1076. https://doi.org/10.1111/evo.12638

Dziallas C, Allgaier M, Monaghan MT, Grossart H (2012) Act together—implications of symbioses in aquatic ciliates. Front Microbiol 3:1–17. https://doi.org/10.3389/fmicb.2012.00288

Edgcomb VP (2016) Marine protist associations and environmental impacts across trophic levels in the twilight zone and below. Curr Opin Microbiol 31:169–175. https://doi.org/10.1016/j.mib.2016.04.001

Ehrsam E, Görtz H-D (1999) Surface proteins of the gram-negative bacterium *Holospora obtusa* bind to macronuclear proteins of its host *Paramecium caudatum*. Eur J Protistol 35:304–308. https://doi.org/10.1016/S0932-4739(99)80008-X

Eschbach E, Pfannkuchen M, Schweikert M et al (2009) "*Candidatus* Paraholospora nucleivisitans", an intracellular bacterium in *Paramecium sexaurelia* shuttles between the cytoplasm and the nucleus of its host. Syst Appl Microbiol 32:490–500. https://doi.org/10.1016/j.syapm.2009.07.004

Fellous S, Quillery E, Duncan AB, Kaltz O (2011) Parasitic infection reduces dispersal of ciliate host. Biol Lett 7:327–329. https://doi.org/10.1098/rsbl.2010.0862

Fels D, Vignon M, Kaltz O (2008) Ecological and genetic determinants of multiple infection and aggregation in a microbial host-parasite system. Parasitology 135:1373–1383. https://doi.org/10.1017/S0031182008004940

Fokin SI (1991) *Holospora recta* sp. nov. - a micronucleus-specific endobiont of the ciliate *Paramecium caudatum*. Tsitologiya 33:135–141

Fokin SI (1989) Bacterial endobionts of the ciliate *Paramecium woodruffi*. I Endobionts of the macronucleus. Tsitologiya 31:839–844

Fokin SI (1998) Strategies of the macronuclear endocytobionts of *Paramecium* during the sexual process of the host. Symbiosis 25:323–342

Fokin SI, Brigge T, Brenner J, Görtz H-D (1996) *Holospora* species infecting the nuclei of *Paramecium* appear to belong into two groups of bacteria. Eur J Protistol 32:19–24. https://doi.org/10.1016/S0932-4739(96)80072-1

Fokin SI, Görtz HD (2009) Diversity of *Holospora* bacteria in *Paramecium* and their characterization. In: Fujishima M (ed) Endosymbionts in *Paramecium*, Microbiology Monographs. Springer, Berlin Heidelberg, pp 161–201

Fokin SI, Ossipov DV, Skoblo I et al (1987) *Nonospora macronucleata* g. n., sp. n. - a vegetative nucleus symbiont of the ciliate *Paramecium caudatum*. Tsitologiia 29:963–970

Fokin SI, Sabaneyeva EV (1993) Bacterial endocytobionts of the ciliate *Paramecium calkinsi*. Eur J Protistol 29:390–395

Fokin SI, Schweikert M, Fujishima M (2005) Recovery of the ciliate *Paramecium multimicronucleatum* following bacterial infection with *Holospora obtusa*. Eur J Protistol 41:129–138. https://doi.org/10.1016/j.ejop.2004.11.007

Fokin SI, Serra V, Ferrantini F et al (2019) "*Candidatus* Hafkinia simulans" gen. nov., sp. nov., a novel *Holospora*-like bacterium from the macronucleus of the rare brackish water ciliate

Frontonia salmastra (Oligohymenophorea, Ciliophora). Microb Ecol 77:1092–1106. https://doi.org/10.1007/s00248-018-1311-0

Fokin SO, Skovorodkin IN, Schweikert M, Görtz H-D (2004) Co-infection of the macronucleus of *Paramecium caudatum* by free-living bacteria together with the infectious *Holospora obtusa*. J Eukaryot Microbiol 51:417–424. https://doi.org/10.1111/j.1550-7408.2004.tb00388.x

Freiburg M (1988) Organization and expression of the nuclear genome. In: Görtz H (ed) *Paramecium*. Springer-Verlag, Berlin, Heidelberg, pp 141–154

Fujishima M (2009) Infection and maintenance of *Holospora* species in *Paramecium caudatum*. In: Fujishima M (ed) Endosymbionts in *Paramecium*. Microbiology Monographs. Springer, Berlin Heidelberg, pp 201–225

Fujishima M, Dohra H, Kawai M (1997) Quantitative changes in periplasmic proteins of the macronucleus-specific bacterium *Holospora obtusa* in the infection process of the ciliate *Paramecium caudatum*. J Eukaryot Microbiol 44:636–642

Fujishima M, Fujita M (1985) Infection and maintenance of *Holospora obtusa*, a macronucleus-specific bacterium of the ciliate *Paramecium caudatum*. J Cell Sci 76:179–187

Fujishima M, Görtz H-D (1983) Infection of macronuclear anlagen of *Paramecium caudatum* with the macronucleus-specific symbiont *Holospora obtusa*. J Cell Sci 64:137–146

Fujishima M, Kawai M, Yamamoto R (2005) *Paramecium caudatum* acquires heat-shock resistance in ciliary movement by infection with the endonuclear symbiotic bacterium *Holospora obtusa*. FEMS Microbiol Lett 243:101–105. https://doi.org/10.1016/j.femsle.2004.11.053

Fujishima M, Kodama Y (2012) Endosymbionts in *Paramecium*. Eur J Protistol 48:124–137. https://doi.org/10.1016/j.ejop.2011.10.002

Fujishima M, Nagahara K, Kojima Y, Sayama Y (1991) Sensitivity of the infectious long form of the macronuclear endosymbiont *Holospora obtusa* of the ciliate *Paramecium caudatum* against chemical and physical factors. Eur J Protistol 27:119–126. https://doi.org/10.1016/S0932-4739(11)80333-0

Garushyants SK, Beliavskaia AY, Malko DB et al (2018) Comparative genomic analysis of *Holospora* spp., intranuclear symbionts of paramecia. Front Microbiol 9:738. https://doi.org/10.3389/fmicb.2018.00738

Gast RJ, Sanders RW, Caron DA (2009) Ecological strategies of protists and their symbiotic relationships with prokaryotic microbes. Trends Microbiol 17:563–569. https://doi.org/10.1016/j.tim.2009.09.001

George EE, Husnik F, Tashyreva D et al (2020) Highly reduced genomes of protist endosymbionts show evolutionary convergence. Curr Biol 30(5):925–933.e3. https://doi.org/10.1016/j.cub.2019.12.070

Gillespie JJ, Kaur SJ, Rahman MS et al (2015) Secretome of obligate intracellular *Rickettsia*. FEMS Microbiol Rev 39:47–80. https://doi.org/10.1111/1574-6976.12084

Görtz H-D (1987) Different endocytobionts simultaneously colonizing ciliate cells. Ann N Y Acad Sci 503:261–268

Görtz H-D, Fokin SI (2009) Diversity of endosymbiotic bacteria in *Paramecium*. In: Fujishima M, (ed) Endosymbionts in *Paramecium*. Microbiology Monographs. Springer, Berlin Heidelberg, pp 131–160

Görtz H-D, Fujishima M (1983) Conjugation and meiosis of *Paramecium caudatum* infected with the micronucleus-specific bacterium *Holospora elegans*. Eur J Cell Biol 32:86–91

Görtz H-D, Schmidt H (2005) Family III. *Holosporaceae* fam. Nov. In: Brenner D, Krieg N, Staley J, Garrity G (eds) Bergey's manual of systematic bacteriology, 2nd edn. Springer, New York, US, pp 146–149

Görtz H-D, Wiemann M (1989) Route of infection of the bacteria *Holospora elegans* and *Holospora obtusa* into the nuclei of *Paramecium caudatum*. Eur J Protistol 24:101–109

Gromov BV, Ossipov DV (1981) *Holospora* (ex Hafkine 1890) nom. rev., a genus of bacteria inhabiting the nuclei of paramecia. Int J Syst Bacteriol 31:348–352. https://doi.org/10.1099/00207713-31-3-348

Grosser K, Ramasamy P, Amirabad A et al (2018) More than the "Killer Trait": Infection with the bacterial endosymbiont *Caedibacter taeniospiralis* causes transcriptomic modulation in *Paramecium* host. Genome Biol Evol 10(2):646–656. https://doi.org/10.1093/gbe/evy024

Hafkine WM (1890) Maladies infectieuses des paramecies. Ann Inst Pasteur 4:148–162

Hori M, Fujii K, Fujishima M (2008) Micronucleus-specific bacterium *Holospora elegans* irreversibly enhances stress gene expression of the host *Paramecium caudatum*. J Eukaryot Microbiol 55:515–521. https://doi.org/10.1111/j.1550-7408.2008.00352.x

Hori M, Fujishima M (2003) The endosymbiotic bacterium *Holospora obtusa* enhances heat-shock gene expression of the host *Paramecium caudatum*. J Eukaryot Microbiol 50:293–298. https://doi.org/10.1111/j.1550-7408.2003.tb00137.x

Husnik F, Keeling PJ (2019) The fate of obligate endosymbionts: reduction, integration, or extinction. Curr Opin Genet Dev 58–59:1–8. https://doi.org/10.1016/j.gde.2019.07.014

Iwamoto M, Osakada H, Mori C et al (2017) Compositionally distinct nuclear pore complexes of functionally distinct dimorphic nuclei in the ciliate *Tetrahymena*. J Cell Sci 130:1822–1834. https://doi.org/10.1242/jcs.199398

Iwatani K, Dohra H, Lang BF et al (2005) Translocation of an 89-kDa periplasmic protein is associated with *Holospora* infection. Biochem Biophys Res Commun 337:1198–1205. https://doi.org/10.1016/j.bbrc.2005.09.175

Kaltz O, Koella JC (2003) Host growth conditions regulate the plasticity of horizontal and vertical transmission in *Holospora undulata*, a bacterial parasite of the protozoan *Paramecium caudatum*. Evolution (N Y) 57:1535–1542. https://doi.org/10.1111/j.0014-3820.2003.tb00361.x

Kawai M, Fujishima M (2000) Invasion of the macronucleus of *Paramecium caudatum* by the bacterium *Holospora obtusa*: fates of the bacteria and timings of invasion steps. Eur J Protistol 36:46–52. https://doi.org/10.1016/S0932-4739(00)80022-X

Kodama Y, Suzuki H, Dohra H et al (2014) Comparison of gene expression of *Paramecium bursaria* with and without *Chlorella variabilis* symbionts. BMC Genomics 15:183. https://doi.org/10.1186/1471-2164-15-183

Konecka E, Olszanowski Z (2019) Detection of a new bacterium of the family *Holosporaceae* (Alphaproteobacteria: *Holosporales*) associated with the oribatid mite *Achipteria coleoptrata*. Biologia (Bratisl) 74:1517–1522. https://doi.org/10.2478/s11756-019-00251-w

Korotaev A, Benken K, Sabaneyeva E (2020) "*Candidatus* Mystax nordicus" aggregates with mitochondria of its host, the ciliate *Paramecium nephridiatum*. Diversity 12(6):251. https://doi.org/10.3390/d12060251

Lambrechts A, Gevaert K, Cossart P et al (2008) *Listeria* comet tails: the actin-based motility machinery at work. Trends Cell Biol 18:220–227. https://doi.org/10.1016/j.tcb.2008.03.001

Lambrechts L, Fellous S, Koella JC (2006) Coevolutionary interactions between host and parasite genotypes. Trends Parasitol 22:12–16. https://doi.org/10.1016/j.pt.2005.11.008

Lanzoni O, Fokin SI, Lebedeva N et al (2016) Rare freshwater ciliate *Paramecium chlorelligerum* Kahl, 1935 and its macronuclear symbiotic bacterium "*Candidatus* Holospora parva". PLoS One 11:e0167928. https://doi.org/10.1371/journal.pone.0167928

Lanzoni O, Sabaneyeva E, Modeo L et al (2019) Diversity and environmental distribution of the cosmopolitan endosymbiont "*Candidatus* Megaira". Sci Rep 9:1–13. https://doi.org/10.1038/s41598-018-37629-w

Lebedeva N, Rodionova G, Skoblo I (1992) Maintenance of *Holospora undulata*, micronucleus-specific symbiotic bacterium of *Paramecium caudatum*, in macronucleus infected by *Holospora obtusa*. Tsitologiia 34:85

Linka N, Hurka H, Lang BF et al (2003) Phylogenetic relationships of non-mitochondrial nucleotide transport proteins in bacteria and eukaryotes. Gene 306:27–35. https://doi.org/10.1016/S0378-1119(03)00429-3

Lion S, Boots M (2010) Are parasites ""prudent"" in space? Ecol Lett 13:1245–1255. https://doi.org/10.1111/j.1461-0248.2010.01516.x

Magalon H, Nidelet T, Martin G, Kaltz O (2010) Host growth conditions influence experimental evolution of life history and virulence of a parasite with vertical and horizontal transmission. Evolution (N Y) 64:2126–2138. https://doi.org/10.1111/j.1558-5646.2010.00974.x

Mergaert P (2018) Role of antimicrobial peptides in controlling symbiotic bacterial populations. Nat Prod Rep 35:336–356. https://doi.org/10.1039/C7NP00056A

Mulisch M (2003) Nuclei and sexual reprodution. In: Hausmann K, Hülsmann N, Radek R (eds) Protistology, 3rd edn. Schweizerbart Science Publishers, Stuttgart, pp 241–255

Muñoz-Gómez SA, Hess S, Burger G et al (2019) An updated phylogeny of the *Alphaproteobacteria* reveals that the parasitic *Rickettsiales* and *Holosporales* have independent origins. elife 8:e42535. https://doi.org/10.7554/eLife.42535

Nakamura Y, Aki M, Aikawa T et al (2004) Differences in gene expression of the ciliate *Paramecium caudatum* caused by endonuclear symbiosis with *Holospora* obtusa, revealed using differential display reverse transcribed PCR. FEMS Microbiol Lett 240:209–213. https://doi.org/10.1016/j.femsle.2004.09.036

Nowack ECM, Melkonian M (2010) Endosymbiotic associations within protists. Philos Trans R Soc B Biol Sci 365:699–712. https://doi.org/10.1098/rstb.2009.0188

Ojogun N, Kahlon A, Ragland SA et al (2012) *Anaplasma phagocytophilum* outer membrane protein a interacts with sialylated glycoproteins to promote infection of mammalian host cells. Infect Immun 80:3748–3760. https://doi.org/10.1128/IAI.00654-12

Ossipov D (1981) Problems of nuclear heteromorphism in the unicellular organisms. Nauka, Leningrad

Ossipov D, Borchsenius O, Skoblo I (1983) Ultrastructural changes in the nuclear apparatus of *Paramecium caudatum* induced by symbiotic bacteria *Holospora undulata*. Tsitologiia 25:33–38

Ossipov DV, Skoblo II, Borchsenius ON, Lebedeva NA (1993) Interactions between *Paramecium bursaria* (protozoa, Ciliophora, Hymenostomatida) and their nuclear symbionts. I. Phenomenon of symbiogenic lysis of the bacterium *Holospora acuminata*. Eur J Protistol 29:61–71. https://doi.org/10.1016/S0932-4739(11)80298-1

Oulhen N, Schulz BJ, Carrier TJ (2016) English translation of Heinrich Anton de Bary's 1878 speech, 'Die Erscheinung der Symbiose' ('De la symbiose'). Symbiosis 69:131–139. https://doi.org/10.1007/s13199-016-0409-8

Parkinson JF, Gobin B, Hughes WOH (2017) The more, the merrier? Obligate symbiont density changes over time under controlled environmental conditions, yet holds no clear fitness consequences. Physiol Entomol 42:163–172. https://doi.org/10.1111/phen.12186

Plotnikov AO, Balkin AS, Gogoleva NE et al (2019) High-throughput sequencing of the 16S rRNA gene as a survey to analyze the microbiomes of free-living ciliates *Paramecium*. Microb Ecol 78:286–298. https://doi.org/10.1007/s00248-019-01321-x

Potekhin A, Brigge T, Rautian M (1999) Genetic consequences of the intranuclear symbiosis *Paramecium-Holospora* analyzed by PFGE. In: Wagner E (ed) Symbiosis to Eukaryotism. University Press, Geneva, pp 169–177

Potekhin A, Schweikert M, Nekrasova I et al (2018) Complex life cycle, broad host range and adaptation strategy of the intranuclear *Paramecium* symbiont *Preeria caryophila* comb. nov. FEMS Microbiol Ecol 94:fiy076. https://doi.org/10.1093/femsec/fiy076

Preer LB (1969) Alpha, an infectious macronuclear symbiont of *Paramecium aurelia*. J Protozool 16:570–578

Rautian M, Vishnyakov A, Makarov S, Ossipov D (1996) Transformation of *Paramecium caudatum* clone resistant to infection by intranuclear symbiotic bacteria of genus *Holospora*. Eur J Protistol 32:135–140. https://doi.org/10.1016/S0932-4739(96)80090-3

Rautian MS, Skoblo II, Lebedeva NA, Ossipov DV (1993) Genetics of symbiotic interactions between *Paramecium bursaria* and the intranuclear bacterium *Holospora acuminata*, natural genetic variability by infectivity and susceptibility. Acta Protozool 32:165–173

Rautian MS, Skoblo II, Lebedeva NA, Ossipov DV (1990) Complementation in the symbiont-host system formation between the clones in the ciliate *Paramecium caudatum* and the stocks of the bacterium *Holospora obtusa*. Tsitologiia 32:584–591

Restif O, Kaltz O (2006) Condition-dependent virulence in a horizontally and vertically transmitted bacterial parasite. Oikos 114:148–158. https://doi.org/10.1111/j.2006.0030-1299.14611.x

Sabaneyeva EV, Derkacheva ME, Benken KA et al (2009) Actin-based mechanism of *Holospora obtusa* trafficking in *Paramecium caudatum*. Protist 160:205–219. https://doi.org/10.1016/j.protis.2008.11.006

Sabaneyeva EV, Fokin SI, Gavrilova EV, Kornilova ES (2005) Nocodazole inhibits macronuclear infection with *Holospora obtusa* in *Paramecium caudatum*. Protoplasma 226:147–153. https://doi.org/10.1007/s00709-005-0121-7

Samba-Louaka A, Delafont V, Rodier M-H et al (2019) Free-living amoebae and squatters in the wild: ecological and molecular features. FEMS Microbiol Rev 43:415–434. https://doi.org/10.1093/femsre/fuz011

Schmidt MA, Riley LW, Benz I (2003) Sweet new world: glycoproteins in bacterial pathogens. Trends Microbiol 11:554–561. https://doi.org/10.1016/j.tim.2003.10.004

Schrallhammer M, Castelli M, Petroni G (2018) Phylogenetic relationships among endosymbiotic R-body producer: bacteria providing their host the killer trait. Syst Appl Microbiol 41 (3):213–220. https://doi.org/10.1016/j.syapm.2018.01.005

Schrallhammer M, Ferrantini F, Vannini C et al (2013) "*Candidatus* Megaira polyxenophila" gen. nov., sp. nov.: considerations on evolutionary history, host range and shift of early divergent rickettsiae. PLoS One 8:e72581. https://doi.org/10.1371/journal.pone.0072581

Schulz F, Horn M (2015) Intranuclear bacteria: inside the cellular control center of eukaryotes. Trends Cell Biol 25:339–346. https://doi.org/10.1016/j.tcb.2015.01.002

Schweikert M, Fujishima M, Görtz H (2013) Symbiotic associations between ciliates and prokaryotes. In: Rosenberg E, DeLong EF, Lory S et al (eds) The prokaryotes. Springer, Berlin, Heidelberg, pp 427–463

Serra V, Fokin S, Castelli M et al (2016) "*Candidatus* Gortzia shahrazadis", a novel endosymbiont of *Paramecium multimicronucleatum* and a revision of the biogeographical distribution of *Holospora*-like bacteria. Front Microbiol 7:1704. https://doi.org/10.3389/fmicb.2016.01704

Skoblo I, Lebedeva NA (1993) The infectiousness of *Holospora acuminata* - an intranuclear bacterial symbiont of *Paramecium bursaria* (Ciliata). I Syngen specificity of infection Tsitologiia 35:92–99

Skoblo I, Makarov S, Ossipov D (2001) The analysis of symbiotic relationships nature diversity in *Paramecium bursaria-Holospora curviuscula* system. Tsitologiia 43:520–527

Skoblo II, Lebedeva NA, Rodionova GV (1996) The formation of the *Paramecium-Holospora* symbiotic system: a study of the compatibility of different symbiont isolates and host clones. Eur J Protistol 32:147–152. https://doi.org/10.1016/S0932-4739(96)80092-7

Skovorodkin IN, Fokin SI, Fujishima M (2001) Fates of the endonuclear symbiotic bacteria *Holospora obtusa* and *Holospora undulata* injected into the macronucleus of *Paramecium caudatum*. Eur J Protistol 37:137–145. https://doi.org/10.1078/0932-4739-00801

Smurov AO, Fokin SI (1998) Resistance of *Paramecium caudatum* infected with endonuclear bacteria *Holospora* against salinity impact. Proc Zool RAS 276:175–178

Stevens JM, Galyov EE, Stevens MP (2006) Actin-dependent movement of bacterial pathogens. Nat Rev Microbiol 4:91–101. https://doi.org/10.1038/nrmicro1320

Szokoli F, Castelli M, Sabaneyeva E et al (2016) Disentangling the taxonomy of *Rickettsiales* and description of two novel symbionts ("*Candidatus* Bealeia paramacronuclearis" and "*Candidatus* Fokinia cryptica") sharing the cytoplasm of the ciliate protist *Paramecium biaurelia*. Appl Environ Microbiol 82:7236–7247. https://doi.org/10.1128/AEM.02284-16

Takeshita K, Yamada T, Kawahara Y et al (2019) Tripartite symbiosis of an anaerobic scuticociliate with two hydrogenosome-associated endosymbionts, a *Holospora*-related Alphaproteobacterium and a methanogenic archaeon. Appl Env Microbiol 85:e00854–e00819. https://doi.org/10.1128/AEM.00854-19

Tashyreva D, Prokopchuk G, Votýpka J et al (2018) Life cycle, ultrastructure, and phylogeny of new diplonemids and their endosymbiotic bacteria. MBio 9:1–20. https://doi.org/10.1128/mBio.02447-17

Wang Z, Wu M (2015) An integrated phylogenomic approach toward pinpointing the origin of mitochondria. Sci Rep 5:7949. https://doi.org/10.1038/srep07949

Wernegreen JJ (2017) In it for the long haul: evolutionary consequences of persistent endosymbiosis. Curr Opin Genet Dev 47:83–90. https://doi.org/10.1016/j.gde.2017.08.006

Wiemann M, Görtz H (1989) Release of the endonucleobiotic bacterium *Holospora elegans* from its host cell *Paramecium caudatum*. Eur J Protistol 25:100–108

Wollman F-A (2016) An antimicrobial origin of transit peptides accounts for early endosymbiotic events. Traffic 17:1322–1328. https://doi.org/10.1111/tra.12446

Chapter 5
Trends in Symbiont-Induced Host Cellular Differentiation

Shelbi L. Russell and Jennie Ruelas Castillo

Abstract Bacteria participate in a wide diversity of symbiotic associations with eukaryotic hosts that require precise interactions for bacterial recognition and persistence. Most commonly, host-associated bacteria interfere with host gene expression to modulate the immune response to the infection. However, many of these bacteria also interfere with host cellular differentiation pathways to create a hospitable niche, resulting in the formation of novel cell types, tissues, and organs. In both of these situations, bacterial symbionts must interact with eukaryotic regulatory pathways. Here, we detail what is known about how bacterial symbionts, from pathogens to mutualists, control host cellular differentiation across the central dogma, from epigenetic chromatin modifications, to transcription and mRNA processing, to translation and protein modifications. We identify four main trends from this survey. First, mechanisms for controlling host gene expression appear to evolve from symbionts co-opting cross-talk between host signaling pathways. Second, symbiont regulatory capacity is constrained by the processes that drive reductive genome evolution in host-associated bacteria. Third, the regulatory mechanisms symbionts exhibit correlate with the cost/benefit nature of the association. And, fourth, symbiont mechanisms for interacting with host genetic regulatory elements are not bound by native bacterial capabilities. Using this knowledge, we explore how the ubiquitous intracellular *Wolbachia* symbiont of arthropods and nematodes may modulate host cellular differentiation to manipulate host reproduction. Our survey of the literature on how infection alters gene expression in *Wolbachia* and its hosts revealed that, despite their intermediate-sized genomes, different strains appear capable of a wide diversity of regulatory manipulations. Given this and *Wolbachia*'s

---

Electronic Supplementary Material The online version of this chapter (https://doi.org/10.1007/978-3-030-51849-3_5) contains supplementary material, which is available to authorized users.

S. L. Russell (✉)
Department of Molecular Cell and Developmental Biology, University of California, Santa Cruz, CA, USA

J. R. Castillo
Johns Hopkins University School of Medicine, Baltimore, MD, USA

M. Kloc (ed.), *Symbiosis: Cellular, Molecular, Medical and Evolutionary Aspects*, Results and Problems in Cell Differentiation 69,
https://doi.org/10.1007/978-3-030-51849-3_5

diversity of phenotypes and eukaryotic-like proteins, we expect that many symbiont-induced host differentiation mechanisms will be discovered in this system.

Keywords *Wolbachia* · *Drosophila* · Symbiosis · Cellular microbiology · Cellular differentiation · Epigenetics · Transcription · Translation · Proteolysis

5.1 The Symbiotic Lifestyle Requires Cellular Remodeling

Bacterial symbionts of eukaryotic hosts form stable associations by colonizing host tissues or cells. This lifestyle requires an added layer of cellular regulation relative to nonsymbiotic lifestyles because symbionts need to integrate with and control the host environment to create a hospitable niche (La et al. 2008; Schwartzman and Ruby 2016; Borges 2017). Without this ability, the bacteria are quickly eliminated by the host's immune system (Medzhitov 2007). Symbionts are benefitted by their ability to control the host environment, as their free-living relatives cannot do much to influence their abiotic environments. However, influencing host cells and tissues is not a trivial task. To do so, symbionts must decode another organism's regulatory pathways and interfere with them without causing too much damage. This is true for costly parasitisms and beneficial mutualisms, as well as extracellular and intracellular lifestyles: in all types of associations, bacteria must subvert host defenses to create a replicative niche (Medzhitov 2007; Mergaert 2018). Furthermore, owing to the deep, 2 billion year divergence between host and symbiont taxonomic domains, the eukaryotic regulatory pathways that need to be subverted are often completely unique from what the bacterial symbiont uses for its own genetic regulation (Cashin et al. 2006).

Nevertheless, bacterial symbionts have repeatedly found ways of controlling host gene expression for their own purposes. In many instances, this means finding ways of integrating with the biology of their multicellular hosts to be recognized as part of the "self" and colonize particular cell types. Naturally, many of the well-known examples of symbiont control of host gene expression involve mechanisms for limiting and modulating immune responses (Grabiec and Potempa 2018), solving the self/nonself issue. While these abilities are fascinating and essential for host-associated bacteria, they have been explored in depth elsewhere (see Hamon and Cossart 2008; Zhong et al. 2013; Silmon de Monerri and Kim 2014; Cheeseman and Weitzman 2015; Pereira et al. 2016; Vilcinskas 2017; Cornejo et al. 2017). Instead, here, we explore the evidence for bacterial symbiont control of host cellular differentiation, which can be used to control the identity of infected host cells, the size of the infection niche, and host reproduction.

In this review, we summarize what is known about how and why symbionts ranging from pathogens to mutualists control host cellular differentiation to create novel cell, tissue, or organ types for their habitation (Fig. 5.1). We focus on cellular, tissue, and organ-levels of differentiation, as different symbiont taxa can target

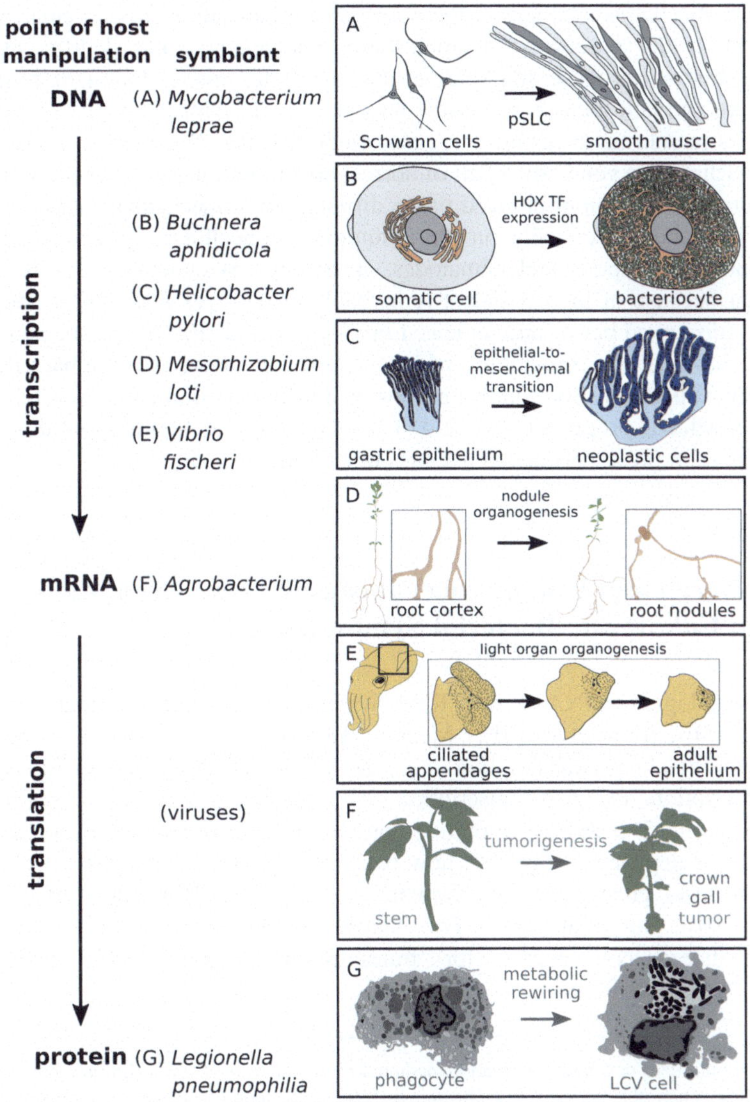

Fig. 5.1 Examples of bacterial symbiont-induced cell, tissue, and organ differentiation across the cost-benefit spectrum of bacterial-eukaryotic symbiotic associations, organized by the point in host genetic regulation that they influence. Interestingly, no bacterial examples of translation-mediated host cellular differentiation were found, making viruses and toxin-secreting lytic bacteria the primary representatives for this strategy. (**a**) *M. leprae* induces dedifferentiation of Schwann cells via altering host epigenetic marks. This produces infected progenitor/stem-like cells (pSLC) that migrate and become new cell types, such as smooth muscle, spreading the infection throughout the host's body (Masaki et al. 2013). (**b**) The fate of host-derived symbiont-housing cells, bacteriocytes, and organs, bacteriomes, is specified through changes in the abundance of host transcription factors (TFs) involved in embryogenesis (Braendle et al. 2003; Matsuura et al. 2015). In the primary aphid endosymbiont, *B. aphidicola,* bacteriocyte formation involves reorganization of the endoplasmic reticulum (orange) to surround dense symbiont (green) aggregates (Simonet et al. 2018). (**c**) Pathogenic *H. pylori* induces host gastric epithelia to dedifferentiate and take on a mesenchymal

regulatory mechanisms at any of these levels of organization. In particular, we are interested in the processes of immortality maintenance and dedifferentiation/redifferentiation, as these strategies enable the stable manipulation of host gene expression and cell identity for symbiont purposes. In parasitisms, these are often viewed as neoplastic structures, i.e., abnormal growths. Whereas, in mutualisms, these structures are generally a part of normal host morphology. After presenting on the diversity of symbiont-induced tissue differentiation mechanisms reported from nature, we focus specifically on the ubiquitous intracellular alphaproteobacterial symbiont of arthropods and nematodes, *Wolbachia*. We focus on *Wolbachia* in particular because of the myriad of remarkable phenotypes it is able to induce in its hosts (discussed below and reviewed in (Werren et al. 2008)) and the tantalizing data that have been accumulating, which suggest that strains of these bacteria have significant capabilities for controlling host cell differentiation pathways. Given the recent growth and progress in the field of *Wolbachia* research, the aim of this review is to inform on the experimental avenues to explore in the future.

5.2 Shared and Unique Mechanisms of Gene Regulation in Eukaryotic-Bacterial Symbioses

The central dogma—DNA encodes RNA, which encodes proteins—holds across the diversity of life (Piras et al. 2012). Meaning, regulation points exist for bacteria and eukaryotes at (1) pretranscription (e.g., epigenetic DNA/histone modifications), (2) transcription, (3) post-transcription (e.g., mRNA processing or regulation), (4) translation, (5) post-translation (e.g., protein modifications), and (6) proteolysis. However, as depicted in Fig. 5.2, how these regulatory mechanisms work in real-time can differ greatly between domains (Kozak 1992; Blumenthal et al. 2002; Belasco 2010; Gur et al. 2011). For example, while both domains of organisms can regulate DNA access for transcription through DNA methylation

Fig. 5.1 (continued) cell fate via effector-mediated influence of host transcription factor retention and binding. Over the course of a chronic infection, this process produces over-proliferative neoplasms that can develop into gastric cancer (Bessède et al. 2014). (**d**) Soil-dwelling rhizobia bacteria localize to legume plant roots, and induce their uptake into root cells and the formation of the root nodule through interacting with host transcription factor signaling (Oldroyd 2013). (**e**) In juvenile bobtail squid, bioluminescent *V. fischeri* colonize a ciliated epithelium on the outside of the nascent light organ, and induce the degradation of the colonization surface's ciliated appendages through interfering with host transcription factor signaling (Nyholm and McFall-Ngai 2004). (**f**) Plant pathogens in genus *Agrobacterium* transfer a mobile element to the host cell, which manipulates host miRNA-based genetic regulation to induce dedifferentiation and tumor formation (Escobar and Dandekar 2003). (**g**) The intracellular pathogen *L. pneumophila* induces the formation of the *Legionella*-containing vacuole (LCV) through co-opting and mimicking host post-translational modifications to inhibit host translation and increase proteolysis of host proteins and peptides (Xu and Luo 2013)

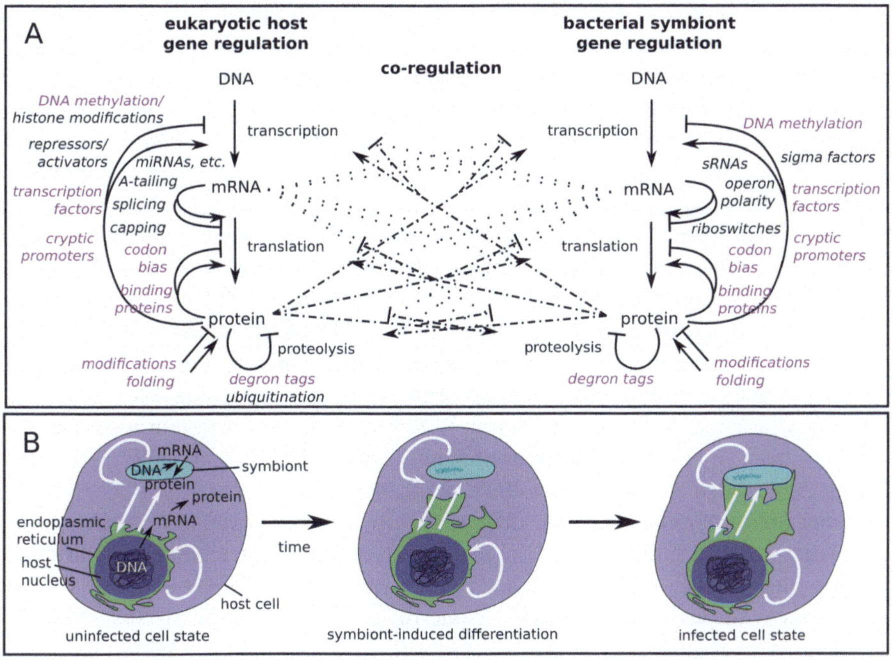

Fig. 5.2 Coordination between host and symbiont gene expression enables host-symbiont interactions. (**a**) Overview of endogenous general mechanisms of eukaryotic and bacterial gene expression from mRNA transcription from DNA, to protein translation from mRNA, to protein turnover (solid lines). Methodological advancements over the past couple of decades have revealed that eukaryotes and bacteria have more mechanisms in common (pink italicized text) than previously estimated (Güell et al. 2011). Interestingly, bacteria can also regulate their mRNA via poly A-tailing, however, in contrast to eukaryotes, this signals for mRNA degradation and represents a small fraction (<<1%) of transcripts (Güell et al. 2011), which is why it is not listed above. Additionally, it should be noted that post-transcriptional regulatory components contained within mRNAs, such as 5′-untranslated regions, influence the access of proteins and other signaling molecules to transcript translation start sites and riboswitches, but are not explicitly listed. Reciprocal control over host/symbiont processes works through endogenous and mimicked mechanisms (dashed lines). (**b**) An example of how host-symbiont interactions (straight white arrows) function with endogenous mechanisms (curved white arrows) to cause phenotypic changes in cell state, such as symbiont-induced formation of an intracellular replicative niche derived from the endoplasmic reticulum membrane, as has been reported for *Wolbachia* (Fattouh et al. 2019) and a variety of other symbionts (see text)

(Sánchez-Romero et al. 2015), eukaryotes also have histones, which can be modified to be more or less permissive to the entry of transcriptional machinery (Verdone et al. 2005). Following transcription, eukaryotes have additional ways to modify their mRNA relative to bacteria, including RNA splicing, poly-A-tailing, and 5′-capping (Belasco 2010), to alter its identity, stability, or accessibility for protein translation, respectively. Although, bacteria do have a range of post-transcriptional regulatory strategies (Dar and Sorek 2018).

In addition to phylogenetic constraints, the different body plans and life histories among hosts and symbionts also underlie their different genetic regulatory capabilities. Multicellular hosts with complex tissue types and body plans require precise mechanisms for controlling gene expression across both space and time to properly control tissue differentiation and maintain stem cell pluripotency. Many plants and animals epigenetically alter their DNA by packaging it into chromatin, which helps maintain differential gene expression in different cell types over the lifespan of the host (Meissner 2010; Li et al. 2011). Interestingly, epigenetic alterations also underlie the transitions between parasite life stages that are evoked by different hosts, both in multicellular (Roquis et al. 2018) and single-cellular (Duraisingh and Horn 2016) eukaryotic parasites. By binding to the DNA promoters and regulatory regions made accessible by epigenetic modifications, transcription factors are also very important to cellular differentiation. This is true for both eukaryotes as well as bacteria, which use transcription factors to differentiate into different metabolic or motility states in response to environmental signals (Laub et al. 2007; Cole and Young 2008; Losick and Desplan 2008; Wolański et al. 2014).

Using these similarities and differences in genetic regulation, many host-associated bacteria have evolved ways to interact with host regulatory pathways. The simplest model for how a bacterium evolves control over its host's gene expression is through the co-option of one of its own pathways. In this situation, the majority of required machinery for the pathway would already be in place, and only modifier components would need to be added for controlling host gene expression. In contrast, it is also possible for bacteria to evolve strategies for interfering with eukaryotic-specific mechanisms of gene expression, such as histone modifications or splicing. In fact, this strategy appears to be quite common among pathogenic bacteria, which can possess proteases, acetyltransferases, kinases, phosphatases, ubiquitin ligases, and deubiquitinases for altering host gene expression (Guven-Maiorov et al. 2017). It is unlikely that genes lacking functions specific to the bacterial cell evolved in concert with the endogenous bacterial gene expression regulatory networks. Thus, their presence implies either introduction via horizontal gene transfer (e.g., Patrick and Blakely 2012) or functional convergence (e.g., Alvarez-Venegas 2014), often resulting in structural mimicry of the host protein (Frank 2019).

The nature of bacterial regulation of host gene expression likely depends on the host cell type and the desired outcome of the interaction. In terms of host cell differentiation, bacterial influence can either cause a host cell to become less differentiated, i.e., more stem-cell-like with pluripotent capabilities, or it can cause a host cell to become more differentiated toward some particular fate. Less-differentiated fates could facilitate bacterial transmission, especially if they are proliferative because bacteria can be inherited by both daughter cells during cell division. For example, the intracellular symbiont *Wolbachia* has been shown to segregate evenly between dividing embryonic cells in *Drosophila melanogaster* (Albertson et al. 2009). More differentiated fates could have a variety of impacts depending on whether the interaction is mutualistic or pathogenic. For example, the differentiation of host cells into bacteriocytes in mutualistic associations (see

Fig. 5.1b) provides an environment for bacterial symbionts to live at high densities and perform metabolic functions necessary to the host (Braendle et al. 2003; Hongo et al. 2013; Matsuura et al. 2015). In pathogenic interactions, bacteria often induce host cell differentiation to reach a metabolic state where more resources are provided to the bacteria for replication, increasing bacterial virulence and infectivity (Cornejo et al. 2017).

5.3 Making a House a Home: Bacterial Symbionts Influence Host Cellular Differentiation During Infection and Establishment

In the sections below, we describe examples from the literature of different ways in which pathogenic and mutualistic symbionts have been found to control host cellular differentiation. These examples are generally organized by their place in the molecular biology hierarchy, from DNA to RNA to protein. Bacterial influence may occur at early points in the hierarchy and have cascading effects on the subsequent stages of gene expression, which are discussed when possible. As regulation becomes circular at the ends of the hierarchy, e.g., post-translational modifications of histone proteins affect DNA accessibility for transcription, this framework serves to organize the discussion.

5.3.1 Epigenetic Control of Host Gene Expression

Multicellular organisms control the differentiation of their cells and tissues through epigenetic modifications put in place during development (Meissner 2010), and bacterial symbionts often use this mechanism to influence host cellular differentiation too (Hamon and Cossart 2008). Indeed, abundant evidence exists that a variety of host-associated bacteria, including *Legionella, Listeria, Clostridium, Streptococcus, Helicobacter,* and *Salmonella,* are able to influence host DNA methylation or histone post-translational modifications to alter chromatin transcriptional accessibility and attenuate the immunological responses their infections solicit (Bierne et al. 2012). Immune responses include the upregulation of inflammatory cytokines, chemokines, toll-like receptors, and antimicrobial peptides, including cationic antimicrobial peptides (CAMPS). Bacterial symbionts can inhibit gene expression underlying these responses by directly altering chromatin packaging with their own enzymes (Alvarez-Venegas 2014). They can also indirectly alter the activities of host proteins such as DNA methyltransferases (DMTs), histone acetyltransferases (HATs), histone deacetylases (HDACs), and histone methyltransferases (HMTs) through protein–protein inhibition or signaling, e.g., through short-chain fatty acids (Grabiec and Potempa 2018). Depending on the molecular specificity of the interactions, epigenetic alterations can be highly targeted to a particular host gene or can be global across the host genome. For example, *Shigella flexneri* produces and secretes an effector protein, OspF, that ultimately prevents histone phosphorylation and NF-B access to transcription binding sites, thus inhibiting an immune response (Arbibe et al. 2007). Importantly, these anti-inflammatory mechanisms are also used by commensal bacteria to a more beneficial effect because chronic inflammation is harmful to hosts (Grabiec and Potempa 2018).

Interestingly, in some instances, pathogen control of the host immune response can also induce developmental effects. For example, in the greater wax moth, *Galleria mellonella*, infection with *Listeria monocytogenes* increases the expression of both HATs and HDACs, resulting in a developmental delay that extends the time until metamorphosis (Mukherjee et al. 2012). Developmental effects such as these could have initially arisen as a side effect of cross-talk between epigenetic mechanisms mediating development and immunity (Vilcinskas 2017) or immune activation being required for nuclear reprogramming (Lee et al. 2012), and been maintained by the pathogen for its benefit. Developmental delays could be beneficial for reallocating resources from the host to the pathogen (Vilcinskas 2017). Thus, influence of host cellular differentiation can be a byproduct of the mechanisms used for infection and virulence (Vilcinskas 2017), and might facilitate the evolution of more intrinsic manipulations that change the identity of the host cell for symbiont purposes.

The known cases of bacterial epigenetic reprogramming of host cellular differentiation are from pathogenic bacteria, potentially due to pathway cross-talk. In an exquisite display of cellular manipulation, the intracellular pathogen that causes leprosy, *Mycobacterium leprae*, has been shown to reprogram the Schwann cells it inhabits to reach a stem cell-like state (Fig. 5.1a). It does this via changes in methylation patterns and gene expression profiles that turn off Schwann cell differentiation genes/transcription factors and turn on developmental and embryonic genes/transcription factors. From this reprogrammed state, these infected stem cell-like cells can then differentiate into different tissue types and migrate from the peripheral nervous system to the surrounding connective tissues and muscles, helping to disseminate the bacteria throughout the host. Interestingly, they also use this reprogrammed state to attract and infect macrophages, further spreading the infection (Masaki et al. 2013).

A more brute-force approach to epigenetic reprogramming of host cells has been reported for the male-killing spirochete parasite of *D. melanogaster*, *Spiroplasma*. While epigenetic regulation via DNA methylation does not occur in *D. melanogaster* because it lacks functional DNA methyltransferase enzymes (Goll and Bestor 2005), it does regulate its gene expression with histone acetylation. In males, acetylation is used to double the expression of X chromosome-linked genes. *Spiroplasma* symbionts are able to interfere with this process to induce male killing, which eliminates these "dead-end" infections from the population so that more resources are available for the females, through which these bacteria will be maternally transmitted (Veneti et al. 2005). These bacteria accomplish male-killing by interfering with the male specific lethal 2 (MSL-2) protein of the dosage compensation complex, which is only active in male embryos, causing it to be randomly mislocalized across the genome. Mislocalization causes randomly elevated transcription across the genome through elevated acetylation, resulting in male lethality (Cheng et al. 2016). Recent work by (Harumoto and Lemaitre 2018) identified the *Spiroplasma* Spaid protein, which contains ankyrin and deubiquitinase domains, as sufficient to induce male lethality through the MSL-2 complex.

While we are still in the early days of characterizing how bacterial symbionts can epigenetically modify host cellular differentiation through chromatin modifications, a number of preliminary data points suggest that this will be a productive area of research in future years. For example, host-pathogen associations have been reported to have long-lasting or transgenerational effects, likely mediated through epigenetic mechanisms, although they have not yet been identified (Fridmann-Sirkis et al. 2014; Mukherjee et al. 2017; Yang et al. 2018; Gegner et al. 2019). Epigenetic-based gene regulation is also implicated in eukaryote-eukaryote mutualisms such as coral-algal symbioses (Li et al. 2018). Furthermore, bioinformatic evidence suggests that many host-associated bacteria contain SET-domain proteins in their genomes (named for their discovery in *Drosophila* proteins **S**u(var)3–9, **E**nhancer-of-zeste and **T**rithorax), which are known to encode lysine histone methyltransferases (Alvarez-Venegas 2014). Given that bacteria do not contain histones, it is highly likely that many of these proteins are used to alter eukaryotic cellular functions.

5.3.2 Symbiont Co-option of Host Signaling Pathways and Transcription Machinery Mimicry

Studying the intertwined and intimate interactions between host and symbiont is often a difficult task, however, the advent of microarrays and next generation sequencing opened up one avenue of investigation significantly: host and symbiont transcriptomics. While these methods enabled the high-throughput collection of gene expression data from hosts and symbionts, challenges continue to this day regarding the amount of mRNA that can be obtained from host-associated bacteria in situ. One of the main issues involves the drastic differences in relative abundance of bacterial versus eukaryotic mRNA. Furthermore, bacterial mRNA only comprises ~4% of total cellular bacterial RNA, with rRNAs and tRNAs making up the bulk of the transcripts. In addition, the half-life of bacterial mRNAs is far shorter than that of eukaryotic mRNAs, making it difficult to accurately capture bacterial gene expression in real-time. On top of all of this, bacteria do not A-tail their transcripts unless they are being marked for degradation. Therefore, while eukaryotic mRNAs can be selected for by poly-dT priming, bacterial mRNAs cannot be directly selected, and instead must be depleted of rRNA (La et al. 2008). Nonetheless, many methodological tricks have been developed over the years to deplete rRNAs and host transcripts or enrich for microbial mRNAs (Güell et al. 2011), and so this is the step of gene expression that we have the most data for presently.

These studies have revealed a few trends in host-symbiont transcriptomics. Importantly, it appears that some, but not all bacterial symbionts are capable of modulating their own or their host's transcription in response to the association. Those that cannot typically exhibit severe degrees of genome erosion, and are discussed later in this section. However, it is worth noting that even the symbionts with extreme levels of genome degradation are able to induce the differentiation of the specific host cells and/or organs they reside in, termed bacteriocytes and bacteriomes, respectively (see Fig. 5.1b). While the mechanism(s) of induction have not been identified, and may involve other elements of host signaling pathways (Smith and Moran 2020), upregulation of the host homeobox transcription factor Ultrabiothorax has been shown to be necessary for bacteriocyte differentiation in seed bug insects (Matsuura et al. 2015) and aphids (Braendle et al. 2003). In general, the symbionts that can influence host transcription do so by either modulating host signaling pathways upstream of transcriptional responses (Rogan et al. 2019) or by mimicking host transcription factors, activators, and suppressors (Saijo and Schulze-Lefert 2008). Examples of these two strategies have been reported for diverse symbiotic systems and are detailed below.

Interaction with host signaling pathways to induce transcriptional changes is the most commonly reported strategy for symbiont-induced modulation of host transcription. Symbionts may be predisposed to evolving this strategy because components of the host signaling cascades that induce immune responses are also used during development (Rogan et al. 2019). This is likely another manifestation of pathway cross-talk discussed above. Of the known signaling pathways, pathogens have been shown to frequently interact with the Notch, Wnt, and STAT3 pathways (Hannemann et al. 2013; Rogan et al. 2019). Wnt signaling is especially fruitful to exploit because its induction through canonical and noncanonical pathways can alter gene expression to manipulate immune responses and increase cell proliferation (Rogan et al. 2019). Additionally, the Wnt signaling-induced transcription factor β-catenin is important for the activation of many genes including ones for adherens junction integrity, which is essential for epithelial integrity. Given that many pathogens are benefited by breaking down epithelial barriers for further dissemination, the ability to target Wnt signaling may be strongly selected for. Thus, by increasing the translocation of transcription factors such as β-catenin to the nucleus, symbionts can simultaneously make a more hospitable and a larger niche for themselves in the host.

Transcription-level bacterial control of host cellular differentiation via the Wnt pathway is also displayed by *Helicobacter pylori*, the leading cause for chronic gastric inflammation and cancer worldwide. This epsilon-proteobacterium colonizes the mammalian stomach epithelium through controlling cell differentiation, proliferation, and apoptosis (see Fig. 5.1c). It accomplishes this via direct interactions with cell adhesion and polarity factors (Amieva 2003; Bagnoli et al. 2005; Wessler and Backert 2008) and indirect interactions with host transcription factors, including β-catenin and (Hatakeyama 2006; Wessler and Backert 2008) and Nuclear factor of activated T-cells (NFAT) (Yokoyama et al. 2005). *H. pylori*-induced transformation of host gastric epithelial cells resembles the process of epithelial-to-mesenchymal cellular transition during embryogenesis, and produces an invasive migratory phenotype through altering the localization and expression of genes involved in controlling cell shape, polarity, and division. While there are several identified mechanisms underlying these abilities, most involve the effector protein cytotoxin-associated gene A (CagA) (e.g., Yokoyama et al. 2005; Bagnoli et al. 2005; Suzuki et al. 2009; Bertaux-Skeirik et al. 2015), which *H. pylori* injects into host epithelial cells with its type IV secretion system. The large, 1200 amino acid CagA protein causes a range of effects due to interactions between host factors and its N- and C-terminal domains, which have different activities in different phosphorylation states (Bagnoli et al. 2005; Hatakeyama 2006; Wessler and Backert 2008). Interestingly, CagA exhibits structural and functional similarity to the eukaryotic Grb-2-associated binder (Gab) adapter proteins, although it does not exhibit any sequence similarity, and likely evolved to mimic Gab interactions with host cellular machinery (Botham et al. 2008).

The nitrogen-fixing mutualistic rhizobia bacteria of leguminous plants (consisting of alpha- and beta-proteobacterial lineages) co-opt host signaling cascades to alter host root tissue differentiation in order to create the nodule structure where the symbionts are housed (see Fig. 5.1d). This structure is essential to the bacteria, as they need an oxygen-free environment to fix atmospheric dinitrogen into biologically available compounds such as ammonium. Nodule development is induced by bacterial colonization from the surrounding soil, and follows an intricate signaling cascade between rhizobia and the root (Oldroyd 2013). The interaction begins when rhizobia encounter legume flavonoids in the soil, which induce the bacteria to produce and secrete nodulation (nod) factors, which bind to host membrane receptors, inducing oscillations in nuclear calcium concentration. The nuclear calcium concentration-dependent transcriptional response is thought to activate the nuclear calcium- and calmodulin-dependent kinase (CCaMK). CCaMK phosphorylates the transcriptional activator CYCLOPS, inducing the expression of genes essential for symbiosis establishment, including infection thread formation and mitotic reactivation at the root cortex. Underscoring the importance of these host genes, CCaMK or CYCLOPS activation alone, without the presence of symbionts, is sufficient to induce nodule formation (Singh et al. 2014). Interestingly, many of the host genes in these pathways have homologs in nonlegumes and are also involved in mycorrhizae establishment, suggesting that they may have evolved for that association first, and were co-opted for the later-evolving rhizobial associations (Singh et al. 2014).

While the full details are not yet available, preliminary evidence suggests that *Vibrio fischeri*-induced development of the squid light organ is also mediated through host transcription factor signaling pathways (Peyer et al. 2017). In this association, bioluminescent gammaproteobacterial *V. fischeri* are lured from the complex community in the surrounding seawater by host production of chitin-like compounds (Mandel et al. 2012). Upon localizing to the juvenile squid's nascent light organ epithelium, general microbe-associated molecular patterns, such as peptidoglycan, induce changes in host gene expression and mucus production. The bacteria then migrate through this mucus to colonize the light organ crypts (Kremer et al. 2013). This process is specific because the bacteria must endure acidic and free radical bombardment by nitric oxide (Nyholm and McFall-Ngai 2004). Once within the crypts, *V. fischeri* induce apoptosis and loss of external appendage structures (see Fig. 5.1e) through interactions with Crumb, the protein regulator of apical-basal polarity and adherens junctions (Peyer et al. 2017). Interestingly, *V. fischeri*-induced tissue differentiation does not end there. In the adult squid, bacterial interactions with genes involved in squid retinal regeneration mediate daily change in light organ epithelial microvilli density (Heath-Heckman et al. 2016; Kremer et al. 2018).

A second mechanism for influencing host transcription has been reported for a range of pathogens and operates through mimicking or influencing host transcription factors, activators, or suppressors. For example, plant pathogenic bacteria, such as *Xanthomonas*, the etiological agent of bacterial blight in rice, synthesize and secrete transcription activator-like effector (TALE) proteins through their type III secretion systems. These proteins cross into the host nucleus and mimic host transcription activators. In susceptible plants, the binding of TALEs to host transcription factors alters transcription start sites and induces the expression of host genes that increase cell size, which facilitates dissemination of the bacteria from the intercellular spaces (Saijo and Schulze-Lefert 2008; Yuan et al. 2016).

Epigenetic and transcriptional control of host differentiation are obviously effective strategies, however, genome erosion in host-associated bacteria has repeatedly limited the capacity for these types of mechanisms in many taxa. Pathogens with no or limited degrees of genome degradation are capable of modulating their gene expression at the transcriptional level (La et al. 2008). Even obligate intracellular pathogens with moderate levels of genome degradation such as the Chlamydiae exhibit evidence of using transcription factors to modulate their own gene expression (de Barsy et al. 2016). In contrast, obligate intracellular pathogens, e.g., *Treponema pallidum* (La et al. 2008), and mutualists, e.g., *Buchnera* (Hansen and Degnan 2014), with extreme levels of genome erosion (genome sizes $\leq$1 Mb) generally have relatively stable transcriptional states, although exceptions do exist (see the *Baumannia* symbiont of the glassy winged sharpshooter (Bennett and Chong 2017)). However, it is clear that some form of post-transcriptional or translational regulation has replaced these mechanisms because, in many associations, differentially expressed mRNA abundances do not correlate with translational abundances (i.e., proteins or "translatomes", which are the ribosome-associated population of mRNAs) (Traubenik et al. 2019).

The loss of reliance on transcriptional regulation for endogenous or host genetic regulation is likely a direct consequence of genome erosion, as many of these bacteria have lost the majority of their transcription factor genes and other genes required for transcriptional regulation (Galán-Vásquez et al. 2016). Indeed, the intracellular pathogen *Mycoplasma pneumoniae* encodes only eight predicted transcription factors in its 0.82 Mb genome (Güell et al. 2009) (compared to *E. coli's* 314 transcription factors (Güell et al. 2011)) and expresses an abundance of antisense RNA and polycistronic operons relative to free-living bacteria (Güell et al. 2009). Interestingly, the substitution of transcriptional regulation with post-transcriptional mechanisms has not resulted in higher transcription errors (Traverse and Ochman 2016). Next, we explore how these restricted regulatory capacities have impacted symbionts' abilities to interact with host biology.

5.3.3 The Pervasiveness of Post-transcriptional Mechanisms for Control of Host Cell State

Control of host gene expression through small RNA (sRNA) pathways is a common feature among symbiotic bacteria, likely because both bacteria and eukaryotes use various types of sRNAs to regulate the turnover of their transcripts. While eukaryotes make a diversity of specific sRNA classes, such as microRNA (miRNA), small interfering RNA (siRNA), and piwi-interacting RNA (piRNA) (Palazzo and Lee 2015), bacteria make more general types of sRNA that are short to long (~50–1000s nt) and highly structured (Bobrovskyy and Vanderpool 2013). Bacterial sRNAs are either cis- or trans-acting, depending on whether they regulate the gene they were transcribed from (in the case of antisense RNAs), or whether they regulate a gene far away, respectively. Trans-acting sRNAs often have multiple targets, making them akin to post-translational transcription factors (Güell et al. 2011). Although eukaryotes and bacteria have very different endogenous mechanisms for sRNA-mediated genetic regulation, the sRNAs themselves have enough similarities to make cross-domain transfer and function possible (reviewed in (Simonov et al. 2016; Zeng et al. 2019)). In some cases, host RNA-processing proteins are even involved in converting bacterial sRNAs into miRNA molecules (Gu et al. 2017).

Bacterial symbionts with extreme levels of genome degradation appear to have converted to an RNA-based strategy of genetic regulation, similar to mitochondrial and plastid organelles (Cognat et al. 2017; Thairu and Hansen 2019). This is an efficient strategy for bacteria with highly degraded or streamlined genomes for three reasons. First, cis-encoded sRNA-based mechanisms of gene regulation are self-contained within the genetic element, making this regulatory approach independent of additional coding sequence, which may not be maintained during genome erosion. As trans-acting sRNAs often require an RNA chaperone protein, e.g., Hfq, for localization, the smallest endosymbiont genomes tend to not have these elements. Second, hosts use sRNA-based gene regulation, making this regulatory mechanism effective for both endogenous and host genetic regulation (Kim et al. 2016). Third, sRNAs have been shown to be critical to bacterial metabolic regulation (Bobrovskyy and Vanderpool 2013). As metabolic functions are often what intracellular mutualists are responsible for in their associations, the retention of their primary regulatory mechanism likely helps to maintain function in the face of coding sequence loss. Consistent with this, the aphid symbiont *Buchnera* has been shown to use its sRNAs to regulate its own arginine biosynthesis (Thairu et al. 2018). Of course, not all expressed sRNAs may be functional, as the often AT-rich sequence content of these genomes may produce spurious promoters (Lloréns-Rico et al. 2016). However, as pointed out by (Thairu and Hansen 2019), this "noise" may produce regulatory raw material for symbionts to select upon.

Pathogens employ sRNAs to regulate their own virulence gene expression as well as host miRNA-mediated immune responses (Sharma and Heidrich 2012; Sesto et al. 2014; Knip et al. 2014; Ortega et al. 2014; Vilcinskas 2017). Bacterial sRNA-based influence of host gene expression is exemplified by the food-borne pathogen *Salmonella*. This intracellular bacterium uses the host Argonaute RNA processing protein to modify double stranded bacterial noncoding RNA derived from the 5'-UTR of its ribosomal transcripts into ~22 bp miRNA, which it uses to promote intracellular survival (Gu et al. 2017) via mechanisms such as inhibiting nitric oxide production (Zhao et al. 2017). Despite these clear functions in host genetic regulation, facultatively host-associated enteric bacteria such as *Escherichia coli* and *Salmonella enterica* exhibit low conservation of antisense RNA expression (Raghavan et al. 2012). Given that pathogens such as *Listeria monocytogenes* do not share sRNAs with their nonpathogenic relatives, these data suggest that sRNAs may be involved in the evolution of virulence (Sesto et al. 2014). Consistent with this notion, similar mechanisms of controlling host gene expression have been reported for eukaryotic pathogens (Knip et al. 2014). For example, the fungal pathogen *Botrytis* secretes its own effector sRNAs into host cells that bind to host Argonaute proteins to inhibit immune gene expression via RNA interference (RNAi) (Weiberg et al. 2013). Bacterial pathogen-produced sRNAs may even mediate an epigenetic memory of the infection, as *Pseudomonas aeruginosa's* sRNAs have been recently shown to induce pathogen avoidance up to four generations after infection (Kaletsky et al. 2019).

Although the majority of reported examples of symbiont-induced host post-transcriptional gene regulation involve modulating immunity or uncharacterized phenotypic effects, one example does exist of a symbiont that uses sRNA to regulate host tissue differentiation. Plant pathogenic bacteria in the genus *Agrobacterium* inhabit soils and, depending on the species, cause neoplastic tumors (galls; see Fig. 5.1f) or excess adventitious roots (hairy roots) when they infect wounded plants (Nilsson and Olsson 1997). Within these new tissue structures, *Agrobacterium* induces host cells to synthesize metabolites (termed opines) that only the bacteria can use, effectively forming a symbiont-specific niche in the host plant (Escobar and Dandekar 2003). All pathogenic *Agrobacterium* species examined to date establish infections via transferring a plasmid-encoded section of their genome called T-DNA. Once within the host cytoplasm, T-DNA-encoded genes are expressed by the host because they contain the required eukaryotic regulatory elements (i.e., TATA box, CAAT box, and polyadenylation signals) (Escobar and Dandekar 2003). Oncogenes encoded by T-DNA are responsible for inducing changes in host cell differentiation by synthesizing auxin and cytokinin plant hormones. Depending on the species' T-DNA content, either undifferentiated tumors or proliferation of differentiated tissues results from these alterations. Also encoded on T-DNA are the opine-producing genes, which synthesize these metabolites for bacterial nutrition (Nilsson and Olsson 1997; Escobar and Dandekar 2003). While it is clear that increased hormone signaling induces host plant tissue differentiation, the precise mechanisms of tumor differentiation are still being elucidated. However, recent high throughput sequencing has made it clear that bacterial factors interact with host RNA silencing pathways to induce tumor formation. Specifically, tumor formation by *Agrobacterium tumefaciens* requires host miRNA pathways, but is inhibited by host siRNA pathways. Over the course of tumorigenesis, dedifferentiation induces an anti-silencing state that inhibits siRNA-based immunity against bacterial T-DNA (Peláez et al. 2017).

In addition to sRNA-mediated mechanisms, some pathogens also modify host immunogenic gene expression post-transcriptionally through RNA-binding proteins and alternate splicing (Svensson and Sharma 2016). The RNA-binding proteins carbon storage regulator (Csr) and regulator of secondary metabolism (Rsm) are produced by a range of pathogenic bacteria, including *Yersinia pseudotuberculosis* and *Legionella pneumophila*, and bind to the translation initiation region of a large diversity of mRNAs, many of which underlie host immune responses, to inhibit their translation (Svensson and Sharma 2016; Kusmierek and Dersch 2018). As splicing is involved in the activation of normal immune responses to infection, (e.g., via release of membrane-bound pre-mRNAs), pathogen-modified splicing has been proposed to be an understudied mechanism for pathogen manipulation of host gene regulation (Chauhan et al. 2019; Rigo et al. 2019). Indeed, coimmunoprecipitation experiments have shown that *Mycobacterium tuberculosis* produces effector proteins that bind to host splicing factors (Chauhan et al. 2019). In *L. pneumophila* infections, the bacteria inhibit the splicing and activation of response regulator mRNAs, which would otherwise activate the host's immunogenic unfolded protein response as a consequence of the bacteria's co-option of endoplasmic reticulum membrane (Treacy-Abarca and Mukherjee 2015). Bacterially induced alternative host gene splicing also appears to have been co-opted by mutualistic root symbionts, as many plant transcripts are alternatively spliced during rhizobia-induced root nodule formation, although the responsible bacterial mechanisms have yet to be identified (Rigo et al. 2019).

Although only a single example of symbiont-induced host cellular differentiation via post-transcriptional gene regulation has been reported (*Agrobacterium*-induced tumors), this mode of host manipulation likely occurs more frequently in nature for a couple of reasons. First, bacterial and fungal pathogens have been shown to use their sRNA to manipulate host RNAi-based gene silencing (Weiberg et al. 2013; Gu et al. 2017). Second, this mechanism is not unique to pathogens. Organellar sRNAs have been found to interact with the nuclear-encoded Argonaute protein, suggesting that bacterially derived organelles have retained the ability to regulate host gene expression through host RNA interference pathways (Cognat et al. 2017).

5.3.4 *Influence of Symbionts on Host Protein Translation*

Eukaryotic translation involves a complex suite of interactions with various protein complexes to bind the 5′ cap and 3′ poly-A tail of mRNA molecules, initiate translation, and elongate the growing peptide. The timing and location of this process influences protein localization and cellular patterning. Initiation is the rate-limiting step of translation because it requires the recruitment of multiple initiation factor proteins to the 5′ cap, recruitment of the poly-A binding protein (PABP) to both the 5′ cap and 3′ poly-A tail, followed by the assembly of elongation factor proteins. Thus, initiation and elongation are the steps most pathogens target to inhibit translation (Mohr and Sonenberg 2012; Jan et al. 2016). In some cases, hosts can overcome translational blocks by overexpressing mRNAs for immune responses, effectively overwhelming the components mediating the block, in a process termed mRNA superinduction (Barry et al. 2017).

Viruses excel at hijacking host translation because all must commandeer it for their own protein synthesis, and many can even induce the host machinery to preferentially translate viral mRNAs (Toribio and Ventoso 2010; Jan et al. 2016; Jaafar and Kieft 2019). One common mechanism for co-opting host ribosomes is through interacting with the cap-initiation complex during translation initiation to inhibit and/or co-opt host factors. For example, picornaviruses such as Poliovirus produce a protease that cleaves the cap-binding domain of host initiation factor protein eIF4G. That protein fragment then binds to viral mRNAs and enables cap-independent translation (Schneider and Mohr 2003). Similarly, RNA viruses like Hepatitis C are able to directly bind host ribosomes with their genome's 5-′-untranslated end and a subset of host initiation factor proteins (i.e., eIF3 and eIF2), enabling translation of the full viral genome (Au and Jan 2014). DsDNA adenoviruses phosphorylate host initiation factor protein eIF4E, which inhibits mRNA cap binding and enables the virus to co-opt the translation machinery for its own mechanism, termed ribosome shunting (Schneider and Mohr 2003). In the previous two examples, viral protein synthesis is accomplished through rendering required host translational components unusable. However, examples also exist in which viral translation is accomplished while host translation is ongoing, such as the human cytomegalovirus (HCMV). Within host cells, HCMV increases the expression of host PABPs, which positively regulate the expression of initiation complexes, resulting in an overall increase in the abundance of translation machinery (Au and Jan 2014). Impressively, these strategies are often robust to host interference, as viruses have evolved counter mechanisms that are enacted in response (Jaafar and Kieft 2019).

A range of bacterially produced toxins and effector proteins target host translation in order to inhibit immune responses and scavenge resources (Mohr and Sonenberg 2012). In intestinal infections, *Pseudomonas aeruginosa*-secreted Exotoxin A is endocytosed by adjacent host cells where it inhibits mRNA translation by ribosylating and inactivating host elongation factor EF2 (Dunbar et al. 2012; McEwan et al. 2012). Interestingly, the exotoxins of *Vibrio cholera* and *Corynebacterium diphtheriae* have been shown to inhibit host translation by EF2 ribosylation as well, suggesting this is a common mechanism (McEwan et al. 2012). The intracellular pathogen *Legionella pneumophila* blocks host translation through modifying host translation machinery using five of its effector proteins (Fontana et al. 2011) that act through at least two distinct mechanisms. Host translation elongation factor eEF1A is inhibited via glycosylation by the secreted *L. pneumophila* glucosyltransferases (Lgts), Lgt1, Lgt2, and Lgt3 (Michard and Doublet 2015). Additionally, phosphorylation of host chaperone protein Hsp70 by the *Legionella* eukaryotic-like gene K4 (LegK4), an effector kinase, causes Hsp70 to stall and further lowers the translation rate (Moss et al. 2019). These mechanisms appear to primarily target the host immune response, but may also potentiate the cell for metabolic rewiring (Michard and Doublet 2015). The rewiring process and *L. pneumophila's* wide diversity of post-translational mechanisms for influencing host gene expression are discussed in the next section. Translation inhibition is essential for the establishment of *L. pneumophila* long-term, as the S-phase of the host's cell cycle is lethal to the bacterium, and blocking translation triggers cell cycle arrest (Sol et al. 2019). Fascinatingly, this attribute may be a side effect of *Legionella's* history of association with free-living amoebae that live in oligotrophic bodies of water, which likely enter S-phase infrequently due to nutrient limitation (de Jesús-Díaz et al. 2017).

From the existing literature, it appears that mutualistic bacteria are unlikely to target host translation for two reasons: first, inhibiting translation induces strong antimicrobial responses and second, the genomes of these bacteria likely do not encode the necessary machinery. Given that all viruses hijack protein translation and many pathogens secrete effector proteins to inhibit translation, hosts have evolved signaling mechanisms to detect this perturbation and induce apoptosis (McEwan et al. 2012; Mohr and Sonenberg 2012; Cornejo et al. 2017). Thus, it is likely in the best interest of a symbiont whose strategy is to live in harmony with its host to not interfere with protein translation. Sensitivity to translational inhibition may also underlie why we were unable to find examples of translation-based symbiont-induced host cellular differentiation. Furthermore, the limited genomic coding capacity of these bacteria suggests that they do not encode the proteins necessary to do so. For example, many of these bacteria have lost a subset of their tRNA genes, and instead rely on codon wobble to pair all 61 codons. Furthermore, the 3'-CCA sequence has been lost from many of the tRNAs that remain in the genome and must be added on post-transcriptionally (Hansen and Moran 2012). Thus, these bacteria are ill equipped to manipulate host translation.

5.3.5 Post-translational Modification of Host Genetic Regulatory Components

In both eukaryotes and bacteria, protein activity, stability, and physical location are easily altered through post-translational modifications such as phosphorylation, acetylation, methylation, and glycosylation (Macek et al. 2019). Eukaryotes have many more modifications, some of which can be applied to bacterial proteins in host cells, such as prenyl groups for lipidation and membrane attachment (Al-Quadan et al. 2011). While the mechanism of protein modification is simple—a functional group is covalently bound to a protein—the downstream impacts of protein modifications can be quite complex. For example, ubiquitination can either lead to proteasomal protein degradation or the induction of signaling cascades, depending on the lysine residue ubiquitinated and how many ubiquitins are added (Haglund and Dikic 2005). Amazingly, despite their differences in endogenous post-translational modification capacities, many bacterial symbionts have evolved their own proteins for adding and removing eukaryotic protein modifications such as ubiquitin (Ribet and Cossart 2010; Rolando and Buchrieser 2014; Zhou and Zhu 2015).

One of the most common reasons for symbionts to manipulate host protein modifications is to alter the metabolic balance of the cell to create a nutritive niche. A straight-forward strategy to accomplish this is to increase protein proteolysis via the host's ubiquitination pathway. Short peptides and amino acids alone can go a long way toward meeting a symbiont's complete nutritional needs because many bacteria can use amino acids as both nitrogen and carbon sources (Zhang and Rubin 2013). Using eukaryotic cellular machinery, three enzymes are needed to ubiquitinate a protein, targeting it for degradation by the proteasome: a ubiquitin-activating enzyme (E1), a ubiquitin-conjugating enzyme (E2), and a ubiquitin-ligating enzyme (E3). These three protein functional classes are not equally represented in eukaryotic genomes, with there being only a few E1 enzymes, several dozen E2 enzymes, and hundreds of E3 enzymes (Zhou and Zhu 2015). Mechanisms of bacterial interference in host ubiquitination have evolved to mirror the host's pattern of protein diversity: the vast majority of mechanisms involve bacterial protein mimics or new versions of E3 enzymes, whereas E1 and E2 inhibitory mechanisms are less common (reviewed in (Zhou and Zhu 2015)). Some pathogens, such as *Legionella*, have even evolved novel mechanisms of ubiquitination that do not involve the E1 or E2 enzymes or ATP (Qiu et al. 2016).

As with many pathways, the ubiquitination pathway overlaps with immune and general signal transduction, making it a large target for bacterial interference. During the infection process, intracellular bacteria first have to deal with host ubiquitination to evade the innate immune system. Direct ubiquitination of intracellular pathogen membranes with host Parkin E3 ligase marks them for xenophagy (Manzanillo et al. 2013). In the event that this mechanism is insufficient, the host perceives symbiont-induced manipulations that interrupt protein synthesis or increase proteolysis, resulting in an excess of ubiquitinated proteins and amino acids in the cytoplasm. General autophagy is induced in this event, if the bacteria do not interrupt the process by reducing the number of ubiquitinated proteins with bacterially encoded deubiquitinating enzymes (Zhou and Zhu 2015). Once the threat of ubiquitin-mediated xenophagy has been ameliorated, symbionts can alter patterns of ubiquitination to trigger changes in host gene expression, which further alter immune responses and shape the cellular niche. This process is illustrated by the obligate intracellular pathogen *Chlamydia*. This bacterium uses its ChlaDub1 effector protein to deubiquitinate β-catenin, preventing its degradation and enabling its transport to the nucleus where it serves as a transcription factor to activate genes invoking cell proliferation, nuclear factor kappa-light-chain-enhancer of activated B cells (NF-κB) signaling, and apoptosis inhibition (Rogan et al. 2019). Given the importance of ubiquitination in normal host biology, it is not surprising that intracellular bacteria have evolved to interact with ubiquitin and its mechanisms for addition and removal.

In addition to ubiquitination, other post-translational modifications are often used by bacterial symbionts to control host gene expression and cellular differentiation (in this case, to create a nutritive niche). The pathogen of amoebas, lung macrophages, and neutrophils, *Legionella pneumophila*, is an excellent example of a bacterium that has become proficient at altering host post-translational protein modifications to metabolically rewire the host cell (see Fig. 5.1g). Within hours of entering a new host cell, *L. pneumophila* induces changes in the cell that cause the *Legionella*-containing vacuole (LCV) to become coated in smooth membrane derived from the endoplasmic reticulum that is lined with mitochondria and ribosomes. It accomplishes these tasks through a diverse array of nearly 300 effector proteins that are able to phosphorylate, alkylate, ubiquitinate, glycosylate, AMPylate, and phosphocholinate host proteins (Michard and Doublet 2015). Furthermore, it is able to co-opt host proteins to perform additional modifications on its own proteins, such as prenylation (e.g., farnesylation) (Al-Quadan et al. 2011). Interestingly, interaction with the endoplasmic reticulum to form a replicative niche is common among pathogens, such as the alphaproteobacterium *Brucella abortus* and the Chlamydiales bacterium *Simkania negevensis* (Cornejo et al. 2017) and is also altered in host-derived bacteriocytes that house mutualistic bacteria (Simonet et al. 2018).

To induce the formation of the LCV, *L. pneumophila* secretes a range of effector proteins into the host cytoplasm to either post-translationally modify host proteins or be post-translationally modified by them. The host-derived membrane surrounding *L. pneumophila* is first altered by the addition of endoplasmic reticulum-derived smooth vesicles, which are directed toward the forming LCV by inactivation of host GTPase Rab1 via adenylation by the effector SidM. Interestingly, *L. pneumophila* secretes two other effectors, SidB and LepD, that antagonize SidM adenylation, as well as one effector, AnkX, that can independently maintain Rab1 in the active state (Michard and Doublet 2015). This genetic redundancy suggests that this step is essential to LCV formation. As this is occurring, the AnkB effector co-opts host machinery to farsynlate AnkB, enabling it to attach to the LCV membrane. Once attached, the F-box E3-ligase interacting domain of AnkB recruits host ubiquitin ligase complexes to the membrane where together they attach ubiquitins to the membrane underlying the bacteria. The dense polyubiquitinated clusters attract the host proteasome, which proceeds to degrade ubiquitinated proteins and provide amino acids for bacterial nutrition (Bruckert et al. 2014). Simultaneously, epigenetic changes are also induced to increase the availability of ribosomes to embed in the LCV membrane. The LegAS4 effector confers increased transcription of host rDNA via functioning as a lysine histone methyltransferase through its SET domain (Rolando et al. 2013). Thus, with *L. pneumophila*, we come full circle in our classification of symbiont-induced host differentiation because through post-translational modification of host histones, these bacteria are able to influence host gene expression at the epigenetic DNA level.

While obligately intracellular mutualists have not yet been reported to influence host post-translational protein modifications, data from symbiotically derived organelles suggest some will have this capability, but may have different functions for it within host cells relative to pathogens. Two pieces of evidence support the idea that obligate mutualists may be able to post-translationally modify host proteins. First, the mitochondrial genome has retained genes capable of making post-translational modifications (Gabaldón and Huynen 2007). Second, both mitochondrial proteins encoded by the mitochondrial genome as well as those transferred to the nuclear genome have been shown to be post-translationally modified via phosphorylation, acetylation, and succinylation, indicating that these processes can occur within and by the organellar genome (Hofer and Wenz 2014). However, there may be striking differences between patterns of symbiont-induced host post-transcriptional modifications between mutualists and pathogens. For example, with regard to ubiquitination, amino acid economies are vastly different between pathogenic infections that usurp them from the host (Zhang and Rubin 2013) and mutualistic infections that synthesize them for the host (Feng et al. 2019). Thus, if mutualists are capable of altering host ubiquitination, they may be more likely to use it to control host signaling cascades than to obtain amino acids.

5.3.6 Trends in Symbiont-Mediated Host Cellular Differentiation Mechanisms

From the examples of symbiont-mediated host cellular differentiation described above, it is clear that bacteria are capable of manipulating host gene expression at every step in the process. Some symbionts can induce host epigenetic alterations that impact the access of transcriptional machinery to chromatin. Many taxa can interfere in transcriptional signaling cascades or transcription factor binding. An abundance of symbionts, including obligate intracellular mutualists, can modify mRNA retention by utilizing the similarities between bacterial sRNA and eukaryotic miRNA pathways. A limited range of pathogens can inhibit translation through the use of toxins and effector proteins. And, lastly, a number of pathogens use effector molecules to post-translationally modify host proteins. Impressively, these host-associated bacteria as a whole are not only able to use their own endogenous regulatory elements to control host gene expression, but they have also repeatedly evolved mechanisms for interacting with elements they do not have in their own genomes, such as histones and ubiquitination machinery.

Looking across this wide diversity of associations, both functionally and taxonomically, a few trends stand out that may reflect shared evolutionary constraints and pressures. First, bacterial symbionts tend to interact with differentiation proteins and pathways that are also involved in innate immune signaling. This may reflect the history of their interactions with their hosts. Symbionts must first evolve strategies to work with the host immune system before they evolve more complex phenotypes.

Given that there is a high degree of overlap between immunological pathways and developmental pathways (Cheng et al. 2010), evasion of the immune system may have exapted, or prepared, symbionts to interact with host cellular differentiation pathways. Thus, symbionts have likely evolved the ability to manipulate new host gene regulatory pathways through cross-talk between pathways (Fig. 5.3a).

The second trend that stands out in these examples is that symbiont genome evolution heavily influences the mechanisms available to the symbiont to control host gene expression (Fig. 5.3b). Some symbionts have evolved mechanisms to interfere with host gene expression at every step, from DNA to mRNA to protein (e.g., *Listeria* (Sesto et al. 2014)). Whereas, other symbionts, especially those with degraded genomes, use only one or a few mechanisms. Genome degradation has proceeded far enough in some bacteria, such as the *Nasuia* and *Sulcia* symbionts of leafhoppers with 0.11 and 0.19 Mb genomes, respectively, that control of essential symbiont cellular processes has been ceded to the host (Mao et al. 2018). In these instances, it seems unlikely that the symbionts retain much capacity to manipulate their hosts. However, as many of the host nuclear genes used to maintain symbiont cellular functions were acquired through ancient horizontal gene transfer events from other bacteria (Husnik et al. 2013; Husnik and McCutcheon 2017), it is clearly not straightforward to say who is in control of who in some of these associations.

The temporal and spatial extent of genetic influence may be a factor in constraining what symbiont-mediated host regulatory mechanisms can evolve - mutualists need to live in their organs/tissues/cells for a long time and form large population sizes (discussed in (Russell and Cavanaugh 2017)), whereas pathogens only need to be there to replicate. Due to the intervening steps, the time to reach a protein-coding effect is much longer for an epigenetic alteration than it is for a post-translational modification, which is nearly instant (Hausser et al. 2013; Sasai et al. 2013; Shamir et al. 2016). Thus, the third trend from the data is that symbiont mechanisms for controlling host gene expression correspond to the organismal scale they are trying to influence (cells, tissues, or organs) and the expected duration of the association (days, weeks, years, or lifetimes) (Fig. 5.3c). Pathogens with highly virulent and acute infection profiles (e.g., *Legionella*, *Salmonella*, *Vibrio,* and *Chlamydia*) implement a diversity of strategies, and are far more dependent on fast-acting, targeted mechanisms such as blocking protein translation or altering post-translational protein modifications within each infected cell. Whereas more chronic types of infection (e.g., *Mycobacterium leprae* and *Helicobacter pylori*) use mechanisms higher up in the gene expression hierarchy, evoking epigenetic and transcriptional control of host gene expression to permanently alter cell fate across tissues. These mechanisms also enable many mutualistic associations (e.g., aphids with *Buchnera*), and the occasional pathogenic association (e.g., *Agrobacterium*) to develop novel symbiont-housing cells, tissues, and organs.

The fourth and final trend from these data is that selection to control host cellular differentiation has driven the evolution of entirely novel proteins and molecular mechanisms. These novel elements conceptually fall in four categories depending on whether bacteria are mimicking host proteins and/or mechanisms to manipulate host gene expression: (1) both host proteins and mechanisms are mimicked, (2) host

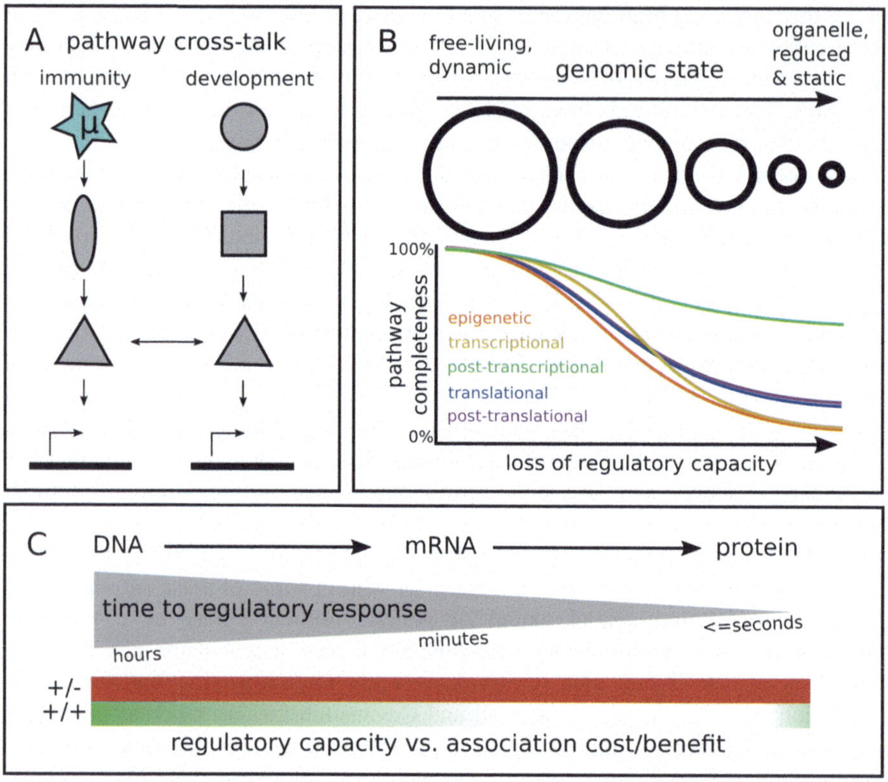

Fig. 5.3 Trends in the distribution of mechanisms for symbiont manipulation of host cellular differentiation. (**a**) Cross-talk between immunological and developmental pathways due to shared components (Cheng et al. 2010) may enable bacterial symbionts (blue star) to develop novel mechanisms of host regulation, such as symbiont-induced cellular differentiation. (**b**) Genetic regulatory capabilities are related to the state of genome erosion in bacterial symbionts. The theory of bacterial endosymbiont genome evolution posits that upon host restriction, bacterial chromosomes begin degrading due to the accumulation of deleterious mutations and the subsequent deletion of pseudogenized regions. This occurs because selection is ineffective in small, host-associated populations. The transmission bottleneck that occurs when a subset of symbionts are transmitted to offspring in vertically transmitted associations further contributes to genetic drift driving the evolution of these genomes (Toft and Andersson 2010). Based upon the reported coding capacities and mechanisms discussed here, we propose this approximate model for the retention/loss of regulatory capacity at each regulatory level during genome erosion. (**c**) Mechanisms of symbiont-induced host differentiation correlate with the cost/benefit trade-off of the association (depicted in red/green above, respectively) potentially due to temporal constraints. For example, virulent pathogens require fast acting mechanisms to circumvent clearance by the host immune system. Protein regulation generates a quicker response than altering host epigenetics or transcription does (Hausser et al. 2013; Sasai et al. 2013; Shamir et al. 2016). Thus, many pathogens likely first evolved to work with these mechanisms. Although, many have subsequently picked up additional mechanisms

mechanisms are mimicked using unique proteins, (3) host protein mimics are used in unique mechanisms, or (4) both the protein and mechanism are novel (Zhou and Zhu 2015). For example, SET domains fall in the first category, as these mimic eukaryotic lysine histone methyltransferase in form and function, but evolved in bacteria (Alvarez-Venegas 2014). The AnkX effector of *Legionella* is an excellent example of the second category, as it contains a conserved FIC protein domain that enacts a novel post-translational modification, phosphocholination, to modulate host Rab protein activity (Mukherjee et al. 2011). The OspF protein, produced and secreted by *Shigella flexneri*, exemplifies the fourth category, as it is a novel protein that irreversibly dephosphorylates mitogen-activated protein kinase (MAPK) via a unique mechanism, which permanently prevents MAPK from phosphorylating histones for immune gene activation (Cornejo et al. 2017).

Interestingly, some hosts are able to induce some symbiotic bacteria to undergo a differentiation-like process that changes their gene expression globally and often permanently. Examples exist from both mutualists and pathogens. In mutualistic rhizobia root infections, some plants induce their symbionts to terminally differentiate, turning them into highly polyploid, often branching cells that cannot divide again. Host plants appear to accomplish this by delivering a diversity of nodule-specific symbiotic peptides, which are similar to antimicrobial peptides, to intracellular rhizobia (Maróti and Kondorosi 2014). In pathogenic *Chlamydia* infections, host cells starve the intracellular bacteria of amino acids while the bacteria replicate in their active form, termed reticulate bodies. Once amino acids become unavailable, reticulate bodies convert into aberrant bodies with low metabolic rates, which cannot always be reactivated (Zhang and Rubin 2013). These two examples suggest that symbiont metabolic activities and cell division rates can be manipulated by host actions. As more data are collected for symbiotic associations, especially from single cell transcriptomes and proteomes, it will be interesting to see if other symbionts enter these or additional types of differentiated states.

5.4 A Natural Aptitude for Host Manipulation: The Intracellular Symbiont *Wolbachia*

The obligately intracellular alphaproteobacterium *Wolbachia* is a ubiquitous infection among arthropod and filarial nematode species. Interest in this group has increased in the past couple of decades due to discoveries that have made it suitable as a biological control agent for mosquito populations (Zheng et al. 2019) and their transmissible viruses (Hedges et al. 2008). This maternally inherited bacterium has achieved high frequencies within and among species through a combination of reproductive manipulation (Werren et al. 2008) and/or mutualism (Gill et al. 2014; Newton and Rice 2019). *Wolbachia's* reproductive phenotypes include feminization, male-killing, cytoplasmic incompatibility, and parthenogenesis, all which manipulate embryogenesis to increase the frequency of infected females in the population (Werren et al. 2008). However, *Wolbachia's* capacity for host manipulation does not end there. Even the cases of apparent "mutualism" in *Wolbachia* may have evolved through the manipulative complementation of host cellular and molecular pathways. In contrast to many mutualistic symbionts that imparted novel functions to the host upon their association, many of *Wolbachia's* mutualistic functions, from apoptosis inhibition (Pannebakker et al. 2007) to oogenesis (Dedeine et al. 2005), involve processes native to the host cell, which the host's ancestors were capable of accomplishing. Thus, *Wolbachia* mutualisms may be more accurately described as "addictive mutualisms" (Sullivan 2017). Clearly, *Wolbachia* is capable of a broad spectrum of host manipulations, which suggests that it encodes a rich diversity of genes and pathways to interact with host gene expression.

5.4.1 Known Wolbachia-Induced Host Reproductive Phenotypes and Mechanisms

Many *Wolbachia*-induced phenotypes occur during host development, and often take place in the germline stem cell, suggesting that this bacterium is able to influence host cellular differentiation. Animal development consists of a series of programmed cell division, migration, and differentiation cascades that create and pattern the adult organism (De Smet and Beeckman 2011). The ability to interact with these processes early-on obviates the need to first dedifferentiate adult host cells, as has been more frequently reported for bacterial pathogens and mutualists acquired from the environment (Wessler and Backert 2008; Masaki et al. 2013; Oldroyd 2013). This is likely due to the differences in transmission mode between these taxa, with vertically inherited *Wolbachia* being present throughout development, opposed to horizontally transmitted pathogens that get taken up by a fully differentiated adult host. Being present in the zygote (Callaini et al. 1994; Albertson et al. 2009; Fast et al. 2011), *Wolbachia* only needs to maintain stem cell status or guide the differentiation process to produce the intended cell type or molecular outcome. This is a skill *Wolbachia* has become adept at, as the following examples illustrate.

Often present in host germline stem cells (Russell et al. 2019), *Wolbachia* has been shown to be capable of rescuing or maintaining this cell lineage in different host taxa. In *D. melanogaster*, the wMel strain of *Wolbachia* can rescue mutations in the germline stem cell maintenance genes *sex lethal* (*sxl*) (Starr and Cline 2002; Sun and Cline 2009) and *bag of marbles* (*bam*) (Flores et al. 2015). In uninfected flies, both of these genes cause sterility in homozygous females due to the loss of germline stem cell maintenance, resulting in tumorous, over-proliferated ovaries. Infection with wMel restores the normal ovary phenotype. While it has not yet been shown whether the rescue of these genes involves one or two bacterially encoded processes, one wMel protein, toxic manipulator of oogenesis (TomO), has been identified that is capable of rescuing part of the phenotype resulting from the loss of *sxl*. TomO is able to maintain host germ cells, preventing their differentiation and loss, by increasing the expression of the germ cell maintenance protein Nanos via binding to *nanos* mRNAs localized within host ribonucleoprotein (RNP) complexes (Ote et al. 2016). Consistent with this mechanism, *Wolbachia* has been reported to interact with other components of host RNPs, such as the protein Gurken (Serbus et al. 2011).

While these germline stem cell maintenance genes are functional in wild-type flies, a scenario could exist in which a *Wolbachia*-infected population goes through a bottleneck and fixes a loss of function allele in the population, converting *Wolbachia* into an "obligate" infection. Wasp species in the genus *Asobara* are potentially an example of this situation. *Asobara tabida* hosts an obligate *Wolbachia* infection that is required for oogenesis, as wasps are unable to reproduce when treated with antibiotics against *Wolbachia*. This appears to have been a very recent occurrence, as all the closely related hymenopteran species do not require *Wolbachia* for reproduction (Dedeine et al. 2005). A similar situation has also been reported for the date stone beetle, *Coccotrypes dactyliperda* (Zchori-Fein et al. 2006). Over time, if a *Wolbachia*-dependent host diversifies and speciates, this process will produce a taxon entirely dependent on these seemingly mutualistic bacteria. This may be what occurred in the filarial nematode lineage. Nearly all of these parasitic worms harbor *Wolbachia* infections that are required for reproduction, development, and survival (Landmann et al. 2011). The requirement for reproduction appears to stem from *Wolbachia's* ability to maintain quiescence in the female germline stem cell, preventing the expression of differentiation-inducing genes, and preserving its totipotency (Foray et al. 2018).

Many *Wolbachia* strains, especially those found in lepidopterans and isopods, are adept at manipulating the sex-determination systems of their hosts, turning genetic males into females (Werren et al. 2008). The induction of sex-specific gene expression across animal cells during development requires two versions of each differentiation pathway that lead to cell types with male or female-specific characteristics. Animals use cell autonomous and hormonal, nonautonomous, mechanisms to control the sex-specific gene expression profiles of their cells. Thus, both mechanisms are targets for *Wolbachia*-control of host sex-specific gene expression (Negri and Pellecchi 2012). Given *Wolbachia's* ability to influence host hormone signaling and the overlap between hormone and epigenetic pathways, it has been suggested that *Wolbachia* may have epigenetic mechanisms for controlling host gene expression (discussed in (Negri 2012)). Consistent with this, *Wolbachia* inhibits the expression of the masculinizing gene *masc* in the adzuki bean borer moth *Ostrinia scapulalis*. As Masc controls both male-specific splicing and activation of dosage compensation in males, inhibition of this gene results in both female features and mortality, respectively (Sugimoto et al. 2010; Fukui et al. 2015). Similarly, in the leafhopper *Zyginidia pullula*, feminized males exhibit female DNA methylation patterns, whereas males with low *Wolbachia* titer exhibit incomplete feminization and male methylation patterns (Negri et al. 2009). While the full mechanisms underlying these phenotypes are not known, it is interesting to note that the *Wolbachia* genome contains a DNA adenine methyltransferase encoded on a prophage (Saridaki et al. 2011). Furthermore, a bacterially induced epigenetic mechanism is reasonable given that many sex-specific differentiation pathways are epigenetically controlled, regardless of the sex-determining mechanism (Piferrer 2013).

In an alternative strategy to feminization, some *Wolbachia* strains kill host males during embryogenesis to alter host sex ratios to favor females. Recent work by (Perlmutter et al. 2019) suggests that in *Drosophila*, the *Wolbachia* infections that cause male-killing may do so via *Wolbachia's* WO phage-encoded *WO-mediated killing* (*wmk*) gene. This DNA-binding gene causes overexpression of the host dosage compensation system at male X chromosomes, resulting in hyperacetylation at histone H4 lysine 16, DNA damage, defects in chromatin remodeling, and altered spindle organization (Riparbelli et al. 2012; Harumoto et al. 2018; Perlmutter et al. 2019). This result is similar yet distinct from the mechanism employed by *Spiroplasma* in *D. melanogaster* (Harumoto and Lemaitre 2018), as *Wolbachia* does not induce alterations the dosage compensation system's localization among chromosomes (Perlmutter et al. 2019). Male-killing exhibits variable penetrance in different hosts, bacterial genomic backgrounds, and environmental contexts. For example, *wmk* does not induce male-killing in natural wMel infections in *D. melanogaster*, despite it causing the phenotype when expressed heterologously in uninfected *D. melanogaster.* Furthermore, the wMel *wmk* sequence is nearly identical to the ortholog from the wRec strain, which causes male-killing when wRec infects the sister species (*Drosophila subquinaria*) of its native host (*Drosophila recens;* (Jaenike 2007)). Regarding environmental variability, the wBif strain that infects *Drosophila bifasciata* exhibits high rates of male-killing at low temperatures and low rates at high temperatures (Hurst and Johnson 2000). Given how costly male-killing is to host fitness (eliminates half of all progeny), the variability in male-killing penetrance described above and the similarity of its mechanism to that of feminization (via the dosage compensation system) suggests that male-killing could be a polygenic phenotype that results when a more fitness-conserving mode of manipulation (e.g., feminization) goes wrong.

The reproductive manipulation termed cytoplasmic incompatibility (CI) involves bacterial modifications of host gamete chromatin packaging, suggesting that this is another example of *Wolbachia* using an epigenetic-like mechanism to control the outcome of host reproduction. CI is a bacterially induced mating incompatibility between infected males and uninfected females, or females with an incompatible strain of *Wolbachia.* Reproduction between these hosts fails during embryogenesis because modifications made to the sperm by *Wolbachia* fail to be compensated for in the eggs. It has been known for some time that the modifications made by *Wolbachia* result in the male pronucleus exhibiting delayed protamine removal and histone deposition in the zygote, which results in mortality at the first mitosis (Landmann et al. 2009). Recent work has revealed the bacterially encoded genes underlying these chromatin modifications. In infected males, *Wolbachia* uses the prophage-encoded deubiquitinase CI-factor (Cif) B and its binding partner CifA (also termed CidA/B) (Beckmann et al. 2017; LePage et al. 2017) to alter sperm chromatin. CifB appears to confer these effects through binding to host nuclear import factor karyopherin-a and P32 protamine-histone exchange factor, which may either prevent histone assembly components from reaching the paternal chromosomes or reduce the efficacy of histone assembly (Beckmann et al. 2019). Expression of CifA in the female germline is necessary and sufficient to compensate for the CifA–CifB induced chromatin alterations made to the male sperm by *Wolbachia* (Shropshire et al. 2018). Thus, CI induction and rescue functions like a toxin-antidote system.

5.4.2 *Other Known Strategies of* Wolbachia*-Mediated Control of Host Gene Expression*

In addition to these bacterial mechanisms of controlling host gene expression that are tied to reproductive manipulations in the host, other mechanisms have been proposed for *Wolbachia's* more general processes of survival and persistence. Compared to the above examples that were primarily focused on epigenetic or post-translational mechanisms of host genetic regulation, the following examples highlight a wider diversity of mechanisms.

To date, two studies suggest that *Wolbachia* can interfere with host translation through using its own as well as the host's transcription factors. The strain of *Wolbachia* found in *Culex molestus* mosquitoes encodes the transcriptional regulator gene *wtrM* that appears to act as a host transcription factor, upregulating the meiotic gene *grauzone*. While *grauzone* expression correlates with CI strength in the *Wolbachia* variants tested, it is not clear how increased *grauzone* expression impacts this phenotype or others (Pinto et al. 2013). In *Aedes aegypti* mosquitoes, *Wolbachia* induces expression of the host transcription factor GATA4, which suppresses expression of the host ovary-specific genes *blastoderm-specific protein 25D* (*bsg25D*) and *imaginal disc growth factor* (*disc*) (Osei-Amo et al. 2018). Given *Wolbachia's* propensity to associate with the germline (Fast et al. 2011), high rates of vertical transmission through oocytes (Narita et al. 2007), and various rescue capabilities in germ stem cells (discussed above), the annotations of these genes suggest that they may be involved in creating or maintaining *Wolbachia's* niche in the female germline.

Abundant evidence exists that *Wolbachia* is able to interact with host post-transcriptional regulation through the host miRNA pathway. In *Aedes aegypti*, *Wolbachia* expresses its own sRNAs that are exported into the host cell and regulate host mRNAs. For example, *Wolbachia's* WsnRNA-46 sRNA has been shown to increase the expression of the host motor protein dynein (Mayoral et al. 2014). Additionally, *Wolbachia* has been shown to alter host miRNA expression in *Aedes aegypti*, which impacts the expression of host protein-coding genes. For example, *Wolbachia* increases the expression of host miRNA aae-miR-2940, causing the upregulation of a host metalloprotease needed for normal infection (Hussain et al. 2011). This miRNA also downregulates host DNA cytosine methyltransferase, AaDnmt2, causing methylation to be reduced genome-wide. Interestingly, while inhibition of this miRNA is necessary for *Wolbachia* infection, its inhibition also confers inhibition of *Flavivirus* replication within infected cells (Zhang et al. 2013). In contrast, and potentially suggesting different mechanisms in different hosts or with different viruses, *Wolbachia*-induced upregulation of *D. melanogaster*

DNA/RNA methyltransferase was shown to inhibit replication and infectivity of the alphavirus, Sindbis virus (Bhattacharya et al. 2017). *Wolbachia* has also been shown to upregulate aae-miR-981, which downregulates the expression of importin β-4, prohibiting AGO1 from entering the nucleus to regulate transcription (Hussain et al. 2013).

To obtain a reliable source of host amino acids, *Wolbachia* appears to have evolved mechanisms to interfere with their sink and their source, i.e., translation and proteolysis, similar to the pathogens discussed above. A recent cell-based genome-wide RNAi screen in *D. melanogaster* cells infected with the wMel strain of *Wolbachia* found that bacterial density, or titer, increases when host ribosomal and translation initiation proteins are knocked down. This suggests that *Wolbachia* interacts with some of these factors in wild-type cells to alter host translation (Grobler et al. 2018). This is fascinating given the trends we reported in the previous section, which found that generally only highly virulent pathogens interfere with host translation. Supporting a role for translation interference in *Wolbachia* nutrition, this (Grobler et al. 2018) and another cell screen (White et al. 2017), found that *Wolbachia* titer decreased when host ubiquitination was inhibited. Furthermore, White et al. (2017) found that *Wolbachia* infection significantly increases ubiquitination levels in the host cell. Thus, *Wolbachia* may alter host protein synthesis as well as ubiquitination-mediated proteolysis to obtain amino acids as their primary source of nutrition. Consistent with using host protein synthesis and degradation pathways for its own nutrition, *Wolbachia* induces the reorganization of host cell endoplasmic reticulum (ER) and surrounds itself with ER-derived membrane (Fattouh et al. 2019), creating a niche near translation and proteolysis machinery. Given that ubiquitination and protein turnover is involved in host cellular differentiation (Kimata 2019), *Wolbachia* may have co-opted its nutrition-provisioning genes for host manipulation. To take the idea of molecular cross-talk in *Wolbachia* associations a step further, it is possible that *Wolbachia's* ability to modify host protein ubiquitination was first co-opted from strategies originally evolved for evading xenophagy (e.g., Manzanillo et al. 2013; Zhou and Zhu 2015).

5.4.3 Exploring Overlooked Mechanisms: Future Prospects in Wolbachia Research

We surveyed the literature for studies that assayed the impact of infection on gene expression in *Wolbachia* and/or its host and found 71 papers published between 2000 and 2019 (Table 10.S1 and Fig. 5.4). These studies characterized gene expression at all stages, from DNA to protein, and suggest that *Wolbachia* has mechanisms to interfere with host gene expression at many points in the process. Transcription-based studies were over-represented relative to the other gene expression stages, which is likely due to how easy generating transcriptomic data has become since the advent of microarrays and RNAseq. Future work should focus on identifying other

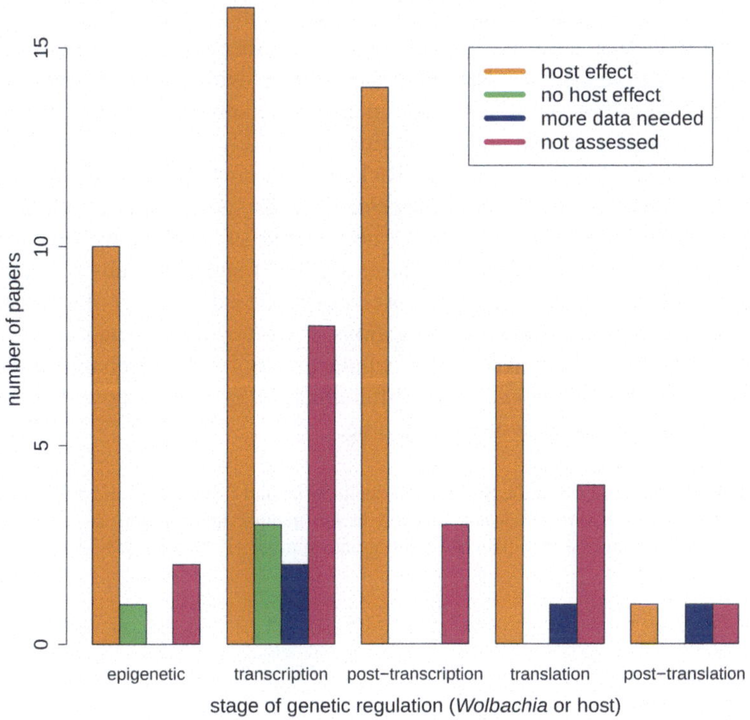

Fig. 5.4 Distribution of existing literature addressing gene expression in *Wolbachia* and/or its hosts. See Table 10.S1 for the full list of papers included here. The excess of papers studying transcription relative to the other stages of regulation reflects the ease with which transcriptomic data can be acquired since the advent of microarrays and Illumina sequencing. Effect = study found *Wolbachia* infection to have an effect on host gene expression; no effect = study found no effect of *Wolbachia* infection on host gene expression; more data needed = results were ambiguous regarding *Wolbachia*'s influence on host gene expression; and not assessed = *Wolbachia's* impact on host gene expression was not assessed by the paper (indicated by "NA" in Table 10.S1)

Wolbachia-mediated post-translational modifications, as these have been studied the least. Furthermore, given the numerous examples of *Wolbachia*-induced miRNA regulation in mosquitoes discussed above, evidence for similar mechanisms should be investigated in other *Wolbachia* infections.

Although the *Wolbachia* field is still in its early days, with complete mechanisms underlying host-symbiont interactions just now being elucidated, the abundance of eukaryotic-like elements in the various *Wolbachia* strain genomes suggest a diversity of mechanisms are waiting to be discovered. These elements include deubiquitinating enzymes (Beckmann et al. 2017), ankyrin repeat proteins (Siozios et al. 2013), and proteins with dynamin domains (Rice et al. 2017). Given *Wolbachia's* known interactions with the host cytoskeleton, including microtubule-dependent motor proteins (Ferree et al. 2005; Serbus and Sullivan 2007; Russell et al. 2018), some of these proteins could mediate these interactions.

Indeed, a *Wolbachia* protein containing a synuclein domain that may mediate interactions with host actin has been characterized (Sheehan et al. 2016).

Wolbachia belongs to the Rickettsiales, a taxon with a long history of host-association, suggesting that it possesses ancient mechanisms for host manipulation. Indeed, the ancestor of the mitochondrion was likely a member of this taxon (Andersson et al. 2003) and today, Rickettsiales contains a wide diversity of pathogens, including species in *Rickettsia, Orientia, Anaplasma,* and *Ehrlichia*. These pathogens have been shown to be capable of modulating host immune responses via epigenetic (Garcia-Garcia et al. 2009) and post-translational (Sahni et al. 2018) modifications, and they themselves encode a diverse set of active sRNAs (Narra et al. 2016). Thus, future investigations of *Wolbachia* associations will likely reveal a wealth of information about the cellular and molecular mechanisms bacterial symbionts use to control host cellular differentiation, as well as how these mechanisms are maintained over evolutionary time.

Acknowledgements We thank the Sullivan, Corbett-Detig, and Cooper labs and Seth Bordenstein for thoughtful feedback on the ideas conveyed in the manuscript. Funding for this work was provided through a National Institutes of Health grant (number 1K99GM135583-01A1).

References

Albertson R, Casper-Lindley C, Cao J et al (2009) Symmetric and asymmetric mitotic segregation patterns influence *Wolbachia* distribution in host somatic tissue. J Cell Sci 122:4570–4583. https://doi.org/10.1242/jcs.054981

Al-Quadan T, Price CT, London N et al (2011) Anchoring of bacterial effectors to host membranes through host-mediated lipidation by prenylation: a common paradigm. Trends Microbiol 19:573–579. https://doi.org/10.1016/j.tim.2011.08.003

Alvarez-Venegas R (2014) Bacterial SET domain proteins and their role in eukaryotic chromatin modification. Front Genet 5:65. https://doi.org/10.3389/fgene.2014.00065

Amieva MR (2003) Disruption of the epithelial apical-junctional complex by helicobacter pylori CagA. Science 300:1430–1434. https://doi.org/10.1126/science.1081919

Andersson GE, Karlberg O, Canbäck B, Kurland CG (2003) On the origin of mitochondria: a genomics perspective. Philos Trans R Soc Lond Ser B Biol Sci 358:165–179. https://doi.org/10.1098/rstb.2002.1193

Arbibe L, Kim DW, Batsche E et al (2007) An injected bacterial effector targets chromatin access for transcription factor NF-κB to alter transcription of host genes involved in immune responses. Nat Immunol 8:47–56. https://doi.org/10.1038/ni1423

Au HH, Jan E (2014) Novel viral translation strategies: novel viral translation strategies. Wiley Interdiscip Rev RNA 5:779–801. https://doi.org/10.1002/wrna.1246

Bagnoli F, Buti L, Tompkins L et al (2005) Helicobacter pylori CagA induces a transition from polarized to invasive phenotypes in MDCK cells. Proc Natl Acad Sci 102:16339–16344. https://doi.org/10.1073/pnas.0502598102

Barry KC, Ingolia NT, Vance RE (2017) Global analysis of gene expression reveals mRNA superinduction is required for the inducible immune response to a bacterial pathogen. eLife 6: e22707. https://doi.org/10.7554/eLife.22707

Beckmann JF, Ronau JA, Hochstrasser M (2017) A *Wolbachia* deubiquitylating enzyme induces cytoplasmic incompatibility. Nat Microbiol 2:17007. https://doi.org/10.1038/nmicrobiol.2017.7

Beckmann JF, Sharma GD, Mendez L et al (2019) The Wolbachia cytoplasmic incompatibility enzyme CidB targets nuclear import and protamine-histone exchange factors. eLife 8:e50026. https://doi.org/10.7554/eLife.50026
Belasco JG (2010) All things must pass: contrasts and commonalities in eukaryotic and bacterial mRNA decay. Nat Rev Mol Cell Biol 11:467–478. https://doi.org/10.1038/nrm2917
Bennett GM, Chong RA (2017) Genome-wide transcriptional dynamics in the companion bacterial symbionts of the glassy-winged sharpshooter (Cicadellidae: *Homalodisca vitripennis*) reveal differential gene expression in bacteria occupying multiple host organs. G3amp58 GenesGenomesGenetics 7:3073–3082. https://doi.org/10.1534/g3.117.044255
Bertaux-Skeirik N, Feng R, Schumacher MA et al (2015) CD44 plays a functional role in helicobacter pylori-induced epithelial cell proliferation. PLoS Pathog 11:e1004663. https://doi.org/10.1371/journal.ppat.1004663
Bessède E, Staedel C, Acuña Amador LA et al (2014) Helicobacter pylori generates cells with cancer stem cell properties via epithelial–mesenchymal transition-like changes. Oncogene 33:4123–4131. https://doi.org/10.1038/onc.2013.380
Bhattacharya T, Newton ILG, Hardy RW (2017) Wolbachia elevates host methyltransferase expression to block an RNA virus early during infection. PLoS Pathog 13:e1006427. https://doi.org/10.1371/journal.ppat.1006427
Bierne H, Hamon M, Cossart P (2012) Epigenetics and bacterial infections. Cold Spring Harb Perspect Med 2:a010272–a010272. https://doi.org/10.1101/cshperspect.a010272
Blumenthal T, Evans D, Link CD et al (2002) A global analysis of *Caenorhabditis elegans* operons. Nature 417:851–854. https://doi.org/10.1038/nature00831
Bobrovskyy M, Vanderpool CK (2013) Regulation of bacterial metabolism by small RNAs using diverse mechanisms. Annu Rev Genet 47:209–232. https://doi.org/10.1146/annurev-genet-111212-133445
Borges RM (2017) Co-niche construction between hosts and symbionts: ideas and evidence. J Genet 96:483–489. https://doi.org/10.1007/s12041-017-0792-9
Botham CM, Wandler AM, Guillemin K (2008) A transgenic Drosophila model demonstrates that the helicobacter pylori CagA protein functions as a eukaryotic gab adaptor. PLoS Pathog 4: e1000064. https://doi.org/10.1371/journal.ppat.1000064
Braendle C, Miura T, Bickel R et al (2003) Developmental origin and evolution of bacteriocytes in the aphid–*Buchnera* symbiosis. PLoS Biol 1:e21. https://doi.org/10.1371/journal.pbio.0000021
Bruckert WM, Price CT, Abu Kwaik Y (2014) Rapid nutritional remodeling of the host cell upon attachment of legionella pneumophila. Infect Immun 82:72–82. https://doi.org/10.1128/IAI.01079-13
Callaini G, Riparbelli MG, Dallai R (1994) The distribution of cytoplasmic bacteria in the early Drosophila embryo is mediated by astral microtubules. J Cell Sci 107:673–682
Cashin P, Goldsack L, Hall D, O'Toole R (2006) Contrasting signal transduction mechanisms in bacterial and eukaryotic gene transcription. FEMS Microbiol Lett 261:155–164. https://doi.org/10.1111/j.1574-6968.2006.00295.x
Chauhan K, Kalam H, Dutt R, Kumar D (2019) RNA splicing: a new paradigm in host–pathogen interactions. J Mol Biol 431:1565–1575. https://doi.org/10.1016/j.jmb.2019.03.001
Cheeseman K, Weitzman JB (2015) Host-parasite interactions: an intimate epigenetic relationship: epigenetics in apicomplexa infections. Cell Microbiol 17:1121–1132. https://doi.org/10.1111/cmi.12471
Cheng P, Zhou J, Gabrilovich D (2010) Regulation of dendritic cell differentiation and function by Notch and Wnt pathways. Immunol Rev 234:105–119. https://doi.org/10.1111/j.0105-2896.2009.00871.x
Cheng B, Kuppanda N, Aldrich JC et al (2016) Male-killing Spiroplasma alters behavior of the dosage compensation complex during *Drosophila melanogaster* embryogenesis. Curr Biol 26:1339–1345. https://doi.org/10.1016/j.cub.2016.03.050

Cognat V, Morelle G, Megel C et al (2017) The nuclear and organellar tRNA-derived RNA fragment population in *Arabidopsis thaliana* is highly dynamic. Nucleic Acids Res 45:3460–3472. https://doi.org/10.1093/nar/gkw1122

Cole MF, Young RA (2008) Mapping key features of transcriptional regulatory circuitry in embryonic stem cells. Cold Spring Harb Symp Quant Biol 73:183–193. https://doi.org/10.1101/sqb.2008.73.027

Cornejo E, Schlaermann P, Mukherjee S (2017) How to rewire the host cell: a home improvement guide for intracellular bacteria. J Cell Biol 216:3931–3948. https://doi.org/10.1083/jcb.201701095

Dar D, Sorek R (2018) Extensive reshaping of bacterial operons by programmed mRNA decay. PLoS Genet 14:e1007354. https://doi.org/10.1371/journal.pgen.1007354

de Barsy M, Frandi A, Panis G et al (2016) Regulatory (pan-)genome of an obligate intracellular pathogen in the PVC superphylum. ISME J 10:2129–2144. https://doi.org/10.1038/ismej.2016.23

de Jesús-Díaz DA, Murphy C, Sol A et al (2017) Host cell S phase restricts *Legionella pneumophila* intracellular replication by destabilizing the membrane-bound replication compartment. mBio 8: e02345-16. https://doi.org/10.1128/mBio.02345-16

De Smet I, Beeckman T (2011) Asymmetric cell division in land plants and algae: the driving force for differentiation. Nat Rev Mol Cell Biol 12:177–188. https://doi.org/10.1038/nrm3064

Dedeine F, Boulétreau M, Vavre F (2005) Wolbachia requirement for oogenesis: occurrence within the genus Asobara (Hymenoptera, Braconidae) and evidence for intraspecific variation in *A. tabida*. Heredity 95:394–400. https://doi.org/10.1038/sj.hdy.6800739

Dunbar TL, Yan Z, Balla KM et al (2012) *C. elegans* detects pathogen-induced translational inhibition to activate immune signaling. Cell Host Microbe 11:375–386. https://doi.org/10.1016/j.chom.2012.02.008

Duraisingh MT, Horn D (2016) Epigenetic regulation of virulence gene expression in parasitic protozoa. Cell Host Microbe 19:629–640. https://doi.org/10.1016/j.chom.2016.04.020

Escobar MA, Dandekar AM (2003) Agrobacterium tumefaciens as an agent of disease. Trends Plant Sci 8:380–386. https://doi.org/10.1016/S1360-1385(03)00162-6

Fast EM, Toomey ME, Panaram K et al (2011) Wolbachia enhance Drosophila stem cell proliferation and target the germline stem cell niche. Science 334:990–992. https://doi.org/10.1126/science.1209609

Fattouh N, Cazevieille C, Landmann F (2019) Wolbachia endosymbionts subvert the endoplasmic reticulum to acquire host membranes without triggering ER stress. PLoS Negl Trop Dis 13:25

Feng H, Edwards N, Anderson CMH et al (2019) Trading amino acids at the aphid–*Buchnera* symbiotic interface. Proc Natl Acad Sci 116:16003–16011. https://doi.org/10.1073/pnas.1906223116

Ferree PM, Frydman HM, Li JM et al (2005) *Wolbachia* utilizes host microtubules and dynein for anterior localization in the *Drosophila* oocyte. PLoS Pathog 1:e14. https://doi.org/10.1371/journal.ppat.0010014

Flores HA, Bubnell JE, Aquadro CF, Barbash DA (2015) The *Drosophila* bag of marbles gene interacts genetically with *Wolbachia* and shows female-specific effects of divergence. PLoS Genet 11:e1005453

Fontana MF, Banga S, Barry KC et al (2011) Secreted bacterial effectors that inhibit host protein synthesis are critical for induction of the innate immune response to virulent legionella pneumophila. PLoS Pathog 7:e1001289. https://doi.org/10.1371/journal.ppat.1001289

Foray V, Perez-Jimenez MM, Fattouh N, Landmann F (2018) Wolbachia control stem cell behavior and stimulate germline proliferation in filarial nematodes. Dev Cell 45:198–211. https://doi.org/10.1101/244046

Frank AC (2019) Molecular host mimicry and manipulation in bacterial symbionts. FEMS Microbiol Lett 366:fnz038. https://doi.org/10.1093/femsle/fnz038

Fridmann-Sirkis Y, Stern S, Elgart M et al (2014) Delayed development induced by toxicity to the host can be inherited by a bacterial-dependent, transgenerational effect. Front Genet 5:27. https://doi.org/10.3389/fgene.2014.00027

Fukui T, Kawamoto M, Shoji K et al (2015) The endosymbiotic bacterium Wolbachia selectively kills male hosts by targeting the masculinizing gene. PLoS Pathog 11:e1005048. https://doi.org/10.1371/journal.ppat.1005048

Gabaldón T, Huynen MA (2007) From endosymbiont to host-controlled organelle: the hijacking of mitochondrial protein synthesis and metabolism. PLoS Comput Biol 3:e219–e219. https://doi.org/10.1371/journal.pcbi.0030219

Galán-Vásquez E, Sánchez-Osorio I, Martínez-Antonio A (2016) Transcription factors exhibit differential conservation in bacteria with reduced genomes. PLoS One 11:e0146901. https://doi.org/10.1371/journal.pone.0146901

Garcia-Garcia JC, Barat NC, Trembley SJ, Dumler JS (2009) Epigenetic silencing of host cell defense genes enhances intracellular survival of the Rickettsial pathogen Anaplasma phagocytophilum. PLoS Pathog 5:e1000488. https://doi.org/10.1371/journal.ppat.1000488

Gegner J, Baudach A, Mukherjee K et al (2019) Epigenetic mechanisms are involved in sex-specific trans-generational immune priming in the lepidopteran model host Manduca sexta. Front Physiol 10:137. https://doi.org/10.3389/fphys.2019.00137

Gill AC, Darby AC, Makepeace BL (2014) Iron necessity: the secret of Wolbachia's success? PLoS Negl Trop Dis 8:e3224. https://doi.org/10.1371/journal.pntd.0003224

Goll MG, Bestor TH (2005) Eukaryotic cytosine methyltransferases. Annu Rev Biochem 74:481–514. https://doi.org/10.1146/annurev.biochem.74.010904.153721

Grabiec AM, Potempa J (2018) Epigenetic regulation in bacterial infections: targeting histone deacetylases. Crit Rev Microbiol 44:336–350. https://doi.org/10.1080/1040841X.2017.1373063

Grobler Y, Yun CY, Kahler DJ et al (2018) Whole genome screen reveals a novel relationship between Wolbachia levels and Drosophila host translation. PLoS Pathog 14:e1007445. https://doi.org/10.1371/journal.ppat.1007445

Gu H, Zhao C, Zhang T et al (2017) Salmonella produce microRNA-like RNA fragment Sal-1 in the infected cells to facilitate intracellular survival. Sci Rep 7:2392. https://doi.org/10.1038/s41598-017-02669-1

Güell M, van Noort V, Yus E et al (2009) Transcriptome complexity in a genome-reduced bacterium. Science 326:1268–1271. https://doi.org/10.1126/science.1176951

Güell M, Yus E, Lluch-Senar M, Serrano L (2011) Bacterial transcriptomics: what is beyond the RNA horiz-ome? Nat Rev Microbiol 9:658–669. https://doi.org/10.1038/nrmicro2620

Gur E, Biran D, Ron EZ (2011) Regulated proteolysis in Gram-negative bacteria—how and when? Nat Rev Microbiol 9:839–848. https://doi.org/10.1038/nrmicro2669

Guven-Maiorov E, Tsai C-J, Nussinov R (2017) Structural host-microbiota interaction networks. PLoS Comput Biol 13:e1005579. https://doi.org/10.1371/journal.pcbi.1005579

Haglund K, Dikic I (2005) Ubiquitylation and cell signaling. EMBO J 24:3353–3359. https://doi.org/10.1038/sj.emboj.7600808

Hamon MA, Cossart P (2008) Histone modifications and chromatin remodeling during bacterial infections. Cell Host Microbe 4:100–109. https://doi.org/10.1016/j.chom.2008.07.009

Hannemann S, Gao B, Galán JE (2013) Salmonella modulation of host cell gene expression promotes its intracellular growth. PLoS Pathog 9:e1003668. https://doi.org/10.1371/journal.ppat.1003668

Hansen AK, Degnan PH (2014) Widespread expression of conserved small RNAs in small symbiont genomes. ISME J 8:2490–2502. https://doi.org/10.1038/ismej.2014.121

Hansen AK, Moran NA (2012) Altered tRNA characteristics and 3′ maturation in bacterial symbionts with reduced genomes. Nucleic Acids Res 40:7870–7884. https://doi.org/10.1093/nar/gks503

Harumoto T, Fukatsu T, Lemaitre B (2018) Common and unique strategies of male killing evolved in two distinct *Drosophila* symbionts. Proc R Soc B Biol Sci 285:20172167. https://doi.org/10.1098/rspb.2017.2167

Harumoto T, Lemaitre B (2018) Male-killing toxin in a bacterial symbiont of Drosophila. Nature 557:252–255. https://doi.org/10.1038/s41586-018-0086-2

Hatakeyama M (2006) The role of helicobacter pylori CagA in gastric carcinogenesis. Int J Hematol 84:301–308. https://doi.org/10.1532/IJH97.06166

Hausser J, Syed AP, Selevsek N et al (2013) Timescales and bottlenecks in miRNA-dependent gene regulation. Mol Syst Biol 9:711. https://doi.org/10.1038/msb.2013.68

Heath-Heckman EAC, Foster J, Apicella MA et al (2016) Environmental cues and symbiont microbe-associated molecular patterns function in concert to drive the daily remodelling of the crypt-cell brush border of the *Euprymna scolopes* light organ: inducers of daily brush-border remodelling. Cell Microbiol 18:1642–1652. https://doi.org/10.1111/cmi.12602

Hedges LM, Brownlie JC, O'Neill SL, Johnson KN (2008) Wolbachia and virus protection in insects. Science 322:702–702

Hofer A, Wenz T (2014) Post-translational modification of mitochondria as a novel mode of regulation. Exp Gerontol 56:202–220. https://doi.org/10.1016/j.exger.2014.03.006

Hongo Y, Nakamura Y, Shimamura S et al (2013) Exclusive localization of carbonic anhydrase in bacteriocytes of the deep-sea clam Calyptogena okutanii with thioautotrophic symbiotic bacteria. J Exp Biol 216:4403–4414. https://doi.org/10.1242/jeb.092809

Hurst GDD, Johnson AP (2000) Male-killing Wolbachia in Drosophila: a temperature-sensitive trait with a threshold bacterial density. Genetics 156:699–709

Husnik F, McCutcheon JP (2017) Functional horizontal gene transfer from bacteria to eukaryotes. Nat Rev Microbiol 16:67–79. https://doi.org/10.1038/nrmicro.2017.137

Husnik F, Nikoh N, Koga R et al (2013) Horizontal gene transfer from diverse bacteria to an insect genome enables a tripartite nested mealybug symbiosis. Cell 153:1567–1578. https://doi.org/10.1016/j.cell.2013.05.040

Hussain M, Frentiu FD, Moreira LA et al (2011) Wolbachia uses host microRNAs to manipulate host gene expression and facilitate colonization of the dengue vector *Aedes aegypti*. Proc Natl Acad Sci 108:9250–9255. https://doi.org/10.1073/pnas.1105469108

Hussain M, O'Neill SL, Asgari S (2013) *Wolbachia* interferes with the intracellular distribution of Argonaute 1 in the dengue vector *Aedes aegypti* by manipulating the host microRNAs. RNA Biol 10:1868–1875. https://doi.org/10.4161/rna.27392

Jaafar ZA, Kieft JS (2019) Viral RNA structure-based strategies to manipulate translation. Nat Rev Microbiol 17:110–123. https://doi.org/10.1038/s41579-018-0117-x

Jaenike J (2007) Spontaneous emergence of a new Wolbachia phenotype. Evolution 61:2244–2252. https://doi.org/10.1111/j.1558-5646.2007.00180.x

Jan E, Mohr I, Walsh D (2016) A cap-to-tail guide to mRNA translation strategies in virus-infected cells. Annu Rev Virol 3:283–307. https://doi.org/10.1146/annurev-virology-100114-055014

Kaletsky R, Moore RS, Parsons LL, Murphy CT (2019) Cross-kingdom recognition of bacterial small RNAs induces transgenerational pathogenic avoidance. bioRxiv:697888. https://doi.org/10.1101/697888

Kim D, Thairu MW, Hansen AK (2016) Novel insights into insect-microbe interactions—role of epigenomics and small RNAs. Front Plant Sci 7. https://doi.org/10.3389/fpls.2016.01164

Kimata Y (2019) APC/C ubiquitin ligase: coupling cellular differentiation to G1/G0 phase in multicellular systems. Trends Cell Biol 29:591–603. https://doi.org/10.1016/j.tcb.2019.03.001

Knip M, Constantin ME, Thordal-Christensen H (2014) Trans-kingdom cross-talk: small RNAs on the move. PLoS Genet 10:e1004602. https://doi.org/10.1371/journal.pgen.1004602

Kozak M (1992) Regulation of translation in eukaryotic systems. Annu Rev Cell Biol 8:197–225

Kremer N, Philipp EER, Carpentier M-C et al (2013) Initial symbiont contact orchestrates host-organ-wide transcriptional changes that prime tissue colonization. Cell Host Microbe 14:183–194. https://doi.org/10.1016/j.chom.2013.07.006

Kremer N, Koch EJ, El Filali A et al (2018) Persistent interactions with bacterial symbionts direct mature-host cell morphology and gene expression in the squid-Vibrio symbiosis. mSystems 3: e00165-18. https://doi.org/10.1128/mSystems.00165-18

Kusmierek M, Dersch P (2018) Regulation of host–pathogen interactions via the post-transcriptional Csr/Rsm system. Curr Opin Microbiol 41:58–67. https://doi.org/10.1016/j.mib.2017.11.022

La M-V, Raoult D, Renesto P (2008) Regulation of whole bacterial pathogen transcription within infected hosts. FEMS Microbiol Rev 32:440–460. https://doi.org/10.1111/j.1574-6976.2008.00103.x

Landmann F, Orsi GA, Loppin B, Sullivan W (2009) Wolbachia-mediated cytoplasmic incompatibility is associated with impaired histone deposition in the male pronucleus. PLoS Pathog 5: e1000343. https://doi.org/10.1371/journal.ppat.1000343

Landmann F, Voronin D, Sullivan W, Taylor MJ (2011) Anti-filarial activity of antibiotic therapy is due to extensive apoptosis after Wolbachia depletion from filarial nematodes. PLoS Pathog 7: e1002351. https://doi.org/10.1371/journal.ppat.1002351

Laub MT, Shapiro L, McAdams HH (2007) Systems biology of *Caulobacter*. Annu Rev Genet 41:429–441. https://doi.org/10.1146/annurev.genet.41.110306.130346

Lee J, Sayed N, Hunter A et al (2012) Activation of innate immunity is required for efficient nuclear reprogramming. Cell 151:547–558. https://doi.org/10.1016/j.cell.2012.09.034

LePage DP, Metcalf JA, Bordenstein SR et al (2017) Prophage WO genes recapitulate and enhance Wolbachia-induced cytoplasmic incompatibility. Nature 543:243–247. https://doi.org/10.1038/nature21391

Li W, Liu H, Cheng ZJ et al (2011) DNA methylation and histone modifications regulate De Novo shoot regeneration in Arabidopsis by modulating WUSCHEL expression and auxin signaling. PLoS Genet 7:e1002243. https://doi.org/10.1371/journal.pgen.1002243

Li Y, Liew YJ, Cui G et al (2018) DNA methylation regulates transcriptional homeostasis of algal endosymbiosis in the coral model Aiptasia. Sci Adv 4(8):eaat2142

Lloréns-Rico V, Cano J, Kamminga T et al (2016) Bacterial antisense RNAs are mainly the product of transcriptional noise. Sci Adv 2:e1501363. https://doi.org/10.1126/sciadv.1501363

Losick R, Desplan C (2008) Stochasticity and cell fate. Science 320:65–68. https://doi.org/10.1126/science.1147888

Macek B, Forchhammer K, Hardouin J et al (2019) Protein post-translational modifications in bacteria. Nat Rev Microbiol 17:651–664. https://doi.org/10.1038/s41579-019-0243-0

Mandel MJ, Schaefer AL, Brennan CA et al (2012) Squid-derived chitin oligosaccharides are a chemotactic signal during colonization by Vibrio fischeri. Appl Environ Microbiol 78:4620–4626. https://doi.org/10.1128/AEM.00377-12

Manzanillo PS, Ayres JS, Watson RO et al (2013) The ubiquitin ligase parkin mediates resistance to intracellular pathogens. Nature 501:512–516. https://doi.org/10.1038/nature12566

Mao M, Yang X, Bennett GM (2018) Evolution of host support for two ancient bacterial symbionts with differentially degraded genomes in a leafhopper host. Proc Natl Acad Sci 115:E11691–E11700. https://doi.org/10.1073/pnas.1811932115

Maróti G, Kondorosi Ã (2014) Nitrogen-fixing Rhizobium-legume symbiosis: are polyploidy and host peptide-governed symbiont differentiation general principles of endosymbiosis? Front Microbiol 5:326. https://doi.org/10.3389/fmicb.2014.00326

Masaki T, Qu J, Cholewa-Waclaw J et al (2013) Reprogramming adult Schwann cells to stem cell-like cells by leprosy bacilli promotes dissemination of infection. Cell 152:51–67. https://doi.org/10.1016/j.cell.2012.12.014

Matsuura Y, Kikuchi Y, Miura T, Fukatsu T (2015) *Ultrabithorax* is essential for bacteriocyte development. Proc Natl Acad Sci 112:9376–9381. https://doi.org/10.1073/pnas.1503371112

Mayoral JG, Hussain M, Joubert DA et al (2014) *Wolbachia* small noncoding RNAs and their role in cross-kingdom communications. Proc Natl Acad Sci 111:18721–18726. https://doi.org/10.1073/pnas.1420131112

McEwan DL, Kirienko NV, Ausubel FM (2012) Host translational inhibition by *Pseudomonas aeruginosa* exotoxin a triggers an immune response in *Caenorhabditis elegans*. Cell Host Microbe 11:364–374. https://doi.org/10.1016/j.chom.2012.02.007

Medzhitov R (2007) Recognition of microorganisms and activation of the immune response. Nature 449:819–826. https://doi.org/10.1038/nature06246

Meissner A (2010) Epigenetic modifications in pluripotent and differentiated cells. Nat Biotechnol 28:1079–1088. https://doi.org/10.1038/nbt.1684

Mergaert P (2018) Role of antimicrobial peptides in controlling symbiotic bacterial populations. Nat Prod Rep 35:336. https://doi.org/10.1039/C7NP00056A

Michard C, Doublet P (2015) Post-translational modifications are key players of the Legionella pneumophila infection strategy. Front Microbiol 6:87. https://doi.org/10.3389/fmicb.2015.00087

Mohr I, Sonenberg N (2012) Host translation at the Nexus of infection and immunity. Cell Host Microbe 12:470–483. https://doi.org/10.1016/j.chom.2012.09.006

Moss SM, Taylor IR, Ruggero D et al (2019) A legionella pneumophila kinase phosphorylates the Hsp70 chaperone family to inhibit eukaryotic protein synthesis. Cell Host Microbe 25:454–462. https://doi.org/10.1016/j.chom.2019.01.006

Mukherjee S, Liu X, Arasaki K et al (2011) Modulation of Rab GTPase function by a protein phosphocholine transferase. Nature 477:103–106. https://doi.org/10.1038/nature10335

Mukherjee K, Fischer R, Vilcinskas A (2012) Histone acetylation mediates epigenetic regulation of transcriptional reprogramming in insects during metamorphosis, wounding and infection. Front Zool 9:25. https://doi.org/10.1186/1742-9994-9-25

Mukherjee K, Grizanova E, Chertkova E et al (2017) Experimental evolution of resistance against *Bacillus thuringiensis* in the insect model host *Galleria mellonella* results in epigenetic modifications. Virulence 8:1618–1630. https://doi.org/10.1080/21505594.2017.1325975

Narita S, Nomura M, Kageyama D (2007) Naturally occurring single and double infection with Wolbachia strains in the butterfly Eurema hecabe: transmission efficiencies and population density dynamics of each Wolbachia strain: Wolbachia densities in single- and double-infected E. hecabe. FEMS Microbiol Ecol 61:235–245. https://doi.org/10.1111/j.1574-6941.2007.00333.x

Narra HP, Schroeder CLC, Sahni A et al (2016) Small regulatory RNAs of rickettsia conorii. Sci Rep 6:36728. https://doi.org/10.1038/srep36728

Negri I (2012) Wolbachia as an "infectious" extrinsic factor manipulating host signaling pathways. Front Endocrinol 2:115. https://doi.org/10.3389/fendo.2011.00115

Negri I, Pellecchi M (2012) Sex steroids in insects and the role of the endosymbiont Wolbachia: a new perspective. Sex Horm:353–374

Negri I, Franchini A, Gonella E et al (2009) Unravelling the *Wolbachia* evolutionary role: the reprogramming of the host genomic imprinting. Proc R Soc B Biol Sci 276:2485–2491. https://doi.org/10.1098/rspb.2009.0324

Newton ILG, Rice DW (2019) The Jekyll and Hyde symbiont: could *Wolbachia* be a nutritional mutualist? J Bacteriol 202(4). https://doi.org/10.1128/JB.00589-19

Nilsson O, Olsson O (1997) Getting to the root: the role of the agrobacterium rhizogenes rol genes in the formation of hairy roots. Physiol Plant 100:463–473. https://doi.org/10.1111/j.1399-3054.1997.tb03050.x

Nyholm SV, McFall-Ngai M (2004) The winnowing: establishing the squid–vibrio symbiosis. Nat Rev Microbiol 2:632–642. https://doi.org/10.1038/nrmicro957

Oldroyd GED (2013) Speak, friend, and enter: signalling systems that promote beneficial symbiotic associations in plants. Nat Rev Microbiol 11:252–263. https://doi.org/10.1038/nrmicro2990

Ortega ÃD, Quereda JJ, Pucciarelli MG, GarcÃ-a-del Portillo F (2014) Non-coding RNA regulation in pathogenic bacteria located inside eukaryotic cells. Front Cell Infect Microbiol 4:162. https://doi.org/10.3389/fcimb.2014.00162

Osei-Amo S, Hussein M, Asad S et al (2018) *Wolbachia*-induced transcription factor *GATA4* suppresses ovary-specific genes *blastoderm-specific protein 25D* and *imaginal disc growth factor*: GATA4 *regulation of ovary-specific genes*. Insect Mol Biol 27:295–304. https://doi.org/10.1111/imb.12371

Ote M, Ueyama M, Yamamoto D (2016) *Wolbachia* protein TomO targets nanos mRNA and restores germ stem cells in *Drosophila* sex-lethal mutants. Curr Biol 26:2223–2232. https://doi.org/10.1016/j.cub.2016.06.054

Palazzo AF, Lee ES (2015) Non-coding RNA: what is functional and what is junk? Front Genet 6:2. https://doi.org/10.3389/fgene.2015.00002

Pannebakker BA, Loppin B, Elemans CP et al (2007) Parasitic inhibition of cell death facilitates symbiosis. Proc Natl Acad Sci 104:213–215

Patrick S, Blakely GW (2012) Crossing the eukaryote-prokaryote divide: a ubiquitin homolog in the human commensal bacterium *Bacteroides fragilis*. Mob Genet Elem 2:149–151. https://doi.org/10.4161/mge.21191

Peláez P, Hernández-López A, Estrada-Navarrete G, Sanchez F (2017) Small RNAs Derived from the T-DNA of Agrobacterium rhizogenes in Hairy Roots of Phaseolus vulgaris. Front Plant Sci:8, 96. https://doi.org/10.3389/fpls.2017.00096

Pereira JM, Hamon MA, Cossart P (2016) A lasting impression: epigenetic memory of bacterial infections? Cell Host Microbe 19:579–582. https://doi.org/10.1016/j.chom.2016.04.012

Perlmutter JI, Bordenstein SR, Unckless RL et al (2019) The phage gene wmk is a candidate for male killing by a bacterial endosymbiont. PLOS Pathog 15:e1007936. https://doi.org/10.1371/journal.ppat.1007936

Peyer SM, Heath-Heckman EAC, McFall-Ngai MJ (2017) Characterization of the cell polarity gene crumbs during the early development and maintenance of the squid–vibrio light organ symbiosis. Dev Genes Evol 227:375–387. https://doi.org/10.1007/s00427-017-0576-5

Piferrer F (2013) Epigenetics of sex determination and gonadogenesis: epigenetics and sex determination. Dev Dyn 242:360–370. https://doi.org/10.1002/dvdy.23924

Pinto SB, Stainton K, Harris S et al (2013) Transcriptional regulation of *Culex pipiens* mosquitoes by Wolbachia influences cytoplasmic incompatibility. PLoS Pathog 9:e1003647. https://doi.org/10.1371/journal.ppat.1003647

Piras V, Tomita M, Selvarajoo K (2012) Is central dogma a global property of cellular information flow? Front Physiol 3:442. https://doi.org/10.3389/fphys.2012.00439

Qiu J, Sheedlo MJ, Yu K et al (2016) Ubiquitination independent of E1 and E2 enzymes by bacterial effectors. Nature 533:120–124. https://doi.org/10.1038/nature17657

Raghavan R, Sloan DB, Ochman H (2012) Antisense transcription is pervasive but rarely conserved in enteric bacteria. mBio 3:e00156-12. https://doi.org/10.1128/mBio.00156-12

Ribet D, Cossart P (2010) Post-translational modifications in host cells during bacterial infection. FEBS Lett 584:2748–2758. https://doi.org/10.1016/j.febslet.2010.05.012

Rice DW, Sheehan KB, Newton ILG (2017) Large-scale identification of Wolbachia pipientis effectors. Genome Biol Evol 9:1925–1937. https://doi.org/10.1093/gbe/evx139

Rigo R, Bazin J, Crespi M, Charon C (2019) Alternative splicing in the regulation of plant–microbe interactions. Plant Cell Physiol 60:1906–1916. https://doi.org/10.1093/pcp/pcz086

Riparbelli MG, Giordano R, Ueyama M, Callaini G (2012) Wolbachia-mediated male killing is associated with defective chromatin remodeling. PLoS ONE 7:e30045. https://doi.org/10.1371/journal.pone.0030045

Rogan MR, Patterson LL, Wang JY, McBride JW (2019) Bacterial manipulation of Wnt signaling: a host-pathogen tug-of-Wnt. Front Immunol 10:2390. https://doi.org/10.3389/fimmu.2019.02390

Rolando M, Buchrieser C (2014) Legionella pneumophila type IV effectors hijack the transcription and translation machinery of the host cell. Trends Cell Biol 24:771–778. https://doi.org/10.1016/j.tcb.2014.06.002

Rolando M, Sanulli S, Rusniok C et al (2013) Legionella pneumophila effector RomA uniquely modifies host chromatin to repress gene expression and promote intracellular bacterial replication. Cell Host Microbe 13:395–405. https://doi.org/10.1016/j.chom.2013.03.004

Roquis D, Taudt A, Geyer KK et al (2018) Histone methylation changes are required for life cycle progression in the human parasite *Schistosoma mansoni*. PLoS Pathog 14:e1007066. https://doi.org/10.1371/journal.ppat.1007066

Russell SL, Cavanaugh CM (2017) Intrahost genetic diversity of bacterial symbionts exhibits evidence of mixed infections and recombinant haplotypes. Mol Biol Evol 34:2747–2761. https://doi.org/10.1093/molbev/msx188

Russell SL, Lemseffer N, White P, Sullivan WT (2018) Wolbachia and host germline components compete for kinesin-mediated transport to the posterior pole of the Drosophila oocyte. PLoS Pathog 14:e1007216. https://doi.org/10.1371/journal.ppat.1007216

Russell SL, Chappell L, Sullivan W (2019) A symbiont's guide to the germline. In: Current topics in developmental biology. Elsevier, pp 315–351

Sahni A, Narra H, Patel J, Sahni S (2018) MicroRNA-regulated Rickettsial invasion into host endothelium via fibroblast growth factor 2 and its receptor FGFR1. Cell 7:240. https://doi.org/10.3390/cells7120240

Saijo Y, Schulze-Lefert P (2008) Manipulation of the eukaryotic transcriptional machinery by bacterial pathogens. Cell Host Microbe 4:96–99. https://doi.org/10.1016/j.chom.2008.07.001

Sánchez-Romero MA, Cota I, Casadesús J (2015) DNA methylation in bacteria: from the methyl group to the methylome. Curr Opin Microbiol 25:9–16. https://doi.org/10.1016/j.mib.2015.03.004

Saridaki A, Sapountzis P, Harris HL et al (2011) Wolbachia prophage DNA adenine methyltransferase genes in different Drosophila-Wolbachia associations. PLoS One 6:e19708. https://doi.org/10.1371/journal.pone.0019708

Sasai M, Kawabata Y, Makishi K et al (2013) Time scales in epigenetic dynamics and phenotypic heterogeneity of embryonic stem cells. PLoS Comput Biol 9:e1003380. https://doi.org/10.1371/journal.pcbi.1003380

Serbus LR, Ferreccio A, Zhukova M et al (2011) A feedback loop between Wolbachia and the Drosophila gurken mRNP complex influences Wolbachia titer. J Cell Sci 124:4299–4308. https://doi.org/10.1242/jcs.092510

Schneider RJ, Mohr I (2003) Translation initiation and viral tricks. Trends Biochem Sci 28:130–136. https://doi.org/10.1016/S0968-0004(03)00029-X

Schwartzman JA, Ruby EG (2016) Stress as a normal cue in the symbiotic environment. Trends Microbiol 24:414–424. https://doi.org/10.1016/j.tim.2016.02.012

Serbus LR, Sullivan W (2007) A cellular basis for *Wolbachia* recruitment to the host germline. PLoS Pathog 3:e190. https://doi.org/10.1371/journal.ppat.0030190

Sesto N, Koutero M, Cossart P (2014) Bacterial and cellular RNAs at work during *Listeria* infection. Future Microbiol 9:1025–1037. https://doi.org/10.2217/fmb.14.79

Shamir M, Bar-On Y, Phillips R, Milo R (2016) SnapShot: timescales in cell biology. Cell 164:1302. https://doi.org/10.1016/j.cell.2016.02.058

Sharma C, Heidrich N (2012) Small RNAs and virulence in bacterial pathogens. RNA Biol 9:361–363. https://doi.org/10.4161/rna.20517

Sheehan KB, Martin M, Lesser CF et al (2016) Identification and characterization of a candidate *Wolbachia pipientis* type IV effector that interacts with the actin cytoskeleton. mBio 7:e00622-16. https://doi.org/10.1128/mBio.00622-16

Shropshire JD, On J, Layton EM et al (2018) One prophage WO gene rescues cytoplasmic incompatibility in *Drosophila melanogaster*. Proc Natl Acad Sci 115:4987–4991. https://doi.org/10.1073/pnas.1800650115

Silmon de Monerri NC, Kim K (2014) Pathogens hijack the epigenome. Am J Pathol 184:897–911. https://doi.org/10.1016/j.ajpath.2013.12.022

Simonet P, Gaget K, Balmand S et al (2018) Bacteriocyte cell death in the pea aphid/*Buchnera* symbiotic system. Proc Natl Acad Sci 115:E1819–E1828. https://doi.org/10.1073/pnas.1720237115

Simonov D, Swift S, Blenkiron C, Phillips AR (2016) Bacterial RNA as a signal to eukaryotic cells as part of the infection process. Discover 4:e70. https://doi.org/10.15190/d.2016.17

Singh S, Katzer K, Lambert J et al (2014) CYCLOPS, a DNA-binding transcriptional activator, orchestrates symbiotic root nodule development. Cell Host Microbe 15:139–152. https://doi.org/10.1016/j.chom.2014.01.011

Siozios S, Ioannidis P, Klasson L et al (2013) The diversity and evolution of Wolbachia Ankyrin repeat domain genes. PLoS One 8:e55390. https://doi.org/10.1371/journal.pone.0055390

Smith TE, Moran NA (2020) Coordination of host and symbiont gene expression reveals a metabolic tug-of-war between aphids and *Buchnera*. Proc Natl Acad Sci 117(4):2113–2121. https://doi.org/10.1073/pnas.1916748117

Sol A, Lipo E, de Jesús-Díaz DA et al (2019) *Legionella pneumophila* translocated translation inhibitors are required for bacterial-induced host cell cycle arrest. Proc Natl Acad Sci 116:3221–3228. https://doi.org/10.1073/pnas.1820093116

Starr DJ, Cline TW (2002) A host–parasite interaction rescues Drosophila oogenesis defects. Nature 418:76

Sugimoto TN, Fujii T, Kayukawa T et al (2010) Expression of a doublesex homologue is altered in sexual mosaics of Ostrinia scapulalis moths infected with Wolbachia. Insect Biochem Mol Biol 40:847–854. https://doi.org/10.1016/j.ibmb.2010.08.004

Sullivan W (2017) *Wolbachia*, bottled water, and the dark side of symbiosis. Mol Biol Cell 28:2343–2346. https://doi.org/10.1091/mbc.E17-02-0132

Sun S, Cline TW (2009) Effects of Wolbachia infection and ovarian tumor mutations on sex-lethal germline functioning in Drosophila. Genetics 181:1291–1301. https://doi.org/10.1534/genetics.108.099374

Suzuki M, Mimuro H, Kiga K et al (2009) Helicobacter pylori CagA phosphorylation-independent function in epithelial proliferation and inflammation. Cell Host Microbe 5:23–34. https://doi.org/10.1016/j.chom.2008.11.010

Svensson SL, Sharma CM (2016) Virulence mechanisms of bacterial pathogens. In: Virulence mechanisms of bacterial pathogens. American Society for Microbiology American Society for Microbiology, Washington, DC, p 44

Thairu MW, Hansen AK (2019) It's a small, small world: unravelling the role and evolution of small RNAs in organelle and endosymbiont genomes. FEMS Microbiol Lett 366:fnz049. https://doi.org/10.1093/femsle/fnz049

Thairu MW, Cheng S, Hansen AK (2018) A sRNA in a reduced mutualistic symbiont genome regulates its own gene expression. Mol Ecol 27:1766–1776. https://doi.org/10.1111/mec.14424

Toft C, Andersson SGE (2010) Evolutionary microbial genomics: insights into bacterial host adaptation. Nat Rev Genet 11:465–475. https://doi.org/10.1038/nrg2798

Toribio R, Ventoso I (2010) Inhibition of host translation by virus infection in vivo. Proc Natl Acad Sci 107:9837–9842. https://doi.org/10.1073/pnas.1004110107

Traubenik S, Reynoso MA, Hobecker KV et al (2019) Reprogramming of root cells during nitrogen-fixing symbiosis involves dynamic polysome association of coding and non-coding RNAs. Plant Cell 32(2):352–373. https://doi.org/10.1105/tpc.19.00647

Traverse CC, Ochman H (2016) Conserved rates and patterns of transcription errors across bacterial growth states and lifestyles. Proc Natl Acad Sci 113:3311–3316. https://doi.org/10.1073/pnas.1525329113

Treacy-Abarca S, Mukherjee S (2015) Legionella suppresses the host unfolded protein response via multiple mechanisms. Nat Commun 6:7887. https://doi.org/10.1038/ncomms8887

Veneti Z, Bentley JK, Koana T et al (2005) A functional dosage compensation complex required for male killing in Drosophila. Science 307:1461–1463. https://doi.org/10.1126/science.1107182

Verdone L, Caserta M, Mauro ED (2005) Role of histone acetylation in the control of gene expression. Biochem Cell Biol 83:344–353. https://doi.org/10.1139/o05-041

Vilcinskas A (2017) The impact of parasites on host insect epigenetics. In: Advances in insect physiology. Elsevier, pp 145–165

Weiberg A, Wang M, Lin F-M et al (2013) Fungal small RNAs suppress plant immunity by hijacking host RNA interference pathways. Science 342:118–123. https://doi.org/10.1126/science.1239705

Werren JH, Baldo L, Clark ME (2008) *Wolbachia*: master manipulators of invertebrate biology. Nat Rev Microbiol 6:741–751. https://doi.org/10.1038/nrmicro1969

Wessler S, Backert S (2008) Molecular mechanisms of epithelial-barrier disruption by Helicobacter pylori. Trends Microbiol 16:397–405. https://doi.org/10.1016/j.tim.2008.05.005

White PM, Serbus LR, Debec A et al (2017) Reliance of *Wolbachia* on high rates of host proteolysis revealed by a genome-wide RNAi screen of *Drosophila* cells. Genetics 205(4):1473–1488. https://doi.org/10.1534/genetics.116.198903

Wolański M, Jakimowicz D, Zakrzewska-Czerwiska J (2014) Fifty years after the replicon hypothesis: cell-specific master regulators as new players in chromosome replication control. J Bacteriol 196:2901–2911. https://doi.org/10.1128/JB.01706-14

Xu L, Luo Z-Q (2013) Cell biology of infection by Legionella pneumophila. Microbes Infect 15:157–167. https://doi.org/10.1016/j.micinf.2012.11.001

Yang H-J, Kim SG, Lim JH et al (2018) Helicobacter pylori-induced modulation of the promoter methylation of Wnt antagonist genes in gastric carcinogenesis. Gastric Cancer 21:237–248. https://doi.org/10.1007/s10120-017-0741-6

Yokoyama K, Higashi H, Ishikawa S et al (2005) Functional antagonism between Helicobacter pylori CagA and vacuolating toxin VacA in control of the NFAT signaling pathway in gastric epithelial cells. Proc Natl Acad Sci 102:9661–9666. https://doi.org/10.1073/pnas.0502529102

Yuan M, Ke Y, Huang R et al (2016) A host basal transcription factor is a key component for infection of rice by TALE-carrying bacteria. eLife 5:e19605. https://doi.org/10.7554/eLife.19605

Zchori-Fein E, Borad C, Harari AR (2006) Oogenesis in the date stone beetle, *Coccotrypes dactyliperda*, depends on symbiotic bacteria. Physiol Entomol 31:164–169. https://doi.org/10.1111/j.1365-3032.2006.00504.x

Zeng J, Gupta VK, Jiang Y et al (2019) Cross-kingdom small RNAs among animals, plants and microbes. Cell 8:371. https://doi.org/10.3390/cells8040371

Zhang YJ, Rubin EJ (2013) Feast or famine: the host-pathogen battle over amino acids: host-pathogen battle over amino acids. Cell Microbiol 15:1079–1087. https://doi.org/10.1111/cmi.12140

Zhang G, Hussain M, O'Neill SL, Asgari S (2013) Wolbachia uses a host microRNA to regulate transcripts of a methyltransferase, contributing to dengue virus inhibition in *Aedes aegypti*. Proc Natl Acad Sci 110:10276–10281. https://doi.org/10.1073/pnas.1303603110

Zhao C, Zhou Z, Zhang T et al (2017) Salmonella small RNA fragment Sal-1 facilitates bacterial survival in infected cells via suppressing iNOS induction in a microRNA manner. Sci Rep 7:16979. https://doi.org/10.1038/s41598-017-17205-4

Zheng X, Zhang D, Li Y et al (2019) Incompatible and sterile insect techniques combined eliminate mosquitoes. Nature 572:56–61. https://doi.org/10.1038/s41586-019-1407-9

Zhong Y, Kanagaratham C, Radzioch D (2013) Chromatin remodelling during host-bacterial pathogen interaction. Chromatin Remodelling.

Zhou Y, Zhu Y (2015) Diversity of bacterial manipulation of the host ubiquitin pathways: diversity of bacterial manipulation of the host ubiquitin pathways. Cell Microbiol 17:26–34. https://doi.org/10.1111/cmi.12384

Part II
Origin, Adaptations and Evolutionary Aspects of Symbiosis

Chapter 6
We're in this Together: Sensation of the Host Cell Environment by Endosymbiotic Bacteria

Cory D. Dunn, Tamara Somborac, and Bala Anı Akpınar

Abstract Bacteria inhabit diverse environments, including the inside of eukaryotic cells. While a bacterial invader may initially act as a parasite or pathogen, a subsequent mutualistic relationship can emerge in which the endosymbiotic bacteria and their host share metabolites. While the environment of the host cell provides improved stability when compared to an extracellular environment, the endosymbiont population must still cope with changing conditions, including variable nutrient concentrations, the host cell cycle, host developmental programs, and host genetic variation. Furthermore, the eukaryotic host can deploy mechanisms actively preventing a bacterial return to a pathogenic state. Many endosymbionts are likely to use two-component systems (TCSs) to sense their surroundings, and expanded genomic studies of endosymbionts should reveal how TCSs may promote bacterial integration with a host cell. We suggest that studying TCS maintenance or loss may be informative about the evolutionary pathway taken toward endosymbiosis, or even toward endosymbiont-to-organelle conversion.

Keywords Mutualism · Endosymbiosis · Signaling · Quorum sensing · Two-component system

6.1 Introduction

Numerically, prokaryotes dominate our planet (Whitman et al. 1998) and display metabolic proficiency and flexibility currently unmatched by eukaryotes (Goyal 2018; Torsvik et al. 2002). To maintain their survival and propagation, all organisms must sense their surroundings. Toward this goal, bacteria have evolved a number of

C. D. Dunn (✉) · T. Somborac · B. A. Akpınar
Institute of Biotechnology, Helsinki Institute of Life Science, University of Helsinki, Helsinki, Finland
e-mail: cory.dunn@helsinki.fi

M. Kloc (ed.), *Symbiosis: Cellular, Molecular, Medical and Evolutionary Aspects*, Results and Problems in Cell Differentiation 69,
https://doi.org/10.1007/978-3-030-51849-3_6

mechanisms that allow reaction to their environments, including responses to other microorganisms of the same or different species.

One specific and peculiar environment that bacteria may inhabit is the inside of a eukaryotic cell. Occasionally, an endosymbiotic partnership can form in which two organisms appear to form a mutually beneficial relationship based upon syntrophy, or the sharing of metabolites (Morris et al. 2013). Upon establishment of an endosymbiont within its host, there is often a contraction of genome size prompted by redundancy of gene products (Bennett and Moran 2015; Moran 2003; Shigenobu et al. 2000) and small population size (Kuo et al. 2009). Primary bacterial endosymbionts have become firmly ensconced within their hosts and are typically engaged in mutual metabolic dependency with their eukaryotic partner. Secondary endosymbionts have typically initiated a more recent relationship with their host, are more often transmitted horizontally, potentially survive outside of the host cell, and closely skirt the line between parasitism and mutualism that may mark the progression to endosymbiosis (Zachar and Boza 2020; McCutcheon et al. 2019; Sullivan 2017; Pérez-Brocal et al. 2013; Sachs et al. 2011). For primary endosymbionts, even full-length host proteins may eventually be put to use (Nakabachi et al. 2014; Nowack and Grossman 2012), and the use of host proteins by the endosymbiont may mark a major transition point that occurs during the rare conversion of an endosymbiont to an organelle (Keeling et al. 2015; McCutcheon and Keeling 2014).

Should the environment inhabited by the endosymbiont be considered simple or complex? On the one hand, to the potential benefit of the endosymbiont, multiple features of the environment are stabilized when compared to the environment outside of the eukaryotic host. Strict maternal transmission can limit exposure to phage (Metcalf and Bordenstein 2012). Moreover, residence inside of a eukaryotic cell may provide protection against predation, consistent with the idea that predation can drive major evolutionary transitions (Herron et al. 2019; Boraas et al. 1998; Stanley 1973). Ion concentration and pH within the eukaryotic host would be maintained within tight boundaries acceptable to the host, and therefore may be particularly suitable for many bacterial guests. In addition, an obligate endosymbiont can harvest any metabolite for which consumption does not lead to fitness costs for the host and selection against the conglomerate. Taken together, an endosymbiotic life strategy might be initially regarded as a simplified and hospitable environment.

On the other hand, the intracellular environment of an endosymbiont is not as uncomplicated as it may first appear. Host and endosymbiont cell cycles are expected to be coordinated with the help of the appropriate bacterial signaling pathways (Catta-Preta et al. 2015), and host-derived antimicrobial peptides (AMPs) can be employed in a delicate dance between host and endosymbiont that prevents re-emergence of pathogenicity (Login et al. 2011). The nutritional status and life stage of its host may fluctuate, and the endosymbiont must regulate its number and behavior accordingly (Darby et al. 2012; Snyder et al. 2012; Stoll et al. 2009; Wilkinson et al. 2007; Fenn and Blaxter 2004), even if the spectrum and scale of endosymbiont responses to its environment may eventually become diminished

(Wilcox et al. 2003). Endosymbionts also regulate their gene expression in a manner concordant with the different tissues in which they may reside or, if ever transmitted between host cells, the extracellular environment (Darby et al. 2012; Bright and Bulgheresi 2010). Moreover, beyond a reactive use of sensing mechanisms, endosymbionts may manipulate the germline and somatic activities of their hosts (Foray et al. 2018; Pietri et al. 2016; Landmann et al. 2014; Fast et al. 2011; Serbus and Sullivan 2007). Divergent host genotypes can present additional variation to which the endosymbiont must adjust its gene expression (Smith and Moran 2020). Consequently, one might expect that robust sensation mechanisms would be maintained by many endosymbionts.

Here, we focus our attention upon two-component systems (TCSs), a versatile set of sensors and effectors used by a wide variety of bacteria to detect and respond to their environment. We highlight the small, but expanding number of studies focused upon endosymbiont sensation, and we argue that knowledge of TCS activities may be informative about the evolutionary histories of, and strategies deployed by, endosymbionts.

6.2 Fundamental Aspects of Two-Component Systems

TCSs are prominently used by bacteria to sense and respond to the environment (Gao et al. 2019; Jacob-Dubuisson et al. 2018; Zschiedrich et al. 2016). Individual bacterial species can encode tens, or even hundreds of TCSs (Borland et al. 2016), allowing responses to divergent signals that include myriad small molecules, temperature, gasses, and light (Krell et al. 2010). Within the context of a TCS, a histidine kinase (HK) component and a response regulator (RR) serve as a minimal set of polypeptides that can sense cellular conditions, yet this arrangement can be markedly elaborated by additional regulatory pathway members (Gao et al. 2019; Jacob-Dubuisson et al. 2018; Zschiedrich et al. 2016). HKs involved in sensation are often membrane-bound, with sensor domains extending into the cytoplasm. Other HKs are membrane-inserted, yet lack periplasmic extensions, or can even be wholly cytoplasmic. HK domains used for signal detection are characterized by significant structural diversity, in accordance with the heterogeneous signals sensed by bacteria, but the catalytic core tends to be well conserved. The cytosolic portion of an HK, encompassing the autokinase domain, consists of a *D*imerization and *H*istidine-*p*hosphotransfer domain (DHp) and a *C*atalytic and *A*TP-binding domain (CA) connected by a short loop of amino acids. A diverse array of additional domains (Krell et al. 2010) contribute to protein–protein interactions and help to modulate autokinase activity.

HKs are typically found as homodimers for which autophosphorylation prompted by activation can occur in either a *cis*- or *trans*-fashion (Casino et al. 2009, 2014). Upon activation by the stimulus, which can be sensed even at relatively low binding affinities (Krell et al. 2010; Cheung and Hendrickson 2009), the epsilon nitrogen of a conserved histidine in the DHp domain is phosphorylated by use of ATP (Bhate et al.

2015). Next, phosphotransfer to the appropriate RR is catalyzed, providing tight control of response to the stimulus. The RR is phosphorylated at a conserved aspartate, and the transfer of the phosphoryl group from the key HK histidine is driven primarily by the receiver (REC) domain of the RR (Zschiedrich et al. 2016) (Fig. 6.1).

Specificity of signaling is mostly encoded at the interaction face between a given HK and its cognate RR (Podgornaia and Laub 2013; Fisher et al. 1996), although specificity is also dependent upon proper stoichiometry of TCS components (Steiner et al. 2018). Not all HKs act exclusively with one RR; several HKs can share a particular RR and phenotypic outcome (Stephenson and Hoch 2002). Hybrid HKs also exist for which the HK and RR are fused within the same polypeptide (Townsend 2nd et al. 2013; Capra et al. 2012), ensuring dedicated phosphorylation of the relevant RR. As well as providing kinase activity, HKs can also act as phosphatases, removing instances of direct RR phosphorylation by cellular acetyl-phosphate (Gao et al. 2019; Podgornaia and Laub 2013; Klein et al. 2007) and blocking pathway activation when signal reception is concluded (Huynh and Stewart 2011). Kinase activity of HKs does not simply correspond with the presence of ligand or other stimuli; kinase activity can instead be prompted by the lack of a signaling molecule or environmental condition (Neiditch et al. 2005; Henke and Bassler 2004). HKs are often, but not always, found in the same operon with their cognate RRs (Capra and Laub 2012). Of note, there can be additional elaboration upon the standard theme of the TCS to include complicated phosphorelay systems (Francis and Porter 2019; Dworkin 2015; Wright and Ulijasz 2014). Recent evidence also suggests TCS crosstalk by HK phosphorylation of other HKs (Francis and Porter 2019; Francis et al. 2018).

After the REC domain is phosphorylated, the conformation of RRs, and their multimerization state, can change (Gao et al. 2019; Galperin 2006). Like HKs, RRs harbor many different functional domains that provide for regulatory complexity under diverse environmental conditions (Galperin 2006). The majority of RRs bind to DNA, and upon DNA binding, these factors can regulate transcription by functioning as activators or repressors, or they may block chromosome replication. These RRs can also manifest enzymatic activity and can regulate downstream processes by protein–protein interactions (Gao et al. 2019). Beyond signal shutoff by the phosphatase activity of HKs or by dedicated RR phosphatases (Zschiedrich et al. 2016; Pazy et al. 2010), RRs also have the ability to autodephosphorylate (Gao et al. 2019).

6.3 Two-Component Systems of Endosymbionts

As the functions carried out by a bacterium become intertwined with that of the host, its genome becomes eroded as a result of reduced selection and population bottlenecks (Bennett and Moran 2015; Kuo et al. 2009; Moran 2003; Shigenobu et al. 2000). Like the more generalized metric of genome size, the number of TCSs may serve as a reflection of the relative duration of endosymbiont association with its host

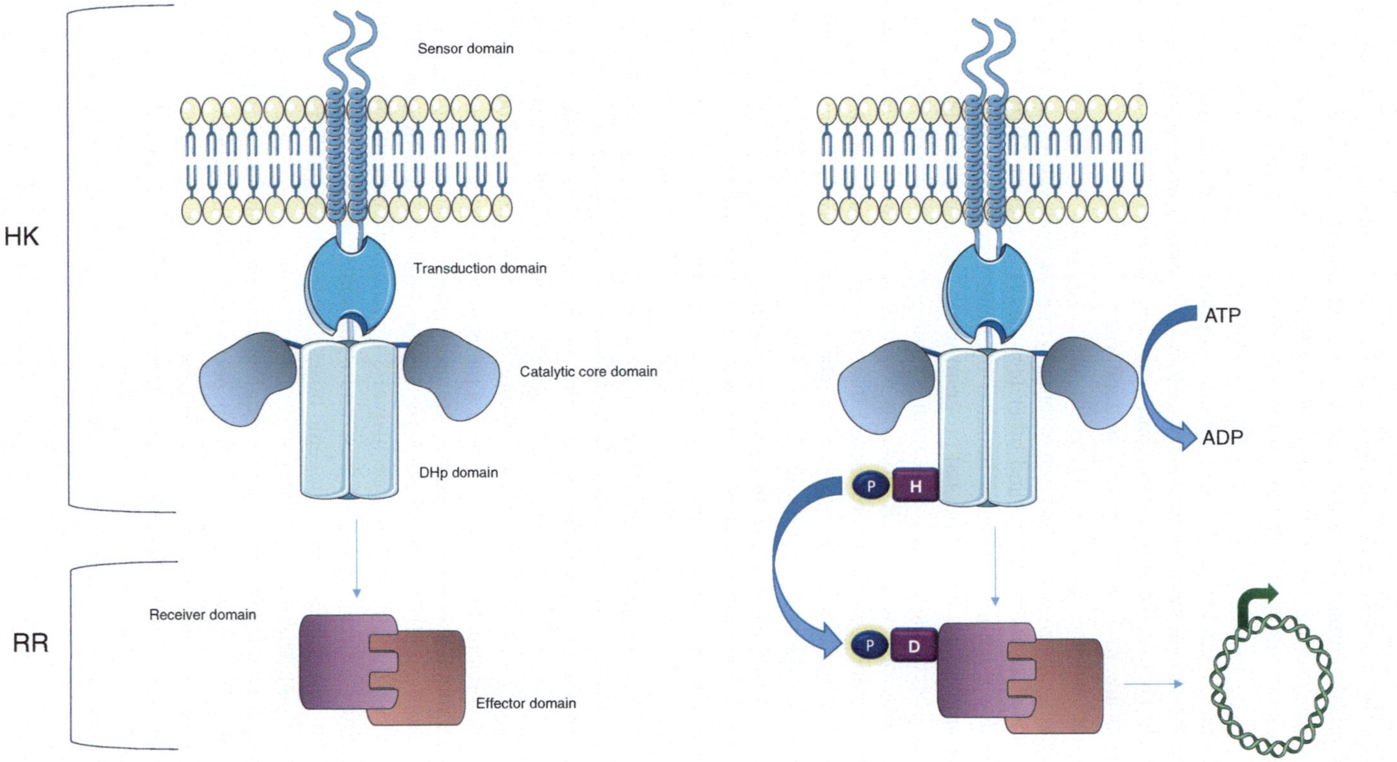

Fig. 6.1 Schematic representation of a typical two-component system structure. A two-component system comprises a histidine kinase (HK) and response regulator (RR). Perception of an extracellular signal by the sensor domain leads to the hydrolysis of ATP by the Catalytic and ATP-Binding Domain, and consequent phosphorylation of the central histidine (H) residue in the DHp domain of the HK. This phosphate is then transferred to the Aspartate (D) residue located in the receiver domain of the RR. Activation of the effector domain of the RR can prompt changes in gene expression to bring about an appropriate cellular response

(Kim et al. 2010). Most evidence does indeed suggest that the number of HKs and RRs can be greatly curtailed in bacteria exclusively localized with a eukaryotic cell (Christensen and Serbus 2015; Rikihisa 2010; Wakeel et al. 2010; Cheng et al. 2006), with some endosymbionts and intracellular pathogens harboring few or no TCS pathways (Capra and Laub 2012; Ashby 2004). Those specific TCSs that are maintained the longest within the degenerating genome may be informative about key aspects of endosymbiont evolutionary history or current aspects of the mutualistic relationship between endosymbiont and host. Yet, the roles of TCSs encoded by endosymbionts are, to date, very poorly understood.

Perhaps the earliest study of TCSs in endosymbionts was focused upon *Bradyrhizobium japonicum*, a facultative endosymbiont that can obtain nitrogen from the atmosphere for soybeans and other legumes (Lardi et al. 2016; Ferguson et al. 2010). By a complex process initiated by plant metabolites, *B. japonicum* activates the appropriate transcriptional program while forming a nodule within the plant root that becomes a suitable location for nitrogen fixation. An operon that includes the HK NodV and the RR NodW is required for nodulation (Göttfert et al. 1990), and subsequent work demonstrated that the vast majority of *B. japonicum* transcriptional targets activated by the soybean nodulation–promoting signal required phosphorylation of NodW by NodV (Lang et al. 2008). *B. japonicum* is not limited to the endosymbiotic lifestyle, but also inhabits the soil. In agreement with this *B. japonicum* life history, its genome is not diminished when compared to other, free-living bacteria, and, along with NodV and NodW this species can encode tens of additional HKs and approximately one hundred RRs (Kaneko et al. 2011).

To illuminate closer genetic and metabolic interdependencies between host and endosymbiont, efforts have been made to understand TCS signaling in Wolbachia, perhaps the most prominent model system for exploration of host–endosymbiont interactions. Wolbachia are intracellular bacteria from the alpha-proteobacterial Rickettsiae family that are widespread among arthropods and nematodes. While some Wolbachia interact with their hosts in a parasitic or pathogenic manner, other Wolbachia are mutualist endosymbionts required by their host for the provision of metabolites (Sullivan 2017; Gutzwiller 2016; Taylor et al. 2013; Darby et al. 2012). Wolbachia is mostly, although not exclusively, transmitted vertically through the female germline (Werren 1997), and these endosymbionts can be tightly associated with the ability of some of their pathogenic hosts to cause disease (Christensen and Serbus 2015; Saint André et al. 2002).

To investigate the landscape of TCS signaling in endosymbionts, a comprehensive search for TCS components has been performed within several Wolbachia species (Christensen and Serbus 2015). Consistent with previous searches within the clade Anaplasmataceae, the number of HKs and RRs recovered by BLAST queries based upon the HK and RR sequences of free-living alpha-proteobacter *Caulobacter crescentus* was exceedingly low. These HKs and RRs were not found within the same operons but were scattered to different chromosomal locations and surrounded by genes for which a functional link to Wolbachia-encoded TCSs was unclear. Specifically, the HK CckA and the RR CtrA were identified in multiple Wolbachia species. These two proteins act within a phosphotransfer cascade controlling cell cycle progression in *C. crescentus* (Biondi et al. 2006; Jacobs et al. 2003), although there appears to be significant divergence among alpha-proteobacterial species when considering targets directly regulated by CtrA (Pini et al. 2015). Additionally, an ortholog of DvlL that lacked a carboxyl-terminal catalytic domain was encoded at a chromosomal location near the *ctrA* locus in several Wolbachia genomes. DvlL is predicted to be a possible potentiator of CckA HK activity, and Wolbachia DvlL harbors multiple Per-Arnt-Sim (PAS) domains, which are common among bacterial polypeptides involved in signal transduction.

Another TCS, consisting of the HK PleC and its target RR PleD, was found among multiple Wolbachia species. While many RRs are DNA-binding proteins (Gao et al. 2019), PleD instead harbors a GGDEF domain, named after a conserved sequence pattern, that may generate bis-(3′-5′)-cyclic diguanylic acid (c-di-GMP) by its diguanylate cyclase activity (Jenal et al. 2017; Lai et al. 2009; Paul et al. 2004). c-di-GMP is an important bacterial second messenger that binds to multiple downstream effectors and controls many processes, including cell polarity, transition to biofilm formation, and virulence (Jenal et al. 2017; Valentini and Filloux 2016; Trampari et al. 2015; Tschowri et al. 2014; Davis et al. 2013; Römling et al. 2013; Moscoso et al. 2012). Interestingly, CckA from *C. crescentus* was found to be directly regulated by c-di-GMP (Lori et al. 2015), suggesting that the coexistence of the CckA/CtrA and PleC/PleD TCSs within the same Wolbachia species is not coincidental and may have functional relevance.

The paucity of Wolbachia TCS components identified in the study described here is quite consistent with a relaxation of selection on, and subsequent loss of, many genes typically required by free-living bacteria. Yet, some TCSs have clearly been maintained, and some evidence supports the idea of positive selection upon the PleD ortholog of the *w*Mel strain of *Wolbachia pipientis* (Brownlie et al. 2007). Moreover, experiments in which gene expression data of *W. pipientis w*Mel were followed during the life cycle of *Drosophila melanogaster* suggest that nearly 8% of Wolbachia genes are differentially expressed in a manner dependent upon sex or developmental stage (Gutzwiller et al. 2015). Intriguingly, one of the genes regulated in a stage-specific manner was CckA (Christensen and Serbus 2015; Gutzwiller et al. 2015), consistent with a role for this HK in responding to developmental cues provided by the host.

A consistent feature of endosymbiont establishment and maintenance within host organisms is likely to be a balance between sensitivity and tolerance to host-synthesized AMPs (Mergaert 2018; Masson et al. 2016). After introduction to the host and the initiation of a mutualistic relationship, endosymbionts may reside within special compartments, such as the bacteriocytes of tsetse flies or the trophosomes of the gutless tube worm *Riftia pachyptila*, and AMPs appear to prevent bacterial escape from some of these special structures (Bing et al. 2017; Masson et al. 2016; Bright et al. 2013; Login et al. 2011). Among other functions, the PhoP-PhoQ TCS, encoded by several Gram-negative pathogens, plays a role in sensation of and response to host-synthesized AMPs (Bader et al. 2005), and the modification of lipopolysaccharide prompted by PhoP-PhoQ activation by AMPs can confer pathogen resistance to these antibacterial agents (Dalebroux and Miller

2014; Groisman and Mouslim 2006). Interestingly, changes to PhoP-PhoQ activity in endosymbionts can correspond with the transition to endosymbiotic mutualism. *Sodalis glossinidius* is a vertically transmitted gamma-proteobacterial endosymbiont that has only recently become established within its tsetse fly host (Chen et al. 1999). *S. glossinidius* appears to have a perpetually activated PhoP-PhoQ TCS that drives high AMP resistance, suggesting that resisting immune functions of the host remains important at an early stage of endosymbiosis (Pontes et al. 2011). A sustained endosymbiotic strategy may correspond with a lack of selection for PhoP-PhoQ and consequent loss of this TCS, consistent with the establishment of confident mutualism less subject to reversion to a state of bacterial pathogenicity.

The PhoP HK and PhoQ RR appear again within the context of a different endosymbiont-mediated phenomenon: resistance of the pea aphid *Acyrthosiphon pisum* to the larvae of parasitic wasp *Aphidius ervi*. Defense of *A. pisum* can be provided by its facultative endosymbionts. To understand the genomic basis of this resistance, the genomic contents of two isolates of the endosymbiont *Regiella insecticola* manifesting disparate capacities to defend against wasp parasites were examined (Hansen et al. 2012). Notably, the PhoP-PhoQ TCS was found to be associated specifically with the isolate that provided parasite resistance. Moreover, the PhoQ transcriptional target PqaA, also encoded by the resistance-conferring *R. insecticola* isolate, has previously been shown to block the activity of parasitoid venom peptides like melittin (Baker et al. 1997), raising the possibility that PhoP-PhoQ-PqaA can act as key modulators of pea aphid resistance. The aphid endosymbiont *Hamiltonella defensa*, which provides protection against parasitoid wasps, also encodes numerous TCS components (Degnan et al. 2009), although their role is not yet characterized.

We performed our own preliminary search for HKs and RRs in *Wigglesworthia glossinidia*, an obligate gamma-bacterial endosymbiont producing B vitamins for its tsetse fly host (Rio et al. 2012; Akman et al. 2002). A BLAST search using PFAM seed sequences revealed only an operon containing the HK CpxA (44% identity to *Escherichia coli* along aligned region) and the RR CpxR (75% identity to *Escherichia coli* along aligned region). CpxA-CpxR signaling can be prompted by protein folding stress in the inner membrane (Mitchell and Silhavy 2019). These findings suggest that changes to conditions within, or demands upon, the endosymbiont within the host may lead to inner membrane proteostasis defects that must be countered by a TCS-mediated transcriptional response.

While a pathway that senses membrane stress may be the last to be maintained by *W. glossinidia*, the genomic sequence of another endosymbiont appears to document the final loss of TCS signaling by destruction of its last HK. An intact, single RR with 93% alignment identity to *E. coli* OmpR is annotated within the genome of the *Cinara cedri* (aphid) endosymbiont *Serratia symbiotica* (*S. symbiotica* SCc), which is almost certainly in the midst of conversion from facultative symbiont to obligate endosymbiont (Lamelas et al. 2011). The OmpR protein is typically partnered in a TCS with the EnvZ protein. However, only a truncated EnvZ protein can be found in the same operon of *S. symbiotica* SCc by BLAST analysis, suggesting that the gene has been pseudogenized and is no longer required by the bacterium. Since this TCS appears to be the last to be lost from *S. symbiotica* SCc, and because OmpR-EnvZ TCS is involved in sensing osmotic stress and acidity, this result suggests that *S. symbiotica* SCc recently circumvented challenges associated with osmotic pressure and/or pH. Of note, the CpxA-CpxR pathway maintained in *W. glossinidia* is functionally connected to the EnvZ-OmpR system (Grabowicz and Silhavy 2017), potentially indicating a need for further experimental emphasis on membrane biogenesis in endosymbionts. Of course, close examination of TCS loss from related endosymbionts making the same transition among similar host species would be necessary to accurately trace the particular stresses and demands encountered by endosymbionts as they become ever more established within their hosts.

6.4 Quorum-Sensing Mechanisms in Endosymbionts

In order to coordinate collective behavior in response to the demands of the local environment, bacteria must sense and respond to members of the same species by use of quorum-sensing mechanisms (Fig. 6.2). Cooperative behavior regulated by

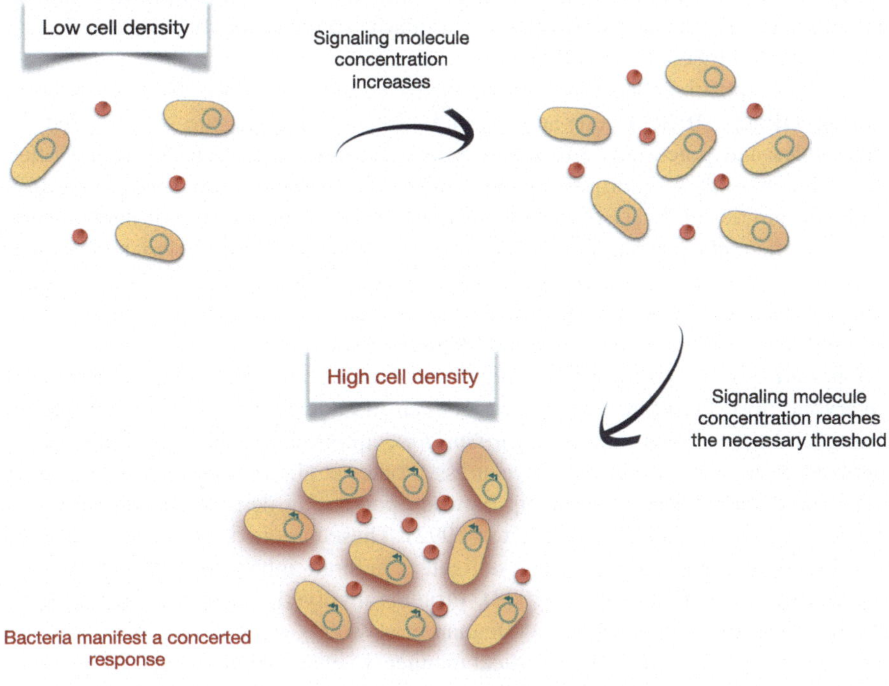

Fig. 6.2 Quorum sensing allows bacteria to change behavior based upon the number of bacteria within the environment. Bacteria produce signaling molecules (denoted here as red spheres). An increase in the signaling molecule concentration allows the population to sense greater numbers. Upon reaching a particular population density, bacteria can respond with concerted group behavior

quorum sensing includes biofilm production, expression of virulence factors, production of antibiotics, and antibiotic resistance (Abisado et al. 2018; Prüß 2017; Rutherford and Bassler 2012). Symbiosis is also modulated by quorum sensing, and indeed, initial efforts to understand quorum sensing focused upon bacterial luminescence by the symbiont *Vibrio fischeri* when it is localized to the light-producing organs of its bobtail squid host (Hastings and Nealson 1977; Nealson et al. 1970).

Mechanisms of quorum sensing differ between Gram-negative and Gram-positive bacteria (Mukherjee and Bassler 2019; Hmelo 2017; Papenfort and Bassler 2016). Gram-negative bacteria synthesize one or more acyl-homoserine lactones or "autoinducer 1" (AI-1) ligands when communicating with one another in a more specific manner. More generalized "autoinducer 2" (AI-2) signals, produced by use of the metabolite 4,5-dihydroxy-2,3-pentanedione, appear to allow communication between different species (Pereira et al. 2013). TCSs can play an important role in the detection of specific and general quorum-sensing signals in Gram-negative bacteria (Papenfort and Bassler 2016). For example, the general quorum-sensing molecule AI-2 binds, at high cell density, to the periplasmic LuxP protein of the bioluminescent *Vibrio harveyi*. Ligand binding ensures that the phosphatase activity of the hybrid HK LuxQ predominates, resulting in activation of hundreds of genes (Ball et al. 2017). The more species-specific AI-1 appears to act through a different HK in *V. harveyi*, called LuxN. This ligand binds directly to its periplasmic domain and promotes its phosphatase activity, similarly resulting in the transcription of genes activated by elevated cell density.

For Gram-positive bacteria, peptide-based ligands are often used for quorum sensing (Bhatt 2018; Lyon and Novick 2004). Ligands are synthesized as propeptides and potentially processed before and after secretion. TCSs are often the mediators of these quorum-sensing peptides. As examples, the AgrC-AgrA TCS binds the processed AgrD peptide to mediate toxin synthesis and virulence in the opportunistic pathogen *Staphylococcus aureus* (Wang and Muir 2016), and competence in *Streptococcus pneumoniae* is promoted when peptide used for quorum sensing activates the ComD-ComE TCS, resulting in the upregulation of genes required for DNA uptake (Shanker and Federle 2017).

Quorum-sensing pathways are not limited to bacteria that live outside of a eukaryotic host. Quorum sensing occurs even in endosymbionts. For example, in the secondary endosymbiont *S. glossinidius*, quorum sensing regulates genes involved in the response to oxidative stress (Pontes et al. 2008), which is intriguing given the demonstrated relationship between population density and resistance to reactive oxygen species (Ma and Eaton 1992). These targets of quorum sensing are also found in the closely related, obligate symbiont inhabiting the rice weevil *Sitophilus oryzae* (Pontes et al. 2008). Proteins involved in quorum sensing have also been identified in *H. defensa* (Degnan et al. 2009), which is mainly, although not exclusively, transmitted in a vertical manner (Li et al. 2018). Targets of quorum-sensing pathways can change significantly upon conversion of a free-living bacteria to an endosymbiont, and while quorum sensing is often associated with virulence, quorum-sensing pathways may also serve as a check upon virulence to promote establishment of a mutualistic relationship between bacterium and host (Enomoto et al. 2017; Papenfort and Bassler 2016; Winzer and Williams 2001). We suggest that the host may even exploit endosymbiont quorum-sensing pathways in order to maintain mutualism. Supporting the idea that eukaryotes can control bacterial pathogenicity by exploiting bacterial quorum-sensing mechanisms, proliferation of the pathogen *Acinetobacter baumannii* can be hindered by its sensation of a fungus-produced farnesol within the context of a co-infection paradigm (Kobayashi and Crouch 2009; Peleg et al. 2008).

So far, to our knowledge, TCSs have not been explicitly linked to quorum sensing by an endosymbiont, and ligand sensation by HKs is certainly not strictly required for quorum sensing (Colton et al. 2015; Urbanowski et al. 2004). However, TCSs should be expected to have a prominent role in intraspecies and interspecies communication by endosymbionts. Given the rapid expansion of endosymbiont genomes available for analysis, and the well-characterized general role of TCSs in quorum sensing, we suggest that TCS involvement in endosymbiont quorum sensing should be a focus of future bioinformatic and experimental attention.

6.5 Two-Component Systems and Endosymbiont-to-Organelle Transitions

The ability of an endosymbiont to sense, respond to, and potentially defend itself against AMPs, likely mediated by TCSs, may be relevant to the frequency at which endosymbiont-to-organelle conversions may take place. What it means to be an "organelle" remains ill-defined (Keeling and Archibald 2008; Theissen and Martin 2006). However, a bright line between mutualist endosymbiont and organelle is almost certainly crossed when the import of key host proteins into the endosymbiont becomes required for host survival. The question of how such an import mechanism can evolve is difficult, and the rarity of extant organelles derived from endosymbionts suggests that development of the required translocation machinery is not trivial (Cavalier-Smith and Lee 1985). Recently, several instances in which host proteins are translocated through endosymbiont membranes have been identified (Bublitz et al. 2019; McCutcheon and Keeling 2014; Nakabachi et al. 2014). Among these intriguing examples, the most prominent may be the import of hundreds of host proteins into the photosynthetic endosymbiont residing within the amoeba *Paulinella chromatophora* (Nowack and Grossman 2012), which seems to have been captured in the midst of an endosymbiont-to-organelle transition.

Of the proteins imported from the host into the *P. chromatophora* endosymbiont, many substrates were reported to harbor amino-terminal sequences similar in structure to AMPs (Singer et al. 2017), although additional support for the idea that these regions are related to AMPs is warranted (Knopp et al. 2020). However, if these amino termini do indeed have AMP-like activity, these findings, as well as others focused upon organelle targeting sequences (Garrido et al. 2020), would raise the possibility that the initial import of host proteins into an endosymbiont may not require pre-existing translocation machinery. Instead, endosymbiont-directed proteins may instead penetrate membrane barriers by utilizing the biophysical properties of membrane-permeable domains mimicking, or derived from, AMPs (Mergaert et al. 2017; Wollman 2016). As described above, TCSs like the PhoP-PhoQ system can play a role in AMP resistance. Consequently, the link between AMP sensation and endosymbiont-to-organelle transitions will remain a topic of high interest for those studying the initial and continuing evolution of the eukaryotic cell.

The level of autonomy that the endosymbiont maintains over its most important activities during the endosymbiont-to-organelle transition may be reflected by the TCSs that it encodes, since any semblance of autonomy would require the ability to respond to the appropriate local signals (Allen 1993, 2017). Interestingly, *P. chromatophora* encodes at least one HK protein clearly related to the NblS protein of the cyanobacterium *Synechococcus elongatus* (50% identity over alignment region). NblS can be involved in the sensation of multiple stressors (Ashby and Houmard 2006), is commonly found in cyanobacteria (Morrison et al. 2005), and is linked to regulation of photosynthetic processes (Hsiao et al. 2004; van Waasbergen et al. 2002). The presence of NblS as one of the few TCSs remaining in the *P. chromatophora* endosymbiont is consistent with the idea that this endosymbiont maintains some regulatory control over its metabolism and photosynthetic capacity.

6.6 Concluding Remarks and Perspective

In this chapter, we have described the current status of research into TCS signaling by endosymbionts. Although a number of endosymbiont TCS pathways have been discovered, most of these pathways remain uncharacterized. Yet, given the incredible pace with which new endosymbiont genomes are acquired and characterized, and taking into account the appropriately increasing interest in endosymbionts, we expect an increase in efforts to understand endosymbiont signal reception in the coming years. Moreover, genomic approaches will reveal which TCSs and downstream transcriptional programs might be most easily lost during integration of endosymbionts into their hosts, thereby revealing the stressors and factors most difficult for endosymbionts to circumvent. Finally, instances of host protein import into endosymbionts, implying potential endosymbiont-to-organelle conversion, continue to be identified. Consequently, the study of endosymbiont TCSs that sense AMPs potentially acting as the precursors of organelle targeting sequences may be informative regarding the evolution of endosymbiont-derived organelles.

Acknowledgments C.D.D. is funded by the Sigrid Jusélius Foundation and by an ERC Starter Grant (RevMito 637649). We thank Gülayşe İnce Dunn, John McCutcheon, and Sven Gould for helpful examination of the manuscript. In figure construction, elements from Servier Medical Art (https://smart.servier.com/) were utilized under Creative Commons License attribution 3.0 Unported License.

References

Abisado RG, Benomar S, Klaus JR, Dandekar AA, Chandler JR (2018) Bacterial quorum sensing and microbial community interactions. mBio 9(3):e02331-17

Akman L, Yamashita A, Watanabe H, Oshima K, Shiba T, Hattori M, Aksoy S (2002) Genome sequence of the endocellular obligate symbiont of tsetse flies, *Wigglesworthia glossinidia*. Nat Genet 32(3):402–407

Allen JF (1993) Control of gene expression by redox potential and the requirement for chloroplast and mitochondrial genomes. J Theor Biol 165(4):609–631

Allen JF (2017) The CoRR hypothesis for genes in organelles. J Theor Biol 434:50–57

Ashby MK (2004) Survey of the number of two-component response regulator genes in the complete and annotated genome sequences of prokaryotes. FEMS Microbiol Lett 231 (2):277–281

Ashby MK, Houmard J (2006) Cyanobacterial two-component proteins: structure, diversity, distribution, and evolution. Microbiol Mol Biol Rev 70(2):472–509

Bader MW, Sanowar S, Daley ME, Schneider AR, Cho U, Xu W, Klevit RE, Le Moual H, Miller SI (2005) Recognition of antimicrobial peptides by a bacterial sensor kinase. Cell 122(3):461–472

Baker SJ, Daniels C, Morona R (1997) PhoP/Q regulated genes in *Salmonella typhi*: identification of melittin sensitive mutants. Microb Pathog 22(3):165–179

Ball AS, Chaparian RR, van Kessel JC (2017) Quorum sensing gene regulation by LuxR/HapR master regulators in vibrios. J Bacteriol 199(19):e00105-17

Bennett GM, Moran NA (2015) Heritable symbiosis: the advantages and perils of an evolutionary rabbit hole. Proc Natl Acad Sci USA 112(33):10169–10176

Bhate MP, Molnar KS, Goulian M, DeGrado WF (2015) Signal transduction in histidine kinases: insights from new structures. Structure 23(6):981–994

Bhatt VS (2018) Quorum sensing mechanisms in gram positive bacteria. In: Bramhachari PV (ed) Implication of quorum sensing system in biofilm formation and virulence. Springer, Singapore, pp 297–311

Bing X, Attardo GM, Vigneron A, Aksoy E, Scolari F, Malacrida A, Weiss BL, Aksoy S (2017) Unravelling the relationship between the tsetse fly and its obligate symbiont *Wigglesworthia*: transcriptomic and metabolomic landscapes reveal highly integrated physiological networks. Proc R Soc B 284(1857):20170360

Biondi EG, Reisinger SJ, Skerker JM, Arif M, Perchuk BS, Ryan KR, Laub MT (2006) Regulation of the bacterial cell cycle by an integrated genetic circuit. Nature 444(7121):899–904

Boraas ME, Seale DB, Boxhorn JE (1998) Phagotrophy by a flagellate selects for colonial prey: a possible origin of multicellularity. Evol Ecol 12(2):153–164

Borland S, Prigent-Combaret C, Wisniewski-Dyé F (2016) Bacterial hybrid histidine kinases in plant-bacteria interactions. Microbiology 162(10):1715–1734

Bright M, Bulgheresi S (2010) A complex journey: transmission of microbial symbionts. Nat Rev Microbiol 8(3):218–230

Bright M, Klose J, Nussbaumer AD (2013) Giant tubeworms. Curr Biol 23(6):R224–R225

Brownlie JC, Adamski M, Slatko B, McGraw EA (2007) Diversifying selection and host adaptation in two endosymbiont genomes. BMC Evol Biol 7:68

Bublitz DC, Chadwick GL, Magyar JS, Sandoz KM, Brooks DM, Mesnage S, Ladinsky MS, Garber AI, Bjorkman PJ, Orphan VJ, McCutcheon JP (2019) Peptidoglycan production by an insect-bacterial mosaic. Cell 179(3):703–712

Capra EJ, Laub MT (2012) Evolution of two-component signal transduction systems. Annu Rev Microbiol 66:325–347

Capra EJ, Perchuk BS, Ashenberg O, Seid CA, Snow HR, Skerker JM, Laub MT (2012) Spatial tethering of kinases to their substrates relaxes evolutionary constraints on specificity. Mol Microbiol 86(6):1393–1403

Casino P, Rubio V, Marina A (2009) Structural insight into partner specificity and phosphoryl transfer in two-component signal transduction. Cell 139(2):325–336

Casino P, Miguel-Romero L, Marina A (2014) Visualizing autophosphorylation in histidine kinases. Nat Commun 5:3258

Catta-Preta CMC, Brum FL, da Silva CC, Zuma AA, Elias MC, de Souza W, Schenkman S, Motta MCM (2015) Endosymbiosis in trypanosomatid protozoa: the bacterium division is controlled during the host cell cycle. Front Microbiol 6:520

Cavalier-Smith T, Lee JJ (1985) Protozoa as hosts for endosymbioses and the conversion of symbionts into organelles. J Protozool 32(3):376–379

Chen X, Li S, Aksoy S (1999) Concordant evolution of a symbiont with its host insect species: molecular phylogeny of genus *Glossina* and its bacteriome-associated endosymbiont, *Wigglesworthia glossinidia*. J Mol Evol 48(1):49–58

Cheng Z, Kumagai Y, Lin M, Zhang C, Rikihisa Y (2006) Intra-leukocyte expression of two-component systems in *Ehrlichia chaffeensis* and *Anaplasma phagocytophilum* and effects of the histidine kinase inhibitor closantel. Cell Microbiol 8(8):1241–1252

Cheung J, Hendrickson WA (2009) Structural analysis of ligand stimulation of the histidine kinase NarX. Structure 17(2):190–201

Christensen S, Serbus LR (2015) Comparative analysis of wolbachia genomes reveals streamlining and divergence of minimalist two-component systems. Genetics 5(5):983–996

Colton DM, Stabb EV, Hagen SJ (2015) Modeling analysis of signal sensitivity and specificity by *Vibrio fischeri* LuxR variants. PloS One 10(5):e0126474

Dalebroux ZD, Miller SI (2014) Salmonellae PhoPQ regulation of the outer membrane to resist innate immunity. Curr Opin Microbiol 17:106–113

Darby AC, Armstrong SD, Bah GS, Kaur G, Hughes MA, Kay SM, Koldkjær P, Rainbow L, Radford AD, Blaxter ML, Tanya VN, Trees AJ, Cordaux R, Wastling JM, Makepeace BL (2012) Analysis of gene expression from the Wolbachia genome of a filarial nematode supports both metabolic and defensive roles within the symbiosis. Genome Res 22(12):2467–2477

Davis NJ, Cohen Y, Sanselicio S, Fumeaux C, Ozaki S, Luciano J, Guerrero-Ferreira RC, Wright ER, Jenal U, Viollier PH (2013) De- and repolarization mechanism of flagellar morphogenesis during a bacterial cell cycle. Genes Dev 27(18):2049–2062

Degnan PH, Yu Y, Sisneros N, Wing RA, Moran NA (2009) *Hamiltonella defensa*, genome evolution of protective bacterial endosymbiont from pathogenic ancestors. Proc Natl Acad Sci U S A 106(22):9063–9068

Dworkin J (2015) Ser/Thr phosphorylation as a regulatory mechanism in bacteria. Curr Opin Microbiol 24:47–52

Enomoto S, Chari A, Clayton AL, Dale C (2017) Quorum sensing attenuates virulence in *Sodalis praecaptivus*. Cell Host Microbe 21(5):629–636

Fast EM, Toomey ME, Panaram K, Desjardins D, Kolaczyk ED, Frydman HM (2011) Wolbachia enhance Drosophila stem cell proliferation and target the germline stem cell niche. Science 334 (6058):990–992

Fenn K, Blaxter M (2004) Quantification of Wolbachia bacteria in *Brugia malayi* through the nematode lifecycle. Mol Biochem Parasitol 137(2):361–364

Ferguson BJ, Indrasumunar A, Hayashi S, Lin M-H, Lin Y-H, Reid DE, Gresshoff PM (2010) Molecular analysis of legume nodule development and autoregulation. J Integr Plant Biol 52 (1):61–76

Fisher SL, Kim S-K, Wanner BL, Walsh CT (1996) Kinetic comparison of the specificity of the Vancomycin resistance kinase VanS for two response regulators, VanR and PhoB. Biochemistry 35(15):4732–4740

Foray V, Pérez-Jiménez MM, Fattouh N, Landmann F (2018) Wolbachia control stem cell behavior and stimulate Germline proliferation in filarial nematodes. Dev Cell 45(2):198–211

Francis VI, Porter SL (2019) Multikinase networks: two-component signaling networks integrating multiple stimuli. Annu Rev Microbiol 73:199–223

Francis VI, Waters EM, Finton-James SE, Gori A, Kadioglu A, Brown AR, Porter SL (2018) Multiple communication mechanisms between sensor kinases are crucial for virulence in *Pseudomonas aeruginosa*. Nat Commun 9(1)

Galperin MY (2006) Structural classification of bacterial response regulators: diversity of output domains and domain combinations. J Bacteriol 188(12):4169–4182

Gao R, Bouillet S, Stock AM (2019) Structural basis of response regulator function. Annu Rev Microbiol 73:175–197

Garrido C, Caspari OD, Choquet Y, Wollman FA, Lafontaine I (2020) Evidence Supporting an Antimicrobial Origin of Targeting Peptidesto Endosymbiotic Organelles. Cells 9(8):E1795

Göttfert M, Grob P, Hennecke H (1990) Proposed regulatory pathway encoded by the *nodV* and *nodW* genes, determinants of host specificity in *Bradyrhizobium japonicum*. Proc Natl Acad Sci U S A 87(7):2680–2684

Goyal A (2018) Metabolic adaptations underlying genome flexibility in prokaryotes. PLoS Genet 14(10):e1007763

Grabowicz M, Silhavy TJ (2017) Envelope stress responses: an interconnected safety net. Trends Biochem Sci 42(3):232–242

Groisman EA, Mouslim C (2006) Sensing by bacterial regulatory systems in host and non-host environments. Nat Rev Microbiol 4(9):705–709

Gutzwiller F (2016) Evolution and gene expression of Wolbachia in *D. melanogaster*. The University of Manchester (United Kingdom)

Gutzwiller F, Carmo CR, Miller DE, Rice DW, Newton ILG, Hawley RS, Teixeira L, Bergman CM (2015) Dynamics of *Wolbachia pipientis* gene expression across the *Drosophila melanogaster* life cycle. Genetics 5(12):2843–2856

Hansen AK, Vorburger C, Moran NA (2012) Genomic basis of endosymbiont-conferred protection against an insect parasitoid. Genome Res 22(1):106–114

Hastings JW, Nealson KH (1977) Bacterial bioluminescence. Annu Rev Microbiol 31:549–595

Henke JM, Bassler BL (2004) Three parallel quorum-sensing systems regulate gene expression in *Vibrio harveyi*. J Bacteriol 186(20):6902–6914

Herron MD, Borin JM, Boswell JC, Walker J, Chen I-CK, Knox CA, Boyd M, Rosenzweig F, Ratcliff WC (2019) De novo origins of multicellularity in response to predation. Sci Rep 9:2328

Hmelo LR (2017) Quorum sensing in marine microbial environments. Annu Rev Mar Sci 9:257–281

Hsiao H-Y, He Q, Van Waasbergen LG, Grossman AR (2004) Control of photosynthetic and high-light-responsive genes by the histidine kinase DspA: negative and positive regulation and interactions between signal transduction pathways. J Bacteriol 186(12):3882–3888

Huynh TN, Stewart V (2011) Negative control in two-component signal transduction by transmitter phosphatase activity. Mol Microbiol 82(2):275–286

Jacob-Dubuisson F, Mechaly A, Betton J-M, Antoine R (2018) Structural insights into the signalling mechanisms of two-component systems. Nat Rev Microbiol 16(10):585–593

Jacobs C, Ausmees N, Cordwell SJ, Shapiro L, Laub MT (2003) Functions of the CckA histidine kinase in *Caulobacter* cell cycle control. Mol Microbiol 47(5):1279–1290

Jenal U, Reinders A, Lori C (2017) Cyclic di-GMP: second messenger extraordinaire. Nat Rev Microbiol 15(5):271–284

Kaneko T, Maita H, Hirakawa H, Uchiike N, Minamisawa K, Watanabe A, Sato S (2011) Complete genome sequence of the soybean symbiont *Bradyrhizobium japonicum* strain USDA6T. Genes 2(4):763–787

Keeling PJ, Archibald JM (2008) Organelle evolution: what's in a name? Curr Biol 18(8):R345

Keeling PJ, McCutcheon JP, Doolittle WF (2015) Symbiosis becoming permanent: survival of the luckiest. Proc Natl Acad Sci USA 112(33):10101–10103

Kim S, Hirakawa H, Muta S, Kuhara S (2010) Identification and classification of a two-component system based on domain structures in bacteria and differences in domain structure between Gram-positive and Gram-negative bacteria. Biosci Biotechnol Biochem:1002261868

Klein AH, Shulla A, Reimann SA, Keating DH, Wolfe AJ (2007) The intracellular concentration of acetyl phosphate in *Escherichia coli* is sufficient for direct phosphorylation of two-component response regulators. J Bacteriol 189(15):5574–5581

Knopp M, Garg SG, Handrich M, Gould SB (2020) Major changes in plastid protein import and the origin of the chloroplastida. iScience 23(3):100896

Kobayashi DY, Crouch JA (2009) Bacterial/fungal interactions: from pathogens to mutualistic endosymbionts. Annu Rev Phytopathol 47:63–82

Krell T, Lacal J, Busch A, Silva-Jiménez H, Guazzaroni M-E, Ramos JL (2010) Bacterial sensor kinases: diversity in the recognition of environmental signals. Annu Rev Microbiol 64:539–559

Kuo C-H, Moran NA, Ochman H (2009) The consequences of genetic drift for bacterial genome complexity. Genome Res 19(8):1450–1454

Lai T-H, Kumagai Y, Hyodo M, Hayakawa Y, Rikihisa Y (2009) The *Anaplasma phagocytophilum* PleC histidine kinase and PleD diguanylate cyclase two-component system and role of cyclic Di-GMP in host cell infection. J Bacteriol 191(3):693–700

Lamelas A, Gosalbes MJ, Manzano-Marín A, Peretó J, Moya A, Latorre A (2011) *Serratia symbiotica* from the aphid *Cinara cedri*: a missing link from facultative to obligate insect endosymbiont. PLoS Genet 7(11):e1002357

Landmann F, Foster JM, Michalski ML, Slatko BE, Sullivan W (2014) Co-evolution between an endosymbiont and its nematode host: Wolbachia asymmetric posterior localization and AP polarity establishment. PLoS Negl Trop Dis 8(8):e3096

Lang K, Lindemann A, Hauser F, Göttfert M (2008) The genistein stimulon of *Bradyrhizobium japonicum*. Mol Gen Genomics 279(3):203–211

Lardi M, Murset V, Fischer H-M, Mesa S, Ahrens CH, Zamboni N, Pessi G (2016) Metabolomic profiling of *Bradyrhizobium diazoefficiens*-induced root nodules reveals both host plant-specific and developmental signatures. Int J Mol Sci 17(6):815

Li Q, Fan J, Sun J, Wang M-Q, Chen J (2018) Plant-mediated horizontal transmission of *Hamiltonella defensa* in the wheat Aphid *Sitobion miscanthi*. J Agric Food Chem 66 (51):13367–13377

Login FH, Balmand S, Vallier A, Vincent-Monégat C, Vigneron A, Weiss-Gayet M, Rochat D, Heddi A (2011) Antimicrobial peptides keep insect endosymbionts under control. Science 334 (6054):362–365

Lori C, Ozaki S, Steiner S, Böhm R, Abel S, Dubey BN, Schirmer T, Hiller S, Jenal U (2015) Cyclic di-GMP acts as a cell cycle oscillator to drive chromosome replication. Nature 523 (7559):236–239

Lyon GJ, Novick RP (2004) Peptide signaling in *Staphylococcus aureus* and other Gram-positive bacteria. Peptides 25(9):1389–1403

Ma M, Eaton JW (1992) Multicellular oxidant defense in unicellular organisms. Proc Natl Acad Sci USA 89(17):7924–7928

Masson F, Zaidman-Rémy A, Heddi A (2016) Antimicrobial peptides and cell processes tracking endosymbiont dynamics. Philos Trans R Soc Lond Ser B Biol Sci 371(1695):20150298

McCutcheon JP, Keeling PJ (2014) Endosymbiosis: protein targeting further erodes the organelle/symbiont distinction. Curr Biol 24(14):R654–R655

McCutcheon JP, Boyd BM, Dale C (2019) The life of an insect endosymbiont from the cradle to the grave. Curr Biol 29(11):R485–R495

Mergaert P (2018) Role of antimicrobial peptides in controlling symbiotic bacterial populations. Nat Prod Rep 35(4):336–356

Mergaert P, Kikuchi Y, Shigenobu S, Nowack ECM (2017) Metabolic integration of bacterial endosymbionts through antimicrobial peptides. Trends Microbiol 25(9):703–712

Metcalf JA, Bordenstein SR (2012) The complexity of virus systems: the case of endosymbionts. Curr Opin Microbiol 15(4):546–552

Mitchell AM, Silhavy TJ (2019) Envelope stress responses: balancing damage repair and toxicity. Nat Rev Microbiol 17(7):417–428

Moran NA (2003) Tracing the evolution of gene loss in obligate bacterial symbionts. Curr Opin Microbiol 6(5):512–518

Morris BEL, Henneberger R, Huber H, Moissl-Eichinger C (2013) Microbial syntrophy: interaction for the common good. FEMS Microbiol Rev 37(3):384–406

Morrison SS, Mullineaux CW, Ashby MK (2005) The influence of acetyl phosphate on DspA signalling in the Cyanobacterium *Synechocystis sp.* PCC6803. BMC Microbiol 5:47

Moscoso JA, Mikkelsen H, Heeb S, Williams P, Filloux A (2012) The *Pseudomonas aeruginosa* sensor RetS switches Type III and Type VI secretion via c-di-GMP signalling. Environ Microbiol 14(4):1088–1089

Mukherjee S, Bassler BL (2019) Bacterial quorum sensing in complex and dynamically changing environments. Nat Rev Microbiol 17(6):371–382

Nakabachi A, Ishida K, Hongoh Y, Ohkuma M, Miyagishima S-Y (2014) Aphid gene of bacterial origin encodes a protein transported to an obligate endosymbiont. Curr Biol 24(14):R640–R641

Nealson KH, Platt T, Hastings JW (1970) Cellular control of the synthesis and activity of the bacterial luminescent system. J Bacteriol 104(1):313–322

Neiditch MB, Federle MJ, Miller ST, Bassler BL, Hughson FM (2005) Regulation of LuxPQ receptor activity by the quorum-sensing signal autoinducer-2. Mol Cell 18(5):507–518

Nowack ECM, Grossman AR (2012) Trafficking of protein into the recently established photosynthetic organelles of *Paulinella chromatophora*. Proc Natl Acad Sci U S A 109(14):5340–5345

Papenfort K, Bassler BL (2016) Quorum sensing signal-response systems in Gram-negative bacteria. Nat Rev Microbiol 14(9):576–588

Paul R, Weiser S, Amiot NC, Chan C, Schirmer T, Giese B, Jenal U (2004) Cell cycle-dependent dynamic localization of a bacterial response regulator with a novel di-guanylate cyclase output domain. Genes Dev 18(6):715–727

Pazy Y, Motaleb MA, Guarnieri MT, Charon NW, Zhao R, Silversmith RE (2010) Identical phosphatase mechanisms achieved through distinct modes of binding phosphoprotein substrate. Proc Natl Acad Sci U S A 107(5):1924–1929

Peleg AY, Tampakakis E, Fuchs BB, Eliopoulos GM, Moellering RC, Mylonakis E (2008) Prokaryote-eukaryote interactions identified by using *Caenorhabditis elegans*. Proc Natl Acad Sci U S A 105(38):14585–14590

Pereira CS, Thompson JA, Xavier KB (2013) AI-2-mediated signalling in bacteria. FEMS Microbiol Rev 37(2):156–181

Pérez-Brocal V, Latorre A, Moya A (2013) Symbionts and pathogens: what is the difference? Curr Top Microbiol Immunol 358:215–243

Pietri JE, DeBruhl H, Sullivan W (2016) The rich somatic life of Wolbachia. Microbiology 5 (6):923–936

Pini F, De Nisco NJ, Ferri L, Penterman J, Fioravanti A, Brilli M, Mengoni A, Bazzicalupo M, Viollier PH, Walker GC, Biondi EG (2015) Cell cycle control by the master regulator CtrA in *Sinorhizobium meliloti*. PLoS Genet 11(5):e1005232

Podgornaia AI, Laub MT (2013) Determinants of specificity in two-component signal transduction. Curr Opin Microbiol 16(2):156–162

Pontes MH, Babst M, Lochhead R, Oakeson K, Smith K, Dale C (2008) Quorum sensing primes the oxidative stress response in the insect endosymbiont, *Sodalis glossinidius*. PloS One 3(10): e3541

Pontes MH, Smith KL, De Vooght L, Van Den Abbeele J, Dale C (2011) Attenuation of the sensing capabilities of PhoQ in transition to obligate insect-bacterial association. PLoS Genet 7(11): e1002349

Prüß BM (2017) Involvement of two-component signaling on bacterial motility and biofilm development. J Bacteriol 199(18):e00259-17

Rikihisa Y (2010) *Anaplasma phagocytophilum* and *Ehrlichia chaffeensis*: subversive manipulators of host cells. Nat Rev Microbiol 8(5):328–339

Rio RVM, Symula RE, Wang J, Lohs C, Wu Y-N, Snyder AK, Bjornson RD, Oshima K, Biehl BS, Perna NT, Hattori M, Aksoy S (2012) Insight into the transmission biology and species-specific functional capabilities of tsetse (Diptera: glossinidae) obligate symbiont *Wigglesworthia*. mBio 3(1)

Römling U, Galperin MY, Gomelsky M (2013) Cyclic di-GMP: the first 25 years of a universal bacterial second messenger. Microbiol Mol Biol Rev 77(1):1–52

Rutherford ST, Bassler BL (2012) Bacterial quorum sensing: its role in virulence and possibilities for its control. Cold Spring Harb Perspect Med 2(11):a012427

Sachs JL, Skophammer RG, Regus JU (2011) Evolutionary transitions in bacterial symbiosis. Proc Natl Acad Sci USA 108(2):10800–10807

Saint André AV, Blackwell NM, Hall LR, Hoerauf A, Brattig NW, Volkmann L, Taylor MJ, Ford L, Hise AG, Lass JH, Diaconu E, Pearlman E (2002) The role of endosymbiotic Wolbachia bacteria in the pathogenesis of river blindness. Science 295(5561):1892–1895

Serbus LR, Sullivan W (2007) A cellular basis for Wolbachia recruitment to the host germline. PLoS Pathogens 3(12):e190

Shanker E, Federle MJ (2017) Quorum sensing regulation of competence and bacteriocins in *Streptococcus pneumoniae* and *mutans*. Genes 8(1)

Shigenobu S, Watanabe H, Hattori M, Sakaki Y, Ishikawa H (2000) Genome sequence of the endocellular bacterial symbiont of aphids *Buchnera sp.* APS. Nature 407(6800):81–86

Singer A, Poschmann G, Mühlich C, Valadez-Cano C, Hänsch S, Hüren V, Rensing SA, Stühler K, Nowack ECM (2017) Massive protein import into the early-evolutionary-stage photosynthetic organelle of the Amoeba *Paulinella chromatophora*. Curr Biol 27(18):2763–2773

Smith TE, Moran NA (2020) Coordination of host and symbiont gene expression reveals a metabolic tug-of-war between aphids and Buchnera. Proc Natl Acad Sci USA

Snyder AK, McLain C, Rio RVM (2012) The tsetse fly obligate mutualist *Wigglesworthia morsitans* alters gene expression and population density via exogenous nutrient provisioning. Appl Environ Microbiol 78(21):7792–7797

Stanley SM (1973) An ecological theory for the sudden origin of multicellular life in the late precambrian. Proc Natl Acad Sci U S A 70(5):1486–1489

Steiner BD, Eberly AR, Hurst MN, Zhang EW, Green HD, Behr S, Jung K, Hadjifrangiskou M (2018) Evidence of cross-regulation in two closely related pyruvate-sensing systems in uropathogenic *Escherichia coli*. J Membr Biol 251(1):65–74

Stephenson K, Hoch JA (2002) Evolution of signalling in the sporulation phosphorelay. Mol Microbiol 46(2):297–304

Stoll S, Feldhaar H, Gross R (2009) Transcriptional profiling of the endosymbiont Blochmannia floridanus during different developmental stages of its holometabolous ant host. Environ Microbiol 11(4):877–888

Sullivan W (2017) Wolbachia, bottled water, and the dark side of symbiosis. Mol Biol Cell 28 (18):2343–2346

Taylor MJ, Voronin D, Johnston KL, Ford L (2013) Wolbachia filarial interactions. Cell Microbiol 15(4):520–526

Theissen U, Martin W (2006) The difference between organelles and endosymbionts. Curr Biol 16 (24):R1016–R1017

Torsvik V, Øvreås L, Thingstad TF (2002) Prokaryotic diversity--magnitude, dynamics, and controlling factors. Science 296(5570):1064–1066

Townsend GE 2nd, Raghavan V, Zwir I, Groisman EA (2013) Intramolecular arrangement of sensor and regulator overcomes relaxed specificity in hybrid two-component systems. Proc Natl Acad Sci U S A 110(2):E161–E169

Trampari E, Stevenson CEM, Little RH, Wilhelm T, Lawson DM, Malone JG (2015) Bacterial rotary export ATPases are allosterically regulated by the nucleotide second messenger cyclic-di-GMP. J Biol Chem 290(40):24470–24483

Tschowri N, Schumacher MA, Schlimpert S, Chinnam NB, Findlay KC, Brennan RG, Buttner MJ (2014) Tetrameric c-di-GMP mediates effective transcription factor dimerization to control *Streptomyces* development. Cell 158(5):1136–1147

Urbanowski ML, Lostroh CP, Greenberg EP (2004) Reversible acyl-homoserine lactone binding to purified *Vibrio fischeri* LuxR protein. J Bacteriol 186(3):631–637

Valentini M, Filloux A (2016) Biofilms and cyclic di-GMP (c-di-GMP) signaling: lessons from *Pseudomonas aeruginosa* and other bacteria. J Biol Chem 291(24):12547–12555 van Waasbergen LG, Dolganov N, Grossman AR (2002) *nblS*, a gene involved in controlling photosynthesis-related gene expression during high light and nutrient stress in *Synechococcus elongatus* PCC 7942. J Bacteriol 184(9):2481–2490

Wakeel A, Zhu B, Yu X-J, McBride JW (2010) New insights into molecular *Ehrlichia chaffeensis*-host interactions. Microbes Infect 12(5):337–345

Wang B, Muir TW (2016) Regulation of virulence in *Staphylococcus aureus*: molecular mechanisms and remaining puzzles. Cell Chem Biol 23(2):214–224

Werren JH (1997) Biology of Wolbachia. Annu Rev Entomol 42:587–609

Whitman WB, Coleman DC, Wiebe WJ (1998) Prokaryotes: the unseen majority. Proc Natl Acad Sci USA 95(12):6578–6583

Wilcox JL, Dunbar HE, Wolfinger RD, Moran NA (2003) Consequences of reductive evolution for gene expression in an obligate endosymbiont. Mol Microbiol 48(6):1491–1500

Wilkinson TL, Koga R, Fukatsu T (2007) Role of host nutrition in symbiont regulation: impact of dietary nitrogen on proliferation of obligate and facultative bacterial endosymbionts of the pea aphid *Acyrthosiphon pisum.* Appl Environ Microbiol 73(4):1362–1366

Winzer K, Williams P (2001) Quorum sensing and the regulation of virulence gene expression in pathogenic bacteria. Int J Med Microbiol 291(2):131–143

Wollman F-A (2016) An antimicrobial origin of transit peptides accounts for early endosymbiotic events. Traffic 17(12):1322–1328

Wright DP, Ulijasz AT (2014) Regulation of transcription by eukaryotic-like serine-threonine kinases and phosphatases in Gram-positive bacterial pathogens. Virulence 5(8):863–885

Zachar I, Boza G (2020) Endosymbiosis before eukaryotes: mitochondrial establishment in protoeukaryotes. Cell Mol Life Sci:1–21

Zschiedrich CP, Keidel V, Szurmant H (2016) Molecular mechanisms of two-component signal transduction. J Mol Biol 428(19):3752–3775

Chapter 7
Phenotype Heritability in Holobionts: An Evolutionary Model

Saúl Huitzil, Santiago Sandoval-Motta, Alejandro Frank, and Maximino Aldana

Abstract Many complex diseases are expressed with high incidence only in certain populations. Genealogy studies determine that these diseases are inherited with a high probability. However, genetic studies have been unable to identify the genomic signatures responsible for such heritability, as identifying the genetic variants that make a population prone to a given disease is not enough to explain its high occurrence within the population. This gap is known as the missing heritability problem. We know that the microbiota plays a very important role in determining many important phenotypic characteristics of its host, in particular the complex diseases for which the missing heritability occurs. Therefore, when computing the heritability of a phenotype, it is important to consider not only the genetic variation in the host but also in its microbiota. Here we test this hypothesis by studying an evolutionary model based on gene regulatory networks. Our results show that the holobiont (the host plus its microbiota) is capable of generating a much larger variability than the host alone, greatly reducing the missing heritability of the

S. Huitzil · M. Aldana (✉)
Instituto de Ciencias Físicas, Universidad Nacional Autónoma de México, Cuernavaca, Morelos, Mexico

Centro de Ciencias de la Complejidad, Universidad Nacional Autónoma de México, Mexico City, Mexico
e-mail: max@icf.unam.mx

S. Sandoval-Motta
Centro de Ciencias de la Complejidad, Universidad Nacional Autónoma de México, Mexico City, Mexico

Instituto Nacional de Medicina Genómica, Mexico City, Mexico

Consejo Nacional de Ciencia y Tecnología, Cátedras CONACyT, Mexico City, Mexico

A. Frank
Centro de Ciencias de la Complejidad, Universidad Nacional Autónoma de México, Mexico City, Mexico

Instituto de Ciencias Nucleares, Universidad Nacional Autónoma de México, Mexico City, Mexico

M. Kloc (ed.), *Symbiosis: Cellular, Molecular, Medical and Evolutionary Aspects*, Results and Problems in Cell Differentiation 69,
https://doi.org/10.1007/978-3-030-51849-3_7

phenotype. This result strongly suggests that a considerably large part of the missing heritability can be attributed to the microbiome.

7.1 Introduction

One of the main objectives of the Human Genome Project, the HapMap Project, and later of the GWAS studies, has been to understand the genetic architecture responsible for the emergence of complex diseases (Collins et al. 1998; International HapMap Consortium 2003; Visscher et al. 2017). For a long time, through genealogy studies, it has been estimated that such complex diseases have a strong heritability determined by genetic components (Manolio et al. 2009). However, this type of study has the inconvenience that it is limited to a small number of individuals. By contrast, thanks to high-throughput techniques such as GWAS, the size of the population under study could be considerably increased because the degree of kinship between individuals can be determined through their genetic variants (SNPs).

The first GWAS study was performed in 2002 (Ozaki et al. 2002). Since then, this method has been very successful in associating thousands of genetic variants with different phenotypic traits (Buniello et al. 2018). However, the effect of these genetic variants on a particular phenotype is very small (Maher 2008) and, consequently, they can explain only a small fraction of the heritability of the associated phenotype. This is because the fraction of phenotypic variance that can be attributed to additive genetic factors is considerably smaller than expected when we consider the genetic variants associated with a given phenotype by GWAS (Gibson 2018).

The gap between the heritability observed through genealogy studies and the one measured from GWAS is known as the *missing heritability* and for some phenotypes can be as large as 60% (Visscher 2008). A typical example is a human height. On the one hand, familial studies have estimated that this phenotype has a heritability larger than 80%. On the other hand, through GWAS about 700 SNPs have been associated with a human height. However, these SNPs can explain only 20% of its heritability (Wood et al. 2014). The most used strategy to fill up the missing heritability gap has been to find more and more genetic variants associated with a particular phenotype in order to increase its estimated heritability (Gibson 2018). In 2010, Yang et al. computed the heritability produced by 29,4831 SNPs, which allowed them to estimate the heritability of human height around 45% (Yang et al. 2010). Nonetheless, if the missing heritability problem were to be solved by increasing the number of SNPs, then we would have to consider so many genetic variants that we would still be unable to understand the genetic architecture behind the emergence of complex diseases and phenotypes. Several solutions have been proposed, such as taking into account epigenetic and epistatic effects, the effect of non-coding RNAs in gene regulation, or an overestimation of the heritability itself (Manolio et al. 2009; Maher 2008; Eichler et al. 2010; Grandjean et al. 2013; Slatkin 2009; Zuk et al.

2014). However, there is no consensus yet in the scientific community as to how to solve this problem.

During the last decade, numerous studies have shown the importance of the microbiome in determining important phenotypic traits of the host organism (Brooks et al. 2016). The microbiota composition and functionality have been correlated with many complex diseases, allergies, neurological disorders, metabolism of antibiotics and other drugs, to mention just a few (Cho and Blaser 2012; Inoue and Shimojo 2015; Kim 2015; Tremlett et al. 2017; Wilson and Nicholson 2017). Additionally, there is evidence suggesting that phenotypic traits can be transmitted (or inherited) from the parents to their offspring through the microbiota (Karunakar et al. 2019). For instance, the phylogenetic congruence (phylosymbiosis) between the host organism and its microbiota suggests that a set of microorganisms have been inherited throughout evolution (Ley et al. 2008; Ochman et al. 2010). Furthermore, it is known that the metabolic response of several host organisms to the microbiota is conserved throughout evolution across different vertebrates, including the zebrafish, mice, and humans (Rawls et al. 2004). The microbiota participates in practically all important biological tasks of the host (Cho and Blaser 2012). Given the strong symbiotic interactions between the host and its microbiota, and the evidence showing that microorganisms can be transmitted both vertically and horizontally from parents to offspring, Rosenberg and his collaborators formulated the hypothesis of the holobiont as a unit of selection in evolution (Zilber-Rosenberg and Rosenberg 2008). This hypothesis states that in order to understand the evolution of the phenotypic traits of a given organism one has to consider the hologenome (the genome of the host and the genomes of all of its microbes) as a unit of selection (Zilber-Rosenberg and Rosenberg 2008). This means that changes in the phenotype of a host organism can be produced by mutations in the host's genome, in the genes of its microbes (the host's microbiome), or in both, and to understand these phenotypic changes one has to consider genetic variation in the entire hologenome. This may be particularly true for complex phenotypes for which missing heritability exists. Therefore, it is reasonable to assume that, when computing the heritability of a phenotype, the genetic variants of the entire hologenome have to be taken into account. This hypothesis was proposed in previous works (Sandoval-Motta et al. 2017; Veigl et al. 2019), but to our knowledge, no measurements have been performed so far validating or refuting this idea. Here we present an evolutionary model, based on Boolean networks, showing quantitatively that a large part of the missing heritability is recovered when the genetic variants are computed using the entire hologenome rather than just the genome of the host.

7.2 Boolean Network Model

We choose to work with Boolean networks as representations of the genomes of both the host and its microbes. These networks were proposed by S. Kauffman in 1969 as models for gene regulation, and have proven to reproduce the gene expression

a

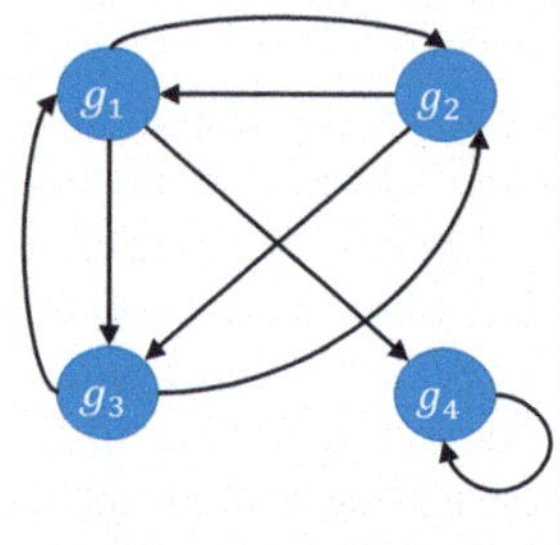

b

g_2	g_3	$F_1(g_2, g_3)$
0	0	0
0	1	1
1	0	1
1	1	0

g_1	g_3	$F_2(g_1, g_3)$
0	0	0
0	1	0
1	0	0
1	1	1

g_1	g_2	$F_3(g_1, g_2)$
0	0	0
0	1	1
1	0	1
1	1	1

g_1	g_4	$F_4(g_1, g_4)$
0	0	1
0	1	0
1	0	0
1	1	1

c

Input configuration			Value
Index	g_2	g_3	$F_1(g_2, g_3)$
1	0	0	$f_1^1 = 0$
2	0	1	$f_1^2 = 1$
3	1	0	$f_1^3 = 1$
4	1	1	$f_1^4 = 0$

d

Boolean DNA

$$BDNA = \{f_1^1 f_1^2 f_1^3 f_1^4 f_2^1 f_2^2 f_2^3 f_2^4 f_3^1 f_3^2 f_3^3 f_3^4 f_4^1 f_4^2 f_4^3 f_4^4\}$$

$$G = \{0110000101111001\}$$

Fig. 7.1 (**a**) Illustrative example of a network with four genes, each one having $K = 2$ inputs. The arrows indicate the regulatory interactions. (**b**) Specific example of the Boolean functions associated with each of the four genes in the network. (**c**) Structure of a Boolean function. The first column represents the index of the input configuration, ranging from $i = 1$ to $i = 4$. The second and third columns show the configurations of the inputs. The last column is the value of the function acquired on the ith configuration of the inputs. (**d**) The Boolean DNA is constructed by concatenating the values of all the Boolean functions over all of their configurations. The genotype G of the network is the particular sequence the Boolean DNA consists of

patterns observed experimentally for several organisms (Bornholdt 2008; Drossel 2008). A Boolean network consists of N nodes representing the genes in the genome. The state of expression of these genes is encoded in a set of variables $\{g_1, g_2, \ldots g_N\}$ such that $g_n = 1$ if the nth gene is expressed, or $g_n = 0$ if it is not expressed. The value of each g_n is determined by a particular set of k_n other genes in the network, which we represent as $\left\{g_{n_1}, g_{n_2}, \ldots, g_{n_{k_n}}\right\}$ (see Fig. 7.1a). We will refer to this set as the *inputs* or *regulators* of g_n, and to k_n as its *input connectivity* which can be different from one gene to another. Once each gene in the networks has been assigned with a set of regulators, the network dynamics are determined by the simultaneous update of the state of all the genes according to the equation

$$g_n(t + \Delta t) = F_n\left(g_{n_1}(t), g_{n_2}(t), \ldots, g_{n_{k_n}}(t)\right), \tag{7.1}$$

where $F_n\left(g_{n_1}, g_{n_2}, \ldots, g_{n_{k_n}}\right)$ is a Boolean function of k_n variables and Δt is the average time it takes for a gene to respond to changes in its regulators. For networks of real organisms, the set of regulators of each gene g_n is determined from experimental observations and its Boolean function is constructed by hand according to the activating or inhibitory nature of its regulators (see Fig. 7.1b). Here we are interested in the general evolutionary properties of holobionts and not on any particular organism. Therefore, we are going to work with Boolean networks constructed in the following way: (a) the number k_n of inputs of each gene g_n will be randomly chosen from a predefined probability function $P_I(k)$, which has average K and variance σ_k^2; (b) the k_n inputs of each gene are chosen randomly from anywhere in the network with uniform probability; (c) each Boolean function F_n will be constructed randomly such that for each of the 2^{k_n} configurations of its k_n arguments, $F_n = 1$ with probability p and $F_n = 0$ with probability $1 - p$.

Once the inputs and Boolean functions have been assigned to every gene in the network, they do not change throughout time. What changes is the state of expression of each gene according to Eq. (7.1). In other words, the inputs to each gene and its Boolean function are randomly assigned at the beginning of the simulation, when the network is constructed. After that, the topological structure of the network (which gene is connected to another) and the Boolean functions do not change in time. This is important since, as the network structure does not change throughout time, we can define a "Boolean DNA sequence" (BDNA) that characterizes the network. For that, let us note that a Boolean function with K inputs (arguments) has 2^K values, one for each configuration of its inputs (Fig. 7.1b). These 2^K configurations can be indexed according to their Boolean value, as Fig. 7.1c shows. Thus, for instance, a function with $K = 3$ inputs has $2^3 = 8$ configurations, ranging from configuration {000} (the three inputs are off) to configuration {111} (the three inputs are on), with all the intermediate configurations in between. As Fig. 7.1c shows, these configurations can be indexed from $i = 1$ to $i = 2^K$. Let us define f_n^i as the value of the nth Boolean function F_n on the ith configuration of its inputs (see Fig. 7.1c). Then, the Boolean DNA of the network is defined as the concatenation of the values of all the Boolean functions in the network for all of their input configurations (see Fig. 7.1d):

$$\text{BDNA} = \left\{ f_1^1, f_1^2, f_1^3, \ldots, f_1^{2^{k_1}}, f_2^1, f_2^2, f_2^3, \ldots, f_2^{2^{k_2}}, \ldots, f_N^1, f_N^2, f_N^3, \ldots, f_N^{2^{k_N}} \right\}. \tag{7.2}$$

A network with N genes, with the nth gene having k_n inputs, has a Boolean DNA with $\Omega = \sum_{n=1}^{N} 2^{k_n}$ digits or loci. If k_n is the same for all the genes, let's say $k_n = K$ for every $n = 1, 2, \ldots, N$, then $\Omega = N \times 2^k$. The dynamics of the network are determined by both its structural topology and its Boolean DNA. However, given a particular network structure, the Boolean DNA completely determines the dynamics of the network. Mutations in the Boolean DNA can change the network dynamics and its gene expression pattern. From now on we will indistinctly refer to the Boolean DNA of the network defined in Eq. (7.2) also as the *genome* of the network, and to a particular Boolean sequence as its *genotype* (see Fig. 7.1d).

7.3 Genetic Variants in a Population

Since we want to study the heritability of phenotypes in a population, we have to define the phenotype in a quantitative way as well as the variants of that phenotype. To do this, let us consider a network with N genes connected in a specific way and with a specific Boolean DNA (see Fig. 7.2). This network, which we will denote as H_0, represents the genome (or part of the genome) of a given organism. Under some

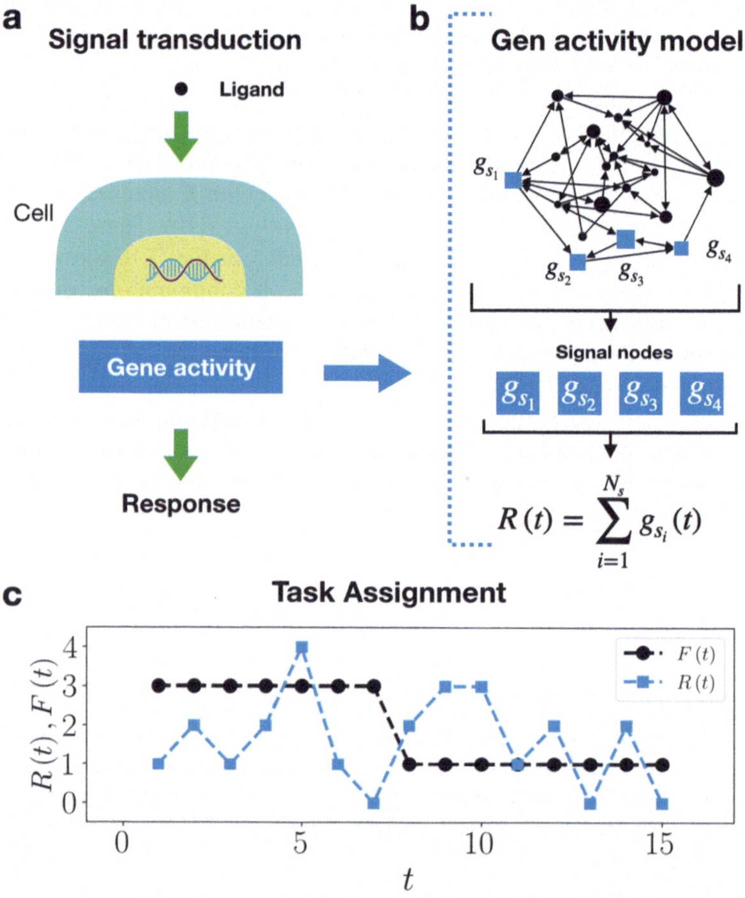

Fig. 7.2 (**a**) Under external signals, a subset of the genes in a cell has to change its expression pattern in order to adapt to the environmental conditions. (**b**) A network representing the genome (or part of the genome) of a cell. The square nodes (in blue) are the transduction nodes that will respond to the external signal (a metabolite, stressor, etc.). These nodes generate a pattern of expression determined by the function $R(t)$, which in this work is just the simple sum of the state of the nodes. (**c**) The desired phenotype $F(t)$ is an arbitrary function defined over an interval of $0 < t \le T$. In general, for a randomly constructed network, its actual phenotype $R(t)$ is quite different from the desired phenotype $F(t)$

environmental conditions a subset of N_S genes of this network, which we will refer to as the *transduction nodes* and represent by $S = \left\{g_1^s, g_2^s, g_3^s, \ldots, g_{N_S}^s\right\}$, has an expression pattern determined by the function $R_0(t)$, defined over a time interval $0 \leq t \leq T$. A gene expression pattern is always associated with a phenotypic trait. Therefore, we will refer to $R_0(t)$ as the *phenotype* of the network. This function may characterize the response of the transduction genes $S = \left\{g_1^s, g_2^s, g_3^s, \ldots, g_{N_S}^s\right\}$ to external metabolites, stressors, or environmental conditions (see Fig. 7.2). It can be, for instance, the phenotype needed to metabolize sugar or fatty acids. Now imagine that the environment permanently changes, and for the organism to adapt to the new environmental challenge, the subset of genes S has to change its expression pattern to a new function $F(t)$. This new function $F(t)$ is the *desired phenotype* for the organism to properly survive in the new environment. Therefore, the *actual* phenotype, encoded in the gene expression pattern $R_0(t)$, has to transform into the desired phenotype $F(t)$. This change will not happen immediately after the environmental change has occurred. Instead, it will be through a series of mutations and partial adaptations that the actual phenotype $R_0(t)$ will transform into the desired phenotype $F(t)$. Therefore, we will have to train the network H_0 to perform the desired function $F(t)$. To do this, we will implement a standard evolutionary algorithm in which a population of networks evolves through mutations and selection so that at each generation, only the networks that better approach the target function $F(t)$ are the ones that survive and pass to the next generations.

The function $R_0(t)$ is the phenotype of the network H_0, and is determined by both its structure and its Boolean DNA (i.e., its genotype). The phenotype $R_0(t)$ has to transform into the desired phenotype $F(t)$. To implement this transformation we will produce mutations in the Boolean DNA by keeping the structural topology of the network fixed. We measure the adaptation of the network to the new environmental conditions through the mean squared error ξ_0 between the actual phenotype $R_0(t)$ and the desired phenotype $F(t)$:

$$\xi_0 = \frac{1}{T} \sum_{t=1}^{T} (R_0(t) - F(t))2. \tag{7.3}$$

If $\xi_0 = 0$, the network is perfectly adapted to the new task, as the actual phenotype $R_0(t)$ of the network is identical to the desired one $F(t)$. By contrast, large values of ξ_0 correspond to a poor adaptation, and the larger the value of ξ_0, the poorer the adaptation.

Since the network H_0 was constructed randomly, its phenotype $R_0(t)$ will, in general, be quite different from the desired phenotype $F(t)$ (see Fig. 7.2c). For $R_0(t)$ to approach $F(t)$ we have to mutate the genome of the network. This allows us to define the *variants* of the network as follows. Let us define G_0 as the genotype of the network H_0, namely, the particular Boolean sequence that characterizes the genome of H_0 (like the particular Boolean sequence in Fig. 7.1d). The variant v_i is a single mutation of G_0 occurring at its ith digit (locus). The variant v_i is thus a single

	Loci																
	1	2	3	4	5	6	7	8	9	10	11	12	13	14	15	16	Phenotype
G_0	0	1	1	0	0	0	0	1	0	1	1	1	1	0	0	1	ξ_0
Variants																	
$G(v_1)$	1	1	1	0	0	0	0	1	0	1	1	1	1	0	0	1	$\xi(v_1)$
$G(v_2)$	0	0	1	0	0	0	0	1	0	1	1	1	1	0	0	1	$\xi(v_2)$
$G(v_3)$	0	1	0	0	0	0	0	1	0	1	1	1	1	0	0	1	$\xi(v_3)$
$G(v_4)$	0	1	1	1	0	0	0	1	0	1	1	1	1	0	0	1	$\xi(v_4)$
$G(v_5)$	0	1	1	0	1	0	0	1	0	1	1	1	1	0	0	1	$\xi(v_5)$
$G(v_6)$	0	1	1	0	0	1	0	1	0	1	1	1	1	0	0	1	$\xi(v_6)$
$G(v_7)$	0	1	1	0	0	0	1	1	0	1	1	1	1	0	0	1	$\xi(v_7)$
$G(v_8)$	0	1	1	0	0	0	0	0	0	1	1	1	1	0	0	1	$\xi(v_8)$
$G(v_9)$	0	1	1	0	0	0	0	1	1	1	1	1	1	0	0	1	$\xi(v_9)$
$G(v_{10})$	0	1	1	0	0	0	0	1	0	0	1	1	1	0	0	1	$\xi(v_{10})$
$G(v_{11})$	0	1	1	0	0	0	0	1	0	1	0	1	1	0	0	1	$\xi(v_{11})$
$G(v_{12})$	0	1	1	0	0	0	0	1	0	1	1	0	1	0	0	1	$\xi(v_{12})$
$G(v_{13})$	0	1	1	0	0	0	0	1	0	1	1	1	0	0	0	1	$\xi(v_{13})$
$G(v_{14})$	0	1	1	0	0	0	0	1	0	1	1	1	1	1	0	1	$\xi(v_{14})$
$G(v_{15})$	0	1	1	0	0	0	0	1	0	1	1	1	1	0	1	1	$\xi(v_{15})$
$G(v_{16})$	0	1	1	0	0	0	0	1	0	1	1	1	1	0	0	0	$\xi(v_{16})$
$G(v_3, v_6, v_{12})$	0	1	0	0	0	1	0	1	0	1	1	0	1	0	0	1	$\xi(v_3, v_6, v_{12})$

Fig. 7.3 Illustrative example of the genetic variants for the genome of the network shown in Fig. 7.1. This genome has $\Omega = 16$ loci, indicated at the top of the matrix. The next row shows the genotype G_0 of the original network H_0 with respect to which the variants are defined. The subsequent rows show the particular genotypes $G(v_i)$ containing the variants v_i, which are indicated in blue. Variant v_i consists of a single mutation (SNP) occurring at the ith position (locus) of the genome G_0 of the original network H_0. The very last row shows a genotype $G(v_3, v_6, v_{12})$ that contains the three variants v_3, v_6 and v_{12}. The column to the far right just illustrates that each variant v_i corresponds to a particular error (phenotype) ξ_i

nucleotide polymorphism (SNP) occurring at the ith locus of the genome. We will denote as $G(v_i)$ the genotype that contains the variant v_i. In other words, $G(v_i)$ is a Boolean sequence almost identical to the genotype G_0 of the original network H_0 except that it has a SNP at the ith locus of the genome (see Fig. 7.3). Clearly, for a network with N genes, each having k_n input connections, there are $\Omega = \sum_{n=1}^{N} 2^{k_n}$ of such variants, $v_1, v_2, \ldots, v_\Omega$ (see Fig. 7.3). In this section and the following one, we will work with networks with $N = 50$ nodes, each with $K = 2$ regulators, which produces $\Omega = 200$ loci in the Boolean DNA and the same number of variants.

We have pointed out before that changes in the Boolean DNA of the network may change its dynamics. Therefore, even one mutation in the genotype of the network can change its phenotype $R_0(t)$. Therefore, to each variant v_i there corresponds a particular phenotype $R_i(t)$. Let us denote as $\xi(v_i)$ the error associated with the variant v_i, which is defined in an analogous way as in Eq. 7.3:

$$\xi(v_i) = \frac{1}{T}\sum_{t=1}^{T}(R_i(t) - F(t))2. \tag{7.4}$$

It should be noted that $\xi(v_i)$ is a quantitative measure of how well the phenotype $R_i(t)$ approximates the desired phenotype $F(t)$. For instance, if $F(t)$ were the phenotype required to metabolize sugar, then $\xi(v_i) \approx 0$ would mean that the organism with the genotype $G(v_i)$ is healthy, whereas a large value of $\xi(v_i)$ would mean that variant v_i is associated with diabetes. Different variants may produce different values of the error $\xi(v_i)$, which could be interpreted as variability in the diabetes phenotype. Therefore, we can consider $\xi(v_i)$ as the quantitative measure of the phenotype $R_i(t)$ that corresponds to the genotype $G(v_i)$, which ultimately is associated with variant v_i. Since the error function $\xi(v)$ is a quantitative measure of the phenotype (or how well the network performs the desired phenotype), we will indistinctly refer to $R(t)$ and $\xi(v)$ as the phenotype of the network.

Analogously, we will represent as $G(v_{i_1}, v_{i_2}, \ldots, v_{i_n})$ the genotype obtained from G_0 by simultaneously implementing the n variants $v_{i_1}, v_{i_2}, \ldots, v_{i_n}$, which are SNPs occurring at positions i_1, i_2, …, i_n of the genome (see Fig. 7.3). The phenotype corresponding to this genotype will be denoted as $R_{i_1,i_2,\ldots,i_n}(t)$, and its corresponding error as $\xi(v_{i_1}, v_{i_2}, \ldots, v_{i_n})$, which is computed in a similar way as in Eq. (7.4).

Some of the variants v_1, v_2, …, v_Ω will have an associated error larger than ξ_0 (the error of the original network H_0), while some other variants will have an error smaller than ξ_0. To measure the effect of the variant v_i on the adaptation of the network to the desired phenotype $F(t)$, we define the error difference $\delta\xi_i$ as

$$\delta\xi_i = \xi(v_i) - \xi_0. \tag{7.5}$$

If $\delta\xi_i < 0$ then the network containing the variant v_i is better adapted to the desired phenotype than the original network H_0, whereas the opposite happens when $\delta\xi_i > 0$. When $\delta\xi_i = 0$ then variant v_i is neutral. Analogously, we can define the effect of the genotype $G(v_{i_1}, v_{i_2}, \ldots, v_{i_n})$ (with n variants) on the adaptation of the network as $\delta\xi_{i_1,i_2,\ldots,i_n} = \xi(v_{i_1}, v_{i_2}, \ldots, v_{i_n}) - \xi_0$. Figure 7.4a shows a plot of the effects $\delta\xi_i$ for all the variants v_1, v_2, …, v_{200} for a network with $N = 50$ and connectivity $K = 2$, while Fig. 7.4b shows the same data ordered in the increasing order of $\delta\xi_i$. Note that, while most of the variants are neutral, some variants produce considerably large effects (up to 50%), both positive and negative. This means that only one SNP can bring the network very close to the desired phenotype $F(t)$, or very far from it.

One of the main arguments trying to explain the missing heritability of phenotypes is that the effects of the different variants on the phenotype are not additive (Eskin 2015; Génin 2019; Zuk et al. 2012). Nonlinear interactions between these variants, known as epistatic effects, may be hiding the heritability of phenotypes. While this may be the case, here we show that nonlinear interactions between the variants are not enough to explain the missing heritability, for even when these epistatic interactions are negligible, the missing heritability persists.

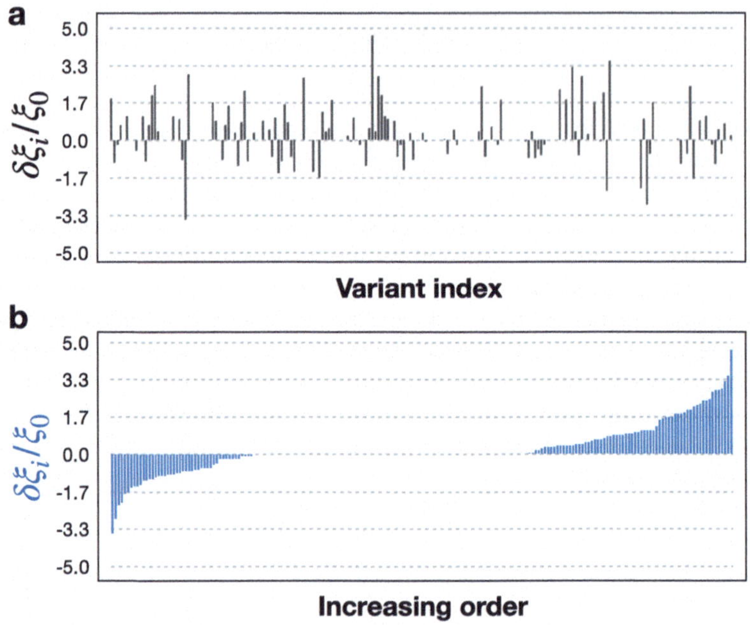

Fig. 7.4 (**a**) Plot of the normalized effect $\delta\xi_i/\xi_0$ of the variants v_i as a function of the variant index i, with $1 \leq i \leq 200$, for a network with $N = 50$ and connectivity $K = 2$. (**b**) Same data as in the previous panel but ordered in increasing order. Note that some variants can have relatively large effects, ranging from -33% to almost 50% of the original error ξ_0. Note also that most of the variants are neutral, with $\delta\xi_i \approx 0$

7.4 Nonlinear (Epistatic) Effects

In the model proposed by Yang et al. (2010) (see also the article by Eskin (2015)), to measure the heritability of a quantitative phenotype, linear regression is performed based on the following equation

$$\xi_j = \overline{\xi} + \sum_{i=1}^{\Omega} g_{ij}\beta_i + e_j, \tag{7.6}$$

where ξ_j represents the value of the phenotype of the jth individual in the population, $\overline{\xi}$ is the average value of the phenotype over the entire population, β_i is the size of the effect of the ith variant v_i on the phenotype, and g_{ij} is a matrix that contains the variants of each individual, so its entries acquire the values 0, 1, or 2 depending on whether none of the alleles of the jth individual has the variant v_i, only one allele has it, or both alleles have the variant. Finally, e_j is an error inherent in the measurement of the phenotype that is associated with the effect of the environment. It is clear that Eq. (7.6) assumes that the effects of the different variants are additive. In this section,

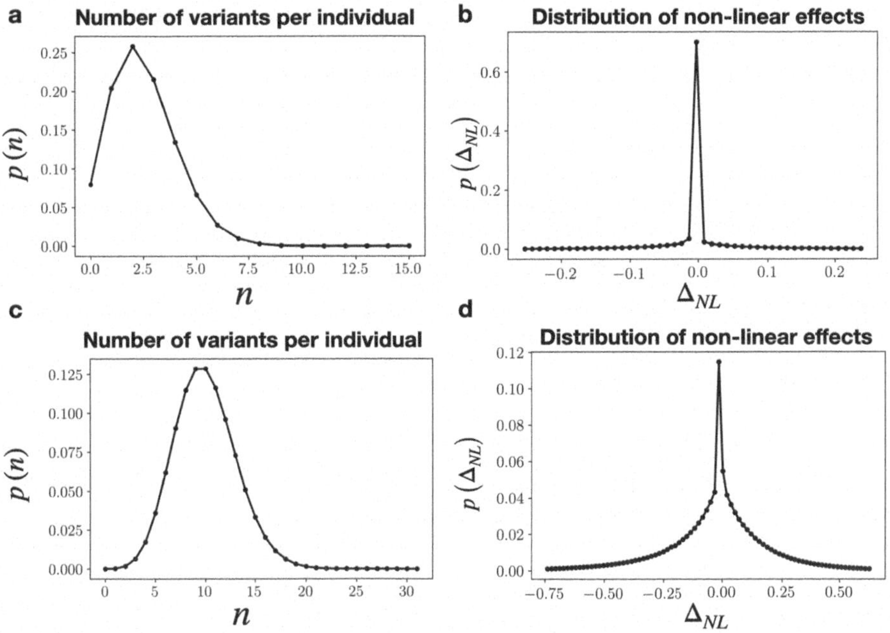

Fig. 7.5 Distribution $P(n)$ of the number of variants n per genome in the population for a variant probability threshold $\delta_v = 0.025$ (panel **a**) and $\delta_v = 0.1$ (panel **c**). In the first case, the average number of variants per individual is $\overline{n} \approx 2.5$, whereas in the second case it is $\overline{n} \approx 10$. Panels **b** and **d** show the distribution $P(\Delta_{NL})$ of nonlinear effects Δ_{NL} for the corresponding populations in the previous two panels. Note that in both cases $P(\Delta_{NL})$ has a sharp maximum at $\Delta_{NL} = 0$, which shows that in our model, the epistatic (nonlinear) interactions between individual variants are not important

we test whether or not the additivity hypothesis is true for the network model we are analyzing.

We construct a population of networks by making $N_P = 10{,}000$ identical copies of H_0, which is a random network with $N = 50$ nodes, each one having $K = 2$ regulators. Therefore, the Boolean DNA has $\Omega = N \times 2^K = 200$ loci. Then, with probability q_i we mutate the ith locus of the genome of each network in the population ($1 \leq i \leq \Omega$). Thus, a mutation in the first position of the genome (variant v_1) will occur in $N_P \times q_1$ networks in the population, a mutation in the second position (variant v_2) will occur in $N_P \times q_2$ networks, and so on. The probabilities q_1, q_2, ..., q_Ω are chosen randomly in the interval $0 < q_i \leq \delta_v$, where δ_v is the upper bound for the mutation probabilities. By increasing the threshold δ_v more mutations will be produced in the genomes of the networks. The reason for choosing different probabilities for the different loci in the Boolean DNA is to allow the different variants $v_1, v_2, \ldots, v_\Omega$ to occur with different frequencies in the population. Note that the genome of one network in the population can accumulate several mutations; that is, it can contain several variants $v_{i_1}, v_{i_2}, \ldots, v_{i_n}$. Figure 7.5a and b show the probability $f(n)$ that the genotype of a randomly chosen network in the population has n variants, for $\delta_v = 0.025$ and $\delta_v = 0.1$, respectively. In the first case the average number of variants in each genome is $\overline{n} = 2.5$, whereas in the second case it is $\overline{n} = 10$.

To determine the nonlinear interaction of the variants we consider a genotype $G(v_{i_1}, v_{i_2}, \ldots, v_{i_n})$ that has n variants. We know the error $\xi(v_{i_1}, v_{i_2}, \ldots, v_{i_n})$ associated with this genotype because we can measure it. Analogously, we know the individual errors $\xi(v_{i_1}), \xi(v_{i_2}), \ldots, \xi(v_{i_n})$ associated with each of the individual variants $v_{i_1}, v_{i_2}, \ldots, v_{i_n}$. To determine whether these individual variants contribute linearly or not to the phenotype $R_{i_1,i_2,\ldots,i_n}(t)$ that corresponds to the genotype $G(v_{i_1}, v_{i_2}, \ldots, v_{i_n})$, we define the *linear contribution* $\xi^L_{i_1,i_2,\ldots,i_n}$ of the individual variants $v_{i_1}, v_{i_2}, \ldots, v_{i_n}$ as

$$\xi^L_{i_1,i_2,\ldots,i_n} = \xi_0 + \sum_{m=1}^{n} \delta\xi_{i_m}. \tag{7.7}$$

The linear contribution $\xi^L_{i_1,i_2,\ldots,i_n}$ is an approximation to the real phenotype $\xi(v_{i_1}, v_{i_2}, \ldots, v_{i_n})$ produced by the genotype $G(v_{i_1}, v_{i_2}, \ldots, v_{i_n})$. If the nonlinear interactions between the individual variants $v_{i_1}, v_{i_2}, \ldots, v_{i_n}$ are indeed very strong, then $\xi^L_{i_1,i_2,\ldots,i_n}$ will be very different from $\xi(v_{i_1}, v_{i_2}, \ldots, v_{i_n})$, whereas if these nonlinear interactions are not so strong, then $\xi^L_{i_1,i_2,\ldots,i_n} \approx \xi(v_{i_1}, v_{i_2}, \ldots, v_{i_n})$. Therefore, a measure of the nonlinearity of the interactions between different variants is determined by the quantity

$$\Delta_{NL} = \xi(v_{i_1}, v_{i_2}, \ldots, v_{i_n}) - \xi^L_{i_1,i_2,\ldots,i_n}. \tag{7.8}$$

Figure 7.5b and d show the probability distribution function $P(\Delta_{NL})$ for the same populations used to generate the data in Fig. 7.5a and c, respectively. Note that in both cases $P(\Delta_{NL})$ has a sharp maximum centered at $\Delta_{NL} = 0$, which means that the nonlinear interactions between individual variants are not quite important in our model. This is particularly true for the case in which there are just a few numbers of SNPs per genome, as in Fig. 7.5a, with an average of $\overline{n} = 2.5$ variants in each Boolean DNA. This represents an SNP occurrence probability of 0.0125 per locus (2.5/200). In the human genome this probability is even lower by at least one order of magnitude, as the SNP occurrence probability in the human genome is 0.0007 per nucleotide (on average 7 SNPs per 10 Kb (Auton et al. 2015; Sachidanandam et al. 2001)). We do not claim that the results reported in Fig. 7.5 can be extrapolated to the human genome (or to any other organism). In our model, and just in our model, the effect of most of the variants can be very well reproduced by a linear contribution of the effects of the corresponding individual variants. This is a remarkable result given that the network dynamics determined by Eq. (7.1) are highly nonlinear.

7.5 GWAS

For the sake of clarity, it is useful to keep in mind the analogy of the desired phenotype $F(t)$ as the capability to metabolize sugars. Taking into account that the error function $\xi(v_{i_1}, v_{i_2}, \ldots, v_{i_n})$ is a quantitative measure of how well the organism with the genotype $G(v_{i_1}, v_{i_2}, \ldots, v_{i_n})$ expresses the desired phenotype, then $\xi = 0$ would correspond to healthy organisms, whereas $\xi \neq 0$ would correspond to diabetic ones. In this analogy, we would be interested in the variants associated with this particular disease, namely, the variants that most increase the value of the phenotype ξ. (From Fig. 7.4 it is clear that there is a symmetric behavior between the positive and negative values of $\delta\xi$; therefore, equivalent results would be obtained if we were looking for the variants that most decrease the value of ξ.) In our model, we can exactly measure the value of the phenotype $\xi(v_i)$ corresponding to each variant v_i, as well as the phenotype $\xi(v_{i_1}, v_{i_2}, \ldots, v_{i_n})$ associated to any combination of variants $v_{i_1}, v_{i_2}, \ldots, v_{i_n}$. However, in real life we cannot know the phenotype $\xi(v_i)$ that each individual variant v_i produces, because such variants generally do not occur isolated in the genome. Therefore, one has to estimate the contribution $\xi(v_i)$ that each individual variant v_i has on the phenotype under analysis. This allows us to identify the variants that are relevant to the phenotype in question. Those are variants that appear with a significantly large frequency among the population that presents the phenotype. This is where the Genome Wide Association Studies (GWAS) enters into play. The main objective of GWAS is to determine the variants that are significantly associated with the phenotype of interest. The main result of GWAS is to associate a p-value $p(v_i)$ to each variant v_i. The variants with the smallest p-values will be the ones that are significantly associated with the phenotype. The natural question arises: how small the p-value has to be for the corresponding variant to be associated with the phenotype? As a rule of thumb, people in the community always choose a threshold $p_T = 0.05$ for the p-value to be significant. If $p(v_i) < p_T$, then the corresponding variant v_i is accepted. Otherwise, it is rejected. Another fact to keep in mind is that, in practice, to perform a GWAS analysis not every individual in the population is tested. It is only a small fraction of the population whose DNA variants are analyzed. The size of the sampled population may have an important influence in the GWAS results, a phenomenon known as the sampling size effect (Eskin 2015; Génin 2019). Some authors claim that by increasing the sampling size (the number of tested individuals), the results GWAS yields would be more trusty.

To simulate a GWAS analysis in our model, we consider a population consisting of $N_P = 10{,}000$ networks, each with $N = 50$ and connectivity $K = 2$. Variant v_i has been implemented with probability q_i in the genotype of the networks, with $0 < q_i < \delta_V = 0.025$ and $1 \leq i \leq 200$. This would result in a population similar to the one used to generate the data reported in Fig. 7.5a and c. To perform GWAS on this population, we extract a subpopulation with $N_{sub} = 500$ networks, which are the ones that have the largest phenotypes ξ (in our analogy, these networks would correspond to clear cases of diabetes). In this subpopulation, the frequency $\widehat{q}_i$ corresponding to the variant v_i may be different from the frequency q_i of this same

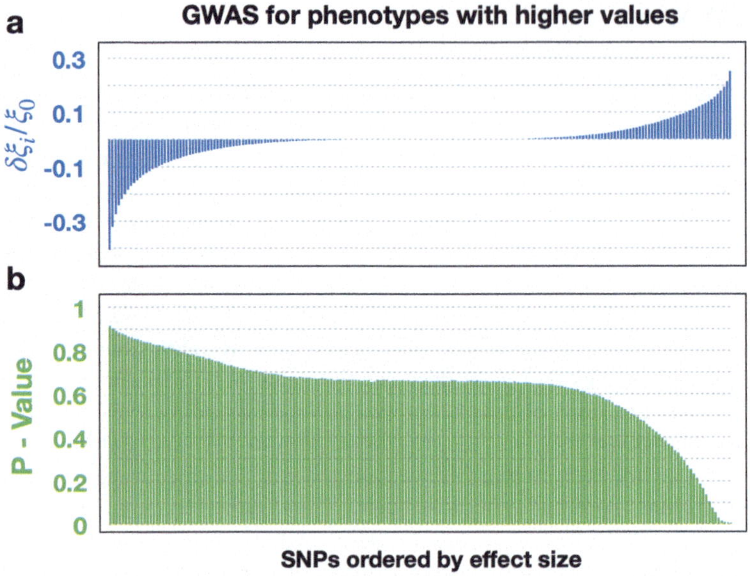

Fig. 7.6 (**a**) Effect size $\delta\xi_i$ for the 200 variants v_i in a population of $N_P = 10{,}000$ networks, each with $N = 50$ nodes and connectivity $K = 2$. The data are plotted in the increasing order of $\delta\xi_i$. (**b**) A subpopulation of $N_{sub} = 500$ neworks (5%) is extracted. These networks are the ones with the largest effect sizes $\delta\xi_i$. The graph shows the p-value, computed through a GWAS analysis, that correlates the occurrence frequency $\widehat{q}_i$ of variant v_i in the subpopulation with the corresponding effect size $\delta\xi_i$. It can be seen that the lowest p-values correspond to the largest effect sizes, which indicates that GWAS is correctly detecting the variants that contributed the most to the phenotype

variant in the original population. If $\widehat{q}_i$ is significantly larger than q_i, then one can think that v_i is a variant associated with the phenotype. By contrast, if $\widehat{q}_i$ is significantly smaller than q_i, then one can conclude that v_i is not associated with the phenotype. GWAS provides, through the p-value, the level of significance of the over-representation (or under-representation), of the variant v_i in the subpopulation that exhibits the phenotype.

Figure 7.6a shows the normalized phenotype $\delta\xi_i/\xi_0 = (\xi(v_i) - \xi_0)/\xi_0$ in increasing order for a subpopulation $N_{\text{sub}} = 500$ networks and $\Omega = 200$ variants, while Fig. 7.6b shows the p-values corresponding to these variants. Clearly, the GWAS analysis in our model is detecting the variants v_i with the highest values of the phenotype $\xi(v_i)$ (the larger the value of $\xi(v_i)$, the smaller the corresponding p-value).

The results presented so far indicate two important aspects of our model: (1) epistatic effects (nonlinear interactions) are negligible when the number of variants in the genome is small, and (2) GWAS effectively detects the variants significantly associated with the phenotype under consideration (in this case, large values of the error $\xi(v_i)$). In the next section we will show that the missing heritability problem is not necessarily a consequence of nonlinear effects or of undersampling in the GWAS analysis. Instead, it may be a consequence of not taking into account the microbiota of the organism in the computation of the heritability, as the microbiota can be fundamental for the occurrence of some phenotypes.

7.6 Holobiont Evolution and the Missing Heritability Problem

So far we have not implemented any evolutionary algorithm to train the network H_0 to perform the desired task $F(t)$. We have just generated genetic and phenotypic variability in a population of networks (initially all of them identical to H_0) by implementing by hand the different variants $v_1, v_2, \ldots, v_\Omega$ with some given probabilities. In this section, in order to compute the heritability of the phenotype, we implement an evolutionary algorithm to actually train (evolve) the network H_0 to perform the desired task $F(t)$. In what follows we will refer to H_0 as the *host network*, in analogy with the host organism in a holobiont.

Following Huitzil et al. (2018), the training of the host network H_0 will be assisted by another network, M, which we will call the *microbial network*. This means that during the evolutionary process, the host network can acquire regulatory connections from the microbial network and vice versa (see Fig. 7.7). The main difference between the host and microbial networks are their mutation rates, with a mutation rate $\mu_h = 0.001$ for the host network and $\mu_m = 0.01$ for the microbial network. The reason for this is that microbes can generate variability at a rate that is at least ten times larger than for the cells in a host organism (plants or animals). In our model, the host network H_0 represents the host organism, whereas the microbial network M represents its microbiota. The regulatory connections between H_0 and

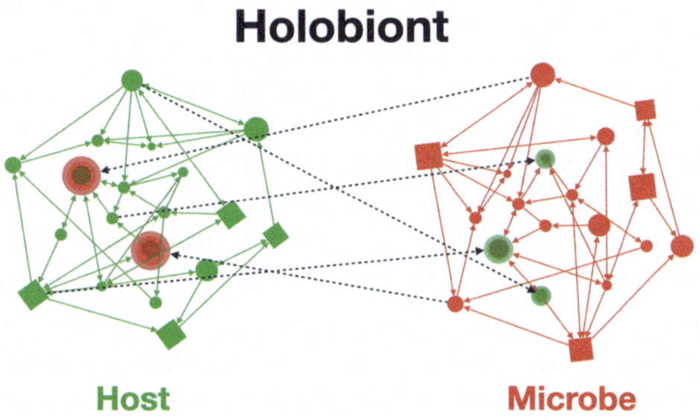

Fig. 7.7 Schematic representation of a holobiont in our model. The holobiont consists of a host network and a microbial network, which can interact through regulatory connections between them (broken lines). In this example, the host network receives two regulatory connections from the microbial network, whereas the microbial network receives three from the host network. The host network still has to evolve to acquire the desired phenotype $F(t)$, as in Fig. 7.2. The difference now is that the microbial network will "help" in the evolution of the host network

M represent the metabolic and genetic interactions between the host and its microbiota. Therefore, we will refer to the entire network made up of H_0 and M as the *holobiont* (Zilber-Rosenberg and Rosenberg 2008). It is important to mention that, although the host network is interacting with the microbial network, the target function $F(t)$ is defined only for the host network and therefore it is just the phenotype $R_0(t)$ of the host network that will be used to train it. Therefore, the microbial network can be considered only as an auxiliary mechanism to help the host network reach its goal. The details of the evolutionary algorithm can be found in (Huitzil et al. 2018). The important thing to mention here is that we have shown that the training of the host network is faster and more efficient with the help of the microbial network than without it, and that most of the variability of the holobiont resides in the microbial network (due in part to its increased mutation rate).

The main idea in this section is to compute the heritability of the phenotype $R(t)$ (which is quantitatively measured through the error function ξ) in two different ways: (i) by computing the genetic variability of the host network only, and (ii) by computing the genetic variability of the whole holobiont (host and microbial networks). As we will see, the missing heritability is considerably larger in the first case than in the second one.

The evolutionary model consists of a population of $N_P = 1000$ holobionts. Initially, the host and microbial networks in each holobiont have $N = 50$ nodes, each node with $K = 2$ regulatory connections. At each generation, each node in each network is mutated with probability μ_h for the host network and μ_m for the microbial network. Once a node has been chosen for mutation, the mutations consist of the following: (i) adding a new regulatory interaction (input connection); (ii) rewiring an existing regulatory interaction; (iii) changing the value of one entry of the Boolean function. Mutations (i) and (ii) can occur between nodes within the same network, or between one node in the host network and the other node in the microbial network. This last possibility is what makes the two networks develop regulatory interactions between them (see Fig. 7.7).

Let us denote as $R_n(t)$ and ξ_n the phenotype of the nth holobiont in the population and its corresponding error, respectively. At each generation, we choose the best 100 holobionts in the population (the ones with the lowest values of ξ_n) to pass to the next generation. Then, we replicate these holobionts by making ten copies of each one in order to restore the population to its original size $N_P = 1000$. Then we repeat the entire process (mutation, selection, replication, etc.). We do this for several generations until the average population error $\overline{\xi} = \frac{1}{N_P}\sum_{n=1}^{N_P}\xi_n$ becomes smaller than a threshold $\varepsilon = 1$. This means that on average most of the holobionts in the population are well adapted to the desired phenotype $F(t)$. We stop the simulation as soon as the condition $\overline{\xi} < \varepsilon$ is satisfied. However, this algorithm eventually produces a population with almost no variability, namely, a population in which almost all the holobionts are identical, a phenomenon known as *purifying selection* (Cvijovic et al. 2018; Huang 2016; Leffler et al. 2012). Therefore, in order to generate variability in the population, we proceed as follows. After the condition $\overline{\xi} < \epsilon$ has been fulfilled, we continue the evolutionary process for five more generations. But now, in each one of these five generations we mutate only the Boolean DNA of the networks, which makes their corresponding errors to change: some errors will increase while some others will decrease. Then, at each one of these five generation, we allow the nth holobiont to pass to the next generation with a probability $p_n = C/\xi_n$ (C is a proportionality constant). Thus, holobionts with a large error (poorly adapted) still have a probability, although small, to continue through the next generation. In this way, the final population will have more variability than if we were always to choose the best holobionts and replicate them.

At the end of this process all the holobionts in the population have the same structural topology but different genotypes, all of the same length. (The genome of a holobiont is the concatenation of the genome of its host network and the genome of its microbial network.) Since through the evolutionary process regulatory connections within and between the host and microbial networks were added or rewired, different genes in the final networks will have a different number of input connections. We will denote as Ω_h, Ω_m and Ω_{HL} the length of the genomes of the host network, the microbial network, and the entire holobiont, respectively, with $\Omega_{HL} = \Omega_h + \Omega_m$. Since all the holobionts in the final population have the same structural topology but different Boolean DNAs, it is possible to align their genotypes and determine a consensus sequence in the same way as one does with the genetic sequences of real organisms (see Fig. 7.8). Once the consensus sequence of the population has been determined, the variants are defined as mutations (SNPs) of the consensus sequence. There are Ω_h variants for the host network, Ω_m variants for the microbial network, and of course $\Omega_{HL} = \Omega_h + \Omega_m$ variants for the holobiont, which we will denote as $\{v_1, v_2, \ldots, v_{\Omega_h}, v_{\Omega_h+1}, v_{\Omega_h+2} \ldots, v_{\Omega_{HL}}\}$. In the final population each variant v_i ($i = 1, \ldots, \Omega_{HL}$) occurs with a frequency q_i. Rare variants in the population are defined as those that satisfy the condition $q_i < \delta_f$, where δ_f is a parameter chosen in the interval [0.001,0.01]. We vary δ_f in this interval to analyze the effect of taking into account (or not) the less common variants in the computation of the heritability, finding no significant changes in the results when δ_f varies in the interval mentioned above.

The strict sense heritability h^2 of the phenotype is defined as (Yang et al. 2010; Eskin 2015).

$$h^2 = \sigma_g^2/\sigma_P^2, \tag{7.9}$$

where σ_g^2 and σ_P^2 are the genotypic and phenotypic variances (or variabilities) of the population, respectively. The phenotypic variance σ_P^2 in our model is easily computed, as it is just the variance of the phenotype ξ over all the holobionts in the population:

$$\sigma_P^2 = \frac{1}{N_P}\sum_{n=1}^{N_P}\left(\xi_n - \overline{\xi}\right)^2. \tag{7.10}$$

a

		Loci										
		1	2	3	4	5	6	7	8	9	10	Phenotype
Boolean DNA	G_1	1	0	1	1	1	0	1	1	1	0	ξ_1
	G_2	0	1	1	0	1	0	0	0	1	0	ξ_2
	G_3	0	1	1	0	1	0	1	0	1	0	ξ_3
	G_4	0	0	1	0	1	1	0	0	1	0	ξ_4
	G_5	1	0	1	1	0	1	1	0	1	1	ξ_5
Consensus		0	0	1	0	1	0	1	0	1	0	

b

Variant	q	Variant	q
v_1	0.4	v_6	0.4
v_2	0.4	v_7	0.4
v_3	0	v_8	0.2
v_4	0.4	v_9	0
v_5	0.2	v_{10}	0.2

c

	Variants in the population	More common variants	Phenotype
G_1	v_1, v_4, v_8	v_1, v_4	ξ_1
G_2	v_2, v_7	v_2, v_7	ξ_2
G_3	v_2	v_2	ξ_3
G_4	v_6, v_7	v_6, v_7	ξ_4
G_5	$v_2, v_4, v_5, v_6, v_{10}$	v_2, v_4, v_6	ξ_5

Fig. 7.8 (**a**) Illustrative example of five Boolean genotypes G_1, G_2, G_3, G_4, and G_5 corresponding to five individuals in a population. Each genome has 10 loci. The consensus sequence, which is shown at the bottom of the table, is constructed using a simple majority rule at each locus. The variants (or SNPs) of each individual are those loci whose value is different from the corresponding one in the consensus sequence. These values are highlighted in blue in the figure. Since the genotypes of the five individuals are different from each other, their phenotypes ξ_i will usually be different from each other as well. (**b**) Table showing the occurrence frequency q_i of each variant v_i in the population. Less frequent variants are highlighted in red. (**c**) Knowing the consensus sequence of the population, all the information in (**a**) can be summarized by just indicating the variants of each individual

The genotypic variance σ_g^2 is more difficult to compute, as it has to do with the variants that are associated with the desired phenotype. Therefore, one has to determine first, through GWAS, the variants that correlate with the phenotype and then compute the variability of those variants throughout the population. Nonetheless, we can see that in the genetic variance σ_g^2 of our holobiont model there are two contributions, one coming from the genome of the host network and the other from the genome of the microbial network. We will denote these contributions as $\sigma_{g,h}^2$ and $\sigma_{g,m}^2$, respectively. Therefore, $\sigma_g^2 = \sigma_{g,h}^2 + \sigma_{g,m}^2$ and the heritability in Eq. (7.9) can be written as

$$h^2 = \left(\sigma_{g,h}^2 + \sigma_{g,m}^2\right)/\sigma_P^2. \tag{7.11}$$

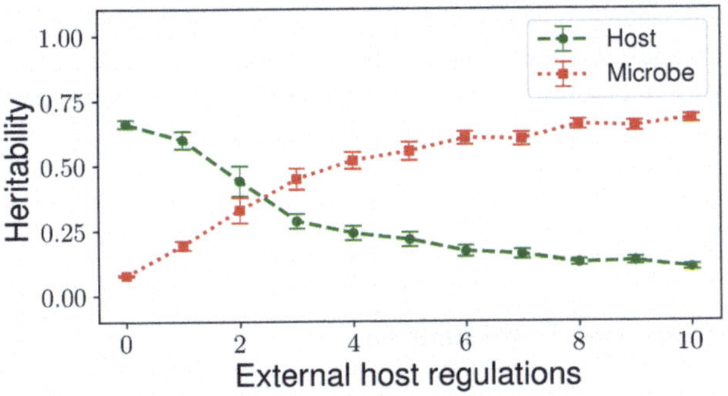

Fig. 7.9 Host and microbial heritabilities, h_h^2 and h_m^2 respectively, as functions of the number K_{ex} of host external regulations. Only for small values of K_{ex} (weak host-microbe interaction) the heritability of the host network is larger than the heritability of the microbial network. However, as K_{ex} increases (strong interaction), h_m^2 increases and very rapidly becomes larger than h_h^2

Defining the host heritability and the microbial heritability as $h_h^2 = \sigma_{g,h}^2/\sigma_P^2$ and $h_m^2 = \sigma_{g,m}^2/\sigma_P^2$, respectively, then the heritability of the holobiont is $h^2 = h_h^2 + h_m^2$. Since both h_h^2 and h_m^2 are positive, it is clear that the two following conditions are satisfied: $h_h^2 \leq h^2$ and $h_m^2 \leq h^2$. If when computing the genetic variance σ_g^2 only the genome of the host organism is considered, as it is usually the case, then we will obtain a value h_h^2 which could be considerably smaller than the real value h^2, unless the microbial heritability h_m^2 is negligibly small compared to the host heritability h_h^2. But this is not a sensible hypothesis, in view of the fact that the microbiota is strongly correlated with the emergence of some phenotypes (like diabetes). In our network model, the microbial network M strongly participates in the training of the host network H_0 to achieve its desired phenotype $F(t)$. The influence of the microbial network M on the training of the host network H_0 is stronger the more regulatory connections occur between them. We will denote as K_{ex} the number of regulatory connections from the microbial network to the host network, and refer to them as the *external host regulations*.

The details of the algorithm to compute the heritability h^2 in our model are presented in the Appendix. It is based in Eq. (7.6) and closely follows the method reported in Yang et al. (2010) and Eskin (2015). The results of this computation are presented in Fig. 7.9, which shows the host and microbial heritabilities, h_h^2 and h_m^2, respectively, as functions of the number of external host regulations, K_{ex}. These results were computed for the final population of holobionts obtained from the evolutionary algorithm described in this section. Note from Fig. 7.9 that when $K_{\text{ex}} = 0$, that is, when there is no interaction between the host and microbial networks, almost all the contribution to the heritability h^2 of the phenotype comes from the heritability h_h^2 of the host network, as should be expected. However, when K_{ex} increases the interaction between the host and microbial networks becomes stronger. In this case, the heritability h_h^2 of the host network decreases while the heritability h_m^2 of the microbial network increases. Even for $K_{\text{ex}} \approx 2.5$, which corresponds to a relatively weak interaction between the host and microbial networks, the microbial heritability h_m^2 starts to surpass the host heritability h_h^2. For $K_{ex} = 10$ (a strong interaction) almost all the contribution to the heritability h^2 comes from the microbial network. In this strong-interaction case h_h^2 is very small with the result that $h^2 \approx h_m^2$.

7.7 Discussion and Conclusions

The missing heritability problem has been a matter of intense debate for more than one decade. The fact that many phenotypes that are transmitted across generations cannot be significantly correlated with a particular set of genetic variants has generated many possible explanations. Among those explanations three stand out (Génin 2019). First, it has been proposed that epistatic (nonlinear) effects emerging from the interactions between these genetic variants may be hindering the identification of significant correlations between the genetic variants and the phenotype under consideration. Second, there is the undersampling problem, which consists in that rare variants that have a strong effect on the phenotype are not being detected due to the combination of two factors: these variants occur with very low frequency in the population and the size of the sampled subpopulation is too small. Third, there are many common variants whose combined effect on the phenotype is very strong, but whose individual effects are too small to be detected individually. While these answers to the missing heritability problem might be true, there is one aspect that has not been considered when measuring the heritability of the phenotype of a given organism, which is the genetic variability of its microbiota.

More than one decade ago, the pioneering work by Turnbaugh and his coworkers (Turnbaugh et al. 2006) showed that the microbiota can have a strong influence on the phenotypic traits of its host organism. This is particularly true for many of the phenotypes that researchers have tried to correlate with genetic variants, such as obesity, diabetes, cancer, and metabolic syndrome. It has also been shown that the microbiota can be transmitted across generations. Furthermore, there is evidence that the microbiota can evolve together with its host organism, strongly influencing (or even substituting) some of the genetic and metabolic functions of the host organism. Therefore, it is natural to assume that, when computing the heritability of some phenotypic trait (particularly one that is strongly influenced by the microbiota), one does have to take into account not only the genotypic variance of the host organism, but also the genotypic variance of its microbiota. This is the approach that we have adopted in this work.

We have presented an evolutionary algorithm based on the Boolean gene regulatory network model proposed by S. Kauffman in 1969, which has proven to accurately reproduce the gene expression patterns experimentally observed for several organisms. The objective is to train a population of networks to perform a predefined task $F(t)$, which represents the phenotype of the networks. To do this training, we mutate the Boolean functions of the networks (their genotypes), which introduce genetic variability in the population (all the networks have the same structural topology but slightly different Boolean functions). The main advantage of working with this model is that we can exactly measure the effect that each variant has on the phenotype, which in turn allows us to simulate a GWAS analysis in order to compute the correlations between the genetic variants and the desired phenotype. From this analysis, three important results are obtained. First, most of the variants have a very small effect on the phenotype, but there are a few variants that have a strong effect (Fig. 7.4). Second, when several variants are present in the genotype of one individual, their effects on the phenotype are mostly additive (Fig. 7.5). Therefore, epistatic (nonlinear) effects can be ignored when computing the contribution of several variants to the phenotype. This is an interesting result given that the dynamics determined by Eq. (7.1) are highly nonlinear. And third, GWAS effectively reveals the variants that have a strong effect on the phenotype (Fig. 7.6). These results indicate that all the conditions are met in our model to obtain a good estimate of the heritability h^2 of the phenotype based on the variability of the Boolean DNA of the networks.

To compute this heritability we evolved a population of holobionts, where each holobiont consists of a host network and a microbial network that can interact through regulatory connections between them (Fig. 7.7). The goal of the evolutionary process is to train the host network to acquire the desired phenotype $F(t)$. It is important to stress that the phenotype is defined only on the host network. However, since the host and microbial networks can interact, the adaptation of the host network to the desired phenotype also depends on the microbial network, and this dependence is greater the more regulatory interactions exist between the host and microbial networks. Therefore, in order to compute the heritability h^2 of the phenotype, one has to take into account the genetic variance of both the host and the microbial networks. The heritability h^2 has two contributions and can thus be written as $h^2 = h_h^2 + h_m^2$, where h_h^2 and h_m^2 are the heritabilities computed by taking into account the genetic variance of only the host and only the microbial networks, respectively. It has happened that when the heritability of a particular phenotypic trait (such as height, cancer, or diabetes) is computed, only the genetic variance of the human genome is taken into account. Therefore, researchers all over the world have been computing h_h^2 instead of h^2. From Fig. 7.9 it can be seen that $h^2 \approx h_h^2$ (i.e., h_h^2 is a good estimate of h^2) only when there are no interactions whatsoever between the host and the microbial networks (this would correspond to an organism with no microbiota). However, as soon as the number of interactions between these two networks starts to increase, very quickly one has $h_h^2 < h^2$. The difference $h_m^2 = h^2 - h_h^2$ is the missing heritability, and is completely attributed to the genetic variability of the microbiota. As Fig. 7.9 shows, this missing heritability becomes larger the stronger the interaction between the host and the microbial networks.

The most important conclusion of the work presented here is the following: even when epistatic (nonlinear) effects between genetic variants can be neglected, and even when GWAS can efficiently detect the most important variants that contribute to a phenotype, if the genetic variance of the microbiota is not taken into account in the computation of the heritability h^2 of some particular phenotype, then this heritability will always be underestimated by a large amount. This is particularly true for those phenotypes that are strongly determined by (or correlated with) the microbiota composition of the host organism. This problem has recently started to be addressed experimentally through Metagenome Wide Association Studies (MWAS), through which the genetic variance of both the host and its microbiota can be measured (Wang and Jia 2016). Our results strongly suggest that MWAS will be essential to fill the missing heritability gap.

Acknowledgments S.H. thanks CONACyT for a PhD scholarship and the Center for Complexity Sciences at UNAM for economic support. This work was partly supported by PAPIIT-UNAM grant IN226917.

Appendix

Following Yang et al. (2010) and Eskin (2015), to measure the heritability h^2 of a given phenotype, we start with Eq. (7.6) of the main text, which determines the value ξ_j of the phenotype of the jth individual in the population as a linear contribution of the effects of the different genetic variants v_i ($i = 1, 2, \ldots, \Omega_{HL}$):

$$\xi_j = \overline{\xi} + \sum_{i=1}^{\Omega_{HL}} g_{ij}\beta_i + e_j.$$

In this equation, $\overline{\xi}$ is the average value of the phenotype in the population; β_i is the contribution of variant v_i to the phenotype; e_j is an error that has to do with the unknown effect of the environment on the phenotype; and g_{ij} is a matrix whose entries acquire the values 1 and 0 depending on whether the jth individual contains variant v_i or not, respectively. For simplicity in the calculation, it is convenient to perform the change of variable

$$x_{ij} = \left(g_{ij} - q_i\right)/\sigma_\xi,$$

where q_i is the occurrence frequency of variant v_i in the population, and $\sigma_\xi = \sqrt{q_i(1 - q_i)}$ is the standard deviation of this quantity. With this change of variable, first equation of Appendix can be written in matrix form as

$$\boldsymbol{\xi} = \overline{\xi} \cdot \boldsymbol{I} + \boldsymbol{X} \cdot \boldsymbol{\beta} + \boldsymbol{e},$$

where $\boldsymbol{\xi} = \left(\xi_1, \xi_2, \ldots \xi_{N_P}\right)$ is the vector containing the phenotypes of the individuals in the population (analogously for the vectors $\boldsymbol{\beta}$ and $\boldsymbol{e}$), and $\boldsymbol{I}$ is the identity matrix. In our model, we know exactly the effect β_i that variant v_i has on the phenotype (it is the error difference $\delta\xi_i$ defined in Eq. (7.5) and reported in Fig. 7.4). However, in a real situation, this effect cannot be known accurately. Instead, it has to be estimated ad $\widehat{\beta}_i = \frac{1}{N_P}\boldsymbol{X}_i^T\boldsymbol{\xi}$, where $\boldsymbol{X}_i$ is the ith row of matrix $\boldsymbol{X}$ and N_P is the number of individuals in the population.

To determine whether the effect of variant v_i is or not correlated with the phenotype, we compute the p-value of the effect of this variant on the phenotype. The p-value $p(v_i)$ corresponding to the variant v_i is computed as

$$p(v_i) = 2(1 - \Phi(1 \leq |t_0(v_i)|)),$$

where $t_0(v_i) = \widehat{\beta}_i/\sqrt{C}$. In this expression, $C = \widehat{\sigma}^2\left(\boldsymbol{X}^T\boldsymbol{X}\right)^{-1}$ is the covariance of the linear regression with $\widehat{\sigma}^2 = \boldsymbol{e}^T\boldsymbol{e}/(N_P - \Omega_{HL})$ being the error of the estimation (Ω_{HL} is the number of different variants occurring in the population). The function $\Phi(x)$ is the (cumulative) normal distribution function.

Once the effect $\widehat{\beta}_i$ of each variant v_i has been estimated, the variants whose effect on the phenotype is significantly small are discarded. Let us denote as S_V the set of variants that are not discarded and remain in the analysis, namely, the set of variants that have a strong effect on the phenotype. These variants can occur in the genome of both the host network and microbial networks. Therefore, the set S_V can be partitioned into two disjoint subsets, S_H and S_M such that $S_V = S_H \cup S_M$, where S_H is the set of relevant variants that occur in the genome of the host network and S_M is the set of relevant variants occurring in the microbial network. The genetic variance of the genotype is then estimated as

$$\sigma_g^2 = \sum\nolimits_{v_i \in S_V}\left(\widehat{\beta}_i\right)^2 = \sum\nolimits_{v_i \in S_H}\left(\widehat{\beta}_i\right)^2 + \sum\nolimits_{v_i \in S_M}\left(\widehat{\beta}_i\right)^2.$$

The heritability h^2 is then computed as

$$h^2 = \sigma_g^2/\sigma_P^2,$$

where σ_P^2 is the phenotypic variability computed as

$$\sigma_P^2 = \frac{1}{N_P}\sum_{i=1}^{N_P}\left(\xi_i - \underline{\xi}\right)^2.$$

References

Auton A, Abecasis GR, Altshuler DM, Durbin RM, David R et al (2015) A global reference for human genetic variation. Nature 526(7571):68–74

Bornholdt S (2008) Boolean network models of cellular regulation: prospects and limitations. J R Soc Interface 5(1):S85–S94

Brooks AW, Kohl KD, Brucker RM, van Opstal EJ, Bordenstein SR (2016) Phylosymbiosis: relationships and functional effects of microbial communities across host evolutionary history. PLoS Biol 14(11):e2000225

Buniello A, MacArthur JAL, Cerezo M, WHarris L, Hayhurst J, Malangone C, McMahon A, Morales J, Mountjoy E, Sollis E et al (2018) The NHGRI-EBI GWAS catalog of published genome-wide association studies, targeted arrays and summary statistics 2019. Nucleic Acids Res 47(D1):D1005–D1012

Cho I, Blaser MJ (2012) The human microbiome: at the interface of health and disease. Nat Rev Genet 13(4):260

Collins FS, Patrinos A, Jordan E, Chakravarti A, Gesteland R, Walters LR et al (1998) New goals for the us human genome project: 1998-2003. Science 282(5389):682–689

Cvijovic I, Good BH, Desai MM (2018) The effect of strong purifying selection on genetic diversity. Genetics 209(4):1235–1278

Drossel B (2008) Random Boolean networks. Reviews of nonlinear dynamics and complexity 1:69–110

Eichler EE, Flint J, Gibson G, Kong A, Leal SM, Moore JH, Nadeau JH (2010) Missing heritability and strategies for finding the underlying causes of complex disease. Nat Rev Genet 11 (6):446–450

Eskin E (2015) Discovering genes involved in disease and the mystery of missing heritability. Commun ACM 58(10):80–87

Génin E (2019) Missing heritability of complex diseases: case solved? Hum Genet 139(1):103–113

Gibson G (2018) Population genetics and gwas: a primer. PLoS Biol 16(3):e2005485

Grandjean V, Badro DA, Kiani J (2013) RNA: a possible contributor to the "missing heritability". Basic Clin Androl 23(1):9

Huang S (2016) New thoughts on an old riddle: what determines genetic diversity within and between species? Genomics 108(1):3–10

Huitzil S, Sandoval-Motta S, Frank A, Aldana M (2018) Modeling the role of the microbiome in evolution. Front Physiol 9:1836

Inoue Y, Shimojo N (2015) Microbiome/microbiota and allergies. In: Seminars in immunopathology, vol 37. Springer, Cham, pp 57–64

International HapMap Consortium et al (2003) The international HapMap project. Nature 426 (6968):789

Karunakar P, Bhalla A, Sharma A (2019) Transgenerational inheritance of cold temperature response in drosophila. FEBS Lett 593(6):594–600

Kim D-H (2015) Gut microbiota-mediated drug-antibiotic interactions. Drug Metab Dispos 43 (10):1581–1589

Leffler EM, Bullaughey K, Matute DR, Meyer WK, Ségurel L, Venkat A, Andolfatto P, Przeworski M (2012) Revisiting an old riddle: what determines genetic diversity levels within species? PLoS Biol 10(9):e1001388

Ley RE, Lozupone CA, Hamady M, Knight R, Gordon JI (2008) Worlds within worlds: evolution of the vertebrate gut microbiota. Nat Rev Microbiol 6(10):776

Maher B (2008) Personal genomes: the case of the missing heritability. Nat News 456(7218):18–21

Manolio TA, Collins FS, Cox NJ, Goldstein DB, Hindorff LA, Hunter DJ, McCarthy MI, Ramos EM, Cardon LR, Chakravarti A et al (2009) Finding the missing heritability of complex diseases. Nature 461(7265):747

Ochman H, Worobey M, Kuo C-H, Ndjango J-BN, Peeters M, Hahn BH, Hugenholtz P (2010) Evolutionary relationships of wild hominids recapitulated by gut microbial communities. PLoS Biol 8(11):1–8

Ozaki K, Ohnishi Y, Iida A, Sekine A, Yamada R, Tsunoda T, Sato H, Sato H, Hori M, Nakamura Y et al (2002) Functional SNPs in the lymphotoxin-gene that are associated with susceptibility to myocardial infarction. Nat Genet 32(4):650

Rawls JF, Samuel BS, Gordon JI (2004) Gnotobiotic zebrafish reveal evolutionarily conserved responses to the gut microbiota. Proc Natl Acad Sci 101(13):4596–4601

Sachidanandam R, Weissman D, Schmidt SC, Kakol JM, Stein LD, Marth G, Sherry S, Mullikin JC, Mortimore BJ, Willey DL et al (2001) A map of human genome sequence variation containing 1.42 million single nucleotide polymorphisms. Nature 409(6822):928–934

Sandoval-Motta S, Aldana M, Martínez-Romero E, Frank A (2017) The human microbiome and the missing heritability problem. Front Genet 8:80

Slatkin M (2009) Epigenetic inheritance and the missing heritability problem. Genetics 182 (3):845–850

Tremlett H, Bauer KC, Appel-Cresswell S, Finlay BB, Waubant E (2017) The gut microbiome in human neurological disease: a review. Ann Neurol 81(3):369–382

Turnbaugh P, Ley R, Mahowald M et al (2006) An obesity-associated gut microbiome with increased capacity for energy harvest. Nature 444:1027–1031

Veigl S, Suárez J, Stencel A. Does inheritance need a rethink? Conceptual tools to extend inheritance beyond the DNA. Extended evolutionary synthesis, 05 November 2019. https://extendedevolutionarysynthesis.com/inheritance-rethink/

Visscher PM (2008) Sizing up human height variation. Nat Genet 40(5):489–490

Visscher PM, Wray NR, Zhang Q, Sklar P, McCarthy MI, Brown MA, Yang J (2017) 10 years of gwas discovery: biology, function, and translation. Am J Hum Genet 101(1):5–22

Wang J, Jia H (2016) Metagenome-wide association studies: fine-mining the microbiome. Nat Rev Microbiol 14(8):508–522

Wilson ID, Nicholson JK (2017) Gut microbiome interactions with drug metabolism, efficacy, and toxicity. Transl Res 179:204–222

Wood AR, Esko T, Yang J, Vedantam S, Pers TH, Gustafsson S, Chu AY, Estrada K, Luan J'a, Kutalik Z et al (2014) Defining the role of common variation in the genomic and biological architecture of adult human height. Nat Genet 46(11):1173

Yang J, Benyamin B, McEvoy BP, Gordon S, Henders AK, Nyholt DR, Madden PA, Heath AC, Martin NG, Montgomery GW et al (2010) Common snps explain a large proportion of the heritability for human height. Nat Genet 42(7):565

Zilber-Rosenberg I, Rosenberg E (2008) Role of microorganisms in the evolution of animals and plants: the hologenome theory of evolution. FEMS Microbiol Rev 32(5):723–735

Zuk O, Hechter E, Sunyaev SR, Lander ES (2012) The mystery of missing heritability: genetic interactions create phantom heritability. Proc Natl Acad Sci 109(4):1193–1198

Zuk O, Schaffner SF, Samocha K, Do R, Hechter E, Kathiresan S, Daly MJ, Neale BM, Sunyaev SR, Lander ES (2014) Searching for missing heritability: designing rare variant association studies. Proc Natl Acad Sci 111(4):E455–E464

Chapter 8
The Role of Constructive Neutral Evolution in the Development of Complexity from Symbioses: A Microbe-Centric View

Ramakrishnan Sitaraman

Abstract Symbiogenesis presents the biologist with very different explanatory issues compared to the lineal and selectionist view of evolution based on individual entities, whether genes, organisms or species. A key question is how the co-existence of two or more partners in close association during a given generation can ultimately be stabilized enough to be transmitted to the next, how the ensuing complexity is maintained and how this arrangement impacts the reproductive fitness of the collective over evolutionary time. In this chapter, we highlight some observations gleaned from the microbial world that could shed light on this problem if viewed within the framework of constructive neutral evolution.

> "The view of evolution as a chronic bloody competition among individuals and species, a popular distortion of Darwin's notion of 'survival of the fittest,' dissolves before a new view of continual cooperation, strong interaction, and mutual dependence among life forms. Life did not take over the globe by combat, but by networking. Life forms multiplied and complexified by co-opting others, not just by killing them."—Lynn Margulis and Dorion Sagan in *Microcosmos: Four billion years of evolution from our microbial ancestors* (1986).

The vernacular usage of the word 'symbiosis' practically assumes the accrual of mutual 'benefits' or 'advantages' for each partner. However, this is only a reflection of the cherished human value of fair dealing that makes the act of symbiosis appear *mutually* beneficial; we would not describe an insufferable co-worker or a habitual sponger as someone with whom we have a 'symbiotic' relationship. Technically, the word 'symbiosis' in essence refers merely to the act of 'living together,' made amply clear by its etymology. Thus, it is applicable with equal force to situations in which the partners could be indifferent (commensalism) or genuinely co-dependent (mutualism) or exploitative (parasitism). Nature observes no niceties. However, we must concede that the exchange of mutual benefits (mutualism) is of considerable importance to the adaptationist-selectionist perspective in ensuring that a given association, once initiated, acquires longevity over lifetimes and, crucially, heritability

---

R. Sitaraman (✉)
Department of Biotechnology, TERI School of Advanced Studies, New Delhi, India

M. Kloc (ed.), *Symbiosis: Cellular, Molecular, Medical and Evolutionary Aspects*, Results and Problems in Cell Differentiation 69,
https://doi.org/10.1007/978-3-030-51849-3_8

across generations. The famous 'plate count anomaly' in microbiology that refers to our inability to isolate only a minor fraction of individual microorganisms as pure cultures from natural microbial communities is a persistent, if underappreciated, commentary on how widespread symbiosis may well be in the microbial world that, after all, represents the majority of life forms on earth. Though the recently developed high-throughput technique of 'culturomics' aims to address this chronic problem, metagenomics remains the method of choice for investigations of the composition of microbial communities. Moreover, we must caution that even the mere act of culturing all the members of a microbial community may not really enlighten us about the kinds and durability of associations prevalent between those members, and how these co-evolve over time.

8.1 A Statement of the Problem

Charles Darwin defined evolution as 'descent with modification,' inadvertently helping reify the idea of vertical transmission of heritable characteristics down given lineages. In the context of symbiosis, it begs the question as to whether associations between organisms can likewise be stable enough to be transmitted from one generation to the next. The idea that an assemblage of organisms can be treated as a unit of natural selection, much like a single organism, is at the root of the word 'holobiont' first used by Adolf Meyer-Abich in 1943[1] (Meyer-Abich 1943; Baedke et al. 2020). Lynn Margulis' insight that the gradualism of Darwinian evolutionary theory is complemented by the saltation inherent in the theory of symbiogenesis, famously exemplified by the case of the transition between prokaryote and eukaryote (Sagan 1967), is crucial to understanding evolutionary transitions.

In this chapter, we discuss how the perspective of constructive neutral evolution (CNE, outlined in the next section) may be seriously considered as an explanation for the origin and stabilization of symbiotic associations, complementing adaptationist and selectionist viewpoints. In particular, CNE is particularly valuable in explaining the eventual fixation of 'gratuitous increases in complexity' (Gray et al. 2010). As Sydney Brenner put it, a mutational change may not have any benefits or disadvantages, and so long as there are no major disadvantages, it may continue to persist in the population. In his words, 'there could be a third value—indifferent—in addition to good and bad, and these "don't care" values immensely complicate the inverse approach . . .We need to remember that whereas mathematics is the art of the perfect and physics the art of the optimal, biology, because of evolution, is only the art of the satisfactory' (Brenner 2010). This immediately begs the question as to why accounts of these 'merely satisfactory' instances are not more widespread in the literature. Are most associations intrinsically less fit in the Darwinian sense of

[1]The first usage of the term 'holobiont' is often mistakenly attributed to Lynn Margulis (1991) for several reasons discussed in detail by Baedke et al. (2020).

leaving behind descendants, as compared to the fitness of their constituents? Or, is this an outcome of our own bias toward analyzing traits that increase fitness by conferring an adaptive value for organisms? After all, the 'dark matter' of negative results in the field of biomedicine is evidence (if any were required) of our tendency to focus on the affirmative and the positive. The acceptance of null hypotheses is generally deemed uninteresting.

A word about terminology is in order here. 'Symbiogenesis' is usually reserved for the evolutionary transition from the ancestral prokaryote to the eukaryote via endosymbiosis. However, it has been suggested recently (and, in our opinion, reasonably) that the term may be used in a general sense to encompass those evolutionary transitions that could have occurred due to symbiosis, and even if the participating organisms are no longer clearly discernible as independent entities (Aanen and Eggleton 2017). For purposes of this chapter, we would like to provisionally accept this reformulation of known terminology, as opposed to new coinage.

Once we posit symbiosis as a source of evolutionary transitions, the quest for 'transitional forms' and the 'inverse problem' haunts the investigator who would venture to reconstruct the sequence of events that could make a case for symbiogenesis. The fossil record has provided several instances of symbiosis indicating that symbiosis has been fairly common in life's history (Bermudes and Back 1991; Yuan et al. 2005; Tapanila 2008; Casazza 2012). Experimental approaches have also indicated the possibility of facultative associations (Shou et al. 2007; Harcombe 2010; Ohkawa et al. 2011; Shapiro and Turner 2018; Mehta et al. 2018). Research on fossils and experimental approaches to observe or induce symbiosis among contemporary organisms are essentially motivated by the inverse problem of evolutionary biology. Much information on the evolutionary history of contemporary life forms has been lost. The experimental and descriptive approaches are the best we can do to convert this fundamentally intractable inverse problem into some semblance of testability and plausibility. A suitable evolutionary model could then provide us with a conceptual framework as well as interpretive and predictive ability to better understand how symbiogenesis came to be. We suggest that constructive neutral evolution provides us with just such a model to better understand symbiogenesis and the unidirectional increase in complexity that it entails.

8.2 Constructive Neutral Evolution (CNE) as a Driving Force for Symbiogenesis?

It must be noted that neutral and unidirectional evolutionary processes were initially proposed to explain the evolution of the mechanistically complex process of RNA editing that requires multi-subunit complexes whose components are incapable of functioning independently (Covello and Gray 1993). This was subsequently detailed and elaborated by Stoltzfus into the constructive neutral evolutionary model to explain the evolution of complexity at the *molecular level*, for example, guide RNA-mediated RNA editing, gene duplication and spliceosomal splicing of RNA (Stoltzfus 1999). Thus, given the complexity of the spliceosome, how could such a mechanism have evolved by incremental mutation and selection? If the subunits of the spliceosome are completely co-dependent for effective functioning and are inactive in isolation, how could the ancestral state with individual but non-associating subunits have performed the essential functions required to survive the sieve of natural selection in the intermediate stages?

As explained by Stolzfus, the usage of the word 'constructive' in the evolutionary sense refers not to advantages associated with variations, but serves as a counterpoint to 'reductive' or 'conservative.' Reductive and conservative refer to decreasing or unchanging complexity respectively, while constructive refers to increasing complexity. Importantly, the word 'neutral' *does not* necessarily mean that the effect of a neutral mutation on the fitness of the organism is exactly zero relative to the wild-type; rather, it simply indicates that the fitness is not appreciably different relative to the wild-type ('approximate parity') (Stoltzfus 1999). There are some crucial features in this theory that recommend it as an alternative to adaptationist or selectionist views (Gray et al. 2010):

(a) A novel mutation may/need not be immediately beneficial in terms of conferring a fitness advantage on its possessor. It is sufficient if it is not deleterious.
(b) Neutral pre-suppression, that is, the occurrence of mutations that can suppress the effects of otherwise deleterious mutations that may arise *later*. However, subsequent mutations elsewhere may render the loss of the original mutation deleterious to the survival of the organism. The problem of the survival of a deleterious intermediate in the progression toward more complex configurations is circumvented. A deleterious mutation need not persist across generations in the face of competition awaiting rescue by a suppressor. Rather, the suppressor exists *before* the occurrence of the deleterious mutation ('pre-suppression'), perhaps serving unrelated functions.
(c) Complex, multi-subunit associations arising in (b) may further accumulate secondary mutations that render the association obligatory for functioning, thereby facilitating a unidirectional increase in complexity and necessitating its continuation for the survival of the organism. In other words, the aggregate undergoes purifying or negative selection.

Can this schema be adapted to microbial symbiogenesis? Proceeding by analogy for the case of a two-member microbial symbiosis (see Fig. 8.1), we may state that:

(a) Independently evolving organisms may physically interact on occasion but without losing the ability to exist autonomously. They continue to independently undergo mutation and elaborate the corresponding phenotypes. We may term this the 'facultative' phase.
(b) One partner elaborates a phenotype that is not immediately 'useful' or 'harmful' to it. This could be, for instance, a metabolic capacity that produces certain beneficial compounds, but is not required by the other. This we may term the

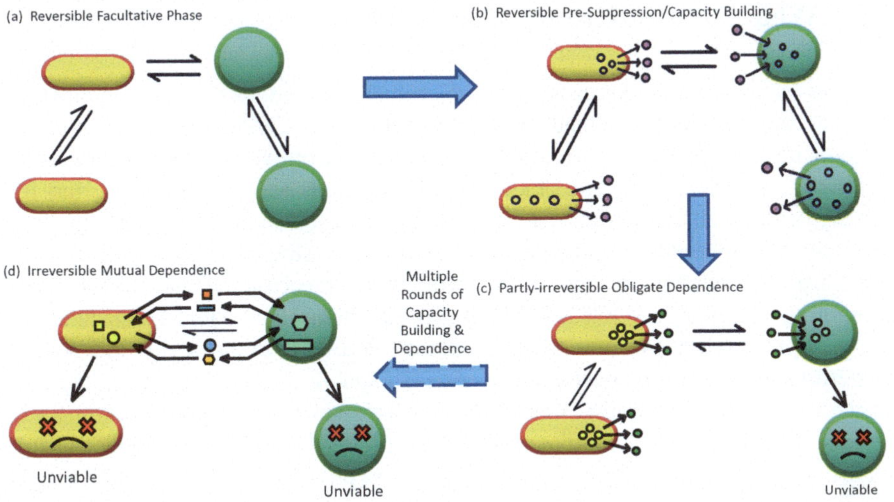

Fig. 8.1 Incremental increase of dependence ensuring irreversibility of a two-microbe symbiosis via neutral changes. The different phases of the association outlined in the text are indicated. The participating microbes are symbolized by the large bacillus- and coccus-like shapes. Metabolites are the smaller shapes, and their associated arrows indicate the source of production and the direction of transfer. (**a**) Both microbes can associate, but can also survive autonomously—the facultative phase. (**b**) In the pre-suppression phase, the bacillus produces a metabolite (circle) that is used by the coccus, as indicated by the direction of the arrows. However, both are still capable of autonomous existence because the coccus can still produce the metabolite if required. (**c**) Obligate dependence arises when the coccus is unable to synthesize the metabolite and becomes completely dependent on the bacillus for survival. (**d**) Multiple rounds of this process result in a mutual dependence for several metabolites between the partners such that neither is able to survive autonomously. The first two phases are reversible in that separation of the partners does not affect viability. Irreversibility starts building up from the third phase onward, with at least one partner rendered unviable without the association

'capacity building' or 'pre-suppression' phase. Note that both this and the earlier phase are perforce neutral insofar as fitness is concerned.

(c) The other partner may lose the capacity to synthesize those beneficial compounds, making it entirely dependent on the first. This commences the phase of 'obligate dependence,' leading to commensalism or parasitism as conventionally stated.

(d) Subsequent rounds of 'capacity-building/pre-suppression' followed by the development of dependencies corresponding to the pre-suppression could lead to 'irreversible mutual dependence' and decisively compromise the capacity for unaided autonomous existence.

Note that this scheme introduces an element of what is termed 'contingent irreversibility' into the model (Szathmáry and Smith 1995; Szathmáry 2015; Maynard Smith and Szathmáry 1995). The exact identity of these excess capacities and the deficiencies they compensate for depends on the nature of the events that occur among specific groups at a given point in their evolutionary histories and, crucially, under particular ecological conditions that may not recur at later times.

8.3 What Natural History Can Tell Us

We now examine particular instances of microbial associations that serve to illustrate the four steps we have outlined for symbiogenesis via CNE above. (The reader is warned that the choice of examples is merely illustrative and not meant as an exhaustive enumeration.) The 'facultative' state may be exemplified by the life history of the well-studied protist 'slime mold' *Dictyostelium discoideum* that can exist either as independent amebae or as a collective 'slug' that eventually matures into a fruiting body that releases spores. Note that in this case the cyclic AMP signal promoting aggregation is released by the amebae themselves. More spectacularly, a sulfonolipid termed 'rosette-inducing factor' (RIF-1[2]) produced by the aquatic bacterium *Algoriphagus machipongonensis* is capable of inducing the aggregation of the cells of the free-living, unicellular, eukaryotic choanoflagellate *Salpingoeca rosetta*, resulting in a rosette-like arrangement (Alegado et al. 2012). Besides being an instance of the recognition of a molecule across phylogenetically distant lineages, the foregoing example also illustrates that our recognition of what we term 'excess capacities' may be highly context-dependent; no amount of genome sequencing of the choanoflagellate would indicate that it might exhibit such a response to a bacterial compound.

A different example is the case of the association between the archaeon *Ignicoccus hospitalis* that is colonized by another archaeon *Nanoarchaeon equitans* (Huber et al. 2002). Detailed investigation of this association in co-culture revealed certain interesting facts (Jahn et al. 2008). This association was found to be obligate in that *N. equitans* is incapable of growth in culture in the absence of its host, and specific in that *N. equitans* cannot proliferate in the presence of other *Ignicoccus* spp. Isotopic labeling experiments strongly suggested the uptake of both lipids and amino acids from *I. hospitalis* by *N. equitans*. But, most interestingly for our case, the presence of *N. equitans* does not seem to be either beneficial or harmful for its host *I. hospitalis*, in contrast to the situation expected of parasitism. Could this be a case of 'excess capacity' in *I. hospitalis* compensating for the diminution of capacity in *N. equitans* resulting in the obligate dependence of the latter on the former, but in a neutral manner? Thus, one partner has retained its autonomy, while the other has become dependent on it.

An instance of complete two-way dependency is observed in phototrophic consortia formed between sulfur bacteria and *Betaproteobacteria* occurring in the chemocline of stratified lakes. Typically, the consortium consists of a single central, colorless betaproteobacterium surrounded by several (up to 69) green sulfur bacterial epibionts. The epibionts are anaerobic and photoautotrophic, besides acting as light sensors. The central bacterium is motile and preferentially moves away from dark areas into lighted ones (scotophobotaxis). Signaling between the epibionts and the central bacterium ensures that cell division between the partners is coordinated. The consortium termed *Chlorochromatium aggregatum* is the only one that has been

[2]Unrelated to eukaryotic RIF1 involved in the process of repairing double-strand DNA breaks.

successfully propagated in the laboratory (reviewed by Müller and Overmann 2011). It consists of 15 cells of *Chlorobium chlorochromatii* strain CaD epibionts surrounding the central betaproteobacterium *Candidatus* Symbiobacter mobilis (Liu et al. 2013). Ca.S. mobilis is entirely dependent on its epibionts and cannot be cultured independently in the laboratory, while the *C. chlorochromatii* can. However, the epibionts have never been detected in isolation in nature, highlighting the importance of ecology in maintaining such associations. From our viewpoint, we note that this is an indication of obligate mutual dependence, with the consortium functioning as a single entity relative to its environment.

The recent isolation and characterization of an Asgard archaeon from deep-sea methane seep sediments incidentally suggests that a neutral process could have been involved in eukaryogenesis (Imachi et al. 2020). This organism belongs to the archaeal phylum Lokiarchaeota and is designated *Candidatus*[3] Prometheoarchaeum syntrophicum strain MK-D1. Initially Ca. P. syntrophicum was enriched under syntrophic (cross-feeding) conditions in combination with a sulfate-reducing bacterium (*Halodesulfovibrio* sp.) and an archaeon (*Methanogenium* sp.). The bacterium was initially included because 16S rRNA analysis of the first successful cultures in a basal medium supplemented with casamino acids and bacteria-suppressing antibiotics revealed a simple microbial community consisting mostly of *Halodesulfovibrio* and small numbers of Lokiarchaeota. Later experiments indicated that Ca. P. syntrophicum could be successfully co-cultured with *Methanogenium* sp. alone. Extensive metabolic analysis of Ca. P. syntrophicum indicated that it could catabolize 10 amino acids through hydrogen/formate exchange during co-culture with *Halodesulfovibrio* and *Methanogenium.* Based on their findings, Imachi et al. proposed a model of eukaryogenesis wherein an ancestral anaerobic archaeon degrading amino acids to short-chain fatty acids and H_2 could have initially partnered with a sulfate-reducing bacterium (SRB) that could scavenge H_2 (and indirectly O_2). An additional partnership with a facultatively aerobic organotrophic bacterium that could also scavenge O_2 could have enabled further development of aerotolerance and its subsequent endosymbiosis with the archaeon, with or without the external SRB partner. Note that, in this model, neither of the partners—archaeon or SRB—*needs* the other in the initial stage. However, it is speculated that a subsequent rise in atmospheric O_2 levels would have stabilized this partnership and also facilitated the acquisition of the facultatively aerobic endosymbiotic bacterium. Here natural selection acts as a negative filter, imposing contingent irreversibility on the association. As an interesting aside, we note recent evidence for the apparent loss of mitochondria (and mitochondrial genes) in two contemporary eukaryotes inhabiting hypoxic environments, namely, *Henneguya salminicola* (Yahalomi et al. 2020) and *Monocercomonoides* sp. PA203 (Karnkowska et al.

---

[3]Imachi et al. opted to retain the *Candidatus* nomenclature even after culturing the organism and sequencing its genome because its very low growth rate and, consequently, low cell mass obtained were insufficient for the kind of biological characterization mandated by current standards of nomenclature. Therefore, the culture was not deposited in any culture collection either. (Supplementary note 3, Imachi et al. 2020).

2016). These two instances illustrate the importance of negative selection in maintaining the irreversibility of a symbiotic association; in these two cases, a long-term change in environmental conditions (hypoxia) seems to have destabilized an established symbiosis.

8.4 The Importance of Neutrality in Evolution and Ecology Vis-à-Vis Symbiogenesis

It is impossible to envision any genuine natural history of living organisms without incorporating the elements of both evolution and ecology, each of which influences the other, into the overall description. Theories of neutrality extend the analytical and predictive spectrum of both ecological and evolutionary models. The importance of neutrality in an evolutionary context was initially enunciated by Motoo Kimura (1977), and in an ecological context by Stephen Hubbell (2001). Note that this section is by no means a comprehensive overview of neutral theories, but only a discussion of their relevance to CNE in the specific case of symbiogenesis by means of suitable examples.

The neutral theory of species diversity emphasizes the importance of stochastic processes in the formation and establishment of community structure and suggests the functional equivalence of species within ecological communities. Its proponents suggest that neutrality is either an approximation or a useful null hypothesis in a strict sense (Hubbell 2005; Rosindell et al. 2012). However, neutrality in the ecological sense does not automatically rule out natural selection. Evolution by natural selection constantly occurs at the phenotypic level—which would imply that function is privileged over specific genotypes (Martínez and Moya 2011). For example, a recent meta-analysis of available microbiota-related datasets across a wide range of animal hosts (sponges, nematodes, mice, hydra, jellyfish and sea anemone) as well as environmental samples illustrated the relative importance of neutral and niche-based assembly (Sieber et al. 2019). Environmental samples tended to be more consistent with a neutral model, while animal microbiota, subjected to selection by host innate and/or adaptive immune systems, deviated to varying extents from expectations of neutrality. Among the animal microbiota analyzed by Sieber et al., the composition of the microbiota of sponges, laboratory populations of jellyfish and, rather surprisingly, that of a wild mouse population followed a neutral model very closely. The microbiota of other animals deviated significantly from neutral expectations, especially across developmental stages in organisms such as the hydra and the sea anemone. Finally, it seems that the process of microbial community assembly *after* passage through an environmental filter in the form of the animal host is also highly consistent with a neutral model. However, for the host to survive and propagate from the evolutionary viewpoint, varied microbiota must eventually elaborate equivalent functions, regardless of the identity of the microbes providing these services (Sitaraman 2018), which incidentally converges to the idea of functional equivalence emphasized by (ecological) neutral theory.

Neutral evolutionary theory, on the other hand, explains the existence of gene variants and, ultimately, phenotypes that do not impose a fitness penalty on the individual organism. Microbes, with their short generation times and large population sizes, can understandably produce a diverse range of variants even if the actual mutation rate per cell per generation is relatively low. To these attributes may be added their biochemical versatility and planet-wide ubiquity in a staggering range of habitats that, we suggest, present immense and unique possibilities for inter- and intra-species interactions waiting to be identified and analyzed. Recent work utilizing computational metabolic modeling of the effect of 'costless'[4] metabolite secretion on hypothetical pairwise associations between 24 well-characterized micro-organisms indicated not only the emergence, but also the stabilization of syntrophic relationships in minimal media under oxic and anoxic conditions (Pacheco et al. 2019). Thus, an essentially neutral phenotype (costless metabolite secretion) can initiate and stabilize inter-microbial interactions.

To reiterate, the theory of CNE can help us better understand the emergence, evolution and stabilization of increasingly complex and irreversible symbiotic relationships over time and in various ecological niches. Most importantly, it helps us conceive an evolutionary trajectory that is *not always* dependent on adaptation and natural selection. Rather, it indicates that, once inter-organismal dependency becomes obligate (e.g. due to loss of biochemical capacity due to mutations in a syntrophic association), the disruption of a partnership amounts to negative selection against the individual partner(s). Thus, the once-fortuitous and facultative association of prokaryotic partners is rendered irreversible, coincidentally leading to an increase in biological complexity.

Acknowledgments This paper is dedicated to my parents, Mr. G. Sitaraman and Mrs. Indubala, for their unstinting support of my studies. I thank my daughter Sukriti Sitaraman for drawing the figure. Mr. Ratan Jha, Assistant Librarian, TERI School of Advanced Studies, is thanked for his diligence in procuring of some of the references in a timely manner via the Developing Library Network (DELNET), India.

References

Aanen DK, Eggleton P (2017) Symbiogenesis: beyond the endosymbiosis theory? J Theor Biol 434:99–103. https://doi.org/10.1016/j.jtbi.2017.08.001

Alegado RA, Brown LW, Cao S et al (2012) A bacterial sulfonolipid triggers multicellular development in the closest living relatives of animals. eLife 1:e00013. https://doi.org/10.7554/eLife.00013

[4]A costless metabolite is defined as one whose secretion does not impose a fitness cost on the organism that produces it.

Baedke J, Fábregas-Tejeda A, Nieves Delgado A (2020) The holobiont concept before Margulis. J Exp Zool (Mol Dev Evol) 334:1–7. https://doi.org/10.1002/jez.b.22931

Bermudes D, Back RC (1991) Symbiosis inferred from the fossil record. In: Margulis L, Fester R (eds) Symbiosis as a source of evolutionary innovation: speciation and morphogenesis. The MIT Press, Cambridge, pp 72–91

Brenner S (2010) Sequences and consequences. Philos Trans R Soc B Biol Sci 365(1537):207–212. https://doi.org/10.1098/rstb.2009.0221

Casazza LR (2012) Symbiosis in the fossil record: Eocene Nummulites and Pleistocene reefs of Egypt. Doctoral thesis, University of California, Berkeley. https://escholarship.org/uc/item/5627w4dd

Covello Patrick S, Gray Michael W (1993) On the evolution of RNA editing. Trends Genet 9 (8):265–268. https://doi.org/10.1016/0168-9525(93)90011-6

Gray MW, Lukes J, Archibald JM et al (2010) Irremediable complexity? Science 330 (6006):920–921. https://doi.org/10.1126/science.1198594

Harcombe W (2010) Novel cooperation experimentally evolved between species. Evol Int J Org Evol 64(7):2166–2172. https://doi.org/10.1111/j.1558-5646.2010.00959.x

Hubbell SP (2001) The unified neutral theory of biodiversity and biogeography (Monographs in population biology – 32). Princeton University Press, Princeton

Hubbell SP (2005) Neutral theory in community ecology and the hypothesis of functional equivalence. Funct Ecol 19:166–172. https://doi.org/10.1111/j.0269-8463.2005.00965.x

Huber H, Hohn MJ, Rachel R et al (2002) A new phylum of Archaea represented by a nanosized hyperthermophilic symbiont. Nature 417(6884):63–67. https://doi.org/10.1038/417063a

Imachi H, Nobu MK, Nakahara N et al (2020) Isolation of an archaeon at the prokaryote–eukaryote interface. Nature 577(7791):519–525. https://doi.org/10.1038/s41586-019-1916-6

Jahn U, Gallenberger M, Paper W et al (2008) *Nanoarchaeum equitans* and *Ignicoccus hospitalis*: new insights into a unique, intimate association of two archaea. J Bacteriol 190(5):1743–1750. https://doi.org/10.1128/JB.01731-07

Karnkowska A, Vacek V, Zubáčová Z et al (2016) A eukaryote without a mitochondrial organelle. Curr Biol 26(10):1274–1284. https://doi.org/10.1016/j.cub.2016.03.053

Kimura M (1977) Preponderance of synonymous changes as evidence for the neutral theory of molecular evolution. Nature 267(5608):275–276. https://doi.org/10.1038/267275a0

Liu Z, Müller J, Li T et al (2013) Genomic analysis reveals key aspects of prokaryotic symbiosis in the phototrophic consortium "Chlorochromatium aggregatum". Genome Biol 14(11):R127. https://doi.org/10.1186/gb-2013-14-11-r127

Margulis L (1991) Symbiogenesis and symbionticism. In: Margulis L, Fester R (eds) Symbiosis as a source of evolutionary innovation: speciation and morphogenesis. The MIT Press, Cambridge, pp 1–14

Martínez M, Moya A (2011) Natural selection and multi-level causation. Philos Theory Biol 3: e202. https://doi.org/10.3998/ptb.6959004.0003.002

Maynard Smith J, Szathmáry E (1995) The major transitions in evolution. W. H. Freeman and Company, Oxford

Mehta AP, Supekova L, Chen J-H et al (2018) Engineering yeast endosymbionts as a step toward the evolution of mitochondria. Proc Natl Acad Sci 115(46):11796–11801. https://doi.org/10.1073/pnas.1813143115

Meyer-Abich A (1943) Beiträge zur Theorie der Evolution der Organismen. I. Das Typologische Grundgesetz und Seine Folgerungen Für Phylogenie und Entwicklungsphysiologie. [Contributions to the evolutionary theory of organisms: I. The basic typological law and its implications for phylogeny and developmental physiology]. Acta Biotheor 7:1–80. https://doi.org/10.1007/BF01603792

Müller J, Overmann J (2011) Close interspecies interactions between prokaryotes from sulfureous environments. Front Microbiol 2:146. https://doi.org/10.3389/fmicb.2011.00146

Ohkawa H, Hashimoto N, Furukawa S et al (2011) Forced symbiosis between *Synechocystis* spp. PCC 6803 and apo-symbiotic *Paramecium bursaria* as an experimental model for evolutionary

emergence of primitive photosynthetic eukaryotes. Plant Signal Behav 6(6):773–776. https://doi.org/10.4161/psb.6.6.15239
Pacheco AR, Moel M, Segrè D (2019) Costless metabolic secretions as drivers of interspecies interactions in microbial ecosystems. Nat Commun 10:103. https://doi.org/10.1038/s41467-018-07946-9
Rosindell J, Hubbell SP, He F et al (2012) The case for ecological neutral theory. Trends Ecol Evol 27(4):203–208. https://doi.org/10.1016/j.tree.2012.01.004
Sagan L (1967) On the origin of mitosing cells. J Theor Biol 14(3):255–274. https://doi.org/10.1016/0022-5193(67)90079-3
Shapiro JW, Turner PE (2018) Evolution of mutualism from parasitism in experimental virus populations. Evolution 72(3):707–712. https://doi.org/10.1111/evo.13440
Shou W, Ram S, Vilar JMG (2007) Synthetic cooperation in engineered yeast populations. Proc Natl Acad Sci U S A 104(6):1877–1882. https://doi.org/10.1073/pnas.0610575104
Sieber M, Pita L, Weiland-Bräuer N et al (2019) Neutrality in the metaorganism. PLoS Biol 17(6): e3000298. https://doi.org/10.1371/journal.pbio.3000298
Sitaraman R (2018) Prokaryotic horizontal gene transfer within the human holobiont: ecological-evolutionary inferences, implications and possibilities. Microbiome 6(1):163. https://doi.org/10.1186/s40168-018-0551-z
Stoltzfus A (1999) On the possibility of constructive neutral evolution. J Mol Evol 49(2):169–181. https://doi.org/10.1007/pl00006540
Szathmáry E (2015) Toward major evolutionary transitions theory 2.0. Proc Natl Acad Sci 112 (33):10104–10111. https://doi.org/10.1073/pnas.1421398112
Szathmáry E, Smith JM (1995) The major evolutionary transitions. Nature 374(6519):227–232. https://doi.org/10.1038/374227a0
Tapanila L (2008) Direct evidence of ancient symbiosis using trace fossils. Paleontol Soc Pap 14:271–287. https://doi.org/10.1017/S1089332600001728
Yahalomi D, Atkinson SD, Neuhof M et al (2020) A cnidarian parasite of salmon (Myxozoa: *Henneguya*) lacks a mitochondrial genome. Proc Natl Acad Sci U S A 117(10):5358–5363. https://doi.org/10.1073/pnas.1909907117
Yuan X, Xiao S, Taylor TN (2005) Lichen-like symbiosis 600 million years ago. Science 308 (5724):1017–1020. https://doi.org/10.1126/science.1111347

Chapter 9
Chemiosmosis, Evolutionary Conflict, and Eukaryotic Symbiosis

Neil W. Blackstone

Abstract Mutualistic symbiosis, in which individuals of different species cooperate and both benefit, has long been an evolutionary puzzle. Why should individuals of two different species cooperate? In this case, as in all others, cooperation is not automatic, but rather requires the mediation of evolutionary conflicts. In chemiosmosis, redox reactions produce a trans-membrane "proton-motive force" that powers energy-requiring reactions in most organisms. Chemiosmosis may also have a role in conflict mediation. Chemiosmosis rapidly produces considerable amounts of products, increasing the risk of end-product inhibition and the formation of dangerous by-products, such as reactive oxygen species. While several mechanisms can modulate chemiosmosis, potential negative effects can also be ameliorated by simply dispersing excess product into the environment. This "free lunch you are forced to make" can attract individuals of other species leading to groups, in which other organisms share the products that are released into the environment by the chemiosmotic cell or organism. Since the time of Darwin, evolutionary biology has recognized that groups are the key to the evolution of cooperation. With many small groups, chance associations of cooperators can arise, even if cooperation is selected against at the individual level. Groups of cooperators can then outcompete groups of defectors, which do not cooperate. Indeed, numerous symbioses may have arisen in this way, perhaps most notably the symbioses of host cells and chemiosmotic bacteria that gave rise to the eukaryotic cell. Other examples in which one partner relies on chemiosmotic products supplied by the other include lichens, corals or other metazoans and dinoflagellates, sap-feeding insects, and plant–rhizobia and plant–mycorrhiza interactions. More problematic are cases of gut microbiomes—for instance, those of termites, ruminants, and even human beings. Under some but not all circumstances, chemiosmosis can be co-opted into punishing defectors and enforcing cooperation, thus leading to mutualistic symbioses.

N. W. Blackstone (✉)
Department of Biological Sciences, Northern Illinois University, DeKalb, IL, USA
e-mail: neilb@niu.edu

M. Kloc (ed.), *Symbiosis: Cellular, Molecular, Medical and Evolutionary Aspects*,
Results and Problems in Cell Differentiation 69,
https://doi.org/10.1007/978-3-030-51849-3_9

9.1 Introduction

Symbiosis, intimate relationships between different organisms, often at the cellular level, can range from parasitism to commensalism to mutualism. The last, in which both the host and the symbiont benefit, has attracted considerable attention. Why should individuals of two different species cooperate? Indeed, such examples seem to be contrary to Darwinian evolution and for some time were supposed to be just that (Blackstone 2016). Currently, however, explicit considerations of evolutionary conflict are recognized as a central issue in understanding symbioses (Bronstein 2015), and cooperation is not an automatic outcome. In other words, as discussed in more detail below, even when a host–symbiont community appears to be dominated by mutualistic interactions, evolutionary conflict can still arise. Cooperation emerges if individuals forgo reproduction to contribute to the group, but selection will inevitably favor the opposite (Nowak 2006). Mechanisms must evolve or more typically be co-opted into mediating these evolutionary conflicts.

In chemiosmosis, redox reactions produce a trans-membrane "proton-motive force" that powers energy-requiring reactions in most organisms. Furthermore, under some circumstances, chemiosmosis may function as a mechanism of conflict mediation. As outlined below, chemiosmotic reactions are extremely fast and can quickly produce large quantities of products. These products can be stored in various ways, but storage mechanisms are slow relative to chemiosmosis and in any event storage capacity is usually limited. When conditions are opportune, chemiosmotic cells and organisms face the possibility of "end-product inhibition" (e.g., Chance and Williams 1956; Goldschmidt and Huber 1992), which can have severe consequences. In some sense, chemiosmosis confronts organisms with the same issues that southern California electric utilities face on sunny, windy days—the need to entice consumers to use more electricity before transmission lines melt (Service 2019). While mechanisms that modulate chemiosmosis are available (e.g., Allen et al. 2011; Malone et al. 2019), an alternative solution is simply to disperse excess product into the environment. Such inadvertent largesse can lead to the formation of symbiotic associations, which are in some ways similar to "byproduct" symbioses (Bronstein 2015). The resulting groups may be the key to the evolution of cooperation. I will suggest that such dynamics can have profound impacts on some mutualistic symbioses. First, however, some background is necessary.

9.2 Energy, the Currency of Biological Evolution

At the outset, it is axiomatic that natural selection favors organisms that devote all available resources into their reproduction. Sharing resources, whether with individuals of the same or different species, will usually be strongly selected against if these resources can be converted into more offspring. Thus, the organism that shares, say, half the available substrate with another organism will be outcompeted by the organism that monopolizes all available substrate and channels this into its reproduction. With regard to symbiosis or any other potentially cooperative interaction, such "defectors" that do not cooperate will usually be favored (Nowak 2006).

9.3 Chemiosmosis

Chemiosmosis describes the ubiquitous process that most living organisms use to convert energy. Notably, it yields considerably more ATPs than energy conversion based on substrate-level phosphorylation (e.g., fermentation). Chemiosmosis also proceeds by mechanisms wholly different from substrate-level phosphorylation. First described in a series of revolutionary papers by Peter Mitchell (e.g., Mitchell 1961), the surviving theory has been substantially modified from that which was originally proposed (e.g., Boyer et al. 1977). Nevertheless, the basic elements of chemiosmosis remain clearly recognizable (e.g., Allen 1993). Electron carriers, embedded in a membrane that is impervious to protons, connect what are ultimately environmental sources and sinks of electrons. As redox reactions proceed, the electron carriers translocate protons across the membrane. These protons then move back across the membrane via ATP synthase triggering the formation of ATP from ADP and inorganic phosphate. While microbial chemiosmotic processes are many and various, here the focus will be on oxidative phosphorylation in mitochondria and oxygenic photosynthesis in chloroplasts. While of course chloroplasts require light, in many ways they function similar to mitochondria. Electrons (from water in the former or coenzymes such as NADH in the latter) power an electron transport chain, producing a proton gradient, which catalyzes the formation of ATP in both chloroplasts and mitochondria (NADPH is also formed in the former) (Fig. 9.1). Cells containing mitochondria can then store ATP as phosphoenolpyruvate or phosphocreatine or something similar, while chloroplasts store the energy in ATP and NADPH by fixing carbon via the soluble enzyme RuBisCO (ribulose-1,5-bisphosphate carboxylase/oxygenase).

At the same time that Mitchell was developing the chemiosmotic theory, other work began to reveal the existence of quantum electron transfer in biological systems (e.g., Chance and Nishimura 1960). In addition, later work showed the existence of "super-complexes" among the membrane-bound electron carriers. Thus, electron transfer within and between membrane-bound complexes in chemiosmosis occurs extremely rapidly (Moser et al. 1992; Dudkina et al. 2005). This rapidity poses problems in linking chemiosmosis to the soluble reactions that store energy. For instance, RuBisCO is perhaps the most abundant protein on Earth because it is "mopping up" the products of chemiosmosis. The linking of chemiosmosis to slower soluble reactions and potentially limited storage capacity has other consequences as well. If an accumulation of products inhibits electron flow, these electrons may divert to molecular oxygen and reactive oxygen species (i.e., partially reduced forms of oxygen, ROS) will form. The chemiosmotic process itself is the cause of ROS formation.

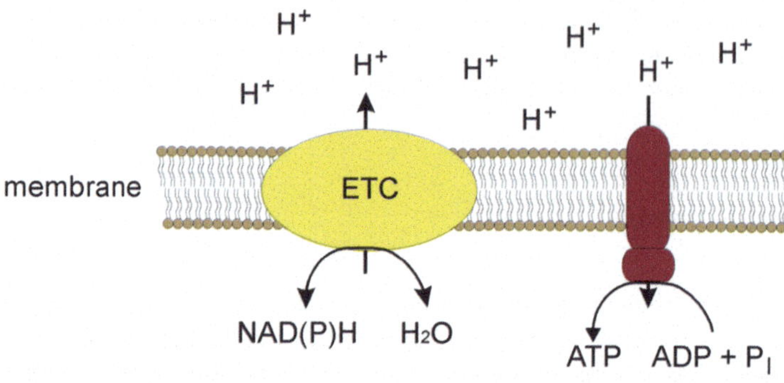

Fig. 9.1 Schematic outlining eukaryotic chemiosmosis. Mitochondria oxidize reduced co-factors such as NADH, run the electrons through an electron transport chain (ETC), and build a transmembrane proton gradient. Protons return via *ATP* synthase, triggering the formation of ATP from ADP and inorganic phosphate (P_i). With the input of light energy, chloroplasts oxidize water, run the electrons through an electron transport chain (which is homologous to that in mitochondria), and build a transmembrane proton gradient. As in mitochondria, protons trigger the formation of *ATP* and electrons reduce $NADP^+$ to NADPH. Mitochondria can store ATP as phosphoenolpyruvate, phosphocreatine, or similar compounds, while chloroplasts store the energy in ATP and NADPH by fixing carbon via the soluble enzyme RuBisCO

9.4 Evolutionary Conflict

In evolutionary terms, cooperation usually involves costs and thus is not an automatic outcome. Hosts and symbionts may respond to divergent selective forces (Bronstein 2015). A defecting symbiont can be selected to sequester resources from the host and symbiont community and to use these resources for its replication. Such defectors may gain a replicatory advantage compared to cooperative symbionts that at least in part forgo reproduction and share resources with the larger community. On the other hand, by sharing resources with the host, the cooperative symbiont community often establishes a more durable environment for their long-term persistence. Despite these long-term advantages, the host and the larger symbiont community (the higher-level unit) remain vulnerable to exploitation by lower-level defectors. Cooperation can emerge only if mechanisms of conflict mediation evolve to suppress defectors. These mechanisms typically decrease the heritable variation at the lower level, thus limiting the evolution of potential defectors, or increase the heritable variation at the higher level, thus potentiating selection against groups of defectors (Michod 1999). While symbiosis is often conceptualized in bilateral terms (e.g., mutualism or parasitism), mechanistically these evolutionary interactions are multilateral and multilevel. In other words, even when a host–symbiont community that appears to be dominated by mutualistic interactions, defecting symbionts can still arise and flourish unless they are controlled by mechanisms of conflict mediation. Population structure often plays a role in conflict mediation, particularly if a population is subdivided into many small groups, in which case groups of cooperators can arise purely by chance. In this way, even if cooperation is selected against at the level of the individual, it can still arise and be favored at the level of the group (Radzvilavicius and Blackstone 2018).

9.5 Chemiosmosis and Conflict Mediation

The hypothesis that chemiosmosis can mediate conflict and lead to associations among organisms is based on the following three premises:

1. Under circumstances that favor chemiosmosis, the energetic needs of a cell or organism may be easily met because this process is extraordinarily fast and efficient.
2. By its biochemical nature—separating hydrogen atoms into component electrons and protons—chemiosmosis can be a potentially fraught process.
3. Given (1), in chemiosmotic cells and organisms too much ATP is more frequently a problem than too little. Given (2), too much ATP can be risky in that it leads to end-product inhibition causing loose electrons to form dangerous by-products.

Cooperation is usually selected against because of the evolutionary costs of sharing, but if chemiosmosis diminishes these costs, or removes them entirely, or even converts them into benefits, cooperation can then be favored and associations among organisms can form. Nevertheless, even in groups of cooperators formed in this manner, defectors may still arise, e.g., by mutations that counteract passive "leakage" of product through a cell wall, or by loss-of-function mutations to genes coding for transporters that would otherwise carry excess product out of the cell or organism. If the excess product can then be diverted into greater reproductive success, defection will be favored. If, however, defection leads to end-product inhibition, it will have costs and will be selected against. Thus, if the origin of life is "the free lunch you are paid to eat" (Lane 2009), then when the lunch is no longer free, you must make your lunch or steal someone else's. With chemiosmosis, however, there may be "the free lunch you are forced to make."

9.6 Chemiosmosis and Symbiosis

Casual observation suggests that many successful symbioses involve an exchange of products produced by chemiosmosis. The symbioses that produced the eukaryotic cell involve chemiosmotic organelles—mitochondria and chloroplasts. Since all eukaryotes primitively contain mitochondria, eukaryotic symbioses typically involve photosynthesis. Secondary symbioses between nonphotosynthetic and photosynthetic eukaryotes have occurred several times. Fungi and algae bond together in lichens; many marine animals rely on photosynthetic dinoflagellates; photosynthetic plants are symbiotic with nitrogen-fixing bacteria and arbuscular mycorrhizal fungi; insects feed on sap from photosynthetic plants; and so on. There is also a large class of associations between gut microbiota and metazoans—e.g., ruminants, termites, and even human beings—which at least in part include fermentative microbes releasing substrate that is utilized by the chemiosmotic host. In all of these symbioses, chemiosmosis figures prominently in the interaction. Is this merely a coincidence? Although existing symbioses cannot provide direct evidence into their formative steps, specific examples can still be instructive in terms of which sort of relationships—e.g., chemiosmotic symbionts or chemiosmotic hosts—favor mutualistic symbioses. By no means is the following intended to be a comprehensive review; rather, these examples are introduced to highlight themes that can then be synthesized in Discussion section.

Coral-dinoflagellate Symbiosis While numerous marine animals form symbiotic associations, clonal and colonial animals such as sponges, ascidians, bryozoans, and cnidarians are particularly likely to do so (Wulff 1985; Burgess et al. 2017; Bang et al. 2018; Blackstone 2020). Notably, all modern reef-building cnidarians contain endosymbiotic dinoflagellates (Hoegh-Guldberg 1999), formerly considered *Symbiodinium* and now classified as the family Symbiodiniaceae (LaJeunesse et al. 2018). Many other colonial cnidarians, whether part of coral reef communities or not, also exhibit similar symbioses. The coral-dinoflagellate symbiosis has attracted considerable study because its breakdown triggers coral bleaching. When environmental stress becomes extreme, these dinoflagellates are lost, and corals bleach (Davy et al. 2012). As elaborated below, chemiosmosis likely not only alleviates evolutionary conflicts but also contributes to the process of coral bleaching in which cooperation breaks down.

Notably, while taxa included in the Symbiodiniaceae form symbioses with corals and many other metazoans, they also remain capable of free-living existence (Fig. 9.2). Given the intense competition for space in the marine benthos, symbiosis is a path by which Symbiodiniaceae can become larger and thus more effective competitors. For the metazoan host, symbiosis is a path to at least partial autotrophy, since these dinoflagellates are photosynthetic and actively export various forms of reduced carbon (Davy et al. 2012).

Despite these mutual benefits, a durable symbiosis requires robust mechanisms of conflict mediation. As suggested above, a population structure of many small groups can often mediate a conflict. No matter how strongly defectors are selected for at the individual level, with many small groups, purely by chance (i.e., genetic drift) some groups will comprise only cooperators. These groups of cooperators will then be strongly selected for at the group level and outcompete groups with more defectors. This sort of scenario likely contributed to the secondary symbioses that gave rise to dinoflagellates among others (Radzvilavicius and Blackstone 2018). Some stages of the life cycle of colonial cnidarians may comprise many small groups, e.g., when small, sexually produced colonies first take up symbionts. Overall, however, colonial cnidarians have a population structure that appears entirely unfavorable in this respect, perhaps best characterized as relatively few very large and very long-lasting groups. In other words, colonies are large, long-lived, and relatively scarce, as compared to, say, insects or nematodes, and a single colony contains many trillions

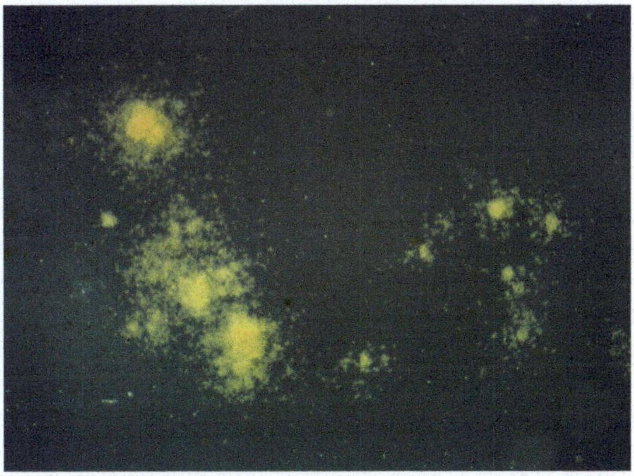

Fig. 9.2 Individuals of *Cladocopium* of the Symbiodiniaceae, formerly clade C of *Symbiodinium*, in laboratory culture. The existence of a free-living population of symbionts has important implications for a symbiosis. *Cladocopium* and other Symbiodiniaceae are capable of a free-living existence and thus can recolonize corals or other metazoan hosts. Because of accessory photosynthetic pigments, these cells absorb green wavelengths and hence appear yellow (each cell is roughly 5–10 micrometers in diameter)

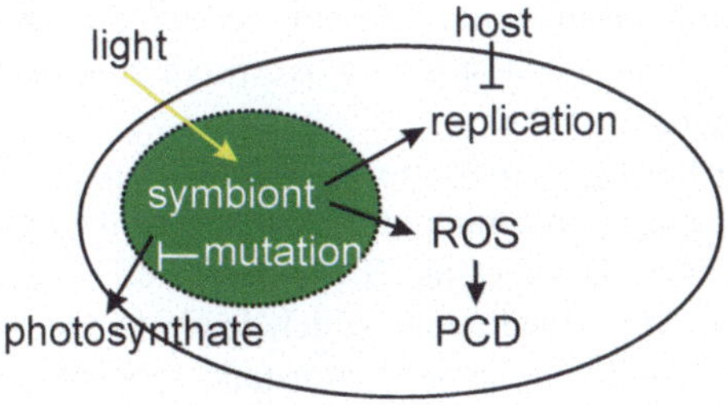

Fig. 9.3 Schematic of evolutionary conflict and its mediation in cnidarian photosymbioses. Via mechanisms that are only partly understood, replication of the symbiont is constrained by the host cell. To avoid product inhibition, the symbiont usually exports reduced carbon. Various mutations to transporter pathway genes can inactivate this export. In the presence of light, product inhibition and other mechanisms result in reactive oxygen species (ROS) formation, which in turn triggers programmed cell death (PCD) of the host cell

of symbionts. Under these conditions, defecting symbionts are strongly selected for within a colony, and additional mechanisms of conflict mediation are a necessity.

Housing symbionts within host cells is a way to create many small groups within a single colony. Furthermore, maintaining symbionts in small groups within cells allows other mechanisms to mediate conflict (Fig. 9.3). Replication of symbionts is typically limited by the host, likely by limiting provisioning with inorganic nutrients (Davy et al. 2012). Combined with limited replication of symbionts, the biophysics of chemiosmosis dictates that excess product must be exported by the symbiont, i.e., shared with the higher-level unit. If the transporter pathways in the symbiont are inactivated by mutation, the redox state of the now-defecting symbiont will shift in the direction of reduction because of product inhibition or other mechanisms (Blackstone and Golladay 2018). High levels of reactive oxygen species will form, in turn triggering programmed cell death.

This hypothesis can illuminate features of the coral-dinoflagellate natural history that otherwise are difficult to explain. Symbiotic corals often release large quantities of reduced carbon (Crossland et al. 1980; Muscatine et al. 1984). Much of this substrate is released in the form of mucus, which includes lipids and polysaccharides and is utilized by many other organisms (Crossland et al. 1980). Coral workers have struggled to explain how this is adaptive; it may, however, simply be a mechanism to disperse excess product into the environment. Indeed, as pointed out by Crossland et al. (1980: p. 89): "Lipid production may provide an alternative to zooxanthellae photorespiratory processes. . .in utilizing excess photosynthetically produced ATP and reducing power [NAD(P)H]. . . .A variety of mechanisms for dispersal of reducing power may be an important feature in maintaining chloroplastic integrity of zooxanthellae contained by the sessile coral in high light environments (e.g., reef flats, shallow reefs)." As was commonplace at the time, Crossland et al. (1980) referred to the Symbiodiniaceae as zooxanthellae.

Furthermore, given such overproduction of a substrate, exposing symbionts to high light levels seems of little value. When considered in terms of conflict mediation, however, exposing symbionts to high light levels, while at the same time limiting their replication, ensures that defectors, which do not export the substrate, essentially self-identify and self-destruct by overproducing reactive oxygen species (Blackstone and Golladay 2018).

By this view, high levels of environmental stress cause photosynthesis to break down, and many symbionts may emit high levels of reactive oxygen species, leading to bleaching (Weis 2008; Davy et al. 2012; Parrin et al. 2017). Bleaching is a by-product of failed conflict mediation, which leads to the failure of the higher-level unit. Nevertheless, under the same circumstances, Symbiodiniaceae themselves do not bleach. Dinoflagellates are a product of a secondary symbiosis between two eukaryotic cells, one of which contained chloroplasts (Janouškovec et al. 2010). This symbiosis is likely based on chemiosmosis as well, but it does not break down, arguably because it exhibits more effective mechanisms of conflict mediation (e.g., genome loss) that entirely prevent this. Adaptation to environmental stress in colonial cnidarians with symbiotic dinoflagellates may require the evolution of more robust mechanisms of conflict mediation.

Lichens By definition, lichens include a variety of terrestrial fungi that form symbiotic associations, usually with cyanobacteria or green algae and sometimes including another fungal partner (Nash 2008). These associations may in many ways parallel those of dinoflagellates and marine metazoans, but with some noteworthy differences. The symbionts are generally not housed in cells but rather in a network of hyphae. The symbionts are taxonomically diverse, and it is not clear whether some symbiont types are capable of a free-living existence. In parallel to coral-dinoflagellate symbiosis, the benefits of the associations include allowing the algal or bacterial symbionts to achieve the larger size and thus better compete in terrestrial systems, while the host can achieve autotrophy at least in part via symbiont-released carbohydrates. In contrast to dinoflagellates, this apparently occurs passively via a permeabilized cell wall. There are some indications that the fungal host forms structures, called haustoria, which allow extraction of reduced carbon from the symbionts as well as perhaps provisioning them with water and inorganic nutrients (Nash 2008).

Given the extra-cellular location of the symbionts, mechanisms of conflict mediation in lichens are unlikely to be based on chemiosmosis, as are those in dinoflagellate–animal symbioses. Rather, it may be that conflict arises when symbiont mutations counteract permeabilization of the cell wall, while mediation occurs simply by the host breaking down and assimilating symbionts that cease to export reduced carbon.

Plant-rhizobia Symbioses In contrast to the previous examples, the host is the photosynthetic partner in these relationships. Provisioning of soil bacteria could thus proceed with little cost, and, in the case of rhizobia, a considerable gain in the form of nitrogen fixation (Spaink et al. 1998). Nevertheless, the provisioning of the host with fixed nitrogen does not involve a chemiosmotic process, so it cannot be facilitated, nor evolutionary conflict mediated, by chemiosmosis. Possibly conflict is mediated by general defenses of plants against parasites, e.g., the hypersensitive response (e.g., Dangol et al. 2019).

Plant-arbuscular Mycorrhizal Fungi Paralleling the previous example, the symbiosis between most land plants and fungi of the phylum Glomeromycota is one of the most consequential for terrestrial ecosystems in general and for cultivated crops in particular (e.g., Jacott et al. 2017; Wipf et al. 2019). As with lichens, the fungal partner supplies inorganic nutrients and water, while the plant supplies reduced carbon. Furthermore, the benefit to the plant increases with available light (Johnson et al. 1997), suggesting that when conditions are favorable for chemiosmosis, the fungal network may serve as a sink for excess photosynthate.

Sap-feeding Insects As with rhizobia and other soil bacteria and fungi, sap-feeding insects can be thought of as utilizing the photosynthetic bounty of terrestrial plants, which may be dispersed at little cost. Thus, these seemly parasitic symbionts may be less harmful than they seem. With a diet of nutrient-poor sap, however, these insects typically require endosymbiotic bacteria, which supply crucial nutrients to their hosts. While these bacteria have been compared to chloroplasts and mitochondria (e.g., McCutcheon 2016), there is at least one crucial difference: it is the host insect that is supplying the chemiosmotic products, albeit second hand, to the bacteria. Thus, as elaborated in Discussion section, these endosymbiotic bacteria parallel, for example, rhizobia much more than true eukaryotic organelles. Symbiont chemiosmosis cannot facilitate the formation of, nor subsequently mediate conflict in, these symbioses.

Termite-flagellate Symbiosis While there is considerable complexity to these interactions (e.g., Brune and Friedrich, 2000), at least in part they involve gut microbes fermenting various ingested material anaerobically into reduced carbon molecules that can be taken up by the host and utilized by mitochondria. Thus, these sorts of symbioses appear to be merely a step in the digestion of food. As elaborated below, chemiosmosis does not drive the association, nor can it reasonably be expected to mediate conflicts. Conflict may be mediated by digestion or excretion of a microbial community that likely includes a heavy burden of defectors.

9.7 Discussion

The biophysics of chemiosmosis, whether of a presumptive host or symbiont, can favor dispersing excess product into the environment. Quantum electron transfer and super-complex formation drive chemiosmotic processes at extremely high rates, rapidly producing large quantities of products. While chemiosmosis can be modulated by several mechanisms, releasing excess product provides an alternative means to protect against end-product inhibition and a buildup of dangerous by-products such as reactive oxygen species. This "no-cost" sharing—the free lunch you are forced to make—facilitates interspecific groups, and such groups can lead to cooperative symbioses. Even after such groups have formed, however, cooperators are always vulnerable to exploitation by defectors, so additional mechanisms of conflict mediation are usually necessary. Under some, but not all, circumstances, chemiosmosis may further mediate the conflict.

The examples briefly summarized above can be used to delineate the likely circumstances under which chemiosmosis can both initiate associations and subsequently provide additional conflict mediation. For instance, in the case of the termite–flagellate symbiosis, and perhaps other gut microbiota, circumstances seem to be unlikely to encourage either form of cooperation. Consider a group of flagellates inhabiting a termite gut. Assuming these protists anaerobically ferment complex polysaccharides and emit small carbon molecules as waste, the flagellate group as a whole may accrue some increased rates of reaction by the host termite taking up their carbon waste. The group of flagellates also clearly benefits from the termite providing a habitat. Meanwhile, the aerobic termite respires the waste emitted by the flagellates. Essentially, cooperation emerges because of anaerobic–aerobic complementation, rather than chemiosmosis. Furthermore, the selection favoring flagellate cooperation is at the group level, and group size is likely quite large, so this selection is expected to be weak compared to individual selection. Thus, while one of the partners in this symbiosis uses chemiosmosis to process the substrate shared by the anaerobic symbionts, this does not meet the criteria outlined above for cooperation to blossom. In other words, the chemiosmotic partner is not dispersing product into the environment and thus incentivizing cooperation. Furthermore, since the lower-level units are not chemiosmotic, they are not at a particularly high risk from the side effects of product inhibition. Indeed, gut microbiomes, in general, may be highly vulnerable to defectors, e.g., lower-level individuals that release toxins to gain a competitive advantage, with consequent negative effects on the higher-level unit, that is, the entire group of symbionts and the host. The host may mediate this conflict simply by digesting or excreting symbionts and periodically repopulating the gut.

More likely to foster long-term cooperation are those symbioses in which chemiosmotic products are dispersed into the environment, although there are differences here as well. Some plants disperse reduced carbon photosynthate to soil bacteria and insects. In the former case, some bacteria engage in a mutualistic symbiosis by providing the plant with fixed nitrogen. Nevertheless, there may be dramatic differences in scale between the plants and the microbes that inhabit root nodules. When conditions are favorable for photosynthesis, a macroscopic plant likely produces far more photosynthate than these microbes can utilize. Thus, these microbes may represent a relatively small sink for the plant's reduced carbon. Perhaps more consequential are potentially parasitic sap-feeding insects. If these insects significantly diminish the surfeit of photosynthate experienced at times by the plant, they are perhaps less parasitic than might otherwise be expected. Meanwhile, the insects can provision their symbiotic bacteria and obtain essential nutrients at little cost. Despite parallels in genome reduction, it would be misleading to characterize these symbiotic bacteria as analogous to common eukaryotic organelles. As discussed below, chemiosmosis is perhaps the key feature of the latter. The symbiotic bacteria of sap-feeding insects, on the other hand, would seem to be more similar to the nitrogen-fixing bacteria of some plants.

The symbiosis between plants and arbuscular mycorrhizal fungi may the most consequential of the plant-based relationships. Not only do plants receive considerable provisioning from the fungal partner but also they reciprocate with up to one-quarter of their photosynthetic products (Wipf et al. 2019). Likely, the benefit to the plant partner increases with available light (Johnson et al. 1997). Possibly, at high light levels the plant avoids end-product inhibition of chemiosmosis by dispersing large quantities of photosynthate at little cost. Indeed, under circumstances favorable to photosynthesis, plants that do not share reduced carbon may inflict a cost on themselves.

In lichens, chemiosmosis may well have driven the initial symbiosis. Photosynthetic microbes may have released excess reduced carbon into the soil to avoid end-product inhibition and maintain redox balance. Fungi took up this substrate and evolved to "farm" these microbes, providing them with water, inorganic nutrients, and shelter. Given the biophysics of chemiosmosis, these benefits were obtained at little cost to the symbionts. Sharing, in this case, seems to be passive via a permeabilized cell wall. Defectors, however, could perhaps evolve a more specialized cell wall, capable of taking up water and inorganic nutrients but limiting the release of the products of photosynthesis. If such defectors could replicate freely, they might endanger the mutualistic symbiosis. The fungal partner could mediate this conflict by perhaps evolving ways to re-permeabilize the cell walls of defectors or by limiting the replication of all symbionts, in which case end-product inhibition would punish defectors.

Limitation of replication seems to be a key feature of marine animal symbioses with representatives of the Symbiodiniaceae. As with lichens, the symbionts are photosynthetic and may have dispersed excess reduced carbon into the environment to escape end-product inhibition. As associations with animals formed, some symbionts may have experienced loss-of-function mutations, limiting the active export of substrate. If these defectors could replicate freely, the mutualistic symbiosis would collapse. At least in the corals, it appears that replication of the symbionts is inhibited. Furthermore, they are housed intracellularly, so if the export of photosynthate is limited, reactive oxygen species triggered by end-product inhibition lead to programmed death of the host cell. In this way, chemiosmotic mechanisms can lead not only to the formation of symbiosis but also to the mediation of evolutionary conflicts that subsequently arise.

With this background, the success of chemiosmotic bacteria in the origin of eukaryotes can be better understood. Indeed, the two most successful symbioses in the history of life involved the chemiosmotic bacteria that became the mitochondrion and the chloroplast. The origin of the mitochondrion remains shrouded in mystery, but the evidence suggests that it occurred concomitantly with the origin of eukaryotes (Lane 2015; Martin et al. 2015). Initially, free-living proto-mitochondria may have passively released high-energy phosphate compounds such as pyrophosphate, perhaps in this way paralleling the photosynthetic symbionts involved in lichen symbioses. By doing so, end-product inhibition was avoided, and redox homeostasis maintained. Furthermore, such largesse may have led to the formation of groups of microbes, perhaps including those that became the host in the partnership that led to eukaryotes. The initial symbiosis may have been based on an exchange of high-energy and low-energy phosphate compounds (Fig. 9.4a). As the eukaryotic cell developed, ADP–ATP carriers evolved, perhaps initially acting as uncouplers to dissipate excess membrane potential (Bertholet et al. 2019). Similarly, when exchanging ATP for ADP, these carriers simulate metabolic demand and alleviate end-product inhibition (Radzvilavicius and Blackstone 2018). Again, while the origin of mitochondria and eukaryotes is an ancient event in the history of life, it is plausible that chemiosmosis had a two-fold impact: triggering the formation of groups that led to a structured population as well as mediating conflict in favor of cooperators once defectors arose in these groups.

The events surrounding the origin chloroplasts are somewhat clearer, particularly in view of the recent reconstruction of ancestral character states (e.g., Sánchez-Baracaldo et al. 2017). Much like lichens or dinoflagellate–animal symbioses, the original symbiosis likely involved the exchange of reduced and oxidized carbon (Fig. 9.4b). Chemiosmosis may have driven the association between early eukaryotes and cyanobacteria. As the symbiosis evolved, mutations that inactivated the export of reduced carbon from chloroplasts may have led to defectors. Coupled with limitations on replication, end-product inhibition, and ROS formation may have punished defectors and enforced cooperation. The secondary symbioses between eukaryotes that primitively possessed chloroplasts and those that lacked them may have proceeded similarly.

Chemiosmosis may thus have had a powerful impact on the two most consequential symbioses in the history of life. It may also impact many modern symbioses particularly those that parallel the origin of mitochondria and chloroplasts:

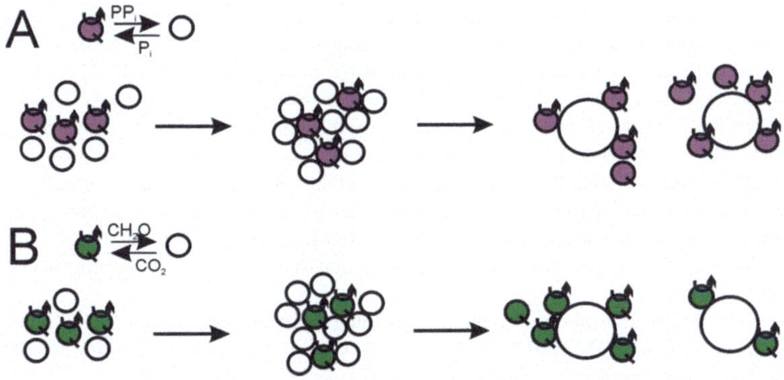

Fig. 9.4 Schematic outlining of the possible role of chemiosmosis in the symbioses that gave rise to the eukaryotic cell. In (**a**), proto-mitochondria (purple circles) carry out oxidative phosphorylation, indicated by curved arrow. At times, these cells emit excess product, indicated by the diagonal line, perhaps in the form of pyrophosphate (PP_i). The proto-host cells (unfilled circles) take up pyrophosphate and emit inorganic phosphate (P_i). The evolutionary dynamics of the proto-mitochondria could have included cells that cease to carry oxidative phosphorylation and take up pyrophosphate as well as cells that continue to carry out oxidative phosphorylation and cease to emit pyrophosphate. At times, the latter could be disadvantaged by end-product inhibition and reactive oxygen formation. In (**b**), proto-chloroplasts (green circles) carry out oxygenic photosynthesis, indicated by the curved arrow. At times, these cells emit excess product, indicated by the diagonal line, in the form of reduced carbon (CH_2O). The eukaryotic proto-host cells (unfilled circles) take up reduced carbon and emit oxidized carbon (CO_2). The evolutionary dynamics of the proto-chloroplasts could include cells that cease to carry oxygenic photosynthesis and take up reduced carbon as well as cells that continue to carry out oxygenic photosynthesis and cease to emit reduced carbon. At times, the latter could be disadvantaged by end-product inhibition and reactive oxygen formation

chemiosmotic symbionts releasing products to maintain redox homeostasis, thus leading to the formation of groups, and as the symbiosis develops, chemiosmosis being co-opted into further mediating evolutionary conflict. While there are many modern eukaryotic symbioses in which one partner is chemiosmotic, many of these do not fit this paradigm, although some do. The powerful effects of chemiosmosis should likely be considered when drawing evolutionary parallels between modern symbioses and events that occurred earlier in the history of life.

Acknowledgments Development of these ideas benefited from the presentations and subsequent discussions at the Wissenschaftskolleg zu Berlin "Evolution of hierarchical organization in adaptive systems" workshop and at the "First international redox symposium: the origin and beyond" sponsored by Knowherebio in Seoul, ROK.

References

Allen JF (1993) Control of gene expression by redox potential and the requirement for chloroplast and mitochondrial genomes. J Theor Biol 165:609–631

Allen JF, Santabarbara S, Allen CA, Puthiyaveetil S (2011) Discrete redox signaling pathways regulate photosynthetic light-harvesting and chloroplast gene transcription. PLoS One 6:e26372
Bang C, Dagan T, Deines P, Dubilier N, Duschl WJ, Fraune S, Hentschel U, Hirt H, Hülter N, Lachnit T, Picazo D, Pita L, Pogoreutz C, Rädecker N, Saad MM, Schmitz RA, Schulenburg H, Voolstra CR, Weiland-Bräuer N, Ziegler M, Bosch TCG (2018) Metaorganisms in extreme environments: do microbes play a role in organismal adaptation. Zoology 127:1–19
Bertholet AM, Chouchani ET, Kazak L, Angelin A, Fedorenko A, Long JZ, Vidoni S, Garrity R, Cho J, Terada N, Wallace DC, Spiegelman BM, Kiricho Y (2019) H+ transport is an integral function of the mitochondrial ADP/ATP carrier. Nature 571:515–520
Blackstone NW (2016) An evolutionary framework for understanding the origin of eukaryotes. Biology 5:18
Blackstone NW (2020) Evolutionary conflict in colonial animals. J Exp Zool MDE (in press)
Blackstone NW, Golladay JM (2018) Why do corals bleach? Conflict and conflict mediation in a host-symbiont community. BioEssays 40:1800021
Boyer PD, Chance B, Ernster L, Mitchell P, Racker E, Slater EC (1977) Oxidative phosphorylation and photophosphorylation. Annu Rev Biochem 46:955–1026
Bronstein JL (ed) (2015) Mutualism. Oxford University Press, Oxford, UK
Brune A, Friedrich M (2000) Microecology of the termite gut: structure and function on a microscale. Curr Opin Microbiol 3:263–269
Burgess SC, Ryan WH, Blackstone NW, Edmunds PJ, Hoogenboom MO, Levitan DR, Wulff JL (2017) Metabolic scaling in modular animals. Invertebr Biol 136:456–472
Chance B, Nishimura M (1960) On the mechanism of chlorophyll-cytochrome interaction: the temperature insensitivity of light-induced cytochrome oxidation in chromatium. Proc Natl Acad Sci U S A 46:19–24
Chance B, Williams GR (1956) The respiratory chain and oxidative phosphorylation. Adv Enzymol Relat Subj Biochem 17:65–134
Crossland CJ, Barnes DJ, Borowitzka MA (1980) Diurnal lipid and mucus production in the staghorn coral *Acropora acuminata*. Mar Biol 60:81–90
Dangol S, Chen Y, Hwang BK, Jwa N-S (2019) Iron- and reactive oxygen species-dependent ferroptotic cell death in rice-*Magnaporthe oryzae* interactions. Plant Cell 31:189–209
Davy SK, Allemand D, Weis VM (2012) Cell biology of cnidarian-dinoflagellate symbiosis. Microbiol Mol Biol Rev 76:229–261
Dudkina NV, Eubel H, Keegstra W, Boekema EJ, Braun H-P (2005) Structure of a mitochondrial supercomplex formed by respiratory-chain complexes I and III. Proc Natl Acad Sci U S A 102:3225–3229
Goldschmidt EE, Huber SC (1992) Regulation of photosynthesis by end-product accumulation in leaves of plants storing starch, sucrose, and hexose sugars. Plant Physiol 99:1443–1448
Hoegh-Guldberg O (1999) Climate change, coral bleaching and the future of the world's coral reefs. Mar Freshw Res 50:839–866
Jacott CN, Murray JD, Ridout CJ (2017) Trade-off in arbuscular mycorrhizal symbiosis: disease resistance, growth responses and perspectives for crop breeding. Agronomy 7:75
Janouškovec J, Horák A, Oborník M, Lukeš J, Keeling PJ (2010) A common red algal origin of the apicomplexan, dinoflagellate, and heterokont plastids. Proc Natl Acad Sci U S A 107:10949–10954
Johnson NC, Graham JH, Smith FA (1997) Functioning of mycorrhizal associations along the mutualism-parasitism continuum. New Phytol 135:575–585
LaJeunesse TC, Parkinson JE, Gabrielson PW, Jeong HJ, Reimer JD, Voolstra CR, Santos SR (2018) Systematic revision of Symbiodiniaceae highlights the antiquity and diversity of coral endosymbionts. Curr Biol 28:2570–2580
Lane N (2009) Life ascending: the ten great inventions of evolution. Norton, New York
Lane N (2015) The vital question: energy, evolution, and the origins of complex life. Norton, New York

Malone LA, Qian P, Mayneord GE, Hitchcock A, Farmer DA, Thompson RF, Swainsbury DJK, Ranson NA, Hunter NA, Johnson MP (2019) Cryo-EM structure of the spinach cytochrome b6f complex at 3.6 Å resolution. Nature 575:535–539
Martin WF, Garg S, Zimorski V (2015) Endosymbiotic theories for eukaryotic origin. Philos Trans R Soc Lond B 370:20140330
McCutcheon JP (2016) From microbiology to cell biology: when an intracellular bacterium becomes part of its host cell. Curr Opin Cell Biol 41:132–136
Michod RE (1999) Darwinian dynamics. Princeton University Press, Princeton, NJ
Mitchell P (1961) Coupling of phosphorylation to electron and hydrogen transfer by a chemi-osmotic type of mechanism. Nature 191:144–148
Moser CC, Keske JM, Warncke K, Farid RS, Dutton PL (1992) Nature of biological electron transfer. Nature 355:796–802
Muscatine L, Falkowski P, Porter J, Dubinsky Z (1984) Fate of photosynthetic-fixed carbon in light and shade-adapted colonies of the symbiotic coral Stylophora pistillata. Proc R Soc Lond B 222:181–202
Nash TH III (ed) (2008) Lichen biology, 2nd edn. Cambridge, Cambridge University Press
Nowak MA (2006) Five rules for the evolution of cooperation. Science 314:1560–1563
Parrin AP, Somova EL, Kern PM, Millet TA, Bross LS, Blackstone NW (2017) The use of in vivo microscopy to image the cnidarian stress response. Invertebr Biol 136:330–344
Radzvilavicius AL, Blackstone NW (2018) The evolution of individuality, revisited. Biol Rev 93:1620–1633
Sánchez-Baracaldo P, Raven JA, Pisani D, Knoll AD (2017) Early photosynthetic eukaryotes inhabited low-salinity habitats. Proc Natl Acad Sci U S A 114:E7737–E7745
Service RF (2019) Renewable bonds. Science 365:1236–1239
Spaink HP, Kondorosi A, Hooykaas PJJ (1998) The Rhizobiaceae. Kluwer Press, Dordrecht, p 566
Weis VM (2008) Cellular mechanisms of cnidarian bleaching: stress causes the collapse of symbiosis. J Exp Biol 211:3059–3066
Wipf D, Krajinski F, van Tuinen D, Recorbet G, Courty P-E (2019) Trading on the arbuscular-mycorrhizal networks. New Phytol 223:1127–1142
Wulff JL (1985) Clonal organisms and the evolution of mutualism. In: Jackson JBC, Buss LW, Cook RE (eds) Population biology and evolution of clonal organisms. Yale University Press, New Haven, CT, pp 437–466

Chapter 10
Symbiotic Origin of Apoptosis

Szymon Kaczanowski

Abstract The progress of evolutionary biology has revealed that symbiosis played a basic role in the evolution of complex eukaryotic organisms, including humans. Mitochondria are actually simplified endosymbiotic bacteria currently playing the role of cellular organelles. Mitochondrial domestication occurred at the very beginning of eukaryotic evolution. Mitochondria have two different basic functions: they produce energy using oxidative respiration, and they initiate different forms of apoptotic programmed/regulated cell death. Apoptotic programmed cell death may have different cytological forms. Mechanisms of apoptotic programmed cell death exist even in the unicellular organisms, and they play a basic role in the development of complex multicellular organisms, such as fungi, green plants, and animals. Multicellularity was independently established many times among eukaryotes. There are indications that apoptotic programmed cell death is a trait required for the establishment of multicellularity. Regulated cell death is initiated by many different parallel biochemical pathways. It is generally accepted that apoptosis evolved during mitochondrial domestication. However, there are different hypothetical models of the origin of apoptosis. The phylogenetic studies of my group indicate that apoptosis probably evolved during an evolutionary arms race between host ancestral eukaryotic predators and ancestral prey mitochondria (named protomitochondria). Protomitochondrial prey produced many different toxins as a defense against predators. From these toxins evolved extant apoptotic factors. There are indications that aerobic respiration and apoptosis co-evolved and are functionally linked in extant organisms. Perturbations of apoptosis and oxidative respiration are frequently observed during neoplastic transition. Our group showed that perturbations of apoptosis in yeasts also cause perturbations of oxidative respiration.

S. Kaczanowski (✉)
Institute of Biochemistry and Biophysics, Polish Academy of Sciences, Warszwa, Poland
e-mail: szymon@ibb.waw.pl

M. Kloc (ed.), *Symbiosis: Cellular, Molecular, Medical and Evolutionary Aspects*,
Results and Problems in Cell Differentiation 69,
https://doi.org/10.1007/978-3-030-51849-3_10

10.1 Introduction

Supported by a great deal of evidence, the endosymbiotic theory of the origin of mitochondria (and chloroplasts) suggests that mitochondria are actually simplified endosymbiotic bacteria. Mitochondria are organelles in which oxidative respiration and mitochondrial domestication occurred at the very beginning of the evolution of eukaryotes. Regulated (or programmed) cell death has been described in the majority of eukaryotic organisms. Regulated cell death is most likely required for the establishment of multicellularity. In most eukaryotes, mitochondria are central players in both cell death and cellular respiration. There are many eukaryotic parallel cell death pathways, which are usually initiated by mitochondrial factors (or at least factors having mitochondrial origins). Traditionally, animal-regulated cell death is called apoptosis, but mechanisms applied by other eukaryotic organisms are very similar. Such mechanisms include mitochondrial membrane permeability transition, the release of mitochondrial factors, and the activation of apoptotic proteases and DNases. The endosymbiotic theory of apoptosis origin was first postulated by Guido Kroemer. Different hypotheses have been proposed to explain how apoptosis evolved during mitochondrial domestication. Kroemer's original hypothesis suggests that ancestral apoptotic cell death was based on addiction modules (toxin-antitoxin systems) similar to those described in plasmid-bacteria interactions. Patrick Fitzgerald and Douglas Green hypothesized that animal apoptosis evolved independently from other forms of regulated cell death and that the main animal apoptotic protease caspase was some kind of immune factor possessed by ancestral eukaryotic host cells. Our group suggests that many different apoptotic mechanisms evolved as a result of either antagonistic host-parasite or predator-prey interactions between ancestral mitochondrial bacteria (protomitochondria) and ancestral hosts (protoeukaryotes). This led to an evolutionary arms race. According to our model, there was a selection pressure favoring protomitochondria producing as many toxins as possible and protoeukaryotic cells producing as many antitoxins as possible. In contrast to Green's model, we suggest that caspase has a protomitochondrial origin. Our phylogenetic studies provide evidence from which we postulate that the model of the evolution of apoptosis is correct. Theoretical predictions of the mathematical model of Nigel Goldenfeld indicate that an evolutionary arms race leads to complexity. The evolution of the complexity of apoptotic mechanisms follows these predictions. The endosymbiotic theory of apoptosis origin suggests that there was a co-evolution between aerobic respiration, regulated cell death, and multicellularity. There are various indications that such co-evolution exists in extant organisms. We tested this hypothesis using the yeast *Saccharomyces cerevisiae* as a model organism. In this chapter, I demonstrate why evolutionary history explains the subtle impact of oxidative respiration on apoptosis. For example, during neoplastic transformation, perturbation of the activity of cancer-suppressing apoptotic machinery co-occurs with perturbations of aerobic respiration. This observation is called the Warburg hypothesis of cancer origin. In contrast, pathological apoptosis during age-related diseases, such as Alzheimer's and Parkinson's, occurs mainly in neuron cells, where oxidative respiration is particularly active (an observation called the "inverse Warburg hypothesis").

10.2 The Endosymbiotic Origin of Multicellular Eukaryotic Organisms

Multicellular organisms are extremely diverse and include animals, green plants, slime molds, and different types of fungi. The progress of evolutionary biology has revealed that multicellularity was established independently at least 25 times (Parfrey and Lahr 2013; Abedin and King 2010; Rokas 2008). In cases of unicellular model lab organisms, the evolutionary transition of yeast from a unicellular to a primordial form of multicellularity or vice versa has been observed frequently during different experiments (Ratcliff et al. 2012; Hope et al. 2017; Soares 2011) (Fig. 10.1).

Although multicellular organisms are extremely diverse, they belong almost exclusively to one systematic group: eukaryotes (Celiker and Gore 2013; Abedin and King 2010; Parfrey and Lahr 2013; Rokas 2008). This observation suggests that eukaryotes have particular traits that are crucial for the development of multicellularity. In this chapter, I will argue that the domestication of mitochondria

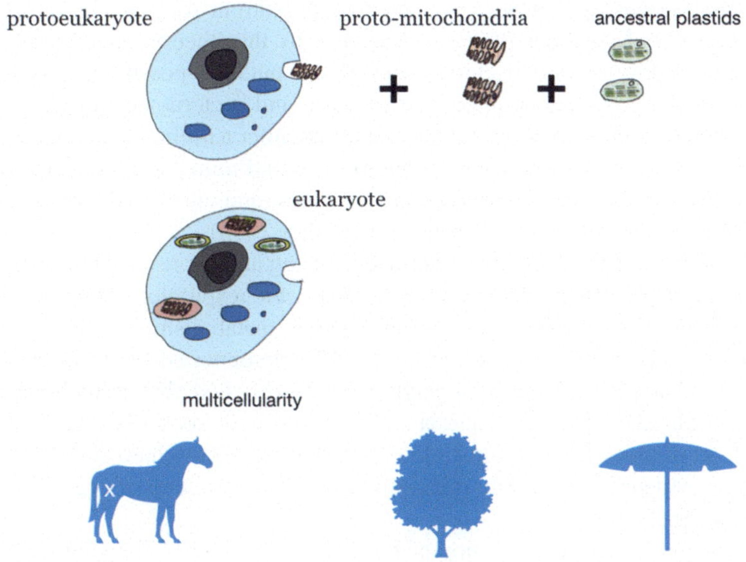

Fig. 10.1 Endosymbiotic theory and the evolution of multicellularity. Mitochondrial and chloroplast domestication lead to the establishment of eukaryotic cells. The multicellularity of eukaryotes has been independently established many times in results involving extremely diverse extant creatures, such as animals, plants, and fungi

and the development of regulated mitochondrial cell death were the preconditions required for the development of multicellularity.

Eukaryotic cells are much more complex than bacterial or archaeal cells (see as a review (O'Connor 2010)). They contain organelles, which are regions of cells that perform different functions. Some are isolated by cellular membranes. Such compartments create different environments for different chemical reactions. They resemble lab tubes in which different chemical reactions occur. As a result, different compartments play different roles in cellular metabolism.

Certain organelles have their own independent DNA, namely, plastids and mitochondria. Plastids are organelles that usually carry out photosynthesis, and they usually contain chlorophyll, the crucial pigment for photosynthesis. They are present in the cells of green plants and green algae. Interestingly, malarial parasites, called *Plasmodium*, originated from unicellular photosynthetic organisms and they contain their own type of plastids called apicoplasts (Moore et al. 2008). Mitochondria are organelles in which oxidative respiration took place. Almost all eukaryotic organisms contain mitochondria, and it is widely accepted that the ancestor of eukaryotes (protoeukaryotes) contained them. It is worth mentioning that there are amitochondrial eukaryotes that lost their mitochondria, such as Giardia (Tovar et al. 2003).

The observation that mitochondria and plastids have their own autonomous DNA was the basis of the theory of symbiogenesis or the endosymbiotic theory of eukaryotic origin. This theory was first formulated by Russian botanist Konstantin Mereschkowski (before the discovery of the function of DNA) using only microscopic observations, and it was presented in its current form by Lynn Margulis (Margulis 1993; Kutschera 2009). According to this theory, mitochondria and plastids are endosymbiotic bacteria. One of the putative possibilities is that the ancestor of eukaryotes was a unicellular predator and bacteria were prey.

Some survived in the body of this predator and, after a million years of evolution, changed into symbiotic organisms: extant plastids and mitochondria. Another putative possibility is that these bacteria were parasites, and current friendly symbiotic organisms evolved from these parasites (O'Connor 2010).

According to this model, only bacteria were using aerobic respiration. In contrast, the ancestral eukaryotic predator was using anaerobic respiration. Due to endosymbiosis evolved the extant complexity of eukaryotic metabolism.

Currently, this theory is generally accepted and supported by a variety of evidence. For example, advances in phylogenetics revealed that mitochondria and plastids are closely related to bacteria and that host cells probably originated from another kind of unicellular organism known as archaea (Kutschera 2009). The most convenient evidence is probably the circular bacterial structure of organellar DNA (O'Connor 2010).

In conclusion, eukaryotes are organisms that are able to establish multicellularity. They have a complex compartmental cellular structure. It is likely that the ancestors of eukaryotes, called protoeukaryotes, were mitochondrial cells. Mitochondria are domesticated endosymbiotic bacteria.

The question arises whether the domestication of mitochondria was a crucial event preceding the evolution of multicellularity. I will argue later in this chapter that the answer is yes because mitochondria fulfill a key function in the eukaryotic regulation of cell death, which performs a crucial function in multicellular organisms.

10.3 Regulated Cell Death Plays a Crucial Role in Establishing Multicellularity

The evolution of multicellularity is an example of kin selection. Kin selection is an altruistic evolutionary strategy. Individuals support their relatives even at the cost of their own reproduction. The mathematical theory describing kin selection was developed in the seminal work of Hamilton (1964a, b).

Social insects, such as ants and honeybees, provide a good example of kin selection. Colonies of such insects are divided unambiguously into different castes. The majority of individuals are sterile and sacrifice their reproductive success. Queens have exclusive reproductive privileges.

As was pointed out in a very recent review of Durand (2019), multicellular organisms are to some extent similar to colonies of such insects. They were established during evolution as colonies, and only some of the cells were capable of reproducing. According to Durand's model, one such altruistic behavior observed among cells of multicellular organisms is regulated cell death.

Regulated cell death (also called programmed cell death [PCD]) was first described in animals. Classical studies were performed using *C. elegans* worms as experimental models. It turns out that these organisms have a very particular and extraordinary feature. An adult worm has predetermined number of cells: an adult hermaphrodite has 959 cells, and an adult male has 1031 cells (Sulston and Horvitz 1977; Sulston et al. 1983). Sulston and Horvitz (Sulston and Horvitz 1977) described the developmental fate of every single cell of *C. elegans* (Sulston and Horvitz 1977). They made a "family tree" of each cell and observed that some of the cells die during development. In the case of *hermaphrodites*, during the generation of the 959 cells of adult hermaphrodites, 131 cells die. Cell death occurs in determined places of the developmental family tree of cells. This is a clear proof that animal cell death can be regulated (or programmed). In animals, apoptosis is involved in development (for example, in neuron systems) and maintaining homeostasis. In animals, it has also been shown that regulated cell death is a primary mechanism involved in tumor suppression. Cellular changes involved in neoplastic transformation induce regulated cell death. Suppression of apoptosis is one of the main hallmarks of cancer (see as a review (Kaczanowski 2016)). Regulated cell death is present even in the most primordial animal organisms, Cnidarians, and plays a crucial role in the establishment of germ layers (Seipp et al. 2006; Seipp et al. 2001).

Regulated cell death has also been described in non-animal multicellular organisms. For example, in green plants, regulated cell death has been described in root development (Drew et al. 2000) and pollen development (Van Durme and Nowack 2016) (see as a review (Fedak et al. 2016)). Regulated cell death has also been described in slime molds, which have a very remote relationship both with animals and plants. During the development of multicellular fruiting bodies, a stalk is generated and stalk cells die in massive numbers (Roisin-Bouffay et al. 2004; Whittingham and Raper 1960).

Regulated cell death has also been described in cases of unicellular organisms. It often plays a role in the regulation of cell density. One of the best examples of such a function is described in *Trypanosoma*, a parasite that causes a disease called sleeping sickness (Duszenko et al. 2006). One of the developmental stages of this parasite is called "stumpy cells." Stumpy cells produce prostaglandin D_2 (PGD_2), which is a putative immune suppressor of the host. It has been shown that this molecule also induces the cell death of parasites. It is likely a kind of "altruistic suicide." Decreasing the cell density prolongs the life of the host. Therefore, parasites have an opportunity to infect other hosts. It is beneficial from the point of view of the entire population, but from the perspective of the dying cells, that it is an altruistic suicide (Duszenko et al. 2006; Figarella et al. 2005). This strategy likely evolved because of kin selection, which is possible as parasites are related. Regulated cell death was also observed in the malarial parasite *Plasmodium* during the infection of mosquitos (Arambage et al. 2009).

The involvement of regulated cell death in the regulation of population size has also been described in cases of free-living organisms. For example, a bloom of the dinoflagellate *Peridinium gatunense* is observed each year in Lake Kinneret (or the Sea of Galilee). After this bloom, the cells rapidly die and there are indications suggesting that this is regulated cell death (Vardi et al. 1999).

The "public goods" hypothesis provides another plausible explanation for why altruistic suicide is beneficial. According to this hypothesis, "public goods" are cellular contents released by dying cells. These contents are used by surviving neighbors. The hypothesis was experimentally tested by Durand's groups using the green algae *Chlamydomonas* as a model (Durand et al. 2014). They demonstrated that the contents liberated during non-programmed cell death are detrimental to other cells, whereas the contents released during apoptotic-like programmed cell death are beneficial. Later, they also demonstrated that this beneficial effect of the liberated contents during programmed cell death is species-specific. They showed that PCD has an inhibitory effect on the growth of other competing species of *Chlamydomonas*.

In fact, regulated cell death has been described in the majority of unicellular eukaryotes and even in bacteria (Kaczanowski 2016; Erental et al. 2012).

In conclusion, eukaryotic cells are able to establish and evolve altruistic traits, such as cell death and multicellularity. Cell death is likely a mechanism required for the evolution of complex multicellular systems. The fact that regulated cell death is widely spread among eukaryotes provokes a question: is the mechanism of cell death required for the establishment of multicellularity?

This hypothesis was experimentally tested using the yeast model organism *S. cerevisiae*. Multicellularity can be easily selected in this organism. As previously mentioned, in nature, yeast strains are often flocculating and establishing primordial multicellular organisms (Soares 2011). It has been shown using experimental evolution that higher rates of cell death occur in such multicellular strains. There are indications suggesting that this is regulated cell death (Ratcliff et al. 2012).

In conclusion, regulated cell death is widely spread among eukaryotes and it is likely that the existence of such a mechanism is a precondition for the evolution of multicellularity. It is tempting to hypothesize that basic mechanisms of regulated cell death appeared in common ancestors of eukaryotes during the domestication of mitochondria. However, as previously mentioned, such mechanisms are also present in bacteria. There are also unicellular eukaryotic organisms that lost mitochondria. In such amitochondrial organisms, regulated cell death has also been described (Kaczanowski et al. 2011). Therefore, it would seem to be difficult to rule out such mechanisms evolving as a result of convergence, which means that different mechanisms regulating cell death evolved independently in different systematic groups.

In the next section, we will see that different mechanisms are applied in the regulation of eukaryotic cell death. However, all of these mechanisms are usually associated with mitochondria. Of course, amitochondrial organisms are an exception.

10.4 Cytology of Eukaryotic Cell Death: Apoptosis, Programmed Necrosis, and Programmed Autophagy Cell Death

Cell death, particularly regulated cell death, may have very different cytological features.

The pioneering paper by Schweichel and Merker described three different types of cell death that occur in rat embryos: Type I cell death was associated with the removal of well-preserved cellular remnants by other cells (“eating of another”). Type II cell death was associated with autophagy (“eating of itself”), and type III cell death did not involve digestion by any cell (Schweichel and Merker 1973).

Today, these types of cell death are called apoptosis, autophagic cell death, and necrosis (Vanden Berghe et al. 2014). Regulated cell death could include all of these forms of cell death (Fig. 10.2).

Apoptosis is an exclusively regulated form of cell death limited to animals. It is also the main form of animal regulated/programmed cell death. Apoptosis was first described by Kerr in 1972 (Kerr et al. 1972). Apoptosis can be easily distinguished from other forms of cell death using cytological and biochemical hallmarks. Apoptosis is initiated by mitochondrial membrane permeability transition, which is characterized by the breakdown of the inner mitochondrial transmembrane potential.

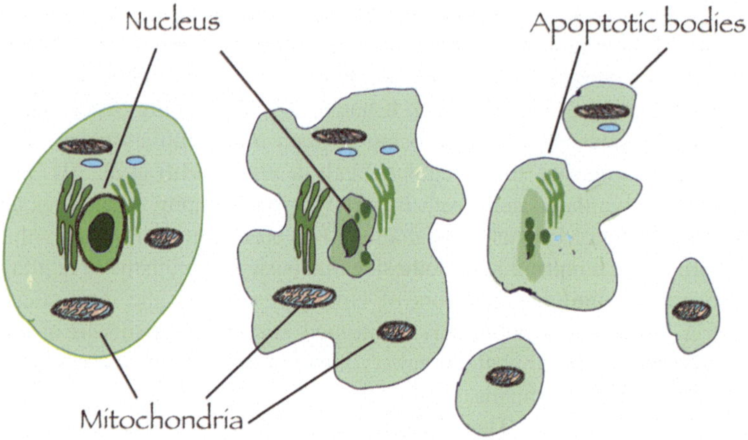

Fig. 10.2 Apoptosis. On the left: a normal cell before apoptosis initiation. In the middle: a dying cell. There is chromatin condensation. The cell shrinks. Fragmentation of the nucleus starts. On the right: the nucleus is fragmented. The cell breaks into membrane-covered, well-preserved fragments (apoptotic bodies) that are ingested by macrophages, which prevents the induction of inflammation

The next stage is characterized by self-degradation of the nucleus. In this stage, chromatin condensation and nuclear fragmentation occur. Then, the cell breaks into membrane-covered, ultrastructural, well-preserved fragments that are ingested by macrophages, which prevents the induction of inflammation. This kind of cell death has been observed in seminal studies about the regulated/programmed cell death of *C. elegans* (Robertson and Thomson 1982). Later, this form of regulated cell death was described many times in animals. Apoptosis is evidently regulated; there are initiation and chronology of events. Mutations causing the perturbation of different stages of apoptosis were first described in pioneering papers about the apoptosis of *C. elegans* (Hedgecock et al. 1983). It later turned out that apoptosis is also a tumor suppressor mechanism (Steller 1995). Cytological studies are widely used for the detection of apoptotic cell death in animals. As a result, the widely accepted definition of apoptosis is that it is programmed cell death, and both terms are used interchangeably. As we see below, this definition can be misleading, as even in the case of animals, there are other cytological forms of regulated cell death. Although apoptosis is exclusive to animals, cell death of other eukaryotes usually has different hallmarks of apoptosis, including self-degradation of DNA, activation of DNA proteases, and mitochondrial membrane permeability transition. This is why this kind of cell death is often referred to as apoptosis or apoptosis-like cell death. These hallmarks of apoptosis were observed, for example, during regulated cell death in yeast (*S. cerevisiae*), slime molds (*Dictyostelium discoideum*), green plants, the malaria parasite *Plasmodium*, trypanosomes, higher plants, and green algae (Kaczanowski 2016) (Fig. 10.3).

Autophagy is a process of self-cannibalism. During autophagy, portions of the cytoplasm are sequestered within double- or multi-membraned vesicles

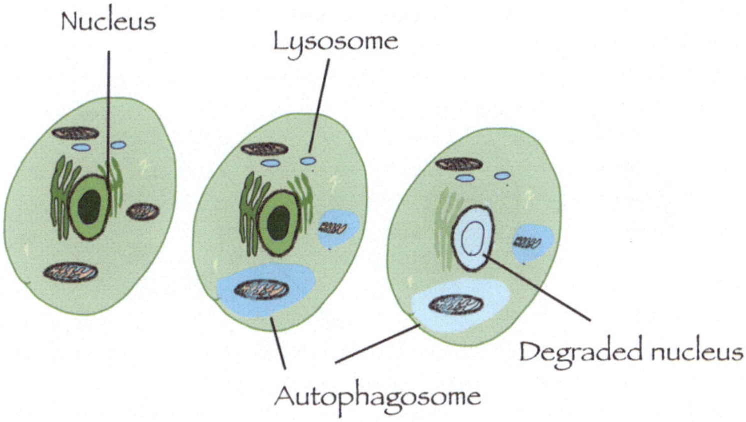

Fig. 10.3 Autophagic cell death. On the left: a normal cell. In the middle: a cell after the initiation of autophagy. During autophagy, fragments of the cytoplasm containing organelles are sequestered within double- or multimembraned vesicles (autophagosomes) and degraded to amino acids by proteases delivered by lysosomes. Autophagosomes and lysosomes are marked in blue. The degraded material is absorbed and reused. On the right: self-cannibalism leads to cellular death. Even the nucleus is degraded. During autophagic cell death, cellular organelles are degraded in autophagosomes (marked in blue)

(autophagosomes), delivered to lysosomes, and degraded (Doherty and Baehrecke 2018). In unicellular organisms, autophagy is a mechanism of energy production during nutrient deprivation. In animals, autophagy is also involved in maintaining cellular homeostasis (Rubinsztein et al. 2011). It is also involved in removing toxic protein aggregates and selectively dysfunctional mitochondria. Indeed, in animals, autophagy can also lead to the selective elimination of pro-apoptotic mechanisms such as apoptotic factors and apoptosis-inducing mitochondria. As a result, autophagy often inhibits the initiation of cell death in animals. There are many cases when the balance between autophagy and apoptosis determines cell death (Mariño et al. 2014).

However, autophagy may also lead to cell death, for example, during starvation. This kind of cell death is accidental. However, in many cases, it has been shown that this is programmed cell death. Even in cases of animals, where apoptosis is the main form of programmed cell death, there are well-described programmed autophagic cell deaths. In *Drosophila*, it is a developmental mechanism involved in the destruction of larval salivary glands as well as in the death of the larval midgut cells of the intestine (Baehrecke 2003). Autophagic cell death has also been observed in cerebral hypoxia–ischemia-induced neuron death (Carloni et al. 2008) in mice and neuronal death in the thalamus following focal cerebral infarction in rats (Zhang et al. 2012), although in this case, it is not clear to what extent this is a programmed cell death. In non-animal eukaryotes, regulated cell death often has different hallmarks of autophagy. This form of regulated cell death has been described in very diverse taxonomic groups, for example, in the slime molds of *Dictyostelium*

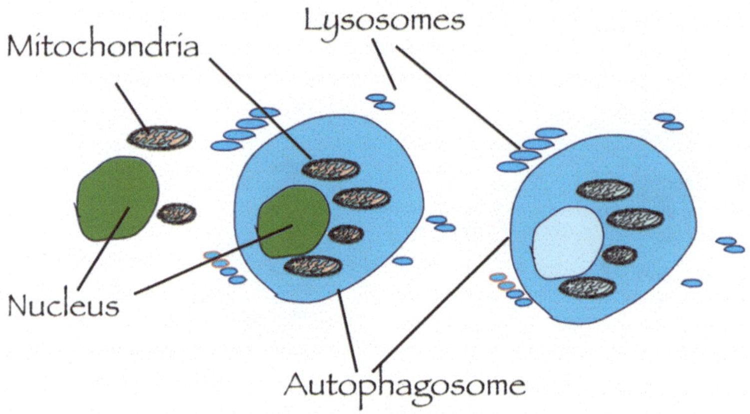

Fig. 10.4 Programmed nuclear death: selective degradation of the nucleus in *Tetrahymena*. The nucleus is green, the autophagosomes and lysosomes are blue, and the mitochondria are pink. On the left: nucleus (green) before cell death. In the middle: large autophagosome (blue) is formed. It contains nucleus and mitochondria that have apoptotic hallmarks (e.g., lost their membrane potential). It is surrounded by lysosomes that deliver proteases. On the right: degraded nucleus and mitochondria before absorption

(Roisin-Bouffay et al. 2004), green plants (Minina et al. 2013, 2014), and during selective nuclear degradation in the unicellular ciliates of *Tetrahymena* (Osada et al. 2014; Liu and Yao 2012) (called programmed nuclear death) (Fig. 10.4).

Regulated cell death in animals could also take the form of necrosis (Vanden Berghe et al. 2014) (Fig 10.5). It is involved in development (e.g., the death of chondrocytes during the development of bones as well as adult tissue homeostasis, such as in intestinal epithelial cells) (Vanden Berghe et al. 2014; Golstein and Kroemer 2007). In non-animal organisms, dying cells do not breaks into membrane-covered, ultrastructural, well-preserved fragments.

Additionally, there are well-described examples of the regulated cell death of unicellular organisms without obvious hallmarks of autophagy (e.g., the regulated cell death of yeasts) (Carmona-Gutierrez et al. 2018). In both animal and non-animal "regulated necrosis," there are frequently observed hallmarks of apoptosis associated with mitochondria, such as the release of mitochondrial apoptotic factors and membrane permeability transition.

In conclusion, eukaryotic regulated cell death can take different forms: apoptosis, autophagic cell death, and necrosis. In non-animal organisms, regulated cell death usually has the hallmarks of different combinations of these three basic types of animal cell death. In all eukaryotes, regulated cell death is usually initiated by mitochondrial permeability transition and the release of mitochondrial apoptotic factors.

In the next section, we will see that many biochemical pathways initiate cell death. However, again, these pathways are usually associated with mitochondria.

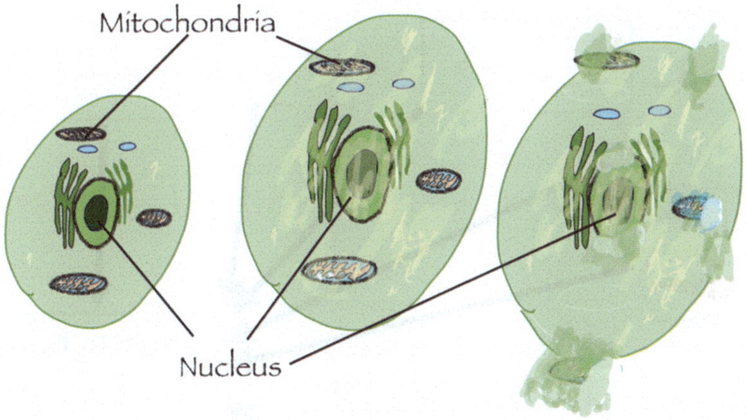

Fig. 10.5 Necrotic cell death. On the left: a normal cell. In the middle: necrosis is initiated. The cell swells. On the right: a membrane rupture. Necrotic cell releases its contents, which are not covered by membranes. In animals, the leaking of the cellular content causes inflammation

10.5 Biochemical Pathways of Programmed Cell Death

Mitochondrial permeability membrane transition is a central event in programmed cell death characterized by the breakdown of the inner mitochondrial transmembrane potential (Marchetti et al. 1996). This hallmark of apoptotic cell death has been described in organisms belonging to different systematic groups, including animals (Marchetti et al. 1996; Bender et al. 2012), plants (Curtis and Wolpert 2002), yeasts (Carraro and Bernardi 2016), malaria parasites (Hurd et al. 2006), trypanosomes, and slime molds (Arnoult et al. 2001).

It has been shown that mitochondrial permeability transition is induced by high concentrations of mitochondrial Ca^{2+} ($[Ca^{2+}]_m$) and to be inhibited by Mg^{2+} and adenosine diphosphate (ADP) (Carraro and Bernardi 2016; Bonora and Pinton 2014). Leaking of mitochondrial apoptotic factors is caused by the formation of the mitochondrial permeability transition (mPT) pore. The mPT pores are megachannels of the inner mitochondrial membrane, with a diameter of 2–3 nm, exhibiting a non-selective conductance of 1–1.3 nm. Still, it is not clear what the structural elements of the pore are. Although there is a controversial hypothesis that animal mPT pores are formed by subunits of the ATP synthase complex (Bernardi 2013; Chinopoulos 2018), results of different studies provide contradicting indications (Walker et al. 2020). Mitochondrial permeability transition activates both cytoplasmatic and mitochondrial apoptotic factors. The mechanism of activation of mitochondrial apoptotic factors is caused by the release of apoptotic factors sequestrated in mitochondria, such as AIF, cytochrome C, or ENDOG (Bonora and Pinton 2014).

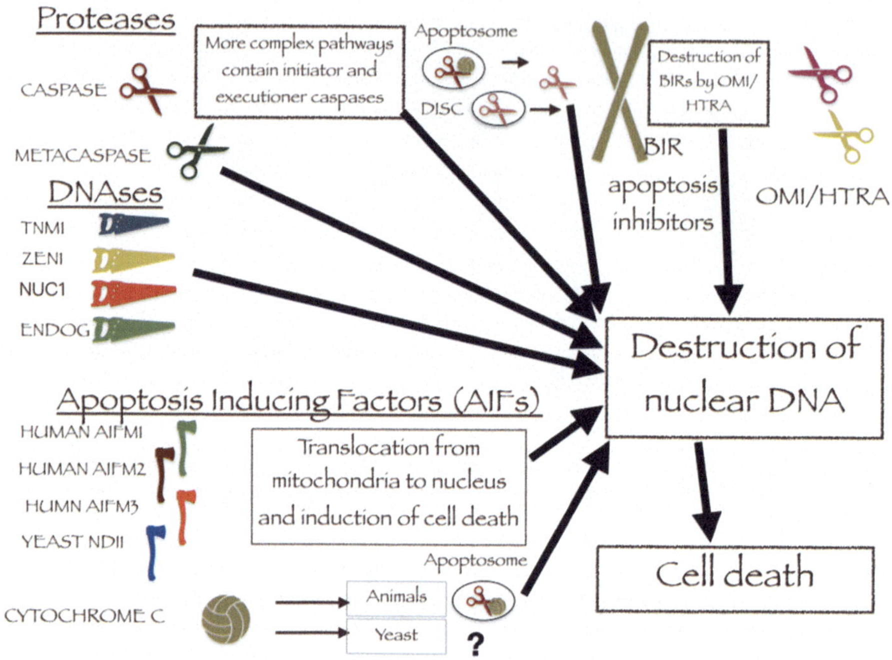

Fig. 10.6 Different factors initiate apoptosis: proteases (scissors), DNases (saws), AIFs (Apoptosis Induction Factors) (axes), and cytochrome C (ball). The more complex caspase (apoptotic cystic aspartic proteases) pathways contain caspase cascades. Initiator caspases activate executioner caspases. Initiator caspases are parts of multiprotein complexes DISC and apoptosome. Proteases HTRA (high-temperature requirement A family) induce apoptosis through the degradation of the apoptosis inhibitors type BIR (*B*aculovirus *I*nhibitor of Apoptosis protein *r*epeat). DNases are enzymes that destroy nuclear DNA. AIFs after translocation from mitochondria to the nucleus induce the self-degradation of DNA

The main ancient "apoptotic" pathways are listed below (Fig 10.6).

10.5.1 Protease Dependent Pathways

Proteases are enzymes that cleave proteins. They have a substrate specificity (i.e., a given protease cuts a given protein). Apoptotic proteases induce cellular suicide (regulated cell death).

10.5.1.1 Caspase-Dependent Pathway

Historically, the main animal caspase pathway was the first described apoptotic pathway, using *C. elegans* as a model organism.

Caspase is an apoptotic cystic aspartic protease. Caspases are involved in the regulation of apoptosis, inflammation, and cellular differentiation (Chowdhury et al.

2008). Phylogenetic studies have revealed that members of this family are encoded by genomes of *Reticulomyxa* (remotely related to animals, the unicellular eukaryotes foraminifera) and bacteria (Klim et al. 2018). In animals, one can distinguish two types of caspases: inflammatory and apoptotic. The activation of these enzymes is tightly controlled by their production as inactive zymogens that gain catalytic activity following signaling events promoting their aggregation into dimers or macromolecular complexes (Chowdhury et al. 2008).

As was already mentioned, the apoptotic function of caspases was discovered using *C. elegans* as a model. In this model, the caspase network is extremely simplified.

Using classic genetics, cell death abnormal (CED) mutations were identified (Ellis and Horvitz 1986). It turns out that proteins with this mutation are involved in caspase-dependent apoptosis. One of them, CED-3, is a caspase (Yuan et al. 1993). This protein is activated by CED-4 (Chen et al. 2000). Both proteins form an apoptosome complex (Chen et al. 2000; Qi et al. 2010). Cells can survive if CED-4 is inactivated. One of the mechanisms of the inactivation of CED-4 is its sequestration at the outer mitochondrial membrane by direct binding with the CED-9 protein (Hengartner and Horvitz 1994; Chen et al. 2000). In flies, the apoptosome has a similar structure (Dorstyn and Kumar 2006). In mammals, the apoptosome contains the cytochrome mitochondrial respiratory protein cytochrome C. There are indications that cytochrome C is part of the apoptosome in different animal systematic groups (Bender et al. 2012). This observation suggests that in its ancestral state, the apoptosome also contained cytochrome C.

The proteolytic cleavage of caspase substrates induces apoptosis. For example, mammalian activation of caspase-activated deoxyribonuclease (CAD) is caused by its inhibitor (ICAD) (Sakahira et al. 1998). Then, CAD performs the self-degradation of DNA.

As was already mentioned, the caspase pathway of *C. elegans* is extremely simplified. In other animals, there are caspase cascades containing different types of caspases. In such pathways, caspases of one kind activate caspases of another type. In mammals, apoptosomes activate executioner caspases that execute cellular death. It is a part of the pathway in which cell death is initiated by mitochondria. This pathway is named the intrinsic pathway.

Additionally, in mammals, there is an extrinsic pathway (for a review, see (D'Arcy 2019)). In this pathway, apoptosis is initiated not by mitochondria but by extracellular signals (death ligands) and death receptors. Death receptors belong to the tumor necrosis factor receptor (TNFR) family. Death ligands activate death receptors. Activated death receptors form a death-inducing signal complex (DISC) containing other proteins and initiator caspases. It is worth mentioning that initiator caspases involved in the extrinsic pathway are encoded by different genes than initiator caspases involved in the "mitochondrial" intrinsic pathway. As a result, initiator caspases activate executioner caspases and again executioner caspases cause cell death. This is an extremely simplified picture of the complex mammalian caspase network.

The caspase pathway usually leads to a classic cytological form of apoptosis but may also induce autophagic cell death (Doherty and Baehrecke 2018) and necrotic cell death (Vanden Berghe et al. 2014; Meurette et al. 2007).

10.5.1.2 Metacaspase-Dependent Pathway

Metacaspase (non-animal apoptotic lysine/arginine specific protease) is the main non-animal protease. This protease is remotely related to caspase and has different specificity. It has been shown that this protein initiates regulated cell death in yeasts (Silva et al. 2005), plants (Minina et al. 2013), and unicellular parasitic protists (Meslin et al. 2011). Both metacaspases and caspases proteatically cleave Tudor staphylococcal nuclease (TNS) (Sundström et al. 2009).

10.5.1.3 OMI/HTRA Dependent Pathway

Proteases from proteins from the high-temperature requirement A (HTRA) family induce apoptosis through the degradation of the apoptosis inhibitors type BIR (*B*aculovirus *I*nhibitor of Apoptosis protein *r*epeat) (such as Survivin) in both animals and fungi (Walter et al. 2006; Suzuki et al. 2001). In animals, Survivin is a direct inhibitor of caspase proteases (Tamm et al. 1998).

10.5.2 DNase-Dependent Pathways

DNases are enzymes that destroy nuclear DNA. The process of self-destruction of DNA is a crucial part of apoptosis.

10.5.2.1 Endonuclease G (EndoG)-Dependent Pathway

Mitochondrial endonuclease G induces cell death via DNA degradation after translocation from mitochondria to the nucleus (Li et al. 2001). This pathway has been described in different animals (Li et al. 2001; Wang et al. 2002), unicellular parasitic protists trypanosomes and *Leishmania* (Gannavaram et al. 2008; Rico et al. 2009), and the yeast *S. cerevisiae* (Oda et al. 2007). This pathway is activated during the development of Parkinson's disease by the aggregation of α-synuclein in neurons, which leads to pathological apoptosis. Interestingly, this mechanism was described using *S. cerevisiae* as a model organism. It has been shown that the expression of this human protein induces EndoG-dependent cell death (Büttner et al. 2013).

10.5.2.2 ZEN1 Nuclease-Dependent Pathway

The ZEN1 nuclease induces programmed cell death in plants via DNA degradation during flower development (Ito and Fukuda 2002; Aoyagi et al. 1998). This cell death forms in a way similar to animal autophagic cell death (Minina et al. 2014).

10.5.2.3 NUC1/DNase II-Dependent Pathway

The apoptotic function of this nuclease has been described in different animals (Lyon et al. 2000).

10.5.2.4 TMN1 Nuclease-Dependent Pathway

The release of the TMN1 mitochondrial nuclease initiates programmed nuclear death in unicellular protists (Osada et al. 2014), as shown in Fig. 10.4.

10.5.3 AIF-Dependent Pathways

Apoptosis-inducing factors (AIFs) are mitochondrial flavoproteins. There are at least four different types of AIFs, which diverged before the origin of eukaryotes. After translocation to the nucleus, they induce the self-degradation of DNA.

10.5.3.1 AIFM1-Dependent Pathway

AIFM1 is one of the human AIFs. The self-degradation of DNA is initiated by the translocation of this protein from mitochondria to the nucleus. The induction of cell death by orthologs of this protein has been demonstrated in animals (Susin et al. 1999) and slime molds (Arnoult et al. 2001).

10.5.3.2 AIFM2 (amid)-Dependent Pathway

AIFM2 is a second human AIF. The self-degradation of DNA is again initiated by the translocation of this factor from mitochondria to the nucleus. It binds directly to DNA. The induction of cell death by AIFM2 has been described in animals and yeasts (Marshall et al. 2005).

10.5.3.3 AIFM3-Dependent Pathway

AIFM3 is an AIF containing the cell death-inducing Rieske domain, which is absent in other AIFs (Xie et al. 2005). This pathway is also involved in inducing apoptosis in mammals as well as the previously mentioned selective nuclear degradation in the unicellular ciliates of *Tetrahymena* (Akematsu and Endoh 2010).

10.5.3.4 Yeast NDI-Dependent Pathway

This is a non-human pathway. Cell death induction by this protein has been described in the yeast *S. cerevisiae* (Li et al. 2006). However, in phylogenetic studies, orthologous factors have been identified in various other organisms, including flowering plants, green algae, and the starlet sea anemone (Klim et al. 2018).

10.5.4 Cytochrome C-dependent Pathway

Cytochrome C is a protein that is directly involved in respiration. When it is released from mitochondria, it activates cell death in animals and yeasts (*S. cerevisiae*). In animals, it forms an apoptosome together with caspases (Zou et al. 1999; Acehan et al. 2002). The mechanism of cell death activation by released cytochrome C in yeasts is unknown. It is obviously different as they do not have caspases (Silva et al. 2005).

10.5.5 Execution of Cell Death

The activation of apoptotic pathways leads to the self-destruction of nuclear DNA. The fragmentation of DNA is also a universal process observed in animals (Petit et al. 1996), plants (Pennell and Lamb 1997), various unicellular organisms such as yeasts (Eisler et al. 2004), malarial parasites (Hurd et al. 2006), and trypanosomes (Welburn et al. 1996) as well as in nuclear death (Mpoke and Wolfe 1996), which was previously mentioned.

The destruction of nuclear DNA and nuclei leads to different cytological forms of cell death (e.g., apoptosis, autophagic cell death, necrosis, and forms that contain hallmarks of different types of animal cell death). In the case of *Tetrahymena*, nuclear death does not lead to the death of the cell and is part of the development of this unicellular organism (Davis et al. 1992).

In conclusion, although there are many eukaryotic cell death pathways and many forms of cell death, all of them are associated with mitochondria.

10.6 Definition: Semantic Confusion Caused by the Term "Apoptosis"

As explained in the previous sections, the term apoptosis is used as the name for three very different phenomena: programmed cell death, a cytological form of animal cell death, and cell death induced by caspases. These definitions can be misleading and confusing. As already mentioned, programmed cell death, even in animals, often features cytological forms of necrosis or autophagic cell death. In animals, an apoptotic cytological form of cell death does not occur exclusively during programmed cell death. It has been shown that pathological apoptosis leads to diseases such as Alzheimer's (LaFerla et al. 1995) and Parkinson's (Mochizuki et al. 1996). Indeed, animal cell death including a cytological form of apoptosis could also be induced by the other pathways described above. In this chapter, apoptosis is defined using a different, broader definition: eukaryotic cell death associated with mitochondria.

This definition is supported by the observation that the regulated cell death of different organisms can be induced by apoptotic factors of other organisms. For example, the BAX protein is an animal apoptotic factor that initiates mitochondrial permeability transition (Petit et al. 1996). Heterologous expression of this protein in the yeast *S. cerevisiae* induces apoptosis (Greenhalf et al. 1996), although yeast has no obvious homologs of the BAX protein.

10.7 Endosymbiotic Theory of the Evolutionary Origin of Apoptosis

Early studies showed that animal apoptosis is associated with mitochondria and mitochondrial permeability transition (Petit et al. 1996). This observation inspired Guido Kroemer, who formulated the endosymbiotic theory of apoptosis origin (Kroemer 1997). In the case of animals, this hypothesis has been supported by phylogenetic analysis (Koonin and Aravind 2002). According to this hypothesis, animal apoptosis appeared during mitochondrial domestication.

In his seminal paper, Kroemer also formulated a putative model of mitochondrial domestication based on the "addiction molecule." According to this model, protomitochondria produced both toxins (apoptotic factors) and antitoxins (antiapoptotic factors). The antitoxins were unstable and played the role of addiction molecules much like the addiction molecules of extant phages. As a result, a protoeukaryotic ancestral host could only survive if it was colonized by ancestral protomitochondria. Later, toxins were transformed into extant apoptotic factors (Fig. 10.7).

A very different model of the origin of animal caspase-based apoptosis was recently suggested by Green and Fitzgerald (Green and Fitzgerald 2016). They were inspired by the observation that proteins from the caspase family also play a

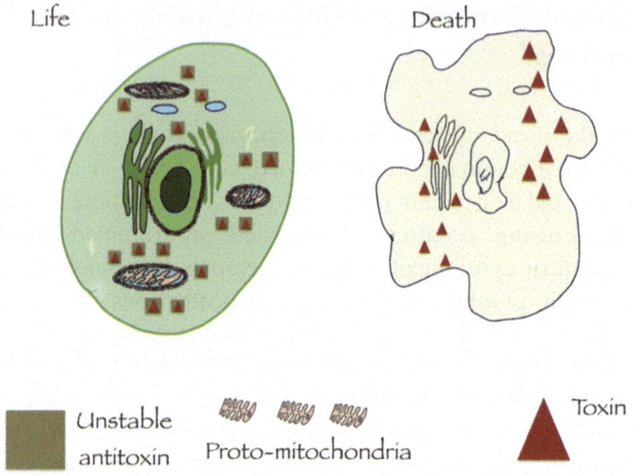

Fig. 10.7 Model of Guido Kroemer's theory of apoptosis evolution. According to this model, protomitochondria released both stable toxins and unstable antitoxins. A toxin was an "addiction" molecule. Protoeukaryotes that killed (or lost) protomitochondria were not protected by unstable protomitochondrial antitoxins, which were degraded. Stable toxins were active and killed protoeukaryotes. As a result, only proto-eukaryotes containing proto-mitochondria survived and apoptotic machinery was maintained

role in the immune response. They assumed that apoptosis evolved as the result of an evolutionary arms race between archaeal cells and bacterial cells. According to this model, caspase was used by archaeal cells (protoeukaryotes) as a kind of immune mechanism against bacteria such as protomitochondria. One of the immunological mechanisms induced by caspases was apoptotic cell death. During infection by bacteria, a cell died, thereby sparing its clone mates' subsequent infection by the replicating pathogen. Caspases also bind and recognize bacterial protomitochondrial cytochrome C. Interactions between caspases and bacterial cytochrome C were transformed into apoptosomes (protein complexes that activate programmed cell death in animals) during evolution (Fig. 10.8).

Recently, our group tested the endosymbiotic hypothesis using a phylogenetic approach and ancestral state reconstruction. In our analysis, we took advantage of recent advances in systematics that have revealed that six to eight major eukaryotic branches appeared very early in evolutionary history. Apoptotic types of cell death were described in different ancient primary eukaryotic systematic groups: Opisthokonta (fungi and animals); Amoebozoa (*Dictyostelium*); SAR–Stramenopiles, Alveolates, and Rhizaria (ciliates, apicomplexan parasites, and *Reticulomyxa*); Excavata (kinetoplastids, *Trichomonas*, *Naegleria*); and Archaeplastida (plants and green algae such as *Volvox*). These groups diverged very early. Therefore, it is likely that the pathway existing in two different ancient eukaryotic systematic groups also existed in protomitochondria. For example, the apoptotic function of the EndoG pathway has been described in Opisthokonts (animals and yeasts), Amoebozoa (*Dictyostelium*), and Excavata (trypanosomes).

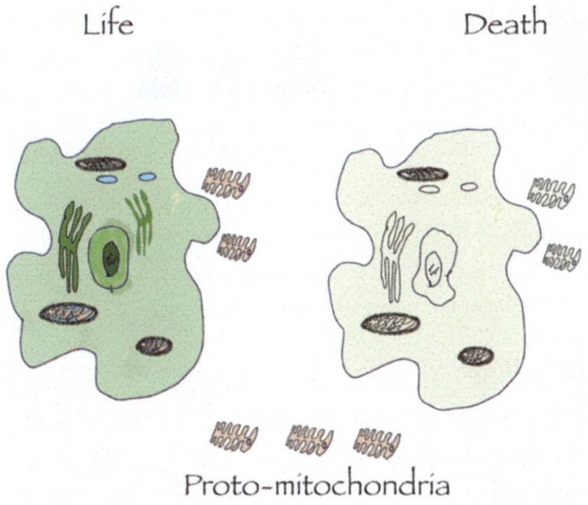

Fig. 10.8 The model developed by Douglas Green and Patrick Fitzgerald. On the left: proto-eukaryote attacked by bacteria (such as proto-mitochondria). On the right: activation of the caspase-dependent primordial immune system caused bacterial death and cellular suicide. Cellular suicide protects clone mates from subsequent infection by the replicating pathogen

One can reason that this pathway had already appeared before the divergence of eukaryotes. We obtained additional information by using less well-described orthologs of apoptotic factors in our analysis. For example, we identified orthologs of caspases belonging to SAR in the genome of the unicellular organism *Reticulomyxa*. This indicates that the common ancestor of SAR and animals (protoeukaryotes) already had genome-encoding caspases. Even if the function of caspases in *Reticulomyxa* is unknown, it is reasonable to hypothesize that the caspase-dependent apoptotic pathway evolved much earlier than animals appeared. Information about homology between different apoptotic factors and non-eukaryotic organisms was also used in our analysis. It turns out that ancient factors inducing apoptotic cell death are very similar to different bacterial proteins and are not similar to archaeal proteins. For example, caspases are similar to caspase-like proteins of the bacterium *Roseibium.* This observation indicates that the caspase-dependent apoptotic pathway probably has bacterial/protomitochondrial origins and was used by these bacteria for killing protoeukaryotes. According to this hypothesis, such cell death induced by bacteria was a primordial form of apoptosis. Interestingly, similar interactions between extant bacteria and eukaryotic cells are known (e.g., bacterial proteases are used as toxins to induce apoptosis) (Rust et al. 2016).

Therefore, phylogenetic analysis suggests a model of the evolution of apoptosis that is opposite to the hypothesis of Green and Fitzgerald, who assumed that caspases were used by protoeukaryotes as an immunological anti-proto-mitochondrial factor (Fig. 10.9).

Using this approach, we demonstrated that it is likely that the apoptosis systems of proto-eukaryotes contained several apoptotic factors/protomitochondrial toxins: apoptotic DNases (ZEN1, EndoG, NUC1), caspase-type proteases, various HTRA/OMI proteases (both fungal and mammalian types), and diverse AIFs. Therefore,

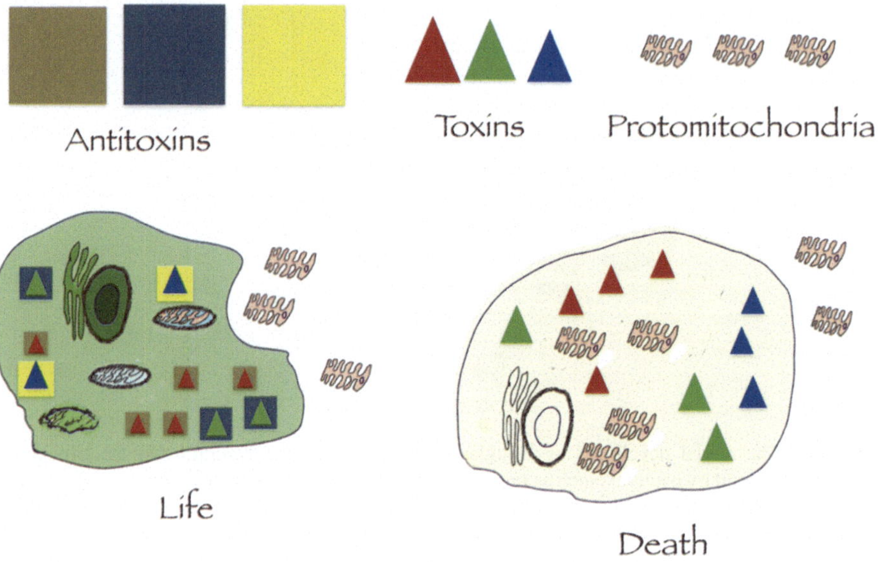

Fig. 10.9 Our model of apoptosis evolution. Due to an evolutionary arms race, proto-mitochondria produced as many toxins/primordial apoptotic factors as possible. The response was the evolution of different protoeukaryotic antitoxins. On the left: protomitochondria lost the evolutionary arms race. Protomitochondrial toxins are inactivated by antitoxins. Proto-mitochondria are eaten and degraded by proto-eukaryotes. On the right: proto-mitochondria won the race. There is a protomitochondrial invasion and the host cell is destroyed by protomitochondrial toxins. This process led to the evolution of complexity in extant apoptotic mechanisms. There are many toxins and antitoxins that are extant apoptotic/antiapoptotic factors

diverse apoptotic pathways appeared very early during evolutionary history. We noticed that protoeukaryotes had several apoptotic factors with redundant apoptotic functions, including apoptotic DNases (ZEN1, EndoG, NUC1), caspase-type proteases, various HTRA/OMI proteases (both fungal and mammalian types), and diverse AIFs. During subsequent evolution, redundant factors were lost in various lineages (for example, caspases and metacaspases, various DNases, and different HTRA/OMI proteases). Such surprising richness of redundant apoptotic factors present in the protomitochondrion suggests that "red queen" co-evolution may have shaped the protomitochondrion to contain as many toxins as possible (Valen 1973).

Phylogenetic analysis suggests, therefore, a putative scenario of apoptosis evolution. Protoeukaryotes were predators. They did not have oxidative respiration and relied on bacterial prey. Bacteria produced as many toxins as possible against enemies. Later, mitochondria were domesticated, and an apoptotic mechanism was used by different organisms for regulated cell death. Regulated cell death was a precondition for establishing multicellularity.

Summarizing this scenario suggests that regulated cell death and oxidative respiration are inherited by eukaryotes from protomitochondria.

10.8 Endosymbiosis as a Source of Complexity

Complexity is one of the main features of living systems. Although there are different definitions of complexity, it is obvious that evolution leads to a selection of complex organisms.

The origin of complexity in biological systems is one of the great questions of theoretical biology. It is not obvious why Darwinian evolution leads to complexity. Unicellular organisms are relatively simple, well adapted to their environment, and capable of establishing huge populations. Complex multicellular organisms have much smaller populations.

One of the potential explanations for the evolution of complexity was proposed by physicists Guttenberg and Goldenfeld (Guttenberg and Goldenfeld 2008). They hypothesized that antagonistic interactions between prey and predator (or host and parasite) are the source of complexity. Such interactions lead to an evolutionary arms race, and both organisms develop complex strategies. They tested this hypothesis using a computer model.

Our model of the endosymbiotic origin of apoptosis follows this theoretical prediction. Complex apoptosis pathways evolved due to interactions between predators (protoeukaryote) and prey (protomitochondrion). However, the impact of mitochondrial domestication on the evolution of complexity is much more complex than in the simplified physical model of Guttenberg and Goldenfeld (Guttenberg and Goldenfeld 2008). During evolution, antagonistic interactions between prey and predator were transformed into a complex strategy of programmed cell death. This strategy is probably required for the establishment of multicellularity. Additionally, more complex metabolism was created. After mitochondrial domestication due to symbiosis, novel organisms gained both aerobic and anaerobic respiration. This observation suggests that co-evolution exists between oxidative respiration machinery and apoptosis.

Additionally, our model explains why different apoptotic pathways are interlinked by one organelle, mitochondria. Such a cross-talk between pathways was probably crucial for establishing multicellularity.

10.9 Co-evolution of Apoptosis and Oxidative Respiration in Extant Organisms

Assuming that multicellularity evolved as a consequence of mitochondrial domestication, one can ask if reversed evolution from multicellular organisms to simplified unicellular organisms is possible. Davila and Zamorano noticed that such a "reversed" evolution occurs during the somatic evolution of cancer cells (Davila and Zamorano 2013). During this rapid evolution, cancer cells lose "altruistic" traits required for multicellularity, which evolved due to keen selection. One of them is apoptosis, an altruistic cell death that is required for the proper function of multicellular organisms. Classical observations indicate that the suppression of apoptosis in cancer cells tends to occur together with perturbations in cellular aerobic metabolism (Warburg hypothesis) (Warburg 1956). Therefore, in extant organisms, during the pathological somatic evolution of cancerous cells, apoptosis and cellular aerobic metabolism co-evolve.

This observation suggests the testable hypothesis that the co-evolution of apoptosis, multicellularity, and oxidative respiration is a general eukaryotic phenomenon. This hypothesis was tested using the yeast model in a previously mentioned experiment. It was shown that during the experimental evolution of multicellularity of yeasts, the frequency of apoptosis increases (Ratcliff et al. 2012). We showed that the deletion of apoptotic factors in yeasts causes a cancer-like phenotype: perturbation of apoptosis and anaerobic respiration (Klim et al. 2018). This observation inspires us to formulate an apoptotic model of the Warburg hypothesis. According to this model, apoptotic factors have a pleiotropic function, and the perturbation of apoptosis often also leads to the perturbation of mitochondrial respiration. Such perturbations occur during the somatic evolution of cancer cells (Kaczanowski et al. 2018).

Interestingly, pathological apoptosis during aging occurs mainly in neuron cells during the development of neurodegenerative diseases, such as Alzheimer's (LaFerla et al. 1995) or Parkinson's (Mochizuki et al. 1996). Interestingly, mitochondrial oxidative respiration is particularly active in neuron cells (Demetrius and Simon 2012).

It has been shown that patients with a history of cancer have experienced a reduced risk of Alzheimer's and Parkinson's diseases (Jansson and Jankovic 1985; Driver et al. 2012; Ibáñez et al. 2014).

This leads to the formulation of the inverse Warburg hypothesis. According to this hypothesis, aerobic respiration accelerates neuronal aging (Demetrius and Simon 2012).

10.10 Conclusions

Mitochondrial domestication leads to the evolution of multicellularity, apoptosis, and anaerobic respiration. Programmed cell death is a complex strategy that evolved as a result of symbiosis and complex interactions between protoeukaryotes and protomitochondria. There are many indications that this strategy is required for the development of multicellularity. Evolutionary history explains the less obvious impact of oxidative respiration on apoptosis. During neoplastic transformation, perturbations of the activity of cancer-suppressing apoptotic machinery co-occur with perturbations of aerobic respiration. This observation is called the Warburg hypothesis of cancer origin. In contrast, pathological apoptosis during the course of age-related diseases, such as Alzheimer's and Parkinson's, occurs mainly in neuron cells, where oxidative respiration is particularly active (an observation called the inverse Warburg hypothesis). Therefore, the evolutionary history of apoptosis has significance for the biology of extant organisms, including humans.

As a result, the endosymbiotic hypothesis of the origin of apoptosis indicates several testable research questions:

1. What is the impact of mitochondrial metabolism on pathological apoptosis during the development of neurodegenerative diseases?
2. What is the impact of mitochondrial metabolism on the development of cancer?
3. What is the impact of diet on the activity of apoptotic machinery in animals, including humans?
4. What is the impact of the evolution of multicellularity on mitochondrial metabolism?
5. What is the impact of kin selection on the evolution of regulated apoptotic cell death in extant unicellular organisms?
6. What is the impact of kin selection on the evolution of the aerobic metabolism of extant unicellular organisms?
7. Do apoptotic factors belonging to the apoptotic pathways described above play similar roles in different eukaryotes? For example, are caspases of *Reticulomyxa* involved in apoptosis in unicellular organisms?

As shown above, these are basic questions of eukaryotic biology. In conclusion, evolutionary history has implications that are fundamental for understanding the main questions of medical science, evolutionary biology, biochemistry, and the cell biology of extant eukaryotes.

Acknowledgments This research was funded by the National Science Centre of Poland NCN grant number 2017/27/B/NZ8/02502.

References

Abedin M, King N (2010) Diverse evolutionary paths to cell adhesion. Trends Cell Biol 20:734–742

Acehan D, Jiang X, Morgan DG, Heuser JE, Wang X, Akey CW (2002) Three-dimensional structure of the apoptosome: implications for assembly, procaspase-9 binding, and activation. Mol Cell 9:423–432

Akematsu T, Endoh H (2010) Role of apoptosis-inducing factor (AIF) in programmed nuclear death during conjugation in Tetrahymena thermophila. BMC Cell Biol 11:13

Aoyagi S, Sugiyama M, Fukuda H (1998) BEN1 and ZEN1 cDNAs encoding S1-type DNases that are associated with programmed cell death in plants. FEBS Lett 429:134–138

Arambage S, Grant K, Pardo I, Ranford-Cartwright L, Hurd H (2009) Malaria ookinetes exhibit multiple markers for apoptosis-like programmed cell death in vitro. Parasit Vectors 2:32

Arnoult D, Tatischeff I, Estaquier J, Girard M, Sureau F, Tissier JP, Grodet A, Dellinger M, Traincard F, Kahn A, Ameisen JC, Petit PX (2001) On the evolutionary conservation of the cell death pathway: mitochondrial release of an apoptosis-inducing factor during Dictyostelium discoideum cell death. Mol Biol Cell 12:3016–3030

Baehrecke EH (2003) Autophagic programmed cell death in Drosophila. Cell Death Differ 10:940–945

Bender CE, Fitzgerald P, Tait SW, Llambi F, GP MS, Tupper DO, Pellettieri J, Sánchez Alvarado A, Salvesen GS, Green DR (2012) Mitochondrial pathway of apoptosis is ancestral in metazoans. Proc Natl Acad Sci U S A 109:4904–4909

Berghe V, Linkermann A, Jouan-Lanhouet S, Walczak H, Vandenabeele P (2014) Regulated necrosis: the expanding network of non-apoptotic cell death pathways. Nat Rev Mol Cell Biol 15:135–147

Bernardi P (2013) The mitochondrial permeability transition pore: a mystery solved? Front Physiol 4:95

Bonora M, Pinton P (2014) The mitochondrial permeability transition pore and cancer: molecular mechanisms involved in cell death. Front Oncol 4:302

Büttner S, Habernig L, Broeskamp F, Ruli D, Vögtle FN, Vlachos M, Macchi F, Küttner V, Carmona-Gutierrez D, Eisenberg T, Ring J, Markaki M, Taskin AA, Benke S, Ruckenstuhl C, Braun R, Van den Haute C, Bammens T, van der Perren A, Fröhlich KU, Winderickx J, Kroemer G, Baekelandt V, Tavernarakis N, Kovacs GG, Dengjel J, Meisinger C, Sigrist SJ, Madeo F (2013) Endonuclease G mediates α-synuclein cytotoxicity during Parkinson's disease. EMBO J 32:3041–3054

Carloni S, Buonocore G, Balduini W (2008) Protective role of autophagy in neonatal hypoxia-ischemia induced brain injury. Neurobiol Dis 32:329–339

Carmona-Gutierrez D, Bauer MA, Zimmermann A, Aguilera A, Austriaco N, Ayscough K, Balzan R, Bar-Nun S, Barrientos A, Belenky P, Blondel M, Braun RJ, Breitenbach M, Burhans WC, Büttner S, Cavalieri D, Chang M, Cooper KF, Côrte-Real M, Costa V, Cullin C, Dawes I, Dengjel J, Dickman MB, Eisenberg T, Fahrenkrog B, Fasel N, Fröhlich KU, Gargouri A, Giannattasio S, Goffrini P, Gourlay CW, Grant CM, Greenwood MT, Guaragnella N, Heger T, Heinisch J, Herker E, Herrmann JM, Hofer S, Jiménez-Ruiz A, Jungwirth H, Kainz K, Kontoyiannis DP, Ludovico P, Manon S, Martegani E, Mazzoni C, Megeney LA, Meisinger C, Nielsen J, Nyström T, Osiewacz HD, Outeiro TF, Park HO, Pendl T, Petranovic D, Picot S, Polčic P, Powers T, Ramsdale M, Rinnerthaler M, Rockenfeller P, Ruckenstuhl C, Schaffrath R, Segovia M, Severin FF, Sharon A, Sigrist SJ, Sommer-Ruck C, Sousa MJ, Thevelein JM, Thevissen K, Titorenko V, Toledano MB, Tuite M, Vögtle FN, Westermann B, Winderickx J, Wissing S, Wölfl S, Zhang ZJ, Zhao RY, Zhou B, Galluzzi L, Kroemer G, Madeo F (2018) Guidelines and recommendations on yeast cell death nomenclature. Microb Cell 5:4–31

Carraro M, Bernardi P (2016) Calcium and reactive oxygen species in regulation of the mitochondrial permeability transition and of programmed cell death in yeast. Cell Calcium 60:102–107

Celiker H, Gore J (2013) Cellular cooperation: insights from microbes. Trends Cell Biol 23:9–15

Chen F, Hersh BM, Conradt B, Zhou Z, Riemer D, Gruenbaum Y, Horvitz HR (2000) Translocation of *C. elegans* CED-4 to nuclear membranes during programmed cell death. Science 287:1485–1489

Chinopoulos C (2018) Mitochondrial permeability transition pore: Back to the drawing board. Neurochem Int 117:49–54

Chowdhury I, Tharakan B, Bhat GK (2008) Caspases - an update. Comp Biochem Physiol B Biochem Mol Biol 151:10–27

Curtis MJ, Wolpert TJ (2002) The oat mitochondrial permeability transition and its implication in victorin binding and induced cell death. Plant J 29:295–312

D'Arcy MS (2019) Cell death: a review of the major forms of apoptosis, necrosis and autophagy. Cell Biol Int 43:582–592

Davila AF, Zamorano P (2013) Mitochondria and the evolutionary roots of cancer. Phys Biol 10:026008

Davis MC, Ward JG, Herrick G, Allis CD (1992) Programmed nuclear death: apoptotic-like degradation of specific nuclei in conjugating Tetrahymena. Dev Biol 154:419–432

Demetrius LA, Simon DK (2012) An inverse-Warburg effect and the origin of Alzheimer's disease. Biogerontology 13:583–594

Doherty J, Baehrecke EH (2018) Life, death and autophagy. Nat Cell Biol 20:1110–1117

Dorstyn L, Kumar S (2006) A cytochrome c-free fly apoptosome. Cell Death Differ 13:1049–1051
Drew MC, He CJ, Morgan PW (2000) Programmed cell death and aerenchyma formation in roots. Trends Plant Sci 5:123–127
Driver JA, Beiser A, Au R, Kreger BE, Splansky GL, Kurth T, Kiel DP, Lu KP, Seshadri S, Wolf PA (2012) Inverse association between cancer and Alzheimer's disease: results from the Framingham Heart Study. BMJ 344:e1442
Durand PM, Choudhury R, Rashidi A, Michod RE (2014) Programmed death in a unicellular organism has species-specific fitness effects. Biol Lett 10:20131088
Durand PM, Barreto Filho MM, Michod RE (2019) Cell death in evolutionary transitions in individuality. Yale J Biol Med 92:651–662
Duszenko M, Figarella K, Macleod E, Welburn S (2006) Death of a trypanosome: a selfish altruism. Trends Parasitol 22:536–542
Eisler H, Fröhlich KU, Heidenreich E (2004) Starvation for an essential amino acid induces apoptosis and oxidative stress in yeast. Exp Cell Res 300:345–353
Ellis HM, Horvitz HR (1986) Genetic control of programmed cell death in the nematode *C. elegans*. Cell 44:817–829
Erental A, Sharon I, Engelberg-Kulka H (2012) Two programmed cell death systems in Escherichia coli: an apoptotic-like death is inhibited by the mazEF-mediated death pathway. PLoS Biol 10: e1001281
Fedak H, Palusinska M, Krzyczmonik K, Brzezniak L, Yatusevich R, Pietras Z, Kaczanowski S, Swiezewski S (2016) Control of seed dormancy in Arabidopsis by a cis-acting noncoding antisense transcript. Proc Natl Acad Sci U S A 113:E7846–E7E55
Figarella K, Rawer M, Uzcategui NL, Kubata BK, Lauber K, Madeo F, Wesselborg S, Duszenko M (2005) Prostaglandin D2 induces programmed cell death in Trypanosoma brucei bloodstream form. Cell Death Differ 12:335–346
Gannavaram S, Vedvyas C, Debrabant A (2008) Conservation of the pro-apoptotic nuclease activity of endonuclease G in unicellular trypanosomatid parasites. J Cell Sci 121:99–109
Golstein P, Kroemer G (2007) Cell death by necrosis: towards a molecular definition. Trends Biochem Sci 32:37–43
Green DR, Fitzgerald P (2016) Just so stories about the evolution of Apoptosis. Curr Biol 26:R620–R627
Greenhalf W, Stephan C, Chaudhuri B (1996) Role of mitochondria and C-terminal membrane anchor of Bcl-2 in Bax induced growth arrest and mortality in *Saccharomyces cerevisiae*. FEBS Lett 380:169–175
Guttenberg N, Goldenfeld N (2008) Cascade of complexity in evolving predator-prey dynamics. Phys Rev Lett 100:058102
Hamilton WD (1964a) The genetical evolution of social behaviour. I. J Theor Biol 7:1–16
Hamilton WD (1964b) The genetical evolution of social behaviour. II. J Theor Biol 7:17–52
Hedgecock EM, Sulston JE, Thomson JN (1983) Mutations affecting programmed cell deaths in the nematode Caenorhabditis elegans. Science 220:1277–1279
Hengartner MO, Horvitz HR (1994) *C. elegans* cell survival gene ced-9 encodes a functional homolog of the mammalian proto-oncogene bcl-2. Cell 76:665–676
Hope EA, Amorosi CJ, Miller AW, Dang K, Heil CS, Dunham MJ (2017) Experimental evolution reveals favored adaptive routes to cell aggregation in yeast. Genetics 206:1153–1167
Hurd H, Grant K, Arambage S (2006) Apoptosis-like death as a feature of malaria infection in mosquitoes. Parasitology 132:S33–S47
Ibáñez K, Boullosa C, Tabarés-Seisdedos R, Baudot A, Valencia A (2014) Molecular evidence for the inverse comorbidity between central nervous system disorders and cancers detected by transcriptomic meta-analyses. PLoS Genet 10:e1004173
Ito J, Fukuda H (2002) ZEN1 is a key enzyme in the degradation of nuclear DNA during programmed cell death of tracheary elements. Plant Cell 14:3201–3211
Jansson B, Jankovic J (1985) Low cancer rates among patients with Parkinson's disease. Ann Neurol 17:505–509

Kaczanowski S (2016) Apoptosis: its origin, history, maintenance and the medical implications for cancer and aging. Phys Biol 13:031001

Kaczanowski S, Sajid M, Reece SE (2011) Evolution of apoptosis-like programmed cell death in unicellular protozoan parasites. Parasit Vectors 4:44

Kaczanowski S, Klim J, Zielenkiewicz U (2018) An apoptotic and endosymbiotic explanation of the Warburg and the inverse Warburg hypotheses. Int J Mol Sci 19:3100

Kerr JF, Wyllie AH, Currie AR (1972) Apoptosis: a basic biological phenomenon with wide-ranging implications in tissue kinetics. Br J Cancer 26:239–257

Klim J, Gładki A, Kucharczyk R, Zielenkiewicz U, Kaczanowski S (2018) Ancestral State Reconstruction of the Apoptosis Machinery in the Common Ancestor of Eukaryotes. G3 (Bethesda) 8:2121–2134

Koonin EV, Aravind L (2002) Origin and evolution of eukaryotic apoptosis: the bacterial connection. Cell Death Differ 9:394–404

Kroemer G (1997) Mitochondrial implication in apoptosis: towards an endosymbiont hypothesis of apoptosis evolution. Cell Death Differ 4:443–456

Kutschera U (2009) Symbiogenesis, natural selection, and the dynamic earth. Theory Biosci 128:191–203

LaFerla FM, Tinkle BT, Bieberich CJ, Haudenschild CC, Jay G (1995) The Alzheimer's A beta peptide induces neurodegeneration and apoptotic cell death in transgenic mice. Nat Genet 9:21–30

Li L, Luo X, Wang X (2001) Endonuclease G is an apoptotic DNase when released from mitochondria. Nature 412:95–99

Li W, Sun L, Liang Q, Wang J, Mo W, Zhou B (2006) Yeast AMID homologue Ndi1p displays respiration-restricted apoptotic activity and is involved in chronological aging. Mol Biol Cell 17:1802–1811

Liu ML, Yao MC (2012) Role of ATG8 and autophagy in programmed nuclear degradation in Tetrahymena thermophila. Eukaryot Cell 11:494–506

Lyon C, Evans C, Bill B, Otsuka A, Aguilera R (2000) The *C. elegans* apoptotic nuclease NUC-1 is related in sequence and activity to mammalian DNase II. Gene 252:147–154

Marchetti P, Castedo M, Susin SA, Zamzami N, Hirsch T, Macho A, Haeffner A, Hirsch F, Geuskens M, Kroemer G (1996) Mitochondrial permeability transition is a central coordinating event of apoptosis. J Exp Med 184:1155–1160

Margulis L (1993) Symbiosis in cell evolution. Freeman, New York

Mariño G, Niso-Santano M, Baehrecke EH, Kroemer G (2014) Self-consumption: the interplay of autophagy and apoptosis. Nat Rev Mol Cell Biol 15:81–94

Marshall KR, Gong M, Wodke L, Lamb JH, Jones DJ, Farmer PB, Scrutton NS, Munro AW (2005) The human apoptosis-inducing protein AMID is an oxidoreductase with a modified flavin cofactor and DNA binding activity. J Biol Chem 280:30735–30740

Meslin B, Beavogui AH, Fasel N, Picot S (2011) *Plasmodium falciparum* metacaspase PfMCA-1 triggers a z-VAD-fmk inhibitable protease to promote cell death. PLoS One 6:e23867

Meurette O, Rebillard A, Huc L, Le Moigne G, Merino D, Micheau O, Lagadic-Gossmann D, Dimanche-Boitrel MT (2007) TRAIL induces receptor-interacting protein 1-dependent and caspase-dependent necrosis-like cell death under acidic extracellular conditions. Cancer Res 67:218–226

Minina EA, Filonova LH, Fukada K, Savenkov EI, Gogvadze V, Clapham D, Sanchez-Vera V, Suarez MF, Zhivotovsky B, Daniel G, Smertenko A, Bozhkov PV (2013) Autophagy and metacaspase determine the mode of cell death in plants. J Cell Biol 203:917–927

Minina EA, Bozhkov PV, Hofius D (2014) Autophagy as initiator or executioner of cell death. Trends Plant Sci 19:692–697

Mochizuki H, Goto K, Mori H, Mizuno Y (1996) Histochemical detection of apoptosis in Parkinson's disease. J Neurol Sci 137:120–123

Moore RB, Oborník M, Janouskovec J, Chrudimský T, Vancová M, Green DH, Wright SW, Davies NW, Bolch CJ, Heimann K, Slapeta J, Hoegh-Guldberg O, Logsdon JM, Carter DA (2008) A photosynthetic alveolate closely related to apicomplexan parasites. Nature 451:959–963
Mpoke S, Wolfe J (1996) DNA digestion and chromatin condensation during nuclear death in Tetrahymena. Exp Cell Res 225:357–365
O'Connor C (2010) Essentials of cell biology. https://www.nature.com/scitable/ebooks/essentials-of-cell-biology-14749010/118237915/
Oda K, Kawasaki N, Fukuyama M, Ikeda S (2007) Ectopic expression of mitochondria endonuclease Pnu1p from Schizosaccharomyces pombe induces cell death of the yeast. J Biochem Mol Biol 40:1095–1099
Osada E, Akematsu T, Asano T, Endoh H (2014) A novel mitochondrial nuclease-associated protein: a major executor of the programmed nuclear death in Tetrahymena thermophila. Biol Cell 106:97–109
Parfrey LW, Lahr DJ (2013) Multicellularity arose several times in the evolution of eukaryotes. Bioessays 35:339–347. https://doi.org/10.1002/bies.201100187
Pennell RI, Lamb C (1997) Programmed cell death in plants. Plant Cell 9:1157–1168
Petit P, Susin S, Zamzami N, Mignotte B, Kroemer G (1996) Mitochondria and programmed cell death: back to the future. FEBS Lett 396:7–13
Qi S, Pang Y, Hu Q, Liu Q, Li H, Zhou Y, He T, Liang Q, Liu Y, Yuan X, Luo G, Wang J, Yan N, Shi Y (2010) Crystal structure of the *Caenorhabditis elegans* apoptosome reveals an octameric assembly of CED-4. Cell 141:446–457
Ratcliff WC, Denison RF, Borrello M, Travisano M (2012) Experimental evolution of multicellularity. Proc Natl Acad Sci U S A 109:1595–1600
Rico E, Alzate JF, Arias AA, Moreno D, Clos J, Gago F, Moreno I, Domínguez M, Jiménez-Ruiz A (2009) Leishmania infantum expresses a mitochondrial nuclease homologous to EndoG that migrates to the nucleus in response to an apoptotic stimulus. Mol Biochem Parasitol 163:28–38
Robertson AM, Thomson J (1982) Morphology of programmed cell death in the ventral nerve cord of Caenorhabditis elegans larvae. Development 67:89–100
Roisin-Bouffay C, Luciani MF, Klein G, Levraud JP, Adam M, Golstein P (2004) Developmental cell death in dictyostelium does not require paracaspase. J Biol Chem 279:11489–11494
Rokas A (2008) The origins of multicellularity and the early history of the genetic toolkit for animal development. Annu Rev Genet 42:235–251
Rubinsztein DC, Mariño G, Kroemer G (2011) Autophagy and aging. Cell 146:682–695
Rust A, Leese C, Binz T, Davletov B (2016) Botulinum neurotoxin type C protease induces apoptosis in differentiated human neuroblastoma cells. Oncotarget 7:33220–33228
Sakahira H, Enari M, Nagata S (1998) Cleavage of CAD inhibitor in CAD activation and DNA degradation during apoptosis. Nature 391:96–99
Schweichel JU, Merker HJ (1973) The morphology of various types of cell death in prenatal tissues. Teratology 7:253–266
Seipp S, Schmich J, Leitz T (2001) Apoptosis--a death-inducing mechanism tightly linked with morphogenesis in Hydractina echinata (Cnidaria, Hydrozoa). Development 128:4891–4898
Seipp S, Wittig K, Stiening B, Böttger A, Leitz T (2006) Metamorphosis of Hydractinia echinata (Cnidaria) is caspase-dependent. Int J Dev Biol 50:63–70
Silva RD, Sotoca R, Johansson B, Ludovico P, Sansonetty F, Silva MT, Peinado JM, Côrte-Real M (2005) Hyperosmotic stress induces metacaspase- and mitochondria-dependent apoptosis in Saccharomyces cerevisiae. Mol Microbiol 58:824–834
Soares EV (2011) Flocculation in *Saccharomyces cerevisiae*: a review. J Appl Microbiol 110:1–18
Steller H (1995) Mechanisms and genes of cellular suicide. Science 267:1445–1449
Sulston JE, Horvitz HR (1977) Post-embryonic cell lineages of the nematode, Caenorhabditis elegans. Dev Biol 56:110–156
Sulston JE, Schierenberg E, White JG, Thomson JN (1983) The embryonic cell lineage of the nematode Caenorhabditis elegans. Dev Biol 100:64–119

Sundström J, Vaculova A, Smertenko A, Savenkov E, Golovko A, Minina E, Tiwari B, Rodriguez-Nieto S, Zamyatnin AJ, Välineva T, Saarikettu J, Frilander M, Suarez M, Zavialov A, Ståhl U, Hussey P, Silvennoinen O, Sundberg E, Zhivotovsky B, Bozhkov P (2009) Tudor staphylococcal nuclease is an evolutionarily conserved component of the programmed cell death degradome. Nat Cell Biol 11:1347–1354

Susin SA, Lorenzo HK, Zamzami N, Marzo I, Snow BE, Brothers GM, Mangion J, Jacotot E, Costantini P, Loeffler M, Larochette N, Goodlett DR, Aebersold R, Siderovski DP, Penninger JM, Kroemer G (1999) Molecular characterization of mitochondrial apoptosis-inducing factor. Nature 397:441–446

Suzuki Y, Imai Y, Nakayama H, Takahashi K, Takio K, Takahashi R (2001) A serine protease, HtrA2, is released from the mitochondria and interacts with XIAP, inducing cell death. Mol Cell 8:613–621

Tamm I, Wang Y, Sausville E, Scudiero DA, Vigna N, Oltersdorf T, Reed JC (1998) IAP-family protein survivin inhibits caspase activity and apoptosis induced by Fas (CD95), Bax, caspases, and anticancer drugs. Cancer Res 58:5315–5320

Tovar J, León-Avila G, Sánchez LB, Sutak R, Tachezy J, van der Giezen M, Hernández M, Müller M, Lucocq JM (2003) Mitochondrial remnant organelles of Giardia function in iron-sulphur protein maturation. Nature 426:172–176

Valen LV (1973) A new evolutionary law. Evol Theory 1:1–30

Van Durme M, Nowack MK (2016) Mechanisms of developmentally controlled cell death in plants. Curr Opin Plant Biol 29:29–37

Vardi A, Berman-Frank I, Rozenberg T, Hadas O, Kaplan A, Levine A (1999) Programmed cell death of the dinoflagellate Peridinium gatunense is mediated by CO(2) limitation and oxidative stress. Curr Biol 9:1061–1064

Walker JE, Carroll J, He J (2020) Reply to Bernardi: the mitochondrial permeability transition pore and the ATP synthase. Proc Natl Acad Sci U S A 117:2745–2746

Walter D, Wissing S, Madeo F, Fahrenkrog B (2006) The inhibitor-of-apoptosis protein Bir1p protects against apoptosis in S. cerevisiae and is a substrate for the yeast homologue of Omi/HtrA2. J Cell Sci 119:1843–1851

Wang X, Yang C, Chai J, Shi Y, Xue D (2002) Mechanisms of AIF-mediated apoptotic DNA degradation in *Caenorhabditis elegans*. Science 298:1587–1592

Warburg O (1956) On the origin of cancer cells. Science 123:309–314

Welburn S, Dale C, Ellis D, Beecroft R, Pearson T (1996) Apoptosis in procyclic Trypanosoma brucei rhodesiense in vitro. Cell Death Differ 3:229–236

Whittingham WF, Raper KB (1960) Non-viability of stalk cells in dictyostelium. Proc Natl Acad Sci U S A 46:642–649

Xie Q, Lin T, Zhang Y, Zheng J, Bonanno JA (2005) Molecular cloning and characterization of a human AIF-like gene with ability to induce apoptosis. J Biol Chem 280:19673–19681

Yuan J, Shaham S, Ledoux S, Ellis HM, Horvitz HR (1993) The *C. elegans* cell death gene ced-3 encodes a protein similar to mammalian interleukin-1 beta-converting enzyme. Cell 75:641–652

Zhang J, Zhang Y, Li J, Xing S, Li C, Li Y, Dang C, Fan Y, Yu J, Pei Z, Zeng J (2012) Autophagosomes accumulation is associated with β-amyloid deposits and secondary damage in the thalamus after focal cortical infarction in hypertensive rats. J Neurochem 120:564–573

Zou H, Li Y, Liu X, Wang X (1999) An APAF-1. cytochrome c multimeric complex is a functional apoptosome that activates procaspase-9. J Biol Chem 274:11549–11556

Chapter 11
The Puzzling Conservation and Diversification of Lipid Droplets from Bacteria to Eukaryotes

Josselin Lupette and Eric Maréchal

Abstract Membrane compartments are amongst the most fascinating markers of cell evolution from prokaryotes to eukaryotes, some being conserved and the others having emerged via a series of primary and secondary endosymbiosis events. Membrane compartments comprise the system limiting cells (one or two membranes in bacteria, a unique plasma membrane in eukaryotes) and a variety of internal vesicular, subspherical, tubular, or reticulated organelles. In eukaryotes, the internal membranes comprise on the one hand the general endomembrane system, a dynamic network including organelles like the endoplasmic reticulum, the Golgi apparatus, the nuclear envelope, etc. and also the plasma membrane, which are linked *via* direct lateral connectivity (e.g. between the endoplasmic reticulum and the nuclear outer envelope membrane) or indirectly *via* vesicular trafficking. On the other hand, semi-autonomous organelles, i.e. mitochondria and chloroplasts, are disconnected from the endomembrane system and request vertical transmission following cell division. Membranes are organized as lipid bilayers in which proteins are embedded. The budding of some of these membranes, leading to the formation of the so-called lipid droplets (LDs) loaded with hydrophobic molecules, most notably triacylglycerol, is conserved in all clades. The evolution of eukaryotes is marked by the acquisition of mitochondria and simple plastids from Gram-positive bacteria by primary endosymbiosis events and the emergence of extremely complex plastids, collectively called secondary plastids, bounded by three to four membranes, following multiple and independent secondary endosymbiosis events. There is currently no consensus view of the evolution of LDs in the Tree of Life. Some features are conserved; others show

Josselin Lupette and Eric Maréchal contributed equally with all other contributors.

J. Lupette (✉)
MSU-DOE Plant Research Laboratory, Michigan State University, East Lansing, MI, USA
e-mail: lupettej@msu.edu

E. Maréchal (✉)
Laboratoire de Physiologie Cellulaire et Végétale, CNRS, CEA, INRAE, Université Grenoble Alpes, Institut de Recherche Interdisciplinaire de Grenoble, Grenoble, France
e-mail: eric.marechal@cea.fr

M. Kloc (ed.), *Symbiosis: Cellular, Molecular, Medical and Evolutionary Aspects*, Results and Problems in Cell Differentiation 69,
https://doi.org/10.1007/978-3-030-51849-3_11

a striking level of diversification. Here, we summarize the current knowledge on the architecture, dynamics, and multitude of functions of the lipid droplets in prokaryotes and in eukaryotes deriving from primary and secondary endosymbiosis events.

Keywords Lipid droplets · Evolution · Architecture · Biogenesis · Catabolism

11.1 Introduction

Lipid droplets (LDs) are conserved structures in prokaryotic and eukaryotic cells (Walther et al. 2017; Zhang and Liu 2017). Their architecture consists of a core, loaded with hydrophobic carbon-rich molecules (polyhydroxyalkanoates or PHAs, triacylglycerol or TAG, steryl esters, isoprenoids such as squalene, etc.) bounded by a monolayer of polar glycerolipids, generally phospholipids that can be associated with sterols. Proteins are transiently or permanently associated with its surface (Walther et al. 2017) (Fig. 11.1). Two main classes have been identified amongst LD closely associated proteins, based on their structure (Kory et al. 2016). Class I proteins have a hydrophobic "hairpin" pattern (Bersuker and Olzmann 2017) and class II proteins have at least one amphipathic helix (Bersuker and Olzmann 2017) (Fig. 11.1). Nevertheless, proteins that do not belong to these two classes are also encountered.

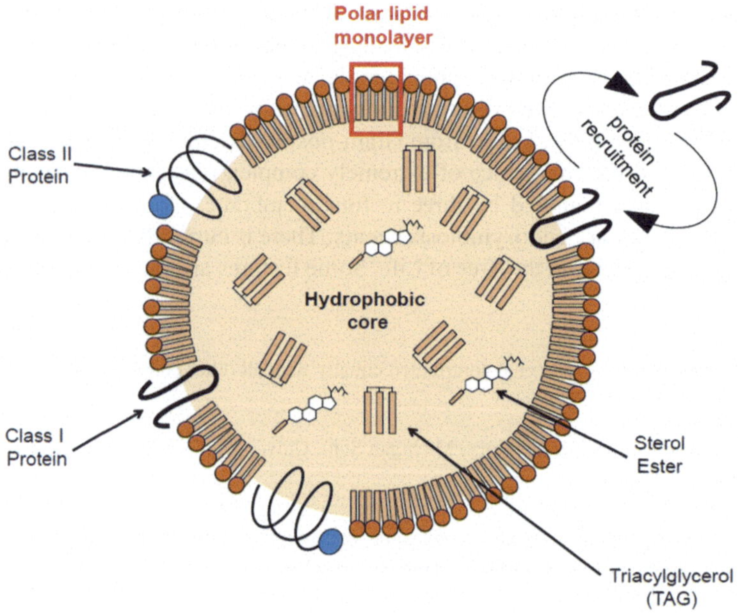

Fig. 11.1 General architecture of a triacylglycerol-containing lipid droplet

LDs derive from the budding of a cell membrane. Membrane compartments are amongst the most fascinating markers of cell evolution from prokaryotes to eukaryotes, some being conserved and the others having emerged via a series of primary and secondary endosymbiosis events. Membrane compartments comprise the system limiting cells (one or two membranes in bacteria, a unique plasma membrane in eukaryotes) and a variety of internal vesicular, subspherical, tubular, or reticulated organelles. In eukaryotes, the internal membranes comprise on the one hand the general endomembrane system, a dynamic network including organelles like the endoplasmic reticulum (ER), the Golgi apparatus, nuclear envelope, etc. and also the plasma membrane, which are linked *via* direct lateral connectivity (e.g. between the endoplasmic reticulum and the nuclear outer envelope membrane) or indirectly *via* vesicular trafficking. On the other hand, semi-autonomous organelles, i.e. mitochondria and chloroplasts, are disconnected from the endomembrane system and request vertical transmission following cell division. Membranes are organized as lipid bilayers in which proteins are embedded. The evolution of eukaryotes is marked by the acquisition of mitochondria and simple plastids from Gram-negative bacteria by primary endosymbiosis events (Maréchal 2018), and the emergence of extremely complex plastids, collectively called secondary plastids, bounded by three to four membranes, following multiple and independent secondary endosymbiosis events (Füssy and Oborník 2018). There is currently no consensus view on the evolution of cell membranes and that of LDs.

Due to their hydrophobic core loaded with carbon-rich molecules, LDs have long been considered as simple carbon and energy storage organelles. Based on the analyses of LD proteomes in various cell models, it is now considered that LDs have other functions that depend on their protein composition (Walther et al. 2017; Den Brok et al. 2018; Henne et al. 2018).

Research on LDs has increased strikingly in recent decades motivated by the multitude of applications ranging from nutrition, health to green chemistry and bioenergy. In 2020, the keyword "lipid droplet" returns as many as 12,500 hits in the Pubmed bibliographic database. Concerning human obesity-related diseases (Faucher and Poitou 2016; Madrigal-Matute and Cuervo 2016), protein actors (CGI-58, SEIPIN, ATGL, LAL) at the surface of LDs have been extensively studied in mammalian models (summarized in Table 11.1). CGI-58-ABHD5 (Comparative Gene Identification 58—α/β hydrolase domain-containing 5; 349 amino acids—45 kDa) is particularly studied because its mutation is responsible for the Chanarin-Dorfman syndrome, an autosomal recessive disease (Missaglia et al. 2014). In mammals, the CGI-58 protein is located on cytosolic LDs interacting with PLIN1 (Subramanian et al. 2004). A second important player in the mechanisms of LD biogenesis in mammals is the SEIPIN protein. A mutation in the human Seipin gene leads to severe forms of generalized Berardinelli-Seip congenital lipodystrophy (Magré et al. 2001). The deletion of ATGL (Adipose triglyceride lipase) in mice reduces the mechanism of lipolysis and promotes the accumulation of lipids in oxidative tissues of the body, leading to the death of mice in 3 months (Zimmermann et al. 2004). The last example is the Wolman disease, which is an autosomal recessive disease affecting young children following a mutation in the

Table 11.1 Human diseases related to LD formation

Diseases	Anatomical pathology	Pathophysiology	References
Atherosclerosis	Accumulation of atheromatous plaques (cholesterol) in the arteries	ACAT1; ABCA1; ADRP	Paul et al. (2008)
Obesity	Accumulation of fat reserves	Multifactorial (genetic, environmental, psychological)	Faucher and Poitou (2016)
Fatty liver	Accumulation of TAGs in the cytoplasm of hepatocytes	Alcohol, hepatitis B and C	Madrigal-Matute and Cuervo (2016)
Chanarin-Dorfman syndrome	Accumulation of lipid droplets in lymphocytes and many tissues	Mutation of CGI-58/ABHD5	Dorfman et al. (1974), Chanarin et al. (1975), Lefèvre et al. (2001), Samuelov et al. (2011), Missaglia et al. (2014), Jordans (1953), Gupta and Kaur (2005), Waheed et al. (2016)
Myopathy		Mutation of PNPLA2	Zimmermann et al. (2004)
Congenital generalized Lipodystrophy (CGL)	Dystrophy of adipose tissue	Mutation of AGPAT2, BSCL2, CAV1 or PTRF	Magré et al. (2001), Agarwal et al. (2002), Kim et al. (2008), Hayashi et al. (2009), Rajab et al. (2010), Quinn and Purcell (2017)
Lysosomal acid lipase deficiency	Lysosomal acid lipase deficiency causing an accumulation of TAGs and cholesterol esters in leukocytes, hepatocytes, and fibroblasts	Mutation of LIPA	Wolman et al. (1961), Onal et al. (2017), Pericleous et al. (2017), Ikari et al. (2018)

LIPA gene (Wolman et al. 1961). The LIPA mutation leads to the synthesis deficiency of lysosomal acid lipase (LAL) (Onal et al. 2017). This disease causes accumulation of cholesterol esters and TAG in leukocyte lysosomes, fibroblasts, and hepatocytes generally leading to the death of the child by liver failure (Pericleous et al. 2017). It is a very rare disease with only 14 cases detected so far, half of which are from the consanguineous union (Ikari et al. 2018).

TAG-rich LDs produced by oleaginous organisms, mainly plants and algae, but sometimes also fungi or animals, are also key to numerous biotechnological applications. Molecules of TAG are composed of a glycerol-3-phosphate backbone on which three fatty acids are esterified (Lupette and Maréchal 2018). Fatty acids (FAs) are carboxylic acids. Their carbon chain length and number of unsaturations (or double bonds, C=C) allow assessing whether they can be used for different applications. Oleaginous crops are an essential resource for human nutrition. Microalgae, whose interest in the scientific community is currently exponential, can also produce TAGs. Microalgae enriched in FAs with short or medium carbon chains without unsaturation are an interesting feedstock for green chemistry or the development of biofuels (Lupette and Maréchal 2018). Microalgae containing high levels of very long-chain fatty acids (carbon number greater than 20) with multiple unsaturations (1–6), called VLC-PUFAs, with unsaturation at the ω-3 position (i.e. eicosapentaenoic acid or EPA, 20:5, and docosahexaenoic acid or DHA, 22:6), are promising for human health applications (Lupette and Maréchal 2018).

There is currently no consensus view of the evolution of LDs in the Tree of Life. Some features are conserved; others show a striking level of diversification. Here, we summarize the current knowledge on the architecture, dynamics, and multitude of functions of the lipid droplets in prokaryotes and eukaryotes deriving from primary and secondary endosymbiosis events.

11.2 Studying Lipid Droplets

11.2.1 Imaging Lipid Droplets

Microscopic observation by confocal or epifluorescence imaging is the main method of detection of LDs in a cell or an organelle. The most commonly used fluorophores are Nile Red (Greenspan et al. 1985) and BODIPY 505/515 (Rumin et al. 2015) or BODIPY 493/503 (Gocze and Freeman 1994). More recently, new fluorophores have been developed (Yang et al. 2012; Gidda et al. 2016). These molecules are compatible with the parallel measurement of the fluorescence of GFP (Green Fluorescent Protein), RFP (Red Fluorescent Protein), and of chlorophyll (Kuntam et al. 2015). Other compounds including AC-202 were recently used in two model species of microalgae *Chlamydomonas reinhardtii* and *Phaeodactylum tricornutum* (Harchouni et al. 2018). These fluorophores make it possible to determine the size and number of LDs in a semi-quantitative manner, as well as their cellular localization.

11.2.2 Purifying Lipid Droplets

The general strategy for studying LD architecture is similar regardless of the organism studied. A culture of cells in a medium promoting the development of LDs is used (e.g. a nutrient deficiency). The LD purification starts by a gentle cell disruption step (French press, cell disruptor, etc.) in a suitable buffer releasing droplets as well as other cellular components. It is then necessary to perform a density gradient purification (Brasaemle and Wolins 2016). Due to their low density, LDs rise to the surface of the gradient during ultracentrifugation (Brasaemle and Wolins 2016). LDs are harvested and washed to limit the presence of contaminants. The study of LD architecture is based first on the determination of all the components of the hydrophobic core and on the composition of the monolayer of polar lipids and proteins on the surface of the droplet (Walther et al. 2017). Proteomic analyses allow the identification of proteins but also the characterization of some post-translational modifications (phosphorylation, nitrosylation, ubiquitinylation, sumoylation, *N* and *O*-glycosylation, farnesylation). The proteome must then be validated by biochemical studies (western blot), the imaging of fusion proteins (with a fluorescent marker to verify the location on the surface of LDs), and functional genetic studies (the study of mutants with altered expression such as knockout, silencing, and overexpression of genes coding for droplet proteins allows their functional characterization).

11.2.3 Biophysical Properties of Lipid Droplets

Biophysical studies of LDs have proven to be critical to advance our understanding of LD biogenesis. After the removal of proteins and polar lipids, it is possible to consider LDs as the product of an emulsion of oil in water (Thiam et al. 2013b). The cytosol of the cell represents the aqueous phase and LDs, the dispersed oily phase. The interface between oil and water generates a surface tension due to the lack of cohesive integrations between the two phases. These emulsions are metastable in the absence of external disturbance. The presence of surfactants makes it possible to reduce the surface tension, thus increasing the (meta)stability of the emulsion and the cohesion energy cost (Georgieva et al. 2009). The polar glycerolipids at the periphery of the droplet then act as surfactants. Mastering this system in vitro is probably one of the most important challenges to understand how such anisotropic hydrophobic cores can be maintained in a cell, where all other components are highly self-assembled and organized (membranes can be considered as two-dimensional fluid crystals, DNA and polymers have three-dimensional architectures, polypeptides form protein structures with rigorous three-dimensional folds). It is also possible to use biophysical methods such as Pulsed Field Gradient-Nuclear Magnetic Resonance (PFG-NMR) to determine the mobility of TAGs inside the volume set by LDs with or without LD-to-LD connections (Gromova et al. 2015). Since the overall structure of LDs appears conserved in the Tree of Life, whereas components may differ, a key to LD conservation may rely on biophysical properties, which now need to be evaluated in different systems.

11.3 Lipid Droplets in the Tree of Life

11.3.1 Lipid Droplets in Prokaryotic Cells

The vast majority of bacterial species have the capacity to accumulate lipid compounds within their cytoplasm, especially during nutritional stress. A distinction is made according to accumulated hydrophobic molecules, with either "lipid droplets" (containing fluid acyl esters, triacylglycerols, or TAGs) or "granules" (containing semi-solid lipopolymers called polyhydroxyalkanoates or PHAs). Here, we describe the formation of LD structures in bacteria, as a possible basis for their evolution in eukaryotes following primary endosymbiosis events. Since primary endosymbiosis events are believed to be facilitated by the presence of pathogenic partners (Maréchal 2018), we also describe how some pathogenic bacteria (and viruses) are known to interact with host cell LDs.

11.3.1.1 Polyhydroxyalkanoate Granules

Polyhydroxyalkanoates (PHAs) are polyesters produced by fermentation of lipids or carbohydrates. They are linear polyesters consisting of hydroxy acid monomers (HA) linked together by an ester bond (Możejko-Ciesielska and Kiewisz 2016). PHAs include poly(3-hydroxybutyrate) (PHB) and polyhydroxyvalerate (PHV) (Murphy 2012). PHAs are also classified into two groups according to the number of carbons per monomer: short-chain (3–5 carbons) PHAs and medium-chain PHAs (6–14 carbons). PHAs are synthesized when the C/N ratio is altered (a nitrogen deficiency coupled with an excess of carbon), stopping growth and division, resulting in an entry into the quiescent phase. Special attention has been given to these polyesters for several decades because they are biodegradable (Pötter and Steinbüchel 2006). PHAs are used in a wide range of applications such as resorbable materials for medical purposes (implants, biodegradable sutures, stents, etc.), materials (paper coating, shape memory gel, etc.), fuel additives, and as metabolic regulators (Możejko-Ciesielska and Kiewisz 2016).

Ralstonia eutropha H16 (new name: *Cupriavidus necator*) is the study model for PHB granules (Reinecke and Steinbüchel 2009). It is a Gram-negative bacterium that can accumulate 10–20 granules per cell, measuring 500 nm in diameter and representing 90% of the dry weight (Anderson and Dawes 1990). A recent study has shown that *R.eutropha* H16 PHB granules do not have a monolayer of polar lipids but only superficial proteins essential for their synthesis and degradation (Bresan et al. 2016). The main proteins detected on their surface are PHB synthase (PhaC), phasin (PhaP), PhaR (PhaR), and PHB depolymerase (PhaZ). The *R. eutropha* H16 PhaC gene has been cloned by three independent laboratories. The localization of the PHB synthase on the surface of the granules was confirmed by immunocytochemical staining with colloidal gold (Gerngross et al. 1993). The study of PHA granules is still too scarce to assess whether they are linked to TAG containing LDs and whether they may also be present in some eukaryotic clades.

11.3.1.2 TAG and Wax Ester Droplets

Bacteria are also able to accumulate TAG in LDs. Bacterial species producing LDs have been described in the following genera: *Mycobacterium*, *Nocardia*,

Rhodococcus, *Micromonospora*, *Dietzia*, and *Gordonia* as well as several Streptomycetes (Alvarez and Steinbüchel 2002; Wältermann and Steinbüchel 2005; Murphy 2012). γ-Proteobacteria (*Marinobacter*, *Alcanivorax*, etc.) are hydrocarbon-based bacteria capable of accumulating LDs (TAGs and wax esters) when entering dormancy (Kalscheuer et al. 2007). These γ-proteobacteria are often found in the oceans associated with microalgae in a system called the phycosphere (Lupette et al. 2016). These bacteria are also able to use petroleum hydrocarbons as a source of carbon raising possible applications in the degradation of hydrocarbons during oil spills (Murphy 2012).

Rhodococcus are oleaginous bacteria containing large amounts of TAGs (Alvarez 2016). Proteomic studies of LDs of several *Rhodococcus* species have been performed: *Rhodococcus opacus* and *Rhodococcus ruber* (Kalscheuer et al. 2001), *Rhodococcus jostii* RHA1 (Ding et al. 2012b), and *Rhodococcus opacus* PD630 (Kalscheuer et al. 2001; Chen et al. 2014). 228 proteins have been identified in *R. jostii* RHA1 including two putative structural proteins representing 15% of LD proteins: Microorganism Lipid Droplet Small (MLDS) and Phage shock protein A (Psp A). Ribosomes and translational regulators have also been isolated in the proteome of the LD of *R. jostii* RHA1 (Ding et al. 2012b). By a functional genetic study, deletion of the MLDS causes the formation of larger LDs (Ding et al. 2012b). A recent study showed that LDs of *R. jostii* RHA1 are bound to genomic DNA via MLDS protein, which increases the survival rate of bacterial cells during nutritional deficiency or genotoxic stress (Zhang et al. 2017).

11.3.1.3 Pathogenic Bacteria

Pathogenic bacteria can divert lipids and even 'hijack' LDs from an infected host. Well-known examples include bacteria of the genera *Mycobacterium* and *Chlamydia*. Mycobacteria are bacilli with pathogenic potency: *Mycobacterium tuberculosis* (Menon et al. 2019), *M. bovis*, and *M. avium* are causative agents of tuberculosis, *M. leprae* is the agent of leprosy, and *M. ulcerans* is responsible for Buruli ulcer. *Mycobacteria* are able to disrupt human lipid homeostasis during infection following the formation of foamy macrophages containing LDs in the cytoplasm (Kim et al. 2010; Caire-Brändli et al. 2014). LDs produced in infected cells serve as a platform for the production of signaling molecules (prostaglandins and leukotrienes eicosanoids) regulating the immune response and inflammation (Melo and Weller 2016). LDs also serve to concentrate and deliver iron via lipophilic siderophores (mycobactins) secreted by the bacteria (Luo et al. 2005). The LD proteome of Bacillus Calmette and Guerin (BCG: a non-replicating hypoxic attenuated strain of *M. bovis*) used in the tuberculosis vaccine allowed the identification of five proteins: two triacylglycerol synthases, Tgs1 (BCG3153c) and Tgs2 (BCG3794c), as well as three other proteins, BCG1169c, BCG1489c, and BCG1721 (Low et al. 2010). BCG1169c is a specific protein of the *Mycobacterium* clade. BCG1489c codes for a putative AGPAT involved in the formation of PA in the Kennedy pathway. A deletion of BCG1489C causes a decrease in the amount of TAG.

Chlamydia trachomatis is an obligate intracellular bacterium responsible for sexually transmitted infections as well as eye diseases. This bacterium has a biphasic life cycle: the elementary body representing the infectious form and the non-infectious reticulate body. *C. trachomatis* is able to translocate an LD from the cytoplasm of the host to the parasitophorous vacuole lumen containing the bacterium *via* an endocytosis process. Internalization occurs after LDs are coated with a family of Lda proteins (Kumar et al. 2006). The proteome of LDs of HeLa human cervical adenocarcinoma epithelial cells infected with *C. trachomatis* LGV-L2 434/Bu showed that they are enriched in PLIN2, PLIN3, ACSL-3, and ACSL-4 proteins (Saka et al. 2015).

11.3.2 Hijacking of Lipid Droplets by Viruses

LDs can serve as a source of energy for the dispersion of viruses such as hepatitis C virus (HCV), dengue virus (DENV), poliovirus (PV), or rotavirus (RV). This section presents several examples of 'hijacking' of lipid metabolism by viruses.

HCV is the most widely used model for studying LD diversion (Roingeard and Melo 2017). The virion initially circulates in a form of lipoviroparticles (rich in cholesterol esters and apolipoproteins apoB and apoE) in the blood of infected patients (Boyer et al. 2014). Its entry into the cell is dependent on LDL receptors (Low Density Lipoproteins). The replication of the viral RNA then induces extensive alterations of the membrane and the formation of vesicular structures that exhibit features similar to lipid rafts (Aizaki et al. 2004). The assembly of HCV requires the localization of some of the HCV proteins in LDs of the host, and the release of virions is strongly associated with the secretion pathway of very low-density lipoproteins (VLDL). Proteins that are important to complete the viral cycle, i.e. the nucleocapsid of HCV and the nonstructural protein NS5A co-localize with the LD in HCV-infected cells and interact with LD proteins, such as DGAT1 (Camus et al. 2013).

Other examples of viruses depending on host cell LDs to complete their cycle include DENV in mosquitoes (Mayer et al. 2017), PV causing poliomyelitis (Nchoutmboube et al. 2013), Flock House Virus (Castorena et al. 2010), Brome mosaic virus (BMV) (Zhang et al. 2016), or RV causing gastroenteritis in infants and young children (Gaunt et al. 2013).

11.3.3 Lipid Droplets in Eukaryotic Cells

Eukaryotes including unicellular and multicellular organisms are characterized by the presence of membrane organelles (endoplasmic reticulum, nucleus, Golgi, trans-Golgi network, peroxisomes, lysosomes, vacuoles, and numerous cytosolic vesicles for the endomembrane system, mitochondria, and chloroplast for semi-autonomous organelles). Primary endosymbiosis events are at the origin of mitochondria and primary plastids. In a very simplistic scheme, mitochondria are considered to derive from Gram-negative alpha-proteobacteria and primary plastids, from Gram-negative cyanobacteria. Based on molecular evidence, other partners have been involved, including pathogenic bacteria (Maréchal 2018). In contrast to prokaryotes, the synthesis of LDs from the ER seems to occur nearly in all eukaryotes studied to date.

In this section, we first consider non-photosynthetic eukaryotes, containing mitochondria as unique semi-autonomous organelles, then photosynthetic eukaryotes containing primary plastids, and finally those containing secondary plastids.

11.3.3.1 *Opisthokonta*: Non-photosynthetic Eukaryotes

Metazoa

Chordata

LDs of mammals are composed of a hydrophobic core consisting of TAGs and cholesteryl esters, generally considered as a form of storage. The monolayer of polar lipids is mainly composed of PC, PE, and PI (Bartz et al. 2007). In mammals, LD surface proteins are grouped into a family formerly called PAT (for Perilipin—Adipophilin—Tail-interacting protein of 47 kDa) (Bickel et al. 2009) but whose nomenclature has evolved since 2010 under the name of Perilipin (PLIN) (Kimmel et al. 2010). PLIN1 has four splice variants PLIN1a, PLIN1b, PLIN1c, and PLIN1d (Kimmel et al. 2010). PLINs contain a hydrophobic PAT domain of 100 amino acids defining an N-terminal region. The study of a mutant PLIN2-N1 (deficient for the PAT domain) reveals that this domain is not involved in lipid binding, but in the stabilization of lipid droplets, and lipid accumulation and degradation of PLIN2 by the proteasome (Najt et al. 2014). The PAT domain is followed by an 11-mer helical pattern of variable size, which might interact with phospholipids (Bussell and Eliezer 2003). Some PLIN proteins can be post-translationally modified by phosphorylations, via PKA, during lipolysis (PLIN3 and PLIN4 do not have a phosphorylation site (Kimmel and Sztalryd 2016)). Recently, the sixth clade of Perilipin, called PLIN6, has been discovered, specific to teleosts (Granneman et al. 2017). Interestingly, PLIN6 is not expressed in tissues associated with lipid metabolism but in the xanthophores of teleost skin. Biochemical analyses have shown that PLIN6 is associated with the surface of droplets enriched in carotenoids and regulates the pigment synthesis pathways (Granneman et al. 2017).

The expression of PLINs (with the exception of PLIN3) is regulated by a family of transcription factors called PPAR (Peroxisome Proliferator Activated Receptor). These transcription factors are activated by the binding of lipid ligands (Poulsen et al. 2012).

Since the early 2000s, at least 25 proteomic studies of the LD of mammalian cells or tissues have been published (Table 11.2). In all these studies, proteomics reveals the presence of at least one of the classes of PLIN on the surface of the mammalian LD. It can also be seen that the distribution of the PLINs varies according to mammalian cells and tissues. PLIN1 and PLIN4 are present on the surface of LDs of adipocytes (Ding et al. 2012a) and adipose tissue (Yu et al. 2015). PLIN2 and PLIN3 are ubiquitous in non-adipose tissues. PLIN2 is strongly expressed on the surface of LDs of the liver and hepatocytes. PLIN5 is present in the oxidative tissues i.e. the heart, brown adipose tissue, and skeletal muscles (Kimmel and Sztalryd 2014).

Functional studies of PLIN have been performed upstream and downstream of proteomic studies. KO mice for PLIN1 have a phenotype of reduction of fat mass, an increase of the lipolytic activity, but also a glucose intolerance and peripheral insulin resistance (Tansey et al. 2001). Stimulation of lipolytic activity by tumor necrosis factor (TNF-α) showed, in a first study, a decrease in the expression of PLIN1 variants. However, the overexpression of PLIN1a and PLIN1b blocks the ability of TNF-α to increase lipolysis in 3T3-L1 cells (Souza et al. 1998). KO mice for PLIN2 present unchanged adipose differentiation, a 60% decrease in hepatic TAGs, but a level of VLDL identical to control mice suggesting the retention of TAGs in the microsomes (Chang et al. 2006). Overexpression of PLIN2 fused to GFP causes an increase in the number and size of LDs in hepatocytes (Imamura et al. 2002). A first antisense study of hepatic PLIN2 causes a decrease in hepatic steatosis, hypertriglyceridemia, and insulin resistance in obese mice without altering the level of expression of PLIN3 and PLIN4 (Imai et al. 2007). A second antisense study of PLIN2 shows the same type of result with a decrease in DAG and TAG in the liver as well as an improvement in insulin production (Varela et al. 2008). Finally, deletion of the PLIN2 exon 5 (Plin2Δ5) in mice causes resistance to obesity induced by a diet rich in fats indicating the role of PLIN2 in obesity and hepatic steatosis (Mcmanaman et al. 2013). The deletion of PLIN3 in mice induces cold tolerance (Lee et al. 2018), probably by regulating beige adipocyte formation and thermogenic activities. The deletion of PLIN4 leads to decreased expression of PLIN5 reducing lipid accumulation in the cardiac muscle (Chen et al. 2013). KO of PLIN6 is responsible for stopping the concentration of carotenoids in the droplets (Granneman et al. 2017).

Arthropoda

Research on LDs of *Arthropoda* was mainly carried out in the fruit fly *Drosophila melanogaster*. LD research began with two RNAi screens (Beller et al. 2008; Guo et al. 2008) showing that approximately 370 genes, or 1.5% of the expressed genome, were involved in LD physiology. A first proteomic study was conducted on the abdominal fat body, the fat tissue of the fly, which allowed the identification of 248 proteins (Beller et al. 2006). A second study was conducted on whole embryos (Cermelli et al. 2006). The proteomes of the LD of *D. melanogaster* have similarities with those of mammals. There are, for example, two members of the

Table 11.2 Analyses of LDs in representative prokaryotic and eukaryotic study models

Phylogeny			Species	LD Proteome/ LD protein characterized	LD lipidome	Other informations	References
Prokaryota	Proteobacteria	β-Proteobacteria	*Cupriavidus necador*	PhaC/PhaP/ PhaR/PhaZ	PHB	PHB granules	Bresan et al. (2016), Peoples and Sinskey (1989), Schubert et al. (1988), Slater et al. (1988), Gerngross et al. (1993), Dennis et al. (1998), Tariq et al. (2015), Han et al. (2007)
		γ-Proteobacteria	*Marinobacter*		TAGs		Kalscheuer et al. (2007)
			Alcanivorax				
	Actinobacteria	Mycobacteriaceae	*Mycobacterium tuberculosis*	PLIN2, PLIN3, ARL8B	TAGs	Pathogenic bacteria	Menon et al. (2019), Kim et al. (2010), Caire-Brändli et al. (2014), Melo and Weller (2016), Luo et al. (2005), Low et al. (2010)
			Mycobacterium bovis/ BCG	Tgs1, Tgs2, BCG1169c, BCG1489c, BCG1721			
			Mycobacterium avium				
		Nocardiaceae	*Rhodococcus jostii* RHA1	MLDS/Psp A	TAGs		Ding et al. (2012b)
			Rhodococcus opacus PD630	LPD06283			Kalscheuer et al. (2001), Chen et al. (2014)
			Rhodococcus ruber	Ribosomal protein L7			Kalscheuer et al. (2001)
	Chlamydiae		*Chlamydia trachomatis*	PLIN2, PLIN2, ACSL-3, ACSL-4, CGI-58	TAGs	Pathogenic bacteria	Saka et al. (2015)

Eukaryota	Unikonta	Opisthokontha	Metazoa	Cnidaria	*Euphyllia glabrescens*	SLDP	TAGs, sterol, WE, PE, PC, Lyso-PC		Peng et al. (2011)
				Chordata	*Cricetulus griseus*	PLINs, CGI-58, ATGL, HSL, CNX	TAGs/SE	Analysis in multiple cell and tissue types	Bartz et al. (2007), Bickel et al. (2009), Kimmel et al. (2010), Greenberg et al. (1991), Orlicky et al. (2008), Najt et al. (2014), Bussell and Eliezer (2003), Hickenbottom et al. (2004), Kimmel and Sztalryd (2016), Granneman et al. (2017), Mandard et al. (2004), Poulsen et al. (2012), Targett-Adams et al. (2005), Dalen et al. (2007), Edvardsson et al. (2006), Chawla et al. (2003), Schmuth et al. (2004), Tobin et al. (2006), Bindesbøll et al. (2013), Stenson et al. (2011), Langhi et al. (2014), Liu et al. (2004)

(continued)

Table 11.2 (continued)

Phylogeny					Species	LD Proteome/ LD protein characterized	LD lipidome	Other informations	References
					Mus musculus				Wu et al. (2000), Brasaemle et al. (2004), Cho et al. (2007), Kanshin et al. (2009), Blouin et al. (2010), Zhang et al. (2011), Ding et al. (2012b), Crunk et al. (2013), Yu et al. (2015), Yamaguchi et al. (2015), Wang et al. (2015), Kramer et al. (2018)
					Rattus norvegicus				Turró et al. (2006), Larsson et al. (2012), Eichmann et al. (2015), Khor et al. (2014)
					Homo sapiens				Fujimoto et al. (2004), Sato et al. (2006), Bouchoux et al. (2011), Moessinger et al. (2011), Beilstein et al. (2013), Dahlhoff et al. (2015), Pataki et al. (2018)

					Danio renio			Granneman et al. (2017)
					Bos taurus			Orban et al. (2011), Talbott and Davis (2017)
				Arthropoda	*Drosophila melanogaster*	PLIN1, PLIN2, Lsdh1, H2Av, CGI-58	TAG/CE	Whole embryos and abdominal fat body	Beller et al. (2006), Cermelli et al. (2006)
					Manduca sexta	LSD-1, LSD-2, ATGL	TAG/DAG		Soulages et al. (2012)
					Anopheles aquasalis	LSD-2, apolipoprotein-o			Dias-Lopes et al. (2016)
				Nematoda	*Caenorhabditis elegans*	PLIN1, DHS-3, MDT-28	TAG/DAG/MAG		Zhang et al. (2012), Na et al. (2015), Vrablik et al. (2015)
			Fungi		*Saccharomyces cerevisiae*	Yes	TAGs/SE/squalene/sterol/PL		Czabany et al. (2008), Leber et al. (1994), Tauchi-sato et al. (2002), Grillitsch et al. (2011), Athenstaedt et al. (1999)
					Pichia pastoris		TAGs/PL/DMPE		Ivashov et al. (2013)
					Yarrowia lipolytica		TAGs/SE		Athenstaedt et al. (2006)
					Mortierella alpina		Neutral and Polar lipids		Yu et al. (2017)
					Trichosporon fermentans				Shen et al. (2016)
					Schizosaccharomyces pombe		TAG/SE		Venkata et al. (2012)
					Rhodosporidium toruloides				Zhu et al. (2015)

(continued)

Table 11.2 (continued)

Phylogeny				Species	LD Proteome/ LD protein characterized	LD lipidome	Other informations	References
		Amoebozoa		*Dyctiostelium discoideum*	plnA	TAGs/FFA		Du et al. (2013)
	Bikonta	Excavata	Euglenozoa	*Trypanozoma cruzi*			LDs induction	D'Avila et al. (2011)
				Leishmania amazonensis				Lecoeur et al. (2013)
				Leishmania major				Rabhi et al. (2012, 2016)
		Viridiplantae	Streptophyta	*Arabidopsis thaliana*	Seeds—SH oleosin, SL oleosin, U oleosin, pollen—U oleosin, Caleosin, Stereoleosin	TAGs	OLDs (seeds, pollen, leaves)	Jolivet et al. (2004), Vermachova et al. (2011)
				Brassica napus				Katavic et al. (2006), Jolivet et al. (2009)
				Sesamum indicum				Lin et al. (2005)
				Jatropha curcas				Popluechai et al. (2011), Liu et al. (2015)
				Madia sativa				Acevedo et al. (2012)
				Camelina sativa				Jolivet et al. (2013)
				Gevuina avellana				Acevedo et al. (2012)
				Zea mays				Tnani et al. (2011)
				Arachis hypogea				Jolivet et al. (2013)
				Nicotiana tabacum				Kretzschmar et al. (2018)

				Brassica tapetum	T-oleosin	TAGs	Tapetosome	Huang (2018)
				Persea americana	LDAP1, LDAP2, M-oleosin	TAGs	OLDs—mesocarp	Horn et al. (2013)
				Allium cepa	U-oleosin/caleosin	TAGs	OLDs—LDs cluster	Huang (2018)
				Vanilla planifolia	U-oleosin			
				Aloe vera	U-oleosin			
				Hevea brasiliensis	SRPP/REF	*cis*-1,4-polyisopropene	Rubber particles—NOLDs	Berthelot et al. (2014), Oh et al. (1999), Sando et al. (2009)
				Capsicum annum	PAP/Fibrillin	Yes	Plastoglobules—NOLDs	Van Wijk and Kessler (2017)
			Chlorophyta	*Chlamydomonas reinhardtii*	MLDP	TAGs		Moellering and Benning (2010), Nguyen et al. (2011), James et al. (2011)
				Dunaliella bardawil		TAGs		Davidi et al. (2012)
				Dunaliella salina				
				Dunaliella parva				
				Scenedesmus quadricauda		TAGs		Javee et al. (2016)
				Haematococcus pluvialis	HOGP	TAGs/DGTS/PC/SQDG/DGDG	Pigments in LDs	Peled et al. (2011)
				Chlorella sp.	Caleosin	TAGs/DAG		Lin et al. (2012)
				Lobosphaera incisa	LiMLDP, LiLBP62, LiLBP36	TAGs		Siegler et al. (2017), Bigogno et al. (2002)

(continued)

Table 11.2 (continued)

Phylogeny							Species	LD Proteome/ LD protein characterized	LD lipidome	Other informations	References
		Chromista	Hacrobia		Haptophyta		*Tisochrysis lutea*	Yes	Alkenone	Alkenone body	Marlowe et al. (1984a), Marlowe et al. (1984b), Song et al. (2013), Shi et al. (2015), Shi (2019)
			SAR super-group/ Harosa	Alvaria	Heterokonta		*Nannochloropsis oceanica*	LDSP	TAGs		Vieler et al. (2012)
							Fistulifera solaris	DOAP1	TAGs		Nojima et al. (2013)
							Phaeodactylum tricornutum	StLDP	TAGs/DGTA/ SQDG/PC		Yoneda et al. (2016), Lupette et al. (2019)
							Auranthiochytrium limacinum	TLDP1	TAGs		Watanabe et al. (2017)
					Alveolata	Apicomplexa	*Toxoplasma gondii*			LDs induction	Hu et al. (2017)
						Dinophyta	*Symbiodinium* sp.	SLDP	TAGs/CE		Pasaribu et al. (2014b), Jiang et al. (2014)
				Rhizaria	Cercozoa		*Plasmodiophora brassicae*	Yes	TAGs		Bi et al. (2016)
					Retaria		*Ammonia tepida*			LDs detection	LeKieffre et al. (2017), Le Cadre et al. (2006)
							Ammonia beccarii				Le Cadre et al. (2006)

PLIN family (*Dm*PLIN1 and *Dm*PLIN2) in the fat body abdominal, as well as the CGI-58 protein in the whole embryo proteome. As a result, *Drosophila* has been considered an interesting model for studying the role of LDs in the context of human pathologies. The conservation of LD-associated proteins is not complete, since PLIN3, PLIN4, and PLIN5 appear to be restricted to vertebrates. *Dm*PLIN1 is present only on the surface of LDs and is involved in promotion/prevention mechanisms for lipolysis (Bi et al. 2012). *Dm*PLIN1 contains four helices in the central region of the protein, capable of binding lipid compounds (Arrese et al. 2008; Lin et al. 2014). Mutant flies deficient of *Dm*PLIN1 have larger LDs. Single and giant LDs within the fat body of these mutants have also been found to confer an obesity phenotype (Beller et al. 2010). *Dm*PLIN2 is present in the cytoplasm and on the surface of LDs (Beller et al. 2010). *Dm*PLIN2 only plays a role in the prevention of lipolysis (Bi et al. 2012). *Dm*PLIN2 mutants have smaller LDs (Li et al. 2012). The double mutant fly *Dm*PLIN1/*Dm*PLIN2 presents a marked reduction of LD size; however, LDs are still present in these mutants suggesting that there is an additional mechanism regulating lipid storage and lipolysis. *Dm*HSL (Hormone-Sensitive-Lipase) is a lipase participating in lipolysis and interacting with *Dm*PLIN1 (Bi et al. 2012). A complementary analysis of the CG2254 protein identified in the proteome of LDs from *Drosophila* abdominal cells (Beller et al. 2006) showed that it was LD subset dehydrogenase 1 (Lsdsh1) (Thul et al. 2017).

Interestingly, the study of *Drophila* LDs has highlighted an unsuspected role in the homeostasis of histones within the cell, a function that may be more frequent in eukaryotes than initially thought. Histones were first identified in the LD proteome of *Drosophila* embryos (Cermelli et al. 2006). These histones are not detected in the fat body abdominal proteome (Beller et al. 2006). These results were confirmed by a secondary study of the Jabba protein that co-immuno-precipitates with histones (Li et al. 2012). The presence of histones H2A, H2B, and H4 was also observed in the tobacco sphinx *Manduca sexta* (Soulages et al. 2012). Finally, it has recently been shown that histone H2Av was dynamically associated with *D. melanogaster* LD during cleavage and syncytial blastoderm stages (Johnson et al. 2018).

A recent study also investigated the protein ABHD4/ABHD5 (CGI-58) in *Drosophila* (Hehlert et al. 2019). The mutation of the *pummelig* (*puml*) gene encoding CGI-58 causes abnormal accumulation of TAG in mutant flies as well as a change in the FA profile of TAGs in Malpighian tubules (kidneys). In contrast to mammals, the Drosophila *puml* does not stimulate ATGL lipase activity (*brummer*) in vitro (Hehlert et al. 2019).

Proteomic studies of LDs have also been performed in other arthropod models such as the tobacco sphinx *Manduca sexta* (Soulages et al. 2012) or *Anopheles aquasalis* (Dias-Lopes et al. 2016), an important vector of *Plasmodium virax*, the main human malarial parasite in the Americas.

Nematoda

The nematode *Caenorhabditis elegans* is a worm constituting a popular study model for the study of apoptosis, embryonic development, and cellular aging. This nematode has LDs measuring 1–1.5 μm in diameter in its intestine and in the hypodermis. Three proteomic studies of *C. elegans* LDs have been performed (Zhang et al. 2012; Na et al. 2015; Vrablik et al. 2015). The first shotgun proteomic analysis allowed the identification of 306 proteins of which 193 were known to be associated with mammalian LDs (Zhang et al. 2012). This first study identified the DHS-3 protein on the surface of the LD *via* a GFP fusion. A second proteome of the LD of *C.elegans* allowed the identification of 154 proteins of which 113 are common with the first proteome (Na et al. 2015). DHS-3 and MDT-28 are the two major proteins in *C. elegans* LD. The deletion of the *dhs-3* gene causes a decrease in the size of LDs as well as the amount of their TAG. The *mdt-28* mutant causes the formation of LD aggregations (Na et al. 2015). A third proteome compared the LD protein composition of a *C. elegans* wild type and high daf-2 (e1370) fat mutant (Vrablik et al. 2015). Using a GFP construct, the ACS-4 protein, an acyl-CoA synthase, was localized at the surface of the *C. elegans* LD. It has long been thought that PLINs were lost in *C. elegans*. However, three isoforms of mammalian PLINs have been identified: PLIN-1a, PLIN-1b, and PLIN-1c. These isoforms have an N-terminal PAT domain, an amphiphilic region with imperfect helices, and four C-terminal helices (Chughtai et al. 2015). The *C. elegans* genome seems to also code for several sequences of the LD protein actors ABHD4 (*Ce*Lid-1) and ABHD5/CGI-58 (*Ce*Abhd5.2) (Lee et al. 2014; Xie and Roy 2015).

Fungi

Saccharomyces cerevisiae is a widely used model for studying lipid biology because the synthetic pathways in the ER are similar to those of plants and animals (Koch et al. 2014). LDs measure about 400 nm in this organism. The hydrophobic core of LDs of *S. cerevisiae* is composed of TAGs grouped in the center and surrounded by the steryl ester molecules (Leber et al. 1994; Czabany et al. 2008). There is also a minor proportion of squalene and sterols. The monolayer of phospholipids consists of PC, PI, PE, PA, and PS (Tauchi-Sato et al. 2002; Grillitsch et al. 2011). The proteome of the LD highlighted proteins that contribute to the synthesis of the hydrophobic core, such as sterol-Δ24-methyltransferase, squalene epoxidase, and lanosterol synthetase (Leber et al. 1994; Athenstaedt et al. 1999) (Table 11.2). In *Pichia pastoris*, the polar lipid monolayer is mainly composed of PC and PE but there is also a lower proportion of PI, PS, PA, cardiolipin, lysophospholipids, and 1,2-dimyristoyl-*sn*-glycero-3-phosphoethanolamine (DMPE) (Ivashov et al. 2013) (Table 11.2). Studies have also been conducted in other models such as *Yarrowia lipolytica* (Athenstaedt et al. 2006), *Mortierella alpina* (Yu et al. 2017), *Cryptococcus albidus* (Shi et al. 2013), *Trichosporon fermentans* (Shen et al. 2016), *Schizosaccharomyces pombe* (Noothalapati Venkata and Shigeto 2012), or *Rhodosporidium toruloides* (Zhu et al. 2015). Identified proteins comprise a majority of orthologues of *S. cerevisiae* and one can deduce from this analysis the enzymatic functions related to the synthesis of ergosterol, phospholipids, sphingolipids, but also proteins involved in the metabolism of fatty acids and degradation of non-polar lipids (Table 11.2).

11.3.3.2 Amoebozoa

Only one study of LD was recorded for *Amoebozoa* in the model species *Dyctyostelium discoideum. D. discoideum* is an amoeba living on dead leaves in forests, phagocyting bacteria, or yeasts (Malchow et al. 1967). LDs of *D. discoideum* are composed of TAG, free fatty acid, and more than 10% of an unknown lipid (Table 11.2). Proteomic analysis of the LD of *D. discoideum* reveals 72 proteins including one perilipin (plnA) (Du et al. 2013). The expression of plnA in CHO cells allowed their localization to the surface of LDs (Miura et al. 2002). Fifteen lipid metabolism enzymes, 31 small GTPases belonging to the Rab family, eleven endoplasmic reticulum component proteins, and six cytoskeletal associated proteins were also identified.

In non-photosynthetic eukaryotes, LD studies support the general conservation of the architecture including some classes of perilipins, but with a striking diversification of proteins associated with the LDs likely associated with specialized functions. A role in histone homeostasis may be an important innovation in eukaryotes. The biogenesis process from the ER seems also conserved, with SEIPIN-associated machineries. Enzymes involved in the biosynthesis of TAGs, also lipases and proteins involved in lipolysis, such as CGI-58, seem to be the markers of LD evolution. However, the molecular function of CGI-58 seems to differ in the various clades of non-photosynthetic eukaryotes studied so far, a functional 'flexibility' which is also observed in photosynthetic eukaryotes (see below).

11.3.3.3 Photosynthetic Eukaryotes Originating from Primary Endosymbiosis

The acquisition of the primary chloroplast occurred when an unknown eukaryotic organism integrated a Gram-negative cyanobacterium (Petroutsos et al. 2014; Maréchal 2018). This event led to the emergence of a photosynthetic organelle with two membranes (inner and outer membranes of the envelope) called the plastid. Based on the machinery of photosynthetic pigments (Archibald and Keeling 2002; Petroutsos et al. 2014; Maréchal 2018), three lineages appeared. The green lineage of primary endosymbionts corresponds to *Viridiplantae*. This lineage includes *Chlorophyta* ("green algae") and *Streptophyta* (commonly called "plants"). The photosynthetic machinery is composed of chlorophyll a and b. The red lineage of primary endosymbionts consists of *Rhodophyta* or "red algae". These organisms have chlorophylls a and c associated with phycobilin. The 'blue' lineage of primary endosymbionts corresponds to Glaucocystophytes (*Cyanophora paradoxa*), having a chloroplast with a residual cell wall rich in peptidoglycans. Chlorophyll a is associated with phycocyanin and allophycocyanin. Primary endosymbiosis has long been considered a unique event during evolution. The study of *Paulinella chromatophora* (a photosynthetic amoeba) has shown that a second, more recent, primary endosymbiosis event (60–100 million years ago) occurred between cyanobacteria and an amoeba (Maréchal 2018). This endosymbiosis led to the formation of an organelle also limited by two membranes, called the chromatophore.

In contrast to non-photosynthetic eukaryotes, in which FA biosynthesis occurs in the cytosol, FAs are synthesized in the stroma of the chloroplast and then exported to the cytosol. On the one hand, chloroplasts contain LD, called plastoglobules. On the other hand, the plastid appears to play a role in the production of TAG and biogenesis of cytosolic LDs in some of the lineages of photosynthetic eukaryotes. Stronger cooperation of the ER and plastid in LD formation may therefore be an important innovation in these primary endosymbionts. This may also be related to the loss of perilipins and the emergence of specific LD-associated proteins. Our understanding of LD evolution in primary endosymbionts is mainly based on analyses performed in Chlorophyta and land plants (Embryophyta, mostly in Angiosperms).

Chlorophyta

Chlamydomonas reinhardtii is a green alga that accumulates oils in the form of LDs following environmental stresses such as a nitrogen deficiency or an increase in salinity. A proteomic study of LD performed in *C. reinhardtii* highlighted a Major Lipid Droplet Protein (MLDP) of 27 kDa (Moellering and Benning 2010; James et al. 2011; Nguyen et al. 2011). The phenotype of a *mldp* mutant suggests that MLDP is involved in the regulation of LD size (Moellering and Benning 2010). A 33 kDa homolog of MLDP was also found in *Haematococcus pluvialis* (Peled et al. 2011) as well as in three *Dunaliella* species (Davidi et al. 2012). MLDP was also detected in *Scenedesmus quadricauda* during salt stress or nitrogen deficiency (Javee et al. 2016). MLDP orthologues are also present in several species of *Chlorophyta*: *Volvox carteri*, *Haematococcus pluvialis*, *Dunaliella salina*, *Coccomyxa* sp., *Chlorella variabilis*, *Polytomella parva*, *Prototheca wickerhamii*, and *Micromonas pusilla* CCMP1545 (Goold et al. 2015). A 28 kDa caleosin protein was shown to be the major protein in the LD of *Chlorella* sp. (Lin et al. 2012). The size of LDs of *Chlorella* can reach 3 μm (Lin et al. 2012). The caleosin localization was specifically determined on the surface of the LD by immunostaining with gold beads (Pasaribu et al. 2014a). In these studies, caleosins first characterized in plants (see below) seem to be conserved LD-associated proteins.

Plantae

Plants accumulate LDs in both vegetative and reproductive tissues (Chapman et al. 2012). The involvement of LDs in the physiology and development of plants are currently little known. The LD proteins of plants are divided into three functional groups: (1) oleosins; (2) caleosins, steroleosins, and dioxygenases; and (3) the proteins associated with the LD.

Oleosins were the first proteins characterized on the surface of LDs of Zea mays seeds (Qu et al. 1986; Vance and Huang 1987). By their small molecular weight (15–26 kDa), oleosins are very abundant proteins on the surface of LDs of plants. Structurally, the oleosins are divided into three portions: (1) a short and amphiphilic N-terminal peptide, (2) a C-terminal amphiphilic peptide of varying length, and (3) a hydrophobic pin of nonpolar amino acids penetrating the monolayer of phospholipids on the surface of the LD (Huang 2018). The N- and C-terminal peptides form receptor binding lipases and other proteins involved in TAG degradation (Huang and Huang 2015). The 72 amino acid pin is a specificity of oleosins (Kory et al. 2016), thus differing from major LD proteins of mammals (PLINs 1-6) or bacteria (Phasin). The pin is also divided into three portions, consisting of two 30-amino acid arms connecting a loop consisting of three prolines (P) and one serine (S) forming a structure called the "Proline Knot" inserted into the hydrophobic core (Chapman et al. 2012; Abell et al. 1997). The secondary structure of the loop has not been defined yet. Seventeen genes code for the oleosins in Arabidopsis: five in the seed, three jointly in the seeds and pollen grains, and nine in the floral cells of the tapetum.

In a bioinformatic study, oleosins could be classified into six major lineages (Huang and Huang 2015): the primitive lineage evolving from green algae to Filicophyta (ferns), the universal lineage (U oleosin) for which genes are present, and Bryophyta (mosses) to higher plants. The universal U line then evolved to specialize in particular structures such as seed-specific oleosins with the Low and High Molecular Weight Seed Oleosin (SH) lines in Angiosperms. Oleosins also specialized in Brassicaceae with the tapetum T line and the M line for oleosins in the Lauraceae mesocarp (avocado) (Kilaru et al. 2015).

Caleosins, stereoleosins, and dioxygenases are grouped into a single cluster because they have a common enzymatic function in the stress response (Huang 2018). Caleosins are enzymes that have been found in microsomes (Frandsen et al. 1996). Caleosins, like oleosins, have a hydrophobic as well as a "proline knot" motif (Huang 2018). A recent study showed that two of the hairpin prolines (P116 and P125) were not essential for LD binding (Müller et al. 2016). Caleosins have an N-terminal EF hand-type calcium binding motif (Chen et al. 1999), a peroxygenase activity (Hanano et al. 2006), and several phosphorylation sites. The genome of *Arabidopsis thaliana* codes for eight caleosins expressed in different structures (Shimada and Hara-Nishimura 2015).

Steroleosins (sterol dehydrogenases) have only two structural domains: a hydrophobic N-terminal region and a C-terminal region having a sequence close to the mammalian hydroxysteroid dehydrogenase (HSD) domain. Steroleosins also have a semi-conserved hydrophobic pin similar to that of oleosin, but of a size similar to that of caleosin (Huang 2018). They are also class I proteins such as oleosins and caleosins (Kory et al. 2016). In contrast to oleosins and caleosins, steroleosins do not have a "proline knot" motif but a "proline knob" (Chapman et al. 2012). Steroleosins are particularly studied because they are capable of converting sterols into brassinosteroids (Baud et al. 2009), a class of phytohormones.

Dioxygenases (α-DOX) have also recently been found to be associated with LDs in leaves of senescent *Arabidopsis thaliana* cells (Shimada et al. 2014). *A. thaliana* has two homologs of these dioxygenases (Atα-DOX1 and Atα-DOX2). Atα-DOX1 is localized on LDs of leaves and Atα-DOX2 is located in the ER (Shimada et al. 2014). These enzymes produce an oxylipin (2-HOT) from α-linolenic acid (18:3). These molecules participate in defense mechanisms in response to biotic and abiotic stresses (Shimada and Hara-Nishimura 2015).

Not all LD proteins listed above belong to the same LDs. Indeed, in plants, cytosolic LDs are divided into two groups: oleosin-based lipid droplets (OLDs) and non-oleosin-based lipid droplets (NOLDs) (Laibach et al. 2015).

Seeds are the most studied structure for the understanding of OLDs because they are able to accumulate TAGs in the form of LDs reserve to support germination after the end of the dormancy phase (Huang 1996). In special cases such as jojoba seed (*Simmondsia chinensis*), LDs can contain cerides (Yermanos 1975). These LDs are small (between 0.5 and 1.5 μm) conferring a large surface area per unit of TAGs, facilitating the binding of lipases during germination (Huang and Huang 2015). Numerous proteomic studies have been performed in plant models: *Brassica napus* (Katavic et al. 2006; Jolivet et al. 2009), *Arabidopsis thaliana* (Jolivet et al. 2004; Vermachova et al. 2011), *Sesamum indicum* (Lin et al. 2005), *Jatropha curcas* (Popluechai et al. 2011; Liu et al. 2015), *Madia sativa* (Acevedo et al. 2012), *Gevuina avellana* (Acevedo et al. 2012), *Zea mays* (Tnani et al. 2011), *Camelina sativa* (Jolivet et al. 2013), or *Arachis hypogaea* (Jolivet et al. 2013). These proteomic analyses show that LDs are covered with oleosins with a minor presence (less than 5%) of caleosin and steroleosin (Chapman et al. 2012; Murphy 2012). Oleosins are involved in regulating the size and stability of LDs of seeds (Chapman et al. 2012). LDs have also been characterized in the tapetum cells of the anther (Hsieh and Huang 2004) and in pollen grains and pollen tubes (Kretzschmar et al. 2018). A focused study on PUX10 (Plant UBX Domain-containing Protein 10) whose localization was confirmed by fusion with enhanced GFP (eGFP) on the surface of LDs during embryonic development, seed germination, and pollen tubes, showed that PUX10 recruits by its UBX domain an AAA-type ATPase Cell Cycle 48 (CDC48) that facilitates the transfer of polyubiquitinated protein to the 26S proteasome (Kretzschmar et al. 2018).

LDs are also present in *Arabidopsis thaliana* leaves. The number of LDs is very low in healthy leaves. LDs accumulate more in the leaves in the senescence phase (Shimada et al. 2015) with a variable size of 1–18 μm (Lersten et al. 2006). In particular, the expression of *A. thaliana* caleosin-3 as well as Atα-DOX1 increases during senescence (Shimada et al. 2014). Proteomic analysis of the LD of aging leaves of *A. thaliana* was performed: 28 proteins including 9 enzymes involved in the secondary defense metabolism of the plant were identified (Brocard et al. 2017). The analysis also revealed the presence of the Small Rubber Particle 1 (AtSRP1) protein. Functional analysis of AtSRP1 reveals that this protein modulates the expression of caleosin-3 in aging leaves. In addition, overexpression of AtSRP1 induces an increase in 18:3 enriched TAG accumulations from galactolipid recycling of thylakoids (Brocard et al. 2017).

LDs of the fruit mesocarp can reach sizes of 10–20 μm, placing them at the top of the ranking of the largest observable LDs in eukaryotic cells (Horn et al. 2013). Proteomic analysis of the mesocarp of the avocado (*Persea americana*) allowed the identification of two LDAP1 and LDAP2 associated proteins (Lipid Droplet-Associated Proteins 1 and 2), which also showed homologies of sequences with Small Rubber Particle Proteins (SRPP) (Horn et al. 2013). Type M oleosins specific to the *Lauraceae* family have been described in avocado (Huang 2018).

NOLDs include two special cases: rubber particles and plastoglobules. More than 20,000 species of higher plants can accumulate rubber particles within their vegetative organs (Hagel et al. 2008). *Hevea brasiliensis* is the main source of latex used by humans. Latex is a colloidal white suspension composed of rubbery and non-rubbery particles, organelles, proteins, lipids, carbohydrates, and minerals. These particles have a hydrophobic core consisting of *cis*-1,4-polyisopropene surrounded by a monolayer of phospholipids in which proteins are bound (Berthelot et al. 2014). The proteome of the rubber particles revealed two major proteins: the SRPP and the Rubber Elongation Factor (Sando et al. 2009).

Finally, plastoglobules are special LDs synthesized inside the chloroplast (Bréhélin et al. 2007). Plastoglobules are continuous with the outer monolayer of thylakoids in higher plants, which is supposed to facilitate the exchange of metabolites (Van Wijk and Kessler 2017). Plastoglobules have been less studied than cytosolic LDs. Analyses support that they have a hydrophobic core containing three classes of molecules: (1) neutral lipids (TAGs, phytol esters, and free fatty acids), (2) tocopherols and quinones (α-tocopherol, plastoquinol-9, plastochromanol-8, and Vitamin K_1), and (3) linear carotenoids (lycopene), cyclic carotenoids (lutein and xanthophylls), and carotenoid esters (Van Wijk and Kessler 2017). These molecules are surrounded by a monolayer of amphiphilic lipids (monogalactosyldiacylglycerol, MGDG; digalactosyldiacylglycerol, DGDG; sulfoquinovosyldiacylglycerol, SQDG) in which proteins are embedded. Proteomic analyses of several study models (the chromoplast of red pepper *Capsicum annuum* and the green microalgae *Dunaliella bardawil* and *Chlamydomonas reinhardtii*) have allowed the identification of about 30 proteins (Kreimer 2009; Davidi et al. 2015). The protein mainly represented on the surface of plastoglobules is Plastid-lipid Associated Protein, Fibrillin (PAP/Fibrillin) (Youssef et al. 2010). This 30-kDa protein does not have transmembrane segments.

11.3.3.4 Photosynthetic Eukaryotes and Non-photosynthetic Relatives Originating from a Secondary Endosymbiosis

Secondary endosymbioses are events that have occurred several times during the evolution of eukaryotes (Petroutsos et al. 2014). Two main types of lineages have emerged as a result of these evolutionary events: (1) green lines resulting from the integration of a green alga within an unknown heterotrophic eukaryotic organism, leading to the appearance of *Euglenozoa* and Chlororarachniophytes and (2) the red lines resulting from the integration of a red alga inside a eukaryotic organism forming the polyphyletic group of *Chromalveolata*.

Organisms from the Green Lineage: *Euglenozoa*

At least two independent events of secondary endosymbiosis between a green alga and an unknown heterotrophic eukaryotic organism led to the appearance of Euglenozoa and Chlororarachniophytes (Petroutsos et al. 2014; Füssy and Oborník 2018). Chlororarachniophytes (*Bigelowiella natans*) contain a four-membrane chloroplast and a residual nucleus, called the nucleomorph, located between the two most internal and external membranes of the chloroplast (Petroutsos et al. 2014). To our knowledge, no data are available on LD formation in Chlororarachniophytes. *Euglenozoa* comprises photosynthetic species with a chloroplast bounded by three membranes but also parasitic species devoid of any chloroplast (Petroutsos et al. 2014). Several studies report that parasitic organisms such as *Trypanosoma cruzi*, the agent of Chagas disease (D'avila et al. 2011), are capable of inducing the formation of large LDs in macrophages. In photosynthetic organisms, *Euglena gracilis* is a microalga living in freshwater, interesting for the research of alternatives to petroleum resources because it accumulates wax esters inside LDs in nitrogen starvation. These wax esters come from the conversion of a crystalline β-1,3-glucan, paramylon. Wax esters can be used as fuel for aviation but also as biofuel after refining (Guo et al. 2017). To our knowledge, there is no detailed study of the structure of the LD of *E. gracilis*.

Organisms from the Red Lineage: *Chromista/Chromalveolates*

Apicomplexa *(Containing a Non-photosynthetic Plastid)*

Apicomplexa is a phylum grouping unicellular parasitic organisms responsible for many diseases in metazoans such as malaria or toxoplasmosis. As noted previously with pathogenic bacteria, parasitic *Euglenozoa* or HCV parasites, these organisms are also able to divert the lipid metabolism of the host by inducing the formation of LDs or by modifying their architecture. One of the most commonly studied *Apicomplexa* models is *Toxoplasma gondii*, the toxoplasmosis agent. *T. gondii* replicates in mammalian cells in a parasitophorous vacuole. *Toxoplasma* induces an increase in the catalytic activity of the host DGAT, which leads to diversion of lipid metabolism toward TAGs and guarantees the import of FAs (Hu et al. 2017).

Dinophyta

Dinophyta or Dinoflagellates are photosynthetic protists, but also mixotrophic and heterotrophic, with two flagella allowing them to move (Sardet 2013). There are symbiotic forms of Dinoflagellates, especially with corals. LDs are present in hosts and symbionts of coral-dinoflagellate endosymbiosis (Kellogg and Patton 1983). The first proteomic analysis of the gastrodermale LDs of *Euphyllia glabrescens* (*Cnidaria*) revealed the presence of 42 proteins involved in the metabolism of lipids and proteins as well as in the response to some stresses (Peng et al. 2011). A second proteomic analysis of LDs of *Symbiodinium* sp. associated with coral tentacles of *Euphyllia glabrescens*, highlighted a 20 kDa Symbiodinium Lipid Droplet Protein (SLDP) that plays a role in the structural and functional stability of the LD (Pasaribu et al. 2014b). Proteomic analysis of LDs of isolated *Symbiodinium* spp. has also been performed (Jiang et al. 2014). This study did not reveal the presence of SLDP but of several proteins involved in lipid metabolism (Sterol transfer family protein), cell signaling (14-3-3 protein, ADP ribolysis factor), stress response (HSP90), and energy metabolism (ATP synthase F1 subunit α, GTP binding protein).

Haptophyta

Haptophyta include photosynthetic species forming extracellular shells of calcium carbonate (coccolithophores' coccospheres) and non-calcified cell walls (Cavalier-Smith 1986). A proteomic study was carried out in the non-calcareous haptophyta *Tisochrysis lutea*, formerly *Isochrysis* aff. *galbana* (Bendif et al. 2013). This microalga is interesting for the aquaculture industry because it is enriched in DHA (Hubert et al. 2017). It is also known to accumulate a particular neutral lipid in place of TAGs: an alkenone, a very long chain ketone (C_{37} to C_{40}), within an alkenone body (AB) (Marlowe et al. 1984a, b). The purification of LDs highlighted the presence of 74.2% of alkenone (C_{37} and C_{38}), 24.6% of other lipids, and 1.2% of alkene. Proteomic analysis revealed the presence of 514 proteins on the surface of *T. lutea* AB, of which three are predominant: a V-ATPase identified previously in a proteomic analysis in nitrogen deficiency (Song et al. 2013), an SPFH (Stomatin/Prohibitin/Flotillin/HflK) domain-containing protein, localized in the endoplasmic reticulum, associated with a lipid raft and a hypothetical protein (Shi et al. 2015). With the exception of SPFH, the transition between a rich medium and a low nitrogen medium causes an increase in the expression of the target genes encoding *T. lutea* AB proteins (Shi 2019).

Heterokonta

Heterokonta (or Stramenopiles) constitute a superphylum in the *Chromista* kingdom (Cavalier-Smith 2018). Five proteomic studies were performed in four model species: *Nannochloropsis oceanica* (Vieler et al. 2012), *Phaeodactylum tricornutum* (Yoneda et al. 2016; Lupette et al. 2019), *Fistulifera solaris* (Nojima et al. 2013), and *Aurantiochytrium limacinum* (Watanabe et al. 2017). The first study conducted with *N. oceanica* (Vieler et al. 2012) revealed a 16.8 kDa Lipid Droplet Surface Protein (LDSP) by proteomic analysis. This protein is present in the six species of *Nannochloropsis* with functional features close to plant oleosins as well as MLDP of *Chlamydomonas reinhardtii*. The second proteomic study in diatoms concerns the pennate diatom *Fistulifera solaris* (Nojima et al. 2013). The authors did not detect any major band and the purification seems to involve protein contaminations. By subtracting the protein purification proteins from the soluble fraction, the authors identified fourteen candidate proteins. By searching for conserved domains, one of the identified proteins has a quinoprotein-alcohol dehydrogenase-like domain detected in the proteome of LDs of *Camelina sativa* seed (Jolivet et al. 2013) but has no hydrophobic domain. This protein of 506 amino acids was named DOAP1 (Diatom-Oleosome-Associated-Protein 1). Based on a fusion between GFP and the N-terminal signal sequence of DOAP1 (Maeda et al. 2014), this protein was initially present in the ER. The third model characterized is the pennate diatom *Phaeodactylum tricornutum*. A first proteomic study allowed the identification of five major proteins on the surface of the LD of *P. tricornutum*, including the Stramenopile LD Protein (StLDP) (Yoneda et al. 2016). This class I protein was localized to the surface of the LD by fusion with a GFP protein (Yoneda et al. 2018). By optimization of the *P. tricornutum* LD purification protocol, a second study was carried out in this organism (Lupette et al. 2019). The proteome of the LD is composed of 86 proteins including most notably the LD-protein StLDP, metabolic actors, organelle membrane-associated proteins, proteins implicated in the treatment of genetic information, or chaperones involved in protein quality control. The hydrophobic core is only made of TAG surrounded by a monolayer of polar lipids consisting of PC, SQDG, and two molecular species (20:5–16:1 and 20:5–16:2) of diacylglycerylhydroxymethyltrimethyl-β-alanine (DGTA), a betaine lipid. A sterol probably located in the polar lipid monolayer, brassicasterol, was also detected and specific enrichment of β-carotene has been observed (Lupette et al. 2019). A last proteomic study was recently conducted on the model *Aurantiochytrium limacinum* F26-b, a non-photosynthetic unicellular microalga classified as *Heterokonta*. This microalga is particularly studied for its biotechnological potential because it is a species rich in DHA (22:6) (Dellero et al. 2018a, b; Morabito et al. 2019). A proteomic study of the LD fraction of *A. limacinum* F26-b identified a Thraustochytrid-specific Lipid Droplet Protein 1 (TLDP1) (Watanabe et al. 2017). A mutant of tldp1 shows a decrease in the amount of TAG and the number of lipid droplets per cell. However, larger and irregular LDs are observed in this mutant. TLDP1 may regulate the accumulation of TAGs as well as the size and number of LDs in *A. limacinum* F26-b (Watanabe et al. 2017).

Based on all these analyses of secondary endosymbionts, it appeared that major LD proteins are specific to distinct clades and that their origin needs to be traced in evolution. Lipids associated with the few LDs analyzed to date sometimes include plastid lipids together with ER lipids, suggesting cooperation of these two organelles in LD biogenesis. Detection of lipases, often associated with mitochondrial and/or lysosomal processes, further suggests a conserved role of mitochondria and lysosomes in the mobilization of FAs deriving from TAG stored in cytosolic LDs.

11.4 Lipid Droplets Are Dynamic Structures

LDs are dynamic subcellular structures, most often transient, rarely stable. Their development includes an intense phase of biosynthesis of the hydrophobic molecules that constitute its core (anabolism) and the biosynthesis and the arrangement of the barrier located at the periphery (a monolayer of polar lipids, most often phosphoglycerolipids, and proteins, either embedded in this monolayer or associated more or less transiently).

11.4.1 Anabolism of PHA and TAG

11.4.1.1 PHA Biosynthesis

PHA biosynthesis routes have been recently reviewed by Możejko-Ciesielska and Kiewisz (2016). Depending on the length of the PHA chain, several pathways exist. For short-chain PHA, a direct pathway involves three enzymes: a β-ketothiolase (PhaA), an acetoacetyl-CoA reductase (PhaB), and a PHA synthase (PhaC). For medium chain PHA, several pathways are possible: (1) a synthetic pathway involving β-oxidation resulting in the formation of R-3-hydroxyacyl-CoA that is converted by PhaC to PHA; (2) a synthetic route involving elongation of acetyl-CoA, via de novo synthesis of fatty acids; and (3) a synthetic route generating precursors of 3-hydroxyacyl-CoA.

11.4.1.2 TAG Biosynthesis

The hydrophobic core of LDs described in eukaryotes contains predominantly TAGs. The synthesis of TAGs can be carried out by two main pathways at the level of the endoplasmic reticulum (the so-called "eukaryotic" pathway): (1) an acyl-CoA dependent pathway commonly referred to as the Kennedy pathway (Fig. 11.2), (2) an acyl-CoA independent pathway involving phosphatidylcholine (PC) as an acyl donor, and (3) a specific route of synthesis of TAGs in the chloroplast.

11.4.2 Biogenesis of the Lipid Droplet

11.4.2.1 Biogenesis of PHA Granules in Prokaryotes

There are currently two models for the in vivo formation of PHA granules in prokaryotes: (1) a model based on the formation of micelles of PHA synthases (PhaC) and (2) a model based on the budding of PHA granules from the cytoplasmic membrane (Pötter and Steinbüchel 2006). The model based on the formation of micelles of PHB synthases is currently the most accepted (Haywood et al. 1989). We do not know if any of these processes has been transferred to a eukaryotic system.

11.4.2.2 Biogenesis of Cytosolic TAG Droplets in Eukaryotes

The general principles of the biogenesis mechanism seem to be similar in non-photosynthetic and photosynthetic organisms (Fig. 11.2—TAG synthesis). The first step consists of the synthesis of TAG at the level of the ER, through the Kennedy pathway (see Sect. 11.3.1.2). Following the synthesis of TAGs, a "lens" of neutral lipids (TAGs and sterol esters) develops within the bilayer of the ER at particular nucleation sites (Fig. 11.2—Nucleation) (Pol et al. 2014; Wilfling et al. 2014). A number of pre-droplets are formed in the ER from freely diffusing TAG molecules, but only a small proportion of them will ultimately form a lipid droplet. The lifespan of these lipid pre-droplets is estimated at a few milliseconds (Khandelia et al. 2010). This phenomenon leads to an energy optimum called a nucleation barrier (Thiam and Forêt 2016). The homogeneity of the phospholipid bilayer of the reticulum is also an important parameter to take into account the success of nucleation. Indeed, on a uniform bilayer, the nucleation of LDs could occur randomly. However, the phospholipid bilayer of the ER is heterogeneous in its biochemical composition. There are therefore specific sites where the nucleation energy is lower. Nucleation can be favored by four different phenomena: (1) the curvature of the membrane developing hydrophobic defects favorable to the accumulation of TAG, (2) the synthesis sites of TAGs defined by the presence of DGATs, or other synthetic enzymes such as PDATs, (3) the presence of proteins (SEIPIN, FIT, PLINs, ACSL) inducing curvature in decreasing the bending energy of the membrane, and (4) the presence of proteins and/or lipids interacting with TAGs (Thiam and Forêt 2016).

Following the formation of this lens, budding (Fig. 11.2—Budding) of the LD occurs on the cytoplasmic side of the ER (Walther et al. 2017). This polarized budding of LDs follows an unknown mechanism. In humans, it has been proposed that an ER protein, FIT2 (Fat-inducing transcript 2), was involved in the polarization of budding (Choudhary et al. 2016). There are currently several models concerning the growth of the LD, including the coalescence and Ostwald ripening (Thiam et al. 2013b). Coalescence is a physical mechanism corresponding to the fusion of two identical substances (in our case, the fusion of two LDs). When an emulsion of oil and water is vigorously mixed, the fusion of LDs with each other is observed to form a giant lipid droplet reflecting the separation between the oil and the water. Ostwald ripening is a destabilization mechanism corresponding to the gradual disappearance of small LDs from an emulsion to give way to larger LDs (Thiam et al. 2013b). This mechanism begins with the transfer of TAG molecules from small LDs to larger ones. The direction of the transfer is dictated by the pressure difference of Laplace (Thiam et al. 2013b; Thiam and Forêt 2016).

An increase in the contact angle between the LD and the ER bilayer causes fission of the LD (Fig. 11.2—Fission). This mechanism also involves proteins on the surface of the reticulum. An interesting candidate is the SEIPIN protein located at the level of the ER with two transmembrane domains and a luminal loop (Lundin et al. 2006). SEIPIN is involved functionally in the initial stages of LD formation into generating nascent structures or participating in their maturation in mature LD (Wang et al. 2016). A morphological screen of LDs of *Saccharomyces cerevisiae* shows that the homolog Fld1 of SEIPIN is necessary for the formation of "normal" droplets (Fei et al. 2008). This protein also participates in the stabilization of the contact sites between the ER and the LD (Salo et al. 2016). A recent study also proposes that the SEIPIN protein and the LDAF1 (Lipid Droplet Assembly Factor 1) protein form an oligomeric complex of approximately 600 kDa in the ER bilayer determining the LD formation sites (Chung et al. 2019). SEIPIN is also involved in the regulation of TAG synthesis: it interacts notably with GPAT, AGAT2, and lipins (phosphatidate phosphatases recruited for the synthesis of DAG, serving as the substrate to the production of TAG). A single Seipin gene is detected in animals and fungi studied so far, whereas three homologs (SEIPIN1, 2, and 3) were reported in higher plant models (Taurino et al. 2017). SEIPINS are divided into two monophyletic groups in plants according to their degree of functional specialization (Cai et al. 2015). A functional genetic study shows that double and triple mutations cause the accumulation of larger LDs in *Arabidopsis* (Taurino et al. 2017). An ortholog of SEIPIN was also detected in the diatom *Phaeodactylum tricornutum* (Lu et al. 2017). The overexpression of SEIPIN in *P. tricornutum* causes an increase in the size of LDs, the amount of neutral lipids (57% increase), and the proportion of saturated fatty acids (16:0) (Lu et al. 2017).

LDs can detach from the ER, forming a population of initial LDs (iLD) measuring 400–800 nm in diameter (Walther et al. 2017). These LDs can be subsequently converted, according to an unknown mechanism, into a new population of LDs called expanding lipid droplets (eLDs) (Walther et al. 2017). This conversion requires the acquisition of an independent TAG synthesis machinery carried out by GPAT4 and DGAT2 (Kuerschner et al. 2007; Stone et al. 2009; Wilfling et al. 2013). These two proteins initially present in the ER are relocated to the LD (Wilfling et al. 2013).

Although the conversion mechanism is unknown, there is still evidence that the vesicular machinery ARF1/COP1 (ADP Ribosylation Factor 1/Coat Proteins type I) is involved in this process. In particular, it has been shown in an in vitro system that the ARF1/COP1 machinery allows the synthesis of nano LDs of 60–80 nm (Thiam et al. 2013a). COP1 is a complex composed of seven subunits (α, β, β', γ, ε, δ, ζ) responsible for the retrograde vesicular transport of the Golgi apparatus to the ER as well as intra-Golgi transport (Beck et al. 2009; Jackson 2014). ARF1 is a member of the family of GTP-binding proteins of the Ras superfamily (small GTPases). ARF1 is localized in the Golgi apparatus and was initially shown to act in intra-Golgi transport. The role of the ARF1/COP1 system in LD dynamics needs therefore to be clarified further compared to that in Golgi dynamics.

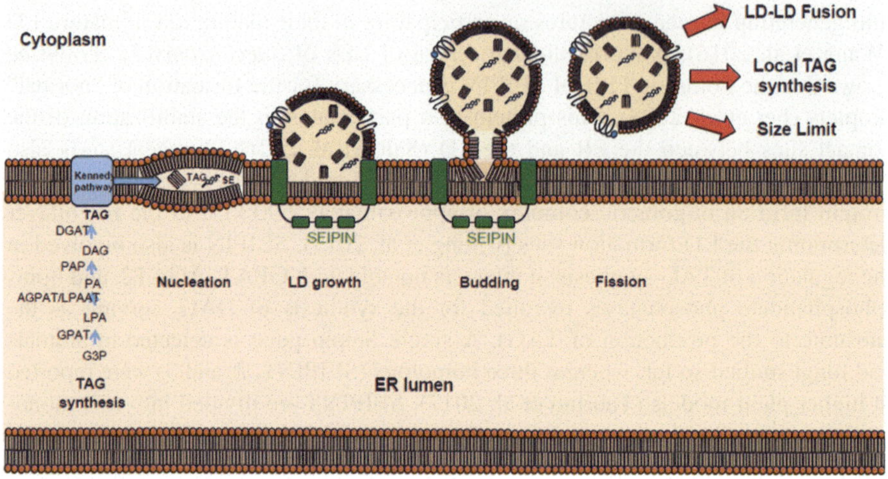

Fig. 11.2 Principles of lipid droplet biogenesis in eukaryotes

11.4.3 Catabolism of Lipid Droplets

The catabolism of LDs corresponds to all the degradation reactions of the LD architecture components, e.g. the lipids and hydrophobic components in the core of the LDs and the proteins especially *via* the ERAD pathway and the 26S proteasome.

11.4.3.1 Degradation of PHA in Prokaryotes

The degradation of PHA granules is carried out by a PHA depolymerase (PhaZ) (Uchino et al. 2008). PhaZ plays a role similar to acyl lipases in eukaryotes (see below). At least two degradation pathways of PHA granules exist: (1) an extracellular pathway in bacteria capable of secreting depolymerases in the environment and (2) an intracellular route, less well known to date. Seven intracellular PHB depolymerases (PhaZ1 to PhaZ7) as well as two PHB hydrolases (PhaY1, PhaY2) have been described in the *R. eutropha* H16 model (Abe et al. 2005; Uchino et al. 2008; Sznajder and Jendrossek 2014). However, PhaZ1 appears to be the only active PHB depolymerase in *R. eutropha* H16 (Sznajder and Jendrossek 2014). PhaZ1 presents a cysteine in the catalytic site of the "box" lipase. The thiolysis of PHB granules into (R)-3-hydroxybutyryl-CoA by PhaZ1 has been demonstrated (Eggers and Steinbüchel 2013). Localization of PhaZ1 has been confirmed on the surface of PHB granules by fusion with eYFP (Uchino et al. 2008). A mechanistic model of PHB granule degradation was proposed by Eggers and Steinbüchel (2013).

11.4.3.2 Degradation of TAG in Eukaryotes

Fatty Acid Mobilization

Multiple processes of TAG hydrolysis have been described in different eukaryotic clades.

In mammals, lipolysis involves different protein actors. The CGI-58 protein interacts with the terminal end of PLIN1a (PLIN1 splice variant) on the surface of the LD in fat cells (Subramanian et al. 2004). The adipose triacylglycerol lipase (ATGL) is also localized at the level of the LD. Before the lipolysis process takes place, the enzymatic activity of ATGL is very low. The majority of ATGL proteins are located in the cytosol before lipolysis occurs. The ATGL activity is inhibited by the G0/G1 switch protein 2 protein (G0S2), attenuating lipolysis in a dose-dependent manner (Schweiger et al. 2012). Lipolysis can be initiated by β-adrenergic stimulation (physical exercise, nutritional deficiency). Adenylate cyclase is activated by a heterotrimeric G protein resulting in the accumulation of cAMP, subsequently activating protein kinase A (PKA) (Viswanadha and Londos 2008). PLIN1 is then phosphorylated by PKA (Granneman et al. 2009). The association of ATGL with LDs is considered to depend on the ARF1/COP1 vesicular machinery (Soni et al. 2009). PKA also phosphorylates CGI-58 thus promoting the binding of CGI-58 to ATGL (Sahu-Osen et al. 2015). This process activates ATGL leading to the hydrolysis of TAGs into DAG and FAs. A Hormone-Sensitive Lipase (HSL) can be activated by phosphorylation: the HSL then binds to the LD, at the level of phosphorylated PLIN1, allowing the conversion of DAG into monoacylglycerol (MAG) and FAs (D'andrea 2016). The HSL/PLIN1 interaction is enhanced by phosphorylation by PKA on several residues of PLIN1: Ser81, Ser222, and Ser276. A monoglyceride lipase (MGL) then completes the release of the last FA and the glycerol backbone. The three released FAs are then transferred to the mitochondria (D'andrea 2016).

LDs are strongly reorganized during lipolysis following a process that reduces their size and increases the surface/volume ratio and accessibility to lipases. Two models exist for the formation of these lipid microdroplets: (1) fragmentation of a giant LD (Moore et al. 2005) and (2) capture of released FAs, which are potentially toxic following lipolysis (Paar et al. 2012). A fat-specific-protein of 27 kDa (FSP27 also known as CIDEC for cell death-inducing DNA fragmentation factor alpha-like effector c) binds to ATGL at the surface of LDs inhibiting the lipolysis mechanism by blocking access of CGI-58 to ATGL. A reduction in FSP27 expression by siRNAs in a 3T3-L1 mouse line causes the formation of multiple small LDs as well as an increase in lipolytic activity (Nishino et al. 2008). FSP27 proteins are involved in lipid transfer mechanisms between LDs favoring mechanisms of fusion, growth, and enlargement (Gong et al. 2011; Jambunathan et al. 2011). FSP27 is notably in direct interaction with PLIN1a in white adipose tissues favoring the formation of LDs (Grahn et al. 2013; Sun et al. 2013). The size of LDs therefore seems regulated also by a balance between PLIN1a and FSP27.

In plants, lipolysis three actors have been described: (1) a Sugar-Dependent Lipase type 1 (SDP1), (2) a homolog of the mammalian protein CGI-58 and (3) a peroxisomal transporter PXA1. These three actors are located at the level of the peroxisome upstream of lipolysis. SDP1 and its SDP1-like homolog (SDP1-L) have been found in the seed of *Arabidopsis thaliana* (Kelly et al. 2011). This enzyme is able to hydrolyze TAG, DAG, and MAG. SDP1 is a papatin-like lipase (Rydel et al. 2003), homologous to mammalian ATGL, *S. cerevisiae* TAG lipases (TGL3, TGL4, TGL5) or *D. melanogaster* BRUMMER protein (Quettier and Eastmond 2009). A homolog of *Arabidopsis thaliana* SDP1 was also detected in the green alga *Lobosphaera incisa* (LiSDP1) (Siegler et al. 2017). A fusion with the fluorescent probe m-Venus showed the localization of this lipase at the surface of LDs of *L. incisa* (Siegler et al. 2017).

A loss of function of the CGI-58 protein in *Arabidopsis thaliana* results in an abnormal accumulation of LDs in the leaves (Ghosh et al. 2008; James et al. 2010). In addition, the seeds of plants deficient in CGI-58 contain equivalent amounts of neutral lipids, indicating that CGI-58 is not involved in the lipolysis mechanisms of the seed (James et al. 2010). Unlike mammals, the CGI-58 protein does not interact with ATGL lipase in plants but with an ATP-binding cassette (ABC) protein (PXA1) (Park et al. 2013). An ortholog of CGI-58 was found in the genome of the green microalga *Chlamydomonas reinhardtii* (Merchant et al. 2007). PXA1 (Peroxysomal ABC-Transporter 1) is localized at the level of the peroxisome and transports FAs upstream of their degradation by β-oxidation (Footitt et al. 2002). It has been estimated that 90% of FAs transported by PXA1 are provided by SDP1 in combination with the retromeric complex, a multiprotein complex involved in the recycling of transmembrane receptors and retrograde transport of cargo proteins from endosomes to *trans*-Golgi (Kelly and Feussner 2016). Physical contact between the LD and the peroxisome is negatively correlated with the presence of sucrose (Cui et al. 2016). PXA1 is also involved in the transfer of lipophilic precursors for jasmonate signaling (Theodoulou et al. 2005) and CGI-58 in the regulation of polyamine metabolism (nitrogen metabolism), interacting with spermidine synthase 1 (SPDS1), in plants (Park et al. 2014).

Knowledge of the catabolism of LDs in secondary endosymbionts, containing secondary plastids, is very poor (Kong et al. 2018). A plant SDP1 lipase homolog, TGL1, was found in the diatom *Phaeodactylum* (Barka et al. 2016). A decrease in its expression causes the accumulation of TAGs. A second actor of catabolism of LDs was discovered in the diatom *Thalassiosira pseudonana*: CGI-58 (Trentacoste et al. 2013). A decrease in the expression of CGI-58 in *Phaeodactylum* (antisense line) causes the accumulation of the amount of TAG (Leterrier et al. 2015). A recent study, based on the search for conserved domains and phylogenetic analysis, identified a putative TAG lipase in *Phaeodactylum*, called OmTGL (Li et al. 2018). By fusion with a fluorescent protein eGFP, this lipase was shown to localize in the third outermost membrane of the chloroplast of *Phaeodactylum*. This localization is puzzling as this enzyme is supposed to operate on TAG originating from cytosolic LDs. It seems that the dynamics of LDs in eukaryotes deriving from secondary endosymbiosis have very complicated links with multiple membrane compartments, including the two outermost membranes of the 4-membrane secondary plastid, the mitochondrial outer envelope membrane, peroxisomes, autophagosomes, etc. (Lupette et al. 2019).

Fatty Acid Degradation Via β-Oxidation

The β-oxidation pathway breaks down FAs and produces acetyl-CoA and reducing power (NADH and $FADH_2$) (Houten and Wanders 2010). This pathway occurs in the mitochondrial matrix in *Mammalia* and/or in the peroxisome (or glyoxysome) in photosynthetic organisms. Short chain fatty acids (less than eight carbons) are degraded in the mitochondria, whereas medium and very long chain fatty acids (greater than eight carbons) are degraded in the peroxisome (Dellero et al. 2018a). Activation and transport of FAs are two preliminary steps required prior β-oxidation initiation. FA activation is catalyzed by specific long chain acyl-CoA ligases. To reach the mitochondria, an acyl-CoA is 'cargoed' via the carnitine shuttle (Houten and Wanders 2010) in two steps: the acyl-CoA is first converted into acyl-carnitine at the outer mitochondrial membrane by carnitine palmitoyltransferase 1 (CPT1). Acylcarnitine then enters the mitochondrial intermembrane space by passing through a porin and then reaches the internal mitochondrial space via a carnitine-acylcarnitine translocase. To reach the peroxisome, transport of acyl-CoA occurs via a specific class of ABC transporter (ATP Binding Cassette).

In both mitochondria and peroxisomes, β-oxidation occurs by an interactive process, via the so-called Lynen helix (Houten and Wanders 2010).

11.4.3.3 Protein Control and Degradation by the ERAD Pathway and the 26S Proteasome

An ER localized pathway also regulates the proteins of cytosolic LDs in eukaryotic cells, called the ER associated protein degradation (ERAD) pathway. During translation, and following binding with the signal recognition protein (SRP), polypeptides carrying a signal peptide are imported by a translocon (Sec61 for example) into ER (Rapoport 2007), where they fold and undergo maturation, co- and post-translational modifications (e.g. cleavage of a signal peptide, *N*-glycosylation, and formation of disulfide bridges). The ERAD machinery is not only a system controlling the quality of proteins targeted to the ER but also a system controlling their quantity (Olzmann et al. 2013a; Stevenson et al. 2016). The presence of mutation, transcription, and translation errors may lead to misfolding, preventing a protein from reaching its functional conformation. The ERAD machinery operates in three stages. First, misfolded or mutated proteins are recognized in the ER. The presence of mismatched cysteine residues, immature glycans, or exposed hydrophobic regions may be targeted by chaperone proteins, such as in the latter case, chaperone proteins of the luminal binding protein (BIP) or cytosolic (Heat Shock Protein 70—Hsp70) (Okuda-Shimizu and Hendershot 2007). Misfolded proteins are then translocated from the

ER to the cytosol. Translocation may involve a Valosin-containing-protein (VCP or p97) in mammals or CDC48 (Cell Cycle Cycle 48 protein) in *Saccharomyces cerevisiae* (Meyer et al. 2012). Finally, misfolded protein can be ubiquitinated and degraded by the 26S proteasome (composed of a 20S barrel-shaped catalytic core in the center and two 19S regulatory complexes at each end) (for review, see Sharma et al. 2016). Multiple LD proteins are controlled by ERAD machinery via a process that still needs to be fully elucidated (Ruggiano et al. 2016).

ERAD machinery is also involved in the regulation of mammalian TAGs (Stevenson et al. 2016). UBXD8 (UBX domain-containing protein 8) is an inhibitor of TAG synthesis when upstream FAs synthesis is very low. An increase in FA synthesis causes the delivery of a portion of UBDX8 from the ER to the LD, where it recruits VCP (Valosin-containing protein) to the surface of the droplet (Suzuki et al. 2012; Olzmann et al. 2013b). This mechanism echoes the recruitment of CDC48 (another name of VCP) by PUX10 at the LD of the pollen tube of *Nicotiana tabacum* (Kretzschmar et al. 2018). UBXD8 inhibits ATGL on the surface of the LD by stimulating the dissociation of ATGL from its cofactor CGI-58 (Olzmann et al. 2013b).

A second checkpoint occurs at the DGAT2 level. Treatment with PS-341, a proteasome inhibitor, reduced the expression of DGAT2 (Oliva et al. 2012). DGAT2 is a relatively unstable protein, degraded by the 26S proteasome after ubiquitination (Choi et al. 2014). Gp78 (Glycoprotein 78) is an E3 ligase involved in the ERAD mechanism (Chen et al. 2012). Functional analysis of this protein by siRNA showed reduction in the ubiquitination of DGAT2 and an increase in its stabilization (Choi et al. 2014). The transfer of the polyubiquitinated DGAT2 to the proteasome is carried out by VCP. The role of UBDX8 in the degradation of DGAT2 is currently unknown (Stevenson et al. 2016).

Overall, the ERAD machinery may be a conserved actor of LD homeostasis in eukaryotes, but our knowledge is even more scarce compared to other regulatory processes.

11.5 Conclusion and Future Prospects

This chapter shows that LDs and oil bodies are similar in their general organization in prokaryotes and eukaryotes. Hydrophobic molecules loaded in their core show a diversity of chemical structures (PHA, TAG, sterols, carotenoids, alkene, alkenone, etc.). The PHA granules seem to be restricted to prokaryotes, whereas TAG-containing LDs are found from bacteria to eukaryotes. The dynamics of LD formation and homeostasis are exquisitely controlled, indicating the importance of LDs in cell physiology and development. LDs are therefore essential for cell survival and development, which is probably one of the reasons for the strong level of conservation of these subcellular structures. Nevertheless, besides their evident function as energy storage and carbon reservoirs, the variety of roles harbored by LDs still needs to be investigated in major eukaryotic clades. It seems that in eukaryotes, the ER plays a conserved role as a platform for LD formation, whereas mitochondria and peroxisome are critical for LD lipid mobilization. The association of proteins such as histones in LDs analyzed from insects to diatoms suggests that the LDs may have unexpected functions. LD biogenesis, dynamics, and role are nevertheless more complex in the organisms deriving from secondary endosymbiosis. Only a few LD-associated proteins seem to be conserved in eukaryotes, and some being involved in LD biogenesis (SEIPIN), TAG biosynthesis, TAG hydrolysis coupled with FA export from LDs (CGI-58). The evolution of CGI-58 is particularly puzzling since this component is involved in TAG mobilization from LDs, whereas CGI-58 molecular activity has apparently diverged between non-photosynthetic and photosynthetic clades. In photosynthetic organisms, the plastid seems to cooperate with the ER in LD biogenesis, possibly in relation to FA synthesis occurring in the stroma of this organelle, whereas FA synthesis occurs in the cytosol of non-photosynthetic eukaryotes. In eukaryotes containing secondary plastids, the role of this organelle may be even more important in LD biogenesis. A conclusion of this chapter may be a frustrating lack of knowledge, especially in important branches of the evolution of eukaryotes. Characterizing the molecular evolution of LD proteins and LD-controlling systems throughout the Tree of Life is a clear challenge for the future. This effort will be necessary to help explore biodiversity and identify oleaginous species, which may be novel promising resources for a multitude of applications from food, health to green chemistry and bioenergy.

Acknowledgements The authors were supported by the French National Research Agency (ANR-10-LABEX-04 GRAL Labex, Grenoble Alliance for Integrated Structural Cell Biology; ANR-11-BTBR-0008 Océanomics; ANR-15-IDEX-02 GlycoAlps) and a Flagship program from the CEA High Commissioner.

References

Abe T, Kobayashi T, Saito T (2005) Properties of a novel intracellular poly(3-hydroxybutyrate) depolymerase with high specific activity (PhaZd) in *Wautersia eutropha* H16. J Bacteriol 187:6982–6990

Abell BM, Holbrook LA, Abenes M, Murphy DJ, Hills MJ, Moloney MM (1997) Role of the proline knot motif in oleosin endoplasmic reticulum topology and oil body targeting. Plant Cell 9(8):1481–1493. https://doi.org/10.1105/tpc.9.8.1481

Acevedo F, Rubilar M, Shene C, Navarrete P, Romero F, Rabert C, Jolivet P, Valot B, Chardot T (2012) Seed oil bodies from *Gevuina avellana* and *Madia sativa*. J Agric Food Chem 60:6994–7004

Agarwal AK et al (2002) AGPAT2 is mutated in congenital generalized lipodystrophy linked to chromosome 9q34. Nat Genet 31:21

Aizaki H, Lee K-J, Sung VMH, Ishiko H, Lai MMC (2004) Characterization of the hepatitis C virus RNA replication complex associated with lipid rafts. Virology 324:450–461

Alvarez HM (2016) Triacylglycerol and wax ester-accumulating machinery in prokaryotes. Biochimie 120:28–39

Alvarez H, Steinbüchel A (2002) Triacylglycerols in prokaryotic microorganisms. Appl Microbiol Biotechnol 60:367–376

Anderson AJ, Dawes EA (1990) Occurrence, metabolism, metabolic role, and industrial uses of bacterial polyhydroxyalkanoates. Microbiol Rev 54:450–472

Archibald JM, Keeling PJ (2002) Recycled plastids: a 'green movement' in eukaryotic evolution. Trends Genet 18:577–584

Arrese EL, Rivera L, Masakazu H, Mirza S, Hartson SD, Weintraub S, Soulages JL (2008) Function and structure of lipid storage droplet protein 1 studied in lipoprotein complexes. Arch Biochem Biophys 473:42–47

Athenstaedt K, Zweytick D, Jandrositz A, Kohlwein SD, Daum G (1999) Identification and characterization of major lipid particle proteins of the yeast *Saccharomyces cerevisiae*. J Bacteriol 181:6441–6448

Athenstaedt K, Jolivet P, Boulard C, Zivy M, Negroni L, Nicaud J-M, Chardot T (2006) Lipid particle composition of the yeast *Yarrowia lipolytica* depends on the carbon source. Proteomics 6:1450–1459

Barka F, Angstenberger M, Ahrendt T, Lorenzen W, Bode HB, Büchel C (2016) Identification of a triacylglycerol lipase in the diatom *Phaeodactylum tricornutum*. Biochim Biophys Acta 1861:239–248

Bartz R, Li W-H, Venables B, Zehmer JK, Roth MR, Welti R, Anderson RGW, Liu P, Chapman KD (2007) Lipidomics reveals that adiposomes store ether lipids and mediate phospholipid traffic. J Lipid Res 48:837–847

Baud S, Dichow NR, Kelemen Z, D'andréa S, To A, Berger N, Canonge M, Kronenberger J, Viterbo D, Dubreucq B, Lepiniec L, Chardot T, Miquel M (2009) Regulation of HSD1 in seeds of *Arabidopsis thaliana*. Plant Cell Physiol 50:1463–1478

Beck R, Ravet M, Wieland FT, Cassel D (2009) The COPI system: molecular mechanisms and function. FEBS Lett 583:2701–2709

Beilstein F et al (2013) Proteomic analysis of lipid droplets from Caco-2/TC7 enterocytes identifies novel modulators of lipid secretion. PLoS One 8(1):e53017

Beller M, Riedel D, Jänsch L, Dieterich G, Wehland J, Jäckle H, Kühnlein RP (2006) Characterization of the *Drosophila* lipid droplet subproteome. Mol Cell Proteomics 5:1082–1094

Beller M, Sztalryd C, Southall N, Bell M, Jäckle H, Auld DS, Oliver B (2008) COPI complex is a regulator of lipid homeostasis. PLoS Biol 6:e292

Beller M, Bulankina AV, Hsiao H-H, Urlaub H, Jäckle H, Kühnlein RP (2010) PERILIPIN-dependent control of lipid droplet structure and fat storage in drosophila. Cell Metab 12:521–532

Bendif EM, Probert I, Schroeder DC, De Vargas C (2013) On the description of *Tisochrysis lutea* gen. nov. sp. nov. and Isochrysis nuda sp. nov. in the Isochrysidales, and the transfer of Dicrateria to the Prymnesiales (Haptophyta). J Appl Phycol 25:1763–1776

Bersuker K, Olzmann JA (2017) Establishing the lipid droplet proteome: mechanisms of lipid droplet protein targeting and degradation. Biochim Biophys Acta 1862:1166–1177

Berthelot K, Lecomte S, Estevez Y, Zhendre V, Henry S, Thévenot J, Dufourc EJ, Alves ID, Peruch F (2014) Rubber particle proteins, HbREF and HbSRPP, show different interactions with model membranes. Biochim Biophys Acta Biomembr 1838:287–299

Bi J, Xiang Y, Chen H, Liu Z, Grönke S, Kühnlein RP, Huang X (2012) Opposite and redundant roles of the two *Drosophila* perilipins in lipid mobilization. J Cell Sci 125:3568

Bi K et al (2016) Integrated omics study of lipid droplets from Plasmodiophora brassicae. Sci Rep 6:36965–36965

Bickel PE, Tansey JT, Welte MA (2009) PAT proteins, an ancient family of lipid droplet proteins that regulate cellular lipid stores. Biochim Biophys Acta 1791:419–440

Bigogno C et al (2002) Biosynthesis of arachidonic acid in the oleaginous microalga Parietochloris incisa (Chlorophyceae): radiolabeling studies. Lipids 37(2):209–216

Bindesbøll C et al (2013) Fatty acids regulate perilipin5 in muscle by activating PPARδ. J Lipid Res 54(7):1949–1963

Blouin CM et al (2010) Lipid droplet analysis in caveolin-deficient adipocytes: alterations in surface phospholipid composition and maturation defects. J Lipid Res 51(5):945–956

Bouchoux J et al (2011) The proteome of cytosolic lipid droplets isolated from differentiated Caco-2/TC7 enterocytes reveals cell-specific characteristics. Biol Cell 103(Pt 11):499–517

Boyer A, Dumans A, Beaumont E, Etienne L, Roingeard P, Meunier J-C (2014) The association of hepatitis C virus glycoproteins with apolipoproteins E and B early in assembly is conserved in lipoviral particles. J Biol Chem 289:18904–18913

Brasaemle DL et al (2004) Proteomic analysis of proteins associated with lipid droplets of basal and lipolytically stimulated 3T3-L1 adipocytes. J Biol Chem 279(45):46835–46842

Brasaemle DL, Wolins NE (2016) Isolation of lipid droplets from cells by density gradient centrifugation. Curr Protoc Cell Biol 72:3.15.11–13.15.13

Bréhélin C, Kessler F, Van Wijk KJ (2007) Plastoglobules: versatile lipoprotein particles in plastids. Trends Plant Sci 12:260–266

Bresan S, Sznajder A, Hauf W, Forchhammer K, Pfeiffer D, Jendrossek D (2016) Polyhydroxyalkanoate (PHA) granules have no phospholipids. Sci Rep 6:26612

Brocard L, Immel F, Coulon D, Esnay N, Tuphile K, Pascal S, Claverol S, Fouillen L, Bessoule J-J, Bréhélin C (2017) Proteomic analysis of lipid droplets from Arabidopsis aging leaves brings new insight into their biogenesis and functions. Front Plant Sci 8:894

Bussell R, Eliezer D (2003) A structural and functional role for 11-mer repeats in α-synuclein and other exchangeable lipid binding proteins. J Mol Biol 329:763–778

Cai Y, Goodman JM, Pyc M, Mullen RT, Dyer JM, Chapman KD (2015) Arabidopsis SEIPIN proteins modulate triacylglycerol accumulation and influence lipid droplet proliferation. Plant Cell 27:2616–2636

Caire-Brändli I, Papadopoulos A, Malaga W, Marais D, Canaan S, Thilo L, De Chastellier C (2014) Reversible lipid accumulation and associated division arrest of mycobacterium avium in lipoprotein-induced foamy macrophages may resemble key events during latency and reactivation of tuberculosis. Infect Immun 82:476–490

Camus G, Herker E, Modi AA, Haas JT, Ramage HR, Farese RV, Ott M (2013) Diacylglycerol acyltransferase-1 localizes hepatitis C virus NS5A protein to lipid droplets and enhances NS5A interaction with the viral capsid core. J Biol Chem 288:9915–9923

Castorena KM, Stapleford KA, Miller DJ (2010) Complementary transcriptomic, lipidomic, and targeted functional genetic analyses in cultured Drosophila cells highlight the role of glycerophospholipid metabolism in Flock House virus RNA replication. BMC Genomics 11:183–183

Cavalier-Smith T (1986) The kingdom Chromista: origin and systematics. Biopress, Bristol

Cavalier-Smith T (2018) Kingdom Chromista and its eight phyla: a new synthesis emphasising periplastid protein targeting, cytoskeletal and periplastid evolution, and ancient divergences. Protoplasma 255:297–357

Cermelli S, Guo Y, Gross SP, Welte MA (2006) The lipid-droplet proteome reveals that droplets are a protein-storage depot. Curr Biol 16:1783–1795

Chanarin I et al (1975) Neutral-lipid storage disease: a new disorder of lipid metabolism. Br Med J 1 (5957):553–555

Chang BH-J, Li L, Paul A, Taniguchi S, Nannegari V, Heird WC, Chan L (2006) Protection against fatty liver but normal adipogenesis in mice lacking adipose differentiation-related protein. Mol Cell Biol 26:1063–1076

Chapman KD, Dyer JM, Mullen RT (2012) Biogenesis and functions of lipid droplets in plants: thematic review series: lipid droplet synthesis and metabolism: from yeast to man. J Lipid Res 53:215–226

Chawla A et al (2003) PPARδ is a very low-density lipoprotein sensor in macrophages. Proc Natl Acad Sci U S A 100(3):1268–1273

Chen JC, Tsai CC, Tzen JT (1999) Cloning and secondary structure analysis of caleosin, a unique calcium-binding protein in oil bodies of plant seeds. Plant Cell Physiol 40:1079–1086

Chen Z, Du S, Fang S (2012) gp78: a multifaceted ubiquitin ligase that integrates a unique protein degradation pathway from the endoplasmic reticulum. Curr Protein Pept Sci 13:414–424

Chen W, Chang B, Wu X, Li L, Sleeman M, Chan L (2013) Inactivation of Plin4 downregulates Plin5 and reduces cardiac lipid accumulation in mice. Am J Physiol-Endocrinol Metab 304: E770–E779

Chen Y, Ding Y, Yang L, Yu J, Liu G, Wang X, Zhang S, Yu D, Song L, Zhang H, Zhang C, Huo L, Huo C, Wang Y, Du Y, Zhang H, Zhang P, Na H, Xu S, Zhu Y, Xie Z, He T, Zhang Y, Wang G, Fan Z, Yang F, Liu H, Wang X, Zhang X, Zhang MQ, Li Y, Steinbüchel A, Fujimoto T, Cichello S, Yu J, Liu P (2014) Integrated omics study delineates the dynamics of lipid droplets in *Rhodococcus opacus* PD630. Nucleic Acids Res 42:1052–1064

Cho SY et al (2007) Identification of mouse Prp19p as a lipid droplet-associated protein and its possible involvement in the biogenesis of lipid droplets. J Biol Chem 282(4):2456–2465

Choi K, Kim H, Kang H, Lee S-Y, Lee SJ, Back SH, Lee SH, Kim MS, Lee JE, Park JY, Kim J, Kim S, Song J-H, Choi Y, Lee S, Lee H-J, Kim JH, Cho S (2014) Regulation of diacylglycerol acyltransferase 2 protein stability by gp78-associated endoplasmic-reticulum-associated degradation. FEBS J 281:3048–3060

Choudhary V, Golden A, Prinz WA (2016) Keeping FIT, storing fat: lipid droplet biogenesis. Worm 5:e1170276

Chughtai AA, Kaššák F, Kostrouchová M, Novotný JP, Krause MW, Saudek V, Kostrouch Z, Kostrouchová M (2015) Perilipin-related protein regulates lipid metabolism in *C. elegans*. PeerJ 3:e1213

Chung J, Wu X, Lambert TJ, Lai ZW, Walther TC, Farese RV (2019) LDAF1 and seipin form a lipid droplet assembly complex. Dev Cell 51:551–563.e557

Crunk AE et al (2013) Dynamic regulation of hepatic lipid droplet properties by diet. PLoS One 8 (7):e67631

Cui S, Hayashi Y, Otomo M, Mano S, Oikawa K, Hayashi M, Nishimura M (2016) Sucrose production mediated by lipid metabolism suppresses physical interaction of peroxisomes and oil bodies during germination of *Arabidopsis thaliana*. J Biol Chem 291:19734–19745

Czabany T, Wagner A, Zweytick D, Lohner K, Leitner E, Ingolic E, Daum G (2008) Structural and biochemical properties of lipid particles from the yeast *Saccharomyces cerevisiae*. J Biol Chem 283:17065–17074

Dahlhoff M et al (2015) Characterization of the sebocyte lipid droplet proteome reveals novel potential regulators of sebaceous lipogenesis. Exp Cell Res 332(1):146–155

Dalen KT et al (2007) LSDP5 is a PAT protein specifically expressed in fatty acid oxidizing tissues. Biochimica et Biophysica Acta (BBA) - Mol Cell Biol Lipids 1771(2):210–227

D'avila H, Freire-De-Lima CG, Roque NR, Teixeira L, Barja-Fidalgo C, Silva AR, Melo RCN, Dosreis GA, Castro-Faria-Neto HC, Bozza PT (2011) Host cell lipid bodies triggered by *Trypanosoma cruzi* infection and enhanced by the uptake of apoptotic cells are associated with prostaglandin E2 generation and increased parasite growth. J Infect Dis 204:951–961

D'andrea S (2016) Lipid droplet mobilization: the different ways to loosen the purse strings. Biochimie 120:17–27

Davidi L, Katz A, Pick U (2012) Characterization of major lipid droplet proteins from Dunaliella. Planta 236:19–33

Davidi L, Levin Y, Ben-Dor S, Pick U (2015) Proteome analysis of cytoplasmatic and plastidic β-carotene lipid droplets in *Dunaliella bardawil*. Plant Physiol 167:60–79

Dellero Y, Cagnac O, Rose S, Seddiki K, Cussac M, Morabito C, Lupette J, Aiese Cigliano R, Sanseverino W, Kuntz M, Jouhet J, Maréchal E, Rébeillé F, Amato A (2018a) Proposal of a new thraustochytrid genus Hondaea gen. nov. and comparison of its lipid dynamics with the closely related pseudo-cryptic genus Aurantiochytrium. Algal Res 35:125–141

Dellero Y, Rose S, Metton C, Morabito C, Lupette J, Jouhet J, Maréchal E, Rébeillé F, Amato A (2018b) Ecophysiology and lipid dynamics of a eukaryotic mangrove decomposer. Environ Microbiol 20:3057–3068

Den Brok MH, Raaijmakers TK, Collado-Camps E, Adema GJ (2018) Lipid droplets as immune modulators in myeloid cells. Trends Immunol 39:380–392

Dennis D et al (1998) Formation of poly(3-hydroxybutyrate-co-3-hydroxyhexanoate) by PHA synthase from Ralstonia eutropha. J Biotechnol 64(2):177–186

Dias-Lopes G, Borges-Veloso A, Saboia-Vahia L, Padrón G, De Faria Castro CL, Guimarães ACR, Britto C, Cuervo P, De Jesus JB (2016) Proteomics reveals major components of oogenesis in the reproductive tract of sugar-fed *Anopheles aquasalis*. Parasitol Res 115:1977–1989

Ding Y, Wu Y, Zeng R, Liao K (2012a) Proteomic profiling of lipid droplet-associated proteins in primary adipocytes of normal and obese mouse. Acta Biochim Biophys Sin 44:394–406

Ding Y, Yang L, Zhang S, Wang Y, Du Y, Pu J, Peng G, Chen Y, Zhang H, Yu J, Hang H, Wu P, Yang F, Yang H, Steinbüchel A, Liu P (2012b) Identification of the major functional proteins of prokaryotic lipid droplets. J Lipid Res 53:399–411

Dorfman ML et al (1974) Ichthyosiform dermatosis with systemic lipidosis. Arch Dermatol 110 (2):261–266

Du X, Barisch C, Paschke P, Herrfurth C, Bertinetti O, Pawolleck N, Otto H, Rühling H, Feussner I, Herberg FW, Maniak M (2013) Dictyostelium lipid droplets host novel proteins. Eukaryot Cell 12:1517–1529

Edvardsson U et al (2006) PPARα activation increases triglyceride mass and adipose differentiation-related protein in hepatocytes. J Lipid Res 47(2):329–340

Eggers J, Steinbüchel A (2013) Poly(3-hydroxybutyrate) degradation in *Ralstonia eutropha* H16 is mediated stereoselectively to (S)-3-hydroxybutyryl coenzyme a (CoA) via crotonyl-CoA. J Bacteriol 195:3213–3223

Eichmann TO et al (2015) ATGL and CGI-58 are lipid droplet proteins of the hepatic stellate cell line HSC-T6. J Lipid Res 56(10):1972–1984

Faucher P, Poitou C (2016) Physiopathologie de l'obésité. Rev Rhum Monogr 83:6–12

Fei W, Shui G, Gaeta B, Du X, Kuerschner L, Li P, Brown AJ, Wenk MR, Parton RG, Yang H (2008) Fld1p, a functional homologue of human seipin, regulates the size of lipid droplets in yeast. J Cell Biol 180:473–482

Footitt S, Slocombe SP, Larner V, Kurup S, Wu Y, Larson T, Graham I, Baker A, Holdsworth M (2002) Control of germination and lipid mobilization by COMATOSE, the Arabidopsis homologue of human ALDP. EMBO J 21:2912–2922

Frandsen G, Müller-Uri F, Nielsen M, Mundy J, Skriver K (1996) Novel plant Ca-binding protein expressed in response to abscisic acid and osmotic stress. J Biol Chem 271:343–348

Fujimoto Y et al (2004) Identification of major proteins in the lipid droplet-enriched fraction isolated from the human hepatocyte cell line HuH7. Biochimica et Biophysica Acta (BBA) - Mol Cell Res 1644(1):47–59

Füssy Z, Oborník M (2018) Complex Endosymbioses I: from primary to complex plastids, multiple independent events. In: Maréchal E (ed) Plastids: methods and protocols. Springer US, New York, NY, pp 17–35

Gaunt ER, Zhang Q, Cheung W, Wakelam MJO, Lever AML, Desselberger U (2013) Lipidome analysis of rotavirus-infected cells confirms the close interaction of lipid droplets with viroplasms. J Gen Virol 94:1576–1586

Georgieva D, Schmitt V, Leal-Calderon F, Langevin D (2009) On the possible role of surface elasticity in emulsion stability. Langmuir 25:5565–5573

Gerngross TU, Reilly P, Stubbe J, Sinskey AJ, Peoples OP (1993) Immunocytochemical analysis of poly-beta-hydroxybutyrate (PHB) synthase in Alcaligenes eutrophus H16: localization of the synthase enzyme at the surface of PHB granules. J Bacteriol 175:5289–5293

Ghosh AK, Ramakrishnan G, Chandramohan C, Rajasekharan R (2008) CGI-58, the causative gene for Chanarin-Dorfman syndrome, mediates acylation of lysophosphatidic acid. J Biol Chem 283:24525–24533

Gidda SK, Park S, Pyc M, Yurchenko O, Cai Y, Wu P, Andrews DW, Chapman KD, Dyer JM, Mullen RT (2016) Lipid droplet-associated proteins (LDAPs) are required for the dynamic regulation of neutral lipid compartmentation in plant cells. Plant Physiol 170:2052–2071

Gocze PM, Freeman DA (1994) Factors underlying the variability of lipid droplet fluorescence in MA-10 leydig tumor cells. Cytometry 17:151–158

Gong J, Sun Z, Wu L, Xu W, Schieber N, Xu D, Shui G, Yang H, Parton RG, Li P (2011) Fsp27 promotes lipid droplet growth by lipid exchange and transfer at lipid droplet contact sites. J Cell Biol 195:953–963
Goold H, Beisson F, Peltier G, Li-Beisson Y (2015) Microalgal lipid droplets: composition, diversity, biogenesis and functions. Plant Cell Rep 34:545–555
Grahn THM, Zhang Y, Lee M-J, Sommer AG, Mostoslavsky G, Fried SK, Greenberg AS, Puri V (2013) FSP27 and PLIN1 interaction promotes the formation of large lipid droplets in human adipocytes. Biochem Biophys Res Commun 432:296–301
Granneman JG, Moore H-PH, Krishnamoorthy R, Rathod M (2009) Perilipin controls lipolysis by regulating the interactions of AB-hydrolase containing 5 (Abhd5) and adipose triglyceride lipase (Atgl). J Biol Chem 284:34538–34544
Granneman JG, Kimler VA, Zhang H, Ye X, Luo X, Postlethwait JH, Thummel R (2017) Lipid droplet biology and evolution illuminated by the characterization of a novel perilipin in teleost fish. elife 6:e21771
Greenspan P, Mayer EP, Fowler SD (1985) Nile red: a selective fluorescent stain for intracellular lipid droplets. J Cell Biol 100:965–973
Greenberg AS et al (1991) Perilipin, a major hormonally regulated adipocyte-specific phosphoprotein associated with the periphery of lipid storage droplets. J Biol Chem 266(17):11341–11346
Grillitsch K, Connerth M, Köfeler H, Arrey TN, Rietschel B, Wagner B, Karas M, Daum G (2011) Lipid particles/droplets of the yeast *Saccharomyces cerevisiae* revisited: lipidome meets proteome. Biochim Biophys Acta 1811:1165–1176
Gromova M, Guillermo A, Bayle P-A, Bardet M (2015) In vivo measurement of the size of oil bodies in plant seeds using a simple and robust pulsed field gradient NMR method. Eur Biophys J 44:121–129
Guo Y, Walther TC, Rao M, Stuurman N, Goshima G, Terayama K, Wong JS, Vale RD, Walter P, Farese RV (2008) Functional genomic screen reveals genes involved in lipid-droplet formation and utilization. Nature 453:657–661
Guo B, Lei C, Ito T, Yaxiaer Y, Kobayashi H, Jiang Y, Tanaka Y, Ozeki Y, Goda K (2017) High-throughput label-free screening of euglena gracilis with optofluidic time-stretch quantitative phase microscopy. SPIE BiOS 12
Gupta P, Kaur G (2005) Chanarin Dorfman syndrome neonatal diagnosis and 3-year follow-up. Indian Pediatr 42:1054–1055
Hagel JM, Yeung EC, Facchini PJ (2008) Got milk? The secret life of laticifers. Trends Plant Sci 13:631–639
Han J et al (2007) Molecular characterization of the phaEC(Hm) genes, required for biosynthesis of Poly(3-Hydroxybutyrate) in the extremely halophilic archaeon haloarcula marismortui. Appl Environ Microbiol 73(19):6058–6065
Hanano A, Burcklen M, Flenet M, Ivancich A, Louwagie M, Garin J, Blée E (2006) Plant seed peroxygenase is an original heme-oxygenase with an EF-hand calcium binding motif. J Biol Chem 281:33140–33151
Harchouni S, Field B, Menand B (2018) AC-202, a highly effective fluorophore for the visualization of lipid droplets in green algae and diatoms. Biotechnol Biofuels 11:120
Hayashi YK et al (2009) Human PTRF mutations cause secondary deficiency of caveolins resulting in muscular dystrophy with generalized lipodystrophy. J Clin Invest 119(9):2623–2633
Haywood GW, Anderson AJ, Dawes EA (1989) The importance of PHB-synthase substrate specificity in polyhydroxyalkanoate synthesis by Alcaligenes eutrophus. FEMS Microbiol Lett 57:1–6
Hehlert P, Hofferek V, Heier C, Eichmann TO, Riedel D, Rosenberg J, Takaćs A, Nagy HM, Oberer M, Zimmermann R, Kühnlein RP (2019) The α/β-hydrolase domain-containing 4- and 5-related phospholipase Pummelig controls energy storage in Drosophila. J Lipid Res 60:1365–1378
Henne WM, Reese ML, Goodman JM (2018) The assembly of lipid droplets and their roles in challenged cells. EMBO J 37

Hickenbottom SJ et al (2004) Structure of a lipid droplet protein: the PAT family member TIP47. Structure 12(7):1199–1207
Horn PJ, James CN, Gidda SK, Kilaru A, Dyer JM, Mullen RT, Ohlrogge JB, Chapman KD (2013) Identification of a new class of lipid droplet-associated proteins in plants. Plant Physiol 162:1926–1936
Houten SM, Wanders RJA (2010) A general introduction to the biochemistry of mitochondrial fatty acid β-oxidation. J Inherit Metab Dis 33:469–477
Hsieh K, Huang AHC (2004) Endoplasmic reticulum, oleosins, and oils in seeds and tapetum cells. Plant Physiol 136:3427–3434
Hu X, Binns D, Reese ML (2017) The coccidian parasites Toxoplasma and Neospora dysregulate mammalian lipid droplet biogenesis. J Biol Chem 292:11009–11020
Huang AH (1996) Oleosins and oil bodies in seeds and other organs. Plant Physiol 110:1055–1061
Huang AHC (2018) Plant lipid droplets and their associated proteins: potential for rapid advances ([OPEN]). Plant Physiol 176:1894–1918
Huang M-D, Huang AHC (2015) Bioinformatics reveal five lineages of oleosins and the mechanism of lineage evolution related to structure/function from green algae to seed plants. Plant Physiol 169:453–470
Hubert F, Poisson L, Loiseau C, Gauvry L, Pencréac'h G, Hérault J, Ergan F (2017) Lipids and lipolytic enzymes of the microalga *Isochrysis galbana*. OCL 24:D407
Ikari N, Shimizu A, Asano T (2018) Lysosomal acid lipase deficiency in Japan: a case report of siblings and a literature review of cases in Japan. J Nippon Med Sch 85:131–137
Imai Y, Varela GM, Jackson MB, Graham MJ, Crooke RM, Ahima RS (2007) Reduction of hepatosteatosis and lipid levels by an adipose differentiation-related protein antisense oligonucleotide. Gastroenterology 132:1947–1954
Imamura M, Inoguchi T, Ikuyama S, Taniguchi S, Kobayashi K, Nakashima N, Nawata H (2002) ADRP stimulates lipid accumulation and lipid droplet formation in murine fibroblasts. Am J Physiol-Endocrinol Metab 283:E775–E783
Ivashov VA, Grillitsch K, Koefeler H, Leitner E, Baeumlisberger D, Karas M, Daum G (2013) Lipidome and proteome of lipid droplets from the methylotrophic yeast *Pichia pastoris*. Biochim Biophys Acta 1831:282–290
Jackson LP (2014) Structure and mechanism of COPI vesicle biogenesis. Curr Opin Cell Biol 29:67–73
Jambunathan S, Yin J, Khan W, Tamori Y, Puri V (2011) FSP27 promotes lipid droplet clustering and then fusion to regulate triglyceride accumulation. PLoS One 6:e28614
James CN, Horn PJ, Case CR, Gidda SK, Zhang D, Mullen RT, Dyer JM, Anderson RGW, Chapman KD (2010) Disruption of the Arabidopsis CGI-58 homologue produces Chanarin–Dorfman-like lipid droplet accumulation in plants. Proc Natl Acad Sci USA 107:17833–17838
James GO, Hocart CH, Hillier W, Chen H, Kordbacheh F, Price GD, Djordjevic MA (2011) Fatty acid profiling of *Chlamydomonas reinhardtii* under nitrogen deprivation. Bioresour Technol 102:3343–3351
Javee A, Sulochana SB, Pallissery SJ, Arumugam M (2016) Major lipid body protein: a conserved structural component of lipid body accumulated during abiotic stress in *S. quadricauda* CASA-CC202. Front Energy Res 4
Jiang P-L, Pasaribu B, Chen C-S (2014) Nitrogen-deprivation elevates lipid levels in Symbiodinium spp. by lipid droplet accumulation: morphological and compositional analyses. PLoS One 9:e87416
Johnson MR, Stephenson RA, Ghaemmaghami S, Welte MA (2018) Developmentally regulated H2Av buffering via dynamic sequestration to lipid droplets in Drosophila embryos. elife 7: e36021
Jolivet P, Roux E, D'andrea S, Davanture M, Negroni L, Zivy M, Chardot T (2004) Protein composition of oil bodies in *Arabidopsis thaliana* ecotype WS. Plant Physiol Biochem 42:501–509

Jolivet P, Boulard C, Bellamy A, Larré C, Barre M, Rogniaux H, D'andréa S, Chardot T, Nesi N (2009) Protein composition of oil bodies from mature *Brassica napus* seeds. Proteomics 9:3268–3284

Jolivet P, Acevedo F, Boulard C, D'andréa S, Faure J-D, Kohli A, Nesi N, Valot B, Chardot T (2013) Crop seed oil bodies: from challenges in protein identification to an emerging picture of the oil body proteome. Proteomics 13:1836–1849

Jordans GHW (1953) The familial occurrence of fat containing vacuoles in the Leukocytes diagnosed in two brothers suffering from Dystrophia Musculorum Progressiva (ERB). Acta Med Scand 145(6):419–423

Kalscheuer R, Wältermann M, Alvarez H, Steinbüchel A (2001) Preparative isolation of lipid inclusions from *Rhodococcus opacus* and *Rhodococcus ruber* and identification of granule-associated proteins. Arch Microbiol 177:20–28

Kalscheuer R, Stöveken T, Malkus U, Reichelt R, Golyshin PN, Sabirova JS, Ferrer M, Timmis KN, Steinbüchel A (2007) Analysis of storage lipid accumulation in *Alcanivorax borkumensis*: evidence for alternative triacylglycerol biosynthesis routes in bacteria. J Bacteriol 189:918–928

Kanshin E et al (2009) The stoichiometry of protein phosphorylation in adipocyte lipid droplets: analysis by N-terminal isotope tagging and enzymatic dephosphorylation. Proteomics 9 (22):5067–5077

Katavic V, Agrawal GK, Hajduch M, Harris SL, Thelen JJ (2006) Protein and lipid composition analysis of oil bodies from two Brassica napus cultivars. Proteomics 6:4586–4598

Kellogg RB, Patton JS (1983) Lipid droplets, medium of energy exchange in the symbiotic anemone *Condylactis gigantea*: a model coral polyp. Mar Biol 75:137–149

Kelly AA, Feussner I (2016) Oil is on the agenda: lipid turnover in higher plants. Biochim Biophys Acta 1861:1253–1268

Kelly AA, Quettier A-L, Shaw E, Eastmond PJ (2011) Seed storage oil mobilization is important but not essential for germination or seedling establishment in Arabidopsis. Plant Physiol 157:866–875

Khandelia H, Duelund L, Pakkanen KI, Ipsen JH (2010) Triglyceride blisters in lipid bilayers: implications for lipid droplet biogenesis and the mobile lipid signal in cancer cell membranes. PLoS One 5:e12811

Khor VK et al (2014) The proteome of cholesteryl-ester-enriched versus triacylglycerolenriched lipid droplets. PLoS One 9(8):e105047–e105047

Kilaru A, Cao X, Dabbs PB, Sung H-J, Rahman MM, Thrower N, Zynda G, Podicheti R, Ibarra-Laclette E, Herrera-Estrella L, Mockaitis K, Ohlrogge JB (2015) Oil biosynthesis in a basal angiosperm: transcriptome analysis of *Persea americana* mesocarp. BMC Plant Biol 15:203–203

Kim CA et al (2008) Association of a homozygous nonsense caveolin-1 mutation with berardinelli-seip congenital lipodystrophy. J Clin Endocrinol Metabol 93(4):1129–1134

Kim M-J, Wainwright HC, Locketz M, Bekker L-G, Walther GB, Dittrich C, Visser A, Wang W, Hsu F-F, Wiehart U, Tsenova L, Kaplan G, Russell DG (2010) Caseation of human tuberculosis granulomas correlates with elevated host lipid metabolism. EMBO Mol Med 2:258–274

Kimmel AR, Sztalryd C (2014) Perilipin 5, a lipid droplet protein adapted to mitochondrial energy utilization. Curr Opin Lipidol 25:110–117

Kimmel AR, Sztalryd C (2016) The perilipins: major cytosolic lipid droplet–associated proteins and their roles in cellular lipid storage, mobilization, and systemic homeostasis. Annu Rev Nutr 36:471–509

Kimmel AR, Brasaemle DL, Mcandrews-Hill M, Sztalryd C, Londos C (2010) Adoption of PERILIPIN as a unifying nomenclature for the mammalian PAT-family of intracellular lipid storage droplet proteins. J Lipid Res 51:468–471

Koch B, Schmidt C, Daum G (2014) Storage lipids of yeasts: a survey of nonpolar lipid metabolism in *Saccharomyces cerevisiae*, *Pichia pastoris*, and *Yarrowia lipolytica*. FEMS Microbiol Rev 38:892–915

Kong F, Romero IT, Warakanont J, Li-Beisson Y (2018) Lipid catabolism in microalgae. New Phytol 218:1340–1348

Kory N, Farese RV, Walther TC (2016) Targeting fat: mechanisms of protein localization to lipid droplets. Trends Cell Biol 26:535–546

Kramer DA et al (2018) Fasting and refeeding induces changes in the mouse hepatic lipid droplet proteome. J Proteome 181:213–224

Kreimer G (2009) The green algal eyespot apparatus: a primordial visual system and more? Curr Genet 55:19–43

Kretzschmar FK, Mengel LF, Müller A, Schmitt K, Blersch KF, Valerius O, Braus G, Ischebeck T (2018) PUX10 is a lipid droplet-localized scaffold protein that interacts with CDC48 and is involved in the degradation of lipid droplet proteins. Plant Cell 30:2137–2160

Kuerschner L, Moessinger C, Thiele C (2007) Imaging of lipid biosynthesis: how a neutral lipid enters lipid droplets. Traffic 9:338–352

Kumar Y, Cocchiaro J, Valdivia RH (2006) The obligate intracellular pathogen *Chlamydia trachomatis* targets host lipid droplets. Curr Biol 16:1646–1651

Kuntam S, Puskás LG, Ayaydin F (2015) Characterization of a new class of blue-fluorescent lipid droplet markers for live-cell imaging in plants. Plant Cell Rep 34:655–665

Laibach N, Post J, Twyman RM, Gronover CS, Prüfer D (2015) The characteristics and potential applications of structural lipid droplet proteins in plants. J Biotechnol 201:15–27

Langhi C et al (2014) Perilipin-5 is regulated by statins and controls triglyceride contents in the hepatocyte. J Hepatol 61(2):358–365

Larsson S et al (2012) Characterization of the lipid droplet proteome of a clonal insulin-producing β-cell line (INS-1 832/13). J Proteome Res 11(2):1264–1273

Le Cadre V, Debenay J-P (2006) Morphological and cytological responses of Ammonia (foraminifera) to copper contamination: implication for the use of foraminifera as bioindicators of pollution. Environ Pollut 143(2):304–317

Leber R, Zinser E, Paltauf F, Daum G, Zellnig G (1994) Characterization of lipid particles of the yeast, *Saccharomyces cerevisiae*. Yeast 10:1421–1428

Lecoeur H et al (2013) Reprogramming neutral lipid metabolism in mouse dendritic leucocytes hosting live Leishmania amazonensis amastigotes. PLoS Negl Trop Dis 7(6):e2276

Lee JH, Kong J, Jang JY, Han JS, Ji Y, Lee J, Kim JB (2014) Lipid droplet protein LID-1 mediates ATGL-1-dependent lipolysis during fasting in *Caenorhabditis elegans*. Mol Cell Biol 34:4165–4176

Lee YK, Sohn JH, Han JS, Park YJ, Jeon YG, Ji Y, Dalen KT, Sztalryd C, Kimmel AR, Kim JB (2018) *Perilipin 3* deficiency stimulates thermogenic beige adipocytes through *PPARα* activation. Diabetes 67:791–804

Lefèvre C et al (2001) Mutations in CGI-58, the gene encoding a new protein of the esterase/lipase/thioesterase subfamily, in chanarin-dorfman syndrome. Am J Hum Genet 69(5):1002–1012

LeKieffre C et al (2017) Surviving anoxia in marine sediments: the metabolic response of ubiquitous benthic foraminifera (Ammonia tepida). PLoS One 12(5):e0177604–e0177604

Lersten NR, Czlapinski AR, Curtis JD, Freckmann R, Horner HT (2006) Oil bodies in leaf mesophyll cells of angiosperms: overview and a selected survey. Am J Bot 93:1731–1739

Leterrier M, Calleja P, Maréchal E (2015) Modified algae strain and method of triacylglycerol accumulation using said strain. Patent. WO2015008160A2

Li Z, Thiel K, Thul PJ, Beller M, Kühnlein RP, Welte MA (2012) Lipid droplets control the maternal histone supply of Drosophila embryos. Curr Biol 22:2104–2113

Li X, Pan Y, Hu H (2018) Identification of the triacylglycerol lipase in the chloroplast envelope of the diatom *Phaeodactylum tricornutum*. Algal Res 33:440–447

Lin L-J, Liao P-C, Yang H-H, Tzen JTC (2005) Determination and analyses of the N-termini of oil-body proteins, steroleosin, caleosin and oleosin. Plant Physiol Biochem 43:770–776

Lin IP, Jiang P-L, Chen C-S, Tzen JTC (2012) A unique caleosin serving as the major integral protein in oil bodies isolated from Chlorella sp. cells cultured with limited nitrogen. Plant Physiol Biochem 61:80–87

Lin P, Chen X, Moktan H, Arrese EL, Duan L, Wang L, Soulages JL, Zhou DH (2014) Membrane attachment and structure models of lipid storage droplet protein 1. Biochim Biophys Acta 1838:874–881

Liu P et al (2004) Chinese hamster ovary K2 cell lipid droplets appear to be metabolic organelles involved in membrane traffic. J Biol Chem 279(5):3787–3792

Liu H, Wang C, Chen F, Shen S (2015) Proteomic analysis of oil bodies in mature *Jatropha curcas* seeds with different lipid content. J Proteome 113:403–414

Low KL, Shui G, Natter K, Yeo WK, Kohlwein SD, Dick T, Rao SPS, Wenk MR (2010) Lipid droplet-associated proteins are involved in the biosynthesis and hydrolysis of triacylglycerol in *Mycobacterium bovis* bacillus Calmette-Guérin. J Biol Chem 285:21662–21670

Lu Y, Wang X, Balamurugan S, Yang W-D, Liu J-S, Dong H-P, Li H-Y (2017) Identification of a putative seipin ortholog involved in lipid accumulation in marine microalga *Phaeodactylum tricornutum*. J Appl Phycol 29:2821–2829

Lundin C, Nordström R, Wagner K, Windpassinger C, Andersson H, Von Heijne G, Nilsson I (2006) Membrane topology of the human seipin protein. FEBS Lett 580:2281–2284

Luo M, Fadeev EA, Groves JT (2005) Mycobactin-mediated iron acquisition within macrophages. Nat Chem Biol 1:149

Lupette J, Maréchal E (2018) Phytoplankton glycerolipids: challenging but promising prospects from biomedicine to green chemistry and biofuels. Blue Biotechnology

Lupette J, Lami R, Krasovec M, Grimsley N, Moreau H, Piganeau G, Sanchez-Ferandin S (2016) Marinobacter dominates the bacterial community of the *Ostreococcus tauri* phycosphere in culture. Front Microbiol 7:1414

Lupette J, Jaussaud A, Seddiki K, Morabito C, Brugière S, Schaller H, Kuntz M, Putaux J-L, Jouneau P-H, Rébeillé F, Falconet D, Couté Y, Jouhet J, Tardif M, Salvaing J, Maréchal E (2019) The architecture of lipid droplets in the diatom *Phaeodactylum tricornutum*. Algal Res 38:101415

Madrigal-Matute J, Cuervo AM (2016) Regulation of liver metabolism by autophagy. Gastroenterology 150:328–339

Maeda Y, Sunaga Y, Yoshino T, Tanaka T (2014) Oleosome-associated protein of the oleaginous diatom *Fistulifera solaris* contains an endoplasmic reticulum-targeting signal sequence. Mar Drugs 12:3892–3903

Magré J, Delépine M, Khallouf E, Gedde-Dahl T Jr, Van Maldergem L, Sobel E, Papp J, Meier M, Mégarbané A, BWG, Lathrop M, Capeau J (2001) Identification of the gene altered in Berardinelli–Seip congenital lipodystrophy on chromosome 11q13. Nat Genet 28:365

Malchow D, Lüderitz O, Westphal O, Gerisch G, Riedeal V (1967) Polysaccharide in vegetativen und aggregationsreifen Amöben von *Dictyostelium discoideum*. Eur J Biochem 2:469–479

Mandard S, Müller M, Kersten S (2004) Peroxisome proliferator-activated receptor α target genes. Cell Mol Life Sci 61(4):393–416

Maréchal E (2018) Primary endosymbiosis: emergence of the primary chloroplast and the chromatophore, two independent events. In: Maréchal E (ed) Plastids: methods and protocols. Springer, New York, pp 3–16

Marlowe IT, Brassell SC, Eglinton G, Green JC (1984a) Long chain unsaturated ketones and esters in living algae and marine sediments. Org Geochem 6:135–141

Marlowe IT, Green JC, Neal AC, Brassell SC, Eglinton G, Course PA (1984b) Long chain (n-C37–C39) alkenones in the Prymnesiophyceae. Distribution of alkenones and other lipids and their taxonomic significance. Br Phycol J 19:203–216

Mayer SV, Tesh RB, Vasilakis N (2017) The emergence of arthropod-borne viral diseases: a global prospective on dengue, chikungunya and zika fevers. Acta Trop 166:155–163

Mcmanaman JL, Bales ES, Orlicky DJ, Jackman M, Maclean PS, Cain S, Crunk AE, Mansur A, Graham CE, Bowman TA, Greenberg AS (2013) Perilipin-2-null mice are protected against diet-induced obesity, adipose inflammation, and fatty liver disease. J Lipid Res 54:1346–1359

Melo RCN, Weller PF (2016) Lipid droplets in leukocytes: organelles linked to inflammatory responses. Exp Cell Res 340:193–197

Menon D, Singh K, Pinto SM, Nandy A, Jaisinghani N, Kutum R, Dash D, Prasad TSK, Gandotra S (2019) Quantitative lipid droplet proteomics reveals mycobacterium tuberculosis induced alterations in macrophage response to infection. ACS Infect Dis 5:559–569

Merchant SS, Prochnik SE, Vallon O, Harris EH, Karpowicz SJ, Witman GB, Terry A, Salamov A, Fritz-Laylin LK, Maréchal-Drouard L, Marshall WF, Qu L-H, Nelson DR, Sanderfoot AA, Spalding MH, Kapitonov VV, Ren Q, Ferris P, Lindquist E, Shapiro H, Lucas SM, Grimwood J, Schmutz J, Cardol P, Cerutti H, Chanfreau G, Chen C-L, Cognat V, Croft MT, Dent R, Dutcher S, Fernández E, Ferris P, Fukuzawa H, González-Ballester D, González-Halphen D, Hallmann A, Hanikenne M, Hippler M, Inwood W, Jabbari K, Kalanon M, Kuras R, Lefebvre PA, Lemaire SD, Lobanov AV, Lohr M, Manuell A, Meier I, Mets L, Mittag M, Mittelmeier T, Moroney JV, Moseley J, Napoli C, Nedelcu AM, Niyogi K, Novoselov SV, Paulsen IT, Pazour G, Purton S, Ral J-P, Riaño-Pachón DM, Riekhof W, Rymarquis L, Schroda M, Stern D, Umen J, Willows R, Wilson N, Zimmer SL, Allmer J, Balk J, Bisova K, Chen C-J, Elias M, Gendler K, Hauser C, Lamb MR, Ledford H, Long JC, Minagawa J, Page MD, Pan J, Pootakham W, Roje S, Rose A, Stahlberg E, Terauchi AM, Yang P, Ball S, Bowler C, Dieckmann CL, Gladyshev VN, Green P, Jorgensen R, Mayfield S, Mueller-Roeber B, Rajamani S, Sayre RT et al (2007) The Chlamydomonas genome reveals the evolution of key animal and plant functions. Science 318:245–250

Meyer H, Bug M, Bremer S (2012) Emerging functions of the VCP/p97 AAA-ATPase in the ubiquitin system. Nat Cell Biol 14:117

Missaglia S, Valadares ER, Moro L, Faguntes EDT, Quintão Roque R, Giardina B, Tavian D (2014) Early onset of Chanarin-Dorfman syndrome with severe liver involvement in a patient with a complex rearrangement of ABHD5 promoter. BMC Med Genet 15:32–32

Miura S, Gan J-W, Brzostowski J, Parisi MJ, Schultz CJ, Londos C, Oliver B, Kimmel AR (2002) Functional conservation for lipid storage droplet association among perilipin, ADRP, and TIP47 (PAT)-related proteins in mammals, Drosophila, and Dictyostelium. J Biol Chem 277:32253–32257

Moellering ER, Benning C (2010) RNA interference silencing of a major lipid droplet protein affects lipid droplet size in *Chlamydomonas reinhardtii*. Eukaryot Cell 9:97–106

Moessinger C et al (2011) Human lysophosphatidylcholine acyltransferases 1 and 2 are located in lipid droplets where they catalyze the formation of phosphatidylcholine. J Biol Chem 286 (24):21330–21339

Moore H-PH, Silver RB, Mottillo EP, Bernlohr DA, Granneman JG (2005) Perilipin targets a novel pool of lipid droplets for lipolytic attack by hormone-sensitive lipase. J Biol Chem 280:43109–43120

Morabito C, Bournaud C, Maës C, Schuler M, Aiese Cigliano R, Dellero Y, Maréchal E, Amato A, Rébeillé F (2019) The lipid metabolism in thraustochytrids. Prog Lipid Res 76:101007

Możejko-Ciesielska J, Kiewisz R (2016) Bacterial polyhydroxyalkanoates: still fabulous? Microbiol Res 192:271–282

Müller AO, Blersch KF, Gippert AL, Ischebeck T (2016) Tobacco pollen tubes – a fast and easy tool for studying lipid droplet association of plant proteins. Plant J 89:1055–1064

Murphy DJ (2012) The dynamic roles of intracellular lipid droplets: from archaea to mammals. Protoplasma 249:541–585

Na H, Zhang P, Chen Y, Zhu X, Liu Y, Liu Y, Xie K, Xu N, Yang F, Yu Y, Cichello S, Mak HY, Wang MC, Zhang H, Liu P (2015) Identification of lipid droplet structure-like/resident proteins in *Caenorhabditis elegans*. Biochim Biophys Acta 1853:2481–2491

Najt CP, Lwande JS, Mcintosh AL, Senthivinayagam S, Gupta S, Kuhn LA, Atshaves BP (2014) Structural and functional assessment of perilipin 2 lipid binding domain(s). Biochemistry 53:7051–7066

Nchoutmboube JA, Viktorova EG, Scott AJ, Ford LA, Pei Z, Watkins PA, Ernst RK, Belov GA (2013) Increased long chain acyl-Coa synthetase activity and fatty acid import is linked to membrane synthesis for development of picornavirus replication organelles. PLoS Pathog 9: e1003401

Nguyen HM, Baudet M, Cuiné S, Adriano J-M, Barthe D, Billon E, Bruley C, Beisson F, Peltier G, Ferro M, Li-Beisson Y (2011) Proteomic profiling of oil bodies isolated from the unicellular green microalga *Chlamydomonas reinhardtii*: with focus on proteins involved in lipid metabolism. Proteomics 11:4266–4273

Nishino N, Tamori Y, Tateya S, Kawaguchi T, Shibakusa T, Mizunoya W, Inoue K, Kitazawa R, Kitazawa S, Matsuki Y, Hiramatsu R, Masubuchi S, Omachi A, Kimura K, Saito M, Amo T, Ohta S, Yamaguchi T, Osumi T, Cheng J, Fujimoto T, Nakao H, Nakao K, Aiba A, Okamura H, Fushiki T, Kasuga M (2008) FSP27 contributes to efficient energy storage in murine white adipocytes by promoting the formation of unilocular lipid droplets. J Clin Invest 118:2808–2821

Nojima D, Yoshino T, Maeda Y, Tanaka M, Nemoto M, Tanaka T (2013) Proteomics analysis of oil body-associated proteins in the oleaginous diatom. J Proteome Res 12:5293–5301

Noothalapati Venkata HN, Shigeto S (2012) Stable isotope-labeled Raman imaging reveals dynamic proteome localization to lipid droplets in single fission yeast cells. Chem Biol 19:1373–1380

Oh SK et al (1999) Isolation, characterization, and functional analysis of a novel cDNA clone encoding a small rubber particle protein from Hevea brasiliensis. J Biol Chem 274(24):17132–17138

Okuda-Shimizu Y, Hendershot LM (2007) Characterization of an ERAD pathway for non-glycosylated BiP substrates which requires Herp. Mol Cell 28:544–554

Oliva J, French SW, Li J, Bardag-Gorce F (2012) Proteasome inhibitor treatment reduced fatty acid, triacylglycerol and cholesterol synthesis. Exp Mol Pathol 93:26–34

Olzmann JA, Kopito RR, Christianson JC (2013a) The mammalian endoplasmic reticulum-associated degradation system. Cold Spring Harb Perspect Biol:5, a013185. https://doi.org/10.1101/cshperspect.a013185

Olzmann JA, Richter CM, Kopito RR (2013b) Spatial regulation of UBXD8 and p97/VCP controls ATGL-mediated lipid droplet turnover. Proc Natl Acad Sci USA 110:1345–1350

Onal G, Kutlu O, Gozuacik D, Dokmeci Emre S (2017) Lipid droplets in health and disease. Lipids Health Dis 16:128

Orban T, Palczewska G, Palczewski K (2011) Retinyl ester storage particles (Retinosomes) from the retinal pigmented epithelium resemble lipid droplets in other tissues. J Biol Chem 286 (19):17248–17258

Orlicky DJ et al (2008) Multiple functions encoded by the N-terminal PAT domain of adipophilin. J Cell Sci 121(Pt 17):2921–2929

Paar M, Jüngst C, Steiner NA, Magnes C, Sinner F, Kolb D, Lass A, Zimmermann R, Zumbusch A, Kohlwein SD, Wolinski H (2012) Remodeling of lipid droplets during lipolysis and growth in adipocytes. J Biol Chem 287:11164–11173

Park S, Gidda SK, James CN, Horn PJ, Khuu N, Seay DC, Keereetaweep J, Chapman KD, Mullen RT, Dyer JM (2013) The α/β hydrolase CGI-58 and peroxisomal transport protein PXA1 coregulate lipid homeostasis and signaling in Arabidopsis. Plant Cell 25:1726–1739

Park S, Keereetaweep J, James CN, Gidda SK, Chapman KD, Mullen RT, Dyer JM (2014) CGI-58, a key regulator of lipid homeostasis and signaling in plants, also regulates polyamine metabolism. Plant Signal Behav 9:e27723

Pasaribu B, Lin IP, Chen C-S, Lu C-Y, Jiang P-L (2014a) Nutrient limitation in Auxenochlorella protothecoides induces qualitative changes of fatty acid and expression of caleosin as a membrane protein associated with oil bodies. Biotechnol Lett 36:175–180

Pasaribu B, Lin IP, Tzen JTC, Jauh G-Y, Fan T-Y, Ju Y-M, Cheng J-O, Chen C-S, Jiang P-L (2014b) SLDP: a novel protein related to caleosin is associated with the endosymbiotic symbiodinium lipid droplets from *Euphyllia glabrescens*. Mar Biotechnol 16:560–571

Pataki CI et al (2018) Proteomic analysis of monolayer-integrated proteins on lipid droplets identifies amphipathic interfacial α-helical membrane anchors. Proc Natl Acad Sci U S A 115 (35):E8172–E8180

Paul A, Chan L, Bickel PE (2008) The PAT family of lipid droplet proteins in heart and vascular cells. Curr Hypertens Rep 10(6):461–466

Peled E, Leu S, Zarka A, Weiss M, Pick U, Khozin-Goldberg I, Boussiba S (2011) Isolation of a novel oil globule protein from the green alga *Haematococcus pluvialis* (Chlorophyceae). Lipids 46:851–861

Peng S-E, Chen W-NU, Chen H-K, Lu C-Y, Mayfield AB, Fang L-S, Chen C-S (2011) Lipid bodies in coral–dinoflagellate endosymbiosis: proteomic and ultrastructural studies. Proteomics 11:3540–3555

Peoples OP, Sinskey AJ (1989) Poly-beta-hydroxybutyrate biosynthesis in Alcaligenes eutrophus H16. Characterization of the genes encoding betaketothiolase and acetoacetyl-CoA reductase. J Biol Chem 264(26):15293–15297

Pericleous M, Kelly C, Wang T, Livingstone C, Ala A (2017) Wolman's disease and cholesteryl ester storage disorder: the phenotypic spectrum of lysosomal acid lipase deficiency. Lancet Gastroenterol Hepatol 2:670–679

Petroutsos D, Amiar S, Abida H, Dolch L-J, Bastien O, Rébeillé F, Jouhet J, Falconet D, Block MA, Mcfadden GI, Bowler C, Botté C, Maréchal E (2014) Evolution of galactoglycerolipid biosynthetic pathways – from cyanobacteria to primary plastids and from primary to secondary plastids. Prog Lipid Res 54:68–85

Pol A, Gross SP, Parton RG (2014) Biogenesis of the multifunctional lipid droplet: lipids, proteins, and sites. J Cell Biol 204:635–646

Popluechai S, Froissard M, Jolivet P, Breviario D, Gatehouse AMR, O'donnell AG, Chardot T, Kohli A (2011) *Jatropha curcas* oil body proteome and oleosins: L-form JcOle3 as a potential phylogenetic marker. Plant Physiol Biochem 49:352–356

Pötter M, Steinbüchel A (2006) Biogenesis and structure of polyhydroxyalkanoate granules. In: Shively JM (ed) Inclusions in prokaryotes. Springer, Berlin, pp 109–136

Poulsen LLC, Siersbæk M, Mandrup S (2012) PPARs: fatty acid sensors controlling metabolism. Semin Cell Dev Biol 23:631–639

Qu R, Wang SM, Lin YH, Vance VB, Huang AH (1986) Characteristics and biosynthesis of membrane proteins of lipid bodies in the scutella of maize (*Zea mays* L.). Biochem J 235:57–65

Quettier A-L, Eastmond PJ (2009) Storage oil hydrolysis during early seedling growth. Plant Physiol Biochem 47:485–490

Quinn K, Purcell SM (2017) Lipodystrophies. StatPearls [Internet]

Rabhi I et al (2012) Transcriptomic signature of leishmania infected mice macrophages: a metabolic point of view. PLoS Negl Trop Dis 6(8):e1763

Rabhi S et al (2016) Lipid droplet formation, their localization and dynamics during leishmania major macrophage infection. PLoS One 11(2):e0148640

Rajab A et al (2010) Fatal cardiac arrhythmia and Long-QT syndrome in a new form of congenital generalized lipodystrophy with muscle rippling (CGL4) due to PTRFCAVIN mutations. PLoS Genet 6(3):e1000874

Rapoport TA (2007) Protein translocation across the eukaryotic endoplasmic reticulum and bacterial plasma membranes. Nature 450:663

Reinecke F, Steinbüchel A (2009) *Ralstonia eutropha* strain H16 as model organism for PHA metabolism and for biotechnological production of technically interesting biopolymers. J Mol Microbiol Biotechnol 16:91–108

Roingeard P, Melo RCN (2017) Lipid droplet hijacking by intracellular pathogens. Cell Microbiol 19:e12688

Ruggiano A, Mora G, Buxó L, Carvalho P (2016) Spatial control of lipid droplet proteins by the ERAD ubiquitin ligase Doa10. EMBO J 35:1644–1655

Rumin J, Bonnefond H, Saint-Jean B, Rouxel C, Sciandra A, Bernard O, Cadoret J-P, Bougaran G (2015) The use of fluorescent Nile red and BODIPY for lipid measurement in microalgae. Biotechnol Biofuels 8:42

Rydel TJ, Williams JM, Krieger E, Moshiri F, Stallings WC, Brown SM, Pershing JC, Purcell JP, Alibhai MF (2003) The crystal structure, mutagenesis, and activity studies reveal that patatin is a lipid acyl hydrolase with a Ser-Asp catalytic dyad. Biochemistry 42:6696–6708

Sahu-Osen A, Montero-Moran G, Schittmayer M, Fritz K, Dinh A, Chang Y-F, Mcmahon D, Boeszoermenyi A, Cornaciu I, Russell D, Oberer M, Carman GM, Birner-Gruenberger R, Brasaemle DL (2015) CGI-58/ABHD5 is phosphorylated on Ser239 by protein kinase a: control of subcellular localization. J Lipid Res 56:109–121

Saka HA, Thompson JW, Chen Y-S, Dubois LG, Haas JT, Moseley A, Valdivia RH (2015) Chlamydia trachomatis infection leads to defined alterations to the lipid droplet proteome in epithelial cells. PLoS One 10:e0124630

Salo VT, Belevich I, Li S, Karhinen L, Vihinen H, Vigouroux C, Magré J, Thiele C, Hölttä-Vuori M, Jokitalo E, Ikonen E (2016) Seipin regulates ER–lipid droplet contacts and cargo delivery. EMBO J 35:2699–2716

Samuelov L et al (2011) An exceptional mutational event leading to Chanarin–Dorfman syndrome in a large consanguineous family. Br J Dermatol 164(6):1390–1392

Sando T, Hayashi T, Takeda T, Akiyama Y, Nakazawa Y, Fukusaki E, Kobayashi A (2009) Histochemical study of detailed laticifer structure and rubber biosynthesis-related protein localization in *Hevea brasiliensis* using spectral confocal laser scanning microscopy. Planta 230:215–225

Sardet C (2013) In: Ulmer E (ed) Plancton - Aux origines du vivant, 218 p

Sato S et al (2006) Proteomic profiling of lipid droplet proteins in hepatoma cell lines expressing hepatitis C virus core protein. J Biochem 139(5):921–930

Schmuth M et al (2004) The effect of LXR activators on AP-1 proteins in keratinocytes. J Investig Dermatol 123(1):41–48

Schubert P, Steinbüchel A, Schlegel HG (1988) Cloning of the Alcaligenes eutrophus genes for synthesis of poly-beta-hydroxybutyric acid (PHB) and synthesis of PHB in Escherichia coli. J Bacteriol 170(12):5837–5847

Schweiger M, Paar M, Eder C, Brandis J, Moser E, Gorkiewicz G, Grond S, Radner FPW, Cerk I, Cornaciu I, Oberer M, Kersten S, Zechner R, Zimmermann R, Lass A (2012) G0/G1 switch gene-2 regulates human adipocyte lipolysis by affecting activity and localization of adipose triglyceride lipase. J Lipid Res 53:2307–2317

Sharma B, Joshi D, Yadav PK, Gupta AK, Bhatt TK (2016) Role of ubiquitin-mediated degradation system in plant biology. Front Plant Sci 7:806

Shen P, Zhao Q, Yao C, Wu S, Meng Y, Zhang L, Xue S (2016) Differential proteome analysis and identification of lipid droplet associated proteins in the marine microalgae *Isochrysis zhangjiangensis* (Haptophyta). In: The 4th Asia-Oceania algae innovation summit. Wuhan

Shi Q (2019) Expression profiling of genes coding for abundant proteins in the alkenone body of marine haptophyte alga *Tisochrysis lutea*. BMC Microbiol 19:56–56

Shi X, Li J, Zou X, Greggain J, Rødkær SV, Færgeman NJ, Liang B, Watts JL (2013) Regulation of lipid droplet size and phospholipid composition by stearoyl-CoA desaturase. J Lipid Res 54:2504–2514

Shi Q, Araie H, Bakku RK, Fukao Y, Rakwal R, Suzuki I, Shiraiwa Y (2015) Proteomic analysis of lipid body from the alkenone-producing marine haptophyte alga *Tisochrysis lutea*. Proteomics 15:4145–4158

Shimada TL, Hara-Nishimura I (2015) Leaf oil bodies are subcellular factories producing antifungal oxylipins. Curr Opin Plant Biol 25:145–150

Shimada TL, Takano Y, Shimada T, Fujiwara M, Fukao Y, Mori M, Okazaki Y, Saito K, Sasaki R, Aoki K, Hara-Nishimura I (2014) Leaf oil body functions as a subcellular factory for the production of a Phytoalexin in Arabidopsis. Plant Physiol 164:105–118

Shimada TL, Takano Y, Hara-Nishimura I (2015) Oil body-mediated defense against fungi: from tissues to ecology. Plant Signal Behav 10:e989036

Siegler H, Valerius O, Ischebeck T, Popko J, Tourasse NJ, Vallon O, Khozin-Goldberg I, Braus GH, Feussner I (2017) Analysis of the lipid body proteome of the oleaginous alga *Lobosphaera incisa*. BMC Plant Biol 17:98

Slater SC, Voige WH, Dennis DE (1988) Cloning and expression in Escherichia coli of the Alcaligenes eutrophus H16 poly-beta-hydroxybutyrate biosynthetic pathway. J Bacteriol 170 (10):4431–4436

Song P, Li L, Liu J (2013) Proteomic analysis in nitrogen-deprived *Isochrysis galbana* during lipid accumulation. PLoS One 8:e82188–e82188

Soni KG, Mardones GA, Sougrat R, Smirnova E, Jackson CL, Bonifacino JS (2009) Coatomer-dependent protein delivery to lipid droplets. J Cell Sci 122:1834

Soulages JL, Firdaus SJ, Hartson S, Chen X, Howard AD, Arrese EL (2012) Developmental changes in the protein composition of *Manduca sexta* lipid droplets. Insect Biochem Mol Biol 42:305–320

Souza SC, De Vargas LM, Yamamoto MT, Lien P, Franciosa MD, Moss LG, Greenberg AS (1998) Overexpression of Perilipin A and B blocks the ability of tumor necrosis factor α to increase lipolysis in 3T3-L1 adipocytes. J Biol Chem 273:24665–24669

Stenson BM et al (2011) Liver X receptor (LXR) regulates human adipocyte lipolysis. J Biol Chem 286(1):370–379

Stevenson J, Huang EY, Olzmann JA (2016) Endoplasmic reticulum–associated degradation and lipid homeostasis. Annu Rev Nutr 36:511–542

Stone SJ, Levin MC, Zhou P, Han J, Walther TC, Farese RV (2009) The endoplasmic reticulum enzyme DGAT2 is found in mitochondria-associated membranes and has a mitochondrial targeting signal that promotes its association with mitochondria. J Biol Chem 284:5352–5361

Subramanian V, Rothenberg A, Gomez C, Cohen AW, Garcia A, Bhattacharyya S, Shapiro L, Dolios G, Wang R, Lisanti MP, Brasaemle DL (2004) Perilipin a mediates the reversible binding of CGI-58 to lipid droplets in 3T3-L1 adipocytes. J Biol Chem 279:42062–42071

Sun Z, Gong J, Wu H, Xu W, Wu L, Xu D, Gao J, Wu J-W, Yang H, Yang M, Li P (2013) Perilipin1 promotes unilocular lipid droplet formation through the activation of Fsp27 in adipocytes. Nat Commun 4:1594

Suzuki M, Otsuka T, Ohsaki Y, Cheng J, Taniguchi T, Hashimoto H, Taniguchi H, Fujimoto T (2012) Derlin-1 and UBXD8 are engaged in dislocation and degradation of lipidated ApoB-100 at lipid droplets. Mol Biol Cell 23:800–810

Sznajder A, Jendrossek D (2014) To be or not to be a poly(3-hydroxybutyrate) (PHB) depolymerase: PhaZd1 (PhaZ6) and PhaZd2 (PhaZ7) of *Ralstonia eutropha*, highly active PHB depolymerases with no detectable role in mobilization of accumulated PHB. Appl Environ Microbiol 80:4936–4946

Talbott H, Davis JS (2017) Lipid droplets and metabolic pathways regulate steroidogenesis in the corpus luteum. In: Meidan R (ed) The life cycle of the corpus luteum. Springer International Publishing, Cham, pp 57–78

Tansey JT, Sztalryd C, Gruia-Gray J, Roush DL, Zee JV, Gavrilova O, Reitman ML, Deng CX, Li C, Kimmel AR, Londos C (2001) Perilipin ablation results in a lean mouse with aberrant adipocyte lipolysis, enhanced leptin production, and resistance to diet-induced obesity. Proc Natl Acad Sci 98:6494

Targett-Adams P et al (2005) A PPAR response element regulates transcription of the gene for human adipose differentiation-related protein. Biochimica et Biophysica Acta (BBA) - Gene Struct Expr 1728(1):95–104

Tariq A et al (2015) Is atomic rearrangement of type IV PHA synthases responsible for increased PHA production? J Biomol Struct Dyn 33(6):1225–1238

Tauchi-Sato K, Ozeki S, Houjou T, Taguchi R, Fujimoto T (2002) The surface of lipid droplets is a phospholipid monolayer with a unique fatty acid composition. J Biol Chem 277:44507–44512

Taurino M, Costantini S, De Domenico S, Stefanelli F, Ruano G, Delgadillo MO, Sanchez-Serrano JJ, Sanmartín M, Santino A, Rojo E (2017) SEIPIN proteins mediate lipid droplet biogenesis to promote pollen transmission and reduce seed dormancy. Plant Physiol 176:1531–1546

Theodoulou FL, Job K, Slocombe SP, Footitt S, Holdsworth M, Baker A, Larson TR, Graham IA (2005) Jasmonic acid levels are reduced in COMATOSE ATP-binding cassette transporter

mutants. Implications for transport of jasmonate precursors into peroxisomes. Plant Physiol 137:835–840
Thiam AR, Forêt L (2016) The physics of lipid droplet nucleation, growth and budding. Biochim Biophys Acta 1861:715–722
Thiam AR, Antonny B, Wang J, Delacotte J, Wilfling F, Walther TC, Beck R, Rothman JE, Pincet F (2013a) COPI buds 60-nm lipid droplets from reconstituted water–phospholipid–triacylglyceride interfaces, suggesting a tension clamp function. Proc Natl Acad Sci USA 110:13244–13249
Thiam AR, Farese RV Jr, Walther TC (2013b) The biophysics and cell biology of lipid droplets. Nat Rev Mol Cell Biol 14:775
Thul PJ, Tschapalda K, Kolkhof P, Thiam AR, Oberer M, Beller M (2017) Targeting of the Drosophila protein CG2254/Ldsdh1 to a subset of lipid droplets. J Cell Sci 130:3141
Tnani H, López I, Jouenne T, Vicient CM (2011) Protein composition analysis of oil bodies from maize embryos during germination. J Plant Physiol 168:510–513
Tobin KAR et al (2006) Regulation of ADRP expression by long-chain polyunsaturated fatty acids in BeWo cells, a human placental choriocarcinoma cell line. J Lipid Res 47(4):815–823
Trentacoste EM, Shrestha RP, Smith SR, Glé C, Hartmann AC, Hildebrand M, Gerwick WH (2013) Metabolic engineering of lipid catabolism increases microalgal lipid accumulation without compromising growth. Proc Natl Acad Sci USA 110:19748–19753
Turró S et al (2006) Identification and characterization of associated with lipid droplet protein 1: a novel membrane-associated protein that resides on hepatic lipid droplets. Traffic 7(9):1254–1269
Uchino K, Saito T, Jendrossek D (2008) Poly(3-hydroxybutyrate) (PHB) depolymerase PhaZa1 is involved in mobilization of accumulated PHB in *Ralstonia eutropha* H16. Appl Environ Microbiol 74:1058–1063
Van Wijk KJ, Kessler F (2017) Plastoglobuli: plastid microcompartments with integrated functions in metabolism, plastid developmental transitions, and environmental adaptation. Annu Rev Plant Biol 68:253–289
Vance VB, Huang AH (1987) The major protein from lipid bodies of maize. Characterization and structure based on cDNA cloning. J Biol Chem 262:11275–11279
Varela GM, Antwi DA, Dhir R, Yin X, Singhal NS, Graham MJ, Crooke RM, Ahima RS (2008) Inhibition of ADRP prevents diet-induced insulin resistance. Am J Physiol Gastrointest Liver Physiol 295:G621–G628
Vermachova M, Purkrtova Z, Santrucek J, Jolivet P, Chardot T, Kodicek M (2011) New protein isoforms identified within *Arabidopsis thaliana* seed oil bodies combining chymotrypsin/trypsin digestion and peptide fragmentation analysis. Proteomics 11:3430–3434
Venkata N, Hemanth N, Shigeto S (2012) Stable isotope-labeled Raman imaging reveals dynamic proteome localization to lipid droplets in single fission yeast cells. Chem Biol 19(11):1373–1380
Vieler A, Wu G, Tsai C-H, Bullard B, Cornish AJ, Harvey C, Reca I-B, Thornburg C, Achawanantakun R, Buehl CJ, Campbell MS, Cavalier D, Childs KL, Clark TJ, Deshpande R, Erickson E, Armenia Ferguson A, Handee W, Kong Q, Li X, Liu B, Lundback S, Peng C, Roston RL, Sanjaya, Simpson JP, Terbush A, Warakanont J, Zäuner S, Farre EM, Hegg EL, Jiang N, Kuo M-H, Lu Y, Niyogi KK, Ohlrogge J, Osteryoung KW, Shachar-Hill Y, Sears BB, Sun Y, Takahashi H, Yandell M, Shiu S-H, Benning C (2012) Genome, functional gene annotation, and nuclear transformation of the heterokont oleaginous alga *Nannochloropsis oceanica* CCMP1779. *PLoS Genet* 8:e1003064
Viswanadha S, Londos C (2008) Determination of lipolysis in isolated primary adipocytes. In: Yang K (ed) Adipose tissue protocols. Humana Press, Totowa, pp 299–306
Vrablik TL, Petyuk VA, Larson EM, Smith RD, Watts JL (2015) Lipidomic and proteomic analysis of *C. elegans* lipid droplets and identification of ACS-4 as a lipid droplet-associated protein. Biochim Biophys Acta 1851:1337–1345

Wältermann M, Steinbüchel A (2005) Neutral lipid bodies in prokaryotes: recent insights into structure, formation, and relationship to eukaryotic lipid depots. J Bacteriol 187:3607–3619

Waheed N et al (2016) Chanarin-Dorfman syndrome. J Coll Physicians Surg Pak 26(9):787–789

Walther TC, Chung J, Farese RV (2017) Lipid droplet biogenesis. Annu Rev Cell Dev Biol 33:491–510

Wang W et al (2015) Proteomic analysis of murine testes lipid droplets. Sci Rep 5:12070

Wang H, Becuwe M, Housden BE, Chitraju C, Porras AJ, Graham MM, Liu XN, Thiam AR, Savage DB, Agarwal AK, Garg A, Olarte M-J, Lin Q, Fröhlich F, Hannibal-Bach HK, Upadhyayula S, Perrimon N, Kirchhausen T, Ejsing CS, Walther TC, Farese RV (2016) Seipin is required for converting nascent to mature lipid droplets. elife 5:e16582

Watanabe T, Sakiyama R, Iimi Y, Sekine S, Abe E, Nomura KH, Nomura K, Ishibashi Y, Okino N, Hayashi M, Ito M (2017) Regulation of TG accumulation and lipid droplet morphology by the novel TLDP1 in *Aurantiochytrium limacinum* F26-b. J Lipid Res 58:2334–2347

Wilfling F, Wang H, Haas JT, Krahmer N, Gould TJ, Uchida A, Cheng J-X, Graham M, Christiano R, Fröhlich F, Liu X, Buhman KK, Coleman RA, Bewersdorf J, Farese RV, Walther TC (2013) Triacylglycerol synthesis enzymes mediate lipid droplet growth by relocalizing from the ER to lipid droplets. Dev Cell 24:384–399

Wilfling F, Haas JT, Walther TC, Farese RV (2014) Lipid droplet biogenesis. Curr Opin Cell Biol 29:39–45

Wolman M, Sterk VV, Gatt S, Frenkel M (1961) Primary familial xanthomatosis with involvement and calcification of the adrenals. Report of two more cases in siblings of a previously described infant. Pediatrics 25:742–757

Wu CC et al (2000) Proteomics reveal a link between the endoplasmic reticulum and lipid secretory mechanisms in mammary epithelial cells. Electrophoresis 21(16):3470–3482

Xie M, Roy R (2015) The causative gene in Chanarian Dorfman syndrome regulates lipid droplet homeostasis in *C. elegans*. PLoS Genet 11:e1005284–e1005284

Yamaguchi T et al (2015) Characterization of lipid droplets in steroidogenic MLTC-1 Leydig cells: protein profiles and the morphological change induced by hormone stimulation. Biochimica et Biophysica Acta (BBA) - Mol Cell Biol Lipids 1851(10):1285–1295

Yang H-J, Hsu C-L, Yang J-Y, Yang WY (2012) Monodansylpentane as a blue-fluorescent lipid-droplet marker for multi-color live-cell imaging. PLoS One 7:e32693

Yermanos DM (1975) Composition of jojoba seed during development. J Am Oil Chem Soc 52:115–117

Yoneda K, Yoshida M, Suzuki I, Watanabe MM (2016) Identification of a major lipid droplet protein in a marine diatom *Phaeodactylum tricornutum*. Plant Cell Physiol 57:397–406

Yoneda K, Yoshida M, Suzuki I, Watanabe MM (2018) Homologous expression of lipid droplet protein-enhanced neutral lipid accumulation in the marine diatom *Phaeodactylum tricornutum*. J Appl Phycol 30:2793–2802

Youssef A, Laizet YH, Block MA, Maréchal E, Alcaraz J-P, Larson TR, Pontier D, Gaffé J, Kuntz M (2010) Plant lipid-associated fibrillin proteins condition jasmonate production under photosynthetic stress. Plant J 61:436–445

Yu J, Zhang S, Cui L, Wang W, Na H, Zhu X, Li L, Xu G, Yang F, Christian M, Liu P (2015) Lipid droplet remodeling and interaction with mitochondria in mouse brown adipose tissue during cold treatment. Biochim Biophys Acta 1853:918–928

Yu Y, Li T, Wu N, Jiang L, Ji X, Huang H (2017) The role of lipid droplets in *Mortierella alpina* aging revealed by integrative subcellular and whole-cell proteome analysis. Sci Rep 7:43896

Zhang C, Liu P (2017) The lipid droplet: a conserved cellular organelle. Protein Cell 8:796–800

Zhang H et al (2011) Proteome of skeletal muscle lipid droplet reveals association with mitochondria and apolipoprotein A-I. J Proteome Res 10(10):4757–4768

Zhang P, Na H, Liu Z, Zhang S, Xue P, Chen Y, Pu J, Peng G, Huang X, Yang F, Xie Z, Xu T, Xu P, Ou G, Zhang SO, Liu P (2012) Proteomic study and marker protein identification of *Caenorhabditis elegans* lipid droplets. Mol Cell Proteomics 11:317–328

Zhang J, Zhang Z, Chukkapalli V, Nchoutmboube JA, Li J, Randall G, Belov GA, Wang X (2016) Positive-strand RNA viruses stimulate host phosphatidylcholine synthesis at viral replication sites. Proc Natl Acad Sci USA 113:E1064–E1073

Zhang C, Yang L, Ding Y, Wang Y, Lan L, Ma Q, Chi X, Wei P, Zhao Y, Steinbüchel A, Zhang H, Liu P (2017) Bacterial lipid droplets bind to DNA via an intermediary protein that enhances survival under stress. Nat Commun 8:15979

Zhu Z, Ding Y, Gong Z, Yang L, Zhang S, Zhang C, Lin X, Shen H, Zou H, Xie Z, Yang F, Zhao X, Liu P, Zhao ZK (2015) Dynamics of the lipid droplet proteome of the oleaginous yeast *Rhodosporidium toruloides*. Eukaryot Cell 14:252–264

Zimmermann R, Strauss JG, Haemmerle G, Schoiswohl G, Birner-Gruenberger R, Riederer M, Lass A, Neuberger G, Eisenhaber F, Hermetter A, Zechner R (2004) Fat mobilization in adipose tissue is promoted by adipose triglyceride lipase. Science 306:1383–1386

Part III
Evolution and Role of Symbiosis in Photosynthesis and Nitrogen Fixation

Chapter 12
Evolution of Photosynthetic Eukaryotes; Current Opinion, Perplexity, and a New Perspective

Shinichiro Maruyama and Eunsoo Kim

Abstract The evolution of eukaryotic photosynthesis marked a major transition for life on Earth, profoundly impacting the atmosphere of the Earth and evolutionary trajectory of an array of life forms. There are about ten lineages of photosynthetic eukaryotes, including Chloroplastida, Rhodophyta, and Cryptophyta. Mechanistically, eukaryotic photosynthesis arose via a symbiotic merger between a host eukaryote and either a cyanobacterial or eukaryotic photosymbiont. There are, however, many aspects of this major evolutionary transition that remain unsettled. The field, so far, has been dominated by proposals formulated following the principle of parsimony, such as the Archaeplastida hypothesis, in which a taxonomic lineage is often conceptually recognized as an individual cell (or a distinct entity). Such an assumption could lead to confusion or unrealistic interpretation of discordant genomic and phenotypic data. Here, we propose that the free-living ancestors to the plastids may have originated from a diversified lineage of cyanobacteria that were prone to symbioses, akin to some modern-day algae such as the Symbiodiniaceae dinoflagellates and *Chlorella*-related algae that associate with a number of unrelated host eukaryotes. This scenario, which assumes the plurality of ancestral form, better explains relatively minor but important differences that are observed in the genomes of modern-day eukaryotic algal species. Such a non-typological (or population-aware) way of thinking seems to better-model empirical data, such as discordant phylogenies between plastid and host eukaryote genes.

S. Maruyama (✉)
Department of Ecological Developmental Adaptability Life Sciences, Graduate School of Life Sciences, Tohoku University, Sendai, Japan
e-mail: maruyama@tohoku.ac.jp

E. Kim
Division of Invertebrate Zoology & Sackler Institute for Comparative Genomics, American Museum of Natural History, New York, NY, USA
e-mail: ekim1@amnh.org

M. Kloc (ed.), *Symbiosis: Cellular, Molecular, Medical and Evolutionary Aspects*, Results and Problems in Cell Differentiation 69,
https://doi.org/10.1007/978-3-030-51849-3_12

12.1 Introduction

> Adaptive radiations have been typically studied using two approaches. Classically, one may observe an established radiation and draw inferences about the past processes, which led to the present-day pattern of ecological and phenotypic diversity, informed by the understanding of the phylogenetic relationships among species. Alternatively, one may identify and study the evolutionary processes operating on a clade of relatively few species (or forms), which may be currently diverging with the assumption that the clade is a representative of the early stages of a forthcoming adaptive radiation. These two approaches represent either end of the process-pattern divide in adaptive radiation research. (Stroud and Losos 2020)

Although not all evolutionary outcomes are the consequences of adaptations, many evolutionary studies are designed based on the process- or pattern-centric approaches. If evolutionary biologists want to know why Roquefort cheese is so special, then they could resort to biochemistry, which helps to uncover how blue cheese fungi (e.g. *Penicillium roqueforti*) can produce a special flavor (process-centric). Alternatively, phylogeny may identify how blue cheese fungi are evolutionarily related or distant from other fungi (pattern-centric) (Dumas et al. 2020). Perhaps comparative biochemical analyses using multiple species and strains sampled from different phylogenetic branches may provide a more complete understanding of the evolutionary transition from "ordinary" to "special" blue cheese fungi, where process- and pattern-centric approaches are synthesized, but often such integration of approaches is not feasible. This is particularly so when an evolutionary event is archaic (e.g. the origins of plants, eukaryotes, or life on the Earth). Processes are often only observable in extant (i.e. modern-day) species, which are often highly diverged from their distant ancestors, thereby limiting their utility in inferring ancient events. Deep phylogenetic patterns are often difficult to reconstruct accurately because 'transitional' species between 'in-group' and 'out-group' taxa become scarce over time due to extinction events. For many ancient evolutionary events, especially when fossil records are scarce, the pattern-based approach using modern-day information is often seen as the sole option.

The theory of endosymbiosis explains that the plastids (chloroplasts), the organelles responsible for photosynthesis, originated via engulfment of photosynthetic microorganisms by host eukaryotic cells. The first photosynthetic eukaryotes arose via 'primary endosymbiosis' in which a eukaryotic host engulfed and retained a cyanobacterial endosymbiont[1] (Cavalier-Smith 1982). Primary plastid-bearing groups include green algae plus their land plant descendants, red algae, and glaucophytes, which are together classified as Archaeplastida (Adl et al. 2019). In contrast, some eukaryotes acquired their plastids via 'secondary' or 'tertiary' events that involved eukaryotic endosymbionts (McFadden 2001). Some researchers argue that the three primary plastid-bearing algal groups arose via a single endosymbiotic

[1]There are more recently identified cases of cyanobacterial integration into the eukaryotic cells, such as in the case of photosynthetic *Paulinella* species (Lhee et al. 2019) and rhopalodiacean diatoms (Nakayama and Inagaki 2017). It is, however, debated as to whether they should be called plastids (Keeling and Archibald 2008).

event that occurred more than a billion years ago, and therefore, they form a monophyletic taxonomic group called Archaeplastida. The proponents of this hypothesis suggest that a permanent integration between two distinct organisms, that is, between a unicellular eukaryotic host and a cyanobacterial endosymbiont, must have been an extremely fortuitous and random evolutionary event. By comparison, the proponents of alternative hypotheses, which often but not necesarrily propose non-monophyletic archaeplastidans, assume that the plastid loss could be as rare or difficult as plastid acquisition founded on the observation that plastid-lacking members are scarce within a well-supported algal group whose common ancestors had a plastid (e.g. haptophytes and cryptophytes).

Such a debate is based on an assumption that the rarity (or frequency) of a plastid's gain or loss events has been constant over time, which may not be true. In very early stages of plastid evolution, loss and regain of photosymbionts by host cells that belong to the same 'lineage' or 'population' might have occurred repeatedly (Fig. 12.1). After such associations formed and ensued over generations, variants such as those that are less competent in symbiosis could have arisen, thereby facilitating sympatric speciation (e.g. symbiont-bearing and color-less species) in a population (Fig. 12.1). A population of the phagotrophic cryptistan biflagellate protist, *Hatena arenicola*, may represent a good example of this process. *H. arenicola*, occurring on a sandy beach, internalizes and retains the green alga *Nephroselmis rotunda* in the cytoplasm (Okamoto and Inouye 2005, 2006). When the green-colored 'parent' *H. arenicola* divides into two daughter cells, only one

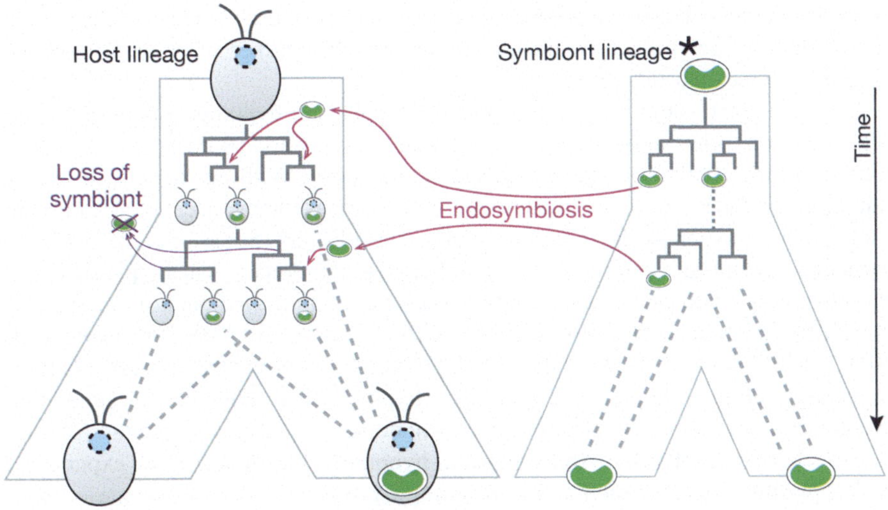

Fig. 12.1 A schematic model showing an early stage of protist–photosymbiont associations. In this model, a population of free-living phototrophs is the source for multiple endosymbioses with a host population. From such multitudinal interactions, those photosymbionts that are kept by the host population are derived from (likely many) different algal cells, but they might be seen collectively as a single entity (shown by asterisk)

inherits the endosymbiont, and the other does not and becomes colorless. What happens if the plastid-lacking cell proliferates further and regains an endosymbiont that is derived from the (ancestral) green algal population with which the parental *H. arenicola* is associated? The host and endosymbiont in the offspring would have genotypes closely related to the ones in the ancestors. In this case, the endosymbiont lineages are not continuous at the cellular level but can be genetically traced to the same green algal population. The bigger the time gap between the ancestor and the offspring generations, the bigger the genotype variations in the 'metapopulation' are expected to be.

Recently, a non-photosynthetic, predatory relative of red algae, named *Rhodelphis*, has been discovered (Gawryluk et al. 2019). This flagellate is the only obligate heterotrophic phagotroph known to date within Archaeplastida. Some early-diverging members of green algae are also phagotrophic, but they are pigmented and photosynthetic; thereby possessing a mixed-mode of nutrition (Maruyama and Kim 2013). Therefore, from the perspective of the trophic mode, *Rhodelphis* represents an oddity considering the Archaeplastida concept, which assumes that their common ancestor was phototrophic. Despite a lack of microscopic evidence for plastids, *Rhodelphis* is suggested to bear plastids based on an in silico identification of a number of putative plastid-targeted peptides. The analyses of putative protein transport machinery and transit peptide sequence motifs did not indicate that *Rhodelphis* had characteristic signals for plastids of secondary origin. The authors, therefore, suggested the plastid of this flagellate is of primary origin. However, in molecular phylogenetic analyses, the majority of *Rhodelphis*' putative plastid-targeted proteins did not branch with red algal proteins, casting doubt on the origin of the *Rhodelphis* plastid and perhaps, even the validity of the existence of the plastid compartment. Therefore, more investigation is needed to see whether *Rhodelphis* indeed possesses plastids or not.

It is also worth pointing out that the Archaeplastida hypothesis is perhaps too dependent on prior knowledge of eukaryotic phylogeny (Baldauf et al. 2000; Moreira et al. 2000; Rodríguez-Ezpeleta et al. 2005). With an increase in taxon sampling and the use of updated phylogenomic matrices, many recent phylogenomic analyses do not support the monophyly of Archaeplastida (Strassert et al. 2019). In particular, Cryptista—comprising cryptophytes, goniomonads, katablepharids, and *Palpitomonas* (Adl et al. 2019)—often branches within Archaeplastida, thereby disrupting its monophyly (Burki et al. 2016; Cenci et al. 2018; Strassert et al. 2019; Gawryluk et al. 2019). If this topology correctly reflects the species' relationships, this may provide evidence against the hypothesis on single plastid-generating event at the ancestry of Chloroplastida, Rhodophyta, and Glaucophyta. Alternatively, some argue that the Archaeplastida hypothesis still holds if we expand the Archaeplastida concept, such as by including Cryptista. Under this scenario, there was a single primary plastid-generating event at the ancestry of Chloroplastida, Rhodophyta, Glaucophyta, and Cryptista, but the complete loss of plastids happened before the common ancestor of Cryptista diverged. This illustrates how the pattern-based (phylogenetic) approach could be limiting in addressing an archaic evolutionary event, such as the origin of plastids.

Here, we propose that 'modern' symbiotic associations may be sources of insight for ancient processes of plastid acquisition. The way by which plastid evolution progressed from free-living bacteria through obligate endosymbionts to permanent cellular organelles remains mysterious. Considering the spectrum of the strength in host–symbiont relationships, facultative associations found in modern-day environments may provide clues about the early stages of plastid evolution. Symbioses also span a spectrum in partner specificity between 'specialists' and 'generalists,' the latter defined here as having a broad host range. While generalist algal symbionts are notable in marine and freshwater environments, this concept has not been taken into account in modeling plastid evolution. In this chapter, we survey those modern associations that involve photosynthetic symbionts, especially those that are generalists in host specificity. Relationship dynamics seen in such modern symbioses may be parallel to the ancient associations that eventually led to the evolution of eukaryotic photosynthesis.

12.2 Generalist Photosymbionts in Modern Aquatic Environments

There are a great variety of photosymbionts in nature, and their diversity has been reviewed extensively elsewhere (Martin et al. 2016). Here, we focus on the three groups of eukaryotic algae that associate with a broad range of host taxa and hence, could be considered as generalist symbiont lineages. Of these, the Symbiodiniaceae dinoflagellates and the *Chlorella*-related algae are symbiotic champions in marine and freshwater ecosystems, respectively.

12.2.1 Symbiodiniaceae Dinoflagellates

Dinoflagellate algae belonging to the family Symbiodiniaceae are known to form stable endosymbiotic relationships with a number of marine eukaryotic hosts, including cnidarians (e.g. coral, sea anemone, and jellyfish), ciliates, and foraminiferans. One of the most ecologically relevant examples is coral–algal symbiosis, which sustains the primary production of coral reefs in oligotrophic oceans.

Based upon the most up-to-date classification proposed by LaJeunesse et al. (2018), Symbiodiniaceae forms a monophyletic taxon in the dinoflagellate phylogeny. Members of the Symbiodiniaceae either (1) associate with only a single host group (e.g. clade I only found in foraminiferans), (2) appear as exclusively free-living in nature (e.g. *Effrenium voratum*), or (3) associate with multiple hosts, which are often colonized by multiple distinct Symbiodiniaceae species/genera. For example, a single species of coral can host a number of symbiont genera, and the composition can vary depending on geographic location, environmental condition,

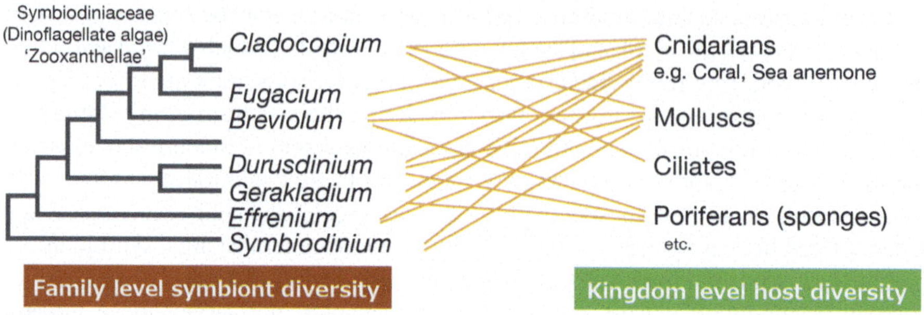

Fig. 12.2 Taxonomic relationships between Symbiodiniaceae dinoflagellates and their various hosts (LaJeunesse et al. 2018; Mies et al. 2017)

and developmental stage (Pochon et al. 2014; Mies et al. 2017). There seems to be no general trend of lineage-specificity in host–dinoflagellate relationships, suggesting that a wide host-range is an ancestral characteristic of the Symbiodiniaceae algae (Fig. 12.2). This 'many-to-many' partnership is likely a factor providing flexibility in the face of changing physiological, developmental, and environmental circumstances.

Flexible 'many-to-many' partnerships appear stable and may be advantageous to both hosts and dinoflagellate symbionts as long as the symbiont population size is large enough to sustain the host population. A disadvantage to the host in such a flexible partnership is the potential for host gastrodermis cells to become occupied by less-beneficial symbiont algae unless a mechanism to selectively recruit the optimally mutualistic algae is already set in place. Similarly, a disadvantage to the endosymbiont is that if a stronger competitor co-occupies the residential space within the host, then it may be forced outside to inhabit the oligotrophic ocean where it could starve due to low nutrient availability.

In the family Symbiodiniaceae, the genus *Cladocopium* (formerly called 'clade C *Symbiodinium*') is known to include generalist members that thrive inside a number of host eukaryotes (e.g. cnidarians, foraminiferans, and acoels) and have an ectosymbiotic partnership with molluscan hosts (LaJeunesse et al. 2018). Conversely, a single host species can accommodate multiple *Cladocopium* species or subspecies. Under a condition where endosymbiosis is destabilized by environmental cues (e.g. thermal stress), host corals expel a portion of residing symbionts (e.g. *Cladocopium*) and take up new symbionts from surrounding environments, including the genus *Durusdinium*, which is also a generalist in host choice (Boulotte et al. 2016). *Durusdinium trenchii* is an opportunistic symbiont of the model sea anemone *Exaiptasia diaphana* (formerly *E. pallida* or *Aiptasia* sp.), but studies based on the transcriptome, proteome, and metabolome suggest that the alga may not be as beneficial as native *Breviolum* spp. symbionts (Matthews et al. 2017, 2018; Sproles et al. 2019). Such versatile, yet selective, many-to-many partnerships are the basis for the stability in cnidarian–dinoflagellate symbioses.

12.2.2 Oophila*: Amphibian Egg-Loving Green Algae*

Symbioses between amphibian embryos and green algae, the latter referred to as *Oophila*, have been known for more than a century (Orr 1888). Symbiont algae live and bloom inside the intracapsular region, rendering a conspicuous green hue to the eggs (Kerney et al. 2011). Previous studies suggest the symbiont algae benefit the host animals by increasing the concentration of oxygen via photosynthesis activity and possibly by producing additional molecules that promote embryonic development (Bachmann et al. 1985; Desnitskiy 2017). Conversely, the algae may flourish living inside the eggs by utilizing ammonia excreted by the embryos (Small et al. 2014). While most of the amphibian embryo–green algal associations appear ectosymbiotic in nature, at least one host salamander (i.e. the spotted salamander *Ambystoma maculatum*) also "allows" algal cells to penetrate into its own tissues and even cells, forming endosymbioses (Kerney et al. 2011, 2019). Globally, amphibian embryo–green algal symbioses have been noted from North America (the USA and Canada), Europe, Russia, and Japan (Desnitskiy 2017). Yet, molecular sequence data for the green algal symbionts are currently limited to those associated with four North American salamander and frog species (Kim et al. 2014; Kerney et al. 2019) and the Japanese black salamander (Muto et al. 2017). The phylogenetic analyses of 18S rDNA sequences suggest the algae that associate with these five amphibian embryos are closely related to each other, forming a clade (together with a few free-living *Chlamydomonas* strains) within the Chlamydomonadales (Kim et al. 2014). While Nema et al. (2019) suggested a polyphyly of *Oophila* species by showing several sequences that fall outside of the *Oophila* clade proposed by Kim et al. (2014), all of those sequences were generated from the isolated and cultured algae and not directly from field materials. Laboratory culturing of algae is typically highly selective, such that there is a good chance of minor, non-symbiont algal species outcompeting *Oophila* algae during the isolation process, as noted previously (Kim et al. 2014). Therefore, it is prudent, and perhaps even necessary, to compare the isolates to the source material(s) by sequencing both.

Within the *Oophila* clade, five subclades have been recognized (Kim et al. 2014; Muto et al. 2017). *Oophila* subclade III has been reported from the eggs of three North American amphibians (Fig. 12.3). Conversely, two *Ambystoma* species were shown to associate with at least two *Oophila* subclades (Fig. 12.3). Even a single amphibian embryo could bear multiple *Oophila* subclades (Kim et al. 2014). Together, these observations suggest that the algal switching of amphibian hosts is (and likely has been) occurring whilst at the same time, *Oophila* has been diversifying into genetically discernable subgroups, without developing host specificity. While *Oophila* has gotten the most attention in the context of their associations with amphibian eggs, they also occur in the water outside the amphibian hosts (Lin and Bishop 2015). This suggests that those free-living chlamydomonad taxa branching within the *Oophila* clade (Kim et al. 2014) may have the capacity to colonize amphibian eggs if given the opportunity. Finally, despite *Oophila*'s apparent loving of amphibian eggs, only a small number of unrelated amphibian taxa are colonized

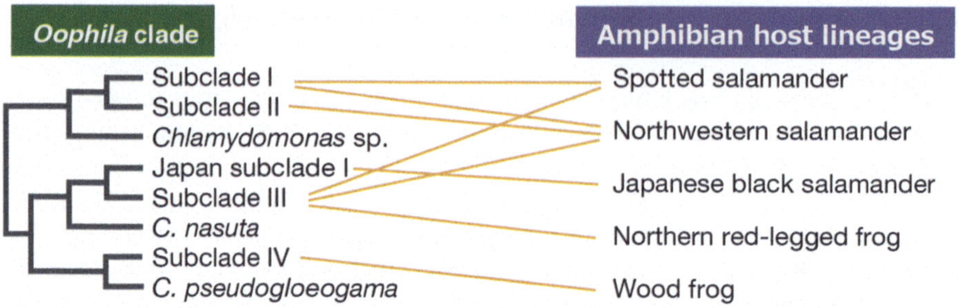

Fig. 12.3 Taxonomic relationships between *Oophila* algae and their amphibian hosts (Kim et al. 2014; Muto et al. 2017; Kerney et al. 2019). Given a limited number of field samples analyzed to date, the actual picture is likely to be more complex than the version presented here

by *Oophila*. To summarize, the amphibian–algal symbioses are affected by the ecology and life history of host animals, whereas the algae appear to have developed lineage-specific traits that enable their proliferation inside the embryos of diverse amphibians.

12.2.3 **Chlorella-*Related Symbionts***

The *Chlorella* clade (Trebouxiophyceae) is well known for its propensity to form symbioses with a wide range of freshwater organisms, including ciliates, (polyphyletic) testate amoebae, the centroheliozoan *Acanthocystis turfacea*, and invertebrates, such as *Hydra* (Pröschold et al. 2011; Pitsch et al. 2017). While there are several genera, including *Chlorella, Micractinium, Didymogenes,* and *Meyerella*, that have been described for the *Chlorella* clade, their SSU rRNA gene sequences are very similar (on average >98%) (Hoshina et al. 2010), suggesting relatively recent divergence times. In contrast, host eukaryotes that associate with members of the *Chlorella* clade are distributed widely and patchily across the eukaryotic tree of life.

The well-studied freshwater ciliate *Paramecium bursaria* associates primarily with *Chlorella variabilis* or *Micractinium reisseri* (Hoshina et al. 2010; Pröschold et al. 2011). Less frequently, *P. bursaria* has been found to form symbioses with *Chlorella vulgaris* or the chlorophycean green alga *Scenedesmus* sp. (Pröschold et al. 2011). Under laboratory conditions, *P. bursaria* can be induced to associate with non-native *Chlorella*-related algae (Summerer et al. 2007). While the algal–*P. bursaria* associations are mutualistic and stable over generations, aposymbiotic *P. bursaria* occurs in nature (Tonooka and Watanabe 2002), and it is possible to experimentally create symbiont-free *P. bursaria* by growing them in the dark (Summerer et al. 2007). To summarize, *P. bursaria* prefers associating with specific *Chlorella*-clade algae, but, in principle, it has not closed its door to other green algae as symbiotic partners.

In oligotrophic *Sphagnum* peatlands of the Northern Hemisphere, green-pigmented testate amoebae are abundant (Jassey et al. 2015). Referred to as the mixotrophic testate amoebae (MTA), they are a polyphyletic assemblage of microbial protists sharing convergently evolved morphological traits, such as the presence of a shell (or test) (Lara and Gomaa 2017). Taxonomically, MTA are distributed in three eukaryotic supergroups, including Amoebozoa, Rhizaria, and Stramenopiles (Lara and Gomaa 2017). All the surveyed MTA cells by Gomaa et al. (2014) harbored the symbionts of the *Chlorella* clade, with the vast majority of them having nearly identical *rbcL* gene sequences, likely representing a single species. This alga, TACS (Testae Amoeba *Chlorella* Symbiont), is most closely related to the *P. bursaria* symbiont *Chlorella variabilis* and appears exceptionally well adapted for living inside (polyphyletic) testate amoebae (Gomaa et al. 2014).

The *Chlorella*-related algae are also found living inside cells of the *viridissima* group of the freshwater cnidarian *Hydra* (Kobayakawa 2017). Their associations are mutualistic; the algae provide photosynthates to the host and in return receive amino acids and possibly additional growth factors that are synthesized by the host (Hamada et al. 2018). Molecular phylogenetic analyses of several strains of *Hydra viridissima* and their respective algal symbionts suggest an intriguing possibility of cospeciation as the host and algal phylogenies largely mirror each other despite the symbiont algae not being monophyletic within the *Chlorella* clade (Kawaida et al. 2013). Nonetheless, aposymbiotic *Hydra* strains can be colonized by non-native *Chlorella*-related algae under laboratory conditions, although questions remain concerning the long-term stability of such non-natural associations (Kessler et al. 1988; Kawaida et al. 2013).

Three algal groups—the *Symbiodiniaceae* dinoflagellates, *Oophila* clade, and *Chlorella* clade—reviewed here are characterized by their apparent propensity to associate with non-photosynthetic organisms of diverse taxonomic origins. A recent study suggests that Symbiodiniaceae originated ~160 mya and since has diversified (LaJeunesse et al. 2018), whereas their hosts are distributed patchily across different eukaryotic supergroups. While comparable molecular clock data are not currently available for *Oophila* and *Chlorella* clades, their respective hosts are also similarly patchy in taxonomic distribution. Such a pattern indicates that the eco-physiological context is a major driving force in host selection, although host–algal cospeciation may be occurring in some sub-lineages. Host switching is common among some members of these generalist algal groups, and laboratory rearing experiments showed host eukaryotes can often be induced to accommodate non-native relatives of algal symbionts. It is also noteworthy that some algal species within these generalist symbiont lineages are found free-living in nature. Overall, these algae appear to be undergoing abundant symbiosis experiments in nature (and experimentally), exemplified by their variability in host choices and stability in associations. The versatile nature seen in modern-day symbioses might also have been a feature of the protist–algal associations that gave rise to eukaryotic photosynthesis.

12.3 Population Thinking for Symbiosis

Considering the pattern of symbiont inheritance in the phagotrophic protist *Hatena*, not a single cell, but a 'population' (an ecological group of individuals) of algal cells is a more realistic representation of an ancestor of the symbionts found in modern-day *Hatena* cells. Here, a 'population' does not necessarily signify the capacity to interbreed with each other within a group of organisms, but it means that all the members share a geological and ecological niche. In this light, the cyanobacterial ancestor of the plastid was not necessarily a distinct individual, but is rather a conceptual framework encompassing a spatiotemporal continuum.

Given the examples of modern-day symbioses as discussed above, the capacity to endocytose symbiont cells was likely a characteristic of the ancestral hosts and their relatives; the algal symbiont, likewise, had the capacity to invade and stay inside a range of host cells. If the cyanobacterial or eukaryotic algal progenitor of the plastid originated from such a population, having a general characteristic of forming plural host–symbiont partnerships—whether the ancestral algal symbiont was a single cell or a group of plural cells (e.g. a few, hundreds, or millions)—would not make a significant impact on our ability to infer its origin. The cumulative effect of mutations and natural selection processes, genome rearrangements, and so on would 'dilute' signals of genetic variation originally present in algal symbionts over time. If, for example, the algal symbiont experienced a reduction in ploidy, the plurality of the allele information would be concealed.

Another challenge of phylogeny is the stability of the out-group, which serves as a reference point for in-group relationships. In a phylogenetic tree, some might see an out-group taxon (taxa) as a 'static' reference in inferring the internal relationships among in-group taxa of interest. However, the out-group is also dynamically evolving and changing over time. If one discusses the origin of plastids, modern cyanobacteria are often seen as out-group taxa, which have evolved independently of the plastids for more than a billion years. Therefore, one needs to be cautious when inferring evolutionary transitions from cyanobacteria to plastids as modern cyanobacteria are likely very different from their >1 billion-year-old ancestors in their genomes.

Cnidarian–algal relationships suggest that genotypic and phenotypic variations of the host animals and algae could provide more stability and flexibility by the formation of consortia in the face of fluctuating environmental conditions, thereby conferring an evolutionary advantage to the combined unit. A flexible many-to-many partnership may allow the cnidarian animals to find their optimally compatible symbiont algae (and vice versa) under a given condition. In an evolutionary timescale, some characteristics are beneficial and others are not in one environmental context, while they may be opposite in another. A subpopulation may evolve through exclusive partnership into an obligate one, similar to the plastids in plant cells. Hundreds of millions of years in the future, if some corals and algae evolve to form an inseparable unit like modern plants with their distinct protist and cyanobacterial ancestors, it may be difficult to imagine their ancestors were a bit loose in their associations and had sustained a variety of many-to-many relationships in nature.

A classical typological view on an ancestor of the plastid does not seem to fit this flexible host–symbiont partnership accommodating substantial genomic variations. Even if a host cell successfully acquires a symbiont, the offspring generation may lose it, but then the next generation may acquire another symbiont related to the one that their parental lineages once had (Fig. 12.1). In a typological view, the 'lost generation' and the discontinuity in the symbiont pedigree might be a problem, but it would be no surprise if such a scenario indeed played out in the early stages of plastid evolution. Rather, in this case, the hosts and the symbionts may better be described as closely associating populations of different organisms.

Population thinking implies the symbiotic spectrum is applicable even within a single population; some members in a symbiont population may be more beneficial to hosts (mutualistic), while others may be less productive or more parasitic (Rueckert et al. 2019). There is merely a conceptual distinction between mutualists and parasites, and it is often extremely difficult to apply this concept to organisms in nature where the boundary is diffusing and changing. The cost–benefit balance in two associating organisms should be condition-dependent, and the degree of their dependence should differ among individuals. Under each distinct environmental or seasonal condition, symbionts can move along an axis of the continuum between mutualism and parasitism. Consequently, the symbionts behave like a population with variations.

12.4 Conclusion

Free-living ancestors of plastids may have been a diversified population of generalist cyanobacteria that include multiple individuals (i.e. cells, for we envisage the ancestors as unicellular organisms) (Fig. 12.4). This hypothesis suggests that the plastids of eukaryotic algae may be traced back to multiple ancestral cyanobacterial cells, which constituted together with their free-living relatives, a coherent ancestral lineage. Over a long period of time, information on the ancestral population and resolution to distinguish between individuals or subpopulations within the population gets inherently lost. Consequently, a phylogenetic lineage is informed from a single or a few sampled individuals, which do not accurately represent the original population structure. Examples of modern-day host–symbiont relationships suggest that ecological, environmental, and developmental conditions, in addition to phylogenetic constraints, have forged various platforms for endosymbioses. Such population thinking may be helpful in resolving issues surrounding the early evolution of plastids. For example, from the perspective of typological thinking, the presence of proteobacterium-type RuBisCO subunits in red algal plastid genomes creates a perplexing situation as green algae and glaucophytes instead have cyanobacterium-type RuBisCO subunits (Delwiche and Palmer 1996). A usual solution to signal conflict such as this is to invoke a lateral gene transfer event, which is based on typological thinking whereby the common ancestor should have a single distinct genotype. Population thinking, however, does not require such a strict

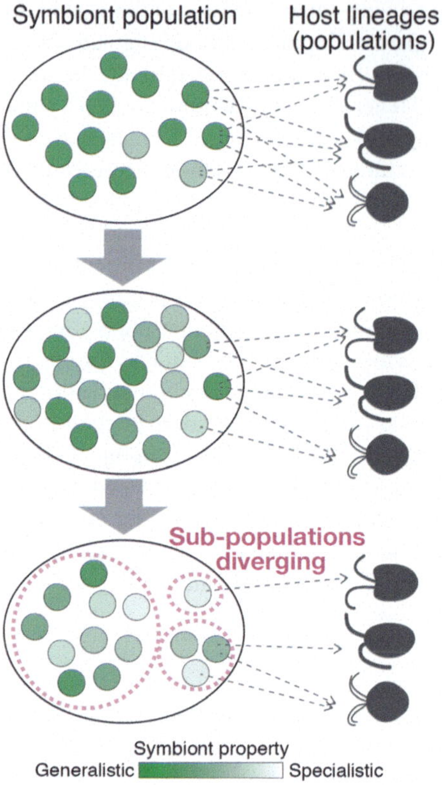

Fig. 12.4 A cartoon illustrating a scenario of plastid evolution. Under this scenario, the plastids originated from a population of a generalist algal symbiont. We hypothesize that such an ancestral generalist symbiont had the capacity to establish partnerships with multiple host lineages. Generalist symbionts are expected to be more stable and persistent through a long time period than specialist symbionts, which could lose their ecological niches more easily (unless their partners proliferate and become abundant). Over time, information on genetic variation within the ancient generalist population is diluted, potentially leading to the fallacy of oversimplification of evolutionary relationships by the observer

assumption, rather it envisages variation and diversity in the genotypes of the common ancestral population. As the research community gathers large-scale genomic and molecular biological data from diverse algal species, we likely face more perplexing cases with our usual pattern-centric approaches. Population thinking may open up a way to change our perspective on how to analyze and interpret data and allow us to draw a more realistic picture of ancient evolutionary processes.

12.5 Acknowledgements

The authors thank Drs. John Burns and Sally Warring for providing helpful comments. This work was supported by JSPS KAKENHI (grant numbers JP17H05713 and JP19H04713), International Activities Support Program of JSPS Grant-in-Aid for Scientific Research on Innovative Areas (grant number JP16K21737), NSF CAREER program (grant number 1453639), and Simons Foundation (grant number 382790).

References

Adl SM, Bass D, Lane CE et al (2019) Revisions to the classification, nomenclature, and diversity of eukaryotes. J Eukaryot Microbiol 66:4–119

Bachmann A, Beard VL, McCarty PL (1985) Performance characteristics of the anaerobic baffled reactor. Water Res 19:99–106

Baldauf SL, Roger AJ, Wenk-Siefert I, Doolittle WF (2000) A kingdom-level phylogeny of eukaryotes based on combined protein data. Science 290:972–977

Boulotte NM, Dalton SJ, Carroll AG et al (2016) Exploring the *Symbiodinium* rare biosphere provides evidence for symbiont switching in reef-building corals. ISME J 10:2693–2701

Burki F, Kaplan M, Tikhonenkov DV et al (2016) Untangling the early diversification of eukaryotes: a phylogenomic study of the evolutionary origins of Centrohelida, Haptophyta and Cryptista. Proc R Soc B Biol Sci 283:20152802

Cavalier-Smith T (1982) The origins of plastids. Biol J Linn Soc 17:289–306

Cenci U, Sibbald SJ, Curtis BA et al (2018) Nuclear genome sequence of the plastid-lacking cryptomonad *Goniomonas avonlea* provides insights into the evolution of secondary plastids. BMC Biol 16:137

Delwiche CF, Palmer JD (1996) Rampant horizontal transfer and duplication of rubisco genes in eubacteria and plastids. Mol Biol Evol 13:873–882

Desnitskiy A (2017) A symbiosis of amphibian embryos and larvae with unicellular green algae. Russ J Herpetol 24:223–227

Dumas E, Feurtey A, Rodríguez de la Vega RC et al (2020) Independent domestication events in the blue-cheese fungus *Penicillium roqueforti*. Mol Ecol (Early View online)

Gawryluk RMR, Tikhonenkov DV, Hehenberger E et al (2019) Non-photosynthetic predators are sister to red algae. Nature 572:240–243

Gomaa F, Kosakyan A, Heger TJ et al (2014) One alga to rule them all: unrelated mixotrophic testate amoebae (Amoebozoa, Rhizaria and Stramenopiles) share the same symbiont (Trebouxiophyceae). Protist 165:161–176

Hamada M, Schröder K, Bathia J et al (2018) Metabolic co-dependence drives the evolutionarily ancient *Hydra–Chlorella* symbiosis. elife 7:e35122

Hoshina R, Iwataki M, Imamura N (2010) *Chlorella variabilis* and *Micractinium reisseri* sp. nov. (Chlorellaceae, Trebouxiophyceae): redescription of the endosymbiotic green algae of *Paramecium bursaria* (Peniculia, Oligohymenophorea) in the 120th year. Phycol Res 58:188–201

Jassey VEJ, Signarbieux C, Hättenschwiler S et al (2015) An unexpected role for mixotrophs in the response of peatland carbon cycling to climate warming. Sci Rep 5:1–10

Kawaida H, Ohba K, Koutake Y et al (2013) Symbiosis between hydra and chlorella: molecular phylogenetic analysis and experimental study provide insight into its origin and evolution. Mol Phylogenet Evol 66:906–914

Keeling PJ, Archibald JM (2008) Organelle evolution: What's in a name? Curr Biol 18:R345–R347

Kerney R, Kim E, Hangarter RP et al (2011) Intracellular invasion of green algae in a salamander host. Proc Natl Acad Sci 108:6497–6502

Kerney R, Leavitt J, Hill E et al (2019) Co-cultures of *Oophila amblystomatis* between *Ambystoma maculatum* and *Ambystoma gracile* hosts show host-symbiont fidelity. Symbiosis 78:73–85

Kessler E, Huss VAR, Rahat M (1988) Species-specific ability of *Chlorella* strains (Chlorophyceae) to form stable symbioses with *Hydra viridis*. Plant Syst Evol 160:241–246

Kim E, Lin Y, Kerney R et al (2014) Phylogenetic analysis of algal symbionts associated with four North American amphibian egg masses. PLoS One 9:e108915

Kobayakawa Y (2017) Symbiosis between green algae and hydra. In: Grube M, Muggia L, Seckenbach J (eds) Algal and cyanobacteria symbioses. World Scientific, London, pp 347–369

LaJeunesse TC, Parkinson JE, Gabrielson PW et al (2018) Systematic revision of Symbiodiniaceae highlights the antiquity and diversity of coral endosymbionts. Curr Biol 28:2570–2580

Lara E, Gomaa F (2017) Symbiosis between testate amoebae and photosynthetic organisms. In: Grube M, Muggia L, Seckenbach J (eds) Algal and cyanobacteria symbioses. World Scientific, London, pp 399–419

Lhee D, Ha J-S, Kim S et al (2019) Evolutionary dynamics of the chromatophore genome in three photosynthetic *Paulinella* species. Sci Rep 9:1–11

Lin Y, Bishop CD (2015) Identification of free-living *Oophila amblystomatis* (Chlorophyceae) from yellow spotted salamander and wood frog breeding habitat. Phycologia 54:183–191

Martin G, Joseph S, Lucia M (2016) Algal and cyanobacteria symbioses. World Scientific, London

Maruyama S, Kim E (2013) A modern descendant of early green algal phagotrophs. Curr Biol 23:1081–1084

Matthews JL, Crowder CM, Oakley CA et al (2017) Optimal nutrient exchange and immune responses operate in partner specificity in the cnidarian-dinoflagellate symbiosis. Proc Natl Acad Sci 114:13194–13199

Matthews JL, Oakley CA, Lutz A et al (2018) Partner switching and metabolic flux in a model cnidarian–dinoflagellate symbiosis. Proc R Soc B Biol Sci 285:20182336

McFadden GI (2001) Primary and secondary endosymbiosis and the origin of plastids. J Phycol 37:951–959

Mies M, Sumida PYG, Rädecker N, Voolstra CR (2017) Marine invertebrate larvae associated with *Symbiodinium*: a mutualism from the start? Front Ecol Evol 5:56

Moreira D, Le Guyader H, Philippe H (2000) The origin of red algae and the evolution of chloroplasts. Nature 405:69–72

Muto K, Nishikawa K, Kamikawa R, Miyashita H (2017) Symbiotic green algae in eggs of *Hynobius nigrescens*, an amphibian endemic to Japan. Phycol Res 65:171–174

Nakayama T, Inagaki Y (2017) Genomic divergence within non-photosynthetic cyanobacterial endosymbionts in rhopalodiacean diatoms. Sci Rep 7:1–8

Nema M, Hanson ML, Müller KM (2019) Phylogeny of the egg-loving green alga *Oophila amblystomatis* (Chlamydomonadales) and its response to the herbicides atrazine and 2,4-D. Symbiosis 77:23–39

Okamoto N, Inouye I (2005) A secondary symbiosis in progress? Science 310:287–287

Okamoto N, Inouye I (2006) *Hatena arenicola* gen. et sp. nov., a katablepharid undergoing probable plastid acquisition. Protist 157:401–419

Orr H (1888) Note on the development of amphibians, chiefly concerning the central nervous system; with additional observations on the hypophysis, mouth, and the appendages and skeleton of the head. Q J Microsc Sci New Ser 29:295–324

Pitsch G, Adamec L, Dirren S et al (2017) The green *Tetrahymena utriculariae* n. sp. (Ciliophora, Oligohymenophorea) with its endosymbiotic algae (*Micractinium* sp.), living in traps of a carnivorous aquatic plant. J Eukaryot Microbiol 64:322–335

Pochon X, Putnam HM, Gates RD (2014) Multi-gene analysis of *Symbiodinium* dinoflagellates: a perspective on rarity, symbiosis, and evolution. PeerJ 2:e394

Pröschold T, Darienko T, Silva PC et al (2011) The systematics of *Zoochlorella* revisited employing an integrative approach. Environ Microbiol 13:350–364

Rodríguez-Ezpeleta N, Brinkmann H, Burey SC et al (2005) Monophyly of primary photosynthetic eukaryotes: green plants, red algae, and glaucophytes. Curr Biol 15:1325–1330

Rueckert S, Betts EL, Tsaousis AD (2019) The symbiotic spectrum: where do the gregarines fit? Trends Parasitol 35:687–694

Small DP, Bennett RS, Bishop CD (2014) The roles of oxygen and ammonia in the symbiotic relationship between the spotted salamander *Ambystoma maculatum* and the green alga *Oophila amblystomatis* during embryonic development. Symbiosis 64:1–10

Sproles AE, Oakley CA, Matthews JL et al (2019) Proteomics quantifies protein expression changes in a model cnidarian colonised by a thermally tolerant but suboptimal symbiont. ISME J 13:2334–2345

Strassert JFH, Jamy M, Mylnikov AP et al (2019) New phylogenomic analysis of the enigmatic phylum Telonemia further resolves the eukaryote tree of life. Mol Biol Evol 36:757–765

Stroud JT, Losos JB (2020) Bridging the process-pattern divide to understand the origins and early stages of adaptive radiation: a review of approaches with insights from studies of *Anolis* lizards. J Hered 111:33–42

Summerer M, Sonntag B, Sommaruga R (2007) An experimental test of the symbiosis specificity between the ciliate *Paramecium bursaria* and strains of the unicellular green alga *Chlorella*. Environ Microbiol 9:2117–2122

Tonooka Y, Watanabe T (2002) A natural strain of *Paramecium bursaria* lacking symbiotic algae. Eur J Protistol 38:55–58

Chapter 13
The Photosynthetic Adventure of *Paulinella* Spp

Przemysław Gagat, Katarzyna Sidorczuk, Filip Pietluch, and Paweł Mackiewicz

Abstract *Paulinella* photosynthetic species are unicellular, silica shell-forming amoebas classified into the supergroup Rhizaria. They crawl at the bottom of freshwater and brackish environments with the help of filose pseudopodia. These protists have drawn the attention of the scientific community because of two photosynthetic bodies, called chromatophores, that fill up their cells permitting fully photoautotrophic existence. *Paulinella* chromatophores, similarly to primary plastids of the Archaeplastida supergroup (including glaucophytes, red algae as well as green algae and land plants), evolved from free-living cyanobacteria in the process of endosymbiosis. Interestingly, these both cyanobacterial acquisitions occurred independently, thereby undermining the paradigm of the rarity of endosymbiotic events. Chromatophores were derived from α-cyanobacteria relatively recently 60–140 million years ago, whereas primary plastids originated from β-cyanobacteria more than 1.5 billion years ago. Since their acquisition, chromatophore genomes have undergone substantial reduction but not to the extent of primary plastid genomes. Consequently, they have also developed mechanisms for transport of metabolites and nuclear-encoded proteins along with appropriate targeting signals. Therefore, chromatophores of *Paulinella* photosynthetic species, similarly to primary plastids, are true cellular organelles. They not only show that endosymbiotic events might not be so rare but also make a perfect model for studying the process of organellogenesis. In this chapter, we summarize the current knowledge and retrace the fascinating adventure of *Paulinella* species on their way to become photoautotrophic organisms.

The original version of this chapter was revised. The funder information was missing from the acknowledgement section and the funding agency asked the author to change this. The correction to this chapter can be found at https://doi.org/10.1007/978-3-030-51849-3_24

P. Gagat (✉) · K. Sidorczuk · F. Pietluch · P. Mackiewicz (✉)
Faculty of Biotechnology, Department of Bioinformatics and Genomics, University of Wrocław, Wrocław, Poland
e-mail: przemyslaw.gagat@uwr.edu.pl; pamac@smorfland.uni.wroc.pl

M. Kloc (ed.), *Symbiosis: Cellular, Molecular, Medical and Evolutionary Aspects*, Results and Problems in Cell Differentiation 69,
https://doi.org/10.1007/978-3-030-51849-3_13

13.1 Introduction

The endosymbiotic theory states that at least some eukaryotic organelles, including mitochondria and plastids, and possibly the eukaryotic cell itself, evolved from free-living prokaryotes (Archibald 2015). This intriguing concept was first mentioned in a footnote of the 1883 article by a German botanist Andreas Schimper. He alluded that plastids could have originated from a symbiosis or fusion of the colorless and chlorophyll-containing organisms (Schimper 1883; Sato 2020). This idea was subsequently picked up at the beginning of the twentieth century by a Russian biologist Constantin Mereschkowsky. He proposed that plastids evolved from cyanobacteria in a process of endosymbiosis based on (1) the continuity of the plastid lineage, i.e. generation by division from existing plastids, (2) high-level of independence, (3) striking resemblance to cyanobacteria, and (4) observations of 'living' cyanobacteria within algal cells, including the endosymbiotic consortium of *Paulinella chromatophora* (Mereschkowsky 1905, 1910; Sato 2020). A similar concept for mitochondria was subsequently put forward by Ivan Wallin (1927).

For half a century, the endosymbiotic theory was hardly mentioned until the microscopic evidence of DNA in plastids and mitochondria was provided by Ris and Plaut (1962) and Nass and Nass (1963), respectively; the formers also suggested a possibility of endosymbiotic origin of plastids citing among others articles of Mereschkowsky. The discovery of extranuclear DNA triggered an American biologist Lynn Margulis (then Lynn Sagan) to formulate her hypothesis about cell evolution. She postulated endosymbiotic origin for mitochondria and plastids, but also incorrectly for the eukaryotic flagella and the mitotic apparatus (Margulis 1967, 1970). Although Margulis was initially lambasted and faced resistance from the wider scientific community, her promotion of endosymbiosis resulted in the accumulation of biochemical, molecular, and phylogenetic data that eventually led to the acceptance of the endosymbiotic theory in the 1980s (Sato 2020).

Already in her 1967 article, Margulis also briefly mentioned multiple plastid origins of different photosynthetic eukaryotes, suggesting that they must have evolved from photosynthetic eukaryotes ingested by heterotrophic protozoans (Margulis 1967). In subsequent years, thinking about the evolution of such plastids, called secondary or complex, was dominated by the principle of minimizing endosymbiotic events (Cavalier-Smith and Lee 1985). This view quickly became a sort of paradigm for cellular evolution, assuming that transformation of endosymbionts, including those of cyanobacterial origin, into cell organelles, is very difficult and consequently unusually rare (Cavalier-Smith 2013; Zimorski et al. 2014; Gould et al. 2015).

The paradigm of minimizing endosymbiotic events (Cavalier-Smith and Lee 1985) has recently been shaken by the discovery that the already mentioned *P. chromatophora* carries not 'living' cyanobacteria but two highly integrated photosynthetic bodies (Bodył et al. 2012; Gagat et al. 2016). It shows that endosymbiotic events might not be so rare after all. Moreover, *Paulinella* represents the third example of prokaryote–eukaryote endosymbiosis (next to mitochondria and primary plastids), and because it is the most recent one, it is expected to shed some more light on the endosymbiotic process in general.

13.2 Endosymbiosis and the Evolution of Photosynthesis in Archaeplastida

It is easier to infer about endosymbiosis from plastids than mitochondria because mitochondria are present in all eukaryotic cells and presumably have always been there since the time of the last eukaryotic common ancestor (Archibald 2015; Karnkowska et al. 2016). We will never know how exactly mitochondria evolved, but probable models postulate an endosymbiotic merger between two prokaryotic cells, an archaeon and a predecessor of α-proteobacterium (Martin and Müller 1998; Roger et al. 2017). The latter, after radical transformation, would provide energy sufficient for expression of thousands of new proteins and consequently evolution of a novel type of cell, i.e. the eukaryotic cell (Lane and Martin 2010; Lane 2011). The first eukaryotes were heterotrophic by nature, but a subsequent endosymbiotic event called cyanobacterial primary endosymbiosis or just primary endosymbiosis resulted in photosynthetic eukaryotes.

13.2.1 The Primary Endosymbiosis

Sometime between 1.9 and 1.5 billion years ago, after the establishment of first eukaryotes, the second bacterial endosymbiosis took place; a phagotrophic, unicellular eukaryote enslaved a *Gloeomargarita*-like cyanobacterium and transformed it into a photosynthetic plastid, the primary plastid (Yoon et al. 2004; Gould et al. 2008; Sánchez-Baracaldo et al. 2017; Ponce-Toledo et al. 2017) (Fig. 13.1a). Before the transformation, however, the eukaryote must have hunted cyanobacteria for sustenance and engulfed them through phagocytosis. The cyanobacteria not only produced glucose, a valuable energy source, through photosynthesis but also fixed nitrogen as well as synthesized amino and fatty acids among other organic molecules (Tetlow et al. 2005; Kneip et al. 2007; Ku et al. 2014). Moreover, they must have been incredibly abundant because the oxygen they had produced as a side effect of photosynthesis changed a weakly reducing Earth's atmosphere into an oxidizing environment (Schirrmeister et al. 2015). Altogether, the cyanobacteria were ideal prey. Yet, through the test of time, they proved more valuable as photosynthesizing endosymbionts rather than prey (McFadden 2014; Ku et al. 2014). Consequently, their relationship with their former eukaryotic predator progressed from transient, through persistent, to obligate, when the hosts could not survive without their cyanobacterial endosymbionts. For the endosymbionts, it was, however, just the beginning of changes. To become true cell organelles, like mitochondria before, they

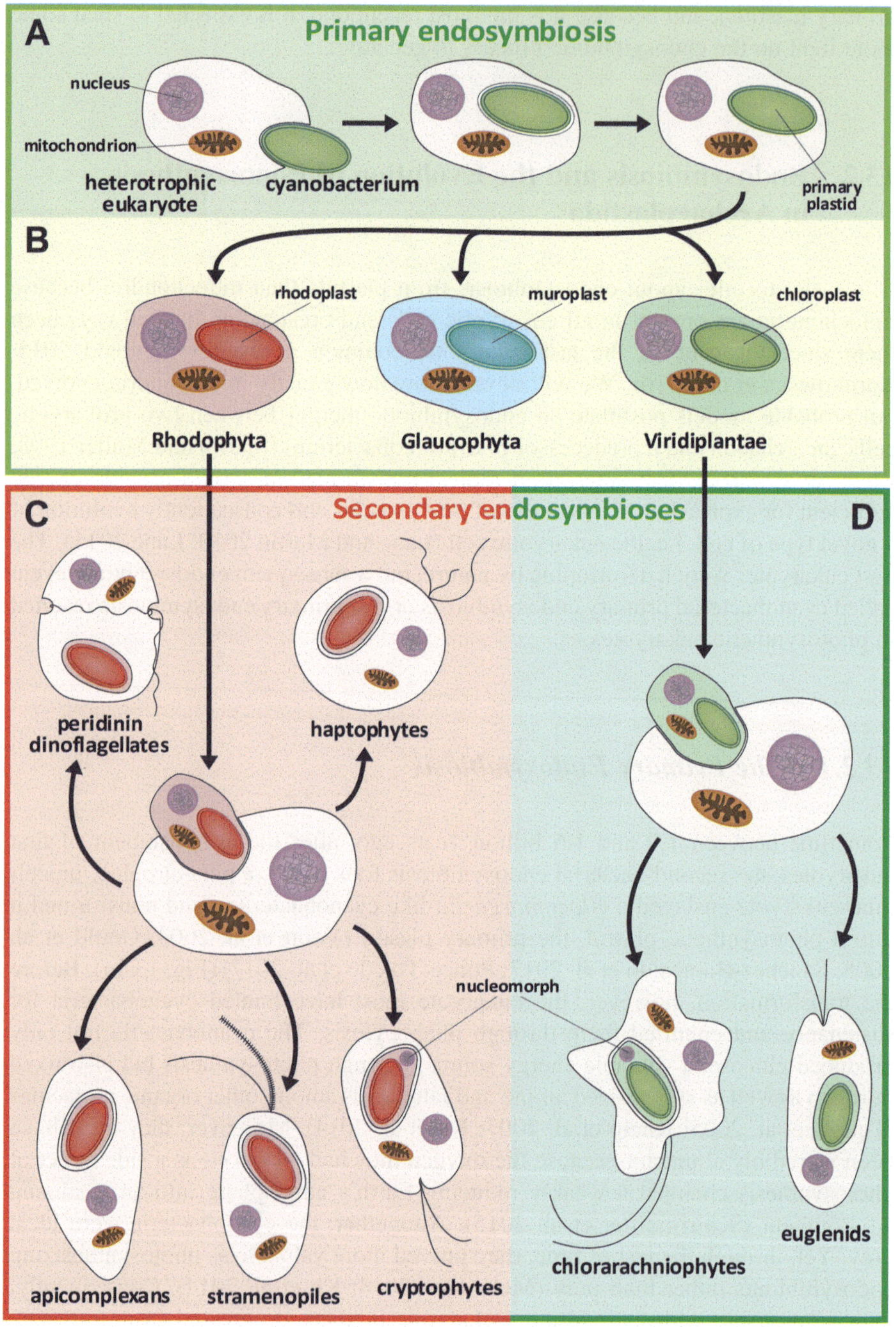

Fig. 13.1 Schematic representation of the evolution of plastids. Primary plastids share a common origin because they evolved from a cyanobacterium engulfed by a heterotrophic eukaryotic cell in a process of primary endosymbiosis (**a**). This endosymbiosis created three autotrophic lineages (**b**): glaucophytes (Glaucophyta), red algae (Rhodophyta), and green algae with their land plant descendants (Viridiplantae). In secondary endosymbioses, red and green algae were horizontally passed to

had to undergo tremendous transformations that involved three key processes: (1) deepening metabolic integration with the host cell, (2) genome reduction by gene loss and endosymbiotic gene transfer (EGT), and (3) evolution of protein import machineries in the endosymbiont membranes for nuclear-encoded proteins along with appropriate targeting signals (Cavalier-Smith and Lee 1985; Bodył et al. 2009a).

Primary plastids, just like cyanobacteria (gram-negative bacteria), are separated from the host (environment) by two membranes, referred to as the outer and inner membranes, and the peptidoglycan layer. Since cyanobacteria were kept and digested within phagosomes, the endosymbionts were initially contained by three membranes. Consequently, there was another precondition for their endosymbiont-to-organelle transformation, i.e. one membrane must have been lost, and very early before the establishment of protein import machinery in the endosymbiont envelope (Bodył et al. 2009a, 2017). Given the fact that the outer membrane of primary plastids has features related both to eukaryotes (e.g. phosphatidylcholine) and prokaryotes (e.g. galactolipids, β-barrel proteins), researchers surmise that the cyanobacterial outer membrane and the phagosome membrane fused and created the chimeric outer membrane seen today. This could have happened during uncoordinated divisions of the endosymbiont while it was escaping the phagosome (Bodył et al. 2009a, 2017).

To fully understand the enormity of cyanobacterium-to-organelle transformation, one should also comprehend the difference between the number of proteins required to keep plastids functional with the number of proteins encoded by their genomes. According to Huang et al. (2003), primary plastids depend on approximately 2000 proteins but their genomes encode only about 130 of them. Moreover, if we compare primary plastid coding capacity with that of their closest, free-living relative, the cyanobacterium *Gloeomargarita lithophora*, with about 3000 potential protein coding genes, the massive EGT becomes even more evident, as well as the loss of numerous bacterial genes that were no longer necessary after the transformation (Ponce-Toledo et al. 2017).

Fig. 13.1 (continued) other eukaryotic lineages; red algae evolved to three membrane plastids of peridinin dinoflagellates and four membrane plastids of cryptophytes, haptophytes, stramenopiles, and apicomplexans (**c**), whereas green algae gave rise to three and four membrane plastids of euglenids and chlorarachniophytes, respectively (**d**). Interestingly, cryptophytes and chlorarachniophytes still retain the reduced nucleus (nucleomorph) of the engulfed red and green alga, respectively. According to the Chromalveolata hypothesis, there was only one secondary red alga endosymbiosis and, consequently, all protists with red alga-derived plastids evolved from a single, red alga-carrying ancestor (Cavalier-Smith 1999, 2013; Zimorski et al. 2014; Gould et al. 2015). However, more and more data support alternative evolutionary scenarios, including hypotheses of serial and multiple endosymbioses, postulating many independent plastid acquisitions (Bodył et al. 2009b; Stiller et al. 2014; Burki 2017; Bodył 2018)

13.2.2 Archaeplastida

The primary endosymbiosis gave rise to three new eukaryotic lineages with three kinds of plastids: Glaucophyta with muroplasts, Rhodophyta with rhodoplasts, and Viridiplantae (green algae and land plants) with chloroplasts, all united into one assemblage: the supergroup of Archaeplastida (Figs. 13.1b and 13.2). Almost all archaeplastidians conduct photosynthesis and only a few lost this ability; however, even the parasitic forms still retain reduced non-photosynthetic plastids (Krause 2008; Wicke and Naumann 2018). The monophyly of Archaeplastida and the single acquisition of their plastid by their common ancestor are widely accepted and seem to be supported by phylogenetic analyses (Mackiewicz and Gagat 2014).

The Glaucophyta is the least numerous phylum of the Archaeplastida supergroup with only 15 described freshwater species (Price et al. 2017). Their plastids were first called cyanelles due to their striking resemblance to cyanobacteria (Pascher 1929), but later the name was changed to more appropriate muroplasts. Similar to other primary plastids, they are surrounded by two membranes, but in contrast to them,

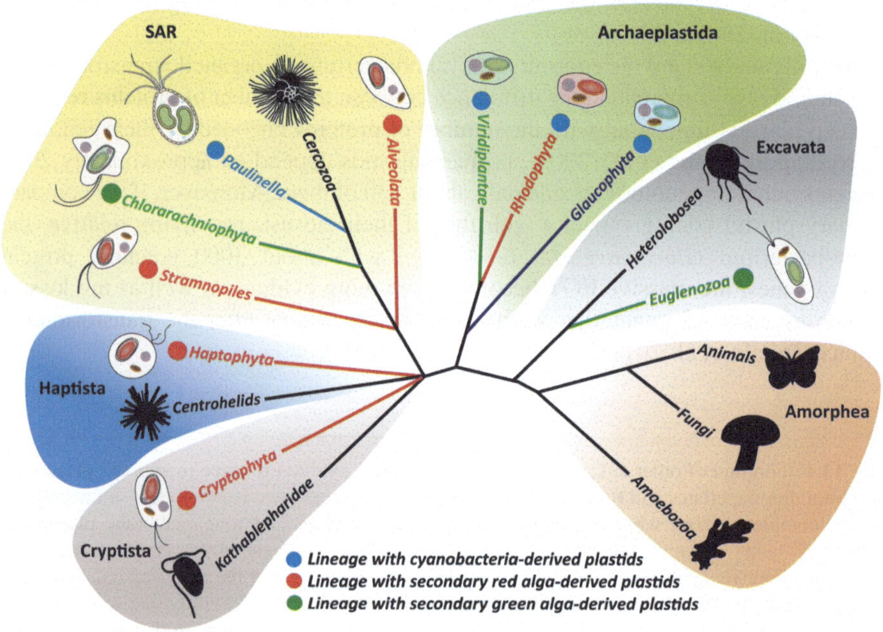

Fig. 13.2 The eukaryotic tree of life. The tree reflects relationships between the large eukaryotic assemblages based on recent phylogenetic analyses (for example, see Brown et al. 2013; Yabuki et al. 2014; Burki et al. 2016; Kern et al. 2020). The black branches correspond to non-photosynthetic eukaryotic clades, whereas the colored ones to photosynthetic lineages; in the case of secondary plastid-bearing lineages, green or red color indicates the type of secondary endosymbiont, green or red alga, respectively. Additionally, plastid-bearing eukaryotes were marked with filled circles, which also correspond to the type of acquired plastid

they preserved more characteristics of their common cyanobacterial ancestor, including a peptidoglycan layer between the outer and inner muroplast membranes. Moreover, they possess phycobilisomes (light-harvesting antennae) and unstacked thylakoids; however, these two features are also shared with rhodoplasts, primary plastids of red algae (Price et al. 2017). The red algae are the second Archaeplastida lineage in terms of abundance with more than 7100 mainly freshwater and marine species reported (Yoon et al. 2017). The most numerous group is Viridiplantae, with possibly more than 500,000 species, divided into two clades: green algae that are mainly aquatic and land plants (Scotland and Wortley 2003; Guiry 2012; Judd et al. 2015). Contrary to other archaeplastidians, which are unicellular and colonial (glaucophytes and red algae) or multicellular (red algae) organisms, land plants have also developed composite tissue organization (Price et al. 2017; Yoon et al. 2017). The Viridiplantae primary plastid, chloroplast, is the most readily recognized photosynthetic organelle. Its thylakoids are arranged in grana stacks and characterized by a high concentration of photosynthetic pigments: chlorophylls a and b. They are also the place of starch synthesis, which was secondarily moved from the cytosol after the Viridiplantae diverged from other Archaeplastida lineages (Ball et al. 2011).

The photosynthetic adventure of eukaryotes has not come to an end with the Archaeplastida because (1) red and green algae were engulfed by other eukaryotes and reduced to multimembrane secondary or complex plastids (Fig. 13.1c, d) and (2) *Paulinella* independently acquired cyanobacteria-derived endosymbionts. Altogether, the establishment of photosynthetic eukaryotes capable of converting light into organic compounds triggered the evolutionary radiation on Earth and resulted in the planet biodiversity we know today (Gould et al. 2008; Archibald 2015) (Fig. 13.2).

13.3 *Paulinella* Photosynthetic Species and Its Chromatophores

Contrary to mitochondria, plastids have been twice established in eukaryotic history. The other known case of cyanobacterial endosymbiosis was discovered within the testate filose amoeba *Paulinella chromatophora*, an organism classified within the supergroup of Rhizaria, which is phylogenetically distant to Archaeplastida (Bhattacharya et al. 1995; Fig. 13.2). Compared to primary plastids, *Paulinella* photosynthetic bodies evolved relatively recently, about 60–140 million years ago (Nowack et al. 2008; Delaye et al. 2016), and therefore represent a very unique opportunity to study cyanobacterial endosymbiosis at a much earlier stage than the one known from Archaeplastida; many similarities and some differences are being discovered and this gives researchers a better understanding of the process of organellogenesis.

13.3.1 Paulinella *Photosynthetic Species*

The person who discovered the species that utterly changed our view on cyanobacterial endosymbiosis and the paradigm of minimizing endosymbiotic events was a German biologist Robert Lauterborn (Lauterborn 1895). Due to his remarkably vast knowledge of zoology, botany, microorganisms, and limnology, he was asked to participate in a long-term project of analyzing water quality of the river Rhine in Germany; the project was motivated by increasing water pollution caused, among other things, by industrial development (Melkonian and Mollenhauer 2005). For many years he studied and described fauna and flora of the river but his most important discovery was made on Christmas Eve 1894. In a sample collected from the riverbed of the Rhine near Neuhofen was a special present for the scientific community, namely *P. chromatophora* (Lauterborn 1895; Melkonian and Mollenhauer 2005). It was a new species of photosynthetic testate amoeba harboring two blue-green, sausage-shaped structures called chromatophores (Lauterborn 1895; Fig. 13.3).

Since Lauterborn's discovery, sightings of *P. chromatophora* have been reported worldwide (Melkonian and Mollenhauer 2005). This filose amoeba lives mainly in freshwater but also brackish environments, where it sluggishly moves at the bottom of water reservoirs (Kepner 1905; Pankow 1982; Melkonian and Mollenhauer 2005). It dwells in eutrophic habitats, characterized by increased salinity, and lowered pH and oxygen content (Lukavský and Cepak 1992; Melkonian and Mollenhauer 2005).

Although *P. chromatophora* was discovered over a hundred years ago, it has not been studied in much detail because it was not isolated into the culture. New insights into this fascinating amoeba were provided by the establishment of its first culture in Germany known as an M0880/a isolate (Marin et al. 2005). The culture was then used to set up a CCAC 0185 strain containing less bacterial contaminants, the first to have its chromatophore genome sequenced (Nowack et al. 2008). Later, another strain, namely FK01, was isolated in Japan and cultured (Yoon et al. 2009). Yoon et al. (2009) performed comparative morphological and phylogenetic analyses of the German and Japanese strains and additionally four other *Paulinella* cell samples collected in Japan from freshwater environments. Their studies led to the conclusion that there are two distinct lineages of *Paulinella*, possibly separate species because the German and Japanese strains differ significantly both in genetic and morphological features (Yoon et al. 2009).

The diversity of *Paulinella* photosynthetic species became evident with the establishment of another photosynthetic *Paulinella* strain KR01, isolated in South Korea, and its comparison with all the available strains (Lhee et al. 2017). The phylogenetic analyses of Lhee et al. (2017) revealed that there are three species of photosynthetic *Paulinella*: *P. chromatophora* (CCAC 0185 strain, fitting the original Lauterborn's description), *P. micropora* (strains FK01 and KR01), and *P. longichromatophora* (Lhee et al. 2017; Table 13.1). In the phylogenetic trees, all these species formed a monophyletic clade that split into two evolutionary

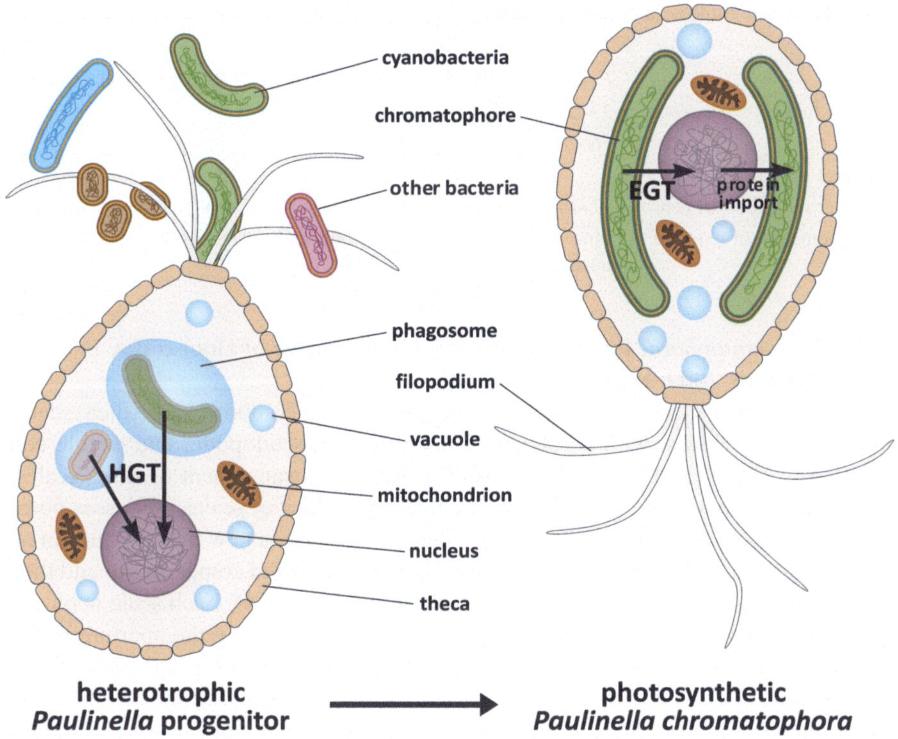

Fig. 13.3 Schematic representation of *Paulinella* before and after cyanobacterial endosymbiosis. *P. chromatophora* is surrounded by the cell wall called theca, which is composed of silica scales. Theca contains an oval aperture through which filopodia enabling the amoeba movement to emerge. Inside the cell, typical eukaryotic organelles are present, including the nucleus, mitochondria, vacuoles, and the photosynthetic bodies of cyanobacterial origin (chromatophores). Before *Paulinella* acquired its chromatophores and became photoautotrophic, it simply fed on bacteria, including cyanobacteria. The cyanobacteria were stored before digestion in phagosomes and consequently separated from the host by three membranes: the phagosomal membrane and a double cyanobacterial envelope. After digestion in phagosomes, some of their genes would get incorporated into the host nuclear genome via endosymbiotic gene transfer (EGT), which is a special case of horizontal gene transfer (HGT) but from an endosymbiont not all available prey. The EGT-, HGT-, and host-derived genes prepared both the cyanobacterium and the amoeba for a more stable relationship between the future endosymbiont and the host. Over time, *Paulinella* learned to benefit from the prolonged upkeep of its cyanobacterial endosymbiont due to their dietary advantages. Consequently, similarly to the primary plastids of Archaeplastida, *Paulinella* cyanobacterial endosymbionts escaped phagosomes and became a permanent feature of the host cell

lineages: (1) *P. micropora* + *P. longichromatophora* and (2) *P. chromatophora* (Lhee et al. 2017). Besides these photosynthetic species, many heterotrophic relatives were also identified: *P. carsoni*, *P. gigantica*, *P. indentata*, *P. intermedia*, *P. lauterborni*, *P. multipora*, *P. ovalis*, and *P. suzukii*. They live in the marine environment, which, interestingly, is also preferred by a photosynthetic species

Table 13.1 Morphological differences between *Paulinella* photosynthetic species (Lhee et al. 2017)

Species	*P. chromatophora*	*P. micropora*	*P. longichromatophora*
Length (μm)	23–29.4	11.5–17	27–35
Width (μm)	16–20	9.1–12.7	14–19
Number of oral scales	3	5	5
Number of scales per row	12–14	8–11	10–12
Ornamentation of external surface	Fine pores and pustules	Fine pores and pustules	Smooth

P. longichromatophora, found in marine sandy sediments of Gomso Bay in South Korea (Nicholls 2009; Kim and Park 2016).

Paulinella cells are surrounded by an oval theca composed of silica scales arranged in five rows, with a collar-like aperture for pseudopodia that enable cell movements (Fig. 13.3). The species differ in size, the arrangement of silica scales, and their ornamentation (Table 13.1). The internal surface of scales possesses pores forming 'sieve-plates' (Kies 1974). Upon the cell division, a new theca is built in an exceptional process. New scales are produced in a specialized compartment and then secreted through the aperture (Kies 1974; Nomura et al. 2014). Each scale is placed in a proper position by pseudopodia to form a new theca facing the aperture of the old one. When it is complete, the cell divides in the old theca and each of the daughter cells possesses one chromatophore (Kies 1974; Nomura et al. 2014). Then, one of the cells is transferred to the new theca through the connection of apertures, using force provided probably by vacuole expansion (Nomura et al. 2014). In the last step of the division, chromatophores of both cells form a U-shaped structure and divide by binary fission (Kies 1974) (Fig. 13.4).

13.3.2 Paulinella *Chromatophores*

The most unusual and unique characteristic of photosynthetic *Paulinella* cells are two chromatophores that allow the amoeba to perform photosynthesis. Although they are larger than their bacterial progenitors and shaped like a sausage, their similarity to cyanobacteria instead of primary plastids was already mentioned in the Lauterborn's original description (Lauterborn 1895; Kies 1974). Mereschkowsky even considered them 'living' cyanobacteria within algal cells in his article introducing the endosymbiotic theory (Sato 2020), but we do know at present that they are highly integrated structures into the *Paulinella* cell and unable to function independently (for review, see Nowack 2014; Bodył et al. 2017).

The early studies by Kies (1974) using electron microscopy revealed that chromatophores are surrounded by two membranes with a peptidoglycan layer located between them. A similar relic of cyanobacterial origin characterizes muroplasts of glaucophytes (Pfanzagl et al. 1996; Price et al. 2017). Chromatophores also possess

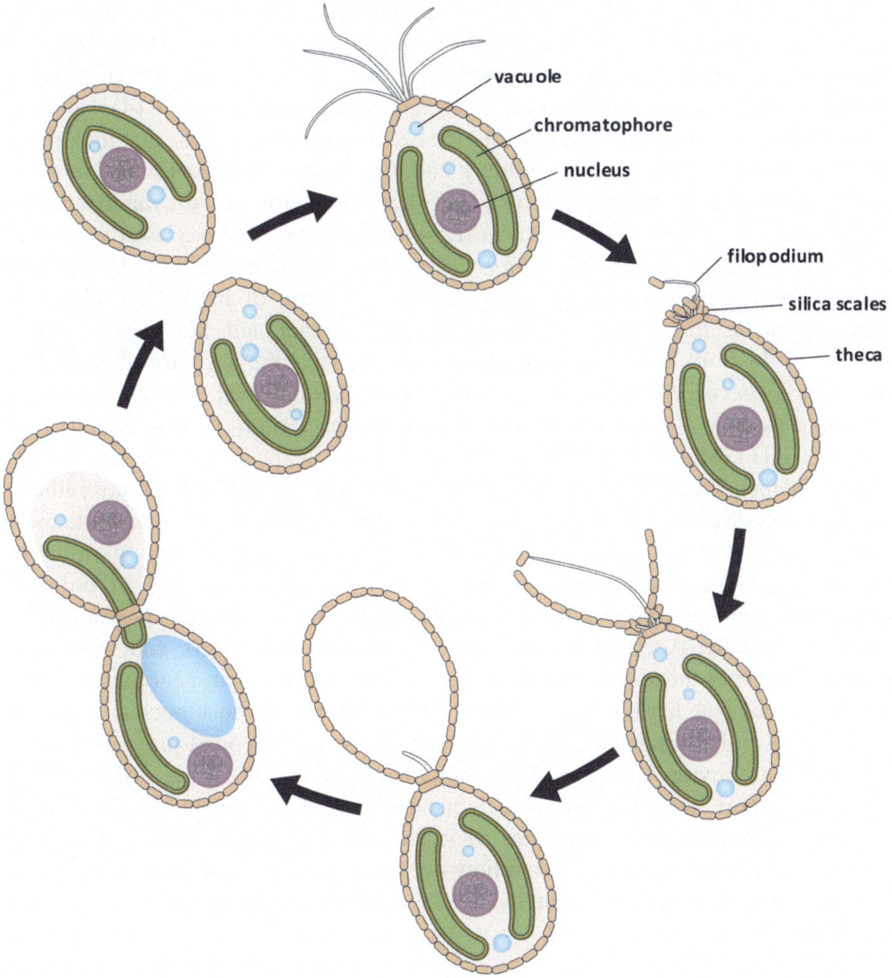

Fig. 13.4 Schematic representation of the cellular division of *Paulinella chromatophora*. Before the division, the mother cell secretes silica scales needed for the construction of the new theca. Scales are produced in specialized vacuoles and accumulated near the aperture. Then, using filopodia, the cell assembles secreted scales into the new theca by placing each scale in a proper place. When the construction is completed, filopodia are withdrawn and the cell division occurs. One of the daughter cells containing one chromatophore is then transferred to the newly assembled theca. Chromatophores of both cells elongate to form a U-shape structure and then divide, yielding two daughter cells with two chromatophores each

carboxysomes, microcompartments involved in CO_2 concentration, and thylakoids covered with cyanobacterial light-harvesting complexes called phycobilisomes (Kies 1974); the latter are also present in glaucophytes and red algae (Gantt 1975; Price et al. 2017; Yoon et al. 2017).

Although chromatophores strongly resemble muroplasts, and indeed at some point were suggested to have evolved from glaucophytes (Raven 2003), phylogenetic analyses clearly indicate that they were acquired independently of primary plastids (Yoon et al. 2009; Burki et al. 2012; Gagat and Mackiewicz 2014). Not only did they evolve in a separate endosymbiotic event, but in the process that took place at least 1 billion years after the primary endosymbiosis (Sánchez-Baracaldo et al. 2017). Moreover, chromatophores are derived from an α-cyanobacterium (containing 1A form of ribulose-1,5-bisphosphate carboxylase/oxygenase, RuBisCo) of *Synechococcus/Prochlorococcus/Cyanobium* clade, whereas primary plastids from a β-cyanobacterium (containing 1B form of RuBisCo) related to *Gloeomargarita lithophora* (Marin et al. 2005; Gagat and Mackiewicz 2014; Ponce-Toledo et al. 2017). Cyanobacteria typically have the 1A form of RuBisCo, but *Synechococcus*, *Prochlorococcus*, and *Cyanobium* replaced their 1A form with 1B form from a *Nitrococcus mobilis*-like proteobacterium via horizontal gene transfer (HGT) (Marin et al. 2007).

The independent cyanobacterial endosymbiosis makes *Paulinella* photosynthetic species a unique system from the evolutionary point of view, but what is the status of these structures? Are they true cell organelles like the primary plastids of Archaeplastida or still prospective organelles, i.e. endosymbionts? In order to answer this question, first, we have to define the difference between an endosymbiont and organelle. According to Cavalier-Smith and Lee (1985), endosymbionts possess all genes necessary for functioning encoded in their genome, whereas organelles have at least some of them transferred to the host nuclear genome. Consequently, the organelles do have to import nuclear-encoded proteins synthesized in the host cytosol, but the endosymbionts do not. In other words, the evolution of protein import machinery seems to define the border between endosymbionts and organelles (Cavalier-Smith and Lee 1985; Theissen and Martin 2006). In the two subsequent chapters, we will present how deeply are the chromatophores integrated into the *Paulinella* cell and that indeed they fulfill the definition of a true cell organelle.

13.4 Metabolic and Genetic Integration of Chromatophores into *Paulinella* Cell

Paulinella chromatophores are well integrated with their host cell because they divide synchronously with it, exchange metabolites with it, and are incapable of independent life (Kies 1974; Kies and Kremer 1979; Marin et al. 2005). The deepening of metabolic integration between the endosymbiont and the host is one of the three key elements of the successful endosymbiont-to-organelle transformation because it provides selective pressure for both parties to tighten the endosymbiotic relationship. In the case of the endosymbiont genomes, the process will be noticeable by loss of genes-encoding proteins: (1) redundant in the host environment and (2) provided by the host, either EGT/HGT-derived or of the host origin.

13.4.1 Chromatophore Genome Evolution

The high level of integration between *Paulinella* chromatophores and their host is reflected in a relatively advanced reduction of the former genomes (Table 13.2, Fig. 13.5). In comparison with their α-cyanobacterial relatives from the *Cyanobium*/*Prochlorococcus*/*Synechococcus* clade (Marin et al. 2005; Gagat and Mackiewicz 2014; Kim and Park 2016; Lhee et al. 2019), their genomes were reduced on average about twofold both in size and coding capacity (Table 13.2, Fig. 13.5). The chromatophore genomes are about 1 Mb, whereas the cyanobacterial genomes usually range from 1.7 to 2.4 Mb; however, the latter can reach even more than 3.34 Mb,

Table 13.2 Comparison of genomes from *Paulinella* chromatophores, primary plastids, and their closest cyanobacterial relatives

Organism Name	Size in Mb	GC%	Protein-coding genes	All genes
Paulinella longichromatophora	0.979	38.8	867	915
Paulinella micropora NZ27	0.977	39.9	863	911
Paulinella micropora KR01	0.977	39.9	860	908
Paulinella micropora FK01	0.977	39.9	860	908
Paulinella chromatophora CCAC0815	1.02	38	878	926
Cyanobacteria related to *Paulinella* (17 genomes)	1.86; 1.64–3.34	36.4; 30.8–68.7	1999; 1785–3255	2059; 1839–3404
Cyanobium gracile PCC6307	3.34	68.7	3255	3404
Prochlorococcus marinus NATL1A	1.86	35.0	1999	2059
Synechococcus sp. WH8102	2.43	59.4	2521	2630
Archaeplastida (4491 genomes)	0.155; 0.011–1.35	37.5; 20.5–57.7	85; 3–273	131; 4–315
Cyanophora paradoxa LB555 (Glaucophyta)	0.136	30.5	149	192
Cyanidioschyzon merolae 10D (Rhodophyta)	0.150	37.6	207	243
Arabidopsis thaliana (Viridiplantae)	0.154	36.3	85	129
Cyanobacterium related to Archaeplastida *Gloeomargarita lithophora* D10	3.05	52.2	2945	3084

The data were obtained from the NCBI Genomes database and Lhee et al. (2019). Summary statistics for Archaeplastida and cyanobacteria contains median and minimum-maximum. The closest relatives of *Paulinella* chromatophores were assumed according to the results by (Marin et al. 2005; Gagat and Mackiewicz 2014; Kim and Park 2016; Lhee et al. 2019)

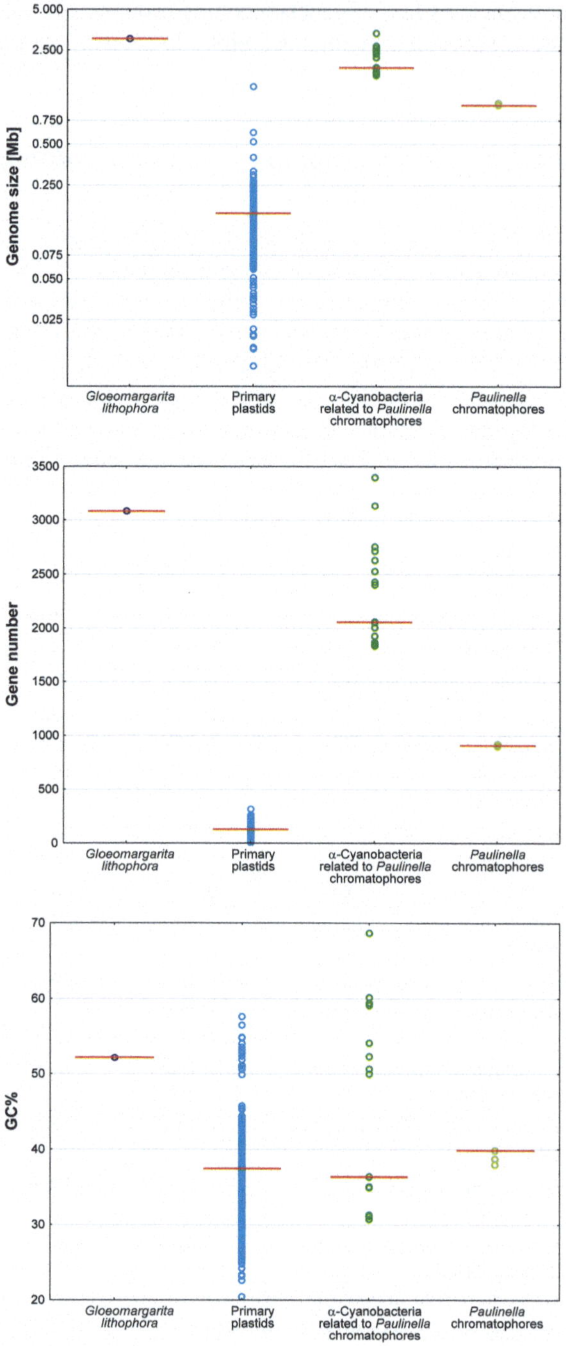

Fig. 13.5 Comparison of genomes from *Paulinella* chromatophores, primary plastids, and their closest cyanobacterial relatives. The data were obtained from the NCBI Genomes database and Lhee et al. (2019). The red horizontal line indicates the median for a given group

e.g. in *Cyanobium gracile* PCC 6307. The total number of genes in chromatophore genomes is about 900, but most cyanobacteria encode from 1900 to 2600 and up to 3404 in *C. gracile* PCC 6307.

The reduction of primary plastid genomes (plastomes) in Archaeplastida is much more substantial, i.e. 20-fold, in comparison with their closest cyanobacterial relative *Gloeomargarita lithophora* (Ponce-Toledo et al. 2017). Assuming that the latter still shares genomic features of the plastid ancestor, the plastomes were reduced from 3.05 Mb to mostly 146–160 kb (Table 13.2, Fig. 13.5), and the number of genes decreased from more than 3000 to only about 130. The most reduced plastome among archaeplastidians, with the size of 11.3 kb and three protein-coding genes, is present in a plant endoparasite *Pilostyles aethiopica* (Bellot and Renner 2015), whereas the largest, a result of expansion of repeated sequences (90% of the genome), is maintained in a green alga *Haematococcus lacustris* (1.35 Mb, 143 protein-coding genes; Bauman et al. 2018; Smith 2018). The highest number of genes encoded by the plastome is 315 and was found in a 117-kb plastid genome of a pine *Pinus koraiensis*. These numbers, including the extreme cases, emphasize the scale of reduction of plastid genomes in Archaeplastida.

Assuming that the primary endosymbiosis involving an archaeplastidial ancestor occurred at least 1.5 billion years ago (Yoon et al. 2004; Parfrey et al. 2011; Zimorski et al. 2014) and that concerning *Paulinella* 90–140 million (Delaye et al. 2016) or 124 million years ago (Lhee et al. 2019), it means that the time of *Paulinella* endosymbionts' integration was about 11–17 times shorter. During this evolutionary period, the primary plastid genomes decreased on average by 2.89 Mb and lost about 2953 genes compared to *G. lithophora*, whereas the chromatophore genomes became smaller by 0.87 Mb and poorer by at least 1145 genes than their closest cyanobacterial relatives. Using these figures, we can estimate that, per 10 million years, the plastid genomes decreased on average by 19 kb and lost on average 20 genes, while the chromatophore genomes decreased by 62–97 kb and were deprived of 82–127 genes. These results indicate that the chromatophore genome reduction is faster than primary plastids' in the whole period of their evolution. It results, however, from the fact that the chromatophores are rather at an early endosymbiotic stage when the genomic changes are more drastic. Since genome reduction is a nonlinear process, we can assume that a similarly high rate of evolution was also typical of primary plastids at the beginning of their endosymbiosis, and then it slowed down.

The genome reduction is usually associated with a decrease in the GC content (or increase in the AT content). This bias was found in the reduced genomes of many microbial endoparasites and endosymbionts (Moran and Wernegreen 2000; Pallen and Wren 2007; McCutcheon and Moran 2010) as well as in mitochondria and plastids (Smith 2012). It is also clearly visible in the data gathered in Table 13.2 and Fig. 13.5 for primary plastid genomes. They are mostly GC-poor, on average 38%, which is much smaller than 52% in the *G. lithophora* genome. Chromatophore genomes have the GC content ranging from 38 to 40%; however, it is not clear what was the nucleotide content in the ancestor of *Paulinella* chromatophores because their close cyanobacterial relatives are either substantially GC-rich (*Cyanobium* and *Synechococcus*) or GC poor (*Prochlorococcus marinus*).

The increase in the genomic AT% is usually explained by enhanced mutation rate (Itoh et al. 2002; Lynch et al. 2006) and Muller's ratchet, which lead to easier fixation of mutations by genetic drift in small asexual populations (Moran 1996; Funk et al. 2001; Woolfit and Bromham 2003). The elevated mutation pressure is coupled with elimination or decrease in the efficiency of genes-encoding DNA repair- and recombination-associated enzymes (Shigenobu et al. 2000; Kang and Hamasaki 2002; Mason et al. 2003; Klasson and Andersson 2004; Moran et al. 2008; Bendich 2010). The biased gene conversion (Khakhlova and Bock 2006), selection for translational efficiency and accuracy (Morton 1998) as well as an adaptation for metabolic efficiency (Jukes and Bhushan 1986; Rocha and Danchin 2002) were also proposed as a process generating the AT bias.

The chromatophore genomes probably also follow the trend associated with the increased mutation rate and AT% because the analyses of their genomes revealed elimination of some genes associated with DNA repair (Nowack et al. 2008). To these genes belong DNA polymerase I (*polA*), involved in DNA repair during synthesis, the genes for UV-excision *uvrABCD*, the repair ATPase *recN*, and the base excision repair gene *mutY*. Moreover, the DNA mismatch repair protein MutS is subjected to severe changes in its sequences compared to *Synechococcus* sp. WH5701 (Nowack et al. 2008). The increase in the AT content was also demonstrated for *Paulinella micropora* strains based on the analysis of mutational pressure (Lhee et al. 2019).

The chromatophore genomes have lost many other genes necessary for free-living organisms (Nowack et al. 2008), which indicates that the chromatophores are quite dependent on their host. It was estimated that 1620 ancestral orthologous gene families were lost leaving 882 such families in the common chromatophore ancestor, which constitutes only 35% of the initial set in the cyanobacterial endosymbiont (Lhee et al. 2019). The chromatophore genomes do not possess genes coding for enzymes involved in: the tricarboxylic acid cycle, several amino acid biosynthetic pathways (Arg, Glu, His, Met, Try), cofactors' syntheses (coenzyme A, NAD, pantothenate, riboflavin, thiamine), and lipopolysaccharides' metabolism. The chromatophore genomes also lost selected genes necessary for biosynthesis of some amino acids (Asn, Asp, Cys, Gln, Lys, Phe, Thr, Tyr), biotin, haem (e.g. *hemD* gene), and peptidoglycan (e.g. *murF* gene), as well as those involved in gene expression (e.g. a DNA ligase *ligA*), lipid metabolism, and Sec-based secretion system. In consequence, these pathways are partially incomplete. A *sulA* gene, playing a role in cell division, and many genes-encoding membrane transporters were also lost, which implies that the host took over the control of the chromatophore division and metabolite transport (Nowack et al. 2008; Nowack 2014). Moreover, the chromatophore genomes do not have many unannotated genes that are characteristic of their cyanobacterial relatives, most probably involved in the response to environmental changes. These genes were lost as they were redundant for permanent endosymbionts of the amoeba cell.

On the other hand, and similarly to the primary plastids, the chromatophore genomes maintained many genes associated with photosynthesis, genetic information processing, biosynthesis of other amino acids (Ala, Ile, Leu, Val), fatty acids, as well as the essential cofactors, such as lipoic acid and folate (Nowack et al. 2008). In comparison to homologs from free-living cyanobacteria, most genes retained in the chromatophore genome have been under relaxed selection, which may be associated with the homogeneous intracellular environment of the host (Lhee et al. 2019). Only genes fulfilling a role in photosynthesis and thus critical for the chromatophore function are subjected to very strong selection.

Despite the drastic reduction of the chromatophore genomes after *Paulinella* cyanobacterial endosymbiosis, the changes slowed down, according to Lhee et al. (2019) about 60 million years ago, when *Paulinella* species diverged from the common ancestor. Since that time, only 40–44 orthologous gene families have been lost and only 3–11 pseudogenes have been reported (Nowack et al. 2008; Lhee et al. 2019). Accordingly, the known chromatophore genomes have a similar size, GC%, and gene content (Table 13.2, Fig. 13.5). More than 90% of orthologous gene families (799 cases) are shared between these genomes (Lhee et al. 2019). Unique chromatophore genes for each species constitute less than 3% (21 orthologues) of the total genes and do not show specific functional features. The genomic comparison also showed significant conservation in chromatophore gene order because only a handful of minor gene rearrangements and up to six inversion events have been reported (Reyes-Prieto et al. 2010; Lhee et al. 2019). It may indicate that the genomes already reached an equilibrium or the process of genomic changes slowed down after the diversification of *Paulinella* photosynthetic species. The inhibition of the genome reorganization after its initial revolutionary changes follows the general trend observed in other intracellular symbionts and parasites (McCutcheon and Moran 2010).

13.4.2 The Role of EGT/HGT and the Host Proteome in Chromatophore Integration

The loss of many genes in the chromatophore genomes was considered to have been triggered by EGT (Martin and Herrmann 1998; Timmis et al. 2004; Huang and Yue 2013). EGT increases the dependence of the endosymbiont on the host and is profitable for both endosymbiotic partners because the transferred genes can be more efficiently regulated in the context of the whole host metabolism. Moreover, the genes located in the nuclear genomes are subjected to lower mutational pressure and influence of free radicals produced by organelles such as mitochondria and plastids. The current studies showed that at least 70 genes (or up to 125) were subjected to EGT in *Paulinella* (Nowack et al. 2011, 2016; Zhang et al. 2017; Singer et al. 2017). They constitute no more than 1% of the host nuclear genome. It is also approximately 10 times less than in the case of the archaeplatidian EGT-derived genes, which fraction was estimated from 6 to 24% depending on the method used and the host nuclear genome considered (Martin et al. 2002; Deusch et al. 2008; Price et al. 2012; Dagan et al. 2013).

Despite the difference in the number of the EGT-derived genes, most of them in *Paulinella* are involved in photosynthesis and photoprotection (Nowack et al. 2011; Zhang et al. 2017) similarly to Archaeplastida (Reyes-Prieto et al. 2006). Many of *Paulinella* EGT-derived genes belong to the family of high-light-inducible proteins. The expansion of this protein family and its host-supervised regulation may be associated with the light/oxidative stress to which *Paulinella* is still subjected, possibly as a consequence of incomplete metabolic integration of both endosymbiotic partners (Zhang et al. 2017). The other EGT-derived genes are responsible for DNA recombination, DNA repair, and protection against oxidative stress (Nowack et al. 2011). Interestingly, the transferred genes in *Paulinella* are small and neutral, which could facilitate their import into chromatophores across their dense and charged peptidoglycan layer (Mackiewicz et al. 2012a).

The EGT process is still ongoing in *Paulinella*, which is exemplified by *csoS4A* (in *P. chromatophora* CCAC 0185) and *psaI* (in *P. micropora* FK01) genes. Their copies are present both in the chromatophore and the nuclear genomes (Reyes-Prieto et al. 2010; Nowack et al. 2011; Nowack 2014). The nuclear version of *csoS4A* is characterized by elevated number of non-synonymous substitutions, suggesting that this copy will be lost or differentiate to gain a new function (Nowack et al. 2011; Mackiewicz et al. 2012a; Nowack 2014). In the case of *psaI*, it is the chromatophore copy that became a pseudogene because it accumulated two nonsense mutations (Reyes-Prieto et al. 2010). Thus, the intact nuclear copy could be functional and an example of a successful transfer according to the general rule favoring the reduction of organelle genomes (Selosse et al. 2001).

Although EGT seems a natural way to acquire genes necessary for the endosymbiont, the analyses of the transcriptome and draft nuclear genome from the first bacteria-free *P. chromatophora* CCAC 0185 culture revealed that there are much more genes of bacterial origin that were obtained through horizontal gene transfer (HGT) from bacteria other than α-cyanobacteria, i.e. via EGT (Nowack 2014; Nowack et al. 2016). Out of 229 genes, only 58 (~25%) arose through EGT from the chromatophore genomes. The rest 171 (~75%) are potential HGT-derived genes from other bacterial sources. It emphasizes the greater importance of HGT than EGT in the integration of *Paulinella* chromatophores. Interestingly, many of the HGT genes have functions that would compensate for gaps in chromatophore metabolic pathways. These genes are, for example, D-Ala-D-Ala ligase, MurF, a DNA polymerase I (PolA), a DNA ligase (LigA), and a serine *O*-acetyltransferase CysE. They were most likely obtained from bacteria on which the phagotrophic ancestor of *Paulinella* photosynthetic species preyed (Nowack et al. 2016) in accordance with the 'you are what you eat' hypothesis (Doolittle 1998) and the 'shopping bag' model (Larkum et al. 2007)—Fig. 13.3. They postulate that a potential host of an endosymbiont regularly acquires genes from distinct food sources and that the genes may be used to establish the endosymbiont as a true cell organelle.

The importance of the HGTs from various bacteria (especially Chlamydiae and Proteobacteria) was also postulated for the establishment of mitochondria and plastids (Ball et al. 2013; Qiu et al. 2013). The contribution of HGT-derived genes to Archaeplastida plastid proteomes was estimated from 7 to 15% (Qiu et al. 2013). They mostly participate in the synthesis of small organic compounds, e.g. amino acids, coenzymes, and vitamins, which are essential in the transformation from phagotrophy to photoautotrophy (Gagat and Mackiewicz 2017).

Interestingly, Singer et al. (2017) suggest that host genes played an even more important role in endosymbiont integration than either EGT- or HGT-derived ones. Out of 433 chromatophore-targeted proteins, only 17 showed a similarity to α-cyanobacteria proteins (EGT candidates), 26 to other bacteria proteins (HGT candidates), whereas as many as 390 were of host- or uncertain-origin. It suggests that the chromatophores are mainly maintained by host proteins that were redirected to chromatophores. The third group of genes-encoding proteins functionally associated with primary plastids constitutes a substantial fraction, i.e. 56–65%, also in Archaeplastida genomes (Qiu et al. 2013).

All these results indicate that both the chromatophore and primary plastid proteomes evolved by evolutionary tinkering using genes from various sources and obtained at different stages of their endosymbioses (Ponce-Toledo et al. 2019).

13.5 Protein Import into Cyanobacteria-Derived Organelles of Archaeplastida and *Paulinella*

13.5.1 Protein Import into Primary Plastids

Given that evolution likes to repeat itself, primary plastids followed in mitochondria footsteps and adapted prokaryotic and eukaryotic components to create their import machineries along with appropriate targeting signals. Both mitochondria and primary plastids evolved two kinds of import pathways for nuclear-encoded proteins: (1) based on transport proteins and (2) on vesicular trafficking. The system based on transport proteins is responsible for the delivery of the vast majority of plastid and mitochondrial nuclear-encoded proteins and it is called Toc-Tic (the translocon at the outer/inner chloroplast membrane; Richardson and Schnell 2019; Chen and Li 2017) and Tom/Tim (the translocon at the outer/inner mitochondrial membrane; Fukasawa et al. 2017; Wiedemann and Pfanner 2017), respectively. They depend on the recognition of N-terminal targeting signals called plastid (pTP) or mitochondrial (mTP) transit peptides and processing proteases that cut off the presequences and degrade them (Kmiec et al. 2018; Ghifari et al. 2019). For greater clarity, only import to primary plastids will be discussed in this section as more relevant for inferring about protein import into *Paulinella* chromatophores.

Nuclear-encoded, plastid-targeted preproteins are maintained in a translocation-competent state by cytosolic chaperones, mostly Hsp70. Some pTPs, which are phosphorylated by cytosolic serine/threonine kinases, are also bound by 14-3-3 protein in order to increase their rate of import (Waegemann and Soil 1991; Sjuts et al. 2017). Preproteins chaperoned by either Hsp70 or 14-3-3 and Hsp70 (so-called guidance complex) are recognized by the outer envelope receptors: Toc159 and Toc34, in a reversible, GTP-dependent manner and transferred to Toc75 (Soll and Schleiff 2004; Kessler and Schnell 2004; Paila et al. 2015). Another receptor of the Toc translocon is Toc64 protein. It contains TPR domains involved in the docking of Hsp90-bound preproteins, which are then passed to Toc34 (Sjuts et al. 2017). The translocation across the outer envelope occurs via β-barrel channel Toc75, a protein belonging to the Omp85 family (Schnell et al. 1994). In the next stage, Tic22 delivers imported preprotein to the Tic translocon (Kouranov et al. 1998; Glaser et al. 2012). Recently, a new Toc-Tic linking component has been identified, Tic236. It is anchored at the inner primary plastid membrane and interacts with both Tic20 and the POTRA domain of Toc75 (Chen et al. 2018; Fig. 13.6).

The exact mechanism of preprotein translocation across the inner primary plastid membrane remains still obscure (Nakai 2018). The classical model of Tic translocon is composed of Tic20, Tic21, Tic110, and Tic40 proteins. The protein-conducting channel can be formed by either Tic20 or Tic21; however, models with Tic110 functioning as a pore have also been proposed (Sjuts et al. 2017). It was recently indicated that Tic110 and Tic20 can form complexes, and accordingly, Tic models with their interactions have also been published (Chen and Li 2017; Bölter 2018) though some argue that there is no direct evidence in support of the claim (Kikuchi et al. 2009, 2013; Nakai 2018). Tic40 interacts with Tic110 and acts as a scaffold for stromal chaperones Hsp70, Hsp90, and Hsp93 that facilitate pulling of imported preproteins into the plastid stroma, where the pTPs are cut off (Sjuts et al. 2017; Fig. 13.6).

The discovery of a novel 1 MDa Tic complex in 2009 resulted in a reconsideration of the classical Tic model (Kikuchi et al. 2009). According to Kikuchi et al. (2009), the 1 MDa translocon consists of Tic20 acting as a protein-conducting channel and three subunits of unclear function: Tic56, Tic100, and Tic214. Tic21 may also associate with this complex but only weak interactions were observed (Kikuchi et al. 2013). Interestingly, Tic110 and Tic40 were not found among components of the new Tic translocon; however, they were suggested to be recruited at later stages of translocation (Nakai 2018) or necessary for the import of only some preproteins (Lee and Hwang 2019; Fig. 13.6).

Regardless of which model is correct, perhaps both are and the Tic complexes complement each other (Sjuts et al. 2017), the protein import through the Tic translocon has been shown to depend on stromal redox state (Stengel et al. 2009; Zhang et al. 2016). Its activity is modulated by additional Tic subunits: Tic32 and Tic62 which form a regulatory redox-sensing complex (Sjuts et al. 2017). For a long time, Tic55 was also suggested to be part of this complex but recently it has been shown to function as a hydroxylase of phyllobilins (Hauenstein et al. 2016).

There is also a small subset of proteins with N-terminal signal peptides targeted to primary plastids in endomembrane vesicles (CAH1, NPP1, αAmy3, and αAmy7; Fig. 13.6; Gagat et al. 2013). Their existence was considered to be evidence for the

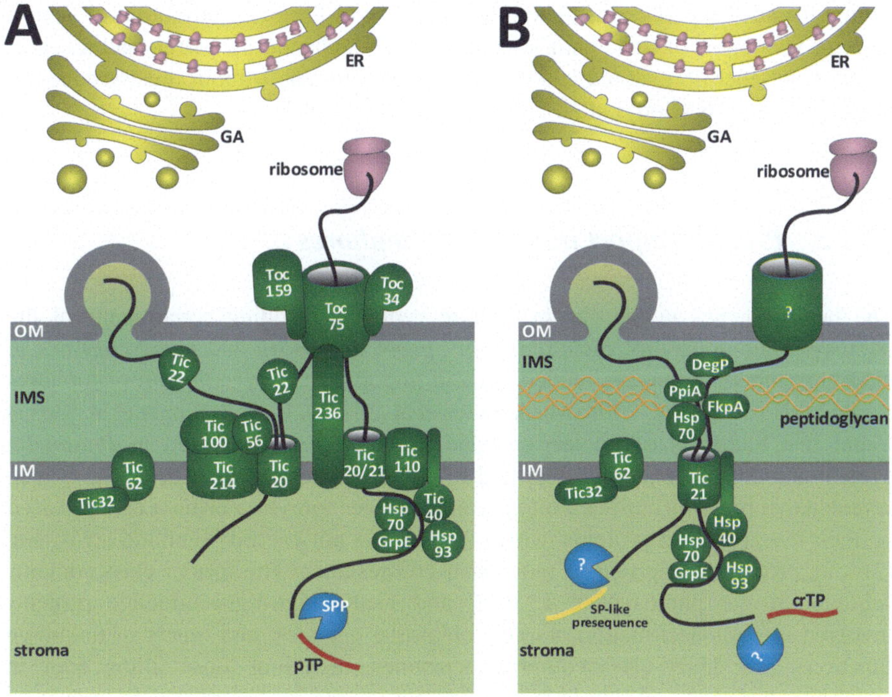

Fig. 13.6 Protein import into primary plastids and *Paulinella* chromatophores. (**a**) In the land plants, the majority of proteins carry transit peptides (pTPs) and are posttranslationally translocated across the outer (OM) and inner (IM) chloroplast membranes with the help of the Toc and Tic translocons, respectively, and their pTPs are cleaved of by stromal processing peptidase (SPP). Only a small subset of proteins is imported in the vesicles of the endomembrane system, including endoplasmic reticulum (ER) and the Golgi apparatus (GA). (**b**) In *Paulinella* chromatophores, some preproteins possess a signal peptide-like (SP-like) presequence and cross the chromatophore envelope with the help of the endomembrane system. Then, chaperones located in the intermembrane space (IMS), such as DegP, PpiA, FkpA, and Hsp70, allow them to get through the peptidoglycan layer. The translocation across the inner membrane occurs via a simplified Tic translocon, similar to chloroplasts, which emphasizes the convergent evolution between both organelles. Interestingly, some proteins imported into chromatophores possess presequences similar to pTPs called chromatophore transit peptides (crTPs). This suggests that a subset of nuclear-encoded, chromatophore-targeted proteins uses yet an unidentified channel to cross the outer chromatophore membrane

so-called 'early endomembrane trafficking' hypothesis, postulating vesicular trafficking to operate in the ancestral plastid of Archaeplastida as the first protein import pathway (Bhattacharya et al. 2007). However, this hypothesis was refuted by multimethod phylogenetic and bioinformatic analyses of proteins that were supposed to support it and analysis of presequences of their homologs (Gagat et al. 2013). Gagat et al. (2013) indicated that vesicular trafficking evolved relatively late in Archaeplastida plastids, contrary to the Tic-Toc translocons, possibly only in higher plants, to permit glycosylation and/or transport to more than one cellular compartment. They also postulated that Toc75 was the most primordial transport protein. In support of this claim, it contains both a channel (β-barrel) and receptor domain and, therefore, could and can translocate proteins without the assistance of other Toc subunits (Bodył et al. 2009a; Gagat et al. 2013).

13.5.2 Protein Import into Chromatophores

The lack of genes in the chromatophore genomes-encoding components of the photosynthetic apparatus, including *psaE*, *psbN*, and *psaK*, and many enzymes of vital biosynthetic pathways renders protein import into *Paulinella* chromatophores essential (Nowack et al. 2008; Mackiewicz et al. 2012b). Indeed, Singer et al. (2017) found that hundreds of nuclear-encoded proteins are synthesized in *Paulinella* cytosol and are subsequently imported to chromatophores. Interestingly, they also revealed that proteins imported into chromatophores form two distinct groups based on their size: (1) short proteins, which length does not exceed 90 amino acids, and (2) long proteins composed of more than 268 residues. The former class contains subunits of the photosystem I: PsaE and PsaK, high-light-inducible proteins involved in photoacclimation, antimicrobial-like peptides, and others of unknown functions. The latter class comprises enzymes, including those filling gaps in metabolic pathways of chromatophores, e.g. pyrroline-5-carboxylate reductase (ProC) and isopropylmalate dehydrogenase (LeuB); they encode proteins involved in proline and leucine biosynthesis, respectively (Singer et al. 2017).

Singer et al. (2017) analyzed both long and short proteins for the presence of typical targeting signals, and short proteins did not seem to contain any. However, Mackiewicz et al. (2012b) previously predicted for *P. micropora* (FK01) PsaE, a signal peptide-like presequence with high confidence (Mackiewicz et al. 2012b). Interestingly, PsaE of *P. chromatophora* CCAC 0185 did not possess such an N-terminal extension but the presence of an alternative, internal, targeting signal was suggested (Mackiewicz et al. 2012a). Mackiewicz et al. (2012b) also found that N-terminal domains of PsaK1 and PsaK2 exhibit increased hydrophobicity in comparison to their cyanobacterial homologs, and therefore could also act as signal peptides.

Unlike the short proteins, the long ones were shown to possess a conserved N-terminal presequence, approximately 200 residues long, that was defined as a chromatophore transit peptide (crTP; Singer et al. 2017). They also used the information about crTPs to construct a model based on support vector machine (SVM) and hidden Markov model (HMM) for in silico prediction of proteins targeted to chromatophores using *P. chromatophora* transcript data. When combined with experimental evidence, the final number of proteins defined as putatively imported to the chromatophores reached 433. The functional analysis of the set revealed that they include proteins involved in: (1) processing of genetic information, (2) metabolism of amino acids, carbohydrates, lipids, cofactors, and nucleotides,

(3) photosynthesis and photoacclimation, (4) oxidative stress, and (5) protein transport (Singer et al. 2017).

The first experimental evidence for protein import into *Paulinella* chromatophores was provided by Nowack and Grossman (2012) for small proteins: PsaE, PsaK1, and PsaK2 of *P. chromatophora* CCAC 0185. These proteins were shown to be synthesized on 80S cytoplasmic ribosomes and then associate with the PSI complex within the chromatophores. Moreover, they used PsaE-specific antibodies for immunogold labeling and demonstrated that the gold particles accumulated not only in chromatophores but also in the Golgi apparatus, thereby providing the first experimental evidence for protein import into chromatophores via the endomembrane system (Nowack and Grossman 2012). Since the vesicular trafficking depends on signal peptides, their results were in accordance with the model published by Mackiewicz et al. (2012a, b; Fig. 13.6).

In contrast, the exact import mechanism for long proteins is still obscure. They are hypothesized to be imported into chromatophores via yet an unidentified translocon located at the outer chromatophore membrane (Fig. 13.6). Its presence is suggested by the crTPs, which are similar to pTPs of Archaeplastida. Moreover, crTPs were shown to allow the import of a yellow fluorescent protein into the chloroplasts of a tobacco *Nicotiana benthamiana* (Singer et al. 2017). This strongly argues for chromatophores containing a similar protein import machinery to the one present in primary plastids. The core of the archaeplastidian Toc translocon at the outer membrane is formed by Toc75 protein-conducting channel (Nakai 2018; Bölter 2018; Fig. 13.6); however, its homolog has been found in *Paulinella* chromatophore genomes or available transcript data so far.

When the preproteins cross the outer chromatophore membrane, there is another obstacle, i.e. the peptidoglycan wall, to overcome (Fig. 13.6). The small proteins could simply diffuse through it as they are characterized by the small size and almost neutral charge (Nowack et al. 2011; Mackiewicz et al. 2012b). The import could also be facilitated by intermembrane space molecular chaperones, such as DegP, FkpA, and PipA, which homologs are present in *Paulinella* chromatophore genomes (Mackiewicz and Bodył 2010; Mackiewicz et al. 2012a, b).

The final step of protein translocation into the chromatophore stroma is the passage across the inner membrane. This stage is suggested to occur via a simplified Tic-like translocon, similar to the one present in plants (Mackiewicz et al. 2012b; Gagat and Mackiewicz 2014; Fig. 13.6). The complex comprises Tic21, Tic32, and Tic62 proteins that were identified in either the *Paulinella* chromatophore or nuclear genome (Mackiewicz et al. 2012b; Gagat and Mackiewicz 2014; Singer et al. 2017). Tic21 constitutes the core of the machinery, i.e. the protein-conducting channel, whereas Tic32 and Tic62 perform the regulatory functions of redox-sensing proteins. Preproteins are pulled into the chromatophore stroma by a molecular motor composed of Hsp40, Hsp70, Hsp93, and GrpE (Mackiewicz et al. 2012b; Gagat and Mackiewicz 2014; Fig. 13.6).

13.6 Antimicrobial Peptides and Evolution of Cyanobacteria-Derived Organelles

13.6.1 Antimicrobial Peptides

Antimicrobial peptides (AMPs) are ancient and evolutionary conserved molecules widespread in all domains of life (Bacteria, Archaea, and Eukaryota) that participate in host defense and/or microbial competition (Kumar et al. 2018). They are short, generally fall within the range of 10–50 amino acids, do not display any consensus sequences, but do share some common features, such as (1) positive charge, (2) hydrophobicity, and (3) amphipathicity. These structural characteristics enable AMPs to preferentially interact with the negatively charged components of the bacterial membranes, e.g. lipopolysaccharides, teichoic acids, some phospholipids (phosphatidylglycerol, cardiolipin, phosphatidylserine) and do not adversely affect the eukaryotic membranes, which are rich in neutral phospholipids (phosphatidylethanolamine, phosphatidylcholine, sphingomyelin) and also stabilizing cholesterol (Bechinger and Gorr 2017). Since AMPs interact with many components of the bacterial membranes, they are also less prone to select for resistance compared to traditional antibiotics; developing resistance to them would require changing the properties of the whole microbial envelope (Andersson et al. 2016; Lázár et al. 2018). AMPs can act on the lipid bilayer in a detergent-like manner and/or penetrate it by forming pores (Marquette and Bechinger 2018). Both interactions lead to membrane disruption resulting in cytoplasmic leakage and finally may trigger cell death. AMPs might also act intracellularly inhibiting, for example, proteases, cell division and biosynthesis of proteins, nucleic acids, and components of the cell wall (Kumar et al. 2018).

13.6.2 A Model for AMP-Dependent Cyanobacterial Endosymbiosis

According to Wollman (2016), AMPs greatly contributed to the establishment of mitochondria and plastids by facilitating two key processes of endosymbiont-to-organelle transformation: (1) endosymbiont gene transfer due to bacterial cell lysis and (2) evolution of efficient protein import machinery by becoming targeting signals (Fig. 13.7). This hypothesis is supported by the fact that AMPs and transit peptides of mitochondria and plastids have similar properties that allow them to interact with membranes of bacterial origin; all show a positive charge and can form amphipathic α-helices in a hydrophobic environment (Bechinger and Gorr 2017; McKinnon and Theg 2019). AMPs have also been demonstrated to keep bacterial endosymbionts in legume plants and insects under control (Login et al. 2011; Tiricz et al. 2013; Stonoha-Arther and Wang 2018). Interestingly, in agreement with Wollman (2016) hypothesis, also in *Paulinella* 39 short chromatophore-targeted,

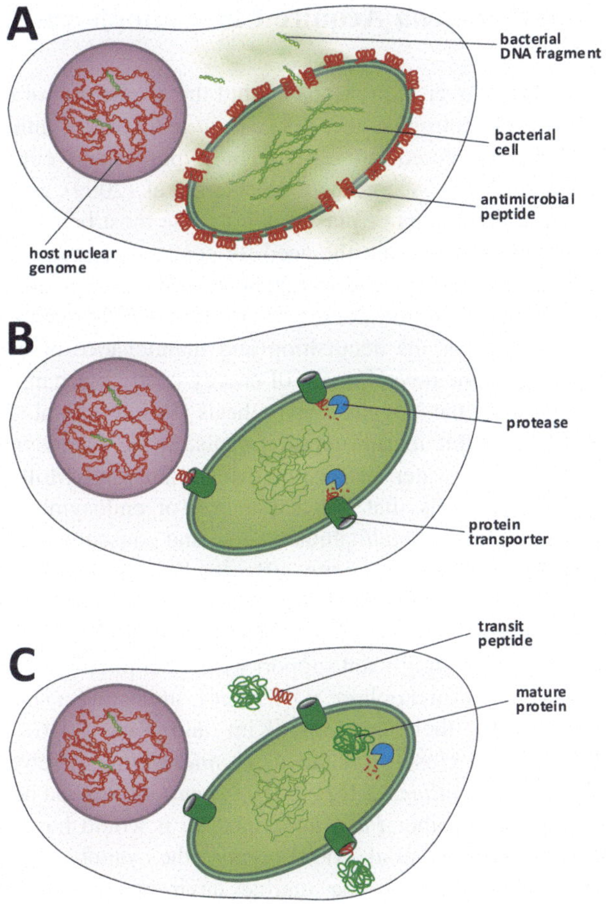

Fig. 13.7 Model for the evolution of protein import into bacteria-derived organelles with antimicrobial peptides (AMPs) playing a central role. Wollman (2016) proposed a three-step scenario explaining the evolutionary transition of AMPs into transit peptides. At first, host AMPs trigger lysis of the bacterial cell, leading to the release of its content into the cytosol, including DNA; some fragments of the bacterial DNA are inserted into the host nuclear genome (**a**). Next, bacteria become resistant to host AMPs by their uptake using a protein transporter and proteolytic degradation by a protease (**b**). In the final step, rearrangements in the host genome lead to the fusion of AMP genes with other nuclear-encoded genes. As a result, nuclear-encoded proteins acquired cleavable transit peptides (TPs) able to deliver them to bacteria-derived organelles using the specific transporters, previously used for AMP uptake. The cell organelles are not in scale, the nucleus is marked in purple and the bacteria-derived organelle in green

nuclear-encoded proteins possess cysteine-rich motifs (CxxC or CxxxxC) and/or positively charged stretches of amino acids characteristic of some AMPs.

13.7 Why Did *Paulinella* Acquire Chromatophores?

It is interesting to ponder over the reasons behind the acquisition of cyanobacterial endosymbionts by the phagotrophic ancestor of *Paulinella* photosynthetic species. A possible explanation is presented in the 'luggage' hypothesis (Wouters et al. 2009; Gagat and Mackiewicz 2017). According to Wouters et al. (2009), as long as the host environment is rich in symbionts or plastid donors, the most beneficial strategy for the host will be eating them without the need for costly investment in their permanent upkeep. However, if the environment changes or the host enters a new habitat, the supply of symbionts or plastid donors might diminish. In such conditions, the selection pressure will favor the acquisition and maintenance of endosymbionts, even though they require the development of energy-costly mechanisms.

Paulinella seems to fit the 'luggage' hypothesis (Wouters et al. 2009) because heterotrophic species inhabit marine environments, while photosynthetic species occur in brackish or freshwater reservoirs (Melkonian and Mollenhauer 2005; Nicholls 2009). This suggests that the acquisition of endosymbionts may have been associated with environmental/habitat change and consequently limited access to food. However, there is one *Paulinella* photosynthetic species, *P. longichromatophora*, that dwells in the marine environment (Kim and Park 2016); but does it disprove Wouters et al. (2009) hypothesis?

This species groups with significant support with other photosynthetic *Paulinella* on both the nuclear and chromatophore rDNA trees, indicating a single acquisition of a cyanobacterium by their ancestor (Kim and Park 2016). Interestingly, *P. longichromatophora* is closely related to *P. micropora*, thereby breaking the monophyly of freshwater *Paulinella*. If *P. longichromatophora* had been in the basal position in the trees to the other *Paulinella* species, it would have disproved the 'luggage' hypothesis (Wouters et al. 2009) because the cyanobacterial endosymbiosis would have taken place in the marine environment. The placement of *P. longichromatophora* among freshwater *Paulinella* species indicates that it must have returned secondarily to the sea (Gagat and Mackiewicz 2017).

Interestingly, the common ancestor of Archaeplastida is also supposed to have acquired their plastids in a freshwater environment as the plastids' closest cyanobacterial relative is a freshwater cyanobacteria *G. lithophora* (Couradeau et al. 2012; Ponce-Toledo et al. 2017). Moreover, this is also corroborated by the fact that glaucophytes, which presumably are the most ancestral Archaeplastida group, exclusively inhabit freshwater ecosystems (Jackson et al. 2015).

The fact that the freshwater habitats seems to be the cradle of both archaeplastidians and *Paulinella* photosynthetic species may also explain why both lineages lost the ability to phagocytize, which is not the case for many other plastid bearing lineages, such as euglenids, chlorarachniophytes, dinoflagellates, haptophytes, and possibly in the case of cryptophytes as well (Bennett et al. 2017; Fig. 13.2). In contrast to them, *Paulinella* and Archaeplastida must have evolved in rather eutrophic (nutrient-rich) environments, where phototrophs can easily acquire simple compounds by endocytic processes and osmotrophy, which are alternative to phagocytosis (Gagat and Mackiewicz 2017). Such conditions have been shown to drive protist diversification into photoautotrophs and heterotrophs with specialized trophic strategies (Troost et al. 2005).

Acknowledgments We are grateful to the reviewers for their helpful comments. We also would like to thank Dr. Malgorzata Kloc for inviting us to write this chapter. This work was supported by the National Science Centre grant UMO-2017/26/D/NZ8/00444 to P. Gagat and UMO-2018/31/N/NZ2/01338 to K. Sidorczuk.

References

Andersson DI, Hughes D, Kubicek-Sutherland JZ (2016) Mechanisms and consequences of bacterial resistance to antimicrobial peptides. Drug Resist Updat 26:43–57

Archibald JM (2015) Endosymbiosis and eukaryotic cell evolution. Curr Biol 25:R911–R921. https://doi.org/10.1016/j.cub.2015.07.055

Ball S, Colleoni C, Cenci U et al (2011) The evolution of glycogen and starch metabolism in eukaryotes gives molecular clues to understand the establishment of plastid endosymbiosis. J Exp Bot 62:1775–1801. https://doi.org/10.1093/jxb/erq411

Ball SG, Subtil A, Bhattacharya D et al (2013) Metabolic effectors secreted by bacterial pathogens: essential facilitators of plastid endosymbiosis? Plant Cell 25:7–21. https://doi.org/10.1105/tpc.112.101329

Bauman N, Akella S, Hann E et al (2018) Next-generation sequencing of *Haematococcus lacustris* reveals an extremely large 1.35-megabase chloroplast genome. Genome Announc 6. https://doi.org/10.1128/genomeA.00181-18

Bechinger B, Gorr S-U (2017) Antimicrobial peptides: mechanisms of action and resistance. J Dent Res 96:254–260

Bellot S, Renner SS (2015) The plastomes of two species in the endoparasite genus pilostyles (Apodanthaceae) each retain just five or six possibly functional genes. Genome Biol Evol 8:189–201. https://doi.org/10.1093/gbe/evv251

Bendich AJ (2010) Mitochondrial DNA, chloroplast DNA and the origins of development in eukaryotic organisms. Biol Direct 5:42. https://doi.org/10.1186/1745-6150-5-42

Bennett RM, Honda D, Beakes GW et al (2017) In: Arch JM, Simpson AGB, Slamovits CH, Margulis Melkonian M, Chapman DJ, Corliss JO (eds) Handbook of the protists. Springer, New York, pp 1–36

Bhattacharya D, Helmchen T, Melkonian M (1995) Molecular evolutionary analyses of nuclear-encoded small subunit ribosomal RNA identify an independent rhizopod lineage containing the Euglyphina and the Chlorarachniophyta. J Eukaryot Microbiol 42:65–69

Bhattacharya D, Archibald JM, Weber APM, Reyes-Prieto A (2007) How do endosymbionts become organelles? Understanding early events in plastid evolution. BioEssays 29:1239–1246. https://doi.org/10.1002/bies.20671

Bodył A (2018) Did some red alga-derived plastids evolve via kleptoplastidy? A hypothesis. Biol Rev 93:201–222

Bodył A, Mackiewicz P, Stiller JW (2009a) Early steps in plastid evolution: current ideas and controversies. BioEssays 31:1219–1232

Bodył A, Stiller JW, Mackiewicz P (2009b) Chromalveolate plastids: direct descent or multiple endosymbioses? Trends Ecol Evol 24:119–121

Bodył A, Mackiewicz P, Gagat P (2012) Organelle evolution: Paulinella breaks a paradigm. Curr Biol 22:R304–R306. https://doi.org/10.1016/j.cub.2012.03.020

Bodył A, Cieśla J, Mackiewicz P, Gagat P (2017) Cyanobacterial endosymbionts of *Paulinella chromatophora* shed new light on the birth of cellular organelles. In: Grube M, Seckbach J, Muggia L (eds) Algal and cyanobacteria symbioses. World Scientific, London, pp 279–328

Bölter B (2018) En route into chloroplasts: preproteins' way home. Photosynth Res 138:263–275. https://doi.org/10.1007/s11120-018-0542-8

Brown MW, Sharpe SC, Silberman JD et al (2013) Phylogenomics demonstrates that breviate flagellates are related to opisthokonts and apusomonads. Proc R Soc B Biol Sci 280:20131755. https://doi.org/10.1098/rspb.2013.1755

Burki F (2017) The convoluted evolution of eukaryotes with complex plastids. In: Hirakawa Y (ed) Advances in botanical research. Academic, New York, pp 1–30

Burki F, Okamoto N, Pombert J-F, Keeling PJ (2012) The evolutionary history of haptophytes and cryptophytes: phylogenomic evidence for separate origins. Proc Biol Sci 279:2246–2254. https://doi.org/10.1098/rspb.2011.2301

Burki F, Kaplan M, Tikhonenkov DV et al (2016) Untangling the early diversification of eukaryotes: a phylogenomic study of the evolutionary origins of Centrohelida, Haptophyta and Cryptista. Proc Biol Sci 283. https://doi.org/10.1098/rspb.2015.2802

Cavalier-Smith T (1999) Principles of protein and lipid targeting in secondary symbiogenesis: euglenoid, dinoflagellate, and sporozoan plastid origins and the eukaryote family tree 1, 2. J Eukaryot Microbiol 46:347–366

Cavalier-Smith T (2013) Symbiogenesis: mechanisms, evolutionary consequences, and systematic implications. Annu Rev Ecol Evol Syst 44:145–172. https://doi.org/10.1146/annurev-ecolsys-110411-160320

Cavalier-Smith T, Lee JJ (1985) Protozoa as hosts for endosymbioses and the conversion of symbionts into organelles. J Protozool 32:376–379. https://doi.org/10.1111/j.1550-7408.1985.tb04031.x

Chen L-J, Li H (2017) Stable megadalton TOC–TIC supercomplexes as major mediators of protein import into chloroplasts. Plant J 92:178–188. https://doi.org/10.1111/tpj.13643

Chen Y-L, Chen L-J, Chu C-C et al (2018) TIC236 links the outer and inner membrane translocons of the chloroplast. Nature 564:125–129. https://doi.org/10.1038/s41586-018-0713-y

Couradeau E, Benzerara K, Gérard E et al (2012) An early-branching microbialite cyanobacterium forms intracellular carbonates. Science 336:459–462

Dagan T, Roettger M, Stucken K et al (2013) Genomes of Stigonematalean cyanobacteria (subsection V) and the evolution of oxygenic photosynthesis from prokaryotes to plastids. Genome Biol Evol 5:31–44. https://doi.org/10.1093/gbe/evs117

Delaye L, Valadez-Cano C, Pérez-Zamorano B (2016) How really ancient is *Paulinella chromatophora*? PLoS Curr 8. https://doi.org/10.1371/currents.tol.e68a099364bb1a1e129a17b4e06b0c6b

Deusch O, Landan G, Roettger M et al (2008) Genes of cyanobacterial origin in plant nuclear genomes point to a heterocyst-forming plastid ancestor. Mol Biol Evol 25:748–761. https://doi.org/10.1093/molbev/msn022

Doolittle WF (1998) You are what you eat: a gene transfer ratchet could account for bacterial genes in eukaryotic nuclear genomes. Trends Genet TIG 14:307–311. https://doi.org/10.1016/s0168-9525(98)01494-2

Fukasawa Y, Oda T, Tomii K, Imai K (2017) Origin and evolutionary alteration of the mitochondrial import system in eukaryotic lineages. Mol Biol Evol 34:1574–1586

Funk DJ, Wernegreen JJ, Moran NA (2001) Intraspecific variation in symbiont genomes: bottlenecks and the aphid-buchnera association. Genetics 157:477–489

Gagat P, Mackiewicz P (2014) Protein translocons in photosynthetic organelles of *Paulinella chromatophora*. Acta Soc Bot Pol 83:399–407. https://doi.org/10.5586/asbp.2014.053

Gagat P, Mackiewicz P (2017) Cymbomonas tetramitiformis – a peculiar prasinophyte with a taste for bacteria sheds light on plastid evolution. Symbiosis 71:1–7. https://doi.org/10.1007/s13199-016-0464-1

Gagat P, Bodył A, Mackiewicz P (2013) How protein targeting to primary plastids via the endomembrane system could have evolved? A new hypothesis based on phylogenetic studies. Biol Direct 8:18

Gagat P, Bodył A, Mackiewicz P (2016) Breaking the endosymbiont-organelle border: the case of *Paulinella chromatophora*. Endocytobiosis Cell Res 27:8

Gantt E (1975) Phycobilisomes: light-harvesting pigment complexes. Bioscience 25:781–788. https://doi.org/10.2307/1297221

Ghifari AS, Huang S, Murcha MW (2019) The peptidases involved in plant mitochondrial protein import. J Exp Bot 70:6005–6018. https://doi.org/10.1093/jxb/erz365

Glaser S, van Dooren GG, Agrawal S et al (2012) Tic22 is an essential chaperone required for protein import into the Apicoplast. J Biol Chem 287:39505–39512. https://doi.org/10.1074/jbc.M112.405100

Gould SB, Waller RF, McFadden GI (2008) Plastid evolution. Annu Rev Plant Biol 59:491–517. https://doi.org/10.1146/annurev.arplant.59.032607.092915

Gould SB, Maier U-G, Martin WF (2015) Protein import and the origin of red complex plastids. Curr Biol 25:R515–R521

Guiry MD (2012) How many species of algae are there? J Phycol 48:1057–1063

Hauenstein M, Christ B, Das A et al (2016) A role for TIC55 as a hydroxylase of phyllobilins, the products of chlorophyll breakdown during plant senescence. Plant Cell 28:2510–2527. https://doi.org/10.1105/tpc.16.00630

Huang J, Yue J (2013) Horizontal gene transfer in the evolution of photosynthetic eukaryotes. J Syst Evol 51:13–29. https://doi.org/10.1111/j.1759-6831.2012.00237.x

Huang CY, Ayliffe MA, Timmis JN (2003) Direct measurement of the transfer rate of chloroplast DNA into the nucleus. Nature 422:72

Itoh T, Martin W, Nei M (2002) Acceleration of genomic evolution caused by enhanced mutation rate in endocellular symbionts. Proc Natl Acad Sci 99:12944–12948. https://doi.org/10.1073/pnas.192449699

Jackson C, Clayden S, Reyes-Prieto A (2015) The Glaucophyta: the blue-green plants in a nutshell. Acta Soc Bot Pol 84:149–165. https://doi.org/10.5586/asbp.2015.020

Judd WS, Campbell CS, Kellogg EA et al (2015) Plant systematics: a phylogenetic approach, 4th edn. Oxford University Press, Oxford

Jukes TH, Bhushan V (1986) Silent nucleotide substitutions and G + C content of some mitochondrial and bacterial genes. J Mol Evol 24:39–44. https://doi.org/10.1007/bf02099949

Kang D, Hamasaki N (2002) Maintenance of mitochondrial DNA integrity: repair and degradation. Curr Genet 41:311–322. https://doi.org/10.1007/s00294-002-0312-0

Karnkowska A, Vacek V, Zubáčová Z et al (2016) A eukaryote without a mitochondrial organelle. Curr Biol 26:1274–1284

Kepner WA (1905) *Paulinella chromatophora*. Biol Bull 9:128–129

Kern R, Facchinelli F, Delwiche C et al (2020) Evolution of photorespiratory glycolate oxidase among Archaeplastida. Plan Theory 9:106. https://doi.org/10.3390/plants9010106

Kessler F, Schnell DJ (2004) Chloroplast protein import: solve the GTPase riddle for entry. Trends Cell Biol 14:334–338. https://doi.org/10.1016/j.tcb.2004.05.004

Khakhlova O, Bock R (2006) Elimination of deleterious mutations in plastid genomes by gene conversion. Plant J Cell Mol Biol 46:85–94. https://doi.org/10.1111/j.1365-313X.2006.02673.x

Kies L (1974) Elektronenmikroskopische Untersuchungen an *Paulinella chromatophora* Lauterborn, einer Thekamobe mit blaugrunen Endosymbionten (Cyanellen). Protoplasma 80:69–89

Kies L, Kremer BP (1979) Function of cyanelles in the thecamoeba *Paulinella chromatophora*. Naturwissenschaften 66:578–579

Kikuchi S, Oishi M, Hirabayashi Y et al (2009) A 1-megadalton translocation complex containing Tic20 and Tic21 mediates chloroplast protein import at the inner envelope membrane. Plant Cell 21:1781–1797. https://doi.org/10.1105/tpc.108.063552

Kikuchi S, Bédard J, Hirano M et al (2013) Uncovering the protein translocon at the chloroplast inner envelope membrane. Science 339:571–574. https://doi.org/10.1126/science.1229262

Kim S, Park MG (2016) Paulinella longichromatophora sp. nov., a new marine photosynthetic testate amoeba containing a chromatophore. Protist 167:1–12. https://doi.org/10.1016/j.protis.2015.11.003

Klasson L, Andersson SGE (2004) Evolution of minimal-gene-sets in host-dependent bacteria. Trends Microbiol 12:37–43. https://doi.org/10.1016/j.tim.2003.11.006

Kmiec B, Branca RMM, Murcha MW et al (2018) A common peptidolytic mechanism for targeting peptide degradation in mitochondria and chloroplasts. Mol Plant 11:342–345. https://doi.org/10.1016/j.molp.2017.11.008

Kneip C, Lockhart P, Voß C, Maier U-G (2007) Nitrogen fixation in eukaryotes–new models for symbiosis. BMC Evol Biol 7:55

Kouranov A, Chen X, Fuks B, Schnell DJ (1998) Tic20 and Tic22 are new components of the protein import apparatus at the chloroplast inner envelope membrane. J Cell Biol 143:991–1002. https://doi.org/10.1083/jcb.143.4.991

Krause K (2008) From chloroplasts to "cryptic" plastids: evolution of plastid genomes in parasitic plants. Curr Genet 54:111

Ku C, Roettger M, Zimorski V et al (2014) Plastid origin: who, when and why? Acta Soc Bot Pol 83:281–289. https://doi.org/10.5586/asbp.2014.045

Kumar P, Kizhakkedathu JN, Straus SK (2018) Antimicrobial peptides: diversity, mechanism of action and strategies to improve the activity and biocompatibility in vivo. Biomol Ther 8:4. https://doi.org/10.3390/biom8010004

Lane N (2011) Energetics and genetics across the prokaryote-eukaryote divide. Biol Direct 6:35. https://doi.org/10.1186/1745-6150-6-35

Lane N, Martin W (2010) The energetics of genome complexity. Nature 467:929–934. https://doi.org/10.1038/nature09486

Larkum AWD, Lockhart PJ, Howe CJ (2007) Shopping for plastids. Trends Plant Sci 12:189–195. https://doi.org/10.1016/j.tplants.2007.03.011

Lauterborn R (1895) Protozoenstudien II. *Paulinella chromatophora* nov. gen., nov. spec., ein beschalter Rhizopode des Sußwassers mit blaugrunen chromatophorenartigen Einschlussen. Z Für Wiss Zool 59:537–544

Lázár V, Martins A, Spohn R et al (2018) Antibiotic-resistant bacteria show widespread collateral sensitivity to antimicrobial peptides. Nat Microbiol 3:718–731

Lee DW, Hwang I (2019) Protein import into chloroplasts via the Tic40-dependent and -independent pathways depends on the amino acid composition of the transit peptide. Biochem Biophys Res Commun 518:66–71. https://doi.org/10.1016/j.bbrc.2019.08.009

Lhee D, Yang EC, Kim JI et al (2017) Diversity of the photosynthetic Paulinella species, with the description of *Paulinella micropora* sp. nov. and the chromatophore genome sequence for strain KR01. Protist 168:155–170. https://doi.org/10.1016/j.protis.2017.01.003

Lhee D, Ha J-S, Kim S et al (2019) Evolutionary dynamics of the chromatophore genome in three photosynthetic Paulinella species. Sci Rep 9:2560. https://doi.org/10.1038/s41598-019-38621-8

Login FH, Balmand S, Vallier A et al (2011) Antimicrobial peptides keep insect endosymbionts under control. Science 334:362–365. https://doi.org/10.1126/science.1209728

Lukavský J, Cepak V (1992) DAPI fluorescent staining of DNA material in cyanelles of the rhizopod *Paulinella chromatophora* LAUTERB. Arch Protistenkd 142:207–212. https://doi.org/10.1016/S0003-9365(11)80086-2

Lynch M, Koskella B, Schaack S (2006) Mutation pressure and the evolution of organelle genomic architecture. Science 311:1727–1730. https://doi.org/10.1126/science.1118884

Mackiewicz P, Bodył A (2010) A hypothesis for import of the nuclear-encoded PsaE protein of *Paulinella chromatophora* (Cercozoa, Rhizaria) into its cyanobacterial endosymbionts/plastids via the endomembrane system. J Phycol 46:847–859. https://doi.org/10.1111/j.1529-8817.2010.00876.x

Mackiewicz P, Gagat P (2014) Monophyly of Archaeplastida supergroup and relationships among its lineages in the light of phylogenetic and phylogenomic studies. Are we close to a consensus? Acta Soc Bot Pol 83:263–280. https://doi.org/10.5586/asbp.2014.044

Mackiewicz P, Bodył A, Gagat P (2012a) Protein import into the photosynthetic organelles of *Paulinella chromatophora* and its implications for primary plastid endosymbiosis. Symbiosis 58:99–107. https://doi.org/10.1007/s13199-012-0202-2

Mackiewicz P, Bodył A, Gagat P (2012b) Possible import routes of proteins into the cyanobacterial endosymbionts/plastids of *Paulinella chromatophora*. Theory Biosci 131:1–18. https://doi.org/10.1007/s12064-011-0147-7

Margulis L (1967) On the origin of mitosing cells. J Theor Biol 14:225–274

Margulis L (1970) Origin of eukaryotic cells: evidence and research implications for a theory of the origin and evolution of microbial, plant and animal cells on the precambrian earth. Yale University Press, New Haven

Marin B, Nowack EC, Melkonian M (2005) A plastid in the making: evidence for a second primary endosymbiosis. Protist 156:425–432. https://doi.org/10.1016/j.protis.2005.09.001

Marin B, Nowack EC, Glöckner G, Melkonian M (2007) The ancestor of the *Paulinella chromatophore* obtained a carboxysomal operon by horizontal gene transfer from a nitrococcus-like γ-proteobacterium. BMC Evol Biol 7:85

Marquette A, Bechinger B (2018) Biophysical investigations elucidating the mechanisms of action of antimicrobial peptides and their synergism. Biomol Ther 8:18

Martin W, Herrmann RG (1998) Gene transfer from organelles to the nucleus: how much, what happens, and why? Plant Physiol 118:9–17. https://doi.org/10.1104/pp.118.1.9

Martin W, Müller M (1998) The hydrogen hypothesis for the first eukaryote. Nature 392:37–41

Martin W, Rujan T, Richly E et al (2002) Evolutionary analysis of Arabidopsis, cyanobacterial, and chloroplast genomes reveals plastid phylogeny and thousands of cyanobacterial genes in the nucleus. Proc Natl Acad Sci 99:12246–12251

Mason PA, Matheson EC, Hall AG, Lightowlers RN (2003) Mismatch repair activity in mammalian mitochondria. Nucleic Acids Res 31:1052–1058. https://doi.org/10.1093/nar/gkg167

McCutcheon JP, Moran NA (2010) Functional convergence in reduced genomes of bacterial symbionts spanning 200 My of evolution. Genome Biol Evol 2:708–718. https://doi.org/10.1093/gbe/evq055

McFadden GI (2014) Origin and evolution of plastids and photosynthesis in eukaryotes. Cold Spring Harb Perspect Biol 6:a016105. https://doi.org/10.1101/cshperspect.a016105

McKinnon L, Theg SM (2019) Determinants of the specificity of protein targeting to chloroplasts or mitochondria. Mol Plant 12:893–895. https://doi.org/10.1016/j.molp.2019.05.004

Melkonian M, Mollenhauer D (2005) Robert Lauterborn (1869-1952) and his *Paulinella chromatophora*. Protist 156:253–262. https://doi.org/10.1016/j.protis.2005.06.001

Mereschkowsky C (1905) Uber natur und ursprung der chromatophoren im pflanzenreiche. Biol Centralbl 25:293–604

Mereschkowsky C (1910) Theorie der zwei Plasmaarten als Grundlage der Symbiogenesis, einer neuen Lehre von der Entstehung der Organismen. Biol Centralbl 30:278–288

Moran NA (1996) Accelerated evolution and Muller's rachet in endosymbiotic bacteria. Proc Natl Acad Sci USA 93:2873–2878. https://doi.org/10.1073/pnas.93.7.2873

Moran NA, Wernegreen JJ (2000) Lifestyle evolution in symbiotic bacteria: insights from genomics. Trends Ecol Evol 15:321–326. https://doi.org/10.1016/s0169-5347(00)01902-9

Moran NA, McCutcheon JP, Nakabachi A (2008) Genomics and evolution of heritable bacterial symbionts. Annu Rev Genet 42:165–190. https://doi.org/10.1146/annurev.genet.41.110306.130119

Morton BR (1998) Selection on the codon bias of chloroplast and cyanelle genes in different plant and algal lineages. J Mol Evol 46:449–459. https://doi.org/10.1007/pl00006325

Nakai M (2018) New perspectives on chloroplast protein import. Plant Cell Physiol 59:1111–1119. https://doi.org/10.1093/pcp/pcy083

Nass MM, Nass S (1963) Intramitochondrial fibers with DNA characteristics: I. Fixation and electron staining reactions. J Cell Biol 19:593–611

Nicholls KH (2009) Six new marine species of the genus Paulinella (Rhizopoda: Filosea, or Rhizaria: Cercozoa). J Mar Biol Assoc UK 89:1415–1425

Nomura M, Nakayama T, Ishida K (2014) Detailed process of shell construction in the photosynthetic testate amoeba *Paulinella chromatophora* (Euglyphid, Rhizaria). J Eukaryot Microbiol 61:317–321. https://doi.org/10.1111/jeu.12102

Nowack ECM (2014) *Paulinella chromatophora*-rethinking the transition from endosymbiont to organelle. Acta Soc Bot Pol 83:387–397

Nowack ECM, Grossman AR (2012) Trafficking of protein into the recently established photosynthetic organelles of *Paulinella chromatophora*. Proc Natl Acad Sci 109:5340–5345. https://doi.org/10.1073/pnas.1118800109

Nowack ECM, Melkonian M, Glöckner G (2008) Chromatophore genome sequence of Paulinella sheds light on acquisition of photosynthesis by eukaryotes. Curr Biol 18:410–418. https://doi.org/10.1016/j.cub.2008.02.051

Nowack ECM, Vogel H, Groth M et al (2011) Endosymbiotic gene transfer and transcriptional regulation of transferred genes in *Paulinella chromatophora*. Mol Biol Evol 28:407–422. https://doi.org/10.1093/molbev/msq209

Nowack ECM, Price DC, Bhattacharya D et al (2016) Gene transfers from diverse bacteria compensate for reductive genome evolution in the chromatophore of *Paulinella chromatophora*. Proc Natl Acad Sci 113:12214–12219. https://doi.org/10.1073/pnas.1608016113

Paila YD, Richardson LGL, Schnell DJ (2015) New insights into the mechanism of chloroplast protein import and its integration with protein quality control, organelle biogenesis and development. J Mol Biol 427:1038–1060. https://doi.org/10.1016/j.jmb.2014.08.016

Pallen MJ, Wren BW (2007) Bacterial pathogenomics. Nature 449:835–842. https://doi.org/10.1038/nature06248

Pankow VH (1982) *Paulinella chromatophora* Lauterb., eine bisher nur im Süßwasser nachgewiesene Thekamöbe, in den Boddengewässern des Darß and des Zingst (südliche Ostsee). Arch Protistenkd 126:261–263. https://doi.org/10.1016/S0003-9365(82)80036-5

Parfrey LW, Lahr DJG, Knoll AH, Katz LA (2011) Estimating the timing of early eukaryotic diversification with multigene molecular clocks. Proc Natl Acad Sci USA 108:13624–13629. https://doi.org/10.1073/pnas.1110633108

Pascher A (1929) Studien uber Symniosen. I. Ueber einige Symbiosen von Blaualgen in Einzellern. Pringsh Jahrb Wiss Bot 71:386–462

Pfanzagl B, Zenker A, Pittenauer E et al (1996) Primary structure of cyanelle peptidoglycan of Cyanophora paradoxa: a prokaryotic cell wall as part of an organelle envelope. J Bacteriol 178:332–339. https://doi.org/10.1128/jb.178.2.332-339.1996

Ponce-Toledo RI, Deschamps P, López-García P et al (2017) An early-branching freshwater cyanobacterium at the origin of plastids. Curr Biol 27:386–391. https://doi.org/10.1016/j.cub.2016.11.056

Ponce-Toledo RI, López-García P, Moreira D (2019) Horizontal and endosymbiotic gene transfer in early plastid evolution. New Phytol 224:618–624. https://doi.org/10.1111/nph.15965

Price DC, Chan CX, Yoon HS et al (2012) Cyanophora paradoxa genome elucidates origin of photosynthesis in algae and plants. Science 335:843–847. https://doi.org/10.1126/science.1213561

Price D, Steiner JM, Yoon HS et al (2017) Glaucophyta. In: Handbook of the protists, 2nd edn. pp 23–87. https://doi.org/10.1007/978-3-319-28149-0_42

Qiu H, Price DC, Weber APM et al (2013) Assessing the bacterial contribution to the plastid proteome. Trends Plant Sci 18:680–687. https://doi.org/10.1016/j.tplants.2013.09.007

Raven JA (2003) Carboxysomes and peptidoglycan walls of cyanelles: possible physiological functions. Eur J Phycol 38:47–53. https://doi.org/10.1080/0967026031000096245

Reyes-Prieto A, Hackett JD, Soares MB et al (2006) Cyanobacterial contribution to algal nuclear genomes is primarily limited to plastid functions. Curr Biol 16:2320–2325. https://doi.org/10.1016/j.cub.2006.09.063

Reyes-Prieto A, Yoon HS, Moustafa A et al (2010) Differential gene retention in plastids of common recent origin. Mol Biol Evol 27:1530–1537. https://doi.org/10.1093/molbev/msq032

Richardson LGL, Schnell DJ (2019) Origins, function, and regulation of the TOC–TIC general protein import machinery of plastids. J Exp Bot. https://doi.org/10.1093/jxb/erz517
Ris H, Plaut W (1962) Ultrastructure of DNA-containing areas in the chloroplast of Chlamydomonas. J Cell Biol 13:383–391
Rocha EPC, Danchin A (2002) Base composition bias might result from competition for metabolic resources. Trends Genet TIG 18:291–294. https://doi.org/10.1016/S0168-9525(02)02690-2
Roger AJ, Muñoz-Gómez SA, Kamikawa R (2017) The origin and diversification of mitochondria. Curr Biol 27:R1177–R1192. https://doi.org/10.1016/j.cub.2017.09.015
Sánchez-Baracaldo P, Raven JA, Pisani D, Knoll AH (2017) Early photosynthetic eukaryotes inhabited low-salinity habitats. Proc Natl Acad Sci 114:E7737. https://doi.org/10.1073/pnas.1620089114
Sato N (2020) Revisiting Endosymbiotic theories of organelles: retrospects and prospects. Springer, New York
Schimper A (1883) Über die Entwicklung der Chlorophyllkörper und Farbkörper, pp 105–162
Schirrmeister BE, Gugger M, Donoghue PCJ (2015) Cyanobacteria and the great oxidation event: evidence from genes and fossils. Palaeontology 58:769–785. https://doi.org/10.1111/pala.12178
Schnell DJ, Kessler F, Blobel G (1994) Isolation of components of the chloroplast protein import machinery. Science 266:1007–1012. https://doi.org/10.1126/science.7973649
Scotland RW, Wortley AH (2003) How many species of seed plants are there? Taxon 52:101–104
Selosse M-A, Albert B, Godelle B (2001) Reducing the genome size of organelles favours gene transfer to the nucleus. Trends Ecol Evol 16:135–141. https://doi.org/10.1016/s0169-5347(00)02084-x
Shigenobu S, Watanabe H, Hattori M et al (2000) Genome sequence of the endocellular bacterial symbiont of aphids Buchnera sp. APS. Nature 407:81–86. https://doi.org/10.1038/35024074
Singer A, Poschmann G, Mühlich C et al (2017) Massive protein import into the early-evolutionary-stage photosynthetic organelle of the amoeba *Paulinella chromatophora*. Curr Biol 27:2763–2773.e5. https://doi.org/10.1016/j.cub.2017.08.010
Sjuts I, Soll J, Bölter B (2017) Import of soluble proteins into chloroplasts and potential regulatory mechanisms. Front Plant Sci 8. https://doi.org/10.3389/fpls.2017.00168
Smith DR (2012) Updating our view of organelle genome nucleotide landscape. Front Genet 3:175. https://doi.org/10.3389/fgene.2012.00175
Smith DR (2018) *Haematococcus lacustris*: the makings of a giant-sized chloroplast genome. AoB Plants 10:ply058. https://doi.org/10.1093/aobpla/ply058
Soll J, Schleiff E (2004) Protein import into chloroplasts. Nat Rev Mol Cell Biol 5:198–208. https://doi.org/10.1038/nrm1333
Stengel A, Benz JP, Buchanan BB et al (2009) Preprotein import into chloroplasts via the Toc and Tic complexes is regulated by redox signals in *Pisum sativum*. Mol Plant 2:1181–1197. https://doi.org/10.1093/mp/ssp043
Stiller JW, Schreiber J, Yue J et al (2014) The evolution of photosynthesis in chromist algae through serial endosymbioses. Nat Commun 5:1–7. https://doi.org/10.1038/ncomms6764
Stonoha-Arther C, Wang D (2018) Tough love: accommodating intracellular bacteria through directed secretion of antimicrobial peptides during the nitrogen-fixing symbiosis. Curr Opin Plant Biol 44:155–163
Tetlow IJ, Rawsthorne S, Raines C, Emes MJ (2005) Plastid metabolic pathways. In: Møller S (ed) Plastids. Blackwell, Oxford, pp 60–109
Theissen U, Martin W (2006) The difference between organelles and endosymbionts. Curr Biol 16: R1016–R1017. https://doi.org/10.1016/j.cub.2006.11.020
Timmis JN, Ayliffe MA, Huang CY, Martin W (2004) Endosymbiotic gene transfer: organelle genomes forge eukaryotic chromosomes. Nat Rev Genet 5:123–135. https://doi.org/10.1038/nrg1271

Tiricz H, Szucs A, Farkas A et al (2013) Antimicrobial nodule-specific cysteine-rich peptides induce membrane depolarization-associated changes in the transcriptome of *Sinorhizobium meliloti*. Appl Environ Microbiol 79:6737–6746. https://doi.org/10.1128/AEM.01791-13

Troost TA, Kooi BW, Kooijman SALM (2005) Ecological specialization of mixotrophic plankton in a mixed water column. Am Nat 166:E45–E61. https://doi.org/10.1086/432038

Waegemann K, Soil J (1991) Characterization of the protein import apparatus in isolated outer envelopes of chloroplasts. Plant J 1:149–158. https://doi.org/10.1111/j.1365-313X.1991.00149.x

Wallin IE (1927) Symbionticism and the origin of species. Рипол Классик

Wicke S, Naumann J (2018) Molecular evolution of plastid genomes in parasitic flowering plants. In: Jansen RK, Chaw S-M (eds) Advances in botanical research. Academic, New York, pp 315–347

Wiedemann N, Pfanner N (2017) Mitochondrial machineries for protein import and assembly. Annu Rev Biochem 86:685–714. https://doi.org/10.1146/annurev-biochem-060815-014352

Wollman F-A (2016) An antimicrobial origin of transit peptides accounts for early endosymbiotic events. Traffic Cph Den 17:1322–1328. https://doi.org/10.1111/tra.12446

Woolfit M, Bromham L (2003) Increased rates of sequence evolution in endosymbiotic bacteria and fungi with small effective population sizes. Mol Biol Evol 20:1545–1555. https://doi.org/10.1093/molbev/msg167

Wouters J, Raven JA, Minnhagen S, Janson S (2009) The luggage hypothesis: comparisons of two phototrophic hosts with nitrogen-fixing cyanobacteria and implications for analogous life strategies for kleptoplastids/secondary symbiosis in dinoflagellates. Symbiosis 49:61–70

Yabuki A, Kamikawa R, Ishikawa SA et al (2014) Palpitomonas bilix represents a basal cryptist lineage: insight into the character evolution in Cryptista. Sci Rep 4:4641. https://doi.org/10.1038/srep04641

Yoon HS, Hackett JD, Ciniglia C et al (2004) A molecular timeline for the origin of photosynthetic eukaryotes. Mol Biol Evol 21:809–818. https://doi.org/10.1093/molbev/msh075

Yoon HS, Nakayama T, Reyes-Prieto A et al (2009) A single origin of the photosynthetic organelle in different Paulinella lineages. BMC Evol Biol 9:98. https://doi.org/10.1186/1471-2148-9-98

Yoon HS, Nelson W, Lindstrom SC et al (2017) Rhodophyta. In: Archibald JM et al (eds) Handbook of the protists, 2nd edn. Springer, New York, pp 89–133. https://doi.org/10.1007/978-3-319-28149-0_33

Zhang D-W, Yuan S, Xu F et al (2016) Light intensity affects chlorophyll synthesis during greening process by metabolite signal from mitochondrial alternative oxidase in Arabidopsis. Plant Cell Environ 39:12–25. https://doi.org/10.1111/pce.12438

Zhang R, Nowack ECM, Price DC et al (2017) Impact of light intensity and quality on chromatophore and nuclear gene expression in *Paulinella chromatophora*, an amoeba with nascent photosynthetic organelles. Plant J 90:221–234. https://doi.org/10.1111/tpj.13488

Zimorski V, Ku C, Martin WF, Gould SB (2014) Endosymbiotic theory for organelle origins. Curr Opin Microbiol 22:38–48. https://doi.org/10.1016/j.mib.2014.09.008

Chapter 14
The Evolutionary Aspects of Legume Nitrogen–Fixing Nodule Symbiosis

Defeng Shen and Ton Bisseling

Abstract Nitrogen-fixing root nodule symbiosis can sustain the development of the host plants under nitrogen-limiting conditions. Such symbiosis occurs only in a clade of angiosperms known as the nitrogen-fixing clade (NFC). It has long been proposed that root nodule symbiosis evolved several times (in parallel) in the NFC. Two recent phylogenomic studies compared the genomes of nodulating and related non-nodulating species across the four orders of the NFC and found that genes essential for nodule formation are lost or pseudogenized in the non-nodulating species. As these symbiosis genes are specifically involved in the symbiotic interaction, it means that the presence of pseudogenes and the loss of symbiosis genes strongly suggest that their ancestor, which still had functional genes, most likely had a symbiosis with nitrogen-fixing bacteria. These findings agree with the hypothesis that nodulation evolved once at the common ancestor of the NFC, and challenge the hypothesis of parallel evolution. In this chapter, we will cover the current understandings on actinorhizal-type and legume nodule development, and discuss the evolution of the legume nodule type.

14.1 Nitrogen-Fixing Clade

Plants can host nitrogen-fixing bacteria by forming novel lateral root organs, named root nodules. Nitrogen-fixing root nodule symbiosis occurs only in the nitrogen-fixing clade (NFC), which is composed of four orders: Fabales, Fagales, Cucurbitales, and Rosales. Legumes (order Fabales) and the non-legume genus *Parasponia* (order Rosales) form nodules with rhizobium bacteria. Other plants that can establish a nitrogen-fixing nodule symbiosis are the actinorhizal plants (orders Fagales, Cucurbitales, and Rosales). They can form nodules with *Frankia*

D. Shen · T. Bisseling (✉)
Laboratory of Molecular Biology, Graduate School Experimental Plant Sciences, Wageningen University, Wageningen, The Netherlands
e-mail: ton.bisseling@wur.nl

M. Kloc (ed.), *Symbiosis: Cellular, Molecular, Medical and Evolutionary Aspects*, Results and Problems in Cell Differentiation 69,
https://doi.org/10.1007/978-3-030-51849-3_14

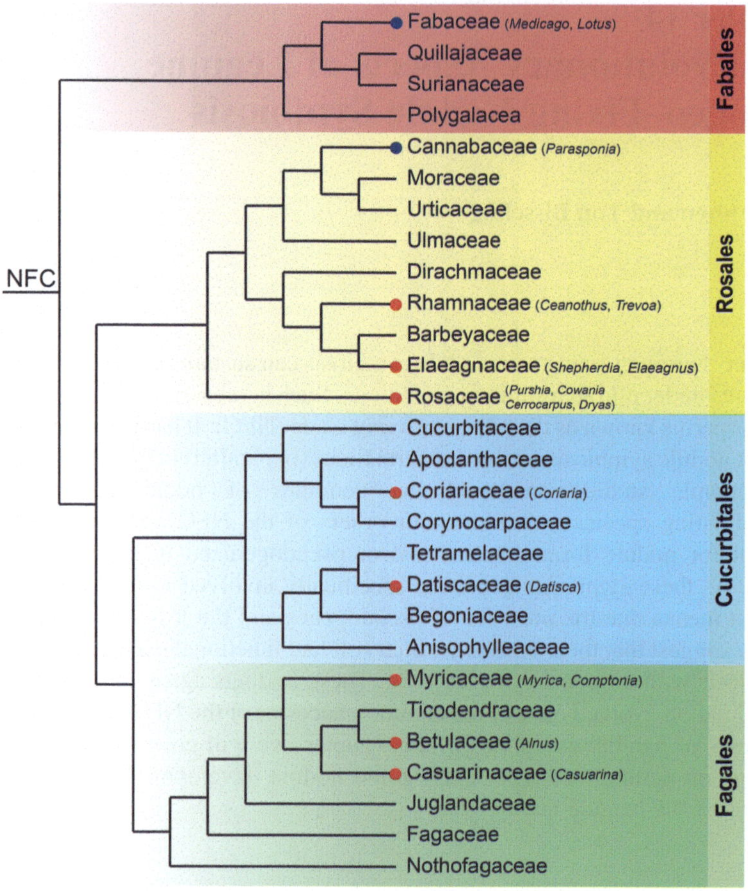

Fig. 14.1 Nitrogen-fixing Clade (NFC). Root nodule symbioses occur only in the NFC, which is composed of four orders: Fabales, Rosales, Cucurbitales, and Fagales. The occurrence of rhizobial symbiosis is indicated with blue circles and *Frankia* symbiosis with red circles. When within a family, symbiosis occurs only in one or a few genera, then these are indicated in parentheses. The exception is Fabaceae (legume) family, in which nodulation is very common. Medicago and Lotus are indicated as model legumes. The phylogenetic tree of the NFC is based on Sun et al. (2016), distribution of nodulating plants based on Tedersoo et al. (2018)

bacteria. It has been hypothesized that nodulation evolved multiple times (in parallel) in the NFC; eight times with *Frankia* and twice with rhizobium (Fig. 14.1). The two major reasons supporting this hypothesis are: first, the ontogeny of legume and actinorhizal nodules is fundamentally different, and it seems unlikely that they could have a common ancestor; second, species forming actinorhizal nodules are rather rare in the three orders in which they occur and independent gain of nodulation seems more parsimonious than massive loss (Soltis et al. 1995; Swensen 1996; Doyle 2011). In this chapter, we will first give an overview of the current knowledge concerning legume and actinorhizal nodule development.

14.2 Legume Nodules

The Leguminosae (Fabaceae) is the third-largest angiosperm family (Legume Phylogeny Working Group, LPWG 2017) and most of its members can form nodules with rhizobium (Tedersoo et al. 2018). Rhizobium is the collective name of Gram-negative bacteria that can form root nodules on legumes and *Parasponia*. They belong to different genera, for example, *Rhizobium*, *Bradyrhizobium*, *Sinorhizobium*, and *Mesorhizobium* (Peter et al. 1996). Legume nodules are generally divided into indeterminate and determinate nodules based on whether they form a persistent nodule meristem or not. In the determinate nodules, a meristem is formed at the periphery of the primordium, leading to a spherical nodule shape. This meristem disappears at an early stage of nodule development (Hirsch 1992; Pawlowski and Bisseling 1996). In indeterminate nodules, a meristem is formed at the apex of the primordium, and it persists to add cells to the different nodule tissues throughout the lifetime of the nodule. By this indeterminate growth, nodules obtain an elongated shape and the central tissue shows a developmental gradient, with the youngest cells adjacent to the nodule meristem and the oldest cells at the proximal region near the root attachment point. The zonation of indeterminate nodules is shown in Fig. 14.2. In the nodule meristem, cells actively divide and are not infected by rhizobia. In the infection zone, cells derived from the nodule meristem become infected by rhizobia. Plant cells and rhizobia gradually enlarge and differentiate, which in both cases involves endoreduplication. In the fixation zone, differentiated rhizobia (bacteroids) start to fix nitrogen. In the senescence zone, nitrogen fixation ceases and the cells are degraded (Pawlowski and Bisseling 1996). The model legumes *Lotus japonicus* (Lotus) and *Medicago truncatula* (Medicago) form determinate and indeterminate nodules, respectively. In this chapter, we will first focus on the model legume species Medicago in the following paragraphs unless otherwise indicated.

14.2.1 Formation of Indeterminate Legume Nodules

The formation of legume nodules involves two coordinated processes: nodule organogenesis and bacterial infection. These two processes are triggered by Nodulation (Nod) factors. Nod factors are lipochitooligosaccharides (LCOs) secreted by Rhizobium bacteria in response to specific compounds, often flavonoids, secreted by plant roots (Limpens et al. 2015). Rhizobia attach to the root hairs and induce root hair curling. In this way, rhizobia become entrapped at the root hair tips in a closed cavity. There, the infection threads are initiated by local cell wall hydrolysis and invagination of the plasma membrane; subsequently, they grow by tip growth. ROS (reactive oxygen species) production is presumed to facilitate the oxidative cross-linking of the infection thread matrix to allow the formation of a tube-like infection thread (Gage 2004; Brewin 2004). The infection threads traverse the root hair and

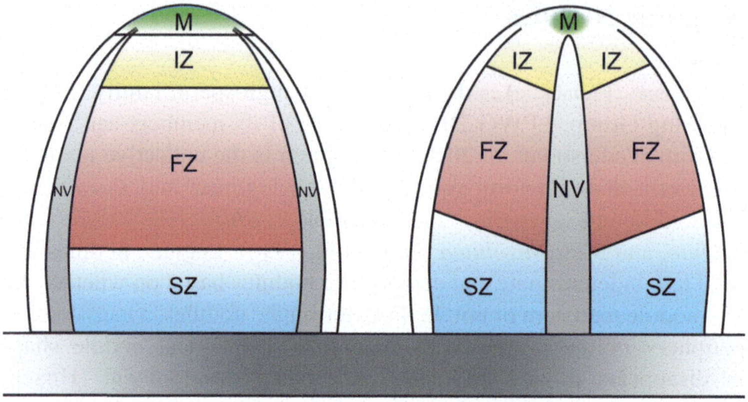

Fig. 14.2 The zonation in legume indeterminate nodules and actinorhizal-type nodules. Left side: the zonation of legume indeterminate nodules. In the nodule meristem (M), cells remain mitotically active and are not infected by rhizobia. In the infection zone (IZ), cells derived from the nodule meristem become infected by rhizobia, which are released from infection threads. Plant and rhizobium cells gradually enlarge and differentiate. In the fixation zone (FZ), the infected cells are fully packed with differentiated rhizobia (bacteroids), which fix nitrogen. In the senescence zone (SZ), nitrogen fixation ceases, and bacteroids disintegrate and plant cells are degraded. Nodule vasculatures (NV) are located at the periphery of the central tissue. Right side: The zonation of actinorhizal-type nodules. Actinorhizal-type nodules are coralloid organs composed of multiple lobes. Shown here is a simplified nodule zonation with one lobe. Actinorhizal-type nodules have a central nodule vasculature (NV), with infected cells in the expanded cortex. In the nodule meristem (M), cells are actively dividing, adding cells to be infected and supporting the growth of the central vasculature. In the infection zone (IZ), cells become gradually filled with branching *Frankia* hyphae. In the fixation zone (FZ) of most host plants, *Frankia* develops vesicles where nitrogen fixation occurs. In the senescence zone (SZ), *Frankia* hyphae and vesicles are degraded. In the nodules formed by actinorhizal Cucurbitales, the infected cells form an uninterrupted domain, on one side of the nodule vasculature. The drawing on the zonation of actinorhizal-type nodules is based on Pawlowski and Demchenko (2012)

outer cortex cells and reach the mitotically activated inner cortical cells (Gage 2004; Brewin 2004; Xiao et al. 2014).

Medicago nodule organogenesis starts with anticlinal cell division in pericycle cells opposite to the protoxylem poles. Subsequently, inner cortical cells (layers C4 and C5) start to divide anticlinally. Cell division is later induced in the endodermis, and it continues in C4 and C5. When the infection thread has reached the middle cortical layer (C3), cells of this layer are mitotically activated. This allows the infection thread to pass the C3 layer, and by continued division C3 forms a multilayered (future) meristem, which is not infected by rhizobia. When C4 and C5 have formed about eight cell layers, and the endodermis and pericycle have formed six to eight cell layers, all these cell layers stop dividing. The C4- and C5-derived cells become penetrated by infection threads from which rhizobia are released (Xiao et al. 2014). During this release, they become surrounded by the plant-derived peribacteroid membrane. In this way, an organelle-like structure, called symbiosome, is formed (Ivanov et al. 2010). The C4- and C5-derived cells at the periphery of the primordium start to differentiate to form peripheral tissues, including vascular bundles. The meristem at the apex starts to add cells to the nodule central tissue as well as the peripheral tissues, including vasculature, by which the nodule grows (Xiao et al. 2014). The process of infection thread penetration and symbiosome formation continues in the nodule and occurs in the cell layer adjacent to the meristem. Upon release, the symbiosomes divide and start to enlarge. Meanwhile, the infected host cells differentiate and enlarge involving endoreduplication, which is mediated by cell cycle regulator *ccs52* (Cebolla et al. 1999). At the transition from infection to fixation zone the expression of nitrogen fixation (*nif*) genes encoding nitrogenase subunits is induced and nitrogen fixation starts (Yang et al. 1991; de Maagd et al. 2011). This transition seems to involve a molecular switch as it is associated with several other changes that occur within one cell layer. This includes the accumulation of starch (Vasse et al. 1990), a major rearrangement of actin cytoskeleton (Gavrin et al. 2015), and the collapse of vacuole (Gavrin et al. 2014). Further, endoreduplication ceases when cells start to fix nitrogen and the expression of *ccs52* is drastically reduced in the fixation zone (Cebolla et al. 1999).

14.2.2 Nod Factor Signaling Pathway

Genetic studies on the model legumes Medicago and Lotus have revealed a set of genes that encode components of the Nod factor signaling pathway. Nod factors are perceived by two types of LysM domain receptor kinases (LysM-RKs) (MtLYK3/MtNFP in Medicago and LjNFR1/LjNFR5 in Lotus) (Madsen et al. 2003; Radutoiu et al. 2003; Limpens et al. 2003; Arrighi et al. 2006; Mulder et al. 2006; Smit et al. 2007). Nod factor receptors trigger nuclear calcium oscillations by activating a signaling cascade, which includes a plasma membrane–localized leucine-rich repeat receptor kinase (MtDMI2 in Medicago, LjSYMRK in Lotus) (Endre et al. 2002; Stracke et al. 2002), nuclear membrane–localized cation channels (potassium-permeable channels: MtDMI1 in Medicago, LjCASTOR and LjPOLLUX in Lotus; and calcium channels: MtCNGCs in Medicago), and several components of the nuclear pore complex (e.g., LjNUP85 and LjNUP133 in Lotus) (Ané et al. 2004; Kanamori et al. 2006; Saito et al. 2007; Charpentier et al. 2008, 2016). Nuclear calcium oscillations are decoded by a nuclear-localized calcium- and calmodulin-dependent protein kinase (MtDMI3 in Medicago; LjCCaMK in Lotus) (Levy 2004; Tirichine et al. 2006). This kinase activates the transcriptional activator MtIPD3/LjCYCLOPS, which subsequently induces the expression of downstream genes (Messinese et al. 2007; Yano et al. 2008).

The above-described part of the Nod factor signaling pathway (common symbiosis signaling pathway) has been proposed to be co-opted from the more ancient arbuscular mycorrhiza (AM) symbiosis. This is because all these Nod factor signaling components, except the Nod factor receptors, are also required for AM symbiosis. As Myc factors and Nod factors are structurally very similar, it has been proposed that the receptors involved in perceiving these factors are similar (Limpens et al. 2015). This hypothesis is in agreement with the fact that Myc factors might be perceived by MtNFP, as the induction of lateral root formation by Myc factors in Medicago is dependent on MtNFP (Maillet et al. 2011). Furthermore, MtNFP is required for the early transcriptional changes induced by Myc factors (Czaja et al. 2012). However, *Mtnfp* and *Ljnfr5* knockout mutants are not impaired in AM colonization (Amor et al. 2003; Radutoiu et al. 2003), suggesting that additional LysM-RK(s) closely related to MtNFP and LjNFR5 is (are) involved in Myc factor perception. In line with this, *Mtlyk3* and *Ljnfr1* mutants have a slightly reduced mycorrhizal infection (Zhang et al. 2015). The involvement of additional Myc factor receptors is also supported by the fact that MtLYR1, a paralog of MtNFP, is upregulated upon mycorrhization (Gomez et al. 2009; Hogekamp et al. 2011; Gaude et al. 2012), suggesting potential functions in the mycorrhizal association. Therefore, it has been hypothesized that duplication of ancestral (LysM-RKs) Myc factor receptor genes enabled neofunctionalization to evolve Nod factor receptors while maintaining Myc factor perception (Op den Camp et al. 2011). Consistent with this, a homolog of MtLYK3/LjNFR1 in *Parasponia andersonii* (Parasponia) has been recently identified that plays a role in AM and root nodule symbioses (R Geurts, pers. comm.).

The primary downstream target of MtIPD3/LjCYCLOPS is the gene encoding the transcription factor NODULE INCEPTION (MtNIN/LjNIN) (Marsh et al. 2007; Singh et al. 2014), which has been shown to be specifically expressed during nodulation and it is essential for infection thread formation and nodule organogenesis (Schauser et al. 1999; Marsh et al. 2007). In *nin* null mutants, excessive root hair curling and deformation are induced by rhizobia, but infection thread, as well as nodule organogenesis, is blocked (Schauser et al. 1999; Marsh et al. 2007). Dominant active forms of the CCaMK protein or a phosphomimetic version of CYCLOPS can induce spontaneous nodule organogenesis, and this is dependent on NIN (Gleason et al. 2006; Tirichine et al. 2006; Singh et al. 2014). In the weak *nin* alleles *daphne* (Lotus) and *daphne-like* (Medicago), the promoter region of *NIN* is mutated. Both weak *nin* alleles have a dramatically increased number of infection threads, but nodule formation is absent (Yoro et al. 2014; Liu et al. 2019). This shows that rhizobial infection and nodule organogenesis can be uncoupled. A detailed study of the promoter region of Medicago *NIN* revealed that the 5 kb region upstream of the *NIN* start codon is sufficient for the epidermal infection process, but nodule organogenesis requires a remote upstream *cis*-regulatory region to induce the expression of *NIN* in the pericycle. This remote region contains putative cytokinin response elements and is conserved in legume species. The gene encoding cytokine receptor 1, which is essential for nodule primordium formation, as well as the B-type cytokine response regulator *RR1*, is expressed in the pericycle cells prior to the induction of *NIN* in the pericycle (Liu et al. 2019). Taken together, it is very likely that *NIN* expression is initially triggered by cytokinin signaling in the pericycle, and this initiates nodule primordium formation. So, at early stages of nodule primordium formation, the expression of *NIN* is induced at two positions (and in different ways): first, induction of *NIN* in the epidermis cells to initiate infection thread formation; second, activation of *NIN* in the pericycle cells by cytokinin signaling to initiate nodule organogenesis. At later stages of primordium formation, the expression of *NIN* extends to the dividing cortical cells. NIN is closely related to NIN-LIKE PROTEINs (NLPs), which are widely present in vascular plants. NIN is orthologous to Arabidopsis NLP1, which is involved in nitrate signaling. NIN has lost the nitrate signaling domain, and its expression came under the control of the Nod factor signaling cascade (Suzuki et al. 2013; Chardin et al. 2014; Van Zeijl et al. 2015). NIN has also been shown to be essential for actinorhizal-type nodules (*Casuarina glauca* and *P. andersonii*) (see below). Therefore, its recruitment into the nodule formation process represents a major step in the evolution of nodulation.

NIN activates the expression of *Nuclear transcription factor Y subunit A-1* (*MtNF-YA1/LjNF-YA1*) (Soyano et al. 2013). In Medicago, MtNF-YA1 is necessary for rhizobial infection and proper formation of nodule meristem (Combier et al. 2006; Xiao et al. 2014; Laporte et al. 2014); In Lotus, LjNF-YA1 is required for normal nodule organogenesis (Hossain et al. 2016). MtNF-YA1 can positively regulate the expression of the *ethylene response factor required for nodulation 1* (*ERN1*), which encodes an AP2/ERF transcription factor and is essential for infection thread formation (Andriankaja et al. 2007; Middleton et al. 2007; Cerri et al. 2012). Nodule formation also requires several other genes. For example, *Rhizobium-directed polar growth* (*RPG*) is required for infection thread progression in root hairs, where it controls the process of polar growth (Arrighi et al. 2008). Two GRAS-domain transcription factors NSP1 and NSP2 are also essential for infection and nodule organogenesis (Kaló et al. 2005; Smit et al. 2005). In Medicago epidermis cells, they promote the expression of *ERN1* and infection marker *Early nodulin 11* (Cerri et al. 2012).

14.2.3 General Aspects of Legume Nodule Formation

Above we described different aspects of legume nodule development and focused on the model plant Medicago. However, most of what we described is in general relevant for most legumes, although variations on the described processes occur.

The infection-by-infection threads that start in the epidermis seem to be the most advanced form of infection. However, rhizobia can enter the roots without forming such infection threads in some legumes. For example, in *Lupinus albus* infections occur between epidermal cells (González-Sama et al. 2004). Further, rhizobia can also enter through natural cracks at the lateral/adventitious root base in an intercellular manner, known as "crack-entry." This is, for example, the case in some *Sesbania* and *Aeschynomene* species (Sprent 2007).

Nodule ontogeny and anatomy are well conserved within the Leguminosae. In all cases, they have central infected tissue and peripheral vascular bundles. The central tissue is derived from the mitotically activated cortical cells. The determinate and indeterminate nodule types are different with respect to the persistence of the meristem.

In all studied legume species, the rhizobia are hosted intracellularly. In most cases, they are in symbiosomes that are not attached to the infection threads. However, in some species of the basal legume genus *Chamaecrista* rhizobia are present in fixation threads. These are extensions of infection threads that can fill a major part of infected cells (Naisbitt et al. 1992). These intracellular fixation threads are very similar to the mode of intracellular infection in actinorhizal nodules (see below). In symbiosomes as well as fixation threads a symbiotic interface is created as no or very little matrix is present between the membrane of the host and that of the rhizobia. In Medicago, it has been shown that the formation of this symbiotic interface involves the recruitment of a symbiosis-specific exocytosis pathway involved in arbuscule formation in the AM symbiosis (Ivanov et al. 2012). In line with the common symbiosis signaling pathway, that is also recruited from the AM symbiosis, we hypothesize that this symbiosis-specific exocytosis pathway is widely used in the legume family during nodule formation.

In all legumes that have been studied the common symbiosis signaling pathway is essential. This pathway is activated after the perception of Nod factors. In the few cases that root nodule formation is activated by rhizobia that do not make Nod factors, for example, some *Aeschynomene* species (Okazaki et al. 2016), members of the common symbiosis signaling pathway are still essential for nodulation (Fabre et al. 2015). This suggests that in these species a new receptor is evolved that can (also) activate the common symbiosis signaling pathway. In addition to the common symbiosis signaling pathway, NIN has been shown to be a key player in nodule organogenesis and infection, in all species that have been studied.

14.3 Actinorhizal Nodules

In contrast to rhizobial symbiosis, the actinorhizal species occur in markedly more plant families, which belong to three different orders (Fagales, Cucurbitales, and Rosales) (Soltis et al. 1995; Swensen 1996). While the vast majority of the legume species can establish a nodule symbiosis, the number of species that can form actinorhizal nodules is rather low. The phylogenetic relationship of genera containing actinorhizal plants is in general rather distant, and genera with actinorhizal species can be part of a family with many genera without actinorhizal species. For example, in the Betulaceae family *Alnus* spp. can form nodules, whereas their close relatives *Betula* spp. do not (Bousquet et al. 1989). This scattered occurrence of actinorhizal species is one of the arguments that was used to support the hypothesis that actinorhizal nodulation is evolved several times independently (Soltis et al. 1995; Swensen 1996; Doyle 2011) (Fig. 14.1).

Nodule formation on actinorhizal plants is induced by *Frankia* bacteria. They are filamentous Gram-positive bacteria, which can be phylogenetically divided into three main clusters. Strains from *Frankia* Cluster I can nodulate most actinorhizal plants of the order Fagales. The strains from Cluster II, which is sister to the other *Frankia* clusters, have a broad range of host plants belonging to four families within the orders Rosales and Cucurbitales. *Frankia* Cluster III strains nodulate plants from two families of the order Rosales, and two genera from the order Fagales. Actinorhizal nodules are coralloid organs composed of multiple lobes. Each lobe has a central vascular system and infected cells in the expanded cortex. Due to the apical meristem of each lobe, the infected cortical cells are arranged in a developmental gradient, similar to indeterminate legume nodules. The cells in the apical meristem remain mitotically active, adding cells to be infected and supporting the growth of the central vasculature. Adjacent to the meristem is the infection zone, where the cells become gradually filled with branching *Frankia* hyphae. Subsequently, in the nitrogen fixation zone of most host plants, *Frankia* develops vesicles where nitrogen fixation can take place. In the infected cells of the senescence zone, *Frankia* hyphae and vesicles are degraded (Pawlowski and Demchenko 2012; Santi et al. 2013) (Fig. 14.2).

14.3.1 Actinorhizal Nodule Formation

Several studies have indicated that actinorhizal nodules originate from root pericycle cells. Upon *Frankia* infection, mitotic activity is induced in the pericycle cells opposite to protoxylem poles. It has been concluded that these cells form nodule primordia from which the nodules develop. These nodules superficially resemble modified lateral roots. However, they are different from lateral roots; for example, they do not form a root cap or an epidermis. The nodule vasculature is at the central position of this lateral root-like nodule, of which cortex cells are infected by *Frankia* in an intracellular manner (Pawlowski and Bisseling 1996; Pawlowski and Demchenko 2012). Similar to nodules formed by some basal legumes (e.g., *Chamaecrista*), actinorhizal infection threads form fixation threads that fill the infected cells and stay attached to the infection threads. *Frankia*, in the fixation threads, forms vesicles. There, the nitrogenase is produced to fix nitrogen, and the vesicles provide a compartment that protects nitrogenase against oxygen damage (Pawlowski and Demchenko 2012).

Although *Parasponia* spp. form nodules with rhizobia, the nodule structure and development resemble those of actinorhizal nodules. Rhizobia enter the root through the intercellular space of epidermis and cortex. Similar to *Frankia* in actinorhizal nodules, rhizobia are hosted in fixation threads (Lancelle and Torrey 1984, 1985; Op den Camp et al. 2011, 2012). Based on the shared features between actinorhizal nodules and *Parasponia* nodules, they are collectively referred to as actinorhizal-type nodules. The common features of actinorhizal-type nodules have a central vasculature and infected cells are located at the periphery. In these cells, nitrogen-fixing bacteria surrounded by a plant-derived membrane are intracellularly accommodated.

14.3.2 Shared Common Signaling Pathway

In legumes, the common symbiosis signaling pathway and NIN play a key role in infection and nodule organogenesis. Therefore, it has been tested whether this is also the case in plants forming actinorhizal-type nodules. These studies focused on *C. glauca* (order Fagales) and *Datisca glomerata* (order Cucurbitales), because these species can be transformed. They belong to different orders and, because of their phylogenetic position within the NFC, they can represent plants in which nodulation evolved independently. The common symbiosis signaling component symbiosis receptor kinase (SYMRK) has been studied in both *C. glauca* and *D. glomerata*. These studies involved knock-down approaches which showed that SYMRK is essential for nodulation in both species (Gherbi et al. 2008; Markmann et al. 2008). This suggests that in both cases the common symbiosis signaling pathway has been recruited to support nodule formation. Further, similar to legumes, an auto-active version of CgCCaMK can induce spontaneous nodules in *C. glauca*. This construct can also induce nodule formation in *Discaria trinervis* (Svistoonoff et al. 2013). As *D. trinervis* belongs to the order Rosales, it suggests that also in this order the common symbiosis signaling pathway is important for nodulation. As mentioned above, *PanNFP* is also required for nodulation in Parasponia (Op den Camp et al. 2011) as this receptor activates the common symbiosis signaling pathway in legumes; this suggests that this is also the case in Parasponia.

Similar to legumes, NIN is also needed for nodule formation in *C. glauca*, as the downregulation of *CgNIN* reduces the nodule number (Clavijo et al. 2015). Further, *NIN* is induced in nodules of *D. glomerata* (Demina et al. 2013), suggesting a role for NIN in nodule formation. The best study of the function of NIN in actinorhizal-type nodules was performed in Parasponia, in which PanNIN is essential for nodule formation. Similar to legumes, PanNIN regulates the expression of *PanNF-YA1* during the early symbiotic response (Bu et al. 2019). Therefore, the common symbiosis signaling pathway and the nodule-specific *NIN* both play an important role in actinorhizal-type nodule formation. This is shared with legume nodule formation.

The involvement of the common symbiosis signaling pathway in actinorhizal nodulation suggests that *Frankia* makes LCO molecules similar to Nod factor. However, the non-characterized signal molecules of Cluster I *Frankia* strains (ACN14a and CcI3) that induce symbiotic responses are hydrophilic and resistant to chitinase degradation, in contrast to the properties of Nod factors that are amphiphilic and chitinase-sensitive (Chabaud et al. 2016). Further, broad host range *Rhizobium* sp. NGR234 strain and its purified Nod factors cannot elicit root hair deformation on *A. glutinosa*, suggesting that *A. glutinosa* does not recognize Nod factors (Cérémonie et al. 1999). Intriguingly, *nodABC*-like genes have been identified in the genome of Cluster II *Candidatus* Frankia datiscae Dg1 strain, a microsymbiont of *D. glomerata* (Persson et al. 2015). However, no strain from Cluster II is culturable, which hinders the identification of *Frankia* LCOs. Although the presence of LCOs has not yet been confirmed, it is tempting to propose that Nod factor-like molecules produced by Dg1 trigger actinorhizal nodule formation in *D. glomerata*. As the *Frankia* strains from Cluster II form the basal group of the symbiotic *Frankia* clusters, it has been hypothesized that the last common ancestor of the symbiotic *Frankia* strains contained the canonical *nod* genes, but these are subsequently lost in the progenitor of *Frankia* Clusters I and III (Persson et al. 2015). This is consistent with the absence of these *nod*-like genes in *Frankia* Clusters I and III. Therefore, similar to legume-rhizobium symbiosis, the genetic components of the common symbiosis signaling pathway and (at least) *NIN* are shared. This is the first indication that legume and actinorhizal-type nodules share a common origin.

14.4 Recent Phylogenomics Studies on Root Nodule Evolution

Nitrogen-fixing root nodule symbiosis exclusively occurs in plant species belonging to four orders that together form the nitrogen-fixing clade (NFC) (Fig. 14.1). As the occurrence of nodulating species is rather scattered within this clade, it was hypothesized that nodulation evolved several times within this clade (Soltis et al. 1995; Swensen 1996). However, two recent phylogenomic studies questioned this hypothesis. These studies revealed that several genes essential for establishing root nodule symbiosis are independently lost or pseudogenized in the non-nodulating relatives of nodulating species in the NFC. The first study compared the genome sequence of *Parasponia* species with that of its closely related non-nodulating sister genus *Trema* (Cannabaceae, order Rosales). This showed that *NFP/NFR5*, *NIN*, and *RPG*, which are essential for nodulation, are lost or pseudogenized in the genomes of *Trema* species. A similar loss or pseudogenization has occurred in more distantly related non-nodulating Rosales species (van Velzen et al. 2018). The second study compared the genomes of multiple nodulating and non-nodulating plants across the four orders of the NFC. This revealed the independent loss or pseudogenization of *NIN* and/or *RPG* in the non-nodulating species (Griesmann et al. 2018). As these pseudogenized/lost symbiosis genes are orthologues of genes specifically involved in nodule symbiosis, these findings suggest that massive loss of nodulation occurred in the NFC and challenged the view that nodulation evolved several times in parallel.

This massive loss of the nitrogen-fixing nodule trait is counterintuitive as nitrogen fixation is considered to be beneficial for the host plant. However, the massive loss of nodulation suggests that this trait became less favorable during the period that loss occurred. The occurrence of widespread loss of nodulation can be best explained by environmental factors that changed on a global scale. Such a factor could be the level of atmospheric CO_2, which is important for photosynthesis. Decreasing CO_2 levels can explain not only the occurrence of massive loss of nodulation in diverse lineages but also the differences in the timing of loss, from very recent (such as Trema) to more ancient (such as Prunus) (van Velzen et al. 2019). During the general decrease of CO_2 levels, there were several geological periods with the particularly steep decline of CO_2 levels. Nitrogen fixation is a high energy–demanding process and therefore it requires a high level of photosynthesis. Therefore, the reduced atmospheric CO_2 levels could have become a limiting factor for plant growth, which made nodulation a less favorable trait (van Velzen et al. 2019). Reduced photosynthesis does block nodulation (Taylor and Menge 2018). So, in the periods with decreasing CO_2 levels nodulation might have been blocked and there would have been no pressure to maintain this trait, by which it could have been massively lost. Decreasing CO_2 levels can explain not only the occurrence of massive loss of nodulation in diverse lineages but also the differences in the timing of loss, from very recent (such as *Trema*) to more ancient (such as *Prunus*) (van Velzen et al. 2019).

The massive loss of nodulation has led to the hypothesis that the common ancestor of the NFC evolved a symbiosis with nitrogen-fixing bacteria. In all nitrogen-fixing root nodule symbiosis, the bacteria are hosted intracellularly and always surrounded by a plant-derived membrane. Based on this, it was proposed that the common ancestor of the NFC can form an intracellular symbiosis in existing root cells. Subsequently, nodulation was proposed to have evolved independently in different lineages (Parniske 2018). The latter was proposed because of the fundamental differences in nodule ontogeny of, for example, legume-type and actinorhizal-type nodules. In contrast, van Velzen et al. (2019) proposed a single gain of nodulation by the common ancestor of the NFC. However, such a single gain of nodulation hypothesis did not take account of the proposed fundamental differences regarding the ontogeny of the two nodule types.

14.5 The Actinorhizal Nodule Type Is Most Likely Ancestral to the Legume Nodule Type

Recently it was shown that actinorhizal-type nodules and legume-type nodules are more similar than previously described. The property that they share is that in both cases cells derived from the mitotically activated root cortex form the infected tissue of the nodule in which bacteria are hosted intracellularly (Shen et al. 2020).

The major difference between the legume-type and actinorhizal-type nodules is the ontogeny of the nodule vascular bundles. In the case of the actinorhizal-type nodules, they are formed from pericycle-derived cells, whereas the legume nodule vasculatures are derived from cortical cells. Furthermore, it was shown that a loss-of-function mutation in *Medicago truncatula* (Medicago) *MtNOOT1* led to the formation of nodule vasculatures from pericycle-derived cells. So, an ontogeny similar to that of actinorhizal-type nodule vasculatures. Therefore, knockout of *MtNOOT1* causes a homeotic switch from a legume-type nodule to an actinorhizal type regarding the ontogeny of nodule vasculature (Shen et al. 2020).

So, on one hand, it was shown that the ontogeny of the two nodule types is more similar as in both cases cortex-derived cells form the infected nodule tissue. Further, as homeotic mutations often cause a reversion to an ancestral phenotype (Garcia-Bellido 1977; Wellmer et al. 2014), this suggests that legume-type nodules evolved from actinorhizal-type nodules. These findings support the hypothesis that the common ancestor of the NFC evolved the nodulation trait (single gain), and this was an actinorhizal-type nodule.

When nodulation evolved in the common ancestor of the NFC, it was most likely induced by *Frankia* bacteria. This implies that there was a switch from *Frankia*-induced nodulation to rhizobium-induced nodulation in *Parasponia* and in legumes. A striking difference between legume-type and actinorhizal-type nodules is that, in general, the rhizobia are released from infection threads in legume nodules, and are present, as the organelle-like structures, in the cytoplasm of host cells. In contrast, bacteria are not released from infection threads in actinorhizal nodules, but stay present in fixation threads that remain connected to the infection threads (reviewed in Pawlowski and Demchenko 2012). This seems not very surprising for the filamentous *Frankia* bacteria. However, also in *Parasponia* nodules, the rhizobia are hosted in fixation threads. Rhizobia cannot be released in *Parasponia* nodules, because the fixation thread is surrounded by a (thin) cell wall (Lancelle and Torrey 1984, 1985). In the nodule cells of most legumes, the cell wall-free droplets are formed at the tip of infection threads, by which the rhizobia can be pinched off and become encapsulated by a host membrane (reviewed in Ivanov et al. 2010). However, rhizobia are not released from infection threads in nodules of some *Chamaecrista* species and fixation threads are formed (Naisbitt et al. 1992). This suggests that this basal legume maintained some characteristics of actinorhizal-type nodules, supporting the hypothesis that legume-type nodules evolved from actinorhizal-type nodules.

14.6 Evolution of Hemoglobin Genes to Facilitate Oxygen Supply in Nodules

All root nodules face a so-called oxygen dilemma of nitrogen fixation. In the infected cells, the oxygen requirement of rhizobia/*Frankia* is high to produce sufficient ATP for the reduction of atmospheric nitrogen to ammonia. However, nitrogenase is sensitive to oxygen, which can irreversibly denature nitrogenase. To solve this dilemma, symbiotic hemoglobin genes have evolved that are specifically expressed in nodules. These genes evolved from non-symbiotic hemoglobin genes that plants use to modulate levels of toxic NO and redox potentials, and oxygen transportation at low levels (Vázquez-Limón et al. 2012). These non-symbiotic hemoglobin genes have been divided into class I and class II types. Legumes (precisely species of Papilionoideae subfamily) use leghemoglobins to facilitate the transportation of oxygen to the rhizobia at low oxygen concentrations. These have evolved from class II hemoglobin genes (Ott et al. 2005). *Chamaecrista fasciculata* is a species from the legume subfamily Caesalpinioideae. It has been suggested that the property of *C. fasciculata* hemoglobin (ppHb) is intermediate between that of class I hemoglobin and leghemoglobin, suggesting that ppHb evolved independently from the leghemoglobins of Papilionoideae species (Gopalasubramaniam et al. 2008). In line with this, the majority of actinorhizal plants (*Alnus firma*, *Myrica gale* (order Fagales), *D. glomerata* (order Cucurbitales) and *Ceanothus thyrsifloru* and Parasponia (order Rosales)) evolved (nodule-specific) hemoglobins from class I hemoglobins for oxygen supply in their nodules (Sasakura et al. 2006; Heckmann et al. 2006; Pawlowski et al. 2007; Sanz-Luque et al. 2015; Salgado et al. 2018). In the actinorhizal species *C. glauca* (order Fagales) a nodule hemoglobin evolved from a class II hemoglobin. Taken together, it shows that there has been pressure to evolve a hemoglobin-based oxygen supply system, and this most likely evolved independently several times.

The *Parasponia* species all have a symbiotic hemoglobin gene (*HB1*) that is derived from a class I hemoglobin. This is the result of *Parasponia*-specific gain-of-function adaptations in HB1, which did not occur in *Trema* hemoglobin genes (Sturms et al. 2010; Kakar et al. 2011; van Velzen et al. 2018). This suggests that the common ancestor of *Parasponia* and *Trema* made nodules, but lacked symbiotic hemoglobins. These did evolve in the *Parasponia* branch, but not in the *Trema* branch. This suggests that nitrogen-fixing efficiency within the *Trema* branch was markedly lower than in the *Parasponia* branch. This might have contributed to the loss of the nodulation trait in the *Trema* branch. It will be interesting to study whether non-nodulating species within the NFC had symbiotic hemoglobin genes. This can provide insight into what extent the lack of symbiotic hemoglobin genes could have contributed to the loss of the nitrogen-fixing trait.

14.7 Tissue Organization of Nodule Vasculatures in Legume- and Actinorhizal-Type Nodules

In actinorhizal-type nodules, vasculatures have a central position and are surrounded by infected cells. In contrast, legume-type nodules have a central infected tissue that is surrounded by peripheral vasculatures. We propose that this different spatial organization of vasculatures and infected tissue led to the evolution of different tissue organization of nodule vasculatures to support efficient nutrient exchange. In actinorhizal-type nodules, the central vasculature consists of multiple xylem and phloem poles. This organization seems well adapted to nutrient exchange (e.g., ammonia and carbohydrates) between the surrounding infected cells and the host plant through this central nodule vasculature (Fig. 14.3a, b). In legume-type nodules, the peripheral vasculature is composed of one xylem pole and one phloem pole (Guinel 2009) (Fig. 14.3c). This relatively simple organization seems sufficient for an efficient exchange of nutrients between the central tissue with infected cells and the host plant through the peripheral vascular system. The ontogeny of the nodule vasculatures in these two nodule types is different, pericycle-derived in actinorhizal-type nodules, cortex-derived in legume-type nodules. Therefore, the different tissue

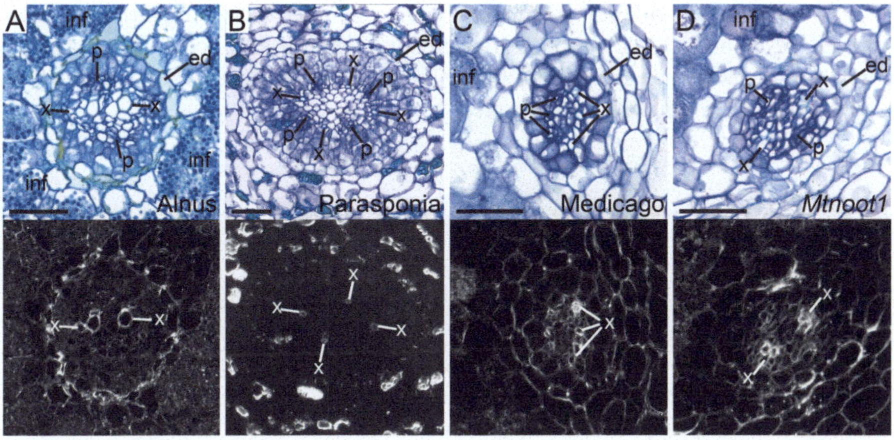

Fig. 14.3 The tissue organization of nodule vasculature in different nodules. (**a–d**) Cross-sections of nodules formed by *Alnus glutinosa* (Alnus) (**a**), *Parasponia andersonii* (Parasponia) (**b**), *Medicago truncatula* (Medicago) (**c**), and *Mtnoot1* mutants (**d**). Below are dark-field images to visualize xylem cells. (**a**, **b**) In actinorhizal-type nodules, the vasculature is centrally localized, surrounded by infected cells (inf). The nodule vasculatures have multiple poles of xylem (x) and phloem (p) cells. (**c**) In Medicago nodules, the peripheral vasculatures show a collateral organization of xylem and phloem tissues, with one phloem pole facing the infected cells and one xylem pole facing the exterior of the nodule. (**d**) In *Mtnoot1* nodules, a file of xylem cells with a xylem pole at both ends is formed in the middle of the vasculature. The xylem cell file is sandwiched by two phloem poles. One of the phloem poles faces the infected cells; the other one faces the exterior of the nodule. ed, endodermis of nodule vasculature. Scale bars: 50 μm

organization between the actinorhizal-type and legume-type nodule vasculatures could be due to their different ontogeny. Medicago *Mtnoot1* mutants provide a possibility to test this hypothesis as the ontogeny of their nodule vasculatures has become more actinorhizal-like.

The nodule vasculature of wild-type Medicago consists of one phloem and one xylem pole, with the xylem pole facing the exterior of nodules and the phloem pole facing the infected cells (Fig. 14.3c). In contrast, in Medicago *Mtnoot1* mutant nodules a file of xylem cells with a xylem pole at both ends is formed in the middle of the vasculature, and this is sandwiched by two phloem poles. This results in a diarchy patterning. One of the phloem poles faces the exterior of the nodule, the other one facing the interior of the nodule (Fig. 14.3d). These results show that the tissue organization of *Mtnoot1* nodule vasculatures becomes more complicated, resembling the vascular patterning in actinorhizal-type nodules. This supports the hypothesis that the ontogeny of nodule vasculatures contributes to their tissue organization. In line with this, the tissue organization of *Lotus japonicus noot-bop-coch-like1* mutant (mutation in the ortholog of *MtNOOT1* in *L. japonicus*) nodule vasculature also becomes more complicated (Magne et al. 2018), similar to that of Medicago *Mtnoot1* and actinorhizal-type nodule vasculatures.

So why would a nodule vasculature with a simpler tissue organization have evolved in legumes? We propose that creating unnecessary xylem and phloem poles causes a waste of energy. Therefore, simpler nodule vasculatures induced by legume NOOT1 might be an advantage.

14.8 Evolution of Legume-Type Nodules

We hypothesize that the evolution of legume-type nodules from actinorhizal-type nodules is (at least) a two-step process: (1) actinorhizal-type vasculatures formed at the periphery of nodules, similar to legume *noot1* nodules; (2) creating legume-type nodule vasculatures by recruiting legume *NOOT1*.

The second step requires the repression of cell division in the pericycle-derived cells. In line with this, we showed that the expression of *MtNOOT1* is induced in the pericycle-derived cells in the Medicago nodule primordia. Combined with the phenotype of *Mtnoot1* mutants, we suggested that Medicago *MtNOOT1* fulfills a cell-autonomous function in repressing cell division in the pericycle-derived cells. In line with this, *PanNOOT1*, an ancestral *NOOT1*, is not expressed in the pericycle-derived cells of Parasponia nodule primordia (Shen et al. 2020). This suggests the neofunctionalization of the *cis*-regulatory elements after the duplication of *NOOT* in legumes. The comparison with Parasponia suggests that it maintained its original/ancestral expression in the meristem of the nodule and it acquired a new expression domain in the pericycle-derived cells. The latter led to the evolution of legume-type nodules.

Then, what could drive the proposed first step of legume-type nodule evolution, which positioned vasculatures at the periphery of the nodule? It has been observed that the vasculatures of *Mtnoot1* nodules can originate at the central basal part but grow toward the peripheral region of the nodule and seem unable to grow in between the cells that will form the infected tissue (Fig. 14.4). This could be (partly) due to the characteristics of legume infected tissue, which is more compact, compared with the infected tissue of actinorhizal-type nodules. We hypothesize that the characteristics (e.g., patterning of cell division) of cells that form the infected tissue can control the positioning (central vs. peripheral) of the nodule vasculature in these two types of nodules, and this is independent of legume *NOOT1*.

Nodulation is very common in legumes; the peripheral positioning of nodule vasculatures has been proposed to be one of the reasons that legume nodulation is so successful (Downie 2014). In the cells of nodule vasculature, oxygen is required to generate ATP to satisfy the energy demands of the vasculature. In the case of central nodule vasculature, it is surrounded by infected cells that consume a lot of oxygen. Further, in these cells, the leghemoglobin facilitates the transport of oxygen to the mitochondria at a low oxygen concentration. As leghemoglobin binds oxygen it even further decreases the availability of oxygen to nodule vascular cells. Therefore, it has been proposed that peripheral vasculature is an advantage as more oxygen will be available (Downie 2014).

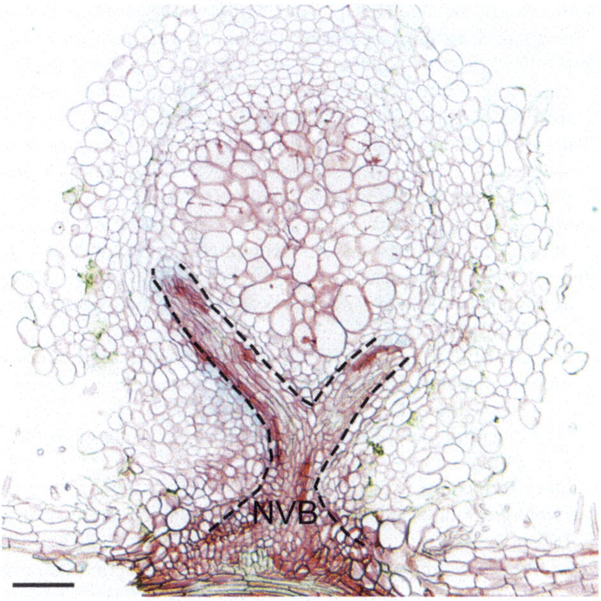

Fig. 14.4 Nodule vasculatures migrate to the peripheral region of Medicago *Mtnoot1* Nodules. Longitudinal section of a transgenic *Mtnoot1* nodule (10 dpi) expressing *pAtCASP1:GUS*. This shows that nodule vasculature originates at the central-basal region of the nodule, then migrates to the peripheral region. The dotted line outlines nodule vasculature (NVB). Scale bar: 100 μm

Acknowledgments TB was supported by European Research Council (grant number ERC-2011-AdG294790), DS was supported by China Scholarship Council (grant number 201306040120).

References

Amor BB, Shaw SL, Oldroyd GED et al (2003) The *NFP* locus of *Medicago truncatula* controls an early step of Nod factor signal transduction upstream of a rapid calcium flux and root hair deformation. Plant J 34:495–506

Andriankaja A, Boisson-Dernier A, Frances L et al (2007) AP2-ERF transcription factors mediate nod factor–dependent Mt *ENOD11* activation in root hairs via a novel *cis*-regulatory motif. Plant Cell 19:2866–2885

Ané J-M, Kiss GB, Riely BK et al (2004) *Medicago truncatula DMI1* required for bacterial and fungal symbioses in legumes. Science 303:1364–1367

Arrighi J-F, Barre A, Ben Amor B et al (2006) The *Medicago truncatula* lysine motif-receptor-like kinase gene family includes *NFP* and new nodule-expressed genes. Plant Physiol 142:265–279

Arrighi J-F, Godfroy O, de Billy F et al (2008) The *RPG* gene of *Medicago truncatula* controls *Rhizobium*-directed polar growth during infection. Proc Natl Acad Sci 105:9817–9822

Bousquet J, Girouard E, Strobeck C et al (1989) Restriction fragment polymorphisms in the rDNA region among seven species of *Alnus* and *Betula papyrifera*. Plant Soil 118:231–240

Brewin NJ (2004) Plant cell wall remodelling in the rhizobium–legume symbiosis. CRC Crit Rev Plant Sci 23:293–316

Bu F, Rutten L, Roswanjaya YP et al (2019) Mutant analysis in the non-legume *Parasponia andersonii* identifies NIN and NF-YA1 transcription factors as a core genetic network in nitrogen-fixing nodule symbioses. New Phytol nph.16386

Cebolla A, Vinardell JM, Kiss E et al (1999) The mitotic inhibitor *ccs52* is required for endoreduplication and ploidy-dependent cell enlargement in plants. EMBO J 18:4476–4484

Cérémonie H, Debellé F, Fernandez MP (1999) Structural and functional comparison of *Frankia* root hair deforming factor and rhizobia Nod factor. Can J Bot 77:1293–1301
Cerri MR, Frances L, Laloum T et al (2012) *Medicago truncatula* ERN transcription factors: regulatory interplay with NSP1/NSP2 GRAS factors and expression dynamics throughout rhizobial infection. Plant Physiol 160:2155–2172
Chabaud M, Gherbi H, Pirolles E et al (2016) Chitinase-resistant hydrophilic symbiotic factors secreted by *Frankia* activate both Ca^{2+} spiking and *NIN* gene expression in the actinorhizal plant *Casuarina glauca*. New Phytol 209:86–93
Chardin C, Girin T, Roudier F et al (2014) The plant RWP-RK transcription factors: key regulators of nitrogen responses and of gametophyte development. J Exp Bot 65:5577–5587
Charpentier M, Bredemeier R, Wanner G et al (2008) *Lotus japonicus* CASTOR and POLLUX are ion channels essential for perinuclear calcium spiking in legume root endosymbiosis. Plant Cell 20:3467–3479
Charpentier M, Sun J, Martins TV et al (2016) Nuclear-localized cyclic nucleotide-gated channels mediate symbiotic calcium oscillations. Science 352:1102–1105
Clavijo F, Diedhiou I, Vaissayre V et al (2015) The *Casuarina NIN* gene is transcriptionally activated throughout *Frankia* root infection as well as in response to bacterial diffusible signals. New Phytol 208:887–903
Combier J-PP, Frugier F, De Billy F et al (2006) *MtHAP2-1* is a key transcriptional regulator of symbiotic nodule development regulated by microRNA169 in *Medicago truncatula*. Genes Dev 20:3084–3088
Czaja LF, Hogekamp C, Lamm P et al (2012) Transcriptional responses toward diffusible signals from symbiotic microbes reveal *MtNFP*- and *MtDMI3*-dependent reprogramming of host gene expression by arbuscular mycorrhizal fungal lipochitooligosaccharides. Plant Physiol 159:1671–1685
de Maagd RA, Yang W-C, Goosen-de Roo L et al (2011) Down-regulation of expression of the *Rhizobium leguminosarum* outer membrane protein gene *ropA* occurs abruptly in interzone ll-lll of pea nodules and can be uncoupled from *nif* gene activation. Mol Plant-Microbe Interact 7:276–281
Demina IV, Persson T, Santos P et al (2013) Comparison of the nodule *vs.* root transcriptome of the actinorhizal plant *Datisca glomerata*: actinorhizal nodules contain a specific class of defensins. PLoS One 8:e72442
Downie JA (2014) Legume nodulation. Curr Biol 24:R184–R190
Doyle JJ (2011) Phylogenetic perspectives on the origins of nodulation. Mol Plant-Microbe Interact 24:1289–1295
Endre G, Kereszt A, Kevei Z et al (2002) A receptor kinase gene regulating symbiotic nodule development. Nature 417:962–966
Fabre S, Gully D, Poitout A et al (2015) The Nod factor-independent nodulation in *Aeschynomene evenia* required the common plant-microbe symbiotic "toolkit." Plant Physiol 169:2654–2664
Gage DJ (2004) Infection and invasion of roots by symbiotic, nitrogen-fixing rhizobia during nodulation of temperate legumes. Microbiol Mol Biol Rev 68:280–300
Garcia-Bellido A (1977) Homoeotic and atavic mutations in insects. Am Zool 17:613–629
Gaude N, Bortfeld S, Duensing N et al (2012) Arbuscule-containing and non-colonized cortical cells of mycorrhizal roots undergo extensive and specific reprogramming during arbuscular mycorrhizal development. Plant J 69:510–528
Gavrin A, Kaiser BN, Geiger D et al (2014) Adjustment of host cells for accommodation of symbiotic bacteria: vacuole defunctionalization, HOPS suppression, and TIP1g retargeting in *Medicago*. Plant Cell 26:3809–3822
Gavrin A, Jansen V, Ivanov S et al (2015) ARP2/3-mediated actin nucleation associated with symbiosome membrane is essential for the development of symbiosomes in infected cells of *Medicago truncatula* root nodules. Mol Plant-Microbe Interact 28:605–614

Gherbi H, Markmann K, Svistoonoff S et al (2008) SymRK defines a common genetic basis for plant root endosymbioses with arbuscular mycorrhiza fungi, rhizobia, and *Frankia* bacteria. Proc Natl Acad Sci USA 105:4928–4932

Gleason C, Chaudhuri S, Yang T et al (2006) Nodulation independent of rhizobia induced by a calcium-activated kinase lacking autoinhibition. Nature 441:1149–1152

Gomez SK, Javot H, Deewatthanawong P et al (2009) *Medicago truncatula* and *Glomus intraradices* gene expression in cortical cells harboring arbuscules in the arbuscular mycorrhizal symbiosis. BMC Plant Biol 9:10

González-Sama A, Lucas MM, De Felipe MR, Pueyo JJ (2004) An unusual infection mechanism and nodule morphogenesis in white lupin (*Lupinus albus*). New Phytol 163:371–380

Gopalasubramaniam SK, Kovacs F, Violante-Mota F et al (2008) Cloning and characterization of a caesalpinoid (*Chamaecrista fasciculata*) hemoglobin: the structural transition from a nonsymbiotic hemoglobin to a leghemoglobin. Proteins Struct Funct Genet 72:252–260

Griesmann M, Chang Y, Liu X et al (2018) Phylogenomics reveals multiple losses of nitrogen-fixing root nodule symbiosis. Science 361:eaat1743

Guinel FC (2009) Getting around the legume nodule: I. The structure of the peripheral zone in four nodule types. Botany 87:1117–1138

Heckmann AB, Hebelstrup KH, Larsen K et al (2006) A single hemoglobin gene in *Myrica gale* retains both symbiotic and non-symbiotic specificity. Plant Mol Biol 61:769–779

Hirsch AM (1992) Developmental biology of legume nodulation. New Phytol 122:211–237

Hogekamp C, Arndt D, Pereira PA et al (2011) Laser microdissection unravels cell-type-specific transcription in arbuscular mycorrhizal roots, including CAAT-box transcription factor gene expression correlating with fungal contact and spread. Plant Physiol 157:2023–2043

Hossain MS, Shrestha A, Zhong S et al (2016) *Lotus japonicus* NF-YA1 plays an essential role during nodule differentiation and targets members of the *SHI/STY* gene family. Mol Plant-Microbe Interact 29:950–964

Ivanov S, Fedorova E, Bisseling T (2010) Intracellular plant microbe associations: secretory pathways and the formation of perimicrobial compartments. Curr Opin Plant Biol 13:372–377

Ivanov S, Fedorova EE, Limpens E et al (2012) *Rhizobium*-legume symbiosis shares an exocytotic pathway required for arbuscule formation. Proc Natl Acad Sci 109:8316–8321

Kakar S, Sturms R, Tiffany A et al (2011) Crystal structures of *Parasponia* and *Trema* hemoglobins: differential heme coordination is linked to quaternary structure. Biochemistry 50:4273–4280

Kaló P, Gleason C, Edwards A et al (2005) Nodulation signaling in legumes requires NSP2, a member of the GRAS family of transcriptional regulators. Science 308:1786–1789

Kanamori N, Madsen LH, Radutoiu S et al (2006) A nucleoporin is required for induction of Ca2+ spiking in legume nodule development and essential for rhizobial and fungal symbiosis. Proc Natl Acad Sci 103:359–364

Lancelle SA, Torrey JG (1984) Early development of *Rhizobium*-induced root-nodules of *Parasponia rigida*. I. Infection and early nodule initiation. Protoplasma 123:26–37

Lancelle SA, Torrey JG (1985) Early development of *Rhizobium*-induced root nodules of *Parasponia rigida*. II. Nodule morphogenesis and symbiotic development. Can J Bot 63:25–35

Laporte P, Lepage A, Fournier J et al (2014) The CCAAT box-binding transcription factor NF-YA1 controls rhizobial infection. J Exp Bot 65:481–494

Levy J (2004) A putative Ca2+ and calmodulin-dependent protein kinase required for bacterial and fungal symbioses. Science 303:1361–1364

Limpens E, Franken C, Smit P et al (2003) LysM domain receptor kinases regulating rhizobial Nod factor-induced infection. Science 302:630–633

Limpens E, van Zeijl A, Geurts R (2015) Lipochitooligosaccharides modulate plant host immunity to enable endosymbioses. Annu Rev Phytopathol 53:311–334

Liu J, Rutten L, Limpens E et al (2019) A remote *cis*-regulatory region is required for *NIN* expression in the pericycle to initiate nodule primordium formation in *Medicago truncatula*. Plant Cell 31:68–83

LPWG (2017) A new subfamily classification of the Leguminosae based on a taxonomically comprehensive phylogeny. Taxon 66:44–77

Madsen EB, Madsen LH, Radutoiu S et al (2003) A receptor kinase gene of the LysM type is involved in legumeperception of rhizobial signals. Nature 425:637–640

Magne K, George J, Berbel Tornero A et al (2018) *Lotus japonicus NOOT-BOP-COCH-LIKE1* is essential for nodule, nectary, leaf and flower development. Plant J 94:880–894

Maillet F, Poinsot V, André O et al (2011) Fungal lipochitooligosaccharide symbiotic signals in arbuscular mycorrhiza. Nature 469:58–64

Markmann K, Giczey G, Parniske M (2008) Functional adaptation of a plant receptor- kinase paved the way for the evolution of intracellular root symbioses with bacteria. PLoS Biol 6:e68

Marsh JF, Rakocevic A, Mitra RM et al (2007) *Medicago truncatula* NIN is essential for rhizobial-independent nodule organogenesis induced by autoactive calcium/calmodulin-dependent protein kinase. Plant Physiol 144:324–335

Messinese E, Mun J-H, Yeun LH et al (2007) A novel nuclear protein interacts with the symbiotic DMI3 calcium- and calmodulin-dependent protein kinase of *Medicago truncatula*. Mol Plant-Microbe Interact 20:912–921

Middleton PH, Jakab J, Penmetsa RV et al (2007) An ERF transcription factor in *Medicago truncatula* that is essential for nod factor signal transduction. Plant Cell 19:1221–1234

Mulder L, Lefebvre B, Cullimore J, Imberty A (2006) LysM domains of *Medicago truncatula* NFP protein involved in Nod factor perception. Glycosylation state, molecular modeling and docking of chitooligosaccharides and Nod factors. Glycobiology 16:801–809

Naisbitt T, James EK, Sprent JI (1992) The evolutionary significance of the legume genus *Chamaecrista*, as determined by nodule structure. New Phytol 122:487–492

Okazaki S, Tittabutr P, Teulet A et al (2016) Rhizobium-legume symbiosis in the absence of Nod factors: two possible scenarios with or without the T3SS. ISME J 10:64–74

Op den Camp R, Streng A, De Mita S et al (2011) LysM-type mycorrhizal receptor recruited for rhizobium symbiosis in nonlegume *Parasponia*. Science 331:909–912

Op den Camp RHM, Polone E, Fedorova E et al (2012) Nonlegume *Parasponia* andersonii deploys a broad rhizobium host range strategy resulting in largely variable symbiotic effectiveness. Mol Plant-Microbe Interact 25:954–963

Ott T, Van Dongen JT, Günther C et al (2005) Symbiotic leghemoglobins are crucial for nitrogen fixation in legume root nodules but not for general plant growth and development. Curr Biol 15:531–535

Parniske M (2018) Uptake of bacteria into living plant cells, the unifying and distinct feature of the nitrogen-fixing root nodule symbiosis. Curr Opin Plant Biol 44:164–174

Pawlowski K, Bisseling T (1996) Rhizobial and actinorhizal symbioses: what are the shared features? Plant Cell 8:1899–1913

Pawlowski K, Demchenko KN (2012) The diversity of actinorhizal symbiosis. Protoplasma 249:967–979

Pawlowski K, Jacobsen KR, Alloisio N et al (2007) Truncated hemoglobins in actinorhizal nodules of *Datisca glomerata*. Plant Biol 9:776–785

Persson T, Battenberg K, Demina IV et al (2015) *Candidatus* Frankia Datiscae Dg1, the actinobacterial microsymbiont of *Datisca glomerata*, expresses the canonical *nod* genes *nodABC* in symbiosis with its host plant. PLoS One 10:e0127630

Peter J, Young W, Haukka KE (1996) Diversity and phylogeny of rhizobia. New Phytol 133:87–94

Radutoiu S, Madsen LH, Madsen EB et al (2003) Plant recognition of symbiotic bacteria requires two LysM receptor-like kinases. Nature 425:585–592

Saito K, Yoshikawa M, Yano K et al (2007) NUCLEOPORIN85 is required for calcium spiking, fungal and bacterial symbioses, and seed production in *Lotus japonicus*. Plant Cell 19:610–624

Salgado MG, van Velzen R, Van Nguyen T et al (2018) Comparative analysis of the nodule transcriptomes of *Ceanothus thyrsiflorus* (Rhamnaceae, Rosales) and *Datisca glomerata* (Datiscaceae, Cucurbitales). Front Plant Sci 9:1629

Santi C, Bogusz D, Franche C (2013) Biological nitrogen fixation in non-legume plants. Ann Bot 111:743–767

Sanz-Luque E, Ocaña-Calahorro F, De Montaigu A et al (2015) THB1, a truncated hemoglobin, modulates nitric oxide levels and nitrate reductase activity. Plant J 81:467–479

Sasakura F, Uchiumi T, Shimoda Y et al (2006) A class 1 hemoglobin gene from *Alnus firma* functions in symbiotic and nonsymbiotic tissues to detoxify nitric oxide. Mol Plant-Microbe Interact 19:441–450

Schauser L, Roussis A, Stiller J, Stougaard J (1999) A plant regulator controlling development of symbiotic root nodules. Nature 402:191–195

Shen D, Xiao TT, van Velzen R et al (2020) A homeotic mutation changes legume nodule ontogeny into actinorhizal-type ontogeny. Plant Cell 32:1868–1885

Singh S, Katzer K, Lambert J et al (2014) CYCLOPS, a DNA-binding transcriptional activator, orchestrates symbiotic root nodule development. Cell Host Microbe 15:139–152

Smit P, Raedts J, Portyanko V et al (2005) NSP1 of the GRAS protein family is essential for rhizobial Nod factor-induced transcription. Science 308:1789–1791

Smit P, Limpens E, Geurts R et al (2007) Medicago LYK3, an entry receptor in rhizobial nodulation factor signaling. Plant Physiol 145:183–191

Soltis DE, Soltis PS, Morgan DR et al (1995) Chloroplast gene sequence data suggest a single origin of the predisposition for symbiotic nitrogen fixation in angiosperms. Proc Natl Acad Sci USA 92:2647–2651

Soyano T, Kouchi H, Hirota A, Hayashi M (2013) Nodule inception directly targets *NF-Y* subunit genes to regulate essential processes of root nodule development in *Lotus japonicus*. PLoS Genet 9:e1003352

Sprent JI (2007) Evolving ideas of legume evolution and diversity: a taxonomic perspective on the occurrence of nodulation: Tansley review. New Phytol 174:11–25

Stracke S, Kistner C, Yoshida S et al (2002) A plant receptor-like kinase required for both bacterial and fungal symbiosis. Nature 417:959–962

Sturms R, Kakar S, Trent J, Hargrove MS (2010) *Trema* and *Parasponia* hemoglobins reveal convergent evolution of oxygen transport in plants. Biochemistry 49:4085–4093

Sun M, Naeem R, Su JX et al (2016) Phylogeny of the Rosidae: a dense taxon sampling analysis. J Syst Evol 54:363–391

Suzuki W, Konishi M, Yanagisawa S (2013) The evolutionary events necessary for the emergence of symbiotic nitrogen fixation in legumes may involve a loss of nitrate responsiveness of the NIN transcription factor. Plant Signal Behav 8:e25975

Svistoonoff S, Benabdoun FM, Nambiar-Veetil M et al (2013) The independent acquisition of plant root nitrogen-fixing symbiosis in fabids recruited the same genetic pathway for nodule organogenesis. PLoS One 8:e64515

Swensen SM (1996) The evolution of actinorhizal symbioses: evidence for multiple origins of the symbiotic association. Am J Bot 83:1503–1512

Taylor BN, Menge DNL (2018) Light regulates tropical symbiotic nitrogen fixation more strongly than soil nitrogen. Nat Plants 4:655–661

Tedersoo L, Laanisto L, Rahimlou S et al (2018) Global database of plants with root-symbiotic nitrogen fixation: NodDB. J Veg Sci 29:560–568

Tirichine L, Imaizumi-Anraku H, Yoshida S et al (2006) Deregulation of a Ca 2+ /calmodulin-dependent kinase leads to spontaneous nodule development. Nature 441:1153–1156

van Velzen R, Holmer R, Bu F et al (2018) Comparative genomics of the nonlegume *Parasponia* reveals insights into evolution of nitrogen-fixing rhizobium symbioses. Proc Natl Acad Sci USA 115:E4700–E4709

van Velzen R, Doyle JJ, Geurts R (2019) A resurrected scenario: single gain and massive loss of nitrogen-fixing nodulation. Trends Plant Sci 24:49–57

van Zeijl A, Op den Camp RHM, Deinum EE et al (2015) Rhizobium lipo-chitooligosaccharide signaling triggers accumulation of cytokinins in *Medicago truncatula* roots. Mol Plant 8:1213–1226

Vasse J, De Billy F, Camut S, Truchet G (1990) Correlation between ultrastructural differentiation of bacteriods and nitrogen fixation in alfalfa nodules. J Bacteriol 172:4295–4306

Vázquez-Limón C, Hoogewijs D, Vinogradov SN, Arredondo-Peter R (2012) The evolution of land plant hemoglobins. Plant Sci 191–192:71–81

Wellmer F, Graciet E, Riechmann JL (2014) Specification of floral organs in *Arabidopsis*. J Exp Bot 65:1–9

Xiao TT, Schilderink S, Moling S et al (2014) Fate map of *Medicago truncatula* root nodules. Development 141:3517–3528

Yang W-C, Horvath B, Hontelez J et al (1991) *In situ* localization of *Rhizobium* mRNAs in pea root nodules: *nif*A and *nif*H localization. Mol Plant-Microbe Interact 4:464–468

Yano K, Yoshida S, Müller J et al (2008) CYCLOPS, a mediator of symbiotic intracellular accommodation. Proc Natl Acad Sci USA 105:20540–20545

Yoro E, Suzaki T, Toyokura K et al (2014) A positive regulator of nodule organogenesis, nodule inception, acts as a negative regulator of rhizobial infection in *Lotus japonicus*. Plant Physiol 165:747–758

Zhang X, Dong W, Sun J et al (2015) The receptor kinase CERK1 has dual functions in symbiosis and immunity signalling. Plant J 81:258–267

Chapter 15
Early Molecular Dialogue Between Legumes and Rhizobia: Why Are They So Important?

Oswaldo Valdés-López, María del Rocío Reyero-Saavedra, Mariel C. Isidra-Arellano, and María del Socorro Sánchez-Correa

Abstract Legume-rhizobia symbiosis has a considerable ecological relevance because it replenishes the soil with fixed-nitrogen (e.g., ammonium) for other plants. Because of this benefit to the environment, the exploitation of the legume-rhizobia symbiosis can contribute to the development of the lower input, sustainable agriculture, thereby, reducing dependency on synthetic fertilizers. To achieve this goal, it is necessary to understand the different levels of regulation of this symbiosis to enhance its nitrogen-fixation efficiency. A different line of evidence attests to the relevance of early molecular events in the establishment of a successful symbiosis between legumes and rhizobia. In this chapter, we will review the early molecular signaling in the legume-rhizobia symbiosis. We will focus on the early molecular responses that are crucial for the recognition of the rhizobia as a potential symbiont.

Keywords Common symbiosis pathway · Nodulation · Calcium spiking · Nodule inception · Root nodule symbiosis

15.1 Introduction

Nitrogen is an essential component in most of the biological molecules, including nucleic acids (i.e., DNA and RNA), amino acids, and proteins. Thus, nitrogen is considered as a vital element for any organism, including plants. Despite being an abundant element in the atmosphere, only limited resources of inorganic nitrogen are available to plants, primarily in the form of nitrate and ammonium. Thus, nitrogen availability is considered as one of the main limitations for the global agriculture

O. Valdés-López (✉) · M. d. R. Reyero-Saavedra · M. C. Isidra-Arellano ·
M. d. S. Sánchez-Correa
Facultad de Estudios Superiores Iztacala, Laboratorio de Genómica Funcional de Leguminosas, Universidad Nacional Autónoma de México, Tlalnepantla, Estado de México, Mexico
e-mail: oswaldovaldesl@unam.mx

M. Kloc (ed.), *Symbiosis: Cellular, Molecular, Medical and Evolutionary Aspects*, Results and Problems in Cell Differentiation 69,
https://doi.org/10.1007/978-3-030-51849-3_15

yield (Pankievicz et al. 2019). To tackle this limitation, different strategies have been used, such as crop rotation, coculture with legumes, and the use of fertilizer mainly in the form of animal waste. At the beginning of the twentieth century, Fritz Haber and Carl Bosh developed a process allowing the production of synthetic nitrogen fertilizers on an industrial scale. The use of synthetic nitrogen fertilizers was the main factor contributing to a drastic increase in crop production and meeting a global food demand (Erisman et al. 2008). However, during their manufacturing, a considerable amount of CO_2 and N_2O, two of the major greenhouse gasses, is released to the atmosphere (Lassaletta et al. 2014). Additionally, the intense use of synthetic fertilizers has also led to the contamination of the groundwater, eutrophication of freshwater, and soil salinization (Galloway et al. 2003; Zhang et al. 2015). Because of all these global sustainability considerations, the use of synthetic fertilizers in global agriculture cannot be considered as a sustainable strategy for food production. Thereby, it is imperative to develop sustainable agriculture with low dependence on synthetic fertilizers.

Unlike most land plants, legumes can grow in nitrogen-deficient soils. This is because legumes can engage in symbiosis with nitrogen-fixing soil bacteria collectively known as rhizobia (Castro-Guerrero et al. 2016). This symbiosis not only allows legume to grow in nitrogen-deficient soils with no synthetic nitrogen fertilizer inputs but also replenishes the soil with fixed-nitrogen (e.g., ammonium) for other plants (Castro-Guerrero et al. 2016; Ferguson et al. 2019). Because of these benefits to the environment, the exploitation of the legume-rhizobia symbiosis can contribute to the development of lower input, sustainable agriculture, thereby, reducing dependency on synthetic fertilizers. To achieve this goal, it is necessary to understand the different levels of regulation of this symbiosis to enhance its nitrogen-fixation efficiency.

To establish the legume-rhizobia symbiosis (hereafter referred to as root nodule symbiosis), two plant genetic programs are required. The first one allows the rhizobia to colonize the roots of the legume host, whereas the second one is required for root nodule development, a new organ where rhizobia are hosted and fix nitrogen (Venkateshwaran et al. 2013). Although both genetic programs are necessary to establish a successful symbiosis with rhizobia, the genetic program controlling the rhizobial infection is considered to be a crucial step because it controls early molecular responses required for the mutual recognition between both partners, and any defect in this program can lead to the abortion of the symbiosis (Venkateshwaran et al. 2013). Hence, in this chapter, we will review the early molecular responses of the legume-rhizobia symbiosis. We will focus on those early molecular responses that are crucial to recognizing rhizobia as a potential symbiont, and we will also discuss how this knowledge can be translated into nonlegume plants.

15.2 Recognizing the Rhizobial Call

Under nitrogen-deficient conditions, legumes secrete flavones and isoflavones into the rhizosphere. These secondary metabolites not only attract compatible rhizobia to the root hairs but also activate rhizobial genes involved in the synthesis of diffusible lipochitooligosaccharides with specific chemical decorations named Nodulation Factors (NFs) (Dénarié et al. 1996). The legume host perceives NFs at the epidermal level through at least three LysM receptor-like kinases named NF Receptor5 (NFR5) and NFR1 in the model legume *Lotus japonicus*, and NF Perception (NFP) and LYK3 in the model legume *Medicago truncatula* (Radutoiu et al. 2003; Broghammer et al. 2012). The third receptor is named as epidermal-NFR (NFRe) and amplifies the NF signal in *L. japonicus* root epidermal cells (Murakami et al. 2018). Mutant plants in any of the NFR5/NFP or NFR1/LYK3 receptors severely affect the communication between legumes and rhizobia (Radutoiu et al. 2003; Broghammer et al. 2012). In contrast, the *nfre* mutant plants are still able to react to NF, but they develop fewer nodules (Murakami et al. 2018). NFRe has an active kinase domain able to phosphorylate NFR5, which, in turn, regulates NFRe downstream signaling (Murakami et al. 2018). This evidence indicates that NFR5/NFP and NFR1/LYK3 are the first receptors to detect the rhizobial call, whereas NFRe is in charge of expanding this symbiotic signal beyond the root hair cells.

Rhizobial access to the legume host roots is controlled by two-stage sequential mechanisms. Both NFR1 and NFR5 participate in the first mechanism by triggering the symbiotic signal transduction in *L. japonicus* (Kawaharada et al. 2015). The first mechanism of rhizobial recognition activates the second level of recognition, in which the bacterial exopolysaccharides (EPS) play a key role (Fraysse et al. 2003). Different lines of evidence indicate that the detection of bacterial EPS by the plant is a crucial step to act positively or negatively in response to compatible or incompatible rhizobia, respectively (Kawaharada et al. 2015, 2017). The detection of EPS is mediated by the EPS receptor3 (EPR3) in *L. japonicus* (Kawaharada et al. 2015). Upon NFs perception, the expression of EPR3 is activated in the rhizobia-infected root hairs (Kawaharada et al. 2015). Comprehensive analyses in *epr3* mutant plants indicate that this receptor regulates both root and nodule infection by the compatible rhizobia (Kawaharada et al. 2015, 2017). Thus, the coordinate interplay between the NF and EPS receptors is crucial to properly detect the compatible rhizobia and initiates the root nodule symbiosis.

15.3 Decoding the Rhizobial Signal Through a Common Symbiosis Pathway

Upon perception of both NFs and EPS, a series of molecular events, including the transcriptional activation and phosphorylation of several symbiosis-related genes and proteins, are activated (Venkateshwaran et al. 2013). The phosphorylation of proteins is a crucial step for deciphering the NFs signal. Several lines of evidence attest to the intricate phosphorylation cascade mechanism required to decode the NFs signal. For instance, a phosphoproteomic study on *M. truncatula* roots revealed that 66 different proteins were differentially phosphorylated upon a one-hour treatment with NFs purified from *Sinorhizobium meliloti* (Rose et al. 2012). A coimmunoprecipitation-based proteomic assay also led to the identification of a receptor-like cytoplasmic kinase (RCLK) that phosphorylates NFR5 (Wong et al. 2019). This RCLK, named as NiCK4 because it does interact with NFR5, is an important link between the perception of NFs by NFR5 and the symbiotic molecular events occurring at the nucleus in *L. japonicus* (Wong et al. 2019).

It is believed that phosphorylation cascades are required to activate an ancient plant common symbiosis pathway (CSP), so-called because it controls the establishment of both root nodule and the arbuscular mycorrhizal symbiosis as well as the endophytic interactions with other beneficial soil-microbes (Venkateshwaran et al. 2013; Skiada et al. 2020). Among the components of this signaling pathway, the leucine-rich repeat receptor-like kinase, Does not Make Infections 2 (DMI2) in *M. truncatula* or SYMRK in *L. japonicus*, which is localized at the plasma membrane of root hair cells, participates in a receptor complex to perceive NFs (Singh and Parniske 2012). It has been demonstrated that DMI2 interacts with the 3-hydroxy-3-methylglutaryl coenzyme A reductase 1 (HMGR1), which participates in the mevalonate biosynthesis in *M. truncatula* (Kevei et al. 2007). Other proteins participating in the CSP include the calcium channels DMI1 in *M. truncatula* and Castor/Pollux in *L. japonicus*, CNGC15, which are localized at the nuclear envelope, as well as different nucleoporins (e.g., NUP85, NUP133, and NENA). This set of ion channels, alongside mevalonate, is required to generate rapid oscillations in the nuclear and perinuclear calcium concentration described as calcium spiking (Kim et al. 2019; Kanamori et al. 2006; Peiter et al. 2007; Saito et al. 2007; Groth et al. 2010; Charpentier et al. 2016). Calcium spiking is an indispensable signal to establish both root nodule and arbuscular mycorrhizal symbiosis. This conclusion is based on the fact that *castor/pollux* and *dmi1* mutant plants are unable to activate calcium spiking and therefore fail to nodulate and form a symbiosis with arbuscular mycorrhizal fungi (Imaizumu-Anraku et al. 2005). Additionally, there is evidence that the Early Phosphorylated Protein1 (EPP1) is a key component required to activate the calcium spiking, thereby, controlling the activation of the CSP in *M. truncatula* (Valdés-López et al. 2019).

Calcium spiking is decoded by a nuclear calcium/calmodulin-dependent protein kinase (DMI3/CCaMK) that further transduces the signal by phosphorylating the transcriptional activator IPD3/CYCLOPS (Lévy et al. 2004; Miller et al. 2013; Singh et al. 2014). In turn, IPD3/CYCLOPS activates the expression of the transcription factor *Nodule INception* (*NIN*), which subsequently promotes the expression of the *Nuclear Factor Y* (*NF-Y*) complexes *NF-YA* and *NF-YB* (Soyano et al. 2013). The coordinated action of these transcription factors alongside the interplay between the transcription factors Nodulation Signaling Pathway2 (NSP2)/NSP1, Ethylene Response Factor Required for Nodulation1 (ERN1), and ERN2 is required to activate the expression of several genes, whose participation is crucial to coordinate

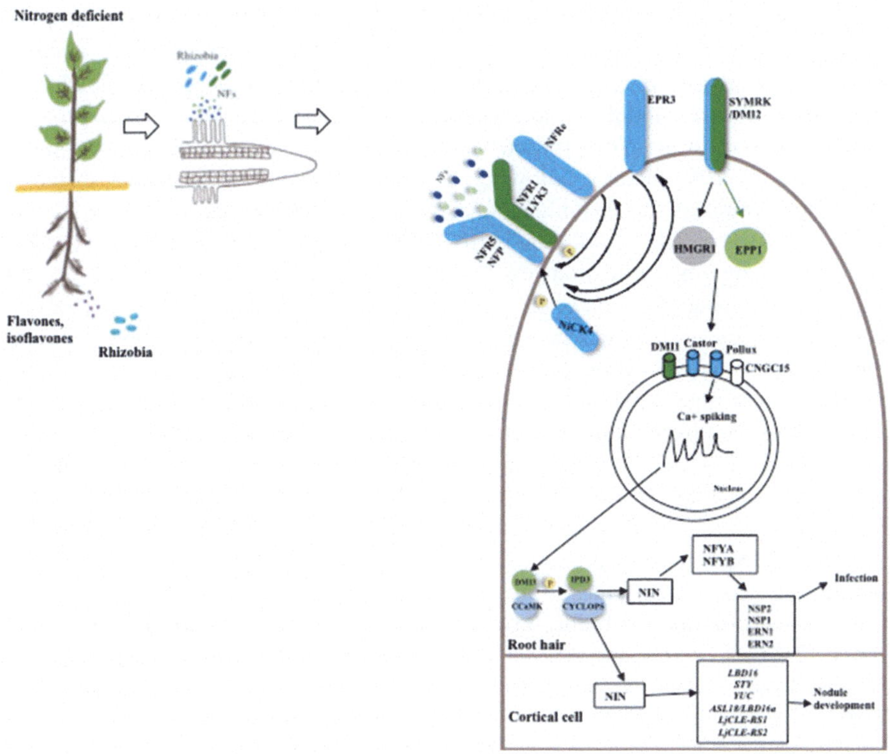

Fig. 15.1 Early molecular events required to establish the root nodule symbiosis. Under nitrogen-deficient conditions, legumes release flavones and isoflavones into the rhizosphere. These secondary metabolites attract the compatible rhizobia to the root hairs. In turn, rhizobia release the so-called NFs, which are detected by the legume host through the LysM receptors NFR5/NFP, NFR1/LYK3, and NFPe. EPS, which are detected by the receptor EPR3, also play a crucial role during the rhizobial recognition process. Upon NFs and EPS detection, a series of molecular events (e.g., gene expression and protein phosphorylation) are activated. The transcription factor NIN plays a crucial role in the establishment of the root nodule symbiosis

the rhizobial process (Genre and Russo 2016) (Fig. 15.1). Comprehensive studies on mutant plants in each of the components of the CSP attest to the relevance of this ancient pathway to properly initiate the root nodule symbiosis.

15.4 NIN, a Master Regulator of the Root Nodule Symbiosis

Although several transcription factors participate in the decoding of the NFs signal, there is an indication that NIN is the most important regulator of the root nodule symbiosis. This assumption is because NIN controls the expression of genes participating in rhizobial infection, nodule development, and in the regulation of nodule number (Soyano et al. 2014; Vernié et al. 2015; Liu et al. 2019a). This versatility in the functions of NIN is explained by the fact that its 20-kb promoter region contains several *cis*-regulatory elements essential for the coordination of its participation in each stage of the root nodule symbiosis (Liu et al. 2019a). Some of the *cis*-regulatory elements present in the *NIN* promoter are crucial for the control of spatiotemporal expression of NIN, thereby, allowing this regulator to coordinate the different genetic programs of this symbiosis occurring at the epidermal and cortical cells (Liu et al. 2019a). The "epidermal" NIN controls the expression of genes participating in the rhizobial infection process, which implies the remodeling of the cell wall, membrane cytoskeleton of the root hairs, and epidermal cells where the mutual recognition between legumes and rhizobia take place (Liu et al. 2019b). A recent transcriptional analysis of root hairs from *nin*, *nf-ya1*, and *ern1 M. truncatula* mutant plants not only reinforced the role of "epidermal" NIN in the activation of genes participating in this cellular reorganization process but also revealed that this transcription factor regulates the expression of genes participating in nutrient uptake and in the biosynthesis and perception of different phytohormones, the processes that are likely important for rhizobial infection (Liu et al. 2019b). In contrast, the "cortical" NIN, whose expression is activated by the phytohormone cytokinin, recruits and coordinates the expression of genes belonging to the lateral root development program, including *LOB-DOMAIN PROTEIN 16* (*LBD16*), *STYLISH* (*STY*), *YUCCAs* (*YUC*), and *ASYMETRIC LEAVES 2-LIKE 18/LATERAL ORGAN BOUNDARIES DOMAIN 16a* (*ASL18/LBD16a*) (Schiessl et al. 2019; Soyano et al. 2019). The recruitment of this program is essential to activate the root nodule development.

NIN not only acts as a positive regulator of the root nodule symbiosis but also as a negative regulator (Soyano et al. 2014). Evidence in *L. japonicus* indicates that NIN positively regulates the expression of rhizobia-induced CLE peptides *LjCLE-RS1* and *LjCLE-RS2*, which belong to the so-called Autoregulation of Nodulation (AON) pathway (Ferguson et al. 2019). To avoid an excessive nodule formation and, thereby, an excessive carbon demand from the rhizobia residing inside the root nodule, the legume host activates the AON pathway (Ferguson et al. 2019). The relevance of the AON pathway resides in the fact that this pathway restricts both rhizobial infection and nodule development (Ferguson et al. 2019). Because NIN is expressed in the different cell types where the root nodule symbiosis occurs, and that controls different genetic programs, NIN is considered to be a master regulator of this symbiosis.

15.5 Early Physiological Responses of the Root Nodule Symbiosis

The molecular responses activated upon NFs and EPS detection are crucial for the coordination of both physiological and morphological modifications in the root hairs. The rhizobia-induced root hair deformation is the ultimate response that allows the legume host to be colonized by the rhizobia (Roy et al. 2020). To allow this modification, a series of cellular rearrangements are required. Genetic analyses on several mutant plants have revealed that modifications in the root hair cell wall and cytoskeleton are determinants for a proper root hair deformation. For instance, mutations in the components of the SCAR/WAVE complex *Nck-associated protein1* (*Nap1*) and *121F-specific p53 inducible RNA* (*Pir1*), which coordinate the actin assembly and, thereby, cell growth, compromise the reorientation of root hair tip growth in response to rhizobia in the model legume *L. japonicus* (Yokota et al. 2009).

The ultimate goal of rhizobia-induced root hair deformation is the rhizobia entrapment and, then, the formation of the infection chamber (Fournier et al. 2015). The formation of the infection chamber is required to form a tubular-like structure named Infection Thread (IT) (Fournier et al. 2015). IT formation and extension require both plant cell wall degradation and membrane trafficking proteins (Roy et al. 2020). Mutation of the *Nodule Pectate Lyase* gene caused a drastic reduction in the number of ITs in *L. japonicus* roots (Xie et al. 2012; Liu et al. 2019b). Likewise, studies on the integral membrane proteins FLOTILLIN2 (FLOT2), FLOT4, and the remorin SYMREM1 demonstrated the role of membrane process in the IT progression and the infection process in *M. truncatula* (Haney and Long 2010; Lefebvre et al. 2010). ITs are crucial structures required to transport rhizobia from the infection chamber to the cortical cell, which will differentiate into the root nodule meristem (Roy et al. 2020).

Another response occurring during the rhizobial infection process is the modulation of plant immunity. There are indications that the modulation of the plant immunity response is a crucial step for successful rhizobial colonization. There is evidence that the plant immunity response is blocked through protein effectors, which are "injected" into the plant cell cytoplasm through the secretion system complex (Cao et al. 2017). Hence, the rhizobia-induced root hair deformation, the IT formation, and the modulation of the plant immunity response are crucial for the rhizobial infection process.

15.6 Is It Possible that Nonlegume Plants Interact with Rhizobia?

One of the golden goals for the international community is to make possible that nonlegume plants fix atmospheric nitrogen in symbiosis with nitrogen-fixing bacteria. One way to achieve this ambitious goal is transferring rhizobial genes, mainly those involved in the nitrogen fixation, to other bacteria able to infect and colonize nonlegume plants. Significant advances in this approach have been made. For instance, it has recently been reported that the transfer of 12 nitrogen fixation-related cluster genes from rhizobia to different bacteria species (e.g., *Azotobacter vinelandii*) resulted in a high nitrogen flux to cereal crops (Ryu et al. 2020).

Another way to achieve this goal is transferring symbiosis-related genes from legumes to nonlegume plants. However, recent phylogenomic studies have revealed that the components of the CSP are conserved in most land plants, including nonlegume plants (Delaux et al. 2014). Indeed, it has been demonstrated that the plant ancestor was "armed" with the majority of the genetic components of the CSP to symbiotically interact with soil-beneficial microbes (Delaux et al. 2015). Likewise, different studies have demonstrated that actinorhizal plants contain most of the components of the CSP (Hocher et al. 2011). This evidence indicates that the transfer of symbiotic-related genes is not necessary because the majority of the land plants have them. Instead, it is necessary to deeply understand how these genes are regulated in legume plants, and then understand why the nonlegume plants "decided" to not interact with rhizobia. Having this knowledge will be critical to achieving the goal that nonlegume plants symbiotically interact with rhizobia and, thereby, reduce the dependency on synthetic nitrogen fertilizers.

15.7 Conclusions and Perspectives

In this chapter, we summarized and discussed the relevance of the early molecular events of the root nodule symbiosis. It is clear that the proper regulation of these molecular responses is crucial for the initiation and establishment of this important symbiosis. Although significant advances have been made in the last two decades, it is clear that we are far from fully understanding how these early stages are regulated. For instance, it is still not clear how the CSP is specifically activated by rhizobia and activates all the genetic programs leading to a successful root nodule symbiosis. Likewise, it is imperative to understand how these early stages are regulated in legumes, and having this knowledge will help us to design experimental strategies oriented to make possible that the nonlegume plants recognize rhizobia as potential symbionts.

Acknowledgments This work was supported by the Programa de Apoyo a Proyectos de Investigación e Inovación Tecnológica (PAPIIT-UNAM grants No. IN213017 and IN201320) and by the Consejo Nacional de Ciencia y Tecnología (CONACyT grant No. A1-S-9454) to OV-L. MDRR-S is a Doctoral Student from Programa de Doctorado en Ciencias Biomédicas, Universidad Nacional Autónoma de México, and receives a doctoral student fellowship from CONACyT (CVU: 347027). MCI-A is a Doctoral Student from Programa de Doctorado en Ciencias Biológicas, Universidad Nacional Autónoma de México, and receives a doctoral student fellowship from CONACyT (CVU: 919676).

References

Broghammer A, Krussel L, Blaise M et al (2012) Legume receptors perceive the rhizobial lipochitin oligosaccharides signal molecules by direct binding. Proc Natl Acad Sci USA 109:13859–13864

Cao Y, Halane MK, Gassmann W, Stacey G (2017) The role of plant innate immunity in the legume-rhizobium symbiosis. Annu Rev Plant Biol 68:535–561

Castro-Guerrero NA, Isidra-Arellano MC, Mendoza-Cozatl DG, Valdés-López O (2016) Common bean: a legume model on the rise for unraveling responses and adaptations to iron, zinc, and phosphate deficiencies. Front Plant Sci 7:600

Charpentier M, Sun J, Martins TV, Radhakrishnan GV et al (2016) Nuclear-localized cyclic nucleotide-gated channels mediate symbiotic calcium oscillations. Science 352:1102–1105

Delaux PM, Varala K, Edger PP, Coruzzi GM, Ané JM (2014) Comparative phylogenomics uncovers the impact of symbiotic associations on host genome evolution. PLoS Genet 10(7): e1004487

Delaux PM, Radhakrishnan GV, Jayaraman D et al (2015) Algal ancestor of land plants was preadapted for symbiosis. Proc Natl Acad Sci USA 112:13390–13395

Dénarié J, Debellé F, Promé JC (1996) Rhizobium lipo-chitooligosaccharide nodulation factors: signaling molecules mediating recognition and morphogenesis. Annu Rev Biochem 65:503–535

Erisman JW, Sutton MA, Galloway J, Klimont Z (2008) How a century of ammonia synthesis changed the world. Nat Geosci 1:636–639

Ferguson BJ, Mens C, Hastwell A, Zhang M, Su H, Jones CH, Chu X, Gresshoff PM (2019) Legume nodulation: the host controls the party. Plant Cell Environ 42:41–51

Fournier J, Teillet A, Chabaud M et al (2015) Remodeling of the infection chamber before infection thread formation reveals two-step mechanism for rhizobial entry into the host legume root hair. Plant Physiol 167:1233–1242

Fraysse N, Couderc F, Poinsot V (2003) Surface polysaccharide involvement in establishing the rhizobium-legume symbiosis. Eur J Biochem 270:1365–1380

Galloway JN, Aber JD, Erisman JW, Seitzinger SP, Howarth RW, Cowling EB, Cosby J (2003) The nitrogen cascade. Bioscience 53:341–356

Genre A, Russo G (2016) Does a common pathway transduce symbiotic signals in plant-microbe interactions? Front Plant Sci 7:96

Groth M, Takeda N, Perry J et al (2010) NENA, a *Lotus japonicus* homolog of Sec13, is required for rhizodermal infection by arbuscular mycorrhizal fungi and rhizobia but dispensable for cortical endosymbiotic development. Plant Cell 22:2509–2526

Haney CH, Long SR (2010) Plant flotillins are required for infection by nitrogen-fixing bacteria. Proc Natl Acad Sci USA 107:478–483

Hocher V, Alloiso N, Auguy F et al (2011) Transcriptomics of Actinorhizal symbiosis reveals homologs of the whole common symbiotic signaling cascade. Plant Physiol 156:700–711

Imaizumu-Anraku H, Takeda N, Charpentier M, Perry J et al (2005) Plastid proteins crucial for symbiotic fungal and bacterial entry into plant roots. Nature 443:527–531

Kanamori N, Madsen LH, Radatoiu S et al (2006) A nucleoporin is required for induction of Ca^{2+} spiking in legume nodule development and essential for rhizobial and fungal symbiosis. Proc Natl Acad Sci USA 103:359–364

Kawaharada Y, Kelly S, Nielsen MW et al (2015) Receptor-mediated exopolysaccharide perception controls bacterial infection. Nature 523:308–312

Kawaharada Y, Nielsen MW, Kelly S et al (2017) Differential regulation of the *Epr3* receptor coordinates membrane-restricted rhizobial colonization of root nodule primordial. Nat Commun 8:14534

Kevei Z, Lougnon G, Mergaert P et al (2007) 3-hydroxy-3-methylglutaryl coenzyme a reductase1 interacts with NORK and is crucial for nodulation in *Medicago truncatula*. Plant Cell 19:3974–3989

Kim S, Zeng W, Bernard S, Liao J, Venkateshwaran M, Ané JM, Jiang Y (2019) Ca^{2+}-regulated Ca^{2+} channels with an RCK gating ring control plant symbiotic associations. Nat Commun 10 (1):3703

Lassaletta L, Billen G, Grizzetti B, Anglade J, Garnier J (2014) 50 year trends in nitrogen use efficiency of world cropping systems: the relationship between yield and nitrogen input to cropland. Environ Res Lett 9:105011

Lefebvre B, Timmers T, Mbengue M et al (2010) A remorin protein interacts with symbiotic receptors and regulates bacterial infections. Proc Natl Acad Sci USA 107:2343–2348

Lévy J, Bres C, Geurts R, Chalhoub B, Kulikova O, Duc G (2004) A putative calcium and calmodulin-dependent protein kinase required for bacterial and fungal symbioses. Science 303:1361–1363

Liu J, Rutten L, Limpens E, van der Molen T et al (2019a) A remote *cis*-regulatory region is required for *NIN* expression in the pericycle to initiate nodule primordium formation in *Medicago truncatula*. Plant Cell 31:68–83

Liu CW, Breakspear A, Guan D et al (2019b) NIN acts as a network hub controlling a growth module required for rhizobial infection. Plant Physiol 179:1704–1722

Miller JB, Pratap A, Miyahara A, Zhou L, Borneman S, Morris RJ, Oldroyd GE (2013) Calcium/calmodulin-dependent protein kinase is negatively and positively regulated by calcium, providing a mechanism for decoding calcium responses during symbiosis signaling. Plant Cell 25:5053–5066

Murakami E, Cheng J, Gysel K et al (2018) Epidermal LysM receptor ensures robust symbiotic signaling in *Lotus japonicus*. elife 7:e33506

Pankievicz VCS, Irving TB, Maia LGS, Ané JM (2019) Are we there yet? The long walk towards the development of efficient symbiotic associations between nitrogen-fixing bacteria and non-leguminous crops. BMC Biol 17:99

Peiter E, Sun J, Heckmann AB et al (2007) The *Medicago truncatula* DMI1 protein modulates cytosolic calcium signaling. Plant Physiol 145:192–203

Radutoiu S, Madsen LH, Madsen EB et al (2003) Plant recognition of symbiotic bacteria requires two LysM receptor-like kinases. Nature 425:585–592

Rose CM, Venkateshwaran M, Volkening JD et al (2012) Rapid phosphoproteomic and transcriptomic changes in the rhizobia-legume symbiosis. Mol Cell Proteomics 11:724–744

Roy S, Liu W, Sekhar R et al (2020) Celebrating 20 years of genetic discoveries in legume nodulation and symbiotic nitrogen fixation. Plant Cell 32:15–41

Ryu MH, Zhang J, Toth T et al (2020) Control of nitrogen fixation in bacteria that associate with cereals. Nat Microbiol 5:314–330

Saito K, Yoshikawa M, Yano K et al (2007) NUCLEOPORIN85 is required for calcium spiking, fungal and bacterial symbioses, and seed production in *Lotus japonicus*. Plant Cell 19:610–624

Schiessl K, Lilley JLS, Lee T et al (2019) *NODULE INCEPTION* recruits the lateral root developmental program for symbiotic nodule organogenesis in *Medicago truncatula*. Curr Biol 29:3657–3668

Singh S, Parniske M (2012) Activation of calcium- and calmodulin-dependent protein kinase (CCaMK), the central regulator of plant root endosymbiosis. Curr Opin Plant Biol 15:444–453

Singh S, Katzer K, Lambert J, Cerri M, Parniske M (2014) CYCLOPS, a DNA-binding transcriptional activator orchestrates symbiotic root nodule development. Cell Host Microbe 15:139–152

Skiada V, Avramidou M, Bonfante P, Genre A, Papadopoulou KK (2020) An endophytic *Fusarium*-legume association is partially dependent on the common symbiotic signaling pathway. New Phytol 226:1429. https://doi.org/10.1111/nph.16457

Soyano T, Kouchi H, Hirota A, Hayashi M (2013) Nodule inception directly targets NF-Y subunit genes to regulate essential processes of root nodule development in Lotus japonicus. PLoS Genet 9:e1003352

Soyano T, Hirakawa H, Sato S, Hayashi M, Kawaguchi M (2014) NODULE INCEPTION creates a long-distance negative feedback loop involved in homeostatic regulation of nodule organ production. Proc Natl Acad Sci USA 111:14607–14612

Soyano T, Shimoda Y, Kawaguchi M, Hayashi M (2019) A shared gene drives lateral root development and root nodule symbiosis pathway in *Lotus*. Science 366:1021–1023

Valdés-López O, Jayaraman D, Maeda J et al (2019) A novel positive regulator of the early stages of root nodule Symbiosis identified by Phosphoproteomics. Plant Cell Physiol 60:575–586
Venkateshwaran M, Volkening JD, Sussman MR, Ané JM (2013) Symbiosis and the social network of higher plants. Curr Opin Plant Biol 16:118.127
Vernié T, Kim J, Frances L et al (2015) The NIN transcription factor coordinates diverse nodulation programs in different tissues of the *Medicago truncatula* root. Plant Cell 27:3410–3424
Wong JEMM, Nadzieja M, Madsen LH et al (2019) A *Lotus japonicus* cytoplasmic kinase connects Nod factor perception by the NFR5 LysM receptor to nodulation. Proc Natl Acad Sci USA 116:14339–14348
Xie F, Murray JD, Kim J, Heckmann AB, Edwards A, Olroyd GE, Downie JA (2012) Legume pectate lyase required for root infection by rhizobia. Proc Natl Acad Sci USA 109:633–638
Yokota K, Fukai E, Madsen LH et al (2009) Rearrangement of actin cytoskeleton mediates invasion of *Lotus japonicus* roots by *Mesorhizobium loti*. Plant Cell 21:267–284
Zhang X, Davidson EA, Mauzerall DL, Searchinger TD, Dumas P, Shen Y (2015) Managing nitrogen for sustainable development. Nature 528:51–59

Part IV
Diversity of Nematode and Insect Symbionts

Chapter 16
The *Wolbachia* Symbiont: Here, There and Everywhere

Emilie Lefoulon, Jeremy M. Foster, Alex Truchon, C. K. S. Carlow, and Barton E. Slatko

I want her everywhere. And if she's beside me, I know I need never care. But to love her is to need her, everywhere. Knowing that love is to share...

(J Lennon, P. McCartney)

Abstract *Wolbachia* symbionts, first observed in the 1920s, are now known to be present in about 30–70% of tested arthropod species, in about half of tested filarial nematodes (including the majority of human filarial nematodes), and some plant-parasitic nematodes. In arthropods, they are generally viewed as parasites while in nematodes they appear to be mutualists although this demarcation is not absolute. Their presence in arthropods generally leads to reproductive anomalies, while in nematodes, they are generally required for worm development and reproduction. In mosquitos, *Wolbachia* inhibit RNA viral infections, leading to populational reductions in human RNA virus pathogens, whereas in filarial nematodes, their requirement for worm fertility and survival has been channeled into their use as drug targets for filariasis control. While much more research on these ubiquitous symbionts is needed, they are viewed as playing significant roles in biological processes, ranging from arthropod speciation to human health.

16.1 Introduction

It is often the case in biological research that discoveries wait for connections to realize their full significance. This was the case for *Wolbachia* endosymbionts, where seemingly disparate observations coalesced to shed light on the fascinating biology of *Wolbachia* and its significance. These *Rickettsia*-like bacteria were first

E. Lefoulon · J. M. Foster · A. Truchon · C. K. S. Carlow · B. E. Slatko (✉)
Molecular Parasitology Group, New England Biolabs, Inc., Ipswich, MA, USA
e-mail: slatko@neb.com

M. Kloc (ed.), *Symbiosis: Cellular, Molecular, Medical and Evolutionary Aspects*, Results and Problems in Cell Differentiation 69,
https://doi.org/10.1007/978-3-030-51849-3_16

observed in mosquitos in 1924 by Marshall Hertig and Simon Wolbach, who observed them in ovarian tissue (Hertig and Wolbach 1924). They were formally named *Wolbachia pipientis* in 1936, in honor of Wolbach (Hertig 1936) who had discovered that lice transmit *Rickettsia prowazekii*, the organism which is the cause of epidemic typhus. A second seemingly isolated discovery was made almost 20 years after that, in the 1950s by S. Ghelelovitch (1952) and H. Laven (1959) who discovered that certain crosses within *Culex* mosquitos were incompatible, being essentially sterile, i.e. they produced few or no progeny. They named the phenomenon cytoplasmic incompatibility (CI), as it appeared to be due to a factor with an inheritance pattern through females but not males. It took another 20 years for these observations to be connected by Janice Yen and A. Ralph Barr in the 1970s (Yen and Barr 1973) who established that CI was associated with the presence of the *Rickettsia*-like organism that could be removed by antibiotic treatments.

Little did we know then, of the full extent of the presence of female-inherited *Wolbachia* in biological systems and their importance. *Wolbachia* are now known to have a wide range of phenotypic effects and have complex interactions with their hosts (Binnington and Hoffmann 1989; Bordenstein and Werren 1998; Bordenstein et al. 2001; Comandatore et al. 2013; Hoerauf et al. 2003a, b; Kampfraath et al. 2019; Landmann 2019; O'Neill 1989; Saul 1961; Serbus et al. 2008; Telschow et al. 2007; Werren et al. 2008; Zimmer 2001). They appear to have evolved as specialists in manipulating reproduction and development (or both) in their eukaryotic hosts as either parasites (arthropods) or mutualists (nematodes), although there can be an overlap, as some systems have components of each (Dedeine et al. 2001; Lefoulon et al. 2016; Nikoh et al. 2014; Werren et al. 2008; Zug and Hammerstein 2015). They have become important in medical applications such as fighting filarial diseases and mosquito-borne human viral diseases (Moreira et al. 2009; Bourtzis et al. 2014; Taylor et al. 2014, 2018). Since their early discovery, a virtual effusion of new examples of *Wolbachia* presence (and effects) has been observed in at least 50% of all tested arthropod species (Table 16.1). PCR, nested PCR, qPCR, microscopy (including immunostaining, fluorescence, and electron microscopy), metagenomic approaches, such as next-generation sequencing or hybridization capture technology can be used to find low levels of *Wolbachia* (Bridgeman et al. 2018; Brown et al. 2016; Gomes et al. 2017; Hartelt et al. 2004; Lefoulon et al. 2019; Noda et al. 1997). As most systems have not been examined in these ways, the frequency of species harboring *Wolbachia* is likely to be significantly higher, further highlighting that *Wolbachia* are the most ubiquitous symbionts on the planet and have been successful in invading many organisms and ecological niches.

16.2 Phenotypes in Arthropod Wolbachia

In some arthropods, CI is observable as sterility or semi-sterility in crosses between certain arthropod strains, and thus CI is demonstrated in reciprocal crosses. The presence of bacteria in ovaries or testes can also be established microscopically

Table 16.1 Diversity of host infected by *Wolbachia*

Wolbachia host					
Phylum	Subphylum	Order/family	Examples	Representative Genomes	References
Arthropod	Hexapods	Blattodea	Termites (*Kalotermes, Microcerotermes, Zootermopsis* spp)	*w*Zoo[a]	Lo et al. (2002), Bordenstein and Rosengaus (2005) and Gerth et al. (2014)
		Coleoptera	Beetles (*Tribolium confusum, Diabrotica virgifera*), weevils (*Hypera postica, Sitophilus oryzae*)	–	Wade and Stevens (1985) and O'Neill et al. (1992), Fialho and Stevens (2000) and Dumler et al. (2001)
		Diptera	Mosquitos (*Culex, Aedes, Anopheles* spp.)	*w*Pip, *w*AlbB	Trpis et al. (1981), Gomes et al. (2017), Carvajal et al. (2019) and WorldMosquitoProgram.org
			Flies (*Drosophila* spp., *Glossina morsitans, Chrysomya megacephala, Haematobia irritans*)	*w*Mel, *w*Ri, *w*IncCu, *w*Mau, *w*Meg, *w*Gmm, *w*Irr	Hoffmann et al. (1986), Wu et al. (2004) and Madhav et al. (2020)
		Hymenoptera	Wasps (*Nasonia* spp., *synergus* spp., *Trichogramma pretiosum, Muscidifurax uniraptor*)	*w*Tpre, *w*OneA1, *w*Uni, *w*VitB, *w*VitA	Perrot-Minnot et al. (1996), Rokas et al. (2002), Lindsey et al. (2016) and Richardson et al. (1987)
			Bee (*Nomada* spp., *Osmia caerulescens*)	*w*Nfla, *w*Nfle, *w*Nleu, *w*Npa, *w*Oc[a]	Lo et al. (2002), Werren and Windsor (2000), Gerth et al. (2014) and Gerth and Bleidorn (2017)
			Ants (*Pheidole* spp., *Myrmica incompleta, Formica exsecta*)	*w*Fex	Wenseleers et al. (1998), Russell et al. (2009), Kautz et al. (2013) and Reeves et al. (2020)
		Lepidoptera	Moths (*Carposina sasakii, Operophtera brumata, Plutella australiana*)	*w*CauA, Ob_Wba, *w*Aus	Brower (1976), Derks et al. (2015) and Ward and Baxter (2017)
			Butterflies (Acraea spp., *Hypolimnas bolina*)	*w*Bol1b	Jiggins et al. (2000), Ahmed et al. (2015) and Duplouy et al. (2013)

(continued)

Table 16.1 (continued)

Wolbachia host					
Phylum	Subphylum	Order/family	Examples	Representative Genomes	References
		Hemiptera	Bedbugs (*Cimex* spp.)	*w*Cle	Sakamoto and Rasgon (2006a, b), Nikoh et al. (2014), Akhoundi et al. (2016) and Siddiqui and Raja (2015)
			Aphid (*Aphis* spp., *Toxoptera aurantii*, *Neophyllaphis podocarpi*, *Pentalonia nigronervosa*)	*w*Pni[a]	Wang et al. (2014) and Brown et al. (2016)
			Cochineals (*Dactylopius coccus*)	*w*DacA, *w*DacB, *w*Coc1	Ramírez-Puebla et al. (2016)
			Planthoppers (*Nilaparvata lugens*, *Laodelphax striatellus*)	*w*Lug, *w*stri	Noda (1984, 1987) and Ju et al. (2020)
			Psyllid (*Diaphorina citri*)	*w*Di	Saha et al. (2012)
			Whiteflies (*Bemisia tabaci*)	*w*Btab	Bing et al. (2014)
		Phthiraptera	Louse (*Bovicola* spp)		Kyei-Poku et al. (2005)
		Odonata	Dragonflies (*Anax guttatus*, *Orthetrum* spp.)	–	Salunkhe et al. (2015)
		Orthoptera	Cricket (Phylopalpus spp., *Hapithus agitator*, *Orocharis saltator*, Gryllus spp.)	–	Werren and Windsor (2000) and Panaram and Marshall (2007)
		Siphonaptera	Fleas (*Ctenocephalides felis*, *Pulex irritans*, *Echidnophaga gallinacea*)	*w*CfeJ, *w*CfeT, *w*Cte[a]	Dittmar and Whiting (2004), Gerth et al. (2014) and Driscoll et al. (2020)
		Strepsiptera	Stylops (*Mengenilla moldrzyki*)	*w*Men[a]	Gerth et al. (2014)
		Collembola	Springtail (*Folsomia candida*, *Mesaphorura italica*, *Megalothorax incertus*)	*w*Fol	Vandekerckhove et al. (1999), Gerth et al. (2014) and Ma et al. (2017)

	Crustacea	Isopoda	Woodlouse (*Armadillidium vulgare*, *Cylisticus convexus*, *Asellus aquaticus*, *Sphaeroma* spp.)	*w*VulC, *w*con	Bouchon et al. (1998), Almerao et al. (2012) and Badawi et al. (2018)
	Arachnids	Acari	Mites (*Syringophilopsis* spp, Torotrogla spp., *Tetranychus urticae*, *Metaseiulus occidentalis*, *Bryobia* spp.)	–	Johanowicz and Hoy (1995), Ros et al. (2009) and Glowska et al. (2015)
		Ixodida	Ticks (*Amblyomma americanum*; *Ixodes ricinus*)	–	Zhang et al. (2011) and Duron et al. (2017)
		Oribatida	Moss mites (*Gustavia microcephala*)	–	Konecka et al. (2019)
		Araneae	Spider (*Diaea* spp., *Misumena* spp., *Nephila plumipes*)	–	Werren and Windsor (2000), Rowley et al. (2004), and Ros et al. (2009)
		Pseudoscorpiones	Pseudoscorpion (*Cordylochernes scorpioides*, *Atemnus politus*)	*w*Apol	Zeh et al. (2005) and Lefoulon et al. (2020a)
		Scorpiones	Scorpions (*Opistophthalmus* spp.)	–	Baldo et al. (2007)
Nematode		Onchocercidae	Filarial nematodes (*Brugia* spp., *Dirofilaria* spp., *Onchocerca* spp., *Wuchereria bancrofti*, *Mansonella* spp., *Dipetalonema* spp., *Madathamugadia hiepei*)	*w*Oo, *w*Ov, *w*Di, *w*Bm, *w*Bp, *w*Wb, *w*Dcau, *w*Mhie	Sironi et al. (1995), Bandi et al. (1998), Foster et al. (2005), Darby et al. (2012), Comandatore et al. (2013), Lefoulon et al. (2020b) and Lebov et al. (2020)
		Pratylenchidae	Plant-parasitic nematodes (*Radopholus similis*, *Pratylenchus penetrans*)	*w*Ppe	Jacob et al. (2008), Haegeman et al. (2009), and Brown et al. (2016)

Some examples of organisms infected by *Wolbachia* and available genomes of *Wolbachia*
[a]Only SRA available on NCBI

and/or their involvement implicated by antibiotic or heat-treatment curing. In addition to CI, *Wolbachia* can also be associated with male killing, feminization, or parthenogenesis in different host species. These phenotypes are the result of *Wolbachia* selfishly attempting to maintain itself in high frequencies in a population (Werren et al. 2008). In addition to the primary effect on reproductive behaviors, *Wolbachia* may also induce evolutionary pressure to select for modifications of their effects, effectively creating reproductive isolation and speciation barriers leading to long-term evolutionary effects on their host lineages (Hoffmann 1988; Hoffmannn et al. 1996; Hurst and Jiggins 2000; Werren et al. 2008; Zug and Hammerstein 2015). In arthropods (namely *Drosophila* and mosquitos), *Wolbachia* confer resistance to their hosts from pathogens, in particular from RNA viruses. This has led to development of strategies to utilize *Wolbachia* to reduce the disease burden of human viral pathogens in natural populations (see below). In addition, some cases of nutritional mutualism have been observed in arthropods with biotin supplementation by *Wolbachia* in the bedbug or planthoppers (Ju et al. 2020; Newton et al. 2020; Nikoh et al. 2014). The biotin synthesis operon seems to appear multiple times independently during *Wolbachia* evolution (Driscoll et al. 2020; Gerth and Bleidorn 2017; Bing et al. 2020; Lefoulon et al. 2020a).

Many arthropod *Wolbachia* species also harbor a WO (named for *Wolbachia*) bacteriophage insertion which appears to be related to some of the reproductive phenotypes associated with arthropod *Wolbachia* (Fujii et al. 2004; Gavotte et al. 2007; Masui et al. 2001; Wright et al. 1978). That these viruses have not been eliminated by evolutionary selective pressure suggests *Wolbachia* bacteriophage play roles in the growth, maintenance, or development of arthropod *Wolbachia* (Tanaka et al. 2009). No intact region or only vestiges of full prophage regions have been observed in studied *Wolbachia* genomes infecting filarial nematodes (Darby et al. 2012; Foster et al. 2005; Martin and Gavotte 2010) even though vestigial remnants of phage-like genes suggest that in their evolutionary lineage, *Wolbachia* associated with these prophages. Recently, the *cifA* and *cifB* genes (*cif*: *c*ytoplasmic *i*ncompatibility *f*actor) contained in the WO region of the *Culex Wolbachia* genome were identified as linked with CI parasitism and the *wmk* gene (*wmk: WO-mediated killing*) was described as a candidate gene involved in male-killing (LePage et al. 2017; Lindsey et al. 2018; Perlmutter et al. 2019). These genes might be diagnostics for *Wolbachia* biological systems where it is unclear if reproductive manipulations are present, due to a lack of the ability to perform appropriate genetic crosses.

In the analysis of phenotypic effects such as reciprocal cross sterility, it is important to separate the effects and or presence of *Wolbachia* from a diverse array of maternally (cytoplasmic) inherited microorganisms that have been discovered that can also alter sex ratio or sex determination in host arthropods. These include protozoa, spiroplasma, other endobacteria, such as *Cardinium* (Giorgini et al. 2009), as well as *Rickettsiae* (Hsiao and Hsiao 1985; Perlman et al. 2006).

16.3 Evolution of *Wolbachia*

Diversity of *Wolbachia* has been described using phylogenetic analysis of *Wolbachia* strains from a monophyletic group comprised of clades, named supergroups. The first designation of "supergroups" was in 1998 (Zhou et al. 1998; Lo et al. 2002). The supergroups have been designated A-S, with two supergroups (G, R) now having been eliminated and merged with preexisting groups. The supergroups A, B, E, H, I, K, M, N, O, P, Q, and S are exclusively composed of symbionts of arthropods (Lo et al. 2002; Glowska et al. 2015; Werren et al. 1995; Ros et al. 2009; Lo et al. 2007; Bing et al. 2014; Bordenstein and Rosengaus 2005; Lefoulon et al. 2020a). *Wolbachia* belonging to supergroups C, D, and J infect exclusively filarial nematodes (Onchocercidae) (Bandi et al. 1998; Casiraghi et al. 2004; Lefoulon et al. 2016). Supergroup L contains exclusively *Wolbachia* of plant-parasitic nematodes (Brown et al. 2016; Haegeman et al. 2009; Jacob et al. 2008). Supergroup F contains some *Wolbachia* strains infecting arthropods and some infecting nematodes (Ferri et al. 2011; Lefoulon et al. 2012, 2020b). It is important to realize the "supergroup" designation only describes the different evolutionary lineages (or clades) of *Wolbachia*, and their limits remain arbitrary.

Recombination events complicate accurate phylogenetics when using single genes or gene regions (Baldo et al. 2005) and thus a multi-locus sequence typing (MLST) system was proposed. However, this was developed based primarily upon analyses of supergroups A and B *Wolbachia*, as they were the majority of genomes sequenced at the time. While early on, molecular phylogeny was based on one single or a few loci, more recently, molecular characterization is based upon multi-locus phylogeny (Baldo et al. 2006; Ferri et al. 2011; Glowska et al. 2015; Lefoulon et al. 2016; Lo et al. 2002, 2007). With the advent of genomic sequencing coupled with phylogenetic analysis, there is an effort to revisit the MLST typing paradigm (Bleidorn and Gerth 2018) or revisit the classification of *Wolbachia* based on phylogenomics (Comandatore et al. 2013; Gerth et al. 2014) or "core genome alignments" (Chung et al. 2018). The notion of *Wolbachia* species remains under debate within the community (Ramirez-Puebla et al. 2015; Lindsey et al. 2016; Chung et al. 2018; Newton and Slatko 2019).

57 draft and 28 complete genomes of *Wolbachia* have been published and while this number seems large, in reality, they do not fully represent the full range and scope of *Wolbachia* biodiversity. Of these, 15 of the 28 completed genome sequences are from insect *Wolbachia* from the A and B supergroups, and 7 are from *Drosophila* species. Only 2 are from insects belonging to supergroups F and E and 4 are symbionts of nematodes (supergroups L, C, or D). Less genome information, if any, is available from the other supergroups, for example, from those present in arachnids (Baldo et al. 2007; Glowska et al. 2015; Johanowicz and Hoy, 1995; Rowley et al. 2004).

The phylogeny of *Wolbachia* from filarial nematodes is in need of more analysis to examine and classify its diversity, as current *Wolbachia* phylogenetics is largely non-nematode based. Due to this underrepresentation of sequence diversity, *Wolbachia* from diverse filarial nematodes are being investigated, as are *Wolbachia* from other under-represented supergroups. For example, Lefoulon et al. (2020a) have sequenced pseudoscorpion *Wolbachia*, of which two species, *Geogarypus minor* and *Chthonius ischnocheles*, are confirmed to be members of clade H. Of interest is that another pseudoscorpion species, *Atemnus politus*, appears to represent a new supergroup S, which also contains *Wolbachia* from a previously described pseudoscorpion *Cordylochernes scorpioides*. Clade S is a sister group to supergroup C (infecting exclusively filarial nematodes) and to supergroup F (infecting filarial nematodes and arthropods). Analysis of the evolutionary history of pseudoscorpions suggests numerous *Wolbachia* transfers/infections may have occurred. Horizontal transmission of *Wolbachia* among insects has been previously documented (O'Neill et al. 1992; Baldo et al. 2008; Gerth et al. 2013) and while the extent is not fully characterized, this is certainly a hallmark of *Wolbachia* evolution.

16.4 The Biology of *Wolbachia* in Filarial Nematodes

Wolbachia were first observed in the filarial nematodes *Dirofilaria immitis*, *Brugia pahangi*, *Onchocerca volvulus*, and *Onchocerca gutturosa*, although not identified as *Wolbachia*, in the 1970s (Harada et al. 1970; Kozek 1977; Kozek and Figueroa 1977; Kozek and Marroquin 1977; Lee 1975; McLaren et al. 1975). Intracellular bacteria were observed in microfilaria (first-stage larvae), female reproductive tissue, and lateral cords[1] (infolded body wall epidermis) of adults. Kozek (1977) suggested that viewed under the electron microscope, they looked *Rickettsia*-like and described their occurrences in various tissues. He also suggested there might be two forms of the endobacteria. In 1995, Sironi and colleagues (Sironi et al. 1995) identified these endobacteria in *Dirofilaria* as *Wolbachia* by 16S PCR analysis. Since then, studies have confirmed this observation in many filarial nematodes in the genera *Onchocerca*, *Bugia*, *Dirofilaria*, *Wuchereria*, *Mansonella*, *Litomosoides*, *Madathamugadia*, *Dipetalonema*, *Yatesia*, *Cruorifilaria*, and interestingly, only one species in the *Cercopithifilaria genus*, *C. japonica*.

In filarial nematodes, *Wolbachia* exist in the hypodermal cells' lateral cords in both males and females. These are tissues in which nutrients are mainly obtained from the host, although some are likely obtained through the digestive system. Presumably, it is in the lateral cords that *Wolbachia* interface with the host and environmental biochemistry. An atypical localization of *Wolbachia* in the intestinal cell wall has been observed in the case of *Mansonella (Cutifilaria) perforata* (Ferri et al. 2011) and *Madathamugadia hiepei* (Lefoulon et al. 2012). Interestingly, they harbored *Wolbachia* closely related to supergroup F members. *Wolbachia* are also found in the ovaries, oocytes, and in a subset of embryonic cells of developing embryos in females (Kramer et al. 2003). They are not present in the male

[1]While the literature generally refers to these as lateral *chords*, we, along with others, feel the proper designation should be lateral *cords*, as they are not harmonic musical sets of notes, but rope-like biological structures.

reproductive tract and thus, as in arthropods, are female transmitted. They are present in every developmental stage of the worms but their titer does not increase during the microfilarial stage or the larval stages in the insect host. After vector transfer to the vertebrate host from the mosquito vector, the *Wolbachia* undergo increased multiplication and their titer rapidly increases as the larvae develop to the adult stages (Kramer et al. 2003; McGarry et al. 2004). Microscopy confirmed that there were few bacteria in mosquito-derived L3 larvae but many, in large groups, in L4 larvae collected 9 and 21 days after infection (McGarry et al. 2004).

In early embryogenesis, *Wolbachia* infect a subset of hypodermal precursors, located in the dorsal-posterior part of the embryo (Landmann et al. 2010, 2014) and from here, cross cellular membranes to inhabit the germline stem cells and their progeny (Fischer et al. 2011, 2014; Landmann et al. 2014). They appear to use the kinesin/actin/dynein cytoskeleton for their movement within cells (Ferree et al. 2005; Serbus and Sullivan 2007), although they are housed inside cytoplasmic vacuoles, in which they may be able to be transported.

In filarial nematodes, *Wolbachia* are transmitted through the female germline but not through the male germline and are first localized in the hypodermis. In early development in the embryo, *Wolbachia* are concentrated at the posterior pole of the egg due to interactions with microtubules and polarized factors (Landmann et al. 2014). During tissue formation, the *Wolbachia* remain in one blastomere which gives rise to the hypodermis where they remain during both embryonic development and in early larval development. From the hypodermis, the *Wolbachia* migrate to the germline by invading distal ovarian syncytial tissue, first in the somatic gonadal cells at the ovarian distal tip. This invasion may be an active engulfment process guided by ovarian-specific signals recognized by *Wolbachia* (Landmann et al. 2010, 2012, 2014). *Wolbachia* thus take a somewhat oblique path to inhabit the female germline by first passing through the hypodermis.

Using the high-pressure freeze substitution microscopic technique, *Wolbachia* appear to be very variable in appearance but usually retain a three-layered double-membrane structure (Fischer et al. 2014). Actin tails, which might be expected if used for movement, are not seen, and thus *Wolbachia* motility in development may be due to movement of vacuoles, as they may co-opt the host cell's secretory pathway to move within and between cells. *Wolbachia* are most often associated with cytoplasmic vacuoles and are often associated with glycogen granules on host cellular membranes (Fischer et al. 2014; Voronin et al. 2016). The vacuoles appear to be of endoplasmic reticulum origin, likely enabling *Wolbachia* to escape from, or reduce the effects of, host primary immune surveillance and response to foreign pathogen infection (Fattouh et al. 2019).

In filarial nematodes, *Wolbachia* have developed mutualistic dependence with the worms as they depend upon this association for their reproduction and survival (Bandi et al. 1999; Bosshardt et al. 1993; Casiraghi et al. 2002; Chirgwin et al. 2003; Hoerauf 2003; Hoerauf et al. 1999, 2001, 2003a, b). This obligate mutualist relationship has taken a different evolutionary trajectory than to that in arthropods where, in general, organisms can survive without *Wolbachia*, albeit, in some cases, with reduced viability. In filarial worms, based upon genomic analysis, many biochemical pathways have been implicated in the *Wolbachia*-filarial host bidirectional association (Foster et al. 2005; Ghedin et al. 2009; Grote et al. 2017; Lentz et al. 2013; Wu et al. 2009). These include heme or riboflavin provisioning, nucleotide biosynthesis, lipoprotein biosynthesis, glycolytic or amino acid metabolism and uptake, bidirectional energy metabolism, potential contributions to immune defense, and the type IV secretion system (T4SS: secretion protein complex able to transport proteins and DNA across the cell membrane) (Darby et al. 2012; Foster et al. 2005; Li and Carlow 2012; Voronin et al. 2016). However, it is important to bear in mind that the *Wolbachia* of filarial nematodes are diverse (Lefoulon et al. 2020b); for example, the *Wolbachia* from *Onchocerca ochengi* have a reduced genome size and numerous pathways are incomplete, including the riboflavin metabolism pathway, suggested as being involved in the mutualism (Darby et al. 2012).

Many symbiotic and pathogenic intracellular bacteria use a T4SS for successful infection, proliferation, and persistence within hosts (Alvarez-Martinez and Christie 2009; Liosa et al. 2009; Zechner et al. 2012). In filarial *Wolbachia*, the T4SS may regulate the riboflavin biosynthesis pathway and interestingly, in some cases, vitamin B2 supplementation partially rescues parasites treated with antibiotic, suggesting that they may supply this essential vitamin to their worm hosts (Li and Carlow 2012). More recent studies have shown that filarial *Wolbachia* T4SS may manipulate eukaryotic membrane traffic to help maintain the essential symbiotic relationship (Carpinone et al. 2018).

In *B. malayi*, the *Wolbachia* surface protein is found in conjunction with six *B. malayi* glycolytic enzymes, including aldolase, which, confirmed by immuno-transmission electron microscopy, are associated with the *Wolbachia* surface (Voronin et al. 2016). The co-localization suggests that *Wolbachia* may utilize host glycogen and its derivatives such as glucose, glycolytic metabolites, and pyruvate as an energy source, as the genome sequence reveals the bacteria are missing two key enzymes in the glycolytic pathway, 6-phosphofructokinase, and pyruvate kinase, and thus are not able to directly convert glucose into pyruvate (Foster et al. 2005; Voronin et al. 2016, 2019). Extra pyruvate added to the media increases the *Wolbachia* population in worms. Inhibition of glycolysis in *Brugia* results in decreasing bacterial population, but the phenotype can be rescued by adding pyruvate, the product of glycolysis (Voronin et al. 2019). This suggests that *Wolbachia* with its reduced glycolytic pathway relies on the worm's glycolysis in order to obtain pyruvate for energy metabolism.

In filarial nematodes that harbor *Wolbachia* (and as mentioned some do not, such as *Loa loa* and *Acanthocheilonema viteae*), they are obligate for worm development, survival, and fertility, which has been confirmed with antibiotic studies. The few studies of filarial nematodes without *Wolbachia* have not revealed major differences in metabolic capabilities, suggesting that one role of *Wolbachia* may be more of a mutually dependent "interactive regulatory" one or that each *Wolbachia*-worm association may be unique.

Wolbachia depletion using antibiotics leads to extensive apoptosis in germline tissue, developing embryos, and developing microfilariae, but not in the hypodermal cords where cytoskeleton defects are nonetheless induced. While the exact mechanisms are unclear, the lack of apoptosis in lateral cord cells and most all other somatic tissues suggests the suppression event is not an "all or none" global consequence of *Wolbachia* depletion. Further, apoptotic suppression occurs in embryonic cells that contain *Wolbachia*, it is not suppressed in those without it (only a few early embryos possess *Wolbachia*) (Landmann et al. 2011).

Since autophagy is a major intracellular defense mechanism, *Wolbachia* appear to have developed mechanisms to evade it, for their benefit in ensuring conveyance to the next generation. A major autophagosomal marker is associated with vacuoles containing *Wolbachia*, but it is also found inside bacteria. Interestingly, the induction of autophagy in worms initiates lysosomal activity and the lysosomes attack *Wolbachia* directly. *Wolbachia* do induce a chain of events in normal worms, similar to that observed in other pathogen systems, suggesting that *Wolbachia* still is recognized as a foreign invader. As the host must surely maintain autophagic systems for defense against other foreign invaders, there must be an evolved mechanism to maintain a balance between autophagy and suppression for maintenance of *Wolbachia*, presumably for the host's own nutritional and biochemical needs (Voronin et al. 2012).

16.5 *Wolbachia* as Filarial Drug Targets

The mutualistic dependency for the presence of *Wolbachia* in filarial nematodes for their development and reproduction suggested their use as drug targets for antifilarial disease elimination (Johnston et al. 2014a, b, 2017; Taylor et al. 2005a, b, 2014, 2018). Human filarial nematodes are responsible for severe disease, affecting over 150 million people in more than 80 countries, with over 1 billion at risk of infection (Molyneux et al. 2003). The nematodes induce lymphatic filariasis (LF) or onchocerciasis (river blindness), depending upon the species. For lymphatic filariasis, the species are *Wuchereria bancrofti*, *Brugia malayi* and *B. timori*, and for onchocerciasis, it is *Onchocerca volvulus*. Once L3 larvae have been delivered to the animal host, they develop into adults after two molts and mate to produce first-stage larvae (microfilaria, mf). Since adult female worms can live for 7–10 years, they continually shed millions of microfilariae which can be picked up during blood meals by the requisite transmission insect vector (mosquitos or black flies) to continue the cycle of infection. LF is characterized by lymphatic system damage and blockage, leading to swellings in body parts associated with the lymphatics, including the limbs, breasts, and scrotal sacs in males, for example. In onchocerciasis (cutaneous filariasis), the worms exist in subcutaneous tissue fibrous nodules (some deep) and give rise to skin lesions, dermatitis and can lead to blindness (river blindness) when microfilariae cross the cornea, causing inflammation reactions. LF and onchocerciasis are rarely fatal but are significant diseases in causing personal and economic devastation.

In every tested filarial nematode harboring *Wolbachia* (including, but not limited to, *Litomosoides sigmodontis*, *Brugia malayi*, *B. pahangi*, *Mansonella perstans*, *Onchocerca ochengi*, *O. lienalis*, *O. gutturosa* and *Dirofilaria immitis*), anti-

Rickettsia antibiotics, such as doxycycline, tetracycline, or rifampicin, result in effects upon the host worm including inhibition of embryogenesis, infertility, inhibition of larval development, stunting of adult worms, and macrofilaricidal (adult worm killing) activity. The observed effects, both in vitro and in vivo, correlate loss of worm functionality with loss or reductions of *Wolbachia* from worm tissues (Aljayyoussi et al. 2017; Bandi et al. 1999; Bazzocchi et al. 2008; Coulibaly et al. 2009; Foster et al. 2013; Genchi et al. 1998; Halliday et al. 2014; Hoerauf et al. 1999, 2000, 2003a, b, 2008; Langworthy et al. 2000; Sharma et al. 2018; Specht et al. 2008; Tamarozzi et al. 2011; Taylor et al. 2005b; Townson et al. 2000; Turner et al. 2017). Furthermore, it has not been possible to produce viable worms that are totally "cured" of their *Wolbachia*. The adulticide aspect is particularly noteworthy, as elimination of adults, not achievable with the current available antifilarial drugs, could stop or reduce transmission by the elimination of the continual source of microfilariae transmitted to the arthropod vectors. A further advantage of the anti-*Wolbachia* drug targeting is the apparent slow onset of antiparasitic activity, thereby avoiding adverse reactions caused by rapid micro- and/or macrofilaricidal activity (Mazzotti reactions: Henson et al. 1979). The antibiotic effects upon *Wolbachia* confirm the findings that indicate the bacterium provides an essential function to the host nematode that enables its development and viability. However, at the time, the existing anti-*Wolbachia* drugs, notably doxycycline, present challenges as a drug for mass administration because of the requirement for long treatment times for effects on worm sterility and viability (4–6 weeks) and contraindications in pregnancy and in children under the age of 8 (Taylor et al. 2014).

Early approaches for drug targeting *Wolbachia* were based upon genomics and/or genome mining for sequence information, first from the sequenced *B. malayi Wolbachia* genome (Foster et al. 2005; Holman et al. 2009). Using a bioinformatic pipeline that eliminated human-related targets, several candidate targets were suggested and researched. These included 2 enzymes involved in *Wolbachia* metabolism. Phosphoglycerate mutases (PGM) interconvert 2- and 3-phosphoglycerate in the glycolytic and gluconeogenic pathways. PGM exists in two distinct forms, cofactor independent phosphoglycerate mutase (iPGM) and cofactor dependent phosphoglycerate mutase (dPGM). The iPGM is the only form identified in filarial *Wolbachia* and their worm hosts. Since iPGM has no sequence or structural similarity to the dPGM form present in mammals, iPGM was pursued as a candidate drug target and specific inhibitors were identified (Foster et al. 2009; Li et al. 2011a; Raverdy et al. 2007; Yu et al. 2017). The genome filtering approach also revealed that filarial *Wolbachia* lack the enzyme pyruvate kinase (PK) and may instead utilize pyruvate phosphate dikinase (PPDK). Most organisms, including mammals, possess PK exclusively. The absence of PPDK in humans and the lack of sequence homology between PPDK and PK suggested that PPDK-specific inhibitors may be identified (Raverdy et al. 2008). The availability of genomic sequences from filarial *Wolbachia* also led to studies on the essential cell division protein FtsZ which has a GTPase activity. The natural plant product berberine was identified in enzyme inhibitor screens and was shown to be effective in reducing the motility and reproduction of filarial parasites in vitro (Li et al. 2011b).

As a more directed approach, anti-*Wolbachia* targeting used mass screening of chemicals, drugs, and biomolecules from varied preexisting diversified or focused molecular libraries to select inhibitors of *Wolbachia* development and reproduction, and thus potentially qualify as antifilarial drugs. This approach first used a developed cell culture system at the Liverpool School of Tropical Medicine (A.WOL project) and then in animal models before initiating human clinical trials. It has involved many industrial and academic partners and has had as a primary goal the identification and characterization of drugs that are adulticides that shorten treatments to less than a week, as there are challenges of patient adherence (Boussinesq et al. 2018; Gualano et al. 2014; Pechère et al. 2007; Taylor et al. 2014), and are safe for the target population. A secondary goal has been the development of treatment protocols which are compatible with the current mass-drug (MDA) procedures.

The A.WOL screening project, using cell culture and animal models, tested previously identified antibiotics and those developed and repurposed, and also developed protocols (for instance, for drug combinations) that have been and are being tested in clinical trials (Bakowski et al. 2019; Boussinesq et al. 2018; Clare et al. 2015; Debrah et al. 2007, 2011, 2015; Hong et al. 2019; Jacobs et al. 2019; Johnston et al. 2014a, b; Mand et al. 2012; Specht et al. 2008; Supali et al. 2008; Tamarozzi et al. 2012; Taylor et al. 2014, 2019; Turner et al. 2006, 2010, 2017; Walker et al. 2015; Wanji et al. 2009; Von Geldern et al. 2019). The candidate drugs have advantages over existing anti-*Wolbachia* compounds by showing higher efficacy than the "gold standard" doxycycline, and which meet their criteria goals. A. WOL is not the only anti-*Wolbachia* drug screening program; other laboratories are also screening for *Wolbachia* inhibitory compounds, as well (Bakowski and McNamara 2019; Serbus et al. 2012; Shrivastava et al. 2015; Xu et al. 2019).

16.6 The Use of *Wolbachia* to Reduce Human RNA Viral Pathogens

In addition to *Wolbachia* being a target for filariasis elimination, *Wolbachia* are also being used as a tool for human disease prevention of other mosquito-borne diseases and as a target for diagnostics (Bourtzis et al. 2014; Slatko et al. 2014; www.worldmosquitoproject.org; www.oxitec.com; www.mosquitomate.com). In terms of disease prevention, *Wolbachia* can protect their arthropod (mostly *Drosophila* and mosquito) hosts from pathogens, in particular from RNA viruses (Bourtzis et al. 2014; Caragata et al. 2016; Cardona-Salgado et al. 2020; Chouin-Carneiro et al. 2019; Ferreira et al. 2020; Ford et al. 2019; Hedges et al. 2008; Moreira et al. 2009; Marcus et al. 2012; Ye et al. 2013). Two different strategies, using CI, have been envisioned and implemented: *Wolbachia* population replacements or population reduction. Population replacement implies adding, driving, or replacing a *Wolbachia* "type" in the insect vector in the population with one that has the requisite antiparasitic attributes (or has more biologically selective "fitness") (Cook et al. 2006, 2008; Ryan et al. 2019; worlmosquitoproject.org).

Aedes aegypti, the insect vector of many viral pathogens, is not naturally infected with *Wolbachia*, but several strains of the *Wolbachia* bacterium have been successfully introduced into it. Containing *Wolbachia*, these mosquito vectors are resistant, and largely fail to transmit, viral pathogens such as dengue, chikungunya, Zika, yellow fever, as well as, malaria. Using this approach for widespread mosquito, antiviral human disease control requires introduction and maintenance of the *Wolbachia* strain in the natural population. Here, the basic biology of arthropod CI, in which the proportion of *Wolbachia*-infected individuals increases in the population and is maintained, plays a significant role.

With CI, males infected with *Wolbachia* (produced from females containing *Wolbachia*) are sterile when they mate with uninfected females; they only give rise to progeny when crossed with *Wolbachia*-infected females. However, *Wolbachia*-infected females produce progeny when mated with males that are either infected or not infected. This gives infected females a reproductive advantage and leads to an increase in the proportion of individuals in the population containing *Wolbachia*. The increase in the percentage of females containing *Wolbachia* leads to decreased frequencies of females able to support and transmit the viral pathogens. For these "populational replacement" biological control approaches, CI can be harnessed to establish and maintain *Wolbachia*-infected mosquito populations in the field.

The second general strategy, population reduction, is to reduce insect populations using CI to create and release *Wolbachia*-containing males which, when mated to females in the population, results in defective embryogenesis within the females (O'Connor et al. 2012; Gilbert and Melton 2018; Mains et al. 2019; Crawford et al. 2020; www.oxitec.com; www.mosquitomate.com). This is a form of the sterile insect technique (SIT), but uses *Wolbachia* to be the sterilizing agent (in the case of the ZAP mosquitos from Mosquitomate or a sterilizing gene from Oxitec). Here, repeated releases of "sterilizing" males (created by CI) are performed to reduce population size and thus limit vector exposure to the population. Since male mosquitoes do not feed on blood and thus do not transmit disease, extensive or repetitive release of male mosquitoes is not a health or nuisance issue. This strategy involves the rearing of large numbers of males, created by CI, to be semi-continually released.

Both strategies are being implemented in natural populations, under strict guidelines where field testing is providing evidence of the effectiveness of these approaches. The World Mosquito Program (https://www.worldmosquitoprogram.org), Mosquitomate (www.mosquitomate.com) with Verily, Inc. (Debug Project, www.verily.com) and Oxitec, Inc. (www.oxitec.com) are currently field-testing the large-scale release of *Wolbachia*-infected *Aedes*, in communities for local mosquito control and in those prone to vector-borne human viral outbreaks, to provide evidence of the effectiveness of these approaches.

It is conceivable that it may be possible in the future to further enhance anti-parasite effects by creating *Wolbachia* strains that express a product (for instance, a microRNA or protein) that would affect pathogen biology or transmission. Using "genome editing" techniques such as CRISPR, Zn-finger nucleases or TALENS or WO phage transduction, it might be feasible to create such *Wolbachia* genomic alterations (Slatko et al. 2010).

16.7 Summary

In summary, *Wolbachia* have morphed from a biological curiosity to a tool for combating human disease. Considerable and substantial information has been accumulated in the last (almost) 100 years of *Wolbachia* research, although much remains to be learned about its host-cellular interactions, immunology, biochemistry, and populational ecology and evolution.

Acknowledgements We wish to thank Don Comb, Jim Ellard, Rich Roberts, Bill Jack, and Tom Evans for their support of the Molecular Parasitology Research Program at New England Biolabs, Inc. We wish to acknowledge former senior members of the group, in particular Fran Perler, Sanjay Kumar, and Larry McReynolds. We also wish to thank Denis Voronin, Frederic Landmann, Bill Sullivan, and reviewers for manuscript editing suggestions.

References

Ahmed M, Li S, Xue X, Yin X, Ren S, Jiggins F, Greeff J, Qiu B (2015) The intracellular bacterium *Wolbachia* uses parasitoid wasps as phoretic vectors for efficient horizontal transmission. PLoS Pathog 10:e1004672. https://doi.org/10.1371/journal.ppat.1004672

Akhoundi M, Cannet A, Loubatier C, Berenger J, Izri A, Marty P, Delaunay P (2016) Molecular characterization of *Wolbachia* infection in bed bugs (*Cimex lectularius*) collected from several localities in France. Parasite 23:31. https://doi.org/10.1051/parasite/2016031

Aljayyoussi G, Tyrer HE, Ford L, Sjoberg H, Pionnier N, Waterhouse D, Davies J, Gamble J, Metugene H, Cook DA, Steven A, Sharma R, Guimaraes AF, Clare RH, Cassidy A, Johnston KL, Myhill L, Hayward L, Wanji S, Turner JD, Taylor MJ, Ward SA (2017) Short-course, high-dose rifampicin achieves *Wolbachia* depletion predictive of curative outcomes in preclinical models of lymphatic filariasis and onchocerciasis. Sci Rep 7:210. https://doi.org/10.1038/s41598-017-00322-5

Almerao M, Fagundes N, de Araujo P, Verne S, Grandjean F, Bouchon D (2012) First record of *Wolbachia* in South American terrestrial isopods: prevalence and diversity in two species of *Balloniscus* (Crustacea, Oniscidea). Genet Mol Biol 35:980–989

Alvarez-Martinez C, Christie P (2009) Biological diversity of prokaryotic type IV secretion systems. Microbiol Mol Biol Rev 73:775–808. https://doi.org/10.1128/MMBR.00023-09

Badawi M, Moumen B, Giraud I, Grève P, Cordaux R (2018) Investigating the molecular genetic basis of cytoplasmic sex determination caused by Wolbachia endosymbionts in terrestrial isopods. Genes 9:290. https://doi.org/10.3390/genes9060290

Bakowski M, McNamara C (2019) Advances in anti-*Wolbachial* drug discovery for treatment of parasitic filarial worm infections. Trop Med Infect Dis 4:108. https://doi.org/10.3390/tropicalmed4030108

Bakowski M, Shiroodi R, Liu R, Olejniczak J, Yang B, Gagaring K, Guo H, White P, Chappell L, Debec A, Landmann F, Dubben B, Lenz F, Struever D, Ehrens A, Frohberger S, Sjoberg H, Pionnier N, Murphy E, Archer A, Steven A, Chunda V, Fombad F, Chounna P, Njouendou A, Metuge H, Ndzeshang B, Gandjui N, Akumtoh D, Kwenti T, Woods A, Joseph S, Hull M, Xiong W, Kuhen K, Taylor M, Wanji S, Turner J, Hübner M, Hoerauf A, Chatterjee A, Roland J, Tremblay M, Schultz P, Sullivan W, Chu X, Petrassi H, McNamara C (2019) Discovery of short-course antiwolbachial quinazolines for elimination of filarial worm infections. Sci Transl Med 11:eaav3523. https://doi.org/10.1126/scitranslmed.aav3523

Baldo L, Lo N, Werren J (2005) Mosaic nature of the *Wolbachia* surface protein. J Bacteriol 187:5406–5418. https://doi.org/10.1128/JB.187.15.5406-5418.2005

Baldo L, Hotopp JD, Jolley K, Bordenstein, S, Biber S, Ray R, Hayashi C, Maiden, C, Tettelin, H, Werren J. (2006) Multilocus sequence typing system for the endosymbiont *Wolbachia pipientis*. Appl Environ Microbiol 72:7098–7110. https://doi.org/10.1128/AEM.00731-06

Baldo L, Prendini L, Corthals A, Werren JH (2007) *Wolbachia* are present in southern African scorpions and cluster with supergroup F. Curr Microbiol 55:367–373. https://doi.org/10.1007/s00284-007-9009-4

Baldo L, Ayoub NA, Hayashi CY, Russell JA, Stahlhut JK, Werren JH (2008) Insight into the routes of *Wolbachia* invasion: high levels of horizontal transfer in the spider genus *Agelenopsis* revealed by *Wolbachia* strain and mitochondrial DNA diversity. Mol Ecol 17:557–569. https://doi.org/10.1111/j.1365-294X.2007.03608.x

Bandi C, Anderson TJ, Genchi C, Blaxter ML (1998) Phylogeny of *Wolbachia* in filarial nematodes. Proc Biol Sci 265:2407–2413. https://doi.org/10.1098/rspb.1998.0591

Bandi C, McCall J, Genchi C, Corona S, Venco L, Luciano L (1999) Effects of tetracycline on the filarial worms *Brugia pahangi* and *Dirofilaria immitis* and their bacterial endosymbionts *Wolbachia*. Int J Parasitol 29:357–364. https://doi.org/10.1016/s0020-7519(98)00200-8

Bazzocchi C, Mortarino M, Grandi G, Kramer LH, Genchi C, Bandi C, Genchi M, Sacchi L, McCall J (2008) Combined ivermectin and doxycycline treatment has microfilaricidal and adulticidal activity against *Dirofilaria immitis* in experimentally infected dogs. Int J Parasitol 38:1401–1410. https://doi.org/10.1016/j.ijpara.2008.03.002

Bing X, Xia W-Q, Gui J-D, Yan G-H, Wang X-W, Liu S-S (2014) Diversity and evolution of the *Wolbachia* endosymbionts of *Bemisia* (Hemiptera: Aleyrodidae) whiteflies. Ecol Evol 4:2714–2737. https://doi.org/10.1002/ece3.1126

Bing X-L, Zhao D-S, Sun J-T, Zhang K-J, Hong X-Y, Sloan D (2020) Genomic analysis of *Wolbachia* from *Laodelphax striatellus* (Delphacidae, Hemiptera) reveals insights into its "Jekyll and Hyde" mode of infection pattern. Genome Biol Evol 12(2):3818–3831

Binnington K, Hoffmann A (1989) *Wolbachia*-like organisms and cytoplasmic incompatibility in *Drosophila simulans*. J Invert Pathol 54:344–352. https://doi.org/10.1016/0022-2011(89)90118-3

Bleidorn C, Gerth M (2018) A critical re-evaluation of multilocus sequence typing (MLST) efforts in *Wolbachia*. FEMS Microbiol Ecol 94(1). https://doi.org/10.1093/femsec/fix163

Bordenstein S, O'Hara F, Werren J (2001) *Wolbachia*-induced incompatibility precedes other hybrid incompatibilities in *Nasonia*. Nature 409:707–710. https://doi.org/10.1038/35055543

Bordenstein S, Rosengaus RB (2005) Discovery of a novel *Wolbachia* supergroup in isoptera. Curr Microbiol 51:393–398. https://doi.org/10.1007/s00284-005-0084-0

Bordenstein S, Werren J (1998) Effects of A and B Wolbachia and host genotype on interspecies cytoplasmic incompatibility in *Nasonia*. Genetics 148:1833–1884

Bosshardt S, McCall J, Coleman S, Jones K, Petit T, Klei T (1993) Prophylactic activity of tetracycline against *Brugia pahangi* infection in jirds (*Meriones unguiculatus*). J Parasitol 79:775–777. https://doi.org/10.2307/3283620

Bouchon D, Rigaud T, Juchault P (1998) Evidence for widespread *Wolbachia* infection in isopod crustaceans: molecular identification and host feminization. Proc Biol Sci 265:1081–1090. https://doi.org/10.1098/rspb.1998.0402

Bourtzis K, Dobson S, Xi Z, Rasgon J, Calvitti M, Moreira L, Bossin H, Moretti R, Baton L, Hughes G, Mavingui P, Gilles J (2014) Harnessing mosquito-*Wolbachia* symbiosis for vector and disease control. Acta Trop 132:S150–S163. https://doi.org/10.1016/j.actatropica.2013.11.004

Boussinesq M, Fobi G, Kuesel A (2018) Alternative treatment strategies to accelerate the elimination of onchocerciasis. Int Health 10:i40–i48. https://doi.org/10.1093/inthealth/ihx054

Bridgeman B, Morgan-Richards M, Wheeler D, Trewick S (2018) First detection of *Wolbachia* in the New Zealand biota. PLoS One 13:e0195517. https://doi.org/10.1371/journal.pone.0195517

Brower J (1976) Cytoplasmic incompatibility: occurrence in a stored-product pest *Ephestia cautella1*. Ann Entomol Soc Am 69:1011–1015

Brown A, Wasala S, Howe D, Peetz A, Zasada I, Denver D (2016) Genomic evidence for plant-parasitic nematodes as the earliest Wolbachia hosts. Sci Rep 6:34955. https://doi.org/10.1038/srep34955

Caragata E, Dutra H, O'Neill S, Moreira L (2016) Zika control through the bacterium *Wolbachia pipientis*. Future Microbiol 11:1499–1502. https://doi.org/10.2217/fmb-2016-0177

Cardona-Salgado D, Campo-Duarte D, Sepulveda-Salcedo L, Vasilieva O (2020) *Wolbachia*-based biocontrol for dengue reduction using dynamic optimization approach. Appl Math Model 82:125–149

Carpinone E, Li Z, Mills M, Foltz C, Brannon E, Carlow C, Starai V (2018) Identification of putative effectors of the Type IV secretion system from the *Wolbachia* endosymbiont of *Brugia malayi*. PLoS One 13:e0204736

Carvajal T, Hashimoto K, Harnandika R, Amalin D, Wantanabe K (2019) Detection of *Wolbachia* in field-collected *Aedes aegypti* mosquitoes in metropolitan Manila, Philippines. Parasit Vectors 12:361. https://doi.org/10.1186/s13071-019-3629-y

Casiraghi M, McCall J, Simonchi L, Kramer L, Genchi C, Werren J, Bandi C (2002) Tetracycline treatment and sex-ratio distortion: a role for *Wolbachia* in the moulting of filarial nematodes? Int J Parasitol 32:1457–1468. https://doi.org/10.1016/S0020-7519(02)00158-3

Casiraghi M, Bain O, Guerrero R, Martin C, Pocacqua V, Gardner S, Franceschi A, Bandi C (2004) Mapping the presence of *Wolbachia pipientis* on the phylogeny of filarial nematodes: evidence for symbiont loss during evolution. Int J Parasitol 34:191–203. https://doi.org/10.1016/j.ijpara.2003.10.004

Chirgwin S, Coleman S, Porthouse K, Nowling J, Punkosdy G, Klei T (2003) Removal of *Wolbachia* from *Brugia pahangi* is closely linked to worm death and fecundity but does not result in altered lymphatic lesion formation in Mongolian gerbils (*Meriones unguiculatus*). Infect Immun 71:6986–6994. https://doi.org/10.1128/IAI.71.12.6986-6994.2003

Chouin-Carneiro T, Ant T, Herd C, Faillou L, Sinkins S (2019). *Wolbachia* strain *w*AlbA blocks Zika virus transmission in *Aedes aegypti*. Med Vet Entomol 34(1):116–119. https://doi.org/10.1111/mve.12384

Chung M, Munro J, Tettelin H, Dunning-Hotopp J (2018) Using core genome alignments to assign bacterial species. mSystems 3:e00236–e00218. https://doi.org/10.1128/mSystems.00236-18

Clare R, Cook D, Johnston K, Ford L, Ward S, Taylor M (2015) Development and validation of a high-throughput anti-*Wolbachia* whole-cell screen: a route to macrofilaricidal drugs against onchocerciasis and lymphatic filariasis. J Biomol Screen 20:64–69. https://doi.org/10.1177/1087057114551518

Comandatore F, Sassera D, Montagna M, Kumar S, Koutsovoulos G, Thomas G, Repton C, Babayan S, Gray N, Cordaux R, Darby A, Makepeace B, Blaxter M (2013) Phylogenomics and analysis of shared genes suggest a single transition to mutualism in *Wolbachia* of nematodes. Genome Biol Evol 5:1668–1674. https://doi.org/10.1093/gbe/evt125

Cook P, Hugo L, Iturbe-Ormaetxe I, Williams C, Chenoweth S, Ritchie S, Ryan P, Kay B, Blows M, O'Neill S (2006) The use of transcriptional profiles to predict adult mosquito age under field conditions. Proc Natl Acad Sci USA 103:18060–18065. https://doi.org/10.1073/pnas.0604875103

Cook P, McMeniman C, O'Neill S (2008) Modifying insect population age structure to control vector-borne disease. Adv Exp Med Biol 627:126–140. https://doi.org/10.1007/978-0-387-78225-6_11

Coulibaly Y, Dembele B, Diallo A, Lipner E, Doumbia S, Coulibaly S, Konate S, Diallo D, Yalcouye D, Kubofcik J (2009) A randomized trial of doxycycline for *Mansonella perstans* infection. N Engl J Med 361:1448–1458. https://doi.org/10.1056/NEJMoa0900863

Crawford J, Clarke D, Criswell V, Desnoyer M, Cornel D, Deegan B, Gong K, Hopkins K, Howell P, Hyde J, Livni J, Behling C, Benza R, Chen W, Dobson K, Eldershaw C, Greeley D,Han Y, Hughes B, Kakani E, Karbowski J, Kitchell A, Lee E, Lin T, Liu J, Lozano,M, MacDonald W,

Mains J, Metlitz M, Mitchell S, Moore D, Ohm J, Parkes K, Porshnikoff A, Robuck C, Sheridan M, Sobecki R, Smith P, Stevenson J, Sullivan J, Wasson B, Weakley A, Wilhelm M, Won J, Yasunaga A, Chan W, Holeman J, Snoad N, Upson L, Zha T, Dobson S, Mulligan F, Massaro P, White B (2020) Efficient production of male *Wolbachia*-infected *Aedes aegypti* mosquitoes enables large-scale suppression of wild populations. Nat Biotechnol 38(4):482–492. https://doi.org/10.1038/s41587-020-0471-x

Darby A, Armstrong S, Bah G, Kaur G, Hughes M, Kay S, Koldkjær P, Rainbow L, Radford A, Blaxter M, Tanya V, Trees A, Cordaux R, Wastling J, Makepeace B (2012) Analysis of gene expression from the *Wolbachia* genome of a filarial nematode supports both metabolic and defensive roles within the symbiosis. Genome Res 22:2467–2477. https://doi.org/10.1101/gr.138420.112

Debrah A, Mand S, Marfo-Debrekyei Y, Batsa L, Pfarr K, Buttner M, Adjei O, Buttner D, Hoerauf A (2007) Macrofilaricidal effect of 4 weeks of treatment with doxycycline on *Wuchereria bancrofti*. Tropical Med Int Health 12(12):1433–1441. https://doi.org/10.1111/j.1365-3156.2007.01949.x

Debrah A, Mand S, Marfo-Debrekyei Y, Batsa L, Albers A, Specht S, Klarmann U, Pfarr K, Adjei O, Hoerauf A (2011) Macrofilaricidal activity in *Wuchereria bancrofti* after 2 weeks treatment with a combination of rifampicin plus doxycycline. J Parasitol Res. https://doi.org/10.1155/2011/201617

Debrah A, Specht S, Klarmann-Schulz U, Batsa L, Mand SY, Marfo-Debrekyei Y, Fimmers R, Dubben B, Kwarteng A, Osei-Atweneboana M, Boakye D, Ricchiuto A, Büttner M, Adjei O, Mackenzie C, Hoerauf A (2015) Doxycycline leads to sterility and enhanced killing of female *Onchocerca volvulus* worms in an area with persistent microfilaridermia after repeated ivermectin treatment—a randomized placebo controlled double-blind trial. Clin Infect Dis 61:517–526. https://doi.org/10.1093/cid/civ363

Dedeine F, Vavre F, Fleury F, Loppin B, Hochberg M, Boulétreau M (2001) Removing symbiotic *Wolbachia* bacteria specifically inhibits oogenesis in a parasitic wasp. Proc Natl Acad Sci USA 98:6247–6252. https://doi.org/10.1073/pnas.101304298

Derks M, Smit S, Salis L, Schijlen E, Bossers A, Mateman C, Pijl A, de Ridder D, Groenen M, Visser M, Megens H (2015) The genome of winter moth (*Operophtera brumata*) provides a genomic perspective on sexual dimorphism and phenology. Genome Biol Evol 7:2321–2332. https://doi.org/10.1093/gbe/evv145

Dittmar K, Whiting M (2004) New *Wolbachia* endosymbionts from nearctic and neotropical fleas (Siphonaptera). J Parasitol 90:953–957. https://www.jstor.org/stable/3286370

Driscoll T, Verhoeve V, Gillespie J, Spencer Johnston J, Guillotte M, Rennoll-Bankert K, Rahman M, Hagen D, Elsik C, Macaluso K, Azad F (2020) A chromosome-level assembly of the cat flea genome uncovers rampant gene duplication and genome size plasticity. BMC Biol 18(1): 70. https://doi.org/10.1186/s12915-020-00802-7

Dumler J, Barbet A, Bekker C, Dasch G, Palmer G, Ray S, Rikihisa Y, Rurangirwa F (2001) Reorganization of genera in the families *Rickettsiaceae* and *Anaplasmataceae* in the order *Rickettsiales*: unification of some species of *Ehrlichi*a with *Anaplasma*, *Cowdria* with *Ehrlichia* and *Ehrlichia* with *Neorickettsia*, descriptions of six new species combinations and designation of *Ehrlichia equi* and 'HGE agent' as subjective synonyms of *Ehrlichia phagocytophila*. Int J Syst Evol Microbiol 51:2145–2165. https://doi.org/10.1099/00207713-51-6-2145

Duplouy A, Iturbe-Ormaetxe I, Beatson S, Szubert J, Brownlie J, McMeniman C, McGraw E, Hurst G, Gregory D, Charlat S, O'Neill S, Woolfit M (2013) Draft genome sequence of the male-killing *Wolbachia* strain wBol1 reveals recent horizontal gene transfers from diverse sources. BMC Genomics 14:20.1–20.13. https://doi.org/10.1186/1471-2164-14-20

Duron O, Binetruy F, Noel V, Cremaschi K, Arnatha C, Chevillon C, Plantar GJ, Perez De Leon A, Heylen D, Estrada-Pena A (2017) Evolutionary changes in symbiont community structure in ticks. Ecol Lett 26:2905–2921. https://doi.org/10.1111/mec.14094

Fattouh N, Cazevieille C, Landmann F (2019) *Wolbachia* endosymbionts subvert the endoplasmic reticulum to acquire host membranes without triggering ER stress. PLoS Negl Trop Dis 13: e0007218. https://doi.org/10.1371/journal.pntd.0007218

Ferree P, Frydman H, Li J, Cao J, Wieschaus E, Sullivan W (2005) *Wolbachia* utilizes host microtubules and dynein for anterior localization in the *Drosophila* oocyte. PLoS Pathog 1: e14. https://doi.org/10.1371/journal.ppat.0010014

Ferreira A, Fairlie S, Moreira L (2020) Insect vectors endosymbionts as solutions against diseases. Curr Opin Insect Sci 40:56–61. https://doi.org/10.1016/j.cois.2020.05.014

Ferri E, Bain O, Barbuto M, Martin C, Lo N, Uni S, Landmann F, Baccei SG, Guerrero R, de Souza LS, Bandi C, Wanji S, Diagne M, Casiraghi M (2011) New insights into the evolution of *Wolbachia* infections in filarial nematodes inferred from a large range of screened species. PLoS One 6:e20843. https://doi.org/10.1371/journal.pone.0020843

Fialho R, Stevens L (2000) Male-killing *Wolbachia* in a flour beetle. Proc Biol Sci 267:1469–1473. https://doi.org/10.1098/rspb.2000.1166

Fischer K, Beatty W, Jiang D, Weil G, Fischer P (2011) Tissue and stage-specific distribution of *Wolbachia* in *Brugia malayi*. PLoS Negl Trop Dis 5:e1174. https://doi.org/10.1371/journal.pone.0086383

Fischer K, Beatty W, Weil G, Fischer P (2014) High pressure freezing/freeze substitution fixation improves the ultrastructural assessment of *Wolbachia* endosymbiont–filarial nematode host interaction. PLoS One 9:e86383. https://doi.org/10.1371/journal.pone.0086383

Ford S, Allen S, Ohm J, Sigle L, Sebastian A, Albert I, Chenoweth S, McGraw E (2019) Selection on *Aedes aegypti* alters *Wolbachia*-mediated dengue virus blocking and fitness. Nat Microbiol 4:1832–1839. https://doi.org/10.1038/s41564-019-0533-3

Foster J, Ganatra M, Kamal I, Ware J, Makarova K, Ivanova N, Bhattacharyya A, Kapatral V, Kumar S, Posfai J, Vincze T, Ingram J, Moran L, Lapidus A, Omelchenko M, Kyrpides N, Ghedin E, Wang S, Goltsman E, Joukov V, Ostrovskaya O, Tsukerman K, Mazur M, Comb D, Koonin E, Slatko B (2005) The *Wolbachia* genome of Brugia malayi: endosymbiont evolution within a human pathogenic nematode. PLoS Biol 3:e121. https://doi.org/10.1371/journal.pbio.0030121

Foster J, Raverdy S, Ganatra M, Colussi P, Taron C, Carlow C (2009) The *Wolbachia* endosymbiont of Brugia malayi has an active phosphoglycerate mutase: a candidate target for anti-filarial therapies. Parasitol Res 104:1047–1052. https://doi.org/10.1007/s00436-008-1287-7

Foster J, Hoerauf A, Slatko B, Taylor M (2013) The molecular biology, immunology and chemotherapy of *Wolbachia* bacterial endosymbionts of filarial nematodes. In: Kennedy M, Harnett W (eds) Parasitic nematodes: molecular biology, biochemistry and immunology, CABI, Wallingford, https://doi.org/10.1079/9781845937591.0308

Fujii Y, Kubo T, Ishikawa H, Sasaki T (2004) Isolation and characterization of the bacteriophage WO from *Wolbachia*, an arthropod endosymbiont. Biochem Biophys Res Commun 317:1183–1188. https://doi.org/10.1016/j.bbrc.2004.03.164

Gavotte L, Henri H, Stouthamer R, Charif D, Charlat S, Bouletreau M, Vavre F (2007) A survey of the bacteriophage WO in the endosymbiotic bacteria *Wolbachia*. Mol Biol Evol 24:427–435. https://doi.org/10.1093/molbev/msl171

Genchi C, Sacchi L, Bandi C, Venco L (1998) Preliminary results on the effect of tetracycline on the embryogenesis and symbiotic bacteria (*Wolbachia*) of *Dirofilaria immitis*. An update and discussion. Parassitologia 40:247–249

Gerth M, Bleidorn C (2017) Comparative genomics provides a timeframe for *Wolbachia* evolution and exposes a recent biotin synthesis operon transfer. Nat Microbiol 2:16241. https://doi.org/10.1038/nmicrobiol.2016.241

Gerth M, Rothe J, Bleidorn C (2013) Tracing horizontal *Wolbachia* movements among bees (*Anthophila*): a combined approach using multilocus sequence typing data and host phylogeny. Mol Ecol 22:6149–6162. https://doi.org/10.1111/mec.12549

Gerth M, Gansauge M, Weigert A, Bleidorn C (2014) Phylogenomic analyses uncover origin and spread of the *Wolbachia* pandemic. Nat Commun 5:5117. https://doi.org/10.1038/ncomms6117

Ghedin E, Hailemariam T, DePasse J, Zhang X, Oskov Y, Unnasch T, Lustigman S (2009) *Brugia malayi* gene expression in response to the targeting of the *Wolbachia* endosymbiont bytetracycline treatment. PLoS Negl Trop Dis 3:e525. https://doi.org/10.1371/journal.pntd.0000525

Ghelelovitch S (1952) Sur le determinisme genetique de la sterlite dans le croisement entre differences souches de *Culex autogenicus* Roubaud. C R Acad Sci 24:2386

Gilbert J, Melton L (2018) Verily project releases millions of factory-reared mosquitoes. Nat Biotechnol 36:781–782. https://doi.org/10.1038/nbt0918-781a

Giorgini M, Monti M, Caprio E, Stouthamer R, Hunter M (2009) Feminization and the collapse of haplodiploidy in an asexual parasitoid wasp harboring the bacterial symbiont *Cardinium*. Heredity 102:365–371. https://doi.org/10.1038/hdy.2008.135

Glowska E, Dragun-Damian A, Dabert M, Gerth M (2015) New *Wolbachia* supergroups detected in quill mites (*Acari*: *Syringophilidae*). Infect Genet Evol 30:140–146. https://doi.org/10.1016/j.meegid.2014.12.019

Gomes F, Hixson B, Tyner M, Ramirez J, Canepa G, Silva T, Molina-Cruz A, Keita M, Kane F, Traoré B, Sogoba N, Barillas-Mury C (2017) Effect of naturally occurring *Wolbachia* in *Anopheles gambiae s.l.* mosquitoes from Mali on *Plasmodium falciparum* malaria transmission. Proc Natl Acad Sci USA 114:12566–12571. https://doi.org/10.1073/pnas.1716181114

Grote A, Voronin D, Ding T, Twaddle A, Unnasch T, Lustigma S, Ghedin E (2017) Defining *Brugia malayi* and *Wolbachia* symbiosis by stage-specific dual RNA-seq. PLoS Negl Trop Dis 11:1–21. https://doi.org/10.1371/journal.pntd.0005357

Gualano M, Gili R, Scaioli G, Bert F, Siliquini R (2014) General population's knowledge and attitudes about antibiotics: a systematic review and meta-analysis. Pharmacoepidemiol Drug Saf 24:2–10. https://doi.org/10.1002/pds.3716

Haegeman A, Vanholme B, Jacob J, Vandekerckhove T, Claeys M, Borgonie G, Gheysen G (2009) An endosymbiotic bacterium in a plant-parasitic nematode: member of a new *Wolbachia* supergroup. Int J Parasitol 39:1045–1054. https://doi.org/10.1016/j.ijpara.2009.01.006

Halliday A, Guimaraes A, Tyrer H, Metuge H, Patrick C, Arnaud K, Kwenti T, Forsbrook G, Steven A, Cook D, Enyong P, Wanji S, Taylor M, Turner J (2014) A murine macrofilaricide pre-clinical screening model for onchocerciasis and lymphatic filariasis. Parasit Vectors 7:472. https://doi.org/10.1186/s13071-014-0472-z

Harada R, Maeda T, Nakashima A, Sadakata Y, Ando M, Yonamine K, Otsuji Y, Sato H (1970) Electronmicroscopical studies on the mechanism of oogénesis and fertilization in Dirofilaria immitis. Recent advances in researches on filariasis and schistosomiasis in Japan. University Park Press, Baltimore, MD, pp. 99–121 Record Number: 19720801359

Hartelt K, Oehme R, Oehme R, Frank H, Brockmann S, Hassle D, Kimmiga P (2004) Pathogens and symbionts in ticks: prevalence of *Anaplasma phagocytophilum* (*Ehrlichia* sp.), *Wolbachia* sp., *Rickettsia* sp., and *Babesia* sp. in Southern Germany. Int J Med Microbiol 293:86–92. https://doi.org/10.1016/S1433-1128(04)80013-5

Hedges LJ, O'Neill S, Johnson K (2008) *Wolbachia* and virus protection in insects. Science 322:702–702. https://doi.org/10.1126/science.1162418

Henson P, Mackenzie C, Spector W (1979) Inflammatory reactions in onchocerciasis: a report on current knowledge and recommendations for further study. Bull World Health Organ 57:667–682

Hertig M (1936) The *Rickettsia*, *Wolbachia pipientis* (gen. et sp.n.) and associated inclusions of the mosquito, *Culex pipiens*. Parasitology 28:453. https://doi.org/10.1017/S0031182000022666

Hertig M, Wolbach SB (1924) Studies on *Rickettsia*-like micro-organisms in insects. J Med Res 44:329

Hoerauf A (2003) Control of filarial infections: not the beginning of the end, but more research is needed. Curr Opin Infect Dis 16:403–410. https://doi.org/10.1097/01.qco.0000092811.64370.2c

Hoerauf A, Nissen-Pahle K, Schmetz C, Henkle-Dührsen K, Blaxter M, Büttner D, Gallin M, Al-Qaoud K, Lucius R, Fleischer B (1999) Tetracycline therapy targets intracellular bacteria in the filarial nematode *Litomosoides sigmodontis* and results in filarial infertility. J Clin Invest 103:11–18. https://doi.org/10.1172/JCI4768

Hoerauf A, Volkmann L, Hamelmann C, Adjei O, Auttennach I, Fleischer B, Buttner D (2000) Endosymbiotic bacteria in worms as targets for a novel chemotherapy in filariasis. Lancet 355:1242–1243. https://doi.org/10.1016/S0140-6736(00)02095-X

Hoerauf A, Volkmann L, Nissen-Paehle K, Schmetz C, Autenrieth I, Buttner D, Fleischer B (2001) Targeting of *Wolbachia* endobacteria in *Litomosoides sigmodontis*: comparison of tetracyclines with chloramphenicol, macrolides and ciprofloxacin. Tropical Med Int Health 5:275–279. https://doi.org/10.1046/j.1365-3156.2000.00544.x

Hoerauf A, Mand S, Volkmann L, Buttner M, Marfo-Debrekyei Y, Taylor M, Adjei O, Buttner DW (2003a) Doxycycline in the treatment of human onchocerciasis: kinetics of Wolbachia endobacteria reduction and of inhibition of embryogenesis in female Onchocerca worms. Microbes Infect 5:261–273. https://doi.org/10.1016/s1286-4579(03)00026-1

Hoerauf A, Mand S, Fischer K, Kruppa T, Marfo-Debrekyei Y, Debrah Y, Pfarr K, Adjei O, Büttner D (2003b) Doxycycline as a novel strategy against bancroftian filariasis—depletion of *Wolbachia* endosymbionts from *Wuchereria bancrofti* and stop of microfilaria production. Med Microbiol Immunol 192:211–216. https://doi.org/10.1007/s00430-002-0174-6

Hoerauf A, Specht S, Buttner M, Pfarr K, Mand S, Fimmers R, Marfo-Debrekyei Y, Konadu P, Debrah A, Bandi C, Brattig N, Albers A, Larbi J, Batsa L, Taylor M, Adjei O, Buttner D (2008) *Wolbachia* endobacteria depletion by doxycycline as antifilarial therapy has macrofilaricidal activity in onchocerciasis: a randomized placebo-controlled study. Med Microbiol Immunol 197:295–311. https://doi.org/10.1007/s00430-007-0062-1

Hoffmann A (1988) Partial cytoplasmic incompatibility between two Australian populations of *Drosophila melanogaster*. Entomol Exp Appl 48:61–67. https://doi.org/10.1111/j.1570-7458.1988.tb02299.x

Hoffmann A, Turelli M, Simmons G (1986) Unidirectional incompatibility between populations of *Drosophila simulans*. Evolution 40:692–701. https://doi.org/10.1111/j.1558-5646.1986.tb00531.x

Hoffmann A, Clancy D, Duncan J (1996) Naturally-occurring *Wolbachia* infection in *Drosophila simulans* that does not cause cytoplasmic incompatibility. Heredity 76:1–8. https://doi.org/10.1038/hdy.1996.1

Holman A, Davis P, Foster J, Carlow C, Kumar S (2009) Computational prediction of essential genes in an unculturable endosymbiotic bacterium, *Wolbachia* of *Brugia malayi*. BMC Microbiol 9:243. https://doi.org/10.1186/1471-2180-9-243

Hong W, Benayoud F, Nixon G, Ford L, Johnston K, Clare R, Cassidy A, Cook D, Siu A, Shiotani M, Webborn P, Kavanagh S, Aljayyoussi G, Murphy E, Steven A, Archer J, Struever D, Frohberger S, Ehrens A, Hubner M, Hoerauf A, Roberts A, Hubbard A, Tate E, Serwa R, Leung S, Qie L, Berry N, Gusovsky F, Hemingway J, Turner J, Taylor M, Ward S, O'Neill P (2019) AWZ1066S, a highly specific anti-Wolbachia drug candidate for a short-course treatment of filariasis. Proc Natl Acad Sci USA 116(4):1414–1419. https://doi.org/10.1073/pnas.1816585116

Hsiao C, Hsiao T (1985) *Rickettsia* as the cause of cytoplasmic incompatibility in the alfalfa weevil *Hypera postica*. J Invert Pathol 45:244–246. https://doi.org/10.1016/0022-2011(85)90016-3

Hurst G, Jiggins F (2000) Male-killing bacteria in insects: mechanisms, incidence, and implications. Emerg Infect Dis 6:329–336. https://doi.org/10.3201/eid0604.000402

Jacob J, Mitreva M, Vanholme B, Ghyssen G (2008) Exploring the transcriptome of the burrowing nematode *Radopholus Similis*. Mol Gen Genomics 280:1–17. https://doi.org/10.1007/s00438-008-0340-7

Jacobs R, Lunde CS, Freund YR, Hernandez V, Li X, Xia Y, Carter D, Berry P, Halladay J, Rock F, Stefanakis R, Easom E, Plattner J, Ford L, Johnston K, Cook D, Clare R, Cassidy A, Myhill L, Tyrer H, Gamble J, Guimaraes A, Steven A, Lenz F, Ehrens A, Frohberge S, Koschel M, Hoerauf A, Hübner M, McNamara C, Bakowski M, Turner J, Taylor M, Ward S (2019) Boron-Pleuromutilins as anti-*Wolbachia* agents with potential for treatment of onchocerciasis and lymphatic filariasis. J Med Chem 62:2521–2540. https://doi.org/10.1021/acs.jmedchem.8b01854

Jiggins F, Hurst G, Dolman C, Majerus M (2000) High-prevalence male-killing *Wolbachia* in the butterfly Acraea encedana. J Evol Biol 13:495–501. https://doi.org/10.1046/j.1420-9101.2000.00180

Johanowicz D, Hoy M (1995) *Wolbachia* in a predator–prey system: 16S ribosomal DNA analysis of two phytoseiids (*Acari: Phytoseiidae*) and their prey (*Acari: Tetranychidae*). Ann Entomol Soc Am 89:435–441. https://doi.org/10.1093/aesa/89.3.435

Johnston K, Ford L, Taylor M (2014a) Overcoming the challenges of drug discovery for neglected tropical diseases: the A·WOL experience. J Biomol Screen 19:335–343. https://doi.org/10.1177/1087057113511270

Johnston K, Umareddy I, Townson S, Specht S, Pfarr K, Hoerauf A, Altmeyer R, Taylor MJ (2014b) Repurposing of approved drugs from the human pharmacopoeia to target *Wolbachia* endosymbionts of onchocerciasis and lymphatic filariasis. Int J Parasitol 4:278–286. https://doi.org/10.1016/j.ijpddr.2014.09.001

Johnston K, Cook D, Berry N, Clare R, Goddard M, Ford L, Nixon G, O'Neill P, Ward S, Taylor M (2017) Identification and prioritization of novel anti-*Wolbachia* chemotypes from screening a 10,000-compound diversity library. Sci Adv 3:1551. https://doi.org/10.1126/sciadv.aao1551

Ju J, Bing X, Zhao D, Guo Y, Xi Z, Hoffmann A, Zhang K, Huang H, Gong J, Zhang X, Hong X (2020) *Wolbachia* supplement biotin and riboflavin to enhance reproduction in planthoppers. ISME J 14:676–687. https://doi.org/10.1038/s41396-019-0559-9

Kampfraath A, Klasson L, Anvar S, Vossen R, Roelofs D, Kraaijeveld K, Ellers J (2019) Genome expansion of an obligate parthenogenesis-associated *Wolbachia* poses an exception to the symbiont reduction model. BMC Genomics 20:106. https://doi.org/10.1186/s12864-019-5492-9

Kautz S, Rubin B, Moreau S (2013) Bacterial infections across the ants: frequency and prevalence of *Wolbachia*, *Spiroplasma*, and *Asaia*. Psyche 2013:1. https://doi.org/10.1155/2013/936341

Konecka E, Olszanowski O, Koczura R (2019) *Wolbachia* of phylogenetic supergroup E identified in oribatid mite *Gustavia microcephala* (Acari: Oribatida). Sci Direct 135:230–235. https://doi.org/10.1016/j.ympev.2019.03.019

Kozek WJ (1977) Transovarially-transmitted intracellular microorganisms in adult and larval stages of *Brugia malayi*. J Parasitol 63:992–1000. https://doi.org/10.2307/3279832. https://www.jstor.org/stable/3279832

Kozek W, Figueroa M (1977) Intracytoplasmic bacteria in *Onchocerca volvulus*. Am J Trop Med Hyg 26:663–678. https://doi.org/10.4269/ajtmh.1977.26.663

Kozek W, Marroquin H (1977) Intracytoplasmic bacteria in *Onchocerca volvulus*. Am J Trop Med Hyg 26(4):663–678. https://doi.org/10.4269/ajtmh.1977.26.663

Kramer L, Passeri B, Corona S, Simoncini L, Casiraghi M (2003) Immunohistochemical/immunogold detection and distribution of the endosymbiont *Wolbachia* of *Dirofilaria immitis* and *Brugia pahangi* using a polyclonal antiserum raised against WSP (*Wolbachia* surface protein). Parasitol Res 89:381–386. https://doi.org/10.1007/s00436-002-0765-6

Kyei-Poku G, Colwell D, Coghlin P, Benkel B, Floate K (2005) On the ubiquity and phylogeny of *Wolbachia* in lice. Mol Ecol 14:285–294. https://doi.org/10.1111/j.1365-294X.2004.02409.x

Landmann F (2019) The *Wolbachia* endosymbionts. Ch 10. In: Cossart P, Roy C, Sansonetti P (eds) Bacteria and intracellularity, pp 139–153. https://doi.org/10.1128/microbiolspec.BAI-0018-2019

Landmann F, Foster J, Slatko B, Sullivan W (2010) Asymmetric Wolbachia segregation during early *Brugia malayi* embryogenesis determines its distribution in adult host tissues. PLoS Negl Trop Dis 4:e758. https://doi.org/10.1371/journal.pntd.0000758

Landmann F, Voronin D, Sullivan W, Taylor M (2011) Anti-filarial activity of antibiotic therapy is due to extensive apoptosis after *Wolbachia* depletion from filarial nematodes. PLoS Pathog 7:e1002351. https://doi.org/10.1371/journal.ppat.1002351

Landmann F, Bain O, Martin C, Uni S, Taylor M, Sullivan W (2012) Both asymmetric mitotic segregation and cell-to-cell invasion are required for stable germline transmission of *Wolbachia* in filarial nematodes. Biol Open 1:536–547. https://doi.org/10.1242/bio.2012737

Landmann F, Foster J, Michalski M, Slatko B, Sullivan W (2014) Co-evolution between an endosymbiont and its nematode host: *Wolbachia* asymmetric posterior localization and AP polarity establishment. PLoS Negl Trop Dis 8(8):e3096. https://doi.org/10.1371/journal.pntd.0003096

Langworthy N, Renz A, Mackenstedt U, Henkle-Dührsen K, Bronsvoort M, Tanya V, Donnelly M, Trees A (2000) Macrofilaricidal activity of tetracycline against the filarial nematode, *Onchocerca ochengi*: elimination of *Wolbachia* preceeds worm death and suggests a dependent relationship. Proc R Soc Lond 267:1063–1069. https://doi.org/10.1098/rspb.2000.1110

Laven H (1959) Speciation in mosquitos: speciation by cytoplasmic isolation in the *Culex Pipiens*-complex. Cold Spring Harb Symp Quant Biol 224:166–173. https://doi.org/10.1101/SQB.1959.024.01.017

Lebov J, Mattick J, Libro S, Sparklin B, Chung M, Bromley R, Nadendla S, Zhao X, Ott S, Sadzewicz L, Tallon L, Michalski M, Foster J, Dunning Hotopp J, Cuomo C (2020) Complete genome sequence of *w*Bp, the endosymbiont of *Brugia pahanghi* FR3. Microbiol Res Announc 9(27):e00480-20. https://doi.org/10.1128/MRA.00480-20

Lee C (1975) *Dirofilaria immitis*: ultrastructural aspects of oocyte development and zygote formation. Exp Parasitol 37:449–468. https://doi.org/10.1016/0014-4894(75)90015-6

Lefoulon E, Gavotte L, Junker K, Barbuto M, Uni S, Landmann F, Laaksonen S, Saari S, Nikander S, de Souza LS, Casiraghi M, Bain O, Martin C (2012) A new type F *Wolbachia* from *Splendidofilariinae* (Onchocercidae) supports the recent emergence of this supergroup. Int J Parasitol 42:1025–1036. https://doi.org/10.1016/j.ijpara.2012.09.004

Lefoulon E, Bain O, Makepeace B, d'Haese C, Uni S, Martin C, Gavotte L (2016) Breakdown of coevolution between symbiotic bacteria *Wolbachia* and their filarial hosts. PeerJ 4:e1840. https://doi.org/10.7717/peerj.1840

Lefoulon E, Vaisman N, Frydman H, Sun L, Voland L, Foster J, Slatko B (2019) Large enriched fragment targeted sequencing (LEFT-SEQ) applied to capture of *Wolbachia* genomes. Sci Rep 9:5939. https://doi.org/10.1038/s41598-019-55305-5

Lefoulon E, Clark T, Borveto F, Perriat-Sanguinet M, Moulia C, Slatko B, Gavotte L (2020a) Pseudoscorpion *Wolbachia* symbionts: diversity and evidence for a new supergroup S. BMC Microbiol 20(1):188. https://doi.org/10.1186/s12866-020-01863-y

Lefoulon E, Clark T, Guerrero R, Canizales I, Cardenas-Collirgos JM, Junker K, Vallarino-Lhermitte N, Makepeace BL, Darby AC, Foster JM, Martin C, Slatko BE (2020b) Diminutive, degraded but dissimilar: *Wolbachia* genomes from filarial nematodes do not conform to a single paradigm. bioRxiv. https://doi.org/10.1101/2020.06.18.160200

Lentz C, Halls V, Hannam J, Niebel B, Strubbing U, Mayer G, Hoerauf A, Famulok M, Pfarr K (2013) A selective inhibitor of heme biosynthesis in endosymbiotic bacteria elicits antifilarial activity *in vitro*. Chem Biol 20:177–187. https://doi.org/10.1016/j.chembiol.2012.11.009

LePage D, Metcalf J, Bordenstein S, On J, Perlmutter J, Shropshire J, Layton E, Funkhouser-Jones L, Beckmann J, Bordenstein S (2017) Prophage WO genes recapitulate and enhance *Wolbachia*-induced cytoplasmic incompatibility. Nature 543:243–247. https://doi.org/10.1038/nature21391

Li Z, Carlow C (2012) Characterization of transcription factors that regulate the type IV secretion system and riboflavin biosynthesis in *Wolbachia* of *Brugia malayi*. PLoS One 7:e51597. https://doi.org/10.1371/journal.pone.0051597

Li Z, Galvin BD, Raverdy S, Carlow CK (2011a) Identification and characterization of the cofactor-independent phosphoglycerate mutases of *Dirofilaria immitis* and its *Wolbachia* endosymbiont. Vet Parasitol 176:350–356. https://doi.org/10.1016/j.vetpar.2011.01.020

Li Z, Garner A, Gloeckne R, Janda K, Carlow C (2011b) Targeting the *Wolbachia* cell division protein *FtsZ* as a new approach for antifilarial therapy. PLoS Negl Trop Dis 5:e1411. https://doi.org/10.1371/journal.pntd.0001411

Lindsey A, Bordenstein S, Newton I, Rasgon J (2016) *Wolbachia pipientis* should not be split into multiple species: a response to Ramírez-Puebla et al., Species in *Wolbachia*? Proposal for the designation of '*Candidatus* Wolbachia bourtzisii', '*Candidatus* Wolbachia onchocercicola', '*Candidatus Wolbachia blaxteri*', '*Candidatus Wolbachia brugii*', '*Candidatus Wolbachia taylori*', '*Candidatus Wolbachia collembolicola*' and '*Candidatus Wolbachia multihospitum*' for the different species within *Wolbachia* supergroups. Syst Appl Microbiol 39:220–222. https://doi.org/10.1016/j.syapm.2016.03.001

Lindsey A, Rice D, Bordenstein S, Brooks A, Bordenstein S, Newton I (2018) Evolutionary genetics of cytoplasmic incompatibility genes *cifA* and *cifB* in prophage WO of *Wolbachia*. Genome Biol Evol 10:434–451. https://doi.org/10.1093/gbe/evy012

Liosa M, Roy C, Dehio C (2009) Bacterial type IV secretion systems in human disease. Mol Microbiol 73:141–151. https://doi.org/10.1111/j.1365-2958.2009.06751.x

Lo N, Casiraghi M, Salati E, Bazzocchi C, Bandi C (2002) How many *Wolbachia* supergroups exist? Mol Biol Evol 19(3):341–346. https://doi.org/10.1093/oxfordjournals.molbev.a004087

Lo N, Paraskevopoulos C, Bourtzis K, O'Neill S, Werren J, Bordenstein S, Bandi C (2007) Taxonomic status of the intracellular bacterium Wolbachia pipientis. Int J Syst Evol Microbiol 57:654–657. https://doi.org/10.1099/ijs.0.64515-0

Ma Y, Chen W, Li Z, Zhang F, Gao Y, Luan Y (2017) Revisiting the phylogeny of *Wolbachia* in Collembola. Ecol Evol 7:2009–2017. https://doi.org/10.1002/ece3.2738

Madhav M, Parry R, Morgan J, James P, Asgari S, Drake H (2020) *Wolbachia* endosymbiont of the horn fly (*Haematobia irritans irritans*): a supergroup A strain with multiple horizontally acquired cytoplasmic incompatibility genes. Appl Environ Microbiol 86(6). https://doi.org/10.1128/AEM.02589-19

Mains J, Kelly PH, Dobson K, Petrie WD, Dobson S (2019) Localized control of *Aedes aegypti* (Diptera: Culicidae) in Miami, FL, via Inundative releases of *Wolbachia*-infected male mosquitoes. J Med Entomol 56:1296–1303. https://doi.org/10.1093/jme/tjz051

Mand S, Debrah A, Klarmann U, Batsa L, Marfo-Debrekyei Y, Kwarteng A, Specht S, Belda-Domene A, Fimmers R, Taylor M, Adjei O, Hoerauf A (2012) Doxycycline improves filarial lymphedema independent of active filarial infection: a randomized controlled trial. Clin Infect Dis 55:621–630. https://doi.org/10.1093/cid/cis486

Marcus S, Blagrove C, Camilo Arias-Goeta C, Failloux A, Sinkins S (2012) *Wolbachia* strain *w*Mel induces cytoplasmic incompatibility and blocks dengue transmission in *Aedes albopictus*. Proc Natl Acad Sci USA 109:255–260. https://doi.org/10.1073/pnas.1112021108

Martin C, Gavotte L (2010) The bacteria *Wolbachia* in filariae, a biological Russian dolls' system: new trends in antifilarial treatments. Parasite 17:79–89. https://doi.org/10.1051/parasite/2010172079

Masui S, Kuroiwa H, Sasaki T, Inui M, Kuroiwa T, Ishikawa H (2001) Bacteriophage WO and virus-like particles in *Wolbachia*, an endosymbiont of arthropods. Biochem Biophys Res Commun 283:1099–1104. https://doi.org/10.1006/bbrc.2001.4906

McGarry H, Egerton G, Taylor M (2004) Population dynamics of *Wolbachia* bacterial endosymbionts in *Brugia malayi*. Mol Biochem Parasitol 135:57–67. https://doi.org/10.1016/j.molbiopara.2004.01.006

Mclaren D, Worms M, Laurence B, Simpson M (1975) Micro-organisms in filarial larvae (Nematoda). Trans R Soc Trop Med Hyg 69:509–514. https://doi.org/10.1016/0035-9203(75)90110-8

Molyneux D, Bradley M, Hoerauf A, Kyelem D, Taylor M (2003) Mass drug treatment for lymphatic filariasis and onchocerciasis. Trends Parasitol 19:516–522. https://doi.org/10.1016/j.pt.2003.09.004

Moreira L, Iturbe-Ormaetxe I, Jeffery J, Lu G, Pyke A, Hedges L, Rocha B, Hall-Mendelin S, Day A, Riegler M (2009) A *Wolbachia* symbiont in *Aedes aegypti* limits infection with dengue, chikungunya, and Plasmodium. Cell 139(7):1268–1278

Newton I, Rice D, Margolin W (2020) The Jekyll and Hyde symbiont: could *Wolbachia* be a nutritional mutualist? J Bacteriol 202:e00589-19. https://doi.org/10.1128/JB.00589-19

Newton I, Slatko B (2019) Symbiosis comes of age at the 10th biennial meeting of *Wolbachia* researchers. Appl Environ Microbiol 85:1–9. https://doi.org/10.1128/AEM.03071-18

Nikoh N, Hosokawa T, Moriyama M, Oshima K, Hattori M, Fukatsu T (2014) Evolutionary origin of insect-*Wolbachia* nutritional mutualism. Proc Natl Acad Sci USA 111(28):10257–10262. https://doi.org/10.1073/pnas.1409284111

Noda H (1984) Cytoplasmic incompatibility in allopatric field populations of the small brown planthopper, *Laodelphax striatellus*, in Japan. Entomol Exp Appl 35:263–267. https://doi.org/10.1111/j.1570-7458.1984.tb03391

Noda H (1987) Further studies of cytoplasmic incompatibility in local populations of *Laodelphax striatellus* in Japan: Homoptera: Delphacidae. Appl Entomol Zool 22:443–448. https://doi.org/10.1303/aez.22.443

Noda H, Munderloh U, Kurtti T (1997) Endosymbionts of ticks and their relationship to *Wolbachia* spp. and tick-borne pathogens of humans and anim. Appl Environ Microbiol 63:3926–3932

O'Connor L, Plichart C, Sang A, Brelsfoard C, Bossin H, Dobson S (2012) Open release of male mosquitoes infected with a wolbachia biopesticide: field performance and infection containment. PLoS Negl Trop Dis 6(11):e1797. https://doi.org/10.1371/journal.pntd.0001797

O'Neill S (1989) Cytoplasmic symbionts in *Tribolium confusum*. J Invert Pathol 53:132–134. https://doi.org/10.1016/0022-2011(89)90087-6

O'Neill S, Giordano R, Colbert A, Karr T, Robertson H (1992) 16S rRNA phylogenetic analysis of the bacterial endosymbionts associated with cytoplasmic incompatibility in insects. Proc Natl Acad Sci USA 89:2699–2702. https://doi.org/10.1073/pnas.89.7.2699

Panaram K, Marshall J (2007) F supergroup *Wolbachia* in bush crickets: what do patterns of sequence variation reveal about this supergroup and horizontal transfer between nematodes and arthropods? Genetica 130:53–60. https://doi.org/10.1007/s10709-006-0020-7

Pechère J, Hughes D, Kardas P, Cornaglia G (2007) Non-compliance with antibiotic therapy for acute community infections: a global survey. Int J Antimicrob Agents 29:245–253. https://doi.org/10.1016/j.ijantimicag.2006.09.026

Perlman S, Hunter M, Zchori-Fein E (2006) The emerging diversity of *Rickettsia*. Proc R Soc B 273:2097–2106. https://doi.org/10.1098/rspb.2006.3541

Perlmutter J, Bordenstein S, Unckless R, LePage D, Metcalf J, Hill T, Martinez J, Jiggins F, Bordenstein S (2019) The phage gene *wmk* is a candidate for male killing by a bacterial endosymbiont. PLoS Pathog 15:e1007936. https://doi.org/10.1371/journal.ppat.1007936

Perrot-Minnot M-J, Guo L, Werren J (1996) Single and double infections with *Wolbachia* in the parasitic Wasp *Nasonia vitripennis* effects on compatibility. Genetics 143:961–972

Ramirez-Puebla S, Servin-Garciduenas L, Ormeno-Orrillo E, Vera-Ponce de León A, Rossenblueth R, Delaye L, Martinez J, Martinez-Romero E (2015) Species in *Wolbachia*? Proposal for the designation of *'Candidatus Wolbachia bourtzisii', 'Candidatus Wolbachia onchocercicola', 'Candidatus Wolbachia blaxteri'*, '*Candidatus Wolbachia brugii*', '*Candidatus Wolbachia taylori*', '*Candidatus Wolbachia collembolicola*' and '*Candidatus Wolbachia multihospitum*' for the different species within *Wolbachia* supergroups. Syst Appl Microbiol 38:390–399. https://doi.org/10.1016/j.syapm.2015.05.005

Ramírez-Puebla S, Ormeño-Orrillo E, Vera-Ponce de León A, Lozano L, Sanchez-Flores A, Rosenblueth M, Martínez-Romero E (2016) Genomes of Candidatus *Wolbachia bourtzisii w*DacA and *Candidatus Wolbachia pipientis w*DacB from the cochineal insect *Dactylopius coccus* (Hemiptera: Dactylopiidae). G3 10:3343–3349. https://doi.org/10.1534/g3.116.031237

Raverdy S, Zhang Y, Foster J, Carlow C (2007) Molecular and biochemical characterization of nematode cofactor independent phosphoglycerate mutases. Mol Biochem Parasitol 156:210. https://doi.org/10.1016/j.molbiopara.2007.08.002

Raverdy S, Foster J, Roopenian E, Carlow CK (2008) The *Wolbachia* endosymbiont of *Brugia malayi* has an active pyruvate phosphate dikinase. Mol Biochem Parasitol 160:163–166. https://doi.org/10.1016/j.molbiopara.2008.04.014

Reeves D, Price S, Ramalho M, Moreau C (2020) The diversity and distribution of *Wolbachia*, *Rhizobiales*, and *Ophiocordyceps* within the widespread neotropical turtle ant, *Cephalotes atratus* (Hymenoptera: Formicidae). Neotrop Entomol 49(1):52–60. https://doi.org/10.1007/s13744-019-00735-z

Richardson P, Holmes W, Saul G (1987) The effect of tetracycline on nonreciprocal cross incompatibility in *Mormoniella [= Nasonia] vitripennis*. J Invertebr Pathol 50:176–183. https://doi.org/10.1016/0022-2011(87)90080-2

Rokas A, Atkinson R, Nieves-Aldrey J, West S, Stone G (2002) The incidence and diversity of *Wolbachia* in gallwasps (*Hymenoptera*; *Cynipidae*) on oak. Mol Ecol 11:1815–1829. https://doi.org/10.1046/j.1365-294X.2002.01556.x

Ros V, Fleming V, Feil E, Breeuwer J (2009) How diverse is the genus *Wolbachia*? Multiple-gene sequencing reveals a putatively new *Wolbachia* supergroup recovered from spider mites (*Acari: Tetranychidae*). Appl Environ Microbiol 75:1036–1043. https://doi.org/10.1128/AEM.01109-08

Rowley S, Raven R, McGraw E (2004) *Wolbachia pipientis* in Australian spiders. Curr Microbiol 49:208–214. https://doi.org/10.1007/s00284-004-4346-z

Russell J, Goldman-Huertas B, Moreau C, Baldo L, Stahlhut J, Werren J, Pierce N (2009) Specialization and geographic isolation among *Wolbachia* symbionts from ants and Lycaenid butterflies. Evolution 63:624–640. https://doi.org/10.1111/j.1558-5646.2008.00579.x

Ryan P, Turley A, Wilson G, Hurst T, Retzki K, Brown-Kenyon J, Hodgson L, Kenny N, Cook H, Montgomery B, Paton C, Ritchie S, Hoffmann A, Jewell N, Tanamas S, Anders K, Simmons C, O'Neill S (2019) Establishment of *w*Mel *Wolbachia* in *Aedes aegypti* mosquitoes and reduction of local dengue transmission in Cairns and surrounding locations in northern Queensland, Australia. Gates Open Res 3:1547. https://doi.org/10.12688/gatesopenres.13061.1

Saha S, Hunter W, Reee J, Morgan J, Marutani-Hert M, Huang H, Lindeberg M (2012) Survey of endosymbionts in the *Diaphorina citri* metagenome and assembly of a *Wolbachia w*Di draft genome. PLoS One 7:e50067. https://doi.org/10.1371/journal.pone.0050067

Sakamoto J, Rasgon J (2006a) Geographic distribution of *Wolbachia* infections in *Cimex lectularius* (Heterotera: Cimicidae). J Med Entomol 43:696–700. https://doi.org/10.1093/jmedent/43.4.696

Sakamoto J, Rasgon J (2006b) Endosymbiotic bacteria of bed bugs: evolution, ecology and genetics. Am Entomol 52:119–122. https://doi.org/10.1093/ae/52.2.119

Salunkhe R, Dhotre D, Salunke B, Patil V, Mahale V, Andrew R, Patole M, Narkhede K, Shouche Y (2015) Distribution and molecular characterization of *Wolbachia* endosymbionts in odonata (insecta) from Central India by multigene approach. Curr Sci 108:971–978. https://www.jstor.org/stable/24216527?seq=1

Saul G (1961) An analysis of non-reciprocal cross incompatibility in *Mormoniella vitripennis* (Walker). Z Vererbungsl 92:28–33. https://doi.org/10.1007/BF018564097

Serbus L, Sullivan W (2007) A cellular basis for *Wolbachia* recruitment to the host germline. PLoS Pathog 3:e190. https://doi.org/10.1371/journal.ppat.0030190

Serbus L, Casper-Lindley C, Landmann F, Sullivan W (2008) The genetics and cell biology of *Wolbachia*-host interactions. Annu Rev Genet 42:683–707. https://doi.org/10.1146/annurev.genet.41.110306.130354

Serbus L, Landmann F, Bray W, White P, Ruybal J, Lokey R, Debec A, Sullivan W (2012) A cell-based screen reveals that the albendazole metabolite, albendazole sulfone, targets Wolbachia. PLoS Pathog 8:e1002922

Sharma R, Aljayyoussi G, Tyrer H, Gamble J, Hayward L, Guimaraes A, Davies J, Waterhouse D, Cook D, Myhill L, Clare R, Cassidy A, Steven A, Johnston K, Ford L, Turner J, Ward S, Taylor M (2018) Corrigendum: minocycline as a re-purposed anti-*Wolbachia* macrofilaricide: superiority compared with doxycycline regimens in a murine infection model of human lymphatic filariasis. Sci Rep 8:46934. https://doi.org/10.1038/srep46934

Shrivastava N, Nag J, Pandey J, Tripathi R, Shah P, Siddiqi M, Misra-Bhattacharya S (2015) Homology modeling of NAD+-dependent DNA ligase of the *Wolbachia* endosymbiont of *Brugia malayi* and its drug target potential using dispiro-cycloalkanones. Animicrob Agents Chemother 59:3736–3717. https://doi.org/10.1128/AAC.03449-14

Siddiqui S, Raja I (2015) Molecular detection of endosymbiont bacteria *Wolbachia* in bedbug species *Cimex lectularius* from Vidarbha region of India. Int J Life Sci 3:200–204

Sironi M, Bandi C, Sacchi L, Sacco B, Damiani G, Genchi C (1995) Molecular evidence for a close relative of the arthropod endosymbiont *Wolbachia* in a filarial worm. Mol Biochem Parasitol 74:223–227. https://doi.org/10.1016/0166-6851(95)02494-8

Slatko B, Taylor M, Foster J (2010) The *Wolbachia* endosymbiont as an anti-filarial nematode target. Symbiosis 51:55–65. https://doi.org/10.1007/s13199-010-0067-1

Slatko B, Luck A, Dobson S, Foster J (2014) *Wolbachia* endosymbionts and human disease control. Mol Biochem Parasitol 195:88–95. https://doi.org/10.1016/j.molbiopara.2014.07.004

Specht S, Mand S, Marfo-Debrekyei Y, Debrah A, Konadu P, Adjei O, Buttner D, Hoerauf A (2008) Efficacy of 2- and 4-week rifampicin treatment on the Wolbachia of Onchocerca volvulus. Parasitol Res 103:1303–1309. https://doi.org/10.1007/s00436-008-1133-y

Supali T, Djuardi Y, Pfarr KM, Wibowo H, Taylor MJ, Hoerauf A, Houwing-Duistermaat JJ, Yazdanbakhsh M, Sartono E (2008) Doxycycline treatment of Brugia malayi-infected persons reduces microfilaremia and adverse reactions after diethylcarbamazine and albendazole treatment. Clin Infect Dis 46(9):1385–1393. https://doi.org/10.1086/586753

Tamarozzi F, Halliday A, Gentil K, Hoerauf A, Pearlman E, Taylor MJ (2011) Onchocerciasis: the role of *Wolbachia* bacterial endosymbionts in parasite biology, disease pathogenesis, and treatment. Clin Microbiol Rev 24(3):459–468. https://doi.org/10.1128/CMR.00057-10

Tamarozzi F, Tendongfor N, Enyong P, Esum M, Faragher B, Wanji S, Taylor M (2012) Long term impact of large scale community-directed delivery of doxycycline for the treatment of onchocerciasis. Parasit Vectors 5:53. https://doi.org/10.1186/1756-3305-5-53

Tanaka K, Furukawa S, Nikoh N, Sasaki T, Fukatsu T (2009) Complete WO phage sequences reveal their dynamic evolutionary trajectories and putative functional elements required for integration into the *Wolbachia* genome. Appl Environ Microbiol 75:5676–5686. https://doi.org/10.1128/AEM.01172-09

Taylor M, Bandi C, Hoerauf A (2005a) *Wolbachia* bacterial endosymbionts of filarial nematodes. Adv Parasitol 60:245–284. https://doi.org/10.1016/s0065-308x(05)60004-

Taylor M, Makunde W, McGarry H, Turner J, Mand S, Hoerauf A (2005b) Macrofilaricidal activity after doxycycline treatment of *Wuchereria bancrofti*: a double-blind, randomised placebo-controlled trial. Lancet 365:2116–2121. https://doi.org/10.1016/S0140-6736(05)66591-9

Taylor M, Hoerauf A, Townson S, Slatko B, Ward S (2014) Anti-*Wolbachia* drug discovery and development: safe macrofilaricides for onchocerciasis and lymphatic filariasis. Parasitology 141 (1):119–127. https://doi.org/10.1017/S0031182013001108

Taylor M, Bordenstein S, Slatko B (2018) *Wolbachia*: a sex selector, a viral protector and a target to treat filarial nematodes. Microbiology 164:1345–1347. https://doi.org/10.1099/mic.0.000724

Taylor M, von Geldern T, Ford L, Hubner M, Marsh K, Johnston K, Sjoberg H, Specht S, Pionnier N, Tyrer H, Clare R, Cook D, Murphy E, Steven A, Archer J, Bloemker D, Lenz F, Koschel M, Ehrens A, Metuge HM, Chunda V, Ndongmo Chounna PW, Njouendou AJ, Fombad FF, Carr R, Morton HE, Aljayyoussi G, Hoerauf A, Wanji S, Kempf DJ, Turner JD, Ward SA (2019) Preclinical development of an oral anti-*Wolbachia* macrolide drug for the treatment of lymphatic filariasis and onchocerciasis. Sci Transl Med 11:eaau2086. https://doi.org/10.1126/scitranslmed.aau2086

Telschow A, Flor M, Kobayashi Y, Hammerstein P, Werren J (2007) *Wolbachia*-induced unidirectional cytoplasmic incompatibility and speciation: mainland-island model. PLoS One 2:e701. https://doi.org/10.1371/journal.pone.0000701

Townson S, Hutton D, Siemienska J, Hollick L, Scanlon T, Tagboto S, Taylor M (2000) Antibiotics and *Wolbachia* in filarial nematodes: antifilarial activity of rifampicin, oxytetracycline and chloramphenicol against *Onchocerca gutturosa*, *Onchocerca lienalis* and *Brugia pahangi*. Ann Trop Med Parasitol 94:801–816. https://doi.org/10.1080/00034983.2000.11813605

Trpis M, Perrone J, Reissig M, Parker K (1981) Control of cytoplasmic incompatibility in the *Aedes scutellaris* complex: incompatible crosses become compatible by treatment of larvae with heat or antibiotics. J Hered 72:313–317. https://doi.org/10.1093/oxfordjournals.jhered.a109513

Turner J, Mand S, Debrah A, Muehlfeld J, Pfarr K, McGarry H, Adjei O, Taylor M, Hoerauf A (2006) A randomized, double-blind clinical trial of a 3-week course of doxycycline plus albendazole and ivermectin for the treatment of *Wuchereria bancrofti* infection. Clin Infect Dis 42:1081–1089. https://doi.org/10.1086/501351

Turner J, Tendongfor N, Esum M, Johnston K, Langley R, Ford L, Faragher B, Specht S, Mand S, Hoerauf A, Enyong P, Wanji S, Taylor M (2010) Macrofilaricidal activity after doxycycline only treatment of Onchocerca volvulus in an area of Loa loa co-endemicity: a randomized controlled trial. PLoS Negl Trop Dis 4:e660. https://doi.org/10.1371/journal.pntd.0000660

Turner J, Sharma R, Al Jayoussi G, Tyrer H, Gamble J, Hayward L, Priestley R, Murphy E, Davies J, Waterhouse D, Cook D, Clare R, Cassidy A, Steven A, Johnston K, McCall J, Ford L, Hemingway J, Ward S, Taylor M (2017) Albendazole and antibiotics synergize to deliver short-course anti-Wolbachia curative treatments in preclinical models of filariasis. Proc Natl Acad Sci USA. https://doi.org/10.1073/pnas.1710845114

Vandekerckhove T, Watteyne S, Willems A, Swings J, Mertens J, Gillis M (1999) Phylogenetic analysis of the 16S rDNA of the cytoplasmic bacterium *Wolbachia* from the novel host *Folsomia candida* (Hexapoda, Collembola) and its implications for *wolbachial* taxonomy. FEMS Microbiol Lett 180:279–286. https://doi.org/10.1111/j.1574-6968.1999.tb08807.x

von Geldern T, Morton H, Clark R, Brown B, Johnston K, Ford L, Specht S, Carr R, Stolarik D, Ma J, Rieser M, Struever D, Frohberger S, Koschel M, Ehrens A, Turner J, Hubner M, Hoerauf A, Taylor M, Ward S, Marsh K, Kempf D (2019) Discovery of ABBV-4083, a novel analog of Tylosin A that has potent anti-*Wolbachia* and anti-filarial activity. PLoS Negl Trop Dis 13:e0007159. https://doi.org/10.1371/journal.pntd.0007159

Voronin D, Cook D, Steven A, Taylor M (2012) Autophagy regulates *Wolbachia* populations across diverse symbiotic associations. Proc Natl Acad Sci USA 109:E1638–E1646. https://doi.org/10.1073/pnas.1203519109

Voronin D, Bachu S, Shlossman M, Unnasch T, Ghedin E, Lustigman S (2016) Glucose and glycogen metabolism in *Brugia malayi* is associated with *Wolbachia* symbiont fitness. PLoS One 11:e0153812. https://doi.org/10.1371/journal.pone.0153812

Voronin D, Schnall E, Grote A, Jawahar S, Ali W, Unnasch T, Ghedin E, Lustigman S (2019) Pyruvate produced by *Brugia* spp. via glycolysis is essential for maintaining the mutualistic association between the parasite and its endosymbiont, *Wolbachia*. PLoS Pathog 15(9): e1008085. https://doi.org/10.1371/journal.ppat.1008085

Wade M, Stevens L (1985) Microorganism mediated reproductive isolation in flour beetles (genus *Tribolium*). Science 227:527–528. https://doi.org/10.1126/science.3966160

Walker M, Specht S, Churcher T, Hoerauf A, Taylor M, Basanez M (2015) Therapeutic efficacy and macrofilaricidal activity of doxycycline for the treatment of river blindness. Clin Infect Dis 60:1199–1207. https://doi.org/10.1093/cid/ciu1152

Wang Z, Su X, Wen J, Jiang L, Qiao G (2014) Widespread infection and diverse infection patterns of *Wolbachia* in Chinese aphids. Insect Sci 21:313–325. https://doi.org/10.1111/1744-7917.12102

Wanji S, Tendongfor N, Nji T, Esum M, Che J, Nkwescheu A, Alassa F, Kamnan G, Enyong P, Taylor M, Hoerauf A, Taylor D (2009) Community-directed delivery of doxycycline for the treatment of onchocerciasis in areas of co-endemicity with loasis in Cameroon. Parasit Vectors 2:39. https://doi.org/10.1186/1756-3305-2-39

Ward C, Baxter S (2017) Draft genome assembly of a *Wolbachia* endosymbiont of *Plutella australiana*. Genome Announc 5:e01134–e01117. https://doi.org/10.1128/genomeA.01134-17

Wenseleers T, Ito F, Van Borm S, Huybrechts R, Volckaert F, Billen J (1998) Widespread occurrence of the micro-organism *Wolbachia* in ants. Proc Biol Sci 265:1447–1452. https://doi.org/10.1098/rspb.1998.0456

Werren J, Windsor D (2000) *Wolbachia* infection frequencies in insects: evidence of a global equilibrium? Proc Biol Sci 267:1277–1285. https://doi.org/10.1098/rspb.2000.1139

Werren J, Zhang W, Guo L (1995) Evolution and phylogeny of *Wolbachia*: reproductive parasites of arthropods. Proc Biol Sci 261:55–63. https://doi.org/10.1098/rspb.1995.0117

Werren J, Baldo L, Clark M (2008) *Wolbachia*: master manipulators of invertebrate biology. Nat Rev Microbiol 6:741–751. https://doi.org/10.1038/nrmicro1969

Wright J, Sjostrand F, Portaro J, Barr A (1978) The ultrastructure of the *Rickettsia*-like microorganism *Wolbachia pipientis* and associated virus-like bodies in the mosquito *Culex pipiens*. J Ultrastruct Res 63:79–85. https://doi.org/10.1016/s0022-5320(78)80046-x

Wu M, Sun L, Vamathevan J, Riegler M, Deboy R, Brownlie J, McGraw E, Martin W, Esser C, Ahmadinejad N, Wiegand C, Madupu R, Beanan M, Brinkac L, Daugherty S, Durkin A, Kolonay J, Nelson W, Mohamoud Y, Lee P, Berry K, Brook M, Young M, Utterback T, Weidman J, Nierman W, Paulsen I, Nelson K, Tettelin H, O'Neill S, Eisen J (2004) Phylogenomics of the reproductive parasite *Wolbachia pipientis w*Mel: a streamlined genome overrun by Mobile genetic elements. PLoS Biol 2:327–341

Wu B, Novelli J, Foster J, Vaisvila R, Conway L, Ingram J, Ganatra M, Rao A, Hamza I, Slatko B (2009) The heme biosynthetic pathway of the obligate *Wolbachia* endosymbiont of *Brugia malayi* as a potential anti-filarial drug target. PLoS Negl Trop Dis 3:1–10. https://doi.org/10.1371/journal.pntd.0000475

Xu Z, Fang S, Bakowski M, Rateb M, Yang D, Zhu X, Huang Y, Zhao L, Jiang Y, Duan Y, Hull M, McNamara C, Shen B (2019) Discovery of kirromycins with anti-*Wolbachia* activity from *Streptomyces* sp. CB00686. ACS Chem Biol 14(6):1174–1182. https://doi.org/10.1021/acschembio.9b00086

Ye Y, Woolfit M, Rances E, O'Neill S, McGraw EA (2013) *Wolbachia*-associated bacterial protection in the mosquito *Aedes aegypti*. PLoS Negl Trop Dis 7(8):e2362. https://doi.org/10.1371/journal.pntd.0002362

Yen J, Barr A (1973) The etiological agent of cytoplasmic incompatibility in *Culex pipiens*. The etiological agent of cytoplasmic incompatibility in *Culex pipiens*. J Invertebr Pathol 22:242–250. https://doi.org/10.1016/0022-2011(73)90141-9

Yu H, Dranchak P, Li Z, MacArthur R, Munson M, Mehzabeen N, Baird N, Battalie K, Ross D, Lovell S, Carlow C, Suga H, Inglese J (2017) Macrocycle peptides delineate locked-open inhibition mechanism for microorganism phosphoglycerate mutases. Nat Commun 8. https://doi.org/10.1038/ncomms14932

Zechner E, Lang S, Schildbach J (2012) Assembly and mechanisms of bacterial type IV secretion machines. Philos Trans R Soc Lond Ser B Biol Sci 367:1073–1097. https://doi.org/10.1098/rstb.2011.0207

Zeh D, Zeh J, Bonilla M (2005) *Wolbachia*, sex ratio bias and apparent male killing in the harlequin beetle riding pseudoscorpion. Heredity 95:41–49. https://doi.org/10.1038/sj.hdy.6800666

Zhang X, Norris D, Rasgon J (2011) Distribution and molecular characterization of *Wolbachia* endosymbionts and filarial nematodes in Maryland populations of the lone star tick (*Amblyomma americanum*). FEMS Microb Ecol 77:50–56. https://doi.org/10.1111/j.1574-6941.2011.01089.x

Zhou W, Rousset W, O'Neill S (1998) Phylogeny and PCR-based classification of *Wolbachia* strains using wsp gene sequences. Proc R Soc Lond B 265:509–515. https://doi.org/10.1098/rspb.1998.0324

Zimmer C (2001) *Wolbachia*: a tale of sex and survival. Science 292:1093–1095. https://doi.org/10.1126/science.292.5519.1093

Zug R, Hammerstein P (2015) Bad guys turned nice? A critical assessment of *Wolbachia* mutualisms in arthropod hosts. Biol Rev Camb Philos Soc 90(1):89–111. https://doi.org/10.1111/brv.12098

Chapter 17
Molecular Regulators of Entomopathogenic Nematode–Bacterial Symbiosis

Ioannis Eleftherianos and Christa Heryanto

Abstract Entomopathogenic nematodes are parasitic organisms with an exceptional capacity to infect rapidly and efficiently a wide range of insect species. Their distinct pathogenic properties have established entomopathogenic nematodes as supreme biocontrol agents of insects as well as excellent models to simulate and dissect the molecular and physiological bases of conserved strategies employed by parasitic nematodes that cause infectious diseases in humans. The extreme infectivity of entomopathogenic nematodes is due in part to the presence of certain species of Gram-negative bacteria that live in mutualistic symbiosis during the infective juvenile stage, which forms the central part of the nematode life cycle. Both nematodes and their mutualistic bacteria are capable of interfering and undermining several aspects of the insect host innate immune system during the infection process. The mutualistic bacteria are also able to modulate other biological functions in their nematode host including growth, development, and reproduction. In this review, we will focus our attention on the mutualistic relationship between entomopathogenic nematodes and their associated bacteria to discuss the nature and distinct characteristics of the regulatory mechanisms, and their molecular as well as physiological components that control this specific biological partnership.

Keywords Entomopathogenic nematodes · Bacteria · Symbiosis · Parasitism · Pathogenicity · Gene expression

I. Eleftherianos (✉) · C. Heryanto
Infection and Innate Immunity Lab, Department of Biological Sciences, Institute for Biomedical Sciences, The George Washington University, Science and Engineering Hall, Washington, DC, USA
e-mail: ioannise@gwu.edu

M. Kloc (ed.), *Symbiosis: Cellular, Molecular, Medical and Evolutionary Aspects*, Results and Problems in Cell Differentiation 69,
https://doi.org/10.1007/978-3-030-51849-3_17

17.1 Introduction

Bacteria–host interactions are ubiquitous in nature (Ruby 2008). They form complex relationships that influence critical biological processes such as the nutrition, development, and immunity of plants and animals (Hentschel et al. 2000; Ochman and Moran 2001). Relationships range from being ancient, stable, and beneficial mutualisms, as exemplified by the origin of the mitochondria and chloroplasts from the endosymbiosis, to those which are more recent, dynamic, and highly pathogenic, such as the evolution of *Yersinia pestis*, the causative agent of plague, from a relatively benign ancestor (Sagan 1967; Parkhill et al. 2001). Although the outcomes of these interactions have very different consequences for their hosts, there is increasing evidence that common mechanisms regulate the ability of bacteria to act as either mutualists or pathogens. Both lifestyles are thought to have evolved from living in close proximity to their hosts and both require the ability to circumvent host immunity and modulate the host environment (Dale and Moran 2006). The diversity of these associations and their importance to medicine and agriculture define them as a key area for research (Ochman and Moran 2001; Maurelli 2007). Recent works have adopted model organisms to determine the overlap between the nature of molecular signaling pathways and their specific genes that are necessary for regulating mutualism and pathogenicity lifestyles in bacteria and their invertebrate hosts. Understanding the genetic mechanisms and dictating the outcomes of bacteria–host interactions will ultimately allow us to determine how microbes switch from one lifestyle to another, thus shedding light on the evolution of complex multi-organism relationships.

Insight into the delicate balance between mutualism and pathogenicity requires a system that allows for the direct study of both interactions (Chaston and Goodrich-Blair 2010). Entomopathogenic nematodes are microscopic worms that target and naturally infect a diverse range of insect hosts, and therefore, they have been implemented in modern agricultural practices as promising biological control agents and alternatives to chemical insecticides for managing destructive insect pests of plants and deleterious vectors of infectious diseases. Due to their remarkable pathogenic properties toward various insect stages, and their unique life cycle that involves mutualistic cooperation with specific bacterial species, entomopathogenic nematodes have been employed in recent years in biomedical research as outstanding and simultaneously environmentally safe tools for unraveling the molecular and physiological basis of mutualistic relationships in animals, and resolving pathogenicity mechanisms in nematode–bacterial complexes, in relation to the host innate immune function. The mutualistic bacteria of entomopathogenic nematodes perform critical biological tasks that are not restricted only to promoting pathogenicity and compromising the insect immune system during infection, but they also protect their nematode host by producing antimicrobial molecules to support the growth of other competitive bacteria. In addition, they promote nematode dispersal and development, growth, and reproduction by supplying nutrients from the bioconverted insect tissues and organs, as well as by acting themselves as a rich food source (Herbert and Goodrich-Blair 2007a).

Application of entomopathogenic nematodes in the field requires careful analysis of various traits of parasites and their associated bacteria acting together as a complex or as a separate one, during the distinct phases of their life cycle. Due to the lack of understanding of the molecular and physiological determinants that control the harmonious coordination between the two mutualistic players, it is considered imperative to invest future efforts and resources on deconstructing the life cycle of different entomopathogenic nematode species to expose the exact elements that enhance or diminish the interaction with the insect host. This approach would, in turn, provide us with the necessary knowledge to validate the attributes that increase the performance of entomopathogenic nematodes, and improve their stability and efficiency in the field. Such information would ultimately enable us to make convenient interventions to refine biocontrol programs that would reduce pesticide use and improve food safety and production.

17.2 Entomopathogenic Nematode–Bacterial Complexes

Nematode–bacterial complexes with insect pathogenic properties are formed specifically in the soil nematodes of the genera *Heterorhabditis* and *Steinernema*, which develop mutualistic relationship with the proteobacteria *Photorhabdus* spp. and *Xenorhabdus* spp., respectively. The nematodes together with their associated bacteria undergo a complex life cycle that comprises two stages; a mutualistic stage that takes place in the nematode gut, during which the bacteria are vectored by their cognate nematode and a pathogenic stage that occurs in the insect host during infection, and involves the manipulation of humoral and cellular innate immune defenses by both partners that lead to the accelerated insect death. Although entomopathogenic nematode life cycles exhibit similar characteristics, variation especially in certain features of nematode reproduction and population growth rate, as well as in host range and phase variants of mutualistic bacteria, can be observed among different genera and species (Forst et al. 1997).

Heterorhabditis nematodes from the Heterorhabditidae family act as 'cruiser' parasites, a behavior that involves active seeking out of suitable insect hosts by burrowing into the soil. *Heterorhabditis* parasitic nematodes form a mutually beneficial symbiotic relationship with the entomopathogenic Gram-negative bacteria from the genus *Photorhabdus*, which belong to the Enterobacteriaceae family (Waterfield et al. 2009). The bacteria are found in the gut of the infective juvenile stage (Ciche et al. 2006; Ciche 2007). The infective juvenile is an obligate stage in the nematode life cycle and is required for the infection of larval stages of mainly lepidopteran insects (Ciche 2007; Kaya and Gaugler 1993) (Fig. 17.1). This stage is analogous to the *C. elegans* dauer stage and the developmentally arrested infective third-stage larva (L3) of many important parasitic nematodes. Infective juveniles gain entry to the insect through natural openings (anus, spiracles, and mouth) or by

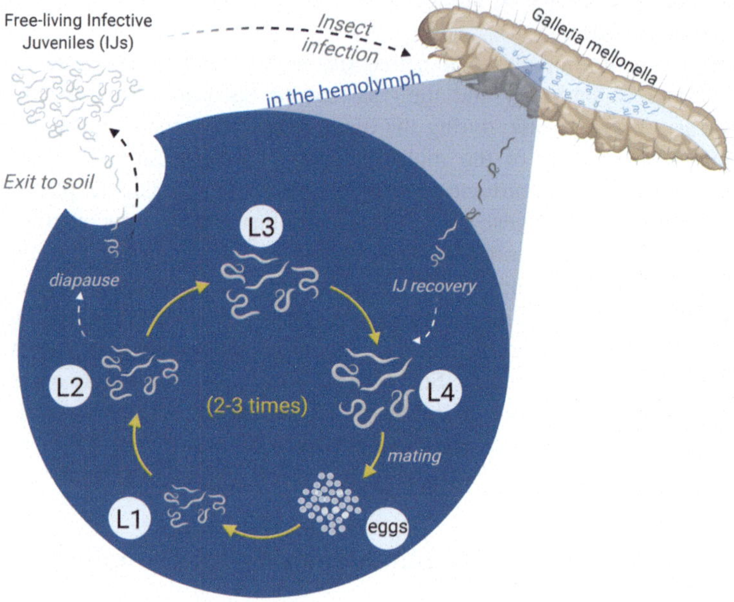

Fig. 17.1 Life cycle of the entomopathogenic nematode *Heterorhabditis bacteriophora*. The infective juveniles (IJs) form the only free-living stage of this parasite. Nematode mating, reproduction, and development occur within the hemolymph of the infected insects (e.g. larvae of the greater wax moth, *Galleria mellonella*) in the presence of high titers of their mutualistic bacteria *Photorhabdus luminescens*. Depending on the amount of resources in the dead insect, two or three generations may take place within the insect cadavers. L1, L2, L3, and L4: Larval molts. Images are made using Biorender graphic software (https://biorender.com)

abrading the insect cuticle using a dorsal tooth (Ciche 2007). Once inside the insect, the infective juveniles expel a small number of *Photorhabdus* cells into the hemolymph where the bacteria begin to divide exponentially. After two to three days of bacterial growth, the insect succumbs to the infection due to septicemia with the concomitant conversion of the internal organs and tissues into bacterial biomass. This bioconversion is facilitated by the production of a wide range of toxins and hydrolytic enzymes by the bacteria (Ffrench-Constant et al. 2007; Eleftherianos 2009; Bode 2009). The worms feed on the bacterial biomass, and subsequent nematodes' growth and development require the presence of high-density *Photorhabdus* bacteria (Ciche and Ensign 2003). The infective juvenile nematodes mature to first-generation hermaphrodite females, which give rise to the second generation of amphimictic males and females (cross-fertilization) and to the self-fertile hermaphrodite females and infective juveniles. Nematodes reproduce and the progeny develops through four juvenile stages (L1, L2, L3, and L4) to adults. Nematode reproduction continues over two to three generations until the nutrient status of the cadaver deteriorates, whereupon adult development is suppressed, and the infective juvenile stage accumulates. These non-feeding infective juveniles enter the soil where they may survive for several months in the absence of a suitable host. The transmission of mutualistic bacteria by infective juveniles is essential for the nematodes to reproduce (Goodrich-Blair and Clarke 2007).

Similar to *Heterorhabditis*, *Steinernema* nematodes are found free in the soil where they gain access to insect larvae through natural body openings but they lack a dorsal tooth that facilitates penetration through the cuticle. However, in contrast to *Heterorhabditis*, the *Steinernema* nematodes exhibit 'ambushing' behavior, which involves waiting and attacking sensitive insect hosts in their vicinity. *Steinernema* nematodes follow a similar life cycle to *Heterorhabditis* that mainly differs in the initial stage of recovery during which amphimictic reproduction occurs. This means that *Steinernema* infective juveniles develop into reproductive males and females. Interestingly, this infective juvenile behavior also takes place during the first- and second-generation offspring. In addition, third-generation females produce eggs all of which develop through the 'endotokia matricida' process that occurs due to the cessation of egg laying and involves intra-uterine birth causing maternal death, a relatively common phenomenon in entomopathogenic nematodes induced in response to the low food supply. As opposed to *Heterorhabditis*, the resulting juvenile stages of *Steinernema* develop into infective juveniles after they exit the mother nematode (Kooliyottil et al. 2013).

17.3 Mutualism Regulators in *Photorhabdus* Bacteria

Recent progress in quantitative proteomic techniques has been started to contribute to the identification and preliminary examination of the factors that control symbiotic processes between animal hosts and microbes, including entomopathogenic nematodes and their related bacteria. To determine the identity of bacterial proteins that underlie symbiotic specificity in the entomopathogenic nematodes *Heterorhabditis*, 2D-gel electrophoresis followed by mass spectrometry were used to analyze and compare the proteomic profiles of two *P. luminescens* subspecies (*P. luminescens* ssp. *laumondii* and *P. luminescens* ssp. *akhurstii*), each occupying a distinct *Heterorhabditis* nematode species (*H. bacteriophora* and *H. indica*, respectively) (Kumar et al. 2016). Results from the proteomic and bioinformatic analyses revealed that either bacterial subspecies expresses several unique proteins, a subset of which (e.g. outer membrane proteins, proteins regulating secondary metabolites, and hypothetical proteins) may define nematode specificity. The functional characterization of certain candidate proteins will undoubtedly provide clues on the evolutionary and mechanistic basis of host–symbiont associations (Fig. 17.2).

Proteomic analysis coupled with bacterial genetics has further explored the role of the *rpoB* gene in the symbiosis between *P. luminescens* LN2 bacteria and their *H. bacteriophora* H06 nematode vectors (Qiu et al. 2012). Gene *rpoB* codes for the bacterial beta subunit of RNA polymerase and interestingly rifampicin prevents the initiation of transcription by repressing the *rpoB* gene. This research showed that certain rifampicin-resistant *P. luminescens* LN2 mutant strains, which surprisingly

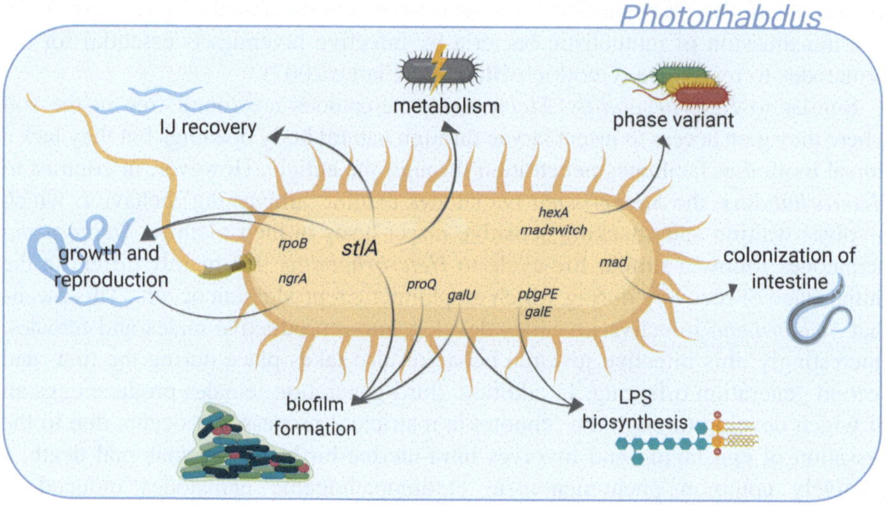

Fig. 17.2 *Photorhabdus* molecular regulators of symbiosis. Growth and reproduction of *Heterorhabditis bacteriophora* nematodes are controlled by genes *rpoB*, *ngrA,* and *stlA* in *P. luminescens*. Recovery of nematode infective juveniles (IJ) and bacterial metabolism are controlled by *P. luminescens* gene *stlA*. The *madswitch* promoter and gene *hexA* regulate the two distinct forms (mutualistic and pathogenic) in *P. luminescens* and *P. temperata* bacteria, respectively. The fimbrial locus *mad* in *P. luminescens* participates in initiating symbiosis through bacterial colonization of the posterior maternal intestinal cells in *H. bacteriophora*. LPS biosynthesis is modulated by genes *galE*, *galU,* and *pbgPE* in *P. luminescens*. Biofilm formation in *P. luminescens* is modulated by genes *galU*, *proQ,* and *stlA*. Images are made using Biorender graphic software (https://biorender.com)

also contained mutations in the *rpoB* gene, were able to support the growth of *H. bacteriophora* H06 infective juveniles. It was further demonstrated that mutations in the *rpoB* gene reconstitute the bacteria as the nutrient source for sustaining nematode reproduction; however, without conferring the ability of the bacteria to colonize the nematode intestines during the infective juvenile stage. Rifampicin selection of *P. luminescens rpoB* mutant strains supporting nematode growth may provide an elegant approach for increasing the production of *H. bacteriophora* in order to achieve more efficient insect pest control in the field.

Genetic analysis of the nematode–bacterial symbiotic relationship using a transposon mutagenesis and screening approach identified a single mutant strain of *P. luminescens* that was deficient in providing growth and reproduction to the *H. bacteriophora* nematode vector (Ciche et al. 2001). Characterization of the mutation localized the transposon insertion into gene *ngrA* encoding the enzyme Ppant transferase, which is involved in the biosynthesis of the siderophore enterobactin. Although this mutation also conferred an inability of the bacteria to produce antibiotics and siderophores, and probably interrupted the biosynthesis of fatty acids or lipids, these deficiencies were not attributed to the nematode defects. Instead, the assumption is that the inactivation of the *ngrA* gene possibly affects the biosynthesis of hormones, polyketides, or other secondary metabolites that are produced by *P. luminescens* when *H. bacteriophora* is also present, and act as signal molecules to promote the nematode's growth and development.

A subsequent study continued this work to examine whether gene *ngrA* encoding a putative phosphopantetheinyl transferase (PPT) that is involved in the biosynthesis of siderophore forms a determining factor for *P. luminescens* to support the growth and reproduction of *H. bacteriohora* nematodes (Ciche et al. 2003). Following a mini-Tn5 mutagenesis approach, *P. luminescens* mutant strain NS414 with a deficiency in producing measurable siderophore activity was first isolated, and then its properties were characterized. The results showed that the mutant bacteria were not able to grow normally in media depleted of iron, but they were capable of promoting the growth and reproduction of their nematode hosts, as well as their transmission by *H. bacteriophora* infective juveniles. Interestingly, the transposon was found to be inserted into gene photobactin synthetase (*phbH*) encoding a putative peptidyl carrier protein, which is covalently modified by PPTase for siderophore production. As *phbH* is not essential for nematode symbiosis, these findings signify that failure of the *ngrA* mutants to support nematode symbiosis was not due to their inability to produce functional siderophore but rather due to their incapacity to synthesize another currently unknown peptide that performs this function.

An interesting feature of *Photorhabdus* is that the bacteria can exist in two forms, the primary and secondary, which are morphologically distinct and are associated with the different phases of the pathogen's lifestyle. Only the primary form bacteria can colonize the intestinal tract of their associated nematode host and promote its growth and development due to the production of extracellular enzymes and antibiotic compounds that support the interaction between the two symbiotic partners during the infection of a suitable insect (Waterfield et al. 2009). Remarkably, it has been previously shown that inactivation through transposon insertion of gene *transcriptional regulator LrhA* (*hexA*) in the secondary phase of *P. temperata* bacteria results in the suppression of nematode colonization, and concomitantly, the mutant bacteria can foster growth and development of the host nematodes *H. downesi* (Joyce and Clarke 2003). These findings provide proof that *hexA* in the secondary phase of *P. temperata* bacteria encodes a molecule that confers direct or indirect repressive effects on symbiotic factors, which are normally expressed in the primary phase variants.

Another library screen of GFP-labeled *P. luminescens* transposon mutants, involving symbiotic assays to examine the qualitative ability of the mutant bacteria to colonize the gut of the infective juvenile stage of *H. bacteriophora* nematodes, further aimed at identifying bacterial genes, and their encoded factors responsible for the symbiotic collaboration between the two organisms (Easom et al. 2010). This work showed that mutations in a subset of genetic loci (e.g. *pbgPE* operon and genes *galE* and *galU*) involved specifically in the biosynthesis of lipopolysaccharide (LPS) and assembly and maintenance of LPS structure, as well as of other bacterial cell surface components, conferred substantially reduced transmission frequency of the mutant bacteria to associate with their nematode host. In addition, the *P. luminescens* mutant for genes *proQ* (encoding an RNA chaperone) and *galU* were also defective in biofilm formation as shown through testing the ability of the mutant bacteria to attach to an abiotic surface. This information highlights further the vital role of cell surface molecules in *P. luminescens*, and probably in other entomopathogenic bacteria, in adjusting the symbiotic outcome of bacterial–nematode partnerships.

Using a similar random transposon mutagenesis screening approach, the transmission ability of GFP-labeled *P. luminescens* mutants in the intestine of *H. bacteriophora* nematode parasites was analyzed in detail. The genetic analysis detected that the *maternal adhesion defective* (*mad*) fimbrial locus in *P. luminescens* has an essential role in initiating symbiosis through the bacterial colonization of the posterior maternal intestinal cells in *H. bacteriophora*. This process facilitates bacterial symbiont transmission from the maternal nematodes to the infective juveniles. Importantly, this is a specialized function because *mad* is required for symbiosis but not for insect pathogenesis, and the effect is regulated by bacterial phase variation in the wild type bacteria but not in the *mad* mutants (Somvanshi et al. 2010). These are the findings of particular significance because although fimbriae are known colonization factors that were previously shown to promote animal tissue or cell colonization by various bacterial pathogens through receptor recognition events, this was the first time that these adhesive organelles were assigned a similar function in modulating nematode–bacteria mutualistic symbiosis.

A previous study identified the production of crystalline inclusion proteins containing high levels of essential amino acids by *P. luminescens* bacteria to assist nematode reproduction (Bintrim and Ensign 1998), and a more recent work linked the two distinct forms of *P. luminescens* (M, initiating nematode Mutualism and P, initiating insect Pathogenicity) with the expression of the *mad* fimbrial locus, which occurs after the inversion of the *madswitch* promoter (Somvanshi et al. 2012). More precisely, it was demonstrated that during the first stage of mutualism in *H. bacteriophora*, *P. luminescens* bacteria switch to the M-form in the posterior intestine of the maternal nematodes, while the P-form bacteria are temporarily present in the intestines. The M-form cells then occupy the intestines of the new generation of infective juveniles before turning into the P-form cells to provide nematodes with bacteria possessing properties that promote infection of susceptible insects. Strikingly, the M-form bacteria are smaller than the P-form bacteria; they grow slower and exhibit decreased bioluminescence, virulence, and ability to secrete secondary metabolic compounds. Therefore, the biological implication of these findings underlines the importance of *madswitch* promoter orientation, which defines not only the phenotypic appearance of *P. luminescens* cells but also influences the lifestyle of the bacteria.

Additional factors have been associated with the persistence of *P. temperata* in the intestine of *H. bacteriophora* nematodes, as demonstrated by the drastic changes in the transcriptional profile of the bacteria during mutualism in the non-feeding infective juvenile stage (An and Grewal 2010). To determine strategies adapted by the bacteria to strengthen their persistence in the parasite through diminishing nutritional reliance on the nematode host, the number and identity of the differentially expressed genes in *P. temperata* was explored using the selective capture of the transcribed sequences technique. Analysis of the results displayed a large number of differentially regulated genes in *P. temperata* when the bacteria reside in the nematode intestine compared to the bacteria being cultured in vitro or being present in the insect host hemolymph. The differentially expressed genes denote modifications in physiological functions when residing in the nematodes' intestine including the activation of the pentose phosphate pathway, alteration in amino acid metabolisms, modification in LPS, induction of intracellular acidification and urea cycle mechanisms, proton transport and biofilm formation, as well as processes involving bacterial replication, transcription, and translation. These findings provide an exciting pool of potential molecular regulators of bacterial symbiosis in parasitic nematodes and future work awaits to dissect the specific mechanistic roles of these symbiosis factors and whether (and how) they are interconnected to enable the close biological link between the two organisms.

P. luminescens is a bacterium that is able to produce the antibiotic 3,5-dihydroxy-4-isopropylstilbene (stilbene, ST) during insect infection. The production of this antibiotic compound eliminates non-symbiotic micro-organisms that colonize insect tissues to compete for resources and nutrients, it prevents decay of the insect carcass and therefore, provides a favorable environment to their *H. bacteriophora* nematode vectors to grow, replicate, and complete their life cycle (Hu and Webster 2000). ST functions as a signal for the nematodes by stimulating the recovery of infective juveniles to adult hermaphrodites. This was demonstrated by the finding that *P. luminescens* mutants deficient in ST production were also unable to support *H. bacteriophora* growth and development (Joyce et al. 2008). The biosynthesis of ST involves the non-oxidative deamination of phenylalanine that leads to the synthesis of cinnamic acid by the enzyme phenylalanine-ammonium lyase, which is encoded by the gene *stlA* (Williams et al. 2005). Interestingly, it was subsequently shown that *stlA* expression is temporally controlled during growth, and it can be regulated by nutrient limitation (Lango-Scholey et al. 2013). This gene regulatory mechanism is further controlled by three transcriptional regulators; LysR-type TyrR, which is absolutely essential for *stlA* expression, as well as *Leucine-responsive regulatory protein* (*Lrp*) and RNA polymerase Sigma factor (*rpoS*), which are also required for normal *stlA* expression under suitable environmental conditions. These findings signify the molecular players that modulate secondary metabolism, and as a consequence, the mutualism of a potent entomopathogenic bacterium with its nematode host. Recently, additional findings provided evidence for the role of *P. luminescens* stilbene in the biology of the bacteria and their association with *H. bacteriophora* nematodes (Hapeshi et al. 2019). Exogenous ectopic addition of stilbene to *P. luminescens stlA* mutants reduces biofilm formation and downregulates the transcriptional expression of genes participating in the secondary metabolism, and basic cellular processes. These findings illustrate that stilbene cannot be only produced but also be detected by *P. luminescens* and plays a modulatory role by possibly acting as a signal for the bacteria to regulate the symbiotic phase of their life cycle through promoting the production of other molecules that are important for the recovery of nematode infective juveniles. This is a crucial process because it facilitates bacterial transmission to the next insect host following parasitic nematode infection.

17.4 Mutualism Regulators in *Xenorhabdus* Bacteria

Similar to *Photorhabdus*, *Xenorhabdus* bacteria are considered to have evolved distinct physiological and metabolic mechanisms that facilitate the close association with their cognate nematode hosts, as revealed by a previous study comparing the genome sequences of the two entomopathogenic bacterial symbionts (Chaston et al. 2011). The current speculation is that the two bacteria most likely share a single progenitor and as a result of multiple selective pressure events following differentiation, they have acquired unique factors to support the close relationship with their related nematode vectors.

Transposon mutagenesis to detect factors in *X. nematophila* that promote symbiosis with *Steinernema carpocapsae* nematodes led to the isolation of a mutant strain that was able to produce certain phospholipases such as lecithinase, but was unable to produce antibiotics and the mutant bacteria failed to grow and emerge normally from their nematode host (Volgyi et al. 2000) (Fig. 17.3). The transposon mutation was detected in the gene *var1* encoding a protein that is involved in the formation of the variant cell type. Deficient growth of the mutant bacteria may suggest a negative effect on the survival of this variant type of cells in the nematode intestines or the inability of the bacteria to swarm properly, which may delay their exit from the nematode. Although the specific physiological and biochemical bases of this

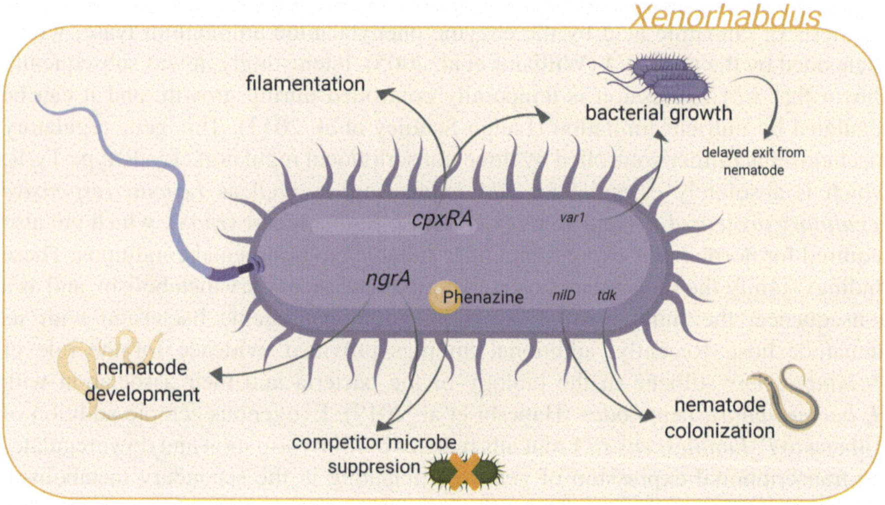

Fig. 17.3 *Xenorhabdus* molecular regulators of symbiosis. Filamentation and replication in *X. nematophila* are regulated by gene *cpxRA*. Bacterial growth and survival are controlled by gene *var1*. Colonization of *Steinernema carpocapsae* infective juveniles by *X. nematophila* is controlled by the bacterial genes *nilD* and *tdk*. Growth of competitor microbes in the dead insect is suppressed by phenazine compounds produced by *X. nematophila* as well as by gene *ngrA*. The latter also modulates the development and emergence of *S. carpocapsae* nematodes. Images are made using Biorender graphic software (https://biorender.com)

phenotype in *X. nematophila* is unclear, further elucidation of the processes and the particular factors modulating the switch from primary to secondary phase cells will enhance our knowledge regarding the type of features that are required to support the symbiotic interactions between entomopathogenic nematodes and their affiliated bacteria.

During the symbiotic relationship with *S. carpocapsae*, the *X. nematophila* uses nutrients from their nematode host; however, the exact identity of these compounds required for bacterial growth during this interaction is not well-determined. Since bacteria can utilize salvaged nucleosides as a supplement to endogenous nucleotides for DNA synthesis, they are able to form nitrogen sources, they can participate in the activation of signal transduction, and they are involved in the construction of cell structures. Therefore, it was previously hypothesized that the regulation of pyrimidine salvage pathways might constitute a process that facilitates the *X. nematophila*-nematode interplay (Orchard and Goodrich-Blair 2005). Curiously, it has been shown that the *X. nematophila* mutant for gene *tdk* encoding the enzyme deoxythymidine kinase, which synthesizes the pyrimidine nucleotide deoxythymidine monophosphate from deoxythymidine, are deficient in nematode colonization in vitro but not when present in the insect host. This defect is also fully restored by the addition of the wild copy *tdk* allele to the mutant strain. Such a mechanism could represent a broader strategy for entomopathogenic bacteria to associate with their nematode hosts.

An important aspect to consider in host–microbe interactions is the ability of the organisms involved in symbiotic relationships to sense and respond to external environmental changes. The two-component regulatory system CpxRA in *X. nematophila* consists of a sensor histidine kinase (CpxA) and a cytoplasmic response regulator (CpxR). It forms a signaling pathway, which in other bacteria, such as *E. coli,* regulates the function of structural components that permit interaction with the host. They might also act as a transducer to transmit internal signals inside the cells for initiating signaling pathways that would generate a cellular response (Herbert and Goodrich-Blair 2007b). Testing the symbiotic competence of *X. nematophila cpxR1* mutant bacteria, in which expression of both *cpxA* and *cpxP* genes is abolished, revealed that their ability to associate with *S. carpocapsae* infective juveniles is markedly impaired, and this effect is not due to a reduced survival ability of the mutants. Alternatively, this effect is primarily associated with modifications in cell morphological features in *cpxR1* mutants that might alter cell division dynamics or cause filamentation, which in turn could lead to interference in the symbiotic partnership between the bacteria and their nematodes, as well as with the decreased expression of genes that participate in nematode colonization. The regulatory role of CpxRA in *X. nematophila* is not unique, as other genes including *Lrp* (Leucine responsive regulatory protein) have also been shown to possess regulatory properties (Cowles et al. 2007). Overall, these findings imply the presence of certain genes in the *X. nematophila* genome performing multiple activities to promote the symbiotic connection of the bacteria with their affiliated entomopathogenic nematode partners.

It is given that molecules with antimicrobial activity may also modulate molecular and physiological processes by acting as signals in certain bacteria, previous efforts have focused on the role of bacterial secondary metabolites in promoting symbiotic relationships. This is because the *P. luminescens ngrA* gene, which encodes various secondary metabolites, is important for the growth and reproduction of *H. bacteriophora* nematodes, the involvement of *X. nematophila ngrA* in establishing or maintaining symbiosis with *S. carpocapsae* nematodes was also analyzed (Singh et al. 2015). Results from this research indicated that the number of nematode progenies from *Manduca sexta* caterpillars inoculated with *S. carpocapsae* infective juveniles containing the *X. nematophila ngrA* mutant were remarkably decreased compared to nematode progeny containing *X. nematophila* wild-type bacteria. These findings support the notion of a dual role for *ngrA*-derived compounds in not only combating competitor microbes in the insect cadaver to facilitate nematode development but also acting as signals for accelerating nematode development and emergence from the infected insect host.

Other molecules with multiple activities in *Xenorhabdus* that might serve as crucial factors for providing efficient bacteria–nematode symbiotic cooperation are the phenazines, which are commonly found in Gram-negative bacteria (Shi et al. 2019). It was recently tested whether phenazine compounds derived from *X. szentirmaii* influence the symbiotic ability between the bacteria and their *S. rarum* nematode vectors. Although the exact function of these molecules in regulating bacteria–nematode interactions is currently obscure because phenazines possess broad-spectrum as well as specific antibiotic activity, the current working hypothesis is that these compounds target a variety of competitor soil microbes present in the insect carcass. Elimination of these competitor microbes enables the *S. rarum* parasitic nematodes to complete and maintain their life cycle together with their closely related *X. szentirmaii* bacteria.

Further efforts to detect bacterial molecular factors promoting nematode colonization have included a signature-tagged mutagenesis screen in *X. nematophila*, an approach that identified a transposon mutant that lost its ability to colonize infective juveniles of *S. carpocapsae* nematodes (Veesenmeyer et al. 2014). Of note, the transposon was found to be inserted into gene *nilD* that codes for a Clustered Regularly Interspaced Short Palindromic Repeats (CRISPR) element. CRISPR associated sequences (Cas) are widely found in bacteria and perform several cellular functions including the regulation of gene expression and DNA repair mechanisms as well as bacterial behavior and resistance to foreign nucleotide sequences (Li and Peng 2019; Hampton et al. 2020). The investigators were able to demonstrate elegantly that the *nilD* CRISPR sequence is sufficient to support the colonization of *S. carpocapsae*, as well as *S. anatoliense* and *S. websteri* nematodes. The *nilD* RNA is expressed in a Cas6e-dependent manner under in vitro growth conditions and during nematode symbiosis, but in the latter case only within a specific genetic background of *X. nematophila*. These exciting new findings open novel avenues of investigation for designing strategies to decode the precise mechanism of CRISPR bacterial systems in modulating symbiotic interdependence with entomopathogenic nematodes.

17.5 Development of Tools for Identifying Nematode Regulators of Mutualism

Although most information currently available on the number and nature of genes and their products acting as regulators of mutualism with entomopathogenic nematodes has been obtained from bacteria, knowledge of analogous molecules in parasites playing a central or complementary role to the evolution, and stability of this process, is missing so far. This, at least till now, was mostly attributed to the lack of availability of genetic and genomic tools in entomopathogenic nematodes, which has impeded progress with dissecting the molecular basis of the mutualistic interaction between the two partners. To this end, recent efforts have mainly focused on developing whole-genome sequencing approaches and molecular procedures in *H. bacteriophora* to genetically manipulate the vector nematode in a similar way to the *C. elegans* model. Earlier work identified species-specific satellite DNA motifs in *H. bacteriophora* and *Steinernema glaseri* nematodes, and employed DNA reassociation kinetics to determine the genome size and complexity in *Heterorhabditis bacteriophora* and *Steinernema carpocapsae* (Grenier et al. 1996, 1997). At the same time, the expressed sequence tags were generated in *H. bacteriophora* to elucidate their involvement in various biological functions and more recently RNA-sequencing studies were carried out to understand the molecular basis of nematode parasitism (Sandhu et al. 2006; Bai et al. 2007, 2009; Vadnal et al. 2017). A tremendous breakthrough in entomopathogenic nematode research, including the nematode–bacterial mutualistic symbiosis, came about with the complete sequencing of the *H. bacteriophora* genome, the annotation which was recently improved (Bai et al. 2013; Vadnal et al. 2018; McLean et al. 2018). In terms of genetic techniques, the development of gene silencing RNAi interference through soaking and microinjection in *H. bacteriophora* has substantially upgraded the research value of this model organism (Ciche and Sternberg 2007; Ratnappan et al. 2016). Such advances are particularly significant because not only they promise to uncover previously unknown players of the interrelationship between the nematodes and their associated bacteria, but also to reveal the exact contribution of the parasites to the infection process of insects.

17.6 Concluding Remarks

Entomopathogenic nematodes are spectacular organisms that have received particular attention due to their complex life cycle that involves the mutualistic interdependence with specific bacteria that act as symbionts for the parasites and potent pathogens for the invaded insects. This extremely efficient relationship provides a fascinating model system for studying the interactions between invertebrates and their mutualistic microbes in relation to the host immune system (Ffrench-Constant et al. 2003; Joyce et al. 2006; Clarke 2008). Appreciating the details of the participation of the molecular processes and the specific genes or their products in fine-tuning the relationship between entomopathogenic nematodes and their associated bacteria is essential for understanding mutualistic interactions in other invertebrate organisms including beneficial insects, and devastating vectors of infectious diseases. It is also important for deciphering the conserved mechanisms that regulate similar types of interconnection between microbes and vertebrate animals, including humans. Such knowledge will significantly advance our biological interpretation of a wide range of host–microbial interplays that occur not only in the lab but also in different environmental settings.

References

An R, Grewal PS (2010) Molecular mechanisms of persistence of mutualistic bacteria *Photorhabdus* in the entomopathogenic nematode host. PLoS One 5:e13154

Bai X, Grewal PS, Hogenhout SA et al (2007) Expressed sequence tag analysis of gene representation in insect parasitic nematode *Heterorhabditis bacteriophora*. J Parasitol 93:1343–1349

Bai X, Adams BJ, Ciche TA et al (2009) Transcriptomic analysis of the entomopathogenic nematode *Heterorhabditis bacteriophora* TTO1. BMC Genomics 10:205

Bai X, Adams BJ, Ciche TA et al (2013) A lover and a fighter: the genome sequence of an entomopathogenic nematode *Heterorhabditis bacteriophora*. PLoS One 8:e69618

Bintrim SB, Ensign JC (1998) Insertional inactivation of genes encoding the crystalline inclusion proteins of *Photorhabdus luminescens* results in mutants with pleiotropic phenotypes. J Bacteriol 180:1261–1269

Bode HB (2009) Entomopathogenic bacteria as a source of secondary metabolites. Curr Opin Chem Biol 13:224–230

Chaston J, Goodrich-Blair H (2010) Common trends in mutualism revealed by model associations between invertebrates and bacteria. FEMS Microbiol Rev 34:41–58

Chaston JM, Suen G, Tucker SL et al (2011) The entomopathogenic bacterial endosymbionts *Xenorhabdus* and *Photorhabdus*: convergent lifestyles from divergent genomes. PLoS One 6: e27909

Ciche T (2007) The biology and genome of *Heterorhabditis bacteriophora*. WormBook:1–9. https://doi.org/10.1895/wormbook.1.135.1

Ciche TA, Ensign JC (2003) For the insect pathogen *Photorhabdus luminescens*, which end of a nematode is out? Appl Environ Microbiol 69:1890–1897

Ciche TA, Sternberg PW (2007) Postembryonic RNAi in *Heterorhabditis bacteriophora*: a nematode insect parasite and host for insect pathogenic symbionts. BMC Dev Biol 7:101

Ciche TA, Bintrim SB, Horswill AR et al (2001) A Phosphopantetheinyl transferase homolog is essential for *Photorhabdus luminescens* to support growth and reproduction of the entomopathogenic nematode *Heterorhabditis bacteriophora*. J Bacteriol 183:3117–3126

Ciche TA, Blackburn M, Carney JR et al (2003) Photobactin: a catechol siderophore produced by *Photorhabdus luminescens*, an entomopathogen mutually associated with *Heterorhabditis bacteriophora* NC1 nematodes. Appl Environ Microbiol 69:4706–4713

Ciche TA, Darby C, Ehlers RU et al (2006) Dangerous liaisons: the symbiosis of entomopathogenic nematodes and bacteria. Biol Control 38:22–46

Clarke DJ (2008) *Photorhabdus*: a model for the analysis of pathogenicity and mutualism. Cell Microbiol 10:2159–2167

Cowles KN, Cowles CE, Richards GR et al (2007) The global regulator Lrp contributes to mutualism, pathogenesis and phenotypic variation in the bacterium *Xenorhabdus nematophila*. Cell Microbiol 9:1311–1323

Dale C, Moran NA (2006) Molecular interactions between bacterial symbionts and their hosts. Cell 126:453–456

Easom CA, Joyce SA, Clarke DJ (2010) Identification of genes involved in the mutualistic colonization of the nematode *Heterorhabditis bacteriophora* by the bacterium *Photorhabdus luminescens*. BMC Microbiol 10:45

Eleftherianos IG (2009) Novel antibiotic compounds produced by the insect pathogenic bacterium *Photorhabdus*. Recent Pat Antiinfect Drug Discov 4:81–89

Ffrench-Constant R, Waterfield N, Daborn P et al (2003) *Photorhabdus*: towards a functional genomic analysis of a symbiont and a pathogen. FEMS Microbiol Rev 26:433–456

Ffrench-Constant RH, Dowling A, Waterfield NR (2007) Insecticidal toxins from *Photorhabdus* bacteria and their potential use in agriculture. Toxicon 49:436–451

Forst S, Dowds B, Boemare N et al (1997) *Xenorhabdus* and *Photorhabdus* spp.: bugs that kill bugs. Annu Rev Microbiol 51:47–72

Goodrich-Blair H, Clarke DJ (2007) Mutualism and pathogenesis in *Xenorhabdus* and *Photorhabdus*: two roads to the same destination. Mol Microbiol 64:260–268

Grenier E, Laumond C, Abad P (1996) Molecular characterization of two species-specific tandemly repeated DNAs from entomopathogenic nematodes *Steinernema* and *Heterorhabditis* (Nematoda: Rhabditida). Mol Biochem Parasitol 83:47–56

Grenier E, Catzeflis FM, Abad P (1997) Genome sizes of the entomopathogenic nematodes *Steinernema carpocapsae* and *Heterorhabditis bacteriophora* (Nematoda: Rhabditida). Parasitology 114:497–501

Hampton HG, Watson BNJ, Fineran PC (2020) The arms race between bacteria and their phage foes. Nature 577:327–336

Hapeshi A, Benarroch JM, Clarke DJ, Waterfield NR (2019) Iso-propyl stilbene: a life cycle signal? Microbiology 165:516–526

Hentschel U, Steinert M, Hacker J (2000) Common molecular mechanisms of symbiosis and pathogenesis. Trends Microbiol 8:226–231

Herbert EE, Goodrich-Blair H (2007a) Friend and foe: the two faces of *Xenorhabdus nematophila*. Nat Rev Microbiol 5:634–646

Herbert EE, Goodrich-Blair H (2007b) CpxRA regulates mutualism and pathogenesis in *Xenorhabdus nematophila*. Appl Environ Microbiol 73:7826–7836

Hu K, Webster JM (2000) Antibiotic production in relation to bacterial growth and nematode development in *Photorhabdus-Heterorhabditis* infected *Galleria mellonella* larvae. FEMS Microbiol Lett 189:219–223

Joyce SA, Clarke DJ (2003) A hexA homologue rom *Photorhabdus* regulates pathogenicity, symbiosis and phenotypic variation. Mol Microbiol 47:1445–1457

Joyce SA, Watson RJ, Clarke DJ (2006) The regulation of pathogenicity and mutualism in *Photorhabdus*. Curr Opin Microbiol 9:127–132

Joyce SA, Brachmann AO, Glazer I et al (2008) Bacterial biosynthesis of a multipotent stilbene. Angew Chem Int Ed Engl 47:1942–1945

Kaya HK, Gaugler R (1993) Entomopathogenic nematodes. Annu Rev Entomol 38:181–206

Kooliyottil R, Upadhyay D, Inman F III et al (2013) A comparative analysis of entomoparasitic nematodes *Heterorhabditis bacteriophora* and *Steinernema carpocapse*. Open J Anim Sci 3:326–333

Kumar R, Kushwah J, Ganguly S et al (2016) Proteomic investigation of *Photorhabdus* Bacteria for nematode-host specificity. Indian J Microbiol 56:361–367

Lango-Scholey L, Brachmann AO, Bode HB et al (2013) The expression of *stlA* in *Photorhabdus luminescens* is controlled by nutrient limitation. PLoS One 8:e82152

Li Y, Peng N (2019) Endogenous CRISPR-Cas system-based genome-editing and antimicrobials: review and prospects. Front Microbiol 10:2471

Maurelli AT (2007) Black holes, antivirulence genes, and gene inactivation in the evolution of bacterial pathogens. FEMS Microbiol Lett 267:1–8

McLean F, Berger D, Laetsch DR, Schwartz HT, Blaxter M (2018) Improving the annotation of the *Heterorhabditis bacteriophora* genome. Gigascience 7:1–12
Ochman H, Moran NA (2001) Genes lost and genes found: evolution of bacterial pathogenesis and symbiosis. Science 292:1096–1098
Orchard SS, Goodrich-Blair H (2005) Pyrimidine nucleoside salvage confers an advantage to *Xenorhabdus nematophila* in its host interactions. Appl Environ Microbiol 71:6254–6259
Parkhill J, Wren BW, Thomson NR et al (2001) Genome sequence of *Yersinia pestis*, the causative agent of plague. Nature 413:523–527
Qiu X, Yan X, Liu M et al (2012) Genetic and proteomic characterization of rpoB mutations and their effect on nematicidal activity in *Photorhabdus luminescens* LN2. PLoS One 7:e43114
Ratnappan R, Vadnal J, Keaney M et al (2016) RNAi-mediated gene knockdown by microinjection in the model entomopathogenic nematode *Heterorhabditis bacteriophora*. Parasit Vectors 9:160
Ruby EG (2008) Symbiotic conversations are revealed under genetic interrogation. Nat Rev Microbiol 6:752–762
Sagan L (1967) On the origin of mitosing cells. J Theor Biol 14:255–274
Sandhu SK, Jagdale GB, Hogenhout SA, Grewal PS (2006) Comparative analysis of the expressed genome of the infective juvenile entomopathogenic nematode, *Heterorhabditis bacteriophora*. Mol Biochem Parasitol 145:239–244
Shi YM, Brachmann AO, Westphalen MA et al (2019) Dual phenazine gene clusters enable diversification during biosynthesis. Nat Chem Biol 15:331–339
Singh S, Orr D, Divinagracia E et al (2015) Role of secondary metabolites in establishment of the mutualistic partnership between *Xenorhabdus nematophila* and the entomopathogenic nematode *Steinernema carpocapsae*. Appl Environ Microbiol 81:754–764
Somvanshi VS, Kaufmann-Daszczuk B, Kim KS et al (2010) *Photorhabdus* phase variants express a novel fimbrial locus, mad, essential for symbiosis. Mol Microbiol 77:1021–1038
Somvanshi VS, Sloup RE, Crawford JM et al (2012) A single promoter inversion switches *Photorhabdus* between pathogenic and mutualistic states. Science 337:88–93
Vadnal J, Ratnappan R, Keaney M et al (2017) Identification of candidate infection genes from the model entomopathogenic nematode *Heterorhabditis bacteriophora*. BMC Genomics 18:8
Vadnal J, Granger OG, Ratnappan R et al (2018) Refined ab initio gene predictions of *Heterorhabditis bacteriophora* using RNA-seq. Int J Parasitol 48:585–590
Veesenmeyer JL, Andersen AW, Lu X et al (2014) NilD CRISPR RNA contributes to *Xenorhabdus nematophila* colonization of symbiotic host nematodes. Mol Microbiol 93:1026–1042
Volgyi A, Fodor A, Forst S (2000) Inactivation of a novel gene produces a phenotypic variant cell and affects the symbiotic behavior of *Xenorhabdus nematophilus*. Appl Environ Microbiol 66:1622–1628
Waterfield NR, Ciche T, Clarke D (2009) *Photorhabdus* and a host of hosts. Annu Rev Microbiol 63:557–574
Williams JS, Thomas M, Clarke DJ (2005) The gene *stlA* encodes a phenylalanine ammonia-lyase that is involved in the production of a stilbene antibiotic in *Photorhabdus luminescens* TT01. Microbiology 151:2543–2550

Chapter 18
The Diversity of Symbiotic Systems in Scale Insects

Teresa Szklarzewicz, Anna Michalik, and Katarzyna Michalik

Abstract Most scale insects, like many other plant sap-sucking hemipterans, harbor obligate symbionts of bacterial or fungal origin, which synthesize and provide the host with substances missing in their restricted diet. Histological, ultrastructural, and molecular analyses have revealed that scale insects differ in the type of symbionts, the localization of symbionts in the host body, and the mode of transmission of symbionts from one generation to the next. Symbiotic microorganisms may be distributed in the cells of the fat body, midgut epithelium, inside the cells of other symbionts, or the specialized cells of a mesodermal origin, termed bacteriocytes. In most scale insects, their symbiotic associates are inherited transovarially, wherein the mode of transmission may have a different course—the symbionts may invade larval ovaries containing undifferentiated germ cells or ovaries of adult females containing vitellogenic or choriogenic oocytes.

18.1 Introduction

Scale insects (coccoids, coccids) are a large group (about 7700 species) of small phloem sap-sucking hemipterans (Gullan and Kosztarab 1997; Gullan and Martin 2003; Kondo et al. 2008). Coccoids are characterized by extreme sexual dimorphism. Adult females are larviform without well-defined body parts and are apterous and usually immobile. Adult males possess one pair of wings and nonfunctional mouthparts. Scale insects are highly diverse in their internal and external morphology, chromosome system, reproductive strategies, egg-protecting methods, and symbiotic systems. Traditionally, coccoids are divided into two informal groups: the ancient archaeococcoids and more advanced neococcoids. The archaeococcoids are made up of 15 extant families (Callipappidae, Carayonemidae, Coelosomidiidae,

T. Szklarzewicz (✉) · A. Michalik · K. Michalik
Department of Developmental Biology and Morphology of Invertebrates, Faculty of Biology, Institute of Zoology and Biomedical Research, Jagiellonian University, Kraków, Poland
e-mail: teresa.szklarzewicz@uj.edu.pl

M. Kloc (ed.), *Symbiosis: Cellular, Molecular, Medical and Evolutionary Aspects*, Results and Problems in Cell Differentiation 69,
https://doi.org/10.1007/978-3-030-51849-3_18

Kuwaniidae, Marchalinidae, Margarodidae, Matsucoccidae, Monophlebidae, Ortheziidae, Phenacoleachiidae, Pityococcidae, Putoidae, Steingeliidae, Stigmacoccidae, and Xylococcidae), whereas the neococcoids consist of 19 extant families (Aclerdidae, Asterolecaniidae, Beesonidae, Cerococcidae, Coccidae, Conchaspididae, Cryptococcidae, Dactylopiidae, Diaspididae, Eriococcidae, Halimococcidae, Kermesidae, Kerridae, Lecanodiaspididae, Micrococcidae, Phoenicococcidae, Pseudococcidae, Rhizoecidae, and Stictococcidae) (García Morales et al. 2016; Kondo et al. 2008).

The majority of scale insects, because of their unbalanced diet, live in obligate symbiotic relationships with different species of bacteria or fungi (Baumann 2005; Buchner 1965; Tremblay 1977). Results of the histological observations (Buchner 1965; Richter 1928; Walczuch 1932), ultrastructural studies (Niżnik and Szklarzewicz 2007; Szklarzewicz et al. 2006, 2010, 2013) and the molecular analyses (e.g., Dhami et al. 2012; Gomez-Polo et al. 2017; Gruwell et al. 2005; Husnik and McCutcheon 2016; López-Madrigal et al. 2015; Matsuura et al. 2009; Michalik et al. 2016, 2018, 2019a, b; Podsiadło et al. 2018; Szklarzewicz et al. 2018; Thao et al. 2002; Vera-Ponce de Leon et al. 2017; von Dohlen et al. 2001) have demonstrated that, in comparison to other hemipterans, scale insects have very diverse symbiotic systems. Their symbiotic associates differ in systematic affiliation and distribution in the host body, as well as in the mode of transmission between generations. It is known that scale insects belonging to the same family (e.g., Eriococcidae, Pseudococcidae) or related families (e.g., Monophlebidae and Marchalinidae) may harbor different symbionts (Buchner 1965). Koteja (1985) suggested that such a great diversity of symbiosis in scale insects is the consequence of the contact of their ancestors with different microorganisms in the forest leaf litter, which was the primary habitat of these insects. According to Koteja (1985), ancestors of scale insects consumed dead organic matter, however, on further evolution, they became plant sap-feeders. Switching to a "new" unbalanced diet required the support of microorganisms, which evolved into the mutualists responsible for the synthesis of missing nutrients. On the other hand, on the basis of cophylogenetic analyses, Rosenblueth et al. (2012, 2018) postulated that the ancestors of scale insects were associated with flavobacteria (phylum Bacteroidetes) which either: (1) coevolved with different lineages of scale insects, (2) were lost and replaced by other microorganisms (also by flavobacteria), or (3) were supported by additional symbionts.

Scale insects may be host to only one obligate symbiont (traditionally termed after Buchner (1965) the primary symbiont) or may harbor additional symbionts (termed the facultative or secondary symbionts). Buchner (1965), based on the observations of paraffin sections, defined primary symbionts as microorganisms, which are: (1) descendants of the bacterium, which infected the ancestor, (2) present in all the individuals of this group, (3) responsible for the synthesis of essential nutrients missing in the diet of the host insect, and (4) vertically (i.e., from mother to progeny) transmitted between generations. More recent molecular analyses supported Buchner's conclusions concerning the nutritional role of the primary symbionts of scale insects (Sabree et al. 2013; Vera-Ponce de Leon et al. 2017).

The functions of the secondary symbionts accompanying scale insects were unknown until recently, however, recent molecular analyses have revealed that in at least some of these insects, they nutritionally complement primary symbionts (Husnik et al. 2013; Husnik and McCutcheon 2016; López-Madrigal et al. 2013, 2014; McCutcheon and von Dohlen 2011; Rosas-Perez et al. 2014; Szabo et al. 2017).

In most scale insects, symbiotic microorganisms (termed bacteriocyte symbionts) are localized in the specialized cells of a mesodermal origin termed the bacteriocytes. However, in some scale insects, the symbionts are distributed in the cells of the fat body (e.g., in Steingeliidae, Kermesidae, Coccidae, and some representatives of Eriococcidae) or midgut epithelium (*Marchalina hellenica*) (Buchner 1965, 1966; Matsuura et al. 2009; Michalik et al. 2016, 2018, 2019a, b; Niżnik and Szklarzewicz 2007; Podsiadło et al. 2018; Szklarzewicz et al. 2006, 2010, 2018; von Dohlen et al. 2001; Walczuch 1932). It is generally accepted that the "bacteriocyte symbiosis" represents the most advanced stage of association of insects and microorganisms, whereas the occurrence of symbionts in the cells of fat body, midgut epithelium, or in the midgut appendages is regarded as an initial stage of symbiotic relationship (Braendle et al. 2003; Kuechler et al. 2011; Michalik et al. 2016). Interestingly, "bacteriocyte" symbiosis and "non-bacteriocyte" symbiosis occur both in archaeococcoids and neococcoids.

It is worth mentioning that some families of scale insects (e.g., Kermesidae, Steingeliidae, Dactylopiidae, Matsucoccidae) had been regarded as the asymbiotic (Buchner 1965; Tremblay 1977) until the ultrastructural and molecular techniques detected the presence of the symbionts (Koteja et al. 2003; Michalik et al. 2019b; Podsiadło et al. 2018; Ramirez-Puebla et al. 2010; Rosenblueth et al. 2018; Szklarzewicz et al. 2014; Vera-Ponce de Leon et al. 2017).

18.2 "Non-bacteriocyte" Symbiosis

In the scale insects, which have not developed specialized bacteriocytes, such as pine bast scales (Matsucoccidae), ensign scales (Ortheziidae), steingeliids (Steingeliidae), giant pine scales (Marchalinidae), and some species of felt scales (Eriococcidae), symbionts may be localized in the fat body, the gut epithelium, and the ovaries (see Table 18.1).

Molecular analyses have revealed that the pine bark scale *Matsucoccus pini* (Matsucoccidae) (Fig. 18.1a), and nettle ensign scale *Orthezia urticae* (Ortheziidae), are host to the alphaproteobacterium of the genus *Wolbachia* (Michalik et al. 2019b). *Wolbachia* commonly occurs in numerous arthropods and nematodes, however, in most cases, its presence in the host body is rather connected with the elimination of males than with the nutritional function (Werren 1997). The role of *Wolbachia* in *Orthezia urticae* and *Matsucoccus pini* remains unclear, however, nobody showed the effect of this "reproductive manipulator" on the frequency of males in these species (Michalik et al. 2019b). Michalik et al. (2019b) reported that in some

Table 18.1 "Non-bacteriocyte" symbiosis in selected species of scale insects—types of symbionts, localization in the host insect, and mode of transmission between generations

Family	Scale insect	Symbionts	Localization in the host insect	Mode of transmission	References
Matsucoccidae (pine bast scales)	*Matsucoccus pini* (Fig. 18.1a)	*Wolbachia* (Alphaproteobacteria)	Fat body cells, gut epithelium, ovaries	Infection of cystocytes in larval ovary	Michalik et al. (2019b)
Ortheziidae (ensign scales)	*Orthezia urticae* (Fig. 18.1b)	*Wolbachia* (Alphaproteobacteria) Enterobacteria related to *Sodalis* (Gammaproteobacteria) (present in some individuals only)	Fat body cells, gut epithelium, ovaries	Infection of cystocytes in larval ovary	Michalik et al. (2019b)
Steingeliidae (steingeliids)	*Steingelia gorodetskia* (Fig. 18.1c, d)	Bacteria related to *Sphingomonas* (Alphaproteobacteria)	Fat body cells, gut epithelium, and ovaries	Infection of cystocytes in larval ovary	Michalik et al. (2019b)
Eriococcidae (felt scales)	*Gossyparia spuria* (Fig. 18.2a–c)	Bacteria related to *Burkholderia* (Betaproteobacteria)	Fat body cells	Infection of neck of ovariole of adult female	Michalik et al. (2016)
Marchalinidae (giant pine scales)	*Marchalina hellenica* (Fig. 18.3a–d)	Large, elongated bacteria of unknown taxonomic identity	Midgut epithelium, "temporary bacteriocytes", and ovaries	Infection of cystocytes in larval ovary	Buchner (1965, 1967) and Szklarzewicz et al. (2013)

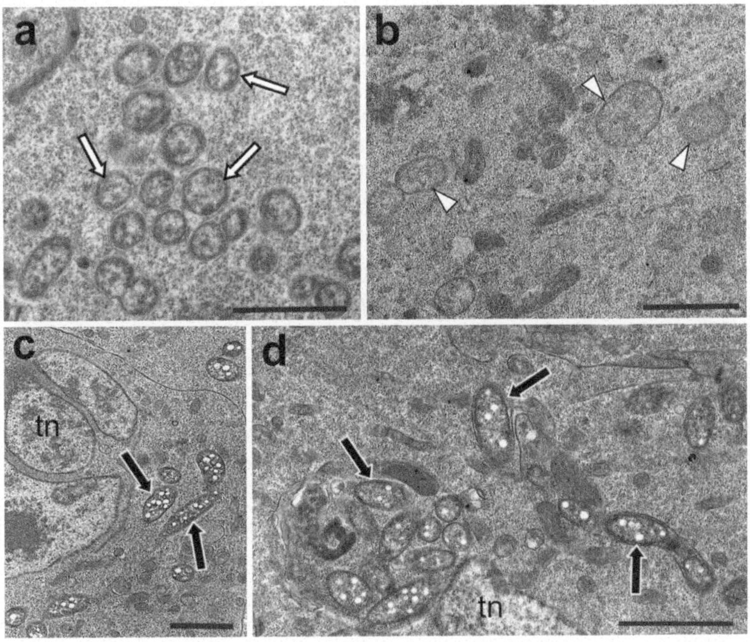

Fig. 18.1 "Non-bacteriocyte" symbiosis in archaeococcoid families Matsucoccidae, Ortheziidae, and Steingeliidae. (**a**) *Matsucoccus pini* (Matsucoccidae). Bacteria *Wolbachia (white arrows)* in the cytoplasm of the trophocyte. (**b**) *Orthezia urticae* (Ortheziidae). Bacteria *Sodalis (white arrowheads)* in the cytoplasm of the oocyte. (**c**, **d**) *Steingelia gorodetskia* (Steingeliidae). (**c**) Bacteria *Sphingomonas (black arrows)* in the trophocyte cytoplasm. (**d**) Bacteria *Sphingomonas* (black arrows) migrate from the trophocytes to the developing oocyte through the trophic core. (**a–d**) TEM, scale bar = 2 μm. *tn* trophocyte nucleus

analyzed individuals of *Orthezia urticae* besides *Wolbachia*, the enterobacterium related to *Sodalis* (Gammaproteobacteria) was found (Fig. 18.1b). *Sodalis*-like bacteria are regarded as common associates of insects because they have been observed in numerous species representing different insect taxa. A wide distribution of *Sodalis*-allied bacteria in insects, their diverse localization in the insect body (in bacteriocytes, in the fat body, and in milk glands of tsetse flies), and different types of relationships with the host insects (mutualism, facultative symbiosis) indicate that this microorganism has a great tendency to colonize the new insect species.

The birch bark scale *Steingelia gorodetskia* (Steingeliidae) was considered an asymbiotic species by Buchner (1966), however, Koteja et al. (2003) observed numerous small rod-shaped bacteria in its ovaries (Fig. 18.1c), and suggested that they may be the symbionts. Molecular phylogenetic analyses based on the sequence of the 16S rRNA gene revealed that these microorganisms are closely related to the soil bacterium *Sphingomonas echinoides* (Alphaproteobacteria) (Michalik et al. 2019b). This finding suggests that the ancestor of *Steingelia gorodetskia* acquired the soil bacterium, which, in further coevolution, became its symbiont. It should be stressed that this suggestion strongly supports the hypothesis of Koteja (1985) that symbionts of scale insects are the descendants of the forest litter bacteria (see Introduction).

The presence of bacteria in the oocytes of *Orthezia urticae*, *Matsucoccus pini*, and *Steingelia gorodetskia* indicates that these microorganisms are transovarially transmitted from mother to progeny (Michalik et al. 2019b). Thus, in spite of the lack of specialized bacteriocytes, *Orthezia urticae*, *Matsucoccus pini*, and *Steingelia gorodetskia* already developed the stable mechanisms ensuring the transfer of symbionts between generations. According to Michalik et al. (2019b), in *Steingelia gorodetskia,* the symbiotic bacteria infect the cystocytes (i.e., undifferentiated germ cells) residing in the larval ovaries (for further details concerning modes of symbiont transmission, see Szklarzewicz and Michalik 2017). Because the cystocytes differentiate into oocytes and trophocytes, in the ovarioles of adult females the symbiotic bacteria reside in both these cell types. In reproductive females, symbionts migrate via the trophic core and nutritive cord (Fig. 18.1d) to the developing oocyte, and after egg fertilization, they are included into the embryo.

The occurrence of the symbionts in the fat body has also been observed in two species of the neococcoid Eriococcidae family, the European elm scale *Gossyparia spuria* (Fig. 18.2a, b) and maple felt scale *Acanthococcus aceris*. Michalik et al. (2016) showed that symbionts of these scale insects are closely related to the widely distributed betaproteobacterium *Burkholderia*. Both in *Gossyparia spuria* and *Acanthococcus aceris*, at the time the oocytes are in the stage of late vitellogenesis, the symbiotic bacteria leave the cells of the fat body and begin to accumulate around the neck of ovariole (the region between the tropharium and vitellarium). The bacteria then cross through the follicular epithelium surrounding the nutritive cord (via endocytic/exocytic pathway) or migrate between neighboring follicular cells (Fig. 18.2c). Finally, symbionts enter the perivitelline space (Fig. 18.2c), where they gather in the deep invagination of the oolemma at the anterior pole of the oocyte (Fig. 18.2d). Such a mode of symbiont transmission relying on the infection of the neck region of the ovariole is unique within insects and has so far been reported only for some scale insects (for further details, see Szklarzewicz and Michalik 2017). The bacteria penetrate the ovariole in its neck region because the oocytes, at this stage, are surrounded by eggshells and the only place on the oocyte surface devoid of the eggshell is the nutritive cord connecting the oocyte with the tropharium. After the nutritive cord had degenerated, the entire oocyte, together with the symbionts located at the invagination of oolemma, becomes covered with egg envelopes (see Fig. 18.9e).

"Typical" bacteriocytes are also absent in the giant pine scale *Marchalina hellenica* (Marchalinidae) (Buchner 1967). Buchner (1967) observed that in this species, the symbiotic bacteria occupy the enlarged cells of the midgut epithelium. In the larvae of the third instar (the last larval stage before imago), the symbiotic bacteria are transported from the midgut to the ovary through "temporary bacteriocytes". The bacteria then infect the cystocytes. Szklarzewicz et al. (2013), who examined the ovaries of the reproductive females of *Marchalina hellenica*, showed that the symbionts (large elongated bacteria of unknown taxonomic identity)

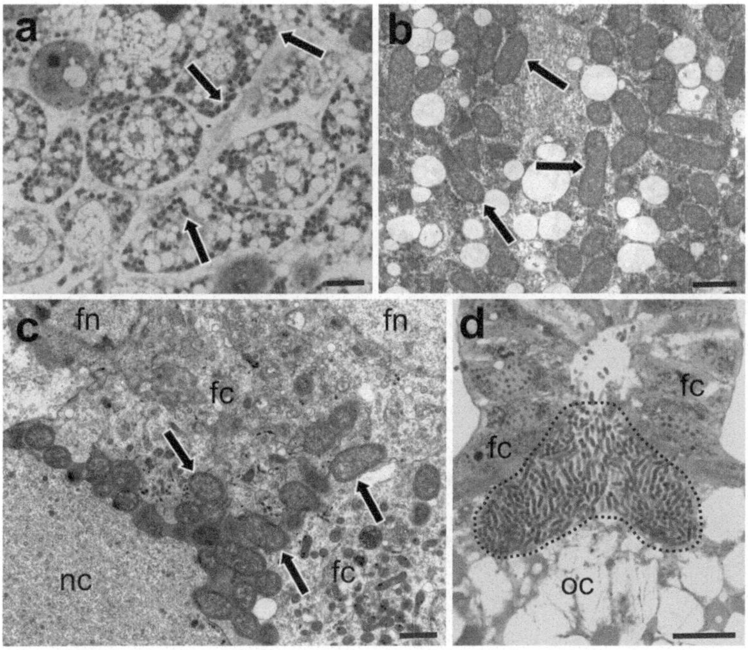

Fig. 18.2 "Non-bacteriocyte" symbiosis in neococcoids *Acanthococcus aceris* and *Gossyparia spuria* (Eriococcidae). (**a–c**) *Gossyparia spuria*. (**a**, **b**) Fragment of the fat body. Cells of the fat body contain numerous rod-shaped bacteria *Burkholderia (black arrows)*. (**c**) Bacteria *Burkholderia (black arrows)* migrate through the follicular epithelium and gather around the nutritve cord. (**d**) *Acanthococcus aceris*. Bacteria *Burkholderia (encircled with black-dotted line)* in the perivitelline space. (**a**, **d**) Methylene blue, scale bar = 20 μm, (**b**, **c**) TEM, scale bar = 2 μm. *fc* follicular cell, *fn* follicular cell nucleus, *nc* nutritive cord, and *oc* oocyte

are transported to the developing oocyte via the trophic core and nutritive cord (Fig. 18.3a–d). Thus, in spite of the lack of specialized bacteriocytes, the abovementioned scale insects already developed stable mechanisms ensuring the transfer of symbionts between generations.

18.3 "Bacteriocyte" Symbiosis

Most scale insects, e.g., giant scales (Monophlebidae), coelostomidiids (Coelostomidiidae), ground pearls (Margarodidae), giant mealybugs (Putoidae), mealybugs (Pseudococcidae), ground mealybugs (Rhizoecidae), cochineal scales (Dactylopiidae), armored scales (Diaspididae), and some species of felt scales (Eriococcidae), like other insect groups living in mutualistic relationships with microorganisms, developed specialized cells of mesodermal origin termed bacteriocytes (the older term "mycetocytes") (see Table 18.2). As a rule,

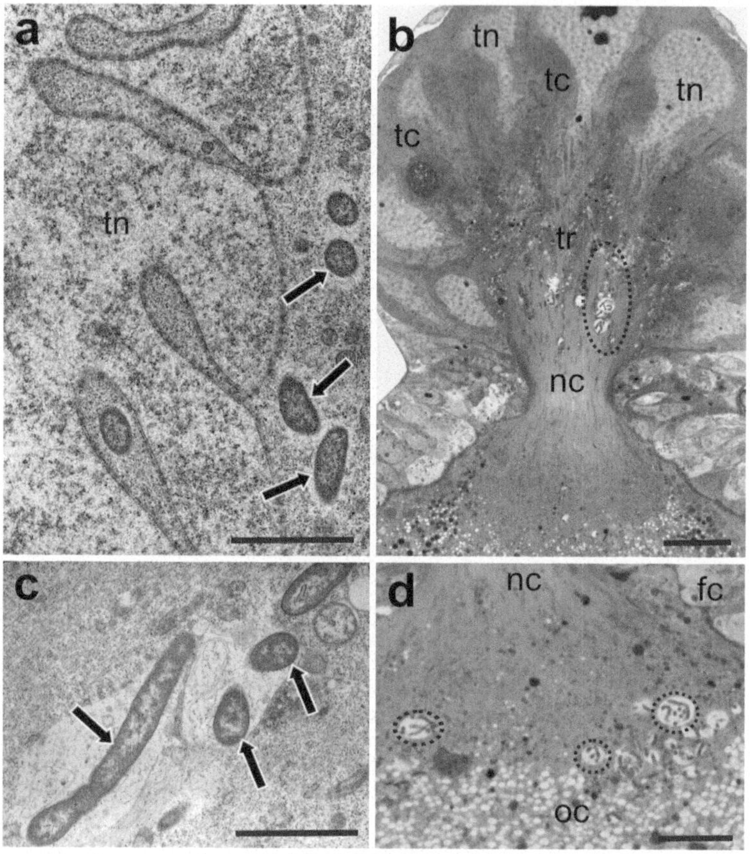

Fig. 18.3 Symbiotic microorganisms in *Marchalina hellenica* (Marchalinidae). (**a**) Fragment of the trophocyte with rod-shaped bacteria (*black arrows*). (**b**) Tropharium (longitudinal section). Note bacteria *(encircled with black-dotted line)* migrating through processes of trophocytes, the trophic core, and the nutritive cord into developing oocyte. (**c**) Bacteria (*black arrows*) in the nutritive cord. (**d**) Anterior pole of the developing oocyte (longitudinal section). Note groups of bacteria *(encircled with black-dotted line)* in the oocyte cytoplasm. (**a**, **c**) TEM, scale bar = 2 μm, (**b**, **d**) Methylene blue, scale bar = 20 μm. *fc* follicular cell, *nc* nutritive cord, *oc* oocyte, *tc* trophocyte, *tr* trophic core, and *tn* trophocyte nucleus

bacteriocytes are grouped into large organs termed the bacteriomes (the older term was "mycetomes"). In the insect body, the bacteriomes are located ventro-laterally between the body wall and gonads (Buchner 1965).

Within the archaeococcoid family Monophlebidae, the symbiotic systems of four genera, *Icerya*, *Palaeococcus*, *Llaveia*, and *Drosicha* were examined by molecular and/or ultrastructural methods (Matsuura et al. 2009; Niżnik and Szklarzewicz 2007; Rosas-Perez et al. 2014; Rosenblueth et al. 2012; Szklarzewicz et al. 2006). Molecular analyses revealed that Monophlebidae harbor two kinds of symbiotic associates:

Table 18.2 "Bacteriocyte symbiosis" in selected species of scale insects—types of symbionts and mode of transmission between generations

Family	Scale insect	Symbionts	Mode of transmission	References
Monophlebidae (giant scales)	*Llaveia axin axin*	*Walczuchella monophlebidarum* (Bacteroidetes) Enterobacteria related to *Sodalis* (Gammaproteobacteria)	Not examined	Rosas-Perez et al. (2014) and Rosenblueth et al. (2012)
	Drosicha piniola	*Walczuchella monophlebidarum* (Bacteroidetes) Enterobacteria related to *Sodalis* (Gammaproteobacteria) *Wolbachia* (Alphaproteobacteria)	Not examined	Matsuura et al. (2009)
	Icerya purchasi (Fig. 18.4a, b, d, e)	*Walczuchella monophlebidarum* (Bacteroidetes) Enterobacteria related to *Sodalis* (Gammaproteobacteria)	*Walczuchella*—infection of posterior end of ovariole of adult female Enterobacteria—infection of cystocytes in larval ovary	Niżnik and Szklarzewicz (2007) and Rosenblueth et al. (2012)
Coelostomidiidae (coelostomidiids)	*Coelostomidia wairoensis*	*Hoataupuhia coelostomidicola* (Bacteroidetes) *Wolbachia* (Alphaproteobacteria) *Erwinia*-like symbionts (Gammaproteobacteria)	Not examined	Dhami et al. (2013)
	Coelostomidia pilosa	*Wolbachia* (Alphaproteobacteria) *Erwinia*-like symbionts (Gammaproteobacteria) Enterobacteria related to *Sodalis* (Gammaproteobacteria)	Not examined	Dhami et al. (2013)
	Ultracoelostoma dracopgylli	*Hoataupuhia coelostomidicola* (Bacteroidetes)	Not examined	Dhami et al. (2013)
	Ultracoelostoma assimile	Enterobacteria related to *Sodalis* (Gammaproteobacteria)	Not examined	Dhami et al. (2013)

(continued)

Table 18.2 (continued)

Family	Scale insect	Symbionts	Mode of transmission	References
Margarodidae (ground pearls)	*Promargarodes australis*	Bacteria of the phylum Bacteroidetes	Not examined	Gruwell et al. (2005)
Putoidae (giant mealybugs)	*Puto superbus* (Fig. 18.5a–e)	Enterobacteria related to *Sodalis* (Gammaproteobacteria) *Wolbachia* (Alphaproteobacteria)	Infection of neck of ovariole of adult female through whole intact bacteriocytes	Buchner (1965) and Szklarzewicz et al. (2018)
	Puto albicans (Fig. 18.5f–h)	Enterobacteria related to *Sodalis* (Gammaproteobacteria)	Infection of cystocytes in larval ovaries	Gruwell et al. (2005) and Szklarzewicz et al. (2010)
Eriococcidae (felt scales)	*Greenisca brachypodii* (Fig. 18.6a–g)	*Kotejella greeniscae* (Gammaproteobacteria) Bacteria related to *Arsenophonus* (Gammaproteobacteria)	Infection of neck of ovariole of the adult female	Michalik et al. (2018)
Diaspididae (armored scales)	*Aspidiotus nerii*	*Uzinura diaspidicola* (Bacteroidetes)	Not examined	Gruwell et al. (2005)
	Leucaspis loewi (Fig. 18.7a, b)	*Uzinura diaspidicola* (Bacteroidetes)	Infection of neck of ovariole of adult female	Szklarzewicz et al. (unpublished data)
Pseudococcidae: Pseudococcinae (mealybugs)	*Planococcus citri*	*Tremblaya princeps* (Betaproteobacteria) *Moranella endobia* (Gammaproteobacteria) inside bacterium *Tremblaya* (nested symbiosis)	Infection of neck of ovariole of adult female	López-Madrigal et al. (2013), McCutcheon and von Dohlen (2011), and von Dohlen et al. (2001)
Pseudococcidae: Phenacoccinae (mealybugs)	*Phenacoccus aceris* (Fig. 18.8d)	*Tremblaya phenacola* (Betaproteobacteria)	Infection of neck of ovariole of adult female	Michalik et al. (2019a)
Rhizoecidae (ground mealybugs)	*Geococcus coffeae*	*Brownia rhizoecola* (Bacteroidetes)	Not examined	Gruwell et al. (2010)
Dactylopiidae (cochineal mealybugs)	*Dactylopius coccus*	*Dactylopiibacterium carminicum* (Betaproteobacteria)	Not examined	Ramirez-Puebla et al. (2010)

flavobacteria (phylum Bacteroidetes), which are considered primary symbionts and enterobacteria related to the bacterium *Sodalis*, which are considered secondary symbionts (Matsuura et al. 2009; Rosas-Perez et al. 2014; Rosenblueth et al. 2012). It should be stressed that the genomic analyses conducted by Rosas-Perez et al. (2014) revealed that flavobacterial symbiont (termed *Walczuchella monophlebidarum*) of *Llaveia axin axin* has strongly reduced genome, in which some genes such as the genes encoding some enzymes involved in the biosynthesis of certain amino acids have been lost or pseudogenized. Based on this finding and the results of genomic analyses of enterobacterial symbiont, which indicated that this bacterium has complete biosynthetic pathways for all essential amino acids, Rosas-Perez et al. (2014) postulated that flavobacteria and enterobacteria complement each other in the synthesis of essential nutrients. In contrast to the highly reduced genome of *Walczuchella*, the size of the genome of the enterobacterial symbiont is similar to the genome of its free-living relatives, which suggests that the secondary symbionts were acquired more recently.

The bacteriomes of the cottony cushion scale *Icerya purchasi* and ancient giant scale *Palaeococcus fuscipennis* consist of numerous bacteriocytes filled with pleomorphic bacteria (Fig. 18.4a, b, d), which are surrounded by the large cells (Fig. 18.4a, b) containing elongated rod-shaped bacteria (Fig. 18.4c) (Niżnik and Szklarzewicz 2007; Szklarzewicz et al. 2006). The ultrastructural observations (Niżnik and Szklarzewicz 2007; Szklarzewicz et al. 2006) and molecular analyses (Rosenblueth et al. 2012) led to the conclusion that the large pleomorphic bacteria present in *Icerya purchasi* represent flavobacteria, whereas small rod-shaped bacteria represent enterobacteria.

In the second larval instar of *Icerya purchasi*, the rod-shaped bacteria infect the cystocytes (Niżnik and Szklarzewicz 2007). In the adult female, these symbionts migrate from the trophocytes into the trophic core (Fig. 18.4e), and next through the nutritive cord into the developing oocyte. Pleomorphic bacteria of *Icerya purchasi* and *Palaeococcus fuscipennis*, as well as the rod-shaped bacteria of *Palaeococcus fuscipennis*, invade the posterior end of the ovariole containing the choriogenic oocyte (Fig. 18.4f). Then, these symbionts migrate through the cytoplasm of follicular cells (Fig. 18.4f, g) to the perivitelline space, where they accumulate in the deep invagination of oolemma, eventually forming a "symbiont ball" (Fig. 18.4g, h).

"Bacteriocyte" symbiosis is also characteristic for the scale insects of the New Zealand family Coelostomidiidae (Dhami et al. 2012), which is closely related to the family Monophlebidae (Gullan and Cook 2007; Hodgson and Hardy 2013). Dhami et al. (2012, 2013) examined the symbionts of several species of two genera (*Coelostomidia* and *Ultracoelostoma*) belonging to the Coelostomidiidae and observed that most of them harbor the flavobacteria (termed *Hoataupuhia coelostomidicola*) (see Table 18.2). Interestingly, species with an absence or low infection density of flavobacterial symbionts were host to an enterobacterial *Sodalis*-like symbiont. Based on this observation, Dhami et al. (2013) postulated that, during the evolution of some species of Coelostomidiidae, the *Sodalis*-like symbionts replaced their ancient symbionts, the flavobacterium *Hoataupuhia*. Apart from the flavobacterial symbiont *Hoataupuhia* and/or *Sodalis*-like symbiont, Dhami et al.

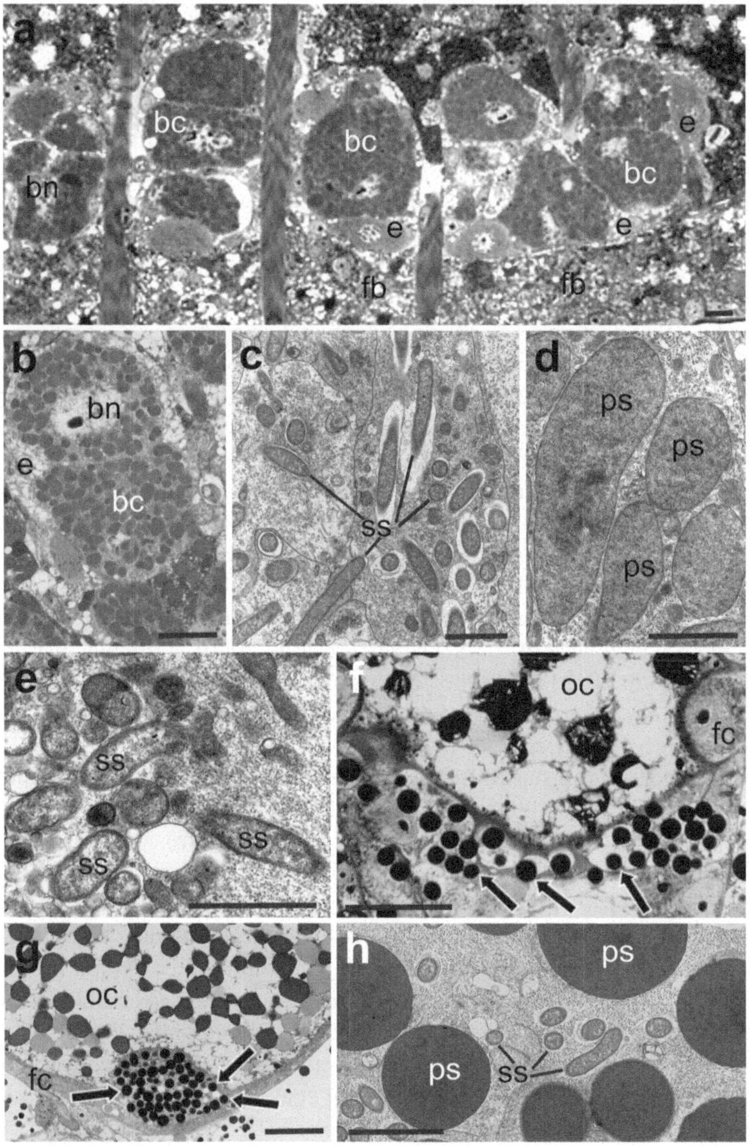

Fig. 18.4 "Bacteriocyte" symbiosis in archaeococcoid family Monophlebidae. (**a**) *Icerya purchasi*. Fragment of the abdomen (longitudinal section) of the last instar larva. Note large bacteriocytes filled with symbiotic bacteria (=primary symbionts). (**b**) *Icerya purchasi*. Bacteriocytes of the young female. (**c**) *Palaeococcus fuscipennis*. Fragment of the epithelium surrounding the bacteriocyte. Note elongated bacteria (=secondary symbionts). (**d**) *Icerya purchasi*. Fragment of the bacteriocyte filled with pleomorphic bacteria (=primary symbionts). (**e**) *Icerya purchasi*. Secondary symbionts migrating from trophocytes to the developing oocyte through the trophic core. (**f**) *Palaeococcus fuscipennis*. Symbionts (*black arrows*) migrate through the follicular epithelium into the perivitelline space. (**g**) *Palaeococcus fuscipennis*. Symbionts (*black arrows*)

(2012) found additional microorganisms in the Coelostomidiidae: the *Erwnia*-related symbionts and *Wolbachia*, the function of these associates, however, remains unknown.

"Bacteriocyte" symbiosis has also been reported to occur in the archaeococcoid family Margarodidae (Buchner 1966), however, in comparison to other scale insects, the symbiotic systems of margarodids have not been extensively studied. Gruwell et al. (2005) detected flavobacteria in two margarodid species, *Eumargarodes laingi* and *Promargarodes australis*. Earlier histological observations of Buchner (1965) and our preliminary ultrastructural studies (unpublished data) on symbionts of the Polish cochineal scale *Porhyrophora polonica* revealed that these microorganisms infest the posterior end of the choriogenic ovariole (as in *Palaeococcus fuscipennis*, see Fig. 18.4f–h).

The other group of archaeococcoids with "bacteriocyte" symbiosis is the family Putoidae containing only a single extant genus *Puto* (Gruwell et al. 2014; Szklarzewicz et al. 2010, 2018). Interestingly, for many years, the genus *Puto* was regarded as a taxon belonging to the neococcoids—the family Pseudococcidae (Kosztarab and Kozár 1988) or to their own family Putoidae (Koteja 1996). Based on the results of molecular analyses, Cook et al. (2002) transferred Putoidae to archaeococcoids. The results of morphological studies on the internal and external organs, as well as on their symbionts, corroborated that the Putoidae are closely related to archaeococcoid families (Gruwell et al. 2014; Hodgson and Hardy 2013; Michalik et al. 2013; Szklarzewicz et al. 2010, 2018).

Gruwell et al. (2005, 2014) found enterobacteria related to *Sodalis* in several American species of *Puto*. Szklarzewicz et al. (2018) observed large elongated bacterial symbionts (Fig. 18.5a, b) in the Palearctic species, *Puto superbus*, and identified them as bacteria related to *Sodalis*. The mode of transmission of symbionts from mother to offspring in *Puto superbus* appeared to be unusual within insects, because, in this scale insect, whole intact bacteriocytes migrate toward the ovaries (Fig. 18.5c, d) (Buchner 1965; Szklarzewicz et al. 2018). In the reproductive females, the bacteriocytes surround the neck of the ovariole containing the choriogenic oocytes and enter the perivitelline space through the gaps between the neighboring follicular cells (Fig. 18.5d). Eventually, the bacteriocytes gather in the perivitelline space where they form the "symbiont ball" (Fig. 18.5e). Interestingly, the North American species, *Puto albicans* developed a completely different mode of transmission of its symbionts from mother to progeny because, in this species, the symbionts (Fig. 18.5f) invade the larval ovaries (Fig. 18.5g, h) (Szklarzewicz et al. 2010).

Fig. 18.4 (continued) gather in the perivitelline space where they form a "symbiont ball". (**h**) *Palaeococcus fuscipennis*. Primary and secondary symbionts in the perivitelline space. (**a**, **b**, **f**, **g**) Methylene blue, scale bar = 20 μm. (**c**–**e**, **h**) TEM, scale bar = 2 μm. *bc* bacteriocyte, *bn* bacteriocyte nucleus, *e* epithelium surrounding the bacteriocyte, *fb* fat body, *fc* follicular epithelium, *oc* oocyte, *ps* primary symbiont, and *ss* secondary symbiont

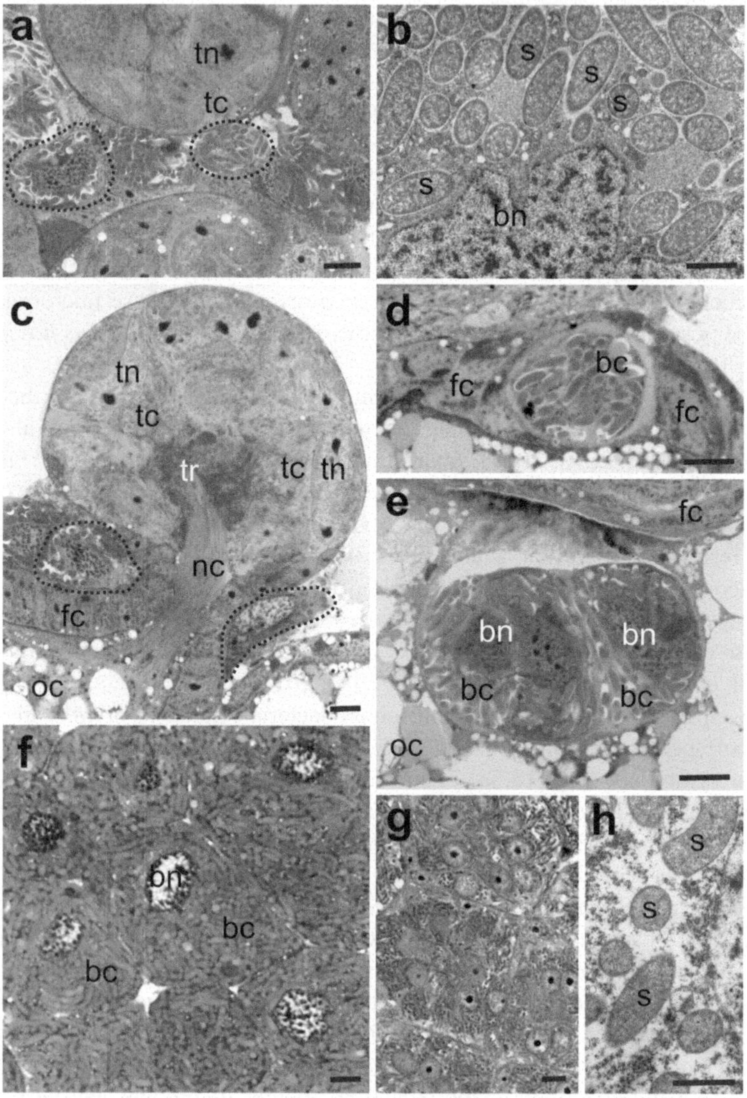

Fig. 18.5 "Bacteriocyte" symbiosis in archaeococcoid family Putoidae. (**a**) *Puto superbus*. Fragment of the abdomen of the adult female. Note bacteriocytes (*encircled with black-dotted line*), which are localized in the close neighborhood to ovaries. (**b**) *Puto superbus*. Fragment of the bacteriocyte packed with elongated bacteria *Sodalis*. (**c**) *Puto superbus*. The anterior end of the ovariole (longitudinal section) containing choriogenic oocyte. Note bacteriocytes (*encircled with black-dotted line*) which surround the neck region of the ovariole. (**d**, **e**) *Puto superbus*. Consecutive stages of infection of the ovariole. (**d**) Whole intact bacteriocyte migrates between neighboring follicular cells. (**e**) Bacteriocytes accumulate in the perivitelline space. (**f**) *Puto albicans*. Bacteriocytes filled with bacteria *Sodalis*. (**g**, **h**) *Puto albicans*. Clusters of cystocytes. Note numerous bacteria *Sodalis* in the cystocyte cytoplasm. (**a**, **c**, **d**–**g**) Methylene blue, scale bar = 20 μm, (**b**, **h**) TEM, scale bar = 2 μm. *bc* bacteriocyte, *bn* bacteriocyte nucleus, *fc* follicular cells, *nc* nutritive cord, *oc* oocyte, *s Sodalis*, *tc* trophocyte, *tn* trophocyte nucleus, and *tr* trophic core

The results of molecular phylogenetic analyses revealed that the symbionts of different species of the genus *Puto* are not closely related, which indicates that they are independently acquired by the ancestors of the extant species (Gruwell et al. 2014; Szklarzewicz et al. 2018). The different organization of symbiotic systems, with respect to the systematic affiliation of symbionts, and mode of their transmission between generations, of *Puto superbus* and scale insects from the family Pseudococccidae, strongly supports the view that Putoidae and Pseudococcidae (see below) represent phylogenetically distant groups.

In contrast to the eriococcids mentioned above, the *Acanthococcus aceris* and *Gossyparia spuria* in which the symbiotic bacteria reside in the fat body, the symbionts of the falsebrome felt scale *Greenisca brachypodii* are harbored in the bacteriocytes (Fig. 18.6a) (Michalik et al. 2018). Using microscopic and molecular techniques, Michalik et al. (2018) showed that *Greenisca brachypodii* is the host to two gammaproteobacterial symbionts: large coccoid bacterium *Kotejella greeniscae* (Fig. 18.6a, b) distantly related to *Sodalis*-like symbionts of Pseudococcinae mealybugs (see below) and elongated bacterium *Arsenophonus* (Fig. 18.6c, d). It should be stressed that neither *Kotejella* nor *Arsenophonus* were detected earlier in other scale insects. Both *Kotejella* and *Arsenophonus* inhabit their bacteriocytes (Fig. 18.6a, c). The symbionts of *Greenisca brachypodii* are transovarially transported between generations (Fig. 18.6e–g) in a way similar to the symbionts of *Acanthococcus aceris* and *Gossyparia spuria* (see Fig. 18.2c, d), i.e., through the infection of the neck of the ovariole.

Since the results of studies on symbionts of eriococcids (Gruwell et al. 2005; Michalik et al. 2016, 2018) revealed that they belong to different taxa, Michalik et al. (2018) concluded that symbioses of eriococcids are the effect of several independent infections of the ancestors of different lineages of these insects with the phylogenetically distant bacteria. Moreover, the hypothesis of the polyphyletic origin of the symbionts of eriococcids corresponds well with the current view that the Eriococcidae family does not represent a monophyletic group (Cook et al. 2002; Gullan and Cook 2007; Hodgson and Hardy 2013).

The symbiotic systems of numerous species of armored scales (Diaspididae) were studied by Gruwell et al. (2007, 2012), who showed that these insects harbor flavobacteria (named *Uzinura diaspidicola*). We observed that in the larvae and young females of *Leucaspis loewi* (unpublished data), the cytoplasm of the bacteriocytes is tightly packed with numerous pleomorphic bacteria *Uzinura* (Fig. 18.7a, b). In older females of armored scales, the bacteria leave the bacteriocytes and disperse between fat body cells (Fig. 18.7c, d). Subsequently, the bacteria invade the neck of the ovariole (Fig. 18.7e).

Bacteriocyte symbionts have also been found in Pseudococcinae (Fig. 18.8a, b) and Phenacoccinae (Fig. 18.8c, d) subfamilies of the Pseudococcidae family. Buchner (1965) reported that almost all members of both mealybug subfamilies live in the symbiotic association with the large pleomorphic bacteria. More recently, these microorganisms have been identified as the betaproteobacteria: bacterium *Tremblaya phenacola* in Phenacoccinae and *Tremblaya princeps* in Pseudococcinae (Downie and Gullan 2005; Fukatsu and Nikoh 2000; Gatehouse et al. 2011; Gil et al.

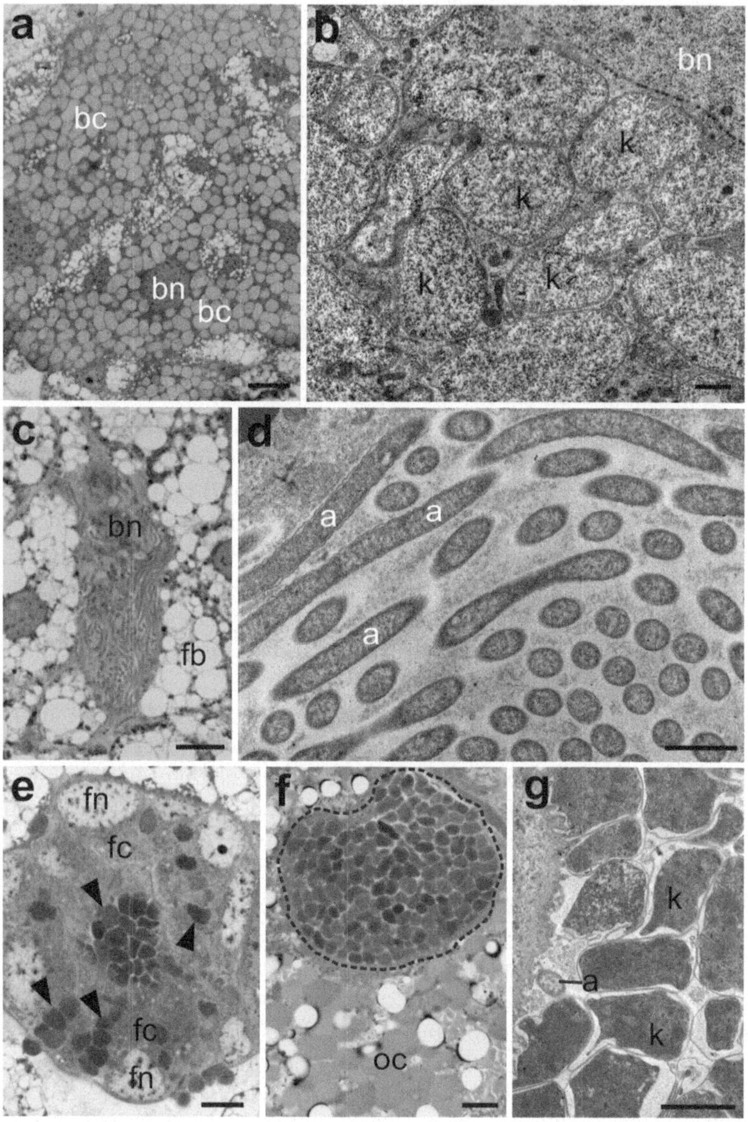

Fig. 18.6 "Bacteriocyte" symbiosis in *Greenisca brachypodii*. (**a**, **b**) Bacteriocytes with bacteria *Kotejella*. (**c**, **d**) Bacteriocytes with bacteria *Arsenophonus*. (**e–g**) Consecutive stages of infection of the ovariole. (**e**) Symbiotic bacteria (*black arrowheads*) enter the ovariole in its neck region (cross section). (**f**) Bacteria accumulate in the perivitelline space in the form of symbiont ball (*encircled with black-dotted line*). (**g**) Bacteria *Kotejella* and *Arsenophonus* in the perivitelline space. (**a**, **c**, **e**, **f**) Methylene blue, scale bar = 20 μm (**b**, **d**, **g**) TEM, scale bar = 2 μm. *a* bacterium *Arsenophonus*, *bc* bacteriocyte, *bn* bacteriocyte nucleus, *fb* fat body, *fc* follicular cells, *fn* follicular cell nucleus, *k* bacterium *Kotejella*, and *oc* oocyte

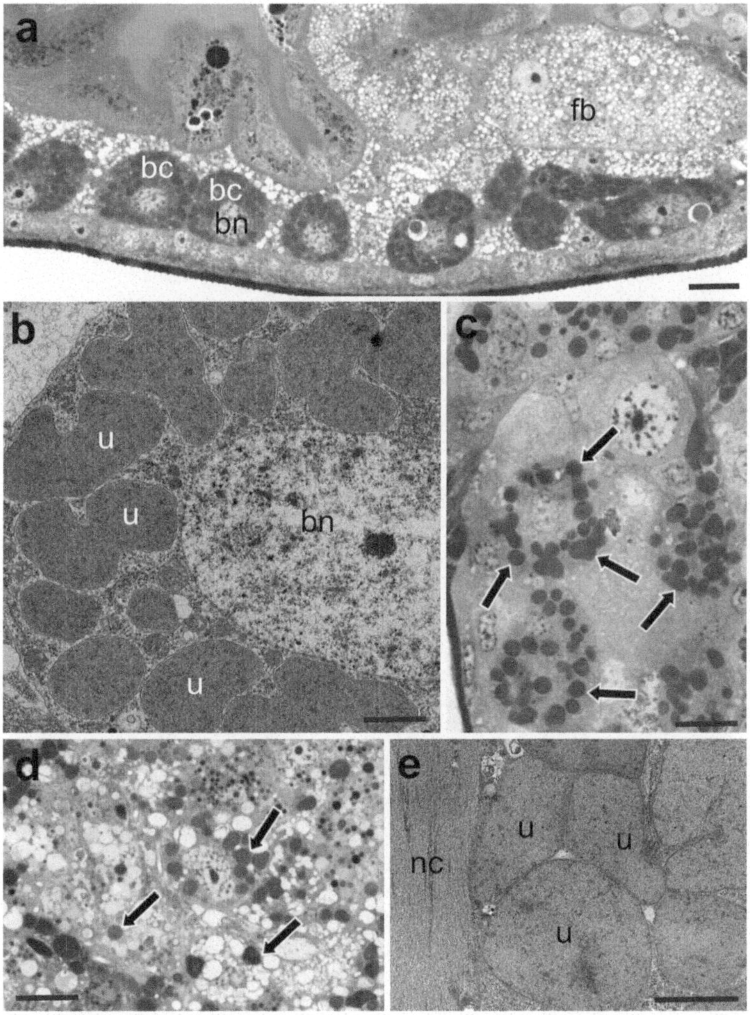

Fig. 18.7 "Bacteriocyte" symbiosis in the neococcoid family Diaspididae. (**a**) *Leucaspis loewi*. Fragment of the abdomen of the last instar larva. Note bacteriocytes tightly packed with bacteria *Uzinura*. (**b**) *Leucaspis loewi*. Bacteriocyte filled with bacteria *Uzinura*. (**c**, **d**) *Diaspidiotus ostreaeformis*. Fragment of the abdomen of the reproductive female. Bacteria *Uzinura* (*black arrows*) leave the bacteriocytes. (**e**) *Diaspidiotus ostreaeformis*. Infection of the ovariole. Bacteria *Uzinura* surround the nutritive cord. (**a**, **c**, **d**) Methylene blue, scale bar = 20 μm, (**b**, **e**) TEM, scale bar = 2 μm. *bc* bacteriocyte, *bn* bacteriocyte nucleus, *fb* fat body, *nc* nutritive cord, and *u* bacterium *Uzinura*

2018; Husnik et al. 2013; Koga et al. 2013; Kono et al. 2008; McCutcheon and von Dohlen 2011; Michalik et al. 2019a; Szabo et al. 2017; Thao et al. 2002; von Dohlen et al. 2001). There is, however, a significant difference in the organization of the symbiotic systems in these subfamilies: in Pseudococcinae all bacteria *Tremblaya*

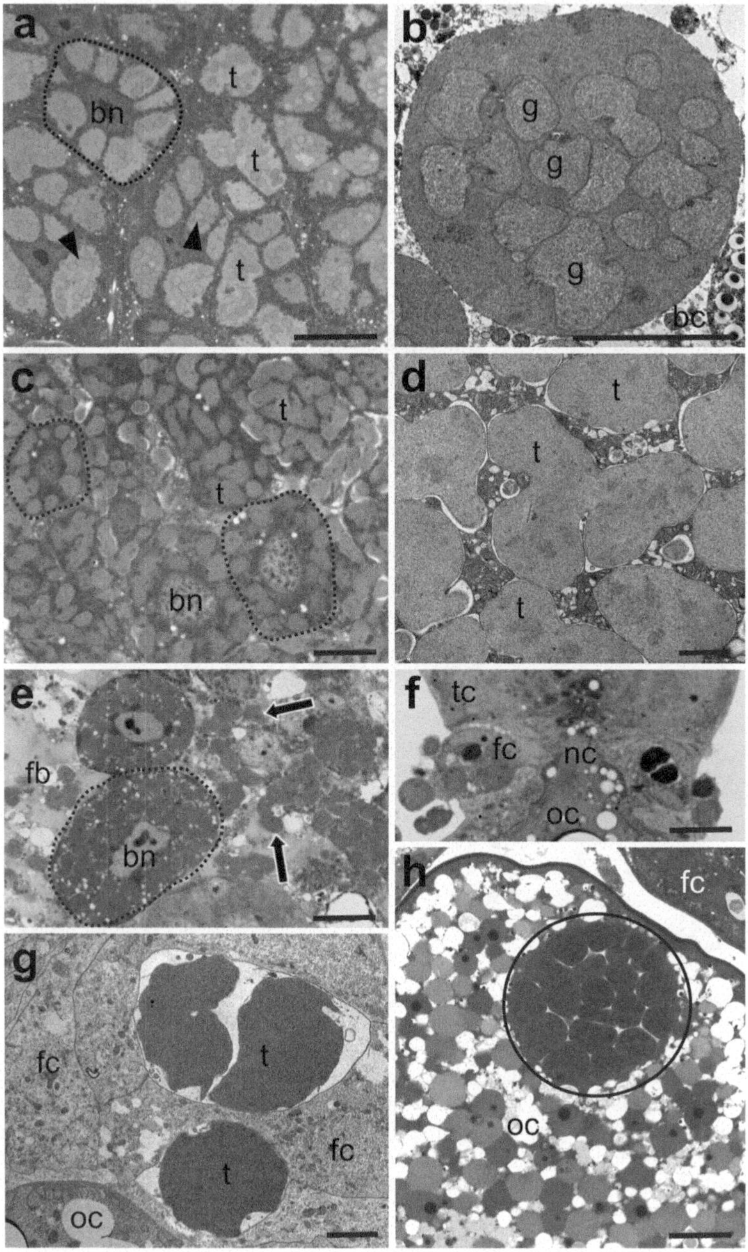

Fig. 18.8 "Bacteriocyte" symbiosis in the neococcoid family Pseudococcidae. (**a**) *Planococcus vovae* (Pseudococcinae). Fragment of the bacteriome. Note bacteriocytes (*encircled with black-dotted line*) filled with bacteria *Tremblaya princes* which contain gammaproteobacteria related to *Sodalis* (*black arrowheads*). (**b**) *Planococcus vovae* (Pseudococcinae). Bacterium *Tremblaya princes* containing numerous gammaproteobacteria. (**c**) *Coccura comari* (Phenacoccinae).

contain *Sodalis*-like bacteria (Fig. 18.8a, b), whereas in the Phenacoccinae subfamily, this phenomenon (termed a "nested symbiosis") does not occur (Fig. 18.8c, d). Bacteria *Tremblaya* were formerly considered to be the primary symbionts of mealybugs, and enterobacteria the secondary symbionts (von Dohlen et al. 2001; Thao et al. 2002), however, recent genomic analyses have shown that both these symbionts are engaged in the synthesis of essential amino acids (Husnik et al. 2013; Husnik and McCutcheon 2016; Kono et al. 2008; López-Madrigal et al. 2011, 2013, 2014; McCutcheon and von Dohlen 2011; Szabo et al. 2017). The presence of bacterium *Tremblaya* in members of both subfamilies indicates that symbioses in Pseudococcidae are the result of an ancient infection of the ancestor of the extant Pseudococcinae and Phenacoccinae before their split (Hardy et al. 2008). In contrast to the ancestral symbiont, i.e., bacterium *Tremblaya*, the enteroteobacterial symbionts accompanying this microorganism in Pseudococcinae mealybugs are of more recent and multiple origins (Fukatsu and Nikoh 2000; Gatehouse et al. 2011; Husnik and McCutcheon 2016; Kono et al. 2008; Thao et al. 2002). Husnik and McCutcheon (2016) revealed that enteroteobacterial associates of different species of Pseudococcinae mealybugs have different sized genome, which is a consequence of its reduction during symbiotic life (more recently, acquired symbionts have larger genomes than more ancestral symbionts). Based on this observation, Husnik and McCutcheon (2016) suggested that the diversity of enterobacterial symbionts in Pseudococcinae mealybugs results from the replacement of one bacterium by another. It has also been shown that the symbiotic systems of the Pseudococcinae mealybugs represent a complex genomic consortium in which: (1) the genome of the host insect contains bacterial genes, which were horizontally transferred and (2) both symbionts (i.e., *Tremblaya princeps* and enterobacteria) complement each other metabolically (Husnik et al. 2013; Husnik and McCutcheon 2016; Szabo et al. 2017).

What is of special interest is that, within the genome of bacterium *Tremblaya phenacola* of several species of Phenacoccinae mealybugs, the gammaproteobacterial DNA has been detected (Gil et al. 2018; López-Madrigal et al. 2014). This finding indicates that during the evolutionary history of

Fig. 18.8 (continued) Fragment of the bacteriome. Note bacteriocytes (*encircled with black-dotted line*) filled with bacteria *Tremblaya phenacola*. (**d**) *Phenacoccus aceris* (Phenacoccinae). Fragment of the bacteriocyte with large lobated bacteria *Tremblaya phenacola*. (**e**) *Ceroputo pilosellae* (Phenacoccinae). Bacteriocytes (*encircled with black-dotted line*) of the reproductive female. Bacteria (*black arrows*) start to leave the bacteriocytes and migrate toward ovaries. (**f**) *Ceroputo pilosellae* (Phenacoccinae). Symbiotic bacteria invade the neck region of the ovariole. (**g**) *Ceroputo pilosellae* (Phenacoccinae). Bacteria *Tremblaya phenacola* migrate through the follicular cells. (**h**) *Trionymus aberrans* (Pseudococcinae). "Symbiont ball" (*encircled with continous line*) in the deep depression of the oolemma at the anterior pole of the oocyte. (**a**, **c**, **e**, **f**, **h**) Methylene blue, scale bar = 20 μm, (**b**, **d**, **g**) TEM, scale bar = 2 μm. *bc* bacteriocyte, *bn* bacteriocyte nucleus, *fb* fat body, *fc* follicular cell, *g* gammaproteiobacteria, *nc* nutritive cord, oc oocyte, *t* bacterium *Tremblaya*, and *tc* trophocyte

Phenacoccinae mealybugs, the horizontal transmission of genes between *Tremblaya* and gammaproteobacteria occurred.

Interestingly, within Pseudococccinae mealybugs, there are two genera (*Hippeococcus* and *Rastrococcus*), which do not contain bacteria *Tremblaya*. The myrmecophilous genus *Hippeococcus* is of special interest because it is devoid of symbionts, although it possesses bacteriocytes (Buchner 1957a, 1965). According to Buchner (1957a, 1965), the loss of symbionts is associated with the fact that scale insects of the genus *Hippeococcus* live in mutualistic relationships with ants, which feed them with juice rich in nutrients and vitamins. Scale insects of the genus *Rastrococcus*, in turn, harbor bacteria of the Bacteroidetes phylum and/or fungal symbionts, which, during the evolutionary history of these insects, replaced bacteria *Tremblaya* (Buchner 1965; Gruwell et al. 2010). The bacterial associates are harbored in the bacteriocytes, whereas the fungal symbionts inhabit the fat body cells.

In both Pseudococcinae and Phenacoccinae mealybugs, symbionts are transovarially transmitted from one generation to the next (Buchner 1965; Michalik et al. 2019a; von Dohlen et al. 2001). The bacteria released from the bacteriocytes (Fig. 18.8e) migrate toward ovaries and then invade the neck region of the ovariole (Fig. 18.8f–h) in a similar mode to Eriococcidae (see Figs. 18.2c, d and 18.6e–g) and Diaspididae (see Fig. 18.7e).

Until recently, the subfamily Rhizoecinae was also included in the family Pseudococcidae, apart from Pseudococcinae and Phenacoccinae (Downie and Gullan 2004), however, Hodgson (2012), based on morphological characters, elevated the subfamily Rhizoecinae to the family status (i.e., Rhizoecidae). What is of special interest is that Gruwell et al. (2010) found that scale insects of the genus *Rhizoecus* do not harbor betaproteobacterium *Tremblaya*, but are the host to flavobacteria (termed by these authors *Brownia rhizoecola*). Thus, the results of the studies on the symbiotic system of Rhizoecidae mealybugs strongly support the classification of the mealybugs proposed by Hodgson (2012).

Interestingly, "bacteriocyte" symbiosis has also been reported in *Dactylopius coccus* (Ramirez-Puebla et al. 2010), a member of the family Dactylopiidae which, until recently, was regarded as the asymbiotic (Tremblay 1977). Ramirez-Puebla (2010) reported that the scale insects of the genus *Dactylopius* harbor the obligate bacterial associate (termed *Dactylopiibacterium carminicum*) representing Betaproteobacteria. Genomic analyses have revealed that *Dactylopiibacterium* has the potential to fix nitrogen, synthesize amino acids, and vitamins (Ramirez-Puebla et al. 2010; Vera-Ponce de Leon et al. 2017). Because the use of the FISH technique indicated the presence of *Dactylopiibacterium* in the ovaries, it seems to be probable that this microorganism is transovarially transmitted from mother to offspring.

18.4 Symbiosis with Fungal Microorganisms

Most insects live in the symbiotic associations with bacteria, however, in some insect taxa, the fungi (traditionally named yeast-like symbionts) also play a nutritional role. So far, the symbionts of the fungal origin have been discovered in members of several unrelated families of scale insects, such as the lac scales (Kerridae), gall-like scales (Kermesidae), soft scales (Coccidae), and some species of mealybugs (Pseudococcidae) (see Table 18.3) (Buchner 1957b, 1965; Gomez-Polo et al. 2017; Michalik et al. unpublished data; Podsiadło et al. 2018; Šulc 1906; Tremblay 1997; Vashishtha et al. 2011).

The molecular phylogenetic analyses, of the 18S rRNA gene of the fungal symbionts accompanying the scale insects from the family Kermesidae, Kerridae, and Coccidae, revealed that: (1) these microorganisms belong to the subphylum Pezizomycotina of the phylum Ascomycota and (2) these symbionts were independently acquired by different lineages of scale insects (Gomez-Polo et al. 2017; Podsiadło et al. 2018; Vashishtha et al. 2011).

What is of special interest is that the fungal symbionts of Kermesidae and Coccidae appeared to be closely related to the entomoparasites such as *Ophiocordyceps* or *Metarhizium* (Gomez-Polo et al. 2017; Michalik et al. unpublished data; Podsiadło et al. 2018). Thus, during the cophylogeny of scale insects and fungi, the transition from the parasitic to the symbiotic relationship between both these organisms occurred.

Microscopic observations (Buchner 1965; Gomez-Polo et al. 2017; Michalik et al. unpublished data; Podsiadło et al. 2018) revealed that the fungal symbionts may occur extracellularly in the hemolymph, and intracellularly in the cells of the fat body (Fig. 18.9a, b). The recent ultrastructural observations of Michalik et al. (unpublished data) of several species of the family Coccidae revealed that the fungal symbionts, like the bacterial symbionts from the family Eriococcidae (see Figs. 18.2c, d, and 18.6e, f), Diaspididae (see Fig. 18.7f), and Pseudococcidae (see Fig. 18.8f, g), infect the neck region of the ovariole containing choriogenic oocytes (Fig. 18.9c). These symbionts migrate to the perivitelline space via follicular cells surrounding the nutritive cord (Fig. 18.9d). After the migration of symbionts had been completed, the oocyte becomes covered by the egg envelopes (Fig. 18.9e) and the tropharium and nutritive cord degenerate (Fig. 18.9e).

18.5 Concluding Remarks and Future Perspectives

The overview presented above shows that during the evolutionary history, the scale insects acquired diverse symbionts, and developed different modes of transmission of symbiotic associates between generations. The use of molecular techniques in two past decades has brought about great progress in the studies of the symbiotic systems of insects, resulting in the determination of the systematic affiliation of symbionts

Table 18.3 Symbiosis with fungi in selected species of scale insects—types of symbionts, localization in the host insect, and mode of transmission between generations

Family	Species	Symbionts	Localization in the host insect	Mode of transmission	References
Kerridae (lac scales)	*Kerria lacca*	Fungi of the subphylum Pezimycotina (phylum Ascomycota)	Fat body cells	Not examined	Vashishtha et al. (2011)
Kermesidae (gall-like scales)	*Kermes quercus* (Fig. 18.9a)	Fungi related to *Ophiocordyceps* (phylum Ascomycota, subphylum Pezimycotina, and class Sordariomycetes)	Fat body cells	Not examined	Podsiadło et al. (2018)
Coccidae (soft scales)	*Coccus hesperidum*	Fungi related to *Ophiocordyceps* (phylum Ascomycota, subphylum *Pezimycotina*, and class Sordariomycetes)	Fat body cells	Infection of neck of ovariole of adult female	Gomez-Polo et al. (2017) and Tremblay (1997)
	Eriopeltis stammeri (Fig. 18.9c–e)	Fungi related to *Ophiocordyceps* (phylum Ascomycota, subphylum Pezimycotina, and class Sordariomycetes)	Fat body cells	Infection of the neck of ovariole of adult female	Michalik et al. (unpublished data)
Pseudococcidae (mealybugs)	*Rastrococcus fransenii*	Unidentified	Fat body cells	Infection of neck of ovariole of adult female	Buchner (1957b, 1965)

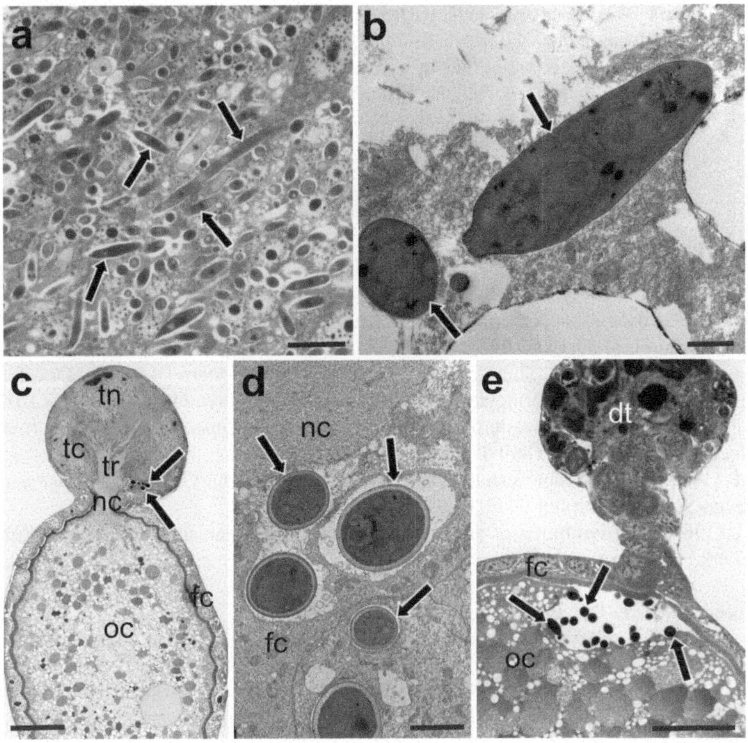

Fig. 18.9 Symbiosis with fungal microorganisms in Kermesidae and Coccidae. (**a**) *Kermes quercus* (Kermesidae). Fungal symbionts (*black arrows*) in cells of the fat body. (**b**) *Parthenolecanium pomeranicum* (Coccidae). Cells of fungal symbionts (*black arrows*) in the fat body cell. (**c**) *Eriopeltis stammeri* (Coccidae). Infection of the neck region of the ovariole (longitudinal section) through fungal symbionts (*black arrows*). (**d**) *Eriopeltis stammeri* (Coccidae). Fungal symbionts (*black arrows*) in the cytoplasm of follicular cells. (**e**) *Eriopeltis stammeri* (Coccidae). Fungal symbionts (*black arrows*) in the deep invagination of the oolemma at the anterior pole of the oocyte. Note degenerating tropharium. (**a**, **c**, **e**) Methylene blue, scale bar = 2 μm, (**b**, **d**) TEM, scale bar = 2 μm. *dt* degenerating tropharium, *fc* follicular cell, *nc* nutritive cord, *oc* oocyte, *tc* trophocyte, *tn* trophocyte nucleus, and *tr* trophic core

and the clarification of the molecular background of their function. There are, however, several families of the scale insects (e.g., Xylococcidae, Kuwaniidae, Phenacoleachiidae, and Carayonemiidae), whose symbionts remain completely unknown or have not been examined by the ultrastructural and molecular techniques. Thus, it seems reasonable to expect that the studies on the above-mentioned scale insects may add new data on the cophylogeny of scale insects and their microbiota. The future studies should also try to explain the role of the particular symbiotic associates within the systems enclosing two or more symbionts, because the situation found in the Pseudococcinae mealybugs or in the monophlebid *Llaveia axin axin* suggests that the functioning of the symbiotic systems may be more complex than previously assumed.

Acknowledgments We are greatly indebted to M.Sc. Ada Jankowska for her skilled technical assistance and Dr. Małgorzata Kalandyk-Kołodziejczyk for the provision and identification of insects. This work was mostly supported by the Iuventus Plus V grant IP2015050374 from the Ministry of Science and Higher Education and grant N18/DBS/000013 from Jagiellonian University.

References

Baumann P (2005) Biology of bacteriocyte-associated endosymbionts of plant sup-sucking insects. Annu Rev Microbiol 59:155–189

Braendle C, Miura T, Bickel R, Shingleton AW, Kambhampati S, Stern DL (2003) Developmental origin and evolution of bacteriocytes in the aphid—*Buchnera* symbiosis. PLoS Biol 1:e21

Buchner P (1957a) Endosymbiosestudien an Schildläusen. IV. *Hippeococcus*, eine myrmekophile Pseudococcine. Z Morphol Ökol Tiere 45:379–410

Buchner P (1957b) Endosymbiosestudien an Schildläusen. V. Die Gattung *Rastrococcus* Ferris (*Ceroputo* Šulc). Z Morphol Ökol Tiere 46:111–148

Buchner P (1965) Endosymbiosis of animals with plant microorganisms. Interscience Publishers, New York

Buchner P (1966) Endosymbiosestudien an Schildläusen. VIII. Die Symbiosen der Palaeococcoidea. 1. Teil. Z Morphol Ökol Tiere 56:275–362

Buchner P (1967) Endosymbiosestudien an Schildläusen. VIII. Die Symbiosen der Palaeococcoidea. 2. Teil. Z Morphol Ökol Tiere 59:211–317

Cook LG, Gullan PJ, Trueman HE (2002) A preliminary phylogeny of the scale insects (Hemiptera: Sternorrhyncha: Coccoidea) based on nuclear small-subunit ribosomal DNA. Mol Phylogenet Evol 25:43–52

Dhami MK, Turner AP, Deines P, Beggs JR, Taylor MW (2012) Ultrastructural and molecular characterization of a bacterial symbiosis in the ecologically important scale insect family Coelostomidiidae. FEMS Microbiol Ecol 81:537–546

Dhami MK, Buckley TR, Beggs JR, Taylor MW (2013) Primary symbiont of the ancient scale insect family Coelostomidiidae exhibits strict cophylogenetic patterns. Symbiosis 61:77–91

Downie DA, Gullan PJ (2004) Phylogenetic analysis of mealybugs (Hemiptera: Coccoidea: Pseudococcidae) based on DNA sequences from three nuclear genes, and a review of the higher classification. Syst Entomol 29:238–259

Downie DA, Gullan PJ (2005) Phylogenetic congruence of mealybugs and their primary endosymbionts. J Evol Biol 18:315–324

Fukatsu T, Nikoh N (2000) Endosymbiotic microbiota of the bamboo pseudococcid *Antonina crawii* (Insecta, Homoptera). Appl Environ Microbiol 66:643–650

García Morales M, Denno BD, Miller DR, Miller GL, Ben-Dov Y, Hardy NB (2016) ScaleNet: a literature-based model of scale insect biology and systematics. Database. https://doi.org/10.1093/database/bav118. http://scalenet.info

Gatehouse LN, Sutherland P, Forgie SA, Kaji R, Christeller JT (2011) Molecular and histological characterization of primary (Betaproteobacteria) and secondary (Gammaproteobacteria) endosymbionts of three mealybug species. Appl Environ Microbiol 78:1187–1197

Gil R, Vargas-Chavez C, López-Madrigal S, Santos-García D, Latorre A, Moya A (2018) *Tremblaya phenacola* PPER: an evolutionary beta-gammaproteobacterium collage. ISME J 12:124–135

Gomez-Polo P, Ballinger MJ, Lalzar M, Malik A, Ben-Dov Y, Mozes-Daube N, Perlman SJ, Iasur-Kruh L, Chief E (2017) An exceptional family: *Ophiocordyceps*-allied fungus dominates the microbiome of soft scale insects (Hemiptera: Sternorrhyncha: Coccidae). Mol Ecol 26:5855–5868

Gruwell ME, von Dohlen CD, Patch K, Normark BB (2005) Preliminary PCR survey of bacteria associated with scale insects (Hemiptera: Coccoidea). In: Erkilic L, Kaydan MB (eds) Proceedings of the X international symposium on scale insect studies. Adana, Turkey, pp 101–116

Gruwell ME, Morse GE, Normark BB (2007) Phylogenetic congruence of armored scale insects (Hemiptera: Diaspididae) and their primary endosymbionts from the phylum Bacteroidetes. Mol Phylogenet Evol 44:267–280

Gruwell ME, Hardy NB, Gullan PJ, Dittmar K (2010) Evolutionary relationships among primary endosymbionts of the mealybug subfamily Phenacoccinae (Hemiptera: Coccoidea: Pseudococcidae). Appl Environ Microbiol 76:7521–7525

Gruwell MM, Flarhety M, Dittmar K (2012) Distribution of the primary endosymbiont (*Candidatus* Uzinura diaspidicola) within host insects from the scale insect family Diaspididae. Insects 3:262–269

Gruwell ME, Duda Z, MacCready J (2014) Investigation of endosymbiotic bacteria associated with scale insects of the family Putoidae (Hemiptera: Coccoidea). Acta Zool Bulg 6:29–34

Gullan PJ, Cook LG (2007) Phylogeny and higher classification of the scale insects (Hemiptera: Sternorrhyncha: Coccoidea). Zootaxa 1668:413–425

Gullan PJ, Kosztarab M (1997) Adaptations in scale insects. Annu Rev Entomol 42:23–50

Gullan PJ, Martin JGH (2003) Sternorrhyncha (jumping plant-lice, whiteflies, aphids and scale insects). In: Cardaé RT, Resh VH (eds) Encyclopedia of insects. Academic Press, New York, pp 1079–1089

Hardy NB, Gullan PJ, Hodgson CJ (2008) A subfamily level classification of mealybugs (Hemiptera: Pseudococcidae) based on integrated molecular and morphological data. Syst Entomol 33:51–71

Hodgson CJ (2012) Comparison of the morphology of the adult males of the rhizoecine, phenacoccine and pseudococcine mealybugs (Hemiptera: Sternorrhyncha: Coccoidea), with the recognition of the family Rhizoecidae Williams. Zootaxa 3291:1–79

Hodgson CJ, Hardy NB (2013) The phylogeny of the superfamily Coccoidea (Hemiptera: Sternorrhyncha) based on the morphology of extant and extinct macropterous males. Syst Entomol 38:794–804

Husnik F, McCutcheon JP (2016) Repeated replacement of an intrabacterial symbiont in the tripartite nested mealybug symbiosis. Proc Natl Acad Sci USA 113:5416–5424

Husnik F, Nikoh R, Koga R, Ross L, Duncan RP, Fujie M, Tanaka M, Satoh N, Bachtrog D, Wilson ACC, von Dohlen CD, Fukatsu T, McCutcheon JP (2013) Horizontal gene transfer from diverse bacteria to an insect genome enables a tripartite nested mealybug symbiosis. Cell 153:1567–1578

Koga R, Nikoh N, Matsuura Y, Meng X-Y, Fukatsu T (2013) Mealybugs with distinct endosymbiotic systems living on the same host plant. FEMS Microbiol Ecol 83:93–100

Kondo T, Gullan PJ, Williams DJ (2008) Coccidology. The study of scale insects (Hemiptera: Sternorrhyncha: Coccoidea). Rev Corp Cienc Tecn Agrop 9:55–61

Kono M, Koga R, Shimada M, Fukatsu T (2008) Infection dynamics of coexisting β and γ-proteobacteria in the nested endosymbiotic system of mealybugs. Appl Environ Microbiol 74:4175–4184

Kosztarab M, Kozár F (1988) Scale insects of Central Europe. Akademiai Kiado, Budapest

Koteja J (1985) Essay on the prehistory of the scale insects (Homoptera, Coccinea). Ann Zool 38:461–503

Koteja J (1996) Scale insects (Homoptera: Coccinea) a day after. In: Schaefer CW (ed) Studies on hemipteran phylogeny. Entomol Soc Am, Lanham, MD, pp 65–88

Koteja J, Pyka-Fosciak G, Vogelgesang M, Szklarzewicz T (2003) Structure of the ovary in *Steingelia* (Sternorrhyncha: Coccinea), and its phylogenetic implications. Arthropod Struct Dev 32:247–256

Kuechler SM, Dettner K, Kehl S (2011) Characterization of an obligate intracellular bacterium in the midgut epithelium of the bulrush bug *Chilacis typhae* (Heteroptera, Lygaeidae, Artheneinae). Appl Environ Microbiol 77:2869–2876

López-Madrigal S, Latorre A, Porcar M, Moya A, Gil R (2011) Complete genome sequence of "*Candidatus* Tremblaya princeps" strain PCVAL, an intriguing translational machine below the living-cell status. J Bacteriol 193:5587–5588

López-Madrigal S, Latorre A, Porcar M, Moya A, Gil R (2013) Mealybugs nested endosymbiosis: going into the 'matryoshka' system in *Planococcus citri* in depth. BMC Microbiol 13:74

López-Madrigal S, Beltra A, Resurrection S, Soto A, Latorre A, Moya A, Gil R (2014) Molecular evidence for ongoing complementarity and horizontal gene transfer in endosymbiotic systems of mealybugs. Front Microbiol 5:449

López-Madrigal S, Latorre A, Moya A, Gil R (2015) The link between independent acquisition of intracellular gamma-endosymbionts and concerted evolution in *Tremblaya princeps*. Front Microbiol 6:642

Matsuura Y, Koga R, Nikoh N, Meng XY, Hanada S, Fukatsu T (2009) Huge symbiotic organs in giant scale insects of the genus *Drosicha* (Coccoidea: Monophlebidae) harbor flavobacterial and enterobacterial endosymbionts. Zool Sci 26:448–456

McCutcheon JP, von Dohlen CD (2011) An interdependent metabolic patchwork in the nested symbiosis of mealybugs. Curr Biol 21:1366–1372

Michalik A, Kalandyk-Kołodziejczyk M, Simon E, Kobiałka M, Szklarzewicz T (2013) Ovaries of *Puto superbus* and *Ceroputo pilosellae* (Hemiptera: Coccoidea): morphology, ultrastructure, phylogenetic and taxonomic implications. Eur J Entomol 110:527–534

Michalik K, Szklarzewicz T, Kalandyk-Kołodziejczyk M, Jankowska W, Michalik A (2016) Bacteria belonging to the genus *Burkholderia* are obligatory symbionts of the eriococcids *Acanthococcus aceris* Signoret, 1875 and *Gossyparia spuria* (Modeer, 1778) (Insecta, Hemiptera, Coccoidea). Arthropod Struct Dev 45:265–272

Michalik A, Schulz F, Michalik K, Wascher F, Horn M, Szklarzewicz T (2018) Coexistence of novel gammaproteobacterial and *Arsenophonus* symbionts in the scale insect *Greenisca brachypodii* (Hemiptera, Coccomorpha: Eriococcidae). Environ Microbiol 20:1148–1157

Michalik A, Michalik K, Grzywacz B, Kalandyk-Kołodziejczyk M, Szklarzewicz T (2019a) Molecular characterization, ultrastructure and transovarial transmission of *Tremblaya phenacola* in six mealybugs of the Phenacoccinae subfamily (Insecta, Hemiptera, Coccomorpha). Protoplasma 256:1597–1608

Michalik K, Szklarzewicz T, Kalandyk-Kołodziejczyk M, Michalik A (2019b) Bacterial associates of *Orthezia urticae*, *Matsucoccus pini* and *Steingelia gorodetskia*—scale insects of archaeoccoid families Ortheziidae, Matsucoccidae and Steingeliidae (Hemiptera, Coccomorpha). Protoplasma 256:1205–1215

Niżnik S, Szklarzewicz T (2007) Structure and development of hermaphroditic gonad in *Icerya purchasi* (Insecta, Hemiptera, Coccinea: Monophlebidae). Zool Polon 52:71–90

Podsiadło E, Michalik K, Michalik A, Szklarzewicz T (2018) Yeast-like microorganisms in the scale insect *Kermes quercus* (Insecta, Hemiptera, Coccomorpha: Kermesidae). Newly acquired symbionts? Arthropod Struct Dev 47:56–63

Ramirez-Puebla ST, Rosenblueth M, Chavez-Moreno CK, Catanho Pereira de Lyra MC, Tecante A, Martinez-Romero A (2010) Molecular phylogeny of the genus *Dactylopius* (Hemiptera: Dactylopiidae) and identification of the symbiotic bacteria. Environ Entomol 39:1178–1118

Richter G (1928) Untersuchungen an Homopterensymbionten. 1. Die Symbionten der Diaspinen und Asterolekaninen. 2. Die Symbionten einiger exotischer Zikaden. Z Morphol Ökol Tiere 10:174–206

Rosas-Perez T, Rosenblueth M, Rincon-Rosales R, Mora J, Martınez-Romero E (2014) Genome sequence of "*Candidatus* Walczuchella monophlebidarum" the flavobacterial endosymbiont of *Llaveia axin axin* (Hemiptera: Coccoidea: Monophlebidae). Genome Biol Evol 6:714–726

Rosenblueth M, Sayavedra L, Sámano-Sánchez H, Roth A, Martínez-Romero E (2012) Evolutionary relationships of flavobacterial and enterobacterial endosymbionts with their scale insect hosts (Hemiptera: Coccoidea). J Evol Biol 25:2357–2368

Rosenblueth M, Martínez-Romero J, Ramírez-Puebla ST, Vera-Ponce de León A, Rosas-Pérez T, Bustamante-Brito R, Rincón-Rosales R, Martínez-Romero E (2018) Endosymbiotic microorganisms of scale insects. Rev Espec Cienc Quím Biol 21:53–69

Sabree ZL, Huang CY, Okusu A, Moran NA, Normark BB (2013) The nutrient supplying capabilities of *Uzinura*, an endosymbiont of armoured scale insects. Environ Microbiol 15:1988–1999

Šulc K (1906) *Kermincola kermesina* n. gn. n. sp., und *Physokermina* n. sp., neue Microendosymbiotiker der Cocciden. Sitzungsber. Königl Böhm Ges Wiss Prag 19:1–6

Szabo G, Schulz F, Toenshoff ER, Volland J-M, Omri M, Finkel OM, Belkin S, Horn M (2017) Convergent patterns in the evolution of mealybug symbioses involving different intrabacterial symbionts. ISME J 11:715–726

Szklarzewicz T, Michalik A (2017) Transovarial transmission of symbionts in insects. In: Kloc M (ed) Molecular mechanisms of cell differentiation in gonad development. Springer, Cham, pp 46–67

Szklarzewicz T, Kędra K, Niżnik S (2006) Ultrastructure and transovarial transmission of endosymbiotic microorganisms in *Palaeococcus fuscipennis* (Burmeister) (Insecta, Hemiptera, Coccinea: Monophlebidae). Folia Biol (Kraków) 54:69–74

Szklarzewicz T, Michalik A, Czaja A, Szydłowska S (2010) Germ cell cluster formation and ovariole structure in *Puto albicans* and *Crypticerya morrilli* (Hemiptera: Coccinea). Phylogenetic implications. Eur J Entomol 107:589–595

Szklarzewicz T, Kalandyk-Kolodziejczyk M, Kot M, Michalik A (2013) Ovary structure and transovarial transmission of endosymbiotic microorganisms in *Marchalina hellenica* (Insecta, Hemiptera, Coccomorpha: Marchalinidae). Acta Zool (Stockholm) 94:184–192

Szklarzewicz T, Michalik A, Kalandyk-Kołodziejczyk M, Kobiałka M, Simon E (2014) Ovary of *Matsucoccus pini* (Insecta, Hemiptera, Coccinea: Matsucoccidae). Morphology, ultrastructure and phylogenetic implications. Microsc Res Tech 77:327–334

Szklarzewicz T, Kalandyk-Kołodziejczyk M, Michalik K, Jankowska W, Michalik A (2018) Symbiotic microorganisms in *Puto superbus* (Leonardi, 1907) (Insecta, Hemiptera, Coccomorpha: Putoidae). Protoplasma 255:129–138

Thao ML, Gullan PJ, Baumann P (2002) Secondary (γ-proteobacteria) endosymbionts infect the primary (β-proteobacteria) endosymbionts of mealybugs multiple times and coevolve with their host. Appl Environ Microbiol 68:3190–3197

Tremblay E (1977) Advances in endosymbiont studies in Coccoidea. VA Polytech Inst State Univ Res Div Bull 127:23–33

Tremblay E (1997) Endosymbionts. In: Ben-Dov Y, Hodgson CJ (eds) Soft scale insects—their biology, natural enemies and control. Elsevier, Amsterdam, pp 257–260

Vashishtha A, Sharama K, Lakhanpaul S (2011) Co-existence, phylogeny and putative role of *Wolbachia* and yeast-like symbiont (YLS) in *Kerria lacca* (Kerr). Curr Microbiol 63:206–212

Vera-Ponce de Leon A, Ormeno-Orrillo E, Ramırez-Puebla ST, Rosenblueth M, Degli Esposti M, Martinez-Romero J, Martinez-Romero E (2017) *Candidatus* Dactylopiibacterium carminicum, a nitrogen-fixing symbiont of *Dactylopius* cochineal insects (Hemiptera: Coccoidea: Dactylopiidae). Genome Biol Evol 9:2237–2250

von Dohlen CD, Kohler S, Alsop ST, McManus WR (2001) Mealybug β-proteobacterial endosymbionts contain γ-proteobacterial symbionts. Nature 412:433–435

Walczuch A (1932) Studien an Coccidensymbionten. Z Morphol Ökol Tiere 41:623–729

Werren JH (1997) Biology of *Wolbachia*. Annu Rev Entomol 42:587–609

Chapter 19
Bacterial Symbionts of Tsetse Flies: Relationships and Functional Interactions Between Tsetse Flies and Their Symbionts

Geoffrey M. Attardo, Francesca Scolari, and Anna Malacrida

Abstract Tsetse flies (*Glossina* spp.) act as the sole vectors of the African trypanosome species that cause Human African Trypanosomiasis (HAT or African Sleeping Sickness) and Nagana in animals. These flies have undergone a variety of specializations during their evolution including an exclusive diet consisting solely of vertebrate blood for both sexes as well as an obligate viviparous reproductive biology. Alongside these adaptations, *Glossina* species have developed intricate relationships with specific microbes ranging from mutualistic to parasitic. These relationships provide fundamental support required to sustain the specializations associated with tsetse's biology. This chapter provides an overview on the knowledge to date regarding the biology behind these relationships and focuses primarily on four bacterial species that are consistently associated with *Glossina* species. Here their interactions with the host are reviewed at the morphological, biochemical and genetic levels. This includes: the obligate symbiont *Wigglesworthia*, which is found in all tsetse species and is essential for nutritional supplementation to the blood-specific diet, immune system maturation and facilitation of viviparous reproduction; the commensal symbiont *Sodalis*, which is a frequently associated symbiont optimized for survival within the fly via nutritional adaptation, vertical transmission through mating and may alter vectorial capacity of *Glossina* for trypanosomes; the parasitic symbiont *Wolbachia*, which can manipulate *Glossina* via cytoplasmic incompatibility and shows unique interactions at the genetic level via horizontal transmission of its genetic material into the genome in two *Glossina* species; finally, knowledge on recently observed relations between *Spiroplasma* and *Glossina* is

G. M. Attardo (✉)
Department of Entomology and Nematology, University of California Davis, Davis, CA, USA
e-mail: gmattardo@ucdavis.edu

F. Scolari
Department of Biology and Biotechnology, University of Pavia, Pavia, Italy

Institute of Molecular Genetics IGM-CNR, National Research Council, Pavia, Italy

A. Malacrida
Department of Biology and Biotechnology, University of Pavia, Pavia, Italy

M. Kloc (ed.), *Symbiosis: Cellular, Molecular, Medical and Evolutionary Aspects*, Results and Problems in Cell Differentiation 69,
https://doi.org/10.1007/978-3-030-51849-3_19

explored and potential interactions are discussed based on knowledge of interactions between this bacterial Genera and other insect species. These flies have a simple microbiome relative to that of other insects. However, these relationships are deep, well-studied and provide a window into the complexity and function of host/ symbiont interactions in an important disease vector.

19.1 Introduction

Human African Trypanosomiasis (HAT, sleeping sickness) and Animal African Trypanosomiasis (AAT, Nagana) are neglected tropical diseases affecting humans and animals throughout countries in sub-Saharan Africa. These diseases are inevitably fatal if left untreated. Recently, coordinated efforts by affected countries have brought case numbers down dramatically (Franco et al. 2018; Meyer et al. 2016). However, these efforts need to be maintained to prevent future resurgences as has occurred in the past when control efforts waned. The animal form of the disease remains a large problem and has tremendous impacts on the economy and people of affected areas (Isaac et al. 2017). The trypanosomes responsible for these diseases are vectored from a reservoir to host or from host to host by tsetse flies (*Glossina spp.*). Flies within the genus *Glossina* are the sole vectors of African trypanosomes in humans and the primary vectors of trypanosomes in animals. Vector control is a primary mechanism for control of trypanosomiasis transmission (Vreysen et al. 2013). This is due to the low natural population numbers of *Glossina* in the wild. This feature differentiates tsetse flies from more prolific vectors such as mosquitoes and is due to their unique biology and unusual adaptations.

Tsetse flies are specialized vectors. They differ from other Dipteran vectors in that both sexes feed on blood alone and derive no nutrition from alternative sources such as floral nectar as mosquitoes do (Buxton 1955; Magnarelli 1978). Tsetse flies also differ in their reproductive biology as females give birth to live fully developed 3rd instar larvae (Mellanby 1937; Tobe 1978). The female fly provides for all of the nutritional requirements of the larvae by secretion of a lactation product (or milk) into the uterus where the developing larvae live (Ma and Denlinger 1974). The evolutionary pressures behind these adaptations are unknown. However, these pressures resulted in extreme adaptations, which required or were facilitated by the presence of symbiotic bacteria. Obligate mutualistic bacterial relationships with insects are well documented and often function to compensate for dietary adaptations to exploit rich but nutrient-deficient dietary sources (Buchner 1965; Douglas 1989). Examples include homopteran insects such as aphids that feed exclusively on plant sap, which lacks essential amino acids. Other insect species with wood-based diets such as carpenter ants (Hymenoptera) (Schroder et al. 1996) and termites (Isoptera) (Jucci 1952) require microbial assistance with the digestion of cellulose. Hematophagous insects subsist entirely on blood, which lacks in B vitamin-associated compounds. In addition to tsetse, examples of other blood feeders with symbiotic relationships include bed bugs (Cimicidae) (Chang and Musgrave 1973) and lice (Anoplura) (Puchta 1955).

Symbiotic relationships can range from mutualistic where the insect and the bacteria are receiving positive effects from their relationship, to parasitic where the bacteria are exploiting the host resulting in negative impacts on fitness and/or reproduction and some relationships are commensal where the bacteria benefit at no apparent cost to the insect host. To date, research shows that tsetse flies have consistent relationships with a few bacterial species. In general, tsetse flies have a relatively simple microbiome consisting primarily of four characterized species. These include *Wigglesworthia glossinidia*, *Sodalis glossinidius*, *Wolbachia pipientis*, and *Spiroplasma* (Fig. 19.1). The presence or absence of all but one of

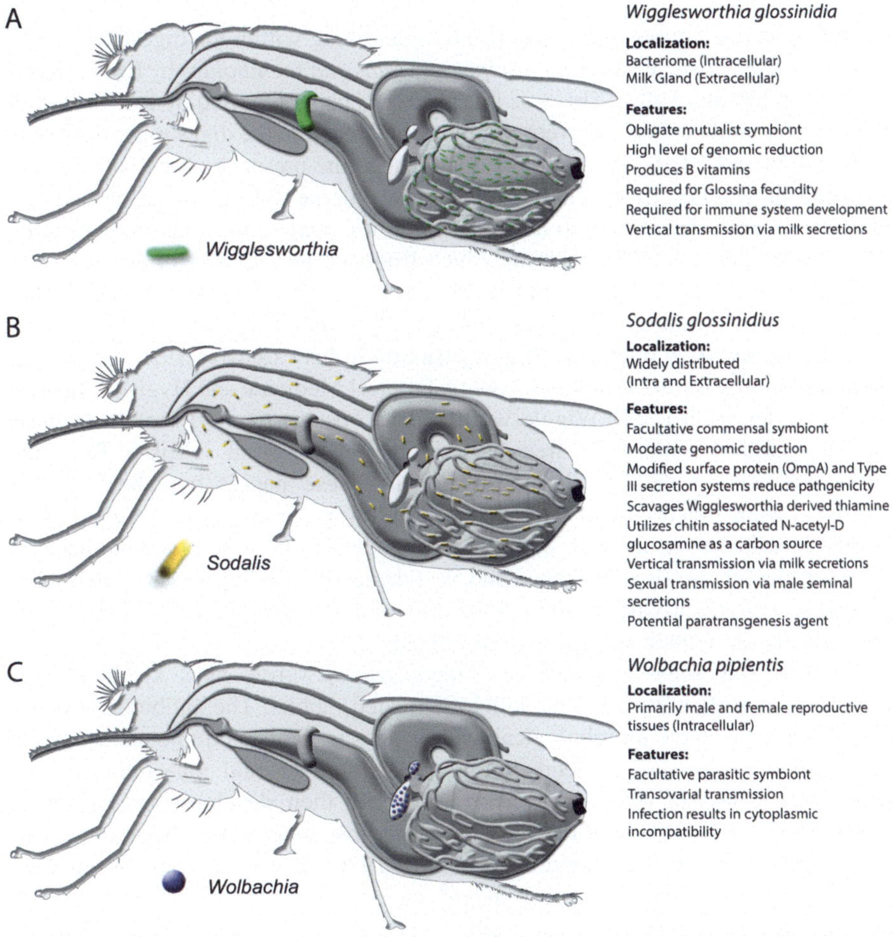

Fig. 19.1 Diagrammatic presentation of the microbiota in *Glossina*. The tissue localization and main features are reported for (**a**) *Wigglesworthia glossinidia*, (**b**) *Sodalis glossinidius*, and (**c**) *Wolbachia pipientis*

these bacterial species is variable. The only bacterial species consistently found in all tsetse fly species is *Wigglesworthia glossinidia,* which functions as an obligate mutualist. The interactions between vectors, their microbiota, and the parasites they vector are intricate. This chapter will explore the biology behind these relationships and the implications for tsetse physiology, vector control, and vectorial capacity.

19.2 *Wigglesworthia glossinidia*

19.2.1 Discovery

The first documentations of bacteria (bacteroids) living within the digestive tract of tsetse flies were made in *Glossina fuscipes* and *Glossina tachinoides* by Robert Koch in 1907 (Stuhlmann 1907). This was followed by a paper reported by Roubaud who hypothesized that the bacteria were symbiotic and aided in the digestion of blood by the fly (Roubaud 1919). These papers documented the presence of the bacteria and observed they were living intracellularly in a tissue (at the time termed the mycetome, currently referred to as the bacteriome), made up of giant hypertrophic cells. These cells contained bacteria, which Roubaud hypothesized that they were released during blood feeding to aid in digestion. The symbionts were further characterized in detail by Wigglesworth in 1929 in his analysis of the physiology of the tsetse digestive tract as well as by Buxton in his extensive 1955 memoir on *Glossina* biology for the London School of Tropical Medicine and Hygiene (Buxton 1955; Wigglesworth 1929). Wigglesworth did not observe free forms of the bacteria and suspected the observed free bacteria may have been released from the bacteriome due to damage during dissection. These papers hypothesized that *Wigglesworthia* functioned to supplement or aid in the digestion of tsetse's blood-specific diet; however, the lack of antibiotics made it difficult to target the bacteria for experimental purposes. Roubaud and Wigglesworth also hypothesized that these bacteria were vertically transmitted to the intrauterine larvae via the milk secretions generated by the female accessory (milk) glands.

The first functional evidence of *Wigglesworthia's* role as a symbiont was described when flies were fed on antibiotic-treated rabbits. The antibiotic fed flies were observed to have morphological disturbances in the bacteriome and a lack of bacteria. These flies showed decreased blood-feeding rates and appeared incapable of developing larvae, which were aborted early in their development (Hill et al. 1973). This observation was repeated by Nogge by direct antibiotic treatment of tsetse flies via an artificial blood meal (Nogge 1976). It was also shown that flies feeding on rabbits inoculated with *Wigglesworthia* become sterile. The antisera against *Wigglesworthia* from the rabbits are specifically bound and killed the bacteria in the bacteriome of flies (Nogge 1978). Nogge also noted that the loss of fecundity in tsetse in which endosymbionts had been removed (aposymbiotic) could be partially complemented by supplementation of blood meals with B-vitamins

(Nogge 1976). While the loss of symbionts has a negative impact on fecundity, it does not seem to influence lifespan (Nogge and Gerresheim 1982).

19.2.2 Localization and Transmission

The localization of *Wigglesworthia* in *Glossina* is specific to two locations. As mentioned previously, they are found living intracellularly within the giant bacteriocyte cells of the bacteriome tissue. The bacteriome is located in the anterior section of the *Glossina* midgut and from the exterior appears as a horseshoe-shaped structure that encircles the dorsal side of the midgut. The bacteriocyte giant cells protrude into the anterior midgut from the basal lamina of the gut and occupy a significant portion of the gut lumen (Wigglesworth 1929). Ultrastructural analysis of the bacteriome by transmission electron microscopy revealed many new observations about the relationship between tsetse and *Wigglesworthia*. Visualization of the bacteriome revealed that *Wigglesworthia* was gram-negative based on their membrane composition and rod-shaped up to 8 microns in length and 1–1.4 microns in width. The analysis also identified the presence of secretory vacuoles within *Wigglesworthia* connected to the plasma membrane (Reinhardt et al. 1972).

Analysis of genes enriched in the bacteriome of *Glossina morsitans* revealed tissue-specific gene expression patterns indicative of the specialization of these cells. The bacteriocytes are enriched in the expression of vesicular transport/exocytosis-related genes, sodium/potassium pumps, a lectin, and the peptidoglycan recognition protein LB (PGRP-LB) (Bing et al. 2017). The lectin is a carbohydrate-binding protein that in some systems is associated with symbiont uptake and localization (Bulgheresi et al. 2006; Chaston and Goodrich-Blair 2010; Kita et al. 2015; Wang et al. 2010). The ion pumps and secretion-associated genes suggest that the bacteriocytes are maintaining an ion gradient that may assist in the exocytosis and secretion of intracellular compounds and may function to export *Wigglesworthia*-derived cofactors. The presence of PGRP-LB is thought to have an immunosuppressive function by binding and cleaving free *Wigglesworthia*-derived peptidoglycan (Wang et al. 2009).

Initial hypotheses by Wigglesworth suggested that *Wigglesworthia* were transmitted vertically from mother to offspring via the glandular "milk" secretions produced by the female accessory gland (milk gland) that flow into the uterus and are imbibed by the developing larvae. Ultrastructural analysis of the milk gland tubules by Denlinger and Ma revealed the presence of bacteria living within the lumen of the gland and often located in close association with the openings of the secretory reservoirs from which milk is released (Ma and Denlinger 1974). The milk gland-associated bacteria showed differences from those in the bacteriome. The bacteria in the milk gland were extracellular, had a thicker cell wall, and were coated in filamentous fimbriae 5–7 nm in diameter and 2 μm in length. These bacteria were later confirmed as *Wigglesworthia* by *in situ* staining of milk gland tissues with a specific 16S ribosomal RNA probe (Attardo et al. 2008; Balmand et al. 2013). The morphological differences between the bacteriocyte and milk gland-derived *Wigglesworthia* appear to reflect differential morphological states required for intra and extracellular life within *Glossina*. Larval *Glossina* are exposed to *Wigglesworthia* only via their gut contents. During the larval stage, the gut is a closed system with larvae lacking an anal opening. A possible mechanism is if bacteria ingested during larvigenesis end up being partitioned, with some invading the cells of the bacteriome and others making their way to the milk gland possibly during metamorphosis. However, the route and mechanism of this colonization process remains an open question.

19.2.3 Phylogeny

Genetic analysis of *Wigglesworthia* by 16S ribosomal DNA analysis revealed that they fall into a discrete lineage close to members of the *Enterobacteriaceae* family within the class *γ Proteobacteria*. The *Enterobacteriaceae* are facultative anaerobes and include many species that function as gastrointestinal pathogens including *Escherichia coli* (*E. coli*) O157:H7, *Salmonella*, and *Shigella*. The *γ Proteobacteria* also contain a lineage of obligate symbionts of aphids in the genus *Buchnera* (Munson et al. 1991; Unterman et al. 1989). Analysis of *Wigglesworthia* derived from five species of *Glossina* representing three subgenera of tsetse flies revealed that these bacteria also form a specific lineage within the *γ*-3 subdivision of the *Proteobacteria*. Phylogenetic comparison of *Wigglesworthia* by 16S rDNA sequencing revealed them as close relatives of *Buchnera*, symbionts identified in other insects and *E. coli* (Aksoy 1995). The phylogeny of *Wigglesworthia* from the different species of *Glossina* closely matched the phylogeny of the flies themselves. This suggests the evolution of *Wigglesworthia* is parallel to that of *Glossina* reflecting an ancient relationship derived from a common ancestor (Aksoy et al. 1995; Chen et al. 1999; Symula et al. 2011). This relationship is estimated to have begun in an ancestral tsetse 50–100 million years ago.

19.2.4 Genetics, Functional Conservation, and Gene Loss

Sequencing of the *Wigglesworthia* genome revealed multiple insights into its biology and functional role as an obligate symbiont. The sequence was obtained from *Wigglesworthia* derived from the tsetse species *Glossina brevipalpis*. The genome is contained in a single chromosome and a plasmid with a total size of 697,742 base pairs containing 621 protein-coding sequences with 89.1% of the genome containing coding regions (Akman et al. 2002). The size of the genome is very small relative to most free-living bacteria and is similar in size to that of the obligate symbiont of the Pea aphid *Buchnera aphidicola*. The small size of these genomes likely reflects the symbiotic nature of these bacteria as they exist in a protected state within the bacteriome tissues of the host (Bennett and Moran 2013; Moran and Bennett 2014). The protective environment and the presence of host-derived nutrients allow for relaxed evolutionary selection on functions required for free-living and competitive environments. In addition, obligate endosymbionts also tend to undergo rapid evolution due to the potential bottlenecks between generations during vertical transmission. To date, attempts to culture *Wigglesworthia* have failed, suggesting that these reductions have compromised their ability to survive outside the parameters of the host environment. *Wigglesworthia's* genome also has a very low guanine/cytosine content of 22%, which is also similar to that of other intracellular bacteria. This is thought to result from the loss of genes coding for repair and recombination enzymes associated with the SOS, base excision, and nucleotide excision repair systems. However, genes with high expression levels such as ribosomal proteins and chaperonins have a bias toward the use of GC-rich amino acid codons relative to the rest of the genome (Herbeck et al. 2003). In addition, *Wigglesworthia* lacks the *DnaA* and *OriC* genes, which are essential components of the chromosomal replication complex, suggesting they are accomplishing this by an alternative method. Another interesting observation is that *Wigglesworthia* also lacks the gene coding for phosphofructokinase (*PfkA*), which is required for energy production via glycolysis and genes required for amino acid biosynthesis (Zientz et al. 2004).

The genes retained by *Wigglesworthia* are informative as to the selective pressures and unique environmental requirements within *Glossina*. These illustrate the biological demands required for maintenance of the vitality and fecundity of the host. While *Wigglesworthia* lacks a key enzyme for glycolysis, it has retained the transketolase and transaldolase enzymes utilized by the nonoxidative branch of the pentose phosphate pathway. This pathway may be facilitating the oxidation of abundant blood meal-derived amino acids for energy. In addition, they have retained the fructose bisphosphatase (*fbp*) gene, which facilitates the synthesis of complex carbohydrates by gluconeogenesis as well as all the enzymatic components required for lipid, phospholipid, and nucleotide biosynthesis (Zientz et al. 2004).

The *Wigglesworthia* genome has maintained genes associated with biosynthetic pathways required for the synthesis of B-vitamins and cofactors such as pantothenate (B5), biotin (B7), riboflavin (B2), folate (B9), thiamine (B1), nicotinamide (B3), and pyridoxine (B6) (Akman et al. 2002). These compounds are deficient in blood and the retention of these biosynthetic pathways appears to be associated with nutritional supplementation of the tsetse host. This observation reinforces the original hypotheses of Roubaud, Wigglesworth, and Nogge. In addition, *Wigglesworthia* maintains all the genes required for the assembly and function of flagella. The function of the flagella in *Wigglesworthia* is unknown. The observation of fimbriae associated with extracellular milk gland *Wigglesworthia* suggests that the fimbriae are flagella and may be involved in motility and cellular invasion of the bacteriome during vertical transfer. Expression of the flagella-associated genes *motA* and *fliC* is specific to maternal milk glands, larvae and early pupal stages of development and their expression are not observed in bacteriome-associated *Wigglesworthia* (Rio et al. 2012).

High-throughput analysis of bacteriome-associated *Wigglesworthia* gene expression revealed enriched expression of functional classes of genes by the symbiont (Bing et al. 2017). The most abundant genes expressed by *Wigglesworthia* code for chaperonins, which aid in protein folding, as well as proteins associated with the degradation of misfolded proteins. These are hypothesized to compensate for the AT-rich nature of *Wigglesworthia's* genome and the absence of rigorous DNA repair systems lost over evolutionary time. The second most highly expressed class of *Wigglesworthia* genes was associated with B-vitamin biosynthesis. Genes associated with thiamine (B1) biosynthesis were most highly expressed, followed by biotin (B7), riboflavin (B2), pantothenate and CoA (B5), nicotinamide (B3), pyridoxine (B6), and folate (B9) (Bing et al. 2017).

19.2.5 Roles in *Glossina* Digestion and Metabolism

The relationship between *Glossina* and *Wigglesworthia* is complex and is essential to multiple aspects of tsetse fly biology including digestion, metabolism, reproduction, and immunity. In the absence of *Wigglesworthia*, female *Glossina* becomes unable to develop intrauterine larval offspring, however, aposymbiotic males do not appear to suffer significant impacts to their fertility (Hill et al. 1973; Nogge 1976; Pais et al. 2008). Females in this state develop and ovulate oocytes. After fertilization, the oocytes appear to undergo embryogenesis, however, females abort their developing larvae early in intrauterine development. This suggests that either the larvae are dying due to malnutrition or that something is initiating premature parturition. Malnutrition could result from the lack of a *Wigglesworthia*-derived compound or the female's inability to effectively transfer nutrients via the milk in the aposymbiotic state. The retention of genes required for B-vitamin compound biosynthesis in the context of *Wigglesworthia's* extreme genomic reduction highlights their significance within the context of the host and their role in dietary supplementation.

Research on the function of these compounds in *Glossina* biology has revealed important insights into this relationship. Dietary supplementation of *Glossina* blood meals with thiamine results in decreased *Wigglesworthia* population density and reduced expression of the *Wigglesworthia thiC* gene. This suggests that *Wigglesworthia* can sense environmental thiamine levels and regulate the expression of its biosynthetic pathway accordingly. Ectopic treatment with thiamine also caused changes in *Wigglesworthia* density suggesting that there is a potential population-regulatory mechanism either via the *Glossina* immune system or a regulatory mechanism within *Wigglesworthia* (Snyder et al. 2012).

In *Glossina*, the amino acid proline is the primary source of ATP production via the tricarboxylic acid (TCA) cycle. The proline is catabolized to alanine to produce ATP. The alanine is then shuttled back to lipid storage tissues where it is restored to proline by the addition of lipid-derived acetyl-CoA. This differs from many other insects that utilize carbohydrates as their primary source of ATP (Bursell 1960,

1963, 1966). A key enzyme required for the conversion of alanine back to proline is alanine-glyoxylate aminotransferase (AGAT). This enzyme requires vitamin B6 (pyridoxal 5′-phosphate) as a cofactor to function. Aposymbiotic flies have significantly lower levels of vitamin B6 as well as lower levels of free proline in their hemolymph. The lack of a key energy-associated metabolite during pregnancy is likely a significant impediment during the energetically demanding process of lactation in females (Michalkova et al. 2014).

The production of folate (Vitamin B9) by *Wigglesworthia* is important for fitness and reproductive function in *Glossina*. Folate functions as a cofactor in many pathways/processes including DNA/RNA synthesis, repair, methylation, and production of the amino acid methionine. *Wigglesworthia* upregulates genes associated with the chorismite and folate biosynthesis in young and pregnant female flies relative to males and virgin females. This suggests that *Wigglesworthia* responds to the requirements of the host. In addition, *Glossina* expresses a folate transporter protein in proportion to the folate level in the bacteriome (Snyder and Rio 2015). There are significant molecular and biochemical interactions occurring at the interface of the symbiont and host that maintain the equilibrium of the system. Female flies fed on glyphosate, an inhibitor of the chorismate and folate biosynthetic pathways, show a number of pathologies associated with folate deprivation. Offspring from folate-deficient mothers had longer larval and pupal development times, weighed less, and had a smaller adult body size (Snyder and Rio 2015).

Comparison of tetracycline-treated aposymbiotic female flies with age-matched pregnant females by untargeted metabolomic analyses revealed dysfunction in multiple metabolic pathways. As predicted by previous work, aposymbiotic females show deficiencies in B-vitamins and associated compounds resulting from the loss of *Wigglesworthia* (Bing et al. 2017). Many pathways with altered metabolite profiles are those with enzymes dependent on B vitamins as enzymatic cofactors. Metabolism of glycogen appears disrupted as aposymbiotic flies show increased levels of unprocessed glycogen metabolites relative to symbiotic flies. This is likely a result of the deficiency in vitamin B6, which is required as a cofactor by the enzyme glycogen phosphorylase. Another disruption appears in a pathway downstream of the glycogen pathway, the pentose phosphate pathway. This pathway processes glucose-6-phosphate derived from glycogen catabolism and converts it into NADPH (utilized in reduction reactions and fatty acid biosynthesis) and 5 carbon sugars (pentoses), which are required for the synthesis of aromatic amino acids and nucleotides. A key enzyme in this pathway is transketolase, which is dependent on vitamin B1 (thiamine). The disruption of this pathway results in a severe deficiency in the metabolite phosphoribosyl pyrophosphate (PRPP). This compound is an essential precursor to nucleotide biosynthesis and aposymbiotic flies appear to be impacted by this as they show significant deficiencies in purine and pyrimidine nucleotide biosynthesis (Bing et al. 2017). The deficiencies in nucleotide biosynthesis in combination with the B-vitamin deficiencies have further impacts downstream in the methionine metabolism pathway. This pathway is responsible for the production of S-adenosylmethionine (SAM), which functions as a universal methyl donor in methylation reactions. Levels of SAM in aposymbiotic flies relative to controls were the second-lowest among all the metabolites identified after PRPP. The biosynthesis of SAM requires adenosine, folate, and vitamin B6 all of which are deficient in aposymbiotic tsetse.

The exact cause behind why aposymbiotic female tsetse is unable to develop intrauterine larvae remains undetermined. However, the observed deficiencies indicate that lipid metabolism could be impacted. The reduced levels of NADPH could negatively impact fatty acid biosynthesis. In addition, the reduced levels of SAM could be impacting the synthesis of phospholipids, which play an important role in lipid storage, metabolism, and mobilization.

19.2.6 Roles in Immunity and Development

The interactions between *Glossina* and *Wigglesworthia* are ancient and complex affecting many aspects of the system. The role of *Glossina* species as vectors of trypanosomes makes the topic of immunity of particular interest. The obligate presence of symbionts requires fine tuning of immune responses such that potential pathogens are selectively targeted to avoid damage of symbiotic populations. Experimental treatment of female flies with the antibiotic ampicillin resulted in selective elimination of *Wigglesworthia* from the milk gland. However, the intracellular population in the bacteriome remained intact due to the inability of the antibiotic to penetrate the bacteriocytes (Pais et al. 2008). This resulted in females that maintain their fecundity; however, they do not pass on *Wigglesworthia* to their offspring. This finding allowed the study of the developmental impacts on tsetse physiology/biology associated with the loss of *Wigglesworthia.*

Glossina that develop in the absence of *Wigglesworthia* shows phenotypic effects that highlight the impacts of development in the absence of their symbionts. Aposymbiotic flies derived from ampicillin-treated mothers show a significant decrease in survival over time relative to symbiotic flies. This phenotype is exacerbated at higher temperatures. These flies also show deficiencies in digestion manifested as a reduced rate of blood meal digestion and the abundant presence of undigested hemoglobin in the gut 2 days after blood feeding. Flies lacking *Wigglesworthia* also appear more susceptible to infection by trypanosomes (Pais et al. 2008).

Analysis of immune gene expression associated with the presence or absence of *Wigglesworthia* showed that the peptidoglycan recognition protein PGRP-LB is upregulated in the bacteriome and milk gland tissues in the presence of *Wigglesworthia* and is downregulated in aposymbiotic flies (Dawadi et al. 2018; Wang and Aksoy 2012). PGRP-LB was first shown to moderate the immune system function in *Drosophila* through degradation of bacterially derived peptidoglycan and buffering activation of the Immune Deficient (IMD) immune pathway (Zaidman-Remy et al. 2006). Flies in which PGRP-LB is knocked down show higher levels of IMD-dependent antimicrobial gene expression. In *Glossina*, this protein is thought to regulate symbiont population numbers and protect the symbionts from the immune system in the bacteriome and the milk gland. A similar role for PGRP-LB in mediating the symbiont/host relationship is also documented in the weevil (*Sitophilus zeamais*) (Anselme et al. 2006). Progeny of female PGRP-LB knockdown flies shows lower densities of *Wigglesworthia* suggesting that this is an important mechanism that ensures safe transfer of the symbiont from mother to offspring (Wang and Aksoy 2012).

Aposymbiotic *Glossina* show higher levels of IMD pathway-associated innate immune gene expression. However, if larvae develop in the absence of symbionts, they display an immunocompromised phenotype resulting from incomplete cellular immune cell development. This manifests as an inability to melanize and clot cuticular wounds and lack of mature hemocytes in their hemolymph (Weiss et al. 2011). Adult flies with this phenotype are very susceptible to hemocelic infections by *E. coli* relative to wild flies. Aposymbiotic *Glossina* appear to be deficient in plasmatocyte and crystal cell type hemocytes. These types of cells are responsible for phagocytosis of foreign bodies (plasmatocytes) and secretion of chemical components required for melanization (crystal cells). The hemocyte deficiency results from an inability of hemocyte precursor cells to differentiate into active immune cells. These findings suggest that *Wigglesworthia* provides stimuli required for proper hemocyte and immune system maturation (Weiss et al. 2011). Deeper analysis of this phenotype revealed that the presence of *Wigglesworthia* in developing larvae stimulates the expression of a gene coding for an odorant-binding protein, *obp6*. Knockdown of *obp6* expression in larval tsetse inhibited differentiation of precursor hemocytes into crystal cells specifically. The lack of crystal cells results in an inability of flies to melanize cuticular wounds or invading pathogens in a manner similar to aposymbiotic flies (Benoit et al. 2017). OBPs are thought to function as transporters for small hydrophobic molecules typically associated with olfaction (Zhou 2010). In this case, OBP6 is hypothesized to mediate the activity of a *Wigglesworthia*-derived compound required for crystal cell hemocyte differentiation. This finding expands the functional role of OBPs (previously thought to primarily function in olfaction) into the realm of development and immunity.

19.2.7 Summary

The relationship between *Wigglesworthia* and *Glossina* is intricate and essential for the survival of both organisms. The dependence of *Glossina* development, immunity, metabolic function, and reproduction on *Wigglesworthia* highlights the intricacy of this partnership. While this partnership is likely unique in some respects, the dependence of other blood-feeding insects suggests that there may be similarities in the relationships between obligate blood feeders and their associated obligate symbiotic relationships.

19.3 *Sodalis glossinidius*

19.3.1 Discovery and Genetic Characterization

Sodalis was first described in 1987 as a Rickettsia-like-organism isolated from the hemolymph of *Glossina* by culturing it in a cell line from the Asian Tiger mosquito *Aedes albopictus* (Welburn et al. 1987). The bacterium was classified as a member of the family *Enterobacteriaceae* within the γ-3 subdivision of the Proteobacteria. This family contains multiple species found as symbionts in other insects (Aksoy et al. 1997; Hosokawa et al. 2015; Novakova et al. 2015). The bacterium was cultured and characterized outside of the fly under microaerobic conditions using a solid-phase culture technique. A new genus, *Sodalis*, was derived from this work to contain secondary symbionts of other tsetse and insect species (Dale and Maudlin 1999). Initial analyses of the *Sodalis* genome by pulse field gel electrophoresis determined it to be ~2 Megabases (Mb) (Akman et al. 2001); however, follow up efforts with more advanced sequencing technologies revealed the genome to be 4.1 Mb (Toh et al. 2006). *Sodalis* also carries multiple extrachromosomal plasmids (Akman et al. 2001). The extrachromosomal material contains genes coding for pilus proteins required for bacterial conjugation and horizontal gene transfer as well as siderophores required for the binding and transport of iron (Darby et al. 2005).

Comparison of the genome relative to *E. coli* (another member of the *Enterobacteriaceae*) via microarray revealed that the two organisms are ~85% orthologous in terms of gene composition. The genome contains three regions encoding Type III secretion systems (SSR-1, SSR-2, and SSR-3), which appear to have independent ancestries and differences in their constitution (Toh et al. 2006). The SSR-1 region is similar in composition to the Type III secretion system *ysa* in the bacteria *Yersinia enterolitica*, while SSR-2+3 bear similarity to SPI-1+2 in *Salmonella*. Many pseudogenes were found in the *Sodalis* genome relative to that of free-living *Salmonella typhi* and *Yersinia pestis*. Genes lacking functional orthologs in *Sodalis* are associated with anaerobic metabolism and carbohydrate transport/metabolism. In addition, there are reductions in genes coding for membrane-associated proteins as well as those involved in cell structure (Rio et al. 2003). Further analysis of the *Sodalis* genome revealed the loss of the arginine biosynthesis pathway in *Sodalis* suggesting that it scavenges this amino acid from *Glossina* (Belda et al. 2010). Another feature of the *Sodalis* genome was the loss of the pathway for synthesis of the B-vitamin thiamine, yet they have retained a thiamine transporter gene. *Wigglesworthia* is capable of thiamine production suggesting that *Sodalis* may scavange *Wigglesworthia*-derived thiamine (Snyder et al. 2010). While *Sodalis* does not show the same level of genomic reduction as *Wigglesworthia*, the loss of these genes suggests it could be at an intermediate stage in the transition from a free-living facultative relationship to that of a symbiont.

An updated annotation of the *Sodalis* genome and development of a new culture system utilizing media with defined ingredients facilitated in-depth analysis of the nutritional and growth requirements of *Sodalis in vitro*. This work determined that

Sodalis does encode an arginine biosynthesis system and is not completely auxotrophic. Growth on media lacking arginine is reduced, but not eliminated. Addition of excess L-glutamate to this system rescued the reduced growth rate associated with arginine depletion suggesting that *Sodalis* can compensate for low environmental arginine with L-glutamate and that this compound functions as an important source of carbon and nitrogen. Another interesting finding is that the carbohydrate N-acetyl-D-glucosamine, a component of insect chitin, is an important dietary factor for *Sodalis*. This suggests that *Sodalis* may have adapted to utilize an abundant carbohydrate associated with the *Glossina* physiological environment (Hall et al. 2019).

19.3.2 Biology, Localization, and Transmission

As opposed to *Wigglesworthia*, the *Sodalis's* range is not limited to specific tissues and are found throughout the fly. During intrauterine larval development, *Sodalis* migrates into the developing larvae via the milk secretions. Upon eclosion from the pupa, a significant increase in the numbers of *Sodalis* relative to *Glossina* host cells is observed over a two-week period. The numbers of *Sodalis* then appear to fluctuate over time in adults (Rio et al. 2006). *Sodalis* is found intra and extracellularly in tissues throughout larval and adult *Glossina* including the midgut, fat body, milk gland, uterus, and oviduct. The ovaries and developing oocytes remain uninfected (Attardo et al. 2008; Balmand et al. 2013). *Sodalis* are also capable of transfer from infected males to females via seminal secretions during mating (De Vooght et al. 2015). Work by Dale and Welburn revealed that treatment of *Glossina* with the antibiotic streptozotocin selectively eliminates *Sodalis* while leaving bacteriome-based W*igglesworthia* intact. This treatment did not impact *Glossina* fecundity, as observed with the elimination of *Wigglesworthia*. However, flies lacking *Sodalis* showed a significant reduction in lifespan and increased susceptibility to trypanosome infection in laboratory settings (Dale and Welburn 2001).

Investigations into the role of Type III secretion systems in the relationship between *Sodalis* and *Glossina* revealed differences relative to orthologous loci in free-living and parasitic bacterial species. These systems are often associated with pathogenicity due to their role in the secretion of toxins and inflammatory agents as well as invasion of host cells. Disruption of this system by transposon-mediated mutagenesis of the *invasion protein C* (*invC*) gene (a component of the type III secretion system) prevented an invasion of insect cells cultured *in vitro* (Dale et al. 2001). SSR-2 seems to have lost gene functionality for the proteins coding for the needle structure of the secretion system. Functional analysis of these pathways suggests that SSR-1 is required for cell invasion while SSR-2 is required for intracellular division (Dale et al. 2005).

Iron acquisition is essential for bacterial survival and proliferation and *Sodalis* has retained protein-coding genes required for iron chelation, transmembrane transport, heme metabolism, and iron storage. These include an inner membrane heme ABC transporter system (*hemTUV*), an outer membrane heme transporter (*hemR*), an iron/

manganese transporter (*sitABCD*), ferritin-like proteins, and an iron-responsive negative transcriptional regulator (*Fur*). Analysis of the regulatory regions from two of these *Sodalis* genes in *E. coli* revealed that the Fur regulatory protein is required for the correct expression of these genes. Under iron-rich environmental conditions, these genes are repressed; however, in the absence of the Fur regulator, they are constitutively expressed. In *Glossina*, these genes appear to be responsive to environmental iron as in unfed teneral flies these genes are upregulated. However, at 48 hours post blood meal, the genes are significantly downregulated (Runyen-Janecky et al. 2010; Smith et al. 2013). The *hemR* outer membrane heme transporter is essential for the survival of *Sodalis* in *Glossina* as strains with mutations in this protein do not establish. The *tonB* gene codes for the protein that supplies energy to *hemR* and is also essential to this system. Strains with mutations in *tonB* show a similar phenotype to that of *hemR* mutants in that they are unable to colonize *Glossina* in its absence (Hrusa et al. 2015).

Analysis of quorum-sensing mechanisms in *Sodalis* revealed that the compound N-(3-oxohexanoyl) homoserine lactone regulates population numbers. The presence of this compound activates the transcription of genes coding for oxidative stress-response proteins that may function to reduce the oxidative burden associated with symbiosis (Pontes et al. 2008). The changes in these systems relative to free-living bacteria may reflect a reduction in mechanisms associated with pathogenesis and adaptations required for invasion by and survival of *Sodalis* in *Glossina* tissues. Transfer of *Sodalis* strains between *Glossina* species revealed that *Sodalis* originating in one species of *Glossina* are capable of colonization and survival in another species indicating that these adaptations are not species-specific. Rather, they allow *Sodalis* to be compatible across the *Glossina* genus (Weiss et al. 2006). *Sodalis* has a truncated lipopolysaccharide (LPS) structure that lacks the O-antigen, which may facilitate its ability to live within *Glossina* with the induction of an immune response (Toh et al. 2006). In depth analysis of outer membrane proteins revealed that the *OmpA* (Outer membrane protein A) gene in *Sodalis* contains polymorphisms not found in other pathogenic bacteria such as *E. coli*. The infection of *Glossina* with *E. coli* is fatal under normal conditions. However, infection with *E. coli* containing an *OmpA* mutation is nonpathogenic while infection with *Sodalis* containing the native *E. coli OmpA* gene results in a lethal infection (Weiss et al. 2008). In *Sodalis*, the *OmpA* surface protein is essential for the establishment of gut infections within *Glossina*. Mutation of the native *Sodalis OmpA* gene prevents biofilm formation that is required for the protection of bacteria from the flies' immune response (Maltz et al. 2012).

19.3.3 Distribution in Wild Populations and Relationship with Vector Competence

Analysis of wild populations of flies revealed that the presence of *Sodalis* is heterogeneous in the field. The field-collected samples of *Glossina austeni* and

Glossina pallidipes from Kenya and South Africa reveal *Sodalis* infection rates to be around 3.7% in *G. austeni* and 16% in *G. pallidipes* (Wamwiri et al. 2013). Analysis of *Glossina* populations from Luambe National Park in Zambia also revealed high levels of variability in the proportions of individuals infected with *Sodalis*. The species with the highest infection level was *Glossina brevipalpis* with 93.7% infection rate followed by *Glossina morsitans* and *G. pallidipes* with 17.5% and 1.4%, respectively (Dennis et al. 2014). llumina sequencing-based analyses of *Glossina* microbiota of fly populations in Uganda have revealed a broader diversity of microbial taxa than previously identified across multiple species of *Glossina* including *Glossina fuscipes fuscipes* (from five distinct populations), *Glossina morsitans morsitans,* and *Glossina pallidipes*. All samples were predominantly occupied by *Wigglesworthia* with it constituting the majority of bacterial sequences. A comparison of the profiles of the remaining bacterial taxa revealed the microbiomes from the different *G. fuscipes* populations showed significant diversity between their microbial constitution. The survey also identified a high prevalence of low-intensity *Sodalis* infections across all the groups tested (Aksoy et al. 2014). Another high-throughput microbiome analysis of *Glossina palpalis palpalis* from Cameroon revealed similar results with *Wigglesworthia* being the predominant species in the flies and the investigators also found low-level infections of *Sodalis* throughout the samples (Tsagmo Ngoune et al. 2019).

Comparison of the *Trypanosoma* infection rates in *Sodalis* infected versus uninfected field collected flies identified a significant positive correlation between the two suggesting that the presence of *Sodalis* may facilitate the establishment of blood meal-derived *Trypanosoma* infections (Farikou et al. 2010; Soumana et al. 2013; Wamwiri et al. 2013). Work on changes in *Sodalis* gene expression in permissive versus nonpermissive *Glossina pallidipes* revealed significant changes in expression profiles between the two groups. A large proportion of the changes associated with the refractory flies is associated with a viral prophage carried by *Sodalis*, which suggests that activation of this phage may be associated with an antitrypanosomal response (Hamidou Soumana et al. 2014a, b). However, recent work studying the correlation between *Sodalis* and *Trypanosoma* coinfections in wild-caught flies suggests that other factors such as geographic location, trypanosomal species, *Glossina* species, and the age and sex of the flies analyzed can be confounding factors (Channumsin et al. 2018).

Selective clearance of *Sodalis* using the antibiotic streptozotocin allowed researchers to generate a *Sodalis*-free line of *Glossina* (Sod-) in the lab for comparison with infected individuals (Sod+) by high-throughput gene expression analysis. An interesting finding from this research is that there were no significant changes in immune-responsive genes between Sod- and Sod+ flies. However, the challenge of these flies with *E. coli* or *Sodalis praecaptivus*, a free-living relative of *S. glossinidius*, resulted in a robust immune response. They also showed that activation of *Glossina*'s innate immune response did not have a significant impact on endogenous *Sodalis* numbers suggesting that the bacterium is resistant to host immune factors. Comparisons of susceptibility to *Trypanosoma brucei brucei* infection between Sod- and Sod+ flies showed no significant difference between the two groups.

19.3.4 Potential for Use in Paratransgenesis

The ability to isolate and culture *Sodalis* outside of *Glossina* provides the opportunity for a reduction of *Glossina*'s vectorial capacity via manipulation of its symbionts. This strategy is called paratransgenesis and is a promising alternative to direct genetic manipulation of the vector species (Coutinho-Abreu et al. 2010). Current genetic transformation technologies, functional in insects such as mosquitoes and *Drosophila*, are not a viable option in *Glossina* due to their low reproductive rate and viviparous physiology. Paratransgenesis works through the genetic transformation of cultured *Sodalis* with a gene encoding an antitrypanosomal factor. The engineered *Sodalis* are then reintroduced into *Glossina* where, in principle, the presence of the modified symbiont would create a hostile environment for invading trypanosomes resulting in reduced or eliminated vectorial capacity. Paratransgenesis was demonstrated in laboratory studies to be an effective strategy in kissing bugs (Order Hemiptera, Family Triatomidae) as a way to reduce their vectorial capacity for the Chagas pathogen, *Trypanosoma cruzi* (Beard et al. 2001, 2002).

Research into the use of *Sodalis* as a paratransgenic agent shows promise. Transformation of *Sodalis* with 9 different cationic antimicrobial peptides toxic to African trypanosomes revealed that *Sodalis* is resistant to 7 of them. This finding opened the door for the utilization of *Sodalis* as a paratrangenesis agent for the delivery of antitrypanosomal compounds in vivo (Haines et al. 2003). A study demonstrated that *Sodalis* transformed with a gene for a trypanolytic nanobody can secrete this factor and that they are capable of invading tissues throughout the fly. The effective establishment of this infection was dependent on the prior treatment of the flies with the antibiotic streptozotocin to deplete native *Sodalis* numbers and reduce competition. These bacteria express the nanobody throughout the tissues of the fly; however, vertical transmission of these bacteria only occurred at a low level (De Vooght et al. 2014). Follow up work shows that the establishment of a stable infection and reliable vertical transmission with engineered bacteria is dependent upon the route of introduction. Flies given intrathoracic injections with recombinant *Sodalis* as adults were able to establish infections, but those infections are not vertically transmitted to offspring. However, the investigators found that injection of larval *Glossina* facilitated disseminated bacterial infection as well as vertical transmission to the resulting offspring (De Vooght et al. 2018).

Sterile Insect Technique (SIT) is another important strategy for population control of *Glossina*. An issue associated with SIT in *Glossina* is that radiation-sterilized males are still capable of functioning as vectors, which is an ethical issue associated with this approach. An alternative approach that minimizes this risk is the release of sterile males infected with engineered *Sodalis* to reduce or eliminate their vectorial capacity. Sterilization of paratransgenic males with gamma radiation revealed that while the treatment caused an initial decline in the number of recombinant *Sodalis*, the bacterial numbers recovered over time making this a potentially practical approach (Demirbas-Uzel et al. 2018).

19.3.5 Summary

The relationship between *Glossina* and *Sodalis* is very different from that of *Glossina* and *Wigglesworthia. Sodalis* appears to be in an evolutionary transition between a free-living organism and a symbiont. Its genomic reductions are not at the level of that observed in *Wigglesworthia*; however, *Sodalis* has accumulated a significant number of pseudogenes associated with functions required for a free-living existence. Analysis of retained features suggests the development of specializations for life within the host. These include alterations to its outer membrane proteins and secretory systems to reduce its pathogenicity and immunogenicity. *Sodalis* has also made nutritional adaptations that optimize it for survival in the fly including the use of N-acetyl-D-glucosamine as a carbon source and the retention of a transporter to scavenge *Wigglesworthia*-derived thiamine. *Glossina* does not require the presence of *Sodalis* for essential functions such as reproduction. However, selective elimination of the bacteria does have a negative impact on *Glossina* lifespan suggesting a beneficial aspect to the relationship. The understanding of the relationship between *Sodalis* and the different *Trypanosoma* types remains uncertain given the conflicting results from field and lab-based studies. The ability to culture and genetically manipulate *Sodalis* makes it an ideal agent for use in paratransgenesis studies. Research on this method of control is ongoing and could be a valuable tool for integration into SIT strategies or on its own to reduce the vectorial capacity of wild *Glossina* populations.

19.4 *Wolbachia*

19.4.1 Discovery

Wolbachia is a genus of obligate intracellular gram-negative bacteria belonging to the Order Rickettsiales and were first identified from *Culex pipiens* in 1924 (Hertig and Wolbach 1924). In 1936, this finding was confirmed and described in detail (Hertig 1936). Since these early discoveries, research has shown that *Wolbachia* is widespread in arthropods (reviewed in (Serbus et al. 2008)), and is probably the most prevalent endosymbiont found in insect germlines. It is found in every insect order (Harris and Braig 2003), and is estimated to infect >65% insect species (de Oliveira et al. 2015; Hilgenboecker et al. 2008).

19.4.2 Basic Biology and Features

The widespread presence of *Wolbachia* is thought to be due to its efficient transmission and its capacity to manipulate host reproduction to favor infected females. *Wolbachia* bacteria are indeed able to colonize female germline cells, through which they are transovarially transmitted to the progeny. *Wolbachia* interspecific horizontal transmission was initially described as a rare phenomenon (O'Neill et al. 1992; Rousset et al. 1992; Turelli et al. 2018; Werren et al. 1995), but studies are increasingly showing that, in a variety of insect species, *Wolbachia* genes have been horizontally transferred to host chromosomes (Aikawa et al. 2009; Dunning Hotopp et al. 2007; Fenn et al. 2006; Klasson et al. 2009; Kondo et al. 2002; Nikoh and Nakabachi 2009; Nikoh et al. 2008; Woolfit et al. 2009).

In different arthropod hosts, *Wolbachia* infection is responsible for several mechanisms that enhance female fertility. These reproductive alterations induce different host phenotypes such as cytoplasmic incompatibility, the feminization of genetically male offspring, male-killing of infected males, and parthenogenesis by infected females (Harris and Braig 2003; Stouthamer et al. 1999; Tram et al. 2003).

Cytoplasmic Incompatibility (CI), the most prevalent and most widely investigated phenomenon, was first observed in the 1970s (Yen and Barr 1973) and results in embryonic mortality in the progeny derived from matings between insects with different *Wolbachia* infection status (Bourtzis et al. 1998; Clark et al. 2003). Unidirectional CI occurs when an infected male mates with an uninfected female; whereas the reciprocal crossing is compatible. Bidirectional CI occurs in crossings between individuals infected with different *Wolbachia* strains (Werren 1997). Embryo lethality has been related to the modifications induced by *Wolbachia* to the paternal chromosomes during spermatogenesis in a way that mitotic synchrony is lost (O'Neill and Karr 1990; Tram and Sullivan 2002). The genetic basis of CI remained unknown for a long time, and only recently LePage and colleagues identified two genes in the eukaryotic association module of prophage WO (Bordenstein and Bordenstein 2016) from *Wolbachia* strain *wMel* that acts as CI factors (Beckmann et al. 2017; LePage et al. 2017). These studies revealed that the mitotic defect is due to a deubiquitinating enzyme encoded by one of the two genes (*i.e.* CidB/CifB) and neutralized by the CidA/CifA product.

In different species, *Wolbachia* has also been proposed to play a key role in sex determination by enhancing female germline development (Cordaux and Gilbert 2017; Kageyama et al. 2017; Kageyama and Traut 2004; Sugimoto et al. 2015). For example, in the parasitoid wasp *Asobara tabida*, *Wolbachia* is essential for oogenesis as its elimination induces apoptosis in the ovaries thus impeding egg maturation (Dedeine et al. 2001; Pannebakker et al. 2007). The role of *Wolbachia* in supporting female germline development in *Drosophila melanogaster* began to be clarified when females carrying mutant alleles of the master gene in female sex determination, *sex-lethal*, rescued their fertility when infected with *Wolbachia* (Starr and Cline 2002). Following these findings, more recent work showed that *Wolbachia* interacts with RNAs encoding proteins involved in the support of germline stem cell maintenance and oocyte polarization (Ote and Yamamoto 2020). These data further support the idea that this bacterium is a genetic manipulator of the infected arthropod hosts (Kozek and Rao 2007).

Wolbachia-mediated manipulations were shown in recent work to affect several other reproduction-related functions. For example, in *Drosophila*, *Wolbachia* affects gene transcription in larval testes (Zheng et al. 2011) as well as the expression of seminal fluid proteins (Yuan et al. 2015). Moreover, *Wolbachia* was shown to impact the expression of immunity genes in a parasitoid wasp (Kremer et al. 2012). The functions *Wolbachia* exerts in mosquitoes are achieved through manipulations of host microRNAs and the production of small RNAs regulating host gene expression (Hussain et al. 2011; Mayoral et al. 2014).

Recent studies expanded knowledge on other effects exerted by *Wolbachia* on the behavior of its hosts. These include influences on sleep, learning, memory, feeding, mating, locomotion, and aggression [for a review see (Bi and Wang 2019)]. For example, it has been suggested that *Wolbachia* is able to affect sleep in *Drosophila* by interactions with the juvenile hormone/sex-determination genes/dopamine pathway. In particular, this bacterium appears to contribute to increased sleep time in order to favor the conservation of resources and energies to support reproductive outputs, thus supporting both the host and its transmission and indicating that the coevolution between *Wolbachia* and its hosts is even more multifaced than so far discovered (Bi and Wang 2019).

19.4.3 Localization in Insect Tissues

Wolbachia primarily resides in the germline tissues of both male and female insects (Dobson et al. 1999); in males, this bacterium is found in the spermatocytes but there is no transmission through the sperm (Bressac and Rousset 1993; Clark et al. 2002; Ijichi et al. 2002). In addition, since its early detection, this bacterium was described to be present also in somatic tissues [(Dobson et al. 1999; Hertig and Wolbach 1924) also see (Pietri et al. 2016) for a review].

In the germline tissue, some *Wolbachia* strains localize at the posterior end of mature oocytes (Ote and Yamamoto 2020) thanks to the presence of RNAs and proteins transported from the nurse cells along microtubules to form a pole plasm (Serbus and Sullivan 2007). Pole cells, together with somatic gonadal cells, form the embryonic gonads, becoming primordial germ cells. During the development of the female pupa, the primordial germ cells initiate their divisions resulting in the germ cell lineage in the ovaries of adult females. *Wolbachia* is found in the germline stem cells, which are localized at the anterior end of each ovariole within the ovaries (Ote et al. 2016; Serbus et al. 2008). In particular, *Wolbachia* tends to be located close to processing bodies in the cytoplasm of these cells that supply maternal factors to the oocyte (Ferree et al. 2005; Franks and Lykke-Andersen 2008; Ote et al. 2016; Serbus et al. 2011), resulting in an association that takes place from oogenesis to embryogenesis. In particular, *Wolbachia* produces a protein able to interact with RNAs encoding proteins that are involved in the support of germline stem cells and oocyte polarization (Ote and Yamamoto 2020).

As far as it concerns *Wolbachia* distribution in somatic tissues, this bacterium is known to be particularly abundant in the nervous system in *Drosophila* and several other insect species (Albertson et al. 2013; Casper-Lindley et al. 2011; Dobson et al. 1999; Mitsuhashi et al. 2002; Moreira et al. 2009; Osborne et al. 2009; Strunov and Kiseleva 2016), as well as in the fat body, gut, salivary glands, hemocytes, and Malphighian tubules, where it has been suggested to play roles related to the host immunity and metabolic regulation (see (Pietri et al. 2016) for a review). The age of the host also impacts *Wolbachia* infection levels and tropism, further suggesting that the physiological relationships between *Wolbachia* and its hosts are extremely complex and require a case-to-case analysis (Binnington and Hoffmann 1989; Bressac and Rousset 1993; Min and Benzer 1997).

19.4.4 **Wolbachia *in* Glossina spp.**

Early hybridization experiments conducted between different species belonging to the *Glossina* genus suggested the presence of incompatibilities resulting in females with reduced fecundity and sterile males. Both bidirectional and unidirectional incompatibility events were reported ((Curtis 1972; Rawlings 1985; Vanderplank 1948) see also (Gooding 1985, 1987, 1989, 1990) for a review). These data, together with light and electron microscopy studies showing gram-negative rods in tsetse ovaries and in the periphery of the yolk of early embryos (Huebner and Davey 1974; Pell and Southern 1975; Pinnock and Hess 1974) and 16s rRNA sequence analyses (Beard et al. 1993), prompted researchers to further investigate the identity of these bacteria. The exploration of 16S rRNA phylogenetic relationships and tissue distribution in tsetse resulted in the identification of *Wolbachia* in laboratory strains of *Glossina* species in 1993 (O'Neill et al. 1993). Initially, *Wolbachia* was detected in *G. m. morsitans* and *G. m. centralis*, while it was found to be absent in *G. p. palpalis* and *G. p. gambiensis*, supporting some of the reported events of reproductive incompatibilities in the genus (O'Neill et al. 1993).

19.4.5 Tissue Localization in Tsetse Tissues

The tissue tropism of *Wolbachia* to the ovarian tissues suggested that transovarial transmission is the primary mode of transmission for this bacterium (O'Neill et al. 1993), which supports previous studies (Pell and Southern 1975). Subsequent work on *G. m. morsitans* laboratory flies confirmed *Wolbachia*'s absence in the milk gland secretions, as well as any other somatic tissues surrounding the uterus, further supporting its transovarial transmission (Balmand et al., 2013). *Wolbachia* was not detected extracellularly, and it was shown to infect only the trophocytes and the oocytes in the ovaries, as well as embryos and larvae (Cheng et al. 2000; Balmand et al. 2013).

However, while in certain tsetse species *Wolbachia* appeared to be restricted to the reproductive tissues, in *G. austeni,* this bacterium was detected also in somatic tissues, in particular the head, salivary glands, milk glands, and fat body (Cheng et al. 2000). These findings stimulate novel studies aimed at understanding whether these differences are due to features of different *Wolbachia* strains infecting the different tsetse species or to insect-specific factors, such as immunity-regulatory mechanisms, controlling the infections. The application of a particularly effective hybridization approach (*i.e.* high-end Stellaris® RNA-FISH) surprisingly revealed that *Wolbachia* is also present in the lumen and secretory cells of the milk glands in *G. m. morsitans* (Schneider et al. 2018). This result opens up new pathways of investigation of a potentially yet undiscovered vertical transmission mechanism for this bacterium.

19.4.6 **Wolbachia** *Role in Tsetse Physiology*

Earlier studies described the existence of bidirectional CI between certain *G. morsitans* subspecies and between *G. palpalis* subspecies, while unidirectional CI has been described only in crossings of certain *G. morsitans* subspecies (Cheng et al. 2000).

The first study showing the functional role of *Wolbachia* in tsetse reared in the laboratory was performed by Alam and colleagues, who showed that the infection of this bacterium is able to support the expression of cytoplasmic incompatibility (Alam et al. 2011). The females expressing CI displayed loss of fecundity due to early embryogenic failure. Currently, there is no way to selectively cure *Wolbachia* infection through antibiotic administration. Aposymbiotic flies lack all endosymbionts, including obligate bacteria responsible for obligate nutritional dialogues, in the absence of which tsetse flies are sterile. Alam and coauthors set up a method to maintain *Wolbachia*-cured *G. m. morsitans* (symbiont-free, Gmm^{Apo}) fertile through dietary provisioning of blood meals supplemented with tetracycline and yeast extract, thus rescuing tsetse fecundity, which is tightly dependent on *Wigglesworthia*. Moreover, cytoplasmic incompatibility in tsetse is particularly strong, differently from many other species that are characterized by incomplete CI (Sinkins and Gould 2006).

Whether *Wolbachia* plays a role in trypanosome infection is still under debate. Alam and colleagues reported the presence of a negative association between *Wolbachia* and trypanosome infections in *G. f. fuscipes*, suggesting that this bacterium could prevent trypanosome infections (Alam et al. 2012). However, the tripartite association between tsetse, trypanosomes, and *Wolbachia* remains unclear. In *G. p. palpalis*, *Wolbachia* infection appeared to have no impact on the establishment of trypanosomes (Kante et al. 2018). This is similar to what was found in *G. tachinoides* and *G. m. submorsitans* (Kame-Ngasse et al. 2018). To shed light on this relevant biological aspect, more extensive data on trypanosome and *Wolbachia* infections, as well as the other symbionts, in different tsetse species and populations are required. Moreover, the potential relationships between specific *Wolbachia* haplotypes and trypanosome infections will be essential to determine the presence and features of this dialogue.

19.4.7 Wolbachia *Distribution in Tsetse Strains and Populations*

Since its discovery, a number of studies have focused on expanding the analysis of the distribution of *Wolbachia* in tsetse species, in both laboratory strains and wild samples. The first extensive work aimed at understanding the presence and infection rates of *Wolbachia* in tsetse was performed by Cheng and colleagues, who analyzed the status of *Wolbachia* infections in laboratory colonies of *G. brevipalpis* and *G. longipinnis* (Fusca group), *G. fuscipes*, *G. tachinoides*, *G. p. palpalis,* and *G. p. gambiensis* (Palpalis group), as well as *G. m. morsitans*, *G. m. centralis*, *G. swynnertoni,* and *G. pallidipes* (Morsitans group) (Cheng et al. 2000). All individuals analyzed from the Morsitans and Fusca groups were positive for *Wolbachia* infection, while none of the flies belonging to the Palpalis group harbored *Wolbachia*, mirroring earlier findings (O'Neill et al. 1993). A 100% prevalence in laboratory strains was confirmed by a more recent study for *G. m. morsitans* and *G. m. centralis*. Similarly, the absence of *Wolbachia* was confirmed in laboratory strains of *G. f. fuscipes* and *G. tachinoides*, whereas in *G. pallidipes*, different colonies displayed different prevalences of *Wolbachia* ranging from the absence (Seibersdorf lab-colony) to low prevalence (KARI-TRC lab-colony, 3%) (Doudoumis et al. 2012). Also in this study, flies from *G. p. palpalis* and *G. p. gambiensis* laboratory colonies showed the absence of *Wolbachia* infection. Differences were reported for *G. brevipalpis*, where prevalence was not complete (41.2%) (Doudoumis et al. 2012).

Significant differences in *Wolbachia* infection frequencies are found between field populations and laboratory strains. In addition, differences are observed within wild populations, which may be dependent on ecological conditions (Mouton et al. 2007; Yun et al. 2011). For example, in the case of *G. m. morsitans*, the *Wolbachia* presence in wild populations ranges from 9.5 to 100% (Doudoumis et al. 2012). In *G. brevipalpis* sampled in South Africa *Wolbachia* appears absent, while, in Kenyan flies, infection levels are reported to be about 30% (Cheng et al. 2000). In *G. austeni*, infection rates ranged from 48 to 98%, in Kenya and South Africa, respectively (Cheng et al. 2000). More recent work also detected variations in *Wolbachia* prevalence in *G. austeni* samples in these two countries, but with the Kenyan population showing higher infection levels (Wamwiri et al. 2013). Low *Wolbachia* prevalence was detected in populations of *G. pallidipes* (below 8.5%) and in *G. gambiensis* (below 8.3%), and the absence of *Wolbachia* infection in *G. p. palpalis* and *G. f. fuscipes* populations was confirmed (Doudoumis et al. 2012). Alam and colleagues conversely showed the presence of *Wolbachia* infections in

G. f. fuscipes from Uganda, although at low density, which may have influenced its detection in previous studies (Alam et al. 2012). A similar result was obtained by Schneider and colleagues, who used sensitive PCR-based methods that allowed the identification of *Wolbachia* in *G. f. fuscipes* (Schneider et al. 2013, 2018). Recent studies identified *Wolbachia* also in *G. p. palpalis* (Kante et al. 2018), as well as in *G. tachinoides* (68.1% prevalence) and in *G. m. submorsitans* (58.5%) (Kame-Ngasse et al. 2018), further supporting the idea that the choices of molecular markers and detection methods play a key role in the sensitivity of determining the infection rates of *Wolbachia*. Indeed, when present at low titers, *Wolbachia* infections can be detected only through the integration of different tools, such as high-sensitivity blot-PCR combined with hybridizations (Schneider et al. 2018).

Finally, in *G. f. quanzensis* (Palpalis group) from Congo, the analysis of midguts revealed that 85% of the analyzed samples were infected by *Wolbachia*, with infection rates varying according to sampling sites (Simo et al. 2019). Moreover, a low number of midguts were naturally coinfected by both *Wolbachia* and *Sodalis*, opening new questions on the potential interactions between these two tsetse symbionts.

19.4.8 Wolbachia *Integrations in Tsetse Genomes*

The genome of *G. m. morsitans* contains large segments of *Wolbachia* that were integrated via horizontal gene transfer (HGT) events. These integrated fragments contain a high degree of nucleotide polymorphisms, as well as insertions and deletions (Brelsfoard et al. 2014).

Subsequent comparative analysis of the genomes of six *Glossina* species, namely *G. morsitans morsitans*, *G. pallidipes*, *G. austeni* (Morsitans group), *G. palpalis* and *G. fuscipes* (Palpalis group), and *G. brevipalpis* (Fusca group), showed that all contain sequences with homology to *Wolbachia*, although the features of these integrations are different. Indeed, in *G. pallidipes*, *G. fuscipes*, *G. palpalis*, and *G. brevipalpis*, the homologous sequences consisted in short fragments and they were initially thought to be artifacts as PCR assays with *Wolbachia*-specific primers on these strains as well as on wild populations of these species resulted in negative results (Doudoumis et al. 2013). However, recent studies suggest that natural *G. p. palpalis* populations from Cameroon do carry *Wolbachia* symbionts (Kante et al. 2018).

However, *G. austeni* contains more extensive *Wolbachia*-derived chromosomal integrations. Both chromosomal and cytoplasmic *Wolbachia* sequences found in *G. austeni* were mapped against the reference *Wolbachia* genomes (*i.e.* *w*Mel and *w*Gmm), as well as the A and B chromosomal insertions found in *G. m. morsitans* (Attardo et al. 2019). The *Wolbachia* insertions in *G. austeni* range in size from 500 to 95,673 bps and display high-sequence homology not only to *w*Mel and *w*Gmm, but also to *G. m. morsitans* A and B insertions. The higher homology (98%) with A and B insertions from *G. m. morsitans* relative to cytoplasmic

Wolbachia sequences suggests they could be derived from an event in a common ancestor, although the absence of comparable insertions in *G. pallidipes* (a closer relative to *G. m. morsitans*) requires further investigations on wild samples from *Glossina* species/subspecies to clarify the true origin of these events.

Whether these *Wolbachia* chromosomal integrations have functional roles in *Glossina* biology is still elusive. Indeed, gene expression analyses of *Wolbachia* insertions in *G. morsitans* found very limited evidence of expression (Brelsfoard et al. 2014), suggesting they may be accidental transfer events associated with the long-term symbiosis between *Wolbachia* and these tsetse species. However, since *Wolbachia* integrations in tsetse chromosomes include genes encoding proteins carrying ankyrin repeat domains, thought to be directly related to *Wolbachia*-host interactions, further research is required to determine their potential involvement in CI (Duron et al. 2007; Iturbe-Ormaetxe et al. 2005; Tram and Sullivan 2002).

19.4.9 Wolbachia *Diversity*

Wolbachia displays a high level of diversity in arthropods and nematodes and it currently comprises 17 phylogenetic clades (or supergroups), named from A to Q (Baldo et al. 2006; Bordenstein and Rosengaus 2005; Bordenstein et al. 2009; Casiraghi et al. 2005; Glowska et al. 2015; Gorham et al. 2003; Lo et al. 2002; Paraskevopoulos et al. 2006; Ros et al. 2009; Rowley et al. 2004). Each supergroup collects strains, most frequently named after their host species (e.g. *w*Pip in *Culex pipiens*, *w*Gff in *G. f. fuscipes*). The most common strategy adopted for strain genotyping relies on multilocus sequence typing (MLST), which includes the sequences of the five conserved genes *fbpA*, *coxA*, *ftsZ*, *gatB*, *coxA*, and *hcpA* and the amino acid sequences of the four hypervariable regions of the WSP protein (Baldo et al. 2006). The diversity displayed in this genus is visible at different levels, including the presence of variants within the same individual host, variation among *Wolbachia* sequences sampled from different individuals belonging to the same species, as well as the molecular changes occurring in the same *Wolbachia* infection in the case of transfer to different host species (*i.e.* in the case of horizontal gene transfer) (Hoffmann et al. 2015). In the case of *Glossina*, *Wolbachia* identified in tsetse species has been regarded to belong to the supergroup A, as determined based on the *Wolbachia* surface protein (*wsp*) gene (Cheng et al. 2000; Zhou et al. 1998). A recent study investigated *Wolbachia* genetic variability in *G. f. fuscipes* from Uganda (Symula et al. 2013). Two *Wolbachia* lineages were identified, suggesting the presence of superinfection in this species, and a high diversity within and between individuals. These data suggest that different *Wolbachia* strains infected this tsetse species multiple times independently.

19.4.10 Wolbachia *as a Tool for Vector Population Control*

Wolbachia has been proposed as a tool for controlling insect disease vectors and agricultural pests through multiple approaches, which are not mutually exclusive. These methods include the release of (1) *Wolbachia*-infected males that are incompatible with females by exploiting cytoplasmic incompatibility phenotypes (O'Connor et al. 2012), (2) *Wolbachia* strains able to induce deleterious fitness effects, in particular under seasonally variable environments (Rasić et al. 2014), and (3) *Wolbachia* strains interfering with pathogen transmission and thus decreasing the ability of vectors to transmit diseases (Kambris et al. 2009; Moreira et al. 2009; Teixeira et al. 2008; Walker et al. 2011). The capacity of rapidly invading insect populations that drive maternally inherited elements into wild insects has made *Wolbachia* a promising tool for inducing genetic manipulations of detrimental species (Beard et al. 1993; Sinkins et al. 1997). The data obtained by Alam and colleagues about CI in tsetse were incorporated into a mathematical model and suggest that *Wolbachia* has the potential for use as a gene-drive mechanism. This could be used to introduce desirable phenotypes, such as resistance to trypanosome infection, into wild *Glossina* populations (Alam et al. 2011).

The exploitation of *Wolbachia*-induced CI was proposed for use in the reduction of population sizes of several insect disease vectors and agricultural pests (Apostolaki et al. 2011; Bourtzis 2008; Stouthamer et al. 1999; Xi et al. 2005; Zabalou et al. 2009). Moreover, *Wolbachia* is able to protect their hosts against viral pathogens (Cook and McGraw 2010). The initial characterization of this phenomenon was performed in *Drosophila* (Hedges et al. 2008; Osborne et al. 2009; Teixeira et al. 2008) and mosquitoes (Bian et al. 2010; Glaser and Meola 2010; Moreira et al. 2009). More recent work demonstrates that *w*AlbB infection in the C6/36 *Ae. albopictus* cell line resulted in reduced titers of several Flaviviruses and Alphaviruses, suggesting a role of *Wolbachia* in reducing transmission of pathogenic RNA viruses (Ekwudu et al. 2020). Moreover, a triple-infected *Ae. albopictus* line carrying, in addition to the two natural symbiotic strains *w*AlbA and *w*AlbB, the *w*Au from *D. simulans*, showed complete resistance to Zika and dengue infections, with moderate fitness costs (Mancini et al. 2020). Pathogen-blocking function has been associated with the upregulation of antimicrobial peptides (e.g. DEFC, defensin c) (Caragata et al. 2019; Pan et al. 2018; Pan et al. 2012).

In the case of tsetse species, as mentioned above, contrasting results have been obtained. In wild populations of *G. f. fuscipes*, the presence of *Wolbachia* has been suggested to be able to prevent trypanosome infection (Alam et al. 2012), while in *G. p. palpalis*, *G. tachinoides,* and *G. m. submorsitans*, *Wolbachia* seems not to impact establishment of trypanosome infection (Kame-Ngasse et al. 2018; Kante et al. 2018). Further studies are thus necessary to understand whether *Wolbachia* may be used in Incompatible Insect Technique (IIT)-based approaches for population control. This strategy is based on the release of *Wolbachia*-infected males that both induce CI when mated with *Wolbachia*-free wild females and refractoriness to trypanosome infection and transmission. This concept is similar to strategies recently designed for mosquitoes (Bourtzis et al. 2016; Zhang et al. 2015a, b, 2016).

19.4.11 Summary

The *Wolbachia* symbiont is present predominantly in tsetse gonadal tissues, and it is transovarially transmitted from females to their progeny. An increasing number of studies now show that this bacterium can colonize also somatic tissues, with still unclarified functions. Horizontal transfer events have been detected in *G. m. morsitans* and *G. austeni*, but whether these integrated sequences play functional roles in tsetse biology is still unclear. *Wolbachia* induces strong cytoplasmic incompatibility in tsetse, supporting the idea that it may be exploited as a tool to control tsetse populations in the field.

19.5 *Spiroplasma*

The innovation of high-throughput 16S ribosomal RNA sequencing has allowed for deeper exploration of the microbiome of *Glossina* species. These studies have identified many other bacterial species in *Glossina* from laboratory colonies and from the field. From these analyses, bacteria from the Genus *Spiroplasma* were found in field-caught and lab colonies of *Glossina f. fuscipes* and *G. tachinoides*. The recent analysis of the genomes of six *Glossina* species also revealed evidence for a significant relationship between *Spiroplasma* and *G. f. fuscipes*. Analysis of the genomic scaffolds from this species revealed the presence of *Spiroplasma* genomic sequences. None of these sequences appear to be the result of a genomic integration and are likely derived from free-living *Spiroplasma* present in the flies used for sequencing (Attardo et al. 2019). Developmental stage and tissue specificity analyses revealed the *Spiroplasma* infections to be highest in the larval gut and male reproductive tissues. In addition, the analysis of flies from a collapsing colony of *G. f. fuscipes* showed that levels of *Spiroplasma* were higher in surviving flies versus recently deceased flies (Doudoumis et al. 2017). Another study found that *Trypanosoma* infection in field-collected *G. f. fuscipes* is negatively correlated with *Spiroplasma* coinfection. This finding was also demonstrated in the laboratory by comparing experimental infection rates between *Spiroplasma* infected and uninfected *G. f. fuscipes* (Schneider et al. 2019).

Spiroplasma species are found to be living in many insect and invertebrate species with relationships ranging from pathological to beneficial. Pathogenic *Spiroplasma* infections leading to death have been observed in aquatic invertebrates, such as shrimps and crabs (Nunan et al. 2005; Wang et al. 2004). In some *Drosophila melanogaster*, infection with *Spiroplasma* results in reproductive manipulation in the form of a male-killing phenotype (Montenegro et al. 2005; Paredes et al. 2015).

However, *Spiroplasma* infections in *Drosophila* can also protect against infection by the parasitic nematode *Howardula aoronymphium*. The nematode infection results in loss of fecundity in female *Drosophila*. However, the coinfection of females with *Spiroplasma* results in the restoration of fecundity and inhibition of nematode development (Haselkorn et al. 2013; Jaenike et al. 2010). The early observations of the relationship between *Spiroplasma* and *Glossina* suggest that this relationship is beneficial in terms of its ability to extend lifespan and reduce the vectorial capacity for *Trypanosoma*. However, relative to the other *Glossina* symbionts, little is known about the details of the interactions between these two species and more work will be required to develop a more comprehensive understanding of this relationship.

19.6 Conclusions

The role bacteria play in the biology of tsetse flies spans the gamut, ranging from parasitic to obligate mutualists. This system is also unique in that the microbiome of species within *Glossina* is limited relative to other insects. This is primarily due to their restricted diet and the protected nature of intrauterine larval development. These limitations simplify the system and provide opportunities to study microbe-host interactions in a way that is impossible in insects with more diverse microbiomes. The relationship between *Wigglesworthia* and *Glossina* is ancient and provides a clear example of obligate symbiosis. It also demonstrates the evolutionary mechanisms and biology behind how an organism can evolve to survive and thrive on a nutrient-rich yet limited diet. The commensal yet nonessential relationship between *Sodalis*/*Glossina* provides a snapshot of two organisms in the early or intermediate stages of a symbiotic relationship and can provide insights as to the adaptations made by both parties to establish a harmonious relationship. In addition, the ability to culture and manipulate *Sodalis* provides opportunities to take advantage of this relationship for purposes of controlling trypanosome transmission by *Glossina*. The nature of the relationship between *Glossina* and *Wolbachia* is more complex and shows aspects of parasitism and manipulation by *Wolbachia* similar to that observed in other insects. However, a deeper understanding of the protective nature of *Wolbachia* infections against viral infections in other insects and the inherent reproductive manipulations driving the spread of these infections may provide opportunities to utilize this relationship to reduce *Glossina* vectorial capacity. The implementation of new technologies such as high-throughput metagenomics, advanced microbial imaging, and novel *in vitro* culture techniques has opened opportunities to investigate these well-studied relationships in more depth. They also provide the ability to identify previously undescribed microbial interactions that were overshadowed by the dominant microbial fauna. Analysis of the diversity of these lesser-explored interactions could provide valuable insights into the ecology, population dynamics, and biology of this unique system.

References

Aikawa T, Anbutsu H, Nikoh N, Kikuchi T, Shibata F, Fukatsu T (2009) Longicorn beetle that vectors pinewood nematode carries many *Wolbachia* genes on an autosome. Proc Biol Sci 276:3791–3798. https://doi.org/10.1098/rspb.2009.1022

Akman L, Rio RV, Beard CB, Aksoy S (2001) Genome size determination and coding capacity of *Sodalis glossinidius*, an enteric symbiont of tsetse flies, as revealed by hybridization to *Escherichia coli* gene arrays. J Bacteriol 183:4517–4525. https://doi.org/10.1128/JB.183.15.4517-4525.2001

Akman L, Yamashita A, Watanabe H, Oshima K, Shiba T, Hattori M, Aksoy S (2002) Genome sequence of the endocellular obligate symbiont of tsetse flies, *Wigglesworthia glossinidia*. Nat Genet 32:402–407. https://doi.org/10.1038/ng986

Aksoy S (1995) *Wigglesworthia* gen. nov. and *Wigglesworthia glossinidia* sp. nov., taxa consisting of the mycetocyte-associated, primary endosymbionts of tsetse flies. Int J Syst Bacteriol 45:848–851. https://doi.org/10.1099/00207713-45-4-848

Aksoy S, Pourhosseini AA, Chow A (1995) Mycetome endosymbionts of tsetse flies constitute a distinct lineage related to Enterobacteriaceae. Insect Mol Biol 4:15–22

Aksoy S, Chen X, Hypsa V (1997) Phylogeny and potential transmission routes of midgut-associated endosymbionts of tsetse (Diptera: Glossinidae). Insect Mol Biol 6:183–190

Aksoy E et al (2014) Analysis of multiple tsetse fly populations in Uganda reveals limited diversity and species-specific gut microbiota. Appl Environ Microbiol 80:4301–4312. https://doi.org/10.1128/AEM.00079-14

Alam U et al (2011) *Wolbachia* symbiont infections induce strong cytoplasmic incompatibility in the tsetse fly *Glossina morsitans*. PLoS Pathog 7:e1002415. https://doi.org/10.1371/journal.ppat.1002415

Alam U et al (2012) Implications of microfauna-host interactions for trypanosome transmission dynamics in *Glossina fuscipes fuscipes* in Uganda. Appl Environ Microbiol 78:4627–4637. https://doi.org/10.1128/AEM.00806-12

Albertson R, Tan V, Leads RR, Reyes M, Sullivan W, Casper-Lindley C (2013) Mapping *Wolbachia* distributions in the adult *Drosophila* brain. Cell Microbiol 15:1527–1544. https://doi.org/10.1111/cmi.12136

Anselme C, Vallier A, Balmand S, Fauvarque MO, Heddi A (2006) Host PGRP gene expression and bacterial release in endosymbiosis of the weevil *Sitophilus zeamais*. Appl Environ Microbiol 72:6766–6772. https://doi.org/10.1128/AEM.00942-06

Apostolaki A, Livadaras I, Saridaki A, Chrysargyris A, Savakis C, Bourtzis K (2011) Transinfection of the olive fruit fly *Bactrocera oleae* with *Wolbachia*: towards a symbiont-based population control strategy. J Appl Entomol 135:546–553. https://doi.org/10.1111/j.1439-0418.2011.01614.x

Attardo GM, Lohs C, Heddi A, Alam UH, Yildirim S, Aksoy S (2008) Analysis of milk gland structure and function in *Glossina morsitans:* milk protein production, symbiont populations and fecundity. J Insect Physiol 54:1236–1242. https://doi.org/10.1016/j.jinsphys.2008.06.008

Attardo GM et al (2019) Comparative genomic analysis of six *Glossina* genomes, vectors of African trypanosomes. Genome Biol 20:187. https://doi.org/10.1186/s13059-019-1768-2

Baldo L et al (2006) Multilocus sequence typing system for the endosymbiont *Wolbachia pipientis*. Appl Environ Microbiol 72:7098–7110. https://doi.org/10.1128/AEM.00731-06

Balmand S, Lohs C, Aksoy S, Heddi A (2013) Tissue distribution and transmission routes for the tsetse fly endosymbionts. J Invertebr Pathol 112. Suppl:S116–S122. https://doi.org/10.1016/j.jip.2012.04.002

Beard CB, O'Neill SL, Tesh RB, Richards FF, Aksoy S (1993) Modification of arthropod vector competence via symbiotic bacteria. Parasitol Today 9:179–183. https://doi.org/10.1016/0169-4758(93)90142-3

Beard CB, Dotson EM, Pennington PM, Eichler S, Cordon-Rosales C, Durvasula RV (2001) Bacterial symbiosis and paratransgenic control of vector-borne Chagas disease. Int J Parasitol 31:621–627. https://doi.org/10.1016/s0020-7519(01)00165-5

Beard CB, Cordon-Rosales C, Durvasula RV (2002) Bacterial symbionts of the triatominae and their potential use in control of Chagas disease transmission. Annu Rev Entomol 47:123–141. https://doi.org/10.1146/annurev.ento.47.091201.145144

Beckmann JF, Ronau JA, Hochstrasser M (2017) A *Wolbachia* deubiquitylating enzyme induces cytoplasmic incompatibility. Nat Microbiol 2:17007. https://doi.org/10.1038/nmicrobiol.2017.7

Belda E, Moya A, Bentley S, Silva FJ (2010) Mobile genetic element proliferation and gene inactivation impact over the genome structure and metabolic capabilities of *Sodalis glossinidius*, the secondary endosymbiont of tsetse flies. BMC Genomics 11:449. https://doi.org/10.1186/1471-2164-11-449

Bennett GM, Moran NA (2013) Small, smaller, smallest: the origins and evolution of ancient dual symbioses in a Phloem-feeding insect. Genome Biol Evol 5:1675–1688. https://doi.org/10.1093/gbe/evt118

Benoit JB et al (2017) Symbiont-induced odorant binding proteins mediate insect host hematopoiesis. Elife 6. https://doi.org/10.7554/eLife.19535

Bi J, Wang YF (2019) The effect of the endosymbiont *Wolbachia* on the behavior of insect hosts. Insect Sci. https://doi.org/10.1111/1744-7917.12731

Bian G, Xu Y, Lu P, Xie Y, Xi Z (2010) The endosymbiotic bacterium *Wolbachia* induces resistance to dengue virus in *Aedes aegypti*. PLoS Pathog 6:e1000833. https://doi.org/10.1371/journal.ppat.1000833

Bing X et al (2017) Unravelling the relationship between the tsetse fly and its obligate symbiont *Wigglesworthia*: transcriptomic and metabolomic landscapes reveal highly integrated physiological networks. Proc Biol Sci 284. https://doi.org/10.1098/rspb.2017.0360

Binnington K, Hoffmann A (1989) *Wolbachia*-like organisms and cytoplasmic incompatibility in *Drosophila simulans*. J Invertebr Pathol 54:344–352

Bordenstein SR, Bordenstein SR (2016) Eukaryotic association module in phage WO genomes from *Wolbachia*. Nat Commun 7:13155. https://doi.org/10.1038/ncomms13155

Bordenstein S, Rosengaus RB (2005) Discovery of a novel *Wolbachia* super group in Isoptera. Curr Microbiol 51:393–398. https://doi.org/10.1007/s00284-005-0084-0

Bordenstein SR et al (2009) Parasitism and mutualism in *Wolbachia*: what the phylogenomic trees can and cannot say. Mol Biol Evol 26:231–241. https://doi.org/10.1093/molbev/msn243

Bourtzis K (2008) *Wolbachia*-based technologies for insect pest population control. In: Aksoy S (ed) Transgenesis and the management of vector-borne disease. Springer, New York, NY, pp 104–113. https://doi.org/10.1007/978-0-387-78225-6_9

Bourtzis K, Dobson SL, Braig HR, O'Neill SL (1998) Rescuing *Wolbachia* have been overlooked. Nature 391:852–853. https://doi.org/10.1038/36017

Bourtzis K, Lees RS, Hendrichs J, Vreysen MJ (2016) More than one rabbit out of the hat: radiation, transgenic and symbiont-based approaches for sustainable management of mosquito and tsetse fly populations. Acta Trop 157:115–130. https://doi.org/10.1016/j.actatropica.2016.01.009

Brelsfoard C et al (2014) *Wolbachia* symbiont genome sequence and extensive chromosomal insertions present in the host *Glossina morsitans morsitans* genome. PLoS Negl Trop Dis 8: e2728. https://doi.org/10.1371/journal.pntd.0002728

Bressac C, Rousset F (1993) The reproductive incompatibility system in *Drosophila simulans*: DAPI-staining analysis of the *Wolbachia* symbionts in sperm cysts. J Invertebr Pathol 61:226–230. https://doi.org/10.1006/jipa.1993.1044

Buchner P (1965) Endosymbiosis of animals with plant microorganisms. Interscience Publishers, New York

Bulgheresi S, Schabussova I, Chen T, Mullin NP, Maizels RM, Ott JA (2006) A new C-type lectin similar to the human immunoreceptor DC-SIGN mediates symbiont acquisition by a marine nematode. Appl Environ Microbiol 72:2950–2956. https://doi.org/10.1128/AEM.72.4.2950-2956.2006

Bursell E (1960) Free amino-acids of the tsetse fly (*Glossina*). Nature 187:778–778

Bursell E (1963) Aspects of the metabolism of amino acids in the tsetse fly, *Glossina* (Diptera). J Insect Physiol 9:439–452

Bursell E (1966) Aspects of flight metabolism of tsetse flies (*Glossina*). Comp Biochem Physiol 19:809–818

Buxton PA (1955) London school of hygiene and tropical medicine memoir 10: the natural history of tsetse flies. H. K. Lewis, London

Caragata EP, Rocha MN, Pereira TN, Mansur SB, Dutra HLC, Moreira LA (2019) Pathogen blocking in *Wolbachia*-infected *Aedes aegypti* is not affected by Zika and dengue virus co-infection. PLoS Negl Trop Dis 13:e0007443. https://doi.org/10.1371/journal.pntd.0007443

Casiraghi M et al (2005) Phylogeny of *Wolbachia pipientis* based on *gltA*, *groEL* and *ftsZ* gene sequences: clustering of arthropod and nematode symbionts in the F supergroup, and evidence for further diversity in the *Wolbachia* tree. Microbiology 151:4015–4022. https://doi.org/10.1099/mic.0.28313-0

Casper-Lindley C, Kimura S, Saxton DS, Essaw Y, Simpson I, Tan V, Sullivan W (2011) Rapid fluorescence-based screening for *Wolbachia* endosymbionts in *Drosophila* germ line and somatic tissues. Appl Environ Microbiol 77:4788–4794. https://doi.org/10.1128/AEM.00215-11

Chang KP, Musgrave AJ (1973) Morphology, histochemistry, and ultrastructure of mycetome and its rickettsial symbiotes in *Cimex lectularius* L. Can J Microbiol 19:1075–1081. https://doi.org/10.1139/m73-171

Channumsin M, Ciosi M, Masiga D, Turner CMR, Mable BK (2018) *Sodalis glossinidius* presence in wild tsetse is only associated with presence of trypanosomes in complex interactions with other tsetse-specific factors. BMC Microbiol 18:163. https://doi.org/10.1186/s12866-018-1285-6

Chaston J, Goodrich-Blair H (2010) Common trends in mutualism revealed by model associations between invertebrates and bacteria. FEMS Microbiol Rev 34:41–58. https://doi.org/10.1111/j.1574-6976.2009.00193.x

Chen X, Li S, Aksoy S (1999) Concordant evolution of a symbiont with its host insect species: molecular phylogeny of genus *Glossina* and its bacteriome-associated endosymbiont, *Wigglesworthia glossinidia*. J Mol Evol 48:49–58. https://doi.org/10.1007/pl00006444

Cheng Q, Ruel TD, Zhou W, Moloo SK, Majiwa P, O'Neill SL, Aksoy S (2000) Tissue distribution and prevalence of *Wolbachia* infections in tsetse flies, *Glossina* spp. Med Vet Entomol 14:44–50. https://doi.org/10.1046/j.1365-2915.2000.00202.x

Clark ME, Veneti Z, Bourtzis K, Karr TL (2002) The distribution and proliferation of the intracellular bacteria *Wolbachia* during spermatogenesis in *Drosophila*. Mech Dev 111:3–15. https://doi.org/10.1016/s0925-4773(01)00594-9

Clark ME, Veneti Z, Bourtzis K, Karr TL (2003) *Wolbachia* distribution and cytoplasmic incompatibility during sperm development: the cyst as the basic cellular unit of CI expression. Mech Dev 120:185–198. https://doi.org/10.1016/s0925-4773(02)00424-0

Cook PE, McGraw EA (2010) *Wolbachia pipientis:* an expanding bag of tricks to explore for disease control. Trends Parasitol 26:373–375. https://doi.org/10.1016/j.pt.2010.05.006

Cordaux R, Gilbert C (2017) Evolutionary significance of *Wolbachia*-to-animal horizontal gene transfer: female sex determination and the f element in the isopod *Armadillidium vulgare*. Genes 8. https://doi.org/10.3390/genes8070186

Coutinho-Abreu IV, Zhu KY, Ramalho-Ortigao M (2010) Transgenesis and paratransgenesis to control insect-borne diseases: current status and future challenges. Parasitol Int 59:1–8. https://doi.org/10.1016/j.parint.2009.10.002

Curtis CF (1972) Sterility from crosses between sub-species of the tsetse fly *Glossina morsitans*. Acta Trop 29:250–268

Dale C, Maudlin I (1999) *Sodalis* gen. nov. and *Sodalis glossinidius* sp. nov., a microaerophilic secondary endosymbiont of the tsetse fly *Glossina morsitans morsitans*. Int J Syst Bacteriol 49 (Pt 1):267–275. https://doi.org/10.1099/00207713-49-1-267

Dale C, Welburn SC (2001) The endosymbionts of tsetse flies: manipulating host-parasite interactions. Int J Parasitol 31:628–631. https://doi.org/10.1016/s0020-7519(01)00151-5

Dale C, Young SA, Haydon DT, Welburn SC (2001) The insect endosymbiont *Sodalis glossinidius* utilizes a type III secretion system for cell invasion. Proc Natl Acad Sci U S A 98:1883–1888. https://doi.org/10.1073/pnas.021450998

Dale C, Jones T, Pontes M (2005) Degenerative evolution and functional diversification of type-III secretion systems in the insect endosymbiont *Sodalis glossinidius*. Mol Biol Evol 22:758–766. https://doi.org/10.1093/molbev/msi061

Darby AC, Lagnel J, Matthew CZ, Bourtzis K, Maudlin I, Welburn SC (2005) Extrachromosomal DNA of the symbiont *Sodalis glossinidius*. J Bacteriol 187:5003–5007. https://doi.org/10.1128/JB.187.14.5003-5007.2005

Dawadi B, Wang X, Xiao R, Muhammad A, Hou Y, Shi Z (2018) PGRP-LB homolog acts as a negative modulator of immunity in maintaining the gut-microbe symbiosis of red palm weevil, *Rhynchophorus ferrugineus* Olivier. Dev Comp Immunol 86:65–77. https://doi.org/10.1016/j.dci.2018.04.021

de Oliveira CD, Gonçalves DS, Baton LA, Shimabukuro PHF, Carvalho FD, Moreira LA (2015) Broader prevalence of *Wolbachia* in insects including potential human disease vectors. Bull Entomol Res 105:305–315. https://doi.org/10.1017/S0007485315000085

De Vooght L, Caljon G, De Ridder K, Van Den Abbeele J (2014) Delivery of a functional anti-trypanosome Nanobody in different tsetse fly tissues via a bacterial symbiont, *Sodalis glossinidius*. Microb Cell Fact 13:156. https://doi.org/10.1186/s12934-014-0156-6

De Vooght L, Caljon G, Van Hees J, Van Den Abbeele J (2015) Paternal transmission of a secondary symbiont during mating in the viviparous tsetse Fly. Mol Biol Evol 32:1977–1980. https://doi.org/10.1093/molbev/msv077

De Vooght L, Van Keer S, Van Den Abbeele J (2018) Towards improving tsetse fly paratransgenesis: stable colonization of *Glossina morsitans morsitans* with genetically modified *Sodalis*. BMC Microbiol 18:165. https://doi.org/10.1186/s12866-018-1282-9

Dedeine F, Vavre F, Fleury F, Loppin B, Hochberg ME, Bouletreau M (2001) Removing symbiotic *Wolbachia* bacteria specifically inhibits oogenesis in a parasitic wasp. Proc Natl Acad Sci U S A 98:6247–6252. https://doi.org/10.1073/pnas.101304298

Demirbas-Uzel G, De Vooght L, Parker AG, Vreysen MJB, Mach RL, Van Den Abbeele J, Abd-Alla AMM (2018) Combining paratransgenesis with SIT: impact of ionizing radiation on the DNA copy number of *Sodalis glossinidius* in tsetse flies. BMC Microbiol 18:160. https://doi.org/10.1186/s12866-018-1283-8

Dennis JW, Durkin SM, Horsley Downie JE, Hamill LC, Anderson NE, MacLeod ET (2014) *Sodalis glossinidius* prevalence and trypanosome presence in tsetse from Luambe National Park, Zambia. Parasit Vectors 7:378. https://doi.org/10.1186/1756-3305-7-378

Dobson SL, Bourtzis K, Braig HR, Jones BF, Zhou W, Rousset F, O'Neill SL (1999) *Wolbachia* infections are distributed throughout insect somatic and germ line tissues. Insect Biochem Mol Biol 29:153–160. https://doi.org/10.1016/s0965-1748(98)00119-2

Doudoumis V et al (2012) Detection and characterization of *Wolbachia* infections in laboratory and natural populations of different species of tsetse flies (genus Glossina). BMC Microbiol 12 (Suppl 1):S3. https://doi.org/10.1186/1471-2180-12-S1-S3

Doudoumis V et al (2013) Tsetse-*Wolbachia* symbiosis: comes of age and has great potential for pest and disease control. J Invertebr Pathol 112. Suppl:S94–S103. https://doi.org/10.1016/j.jip.2012.05.010

Doudoumis V et al (2017) Challenging the *Wigglesworthia, Sodalis, Wolbachia* symbiosis dogma in tsetse flies: *Spiroplasma* is present in both laboratory and natural populations. Sci Rep 7:4699. https://doi.org/10.1038/s41598-017-04740-3

Douglas AE (1989) Mycetocyte symbiosis in insects. Biol Rev Camb Philos Soc 64:409–434. https://doi.org/10.1111/j.1469-185x.1989.tb00682.x

Dunning Hotopp JC et al (2007) Widespread lateral gene transfer from intracellular bacteria to multicellular eukaryotes. Science 317:1753–1756. https://doi.org/10.1126/science.1142490

Duron O, Boureux A, Echaubard P, Berthomieu A, Berticat C, Fort P, Weill M (2007) Variability and expression of ankyrin domain genes in *Wolbachia* variants infecting the mosquito *Culex pipiens*. J Bacteriol 189:4442–4448. https://doi.org/10.1128/JB.00142-07

Ekwudu O, Devine GJ, Aaskov JG, Frentiu FD (2020) *Wolbachia* strain wAlbB blocks replication of flaviviruses and alphaviruses in mosquito cell culture. Parasit Vectors 13:54. https://doi.org/10.1186/s13071-020-3936-3

Farikou O et al (2010) Tripartite interactions between tsetse flies, *Sodalis glossinidius* and trypanosomes – an epidemiological approach in two historical human African trypanosomiasis foci in Cameroon. Infect Genet Evol 10:115–121. https://doi.org/10.1016/j.meegid.2009.10.008

Fenn K, Conlon C, Jones M, Quail MA, Holroyd NE, Parkhill J, Blaxter M (2006) Phylogenetic relationships of the *Wolbachia* of nematodes and arthropods. PLoS Pathog 2:e94. https://doi.org/10.1371/journal.ppat.0020094

Ferree PM, Frydman HM, Li JM, Cao J, Wieschaus E, Sullivan W (2005) *Wolbachia* utilizes host microtubules and Dynein for anterior localization in the *Drosophila* oocyte. PLoS Pathog 1:e14. https://doi.org/10.1371/journal.ppat.0010014

Franco JR et al (2018) Monitoring the elimination of human African trypanosomiasis: Update to 2016. PLoS Negl Trop Dis 12:e0006890. https://doi.org/10.1371/journal.pntd.0006890

Franks TM, Lykke-Andersen J (2008) The control of mRNA decapping and P-body formation. Mol Cell 32:605–615. https://doi.org/10.1016/j.molcel.2008.11.001

Glaser RL, Meola MA (2010) The native *Wolbachia* endosymbionts of *Drosophila melanogaster* and *Culex quinquefasciatus* increase host resistance to West Nile virus infection. PLoS One 5: e11977. https://doi.org/10.1371/journal.pone.0011977

Glowska E, Dragun-Damian A, Dabert M, Gerth M (2015) New *Wolbachia* supergroups detected in quill mites (Acari: Syringophilidae). Infect Genet Evol 30:140–146. https://doi.org/10.1016/j.meegid.2014.12.019

Gooding R (1985) Electrophoretic and hybridization comparisons of *Glossina morsitans morsitans, G. m. centralis* and *G. m. submorsitans* (Diptera:Glossinidae). Can J Zool 63:2694–2702

Gooding R (1987) Genetic basis of sterility in hybrid males from crosses of *Glossina morsitans morsitans* and *Glossina morsitans centralis*. Can J Zool 65:640–646

Gooding R (1989) Genetic basis of sterility in hybrids from crosses of *Glossina morsitans submorsitans* and *Glossina morsitans morsitans* (Diptera:Glossinidae). Genome 32:479–485

Gooding RH (1990) Postmating barriers to gene flow among species and subspecies of tsetse flies (Diptera: Glossinidae). Can J Zool 68:1727–1734

Gorham CH, Fang QQ, Durden LA (2003) *Wolbachia* endosymbionts in fleas (Siphonaptera). J Parasitol 89:283–289. https://doi.org/10.1645/0022-3395(2003)089[0283:WEIFS]2.0.CO;2

Haines LR, Hancock RE, Pearson TW (2003) Cationic antimicrobial peptide killing of African trypanosomes and *Sodalis glossinidius,* a bacterial symbiont of the insect vector of sleeping sickness. Vector Borne Zoonotic Dis 3:175–186. https://doi.org/10.1089/153036603322662165

Hall RJ et al (2019) A tale of three species: adaptation of *Sodalis glossinidius* to tsetse biology, *Wigglesworthia* metabolism, and host Diet. mBio 10:e02106–e02118. https://doi.org/10.1128/mBio.02106-18

Hamidou Soumana I, Loriod B, Ravel S, Tchicaya B, Simo G, Rihet P, Geiger A (2014a) The transcriptional signatures of *Sodalis glossinidius* in the *Glossina palpalis gambiensis* flies negative for *Trypanosoma brucei gambiense* contrast with those of this symbiont in tsetse flies positive for the parasite: possible involvement of a *Sodalis*-hosted prophage in fly *Trypanosoma* refractoriness? Infect Genet Evol 24:41–56. https://doi.org/10.1016/j.meegid.2014.03.005

Hamidou Soumana I, Tchicaya B, Loriod B, Rihet P, Geiger A (2014b) Identification of overexpressed genes in *Sodalis glossinidius* inhabiting trypanosome-infected self-cured tsetse flies. Front Microbiol 5:255. https://doi.org/10.3389/fmicb.2014.00255

Harris HL, Braig HR (2003) Sperm chromatin remodelling and *Wolbachia*-induced cytoplasmic incompatibility in *Drosophila*. Biochem Cell Biol 81:229–240. https://doi.org/10.1139/o03-053

Haselkorn TS, Cockburn SN, Hamilton PT, Perlman SJ, Jaenike J (2013) Infectious adaptation: potential host range of a defensive endosymbiont in *Drosophila*. Evolution 67:934–945. https://doi.org/10.1111/evo.12020

Hedges LM, Brownlie JC, O'Neill SL, Johnson KN (2008) *Wolbachia* and virus protection in insects. Science 322:702. https://doi.org/10.1126/science.1162418

Herbeck JT, Wall DP, Wernegreen JJ (2003) Gene expression level influences amino acid usage, but not codon usage, in the tsetse fly endosymbiont *Wigglesworthia*. Microbiology 149:2585–2596. https://doi.org/10.1099/mic.0.26381-0

Hertig M (1936) The *Rickettsia*, *Wolbachia pipientis* (gen. et sp.n.) and associated inclusions of the mosquito, *Culex pipiens*. Parasitology 28:453–486. https://doi.org/10.1017/S0031182000022666

Hertig M, Wolbach SB (1924) Studies on *Rickettsia*-like micro-organisms in insects. J Med Res 44:329–374.327

Hilgenboecker K, Hammerstein P, Schlattmann P, Telschow A, Werren JH (2008) How many species are infected with *Wolbachia*? – a statistical analysis of current data. FEMS Microbiol Lett 281:215–220. https://doi.org/10.1111/j.1574-6968.2008.01110.x

Hill P, Saunders DS, Campbell JA (1973) Letter: The production of "symbiont-free" *Glossina morsitans* and an associated loss of female fertility. Trans R Soc Trop Med Hyg 67:727–728. https://doi.org/10.1016/0035-9203(73)90051-5

Hoffmann AA, Ross PA, Rašić G (2015) *Wolbachia* strains for disease control: ecological and evolutionary considerations. Evol Appl 8:751–768. https://doi.org/10.1111/eva.12286

Hosokawa T, Kaiwa N, Matsuura Y, Kikuchi Y, Fukatsu T (2015) Infection prevalence of *Sodalis* symbionts among stinkbugs. Zoological Lett 1:5. https://doi.org/10.1186/s40851-014-0009-5

Hrusa G et al (2015) TonB-dependent heme iron acquisition in the tsetse fly symbiont *Sodalis glossinidius*. Appl Environ Microbiol 81:2900–2909. https://doi.org/10.1128/AEM.04166-14

Huebner E, Davey KG (1974) Bacteroids in the ovaries of a tsetse fly. Nature 249:260–261

Hussain M, Frentiu FD, Moreira LA, O'Neill SL, Asgari S (2011) *Wolbachia* uses host microRNAs to manipulate host gene expression and facilitate colonization of the dengue vector *Aedes aegypti*. Proc Natl Acad Sci U S A 108:9250–9255. https://doi.org/10.1073/pnas.1105469108

Ijichi N, Kondo N, Matsumoto R, Shimada M, Ishikawa H, Fukatsu T (2002) Internal spatiotemporal population dynamics of infection with three *Wolbachia* strains in the adzuki bean beetle, *Callosobruchus chinensis* (Coleoptera: Bruchidae). Appl Environ Microbiol 68:4074–4080. https://doi.org/10.1128/aem.68.8.4074-4080.2002

Isaac C, Ohiolei JA, Ebhodaghe F, Igbinosa IB, Eze AA (2017) Animal African Trypanosomiasis in Nigeria: A long way from elimination/eradication. Acta Trop 176:323–331. https://doi.org/10.1016/j.actatropica.2017.08.032

Iturbe-Ormaetxe I, Burke GR, Riegler M, O'Neill SL (2005) Distribution, expression, and motif variability of ankyrin domain genes in *Wolbachia pipientis*. J Bacteriol 187:5136–5145. https://doi.org/10.1128/JB.187.15.5136-5145.2005

Jaenike J, Unckless R, Cockburn SN, Boelio LM, Perlman SJ (2010) Adaptation via symbiosis: recent spread of a *Drosophila* defensive symbiont. Science 329:212–215. https://doi.org/10.1126/science.1188235

Jucci C (1952) Symbiosis and phylogenesis in the isoptera. Nature 169:837. https://doi.org/10.1038/169837a0

Kageyama D, Traut W (2004) Opposite sex-specific effects of *Wolbachia* and interference with the sex determination of its host *Ostrinia scapulalis*. Proc Biol Sci 271:251–258. https://doi.org/10.1098/rspb.2003.2604

Kageyama D et al (2017) Feminizing *Wolbachia* endosymbiont disrupts maternal sex chromosome inheritance in a butterfly species. Evol Lett 1:232–244. https://doi.org/10.1002/evl3.28

Kambris Z, Cook PE, Phuc HK, Sinkins SP (2009) Immune activation by life-shortening *Wolbachia* and reduced filarial competence in mosquitoes. Science 326:134–136. https://doi.org/10.1126/science.1177531

Kame-Ngasse GI, Njiokou F, Melachio-Tanekou TT, Farikou O, Simo G, Geiger A (2018) Prevalence of symbionts and trypanosome infections in tsetse flies of two villages of the "Faro and Deo" division of the Adamawa region of Cameroon. BMC Microbiol 18:159. https://doi.org/10.1186/s12866-018-1286-5

Kante ST, Melachio T, Ofon E, Njiokou F, Simo G (2018) Detection of *Wolbachia* and different trypanosome species in *Glossina palpalis palpalis* populations from three sleeping sickness foci of southern Cameroon. Parasit Vectors 11:630. https://doi.org/10.1186/s13071-018-3229-2

Kita A, Jimbo M, Sakai R, Morimoto Y, Miki K (2015) Crystal structure of a symbiosis-related lectin from octocoral. Glycobiology 25:1016–1023. https://doi.org/10.1093/glycob/cwv033

Klasson L, Kambris Z, Cook PE, Walker T, Sinkins SP (2009) Horizontal gene transfer between *Wolbachia* and the mosquito *Aedes aegypti*. BMC Genomics 10:33. https://doi.org/10.1186/1471-2164-10-33

Kondo N, Nikoh N, Ijichi N, Shimada M, Fukatsu T (2002) Genome fragment of *Wolbachia* endosymbiont transferred to X chromosome of host insect. Proc Natl Acad Sci U S A 99:14280–14285. https://doi.org/10.1073/pnas.222228199

Kozek W, Rao R (2007) The Discovery of *Wolbachia* in arthropods and nematodes – a historical perspective. In: Hoerauf AaR RU (ed) *Wolbachia*: A Bug's Life in another Bug, vol 5. Wolbachia. Issues Infect Dis. Karger, Basel, pp 1–14. https://doi.org/10.1159/000104228

Kremer N, Charif D, Henri H, Gavory F, Wincker P, Mavingui P, Vavre F (2012) Influence of *Wolbachia* on host gene expression in an obligatory symbiosis. BMC Microbiol 12(Suppl 1):S7. https://doi.org/10.1186/1471-2180-12-S1-S7

LePage DP et al (2017) Prophage WO genes recapitulate and enhance *Wolbachia*-induced cytoplasmic incompatibility. Nature 543:243–247. https://doi.org/10.1038/nature21391

Lo N, Casiraghi M, Salati E, Bazzocchi C, Bandi C (2002) How many *Wolbachia* supergroups exist? Mol Biol Evol 19:341–346. https://doi.org/10.1093/oxfordjournals.molbev.a004087

Ma WC, Denlinger DL (1974) Secretory discharge and microflora of milk gland in tsetse flies. Nature 247:301–303

Magnarelli LA (1978) Nectar-feeding by female mosquitoes and its relation to follicular development and parity. J Med Entomol 14:527–530. https://doi.org/10.1093/jmedent/14.5.527

Maltz MA, Weiss BL, O'Neill M, Wu Y, Aksoy S (2012) OmpA-mediated biofilm formation is essential for the commensal bacterium *Sodalis glossinidius* to colonize the tsetse fly gut. Appl Environ Microbiol 78:7760–7768. https://doi.org/10.1128/AEM.01858-12

Mancini MV, Herd CS, Ant TH, Murdochy SM, Sinkins SP (2020) *Wolbachia* strain *w*Au efficiently blocks arbovirus transmission in *Aedes albopictus*. PLoS Negl Trop Dis 14: e0007926. https://doi.org/10.1371/journal.pntd.0007926

Mayoral JG, Hussain M, Joubert DA, Iturbe-Ormaetxe I, O'Neill SL, Asgari S (2014) *Wolbachia* small noncoding RNAs and their role in cross-kingdom communications. Proc Natl Acad Sci U S A 111:18721–18726. https://doi.org/10.1073/pnas.1420131112

Mellanby H (1937) Experimental work on reproduction in the tsetse fly, *Glossina palpalis*. Parasitology 1:131–141

Meyer A, Holt HR, Selby R, Guitian J (2016) Past and ongoing tsetse and animal trypanosomiasis control operations in five African Countries: a systematic review. PLoS Negl Trop Dis 10: e0005247. https://doi.org/10.1371/journal.pntd.0005247

Michalkova V, Benoit JB, Weiss BL, Attardo GM, Aksoy S (2014) Vitamin B6 generated by obligate symbionts is critical for maintaining proline homeostasis and fecundity in tsetse flies. Appl Environ Microbiol 80:5844–5853. https://doi.org/10.1128/AEM.01150-14

Min K-T, Benzer S (1997) *Wolbachia*, normally a symbiont of *Drosophila*, can be virulent, causing degeneration and death. Proc Natl Acad Sci U S A 94:10792–10796

Mitsuhashi W, Saiki T, Wei W, Kawakita H, Sato M (2002) Two novel strains of *Wolbachia* coexisting in both species of mulberry leafhoppers. Insect Mol Biol 11:577–584. https://doi.org/10.1046/j.1365-2583.2002.00368.x

Montenegro H, Solferini VN, Klaczko LB, Hurst GDD (2005) Male-killing *Spiroplasma* naturally infecting *Drosophila melanogaster*. Insect Mol Biol 14:281–287. https://doi.org/10.1111/j.1365-2583.2005.00558.x

Moran NA, Bennett GM (2014) The tiniest tiny genomes. Annu Rev Microbiol 68:195–215. https://doi.org/10.1146/annurev-micro-091213-112901

Moreira LA et al (2009) A *Wolbachia* symbiont in *Aedes aegypti* limits infection with dengue, Chikungunya, and Plasmodium. Cell 139:1268–1278. https://doi.org/10.1016/j.cell.2009.11.042

Mouton L, Henri H, Charif D, Boulétreau M, Vavre F (2007) Interaction between host genotype and environmental conditions affects bacterial density in *Wolbachia* symbiosis. Biol Lett 3:210–213. https://doi.org/10.1098/rsbl.2006.0590

Munson MA, Baumann P, Kinsey MG (1991) *Buchnera* gen. nov. and *Buchnera aphidicola* sp. nov., a taxon consisting of the mycetocyte-associated, primary endosymbionts of aphids. Int J Syst Evol Micr 41:566–568

Nikoh N, Nakabachi A (2009) Aphids acquired symbiotic genes via lateral gene transfer. BMC Biol 7:12. https://doi.org/10.1186/1741-7007-7-12

Nikoh N, Tanaka K, Shibata F, Kondo N, Hizume M, Shimada M, Fukatsu T (2008) *Wolbachia* genome integrated in an insect chromosome: evolution and fate of laterally transferred endosymbiont genes. Genome Res 18:272–280. https://doi.org/10.1101/gr.7144908

Nogge G (1976) Sterility in tsetse flies (*Glossina morsitans* Westwood) caused by loss of symbionts. Experientia 32:995–996

Nogge G (1978) Aposymbiotic tsetse flies, *Glossina morsitans morsitans* obtained by feeding on rabbits immunized specifically with symbionts. J Insect Physiol 24:299–304

Nogge G, Gerresheim A (1982) Experiments on the elimination of symbionts from the tsetse-fly, *Glossina morsitans morsitans* (Diptera, Glossinidae), by antibiotics and lysozyme. J Invertebr Pathol 40:166–179

Novakova E, Husnik F, Sochova E, Hypsa V (2015) *Arsenophonus* and *Sodalis* symbionts in louse flies: an analogy to the *Wigglesworthia* and *Sodalis* system in tsetse flies. Appl Environ Microbiol 81:6189–6199. https://doi.org/10.1128/AEM.01487-15

Nunan LM, Lightner DV, Oduori MA, Gasparich GE (2005) *Spiroplasma penaei* sp. nov., associated with mortalities in *Penaeus vannamei*, Pacific white shrimp. Int J Syst Evol Micr 55:2317–2322

O'Connor L, Plichart C, Sang AC, Brelsfoard CL, Bossin HC, Dobson SL (2012) Open release of male mosquitoes infected with a *Wolbachia* biopesticide: field performance and infection containment. PLoS Negl Trop Dis 6:e1797. https://doi.org/10.1371/journal.pntd.0001797

O'Neill SL, Karr TL (1990) Bidirectional incompatibility between conspecific populations of *Drosophila simulans*. Nature 348:178–180. https://doi.org/10.1038/348178a0

O'Neill SL, Giordano R, Colbert AM, Karr TL, Robertson HM (1992) 16S rRNA phylogenetic analysis of the bacterial endosymbionts associated with cytoplasmic incompatibility in insects. Proc Natl Acad Sci U S A 89:2699–2702. https://doi.org/10.1073/pnas.89.7.2699

O'Neill SL, Gooding RH, Aksoy S (1993) Phylogenetically distant symbiotic microorganisms reside in *Glossina* midgut and ovary tissues. Med Vet Entomol 7:377–383. https://doi.org/10.1111/j.1365-2915.1993.tb00709.x

Osborne SE, Leong YS, O'Neill SL, Johnson KN (2009) Variation in antiviral protection mediated by different *Wolbachia* strains in *Drosophila simulans*. PLoS Pathog 5:e1000656. https://doi.org/10.1371/journal.ppat.1000656

Ote M, Yamamoto D (2020) Impact of *Wolbachia* infection on *Drosophila* female germline stem cells. Curr Opin Insect Sci 37:8–15. https://doi.org/10.1016/j.cois.2019.10.001

Ote M, Ueyama M, Yamamoto D (2016) *Wolbachia* protein TomO targets nanos mRNA and restores germ stem cells in *Drosophila* sex-lethal mutants. Curr Biol 26:2223–2232. https://doi.org/10.1016/j.cub.2016.06.054

Pais R, Lohs C, Wu Y, Wang J, Aksoy S (2008) The obligate mutualist *Wigglesworthia glossinidia* influences reproduction, digestion, and immunity processes of its host, the tsetse fly. Appl Environ Microbiol 74:5965–5974. https://doi.org/10.1128/AEM.00741-08

Pan X, Zhou G, Wu J, Bian G, Lu P, Raikhel AS, Xi Z (2012) *Wolbachia* induces reactive oxygen species (ROS)-dependent activation of the Toll pathway to control dengue virus in the mosquito *Aedes aegypti*. Proc Natl Acad Sci U S A 109:E23–E31. https://doi.org/10.1073/pnas.1116932108

Pan X et al (2018) The bacterium *Wolbachia* exploits host innate immunity to establish a symbiotic relationship with the dengue vector mosquito *Aedes aegypti*. ISME J 12:277–288. https://doi.org/10.1038/ismej.2017.174

Pannebakker BA, Loppin B, Elemans CP, Humblot L, Vavre F (2007) Parasitic inhibition of cell death facilitates symbiosis. Proc Natl Acad Sci U S A 104:213–215. https://doi.org/10.1073/pnas.0607845104

Paraskevopoulos C, Bordenstein SR, Wernegreen JJ, Werren JH, Bourtzis K (2006) Toward a *Wolbachia* multilocus sequence typing system: discrimination of *Wolbachia* strains present in *Drosophila* species. Curr Microbiol 53:388–395. https://doi.org/10.1007/s00284-006-0054-1

Paredes JC et al (2015) Genome sequence of the *Drosophila melanogaster* male-killing *Spiroplasma* strain MSRO endosymbiont. mBio 6:e02437-14. https://doi.org/10.1128/mBio.02437-14

Pell PE, Southern DI (1975) Symbionts in the female tsetse fly *Glossina morsitans morsitans*. Experientia 31:650–651. https://doi.org/10.1007/BF01944608

Pietri JE, DeBruhl H, Sullivan W (2016) The rich somatic life of *Wolbachia*. Microbiologyopen 5:923–936. https://doi.org/10.1002/mbo3.390

Pinnock DE, Hess RT (1974) The occurrence of intracellular rickettsia-like organisms in the tsetse flies, *Glossina morsitans, G. fuscipes, G. brevipalpis* and *G. pallidipes*. Acta Trop 31:70–79

Pontes MH, Babst M, Lochhead R, Oakeson K, Smith K, Dale C (2008) Quorum sensing primes the oxidative stress response in the insect endosymbiont, *Sodalis glossinidius*. PLoS One 3:e3541. https://doi.org/10.1371/journal.pone.0003541

Puchta O (1955) Experimental studies on the significance of symbiosis in the clothes louse *Pediculus vestimenti* Burm. Z Parasitenkd 17:1–40. https://doi.org/10.1007/bf00260226

Rasić G, Endersby NM, Williams C, Hoffmann AA (2014) Using *Wolbachia*-based release for suppression of *Aedes* mosquitoes: insights from genetic data and population simulations. Ecol Appl 24:1226–1234. https://doi.org/10.1890/13-1305.1

Rawlings P (1985) The genetics of hybrid sterility between subspecies of the complex of *Glossina morsitans* Westwood (Diptera: Glossinidae). B Entomol Res 75:689–699

Reinhardt C, Steiger R, Hecker H (1972) Ultrastructural study of the midgut mycetome-bacteroids of the tsetse flies *Glossina morsitans, G. fuscipes,* and *G. brevipalpis* (Diptera, Brachycera). Acta Trop 29:280–288

Rio RV, Lefevre C, Heddi A, Aksoy S (2003) Comparative genomics of insect-symbiotic bacteria: influence of host environment on microbial genome composition. Appl Environ Microbiol 69:6825–6832. https://doi.org/10.1128/aem.69.11.6825-6832.2003

Rio RV, Wu YN, Filardo G, Aksoy S (2006) Dynamics of multiple symbiont density regulation during host development: tsetse fly and its microbial flora. Proc Biol Sci 273:805–814. https://doi.org/10.1098/rspb.2005.3399

Rio RV et al (2012) Insight into the transmission biology and species-specific functional capabilities of tsetse (Diptera: glossinidae) obligate symbiont *Wigglesworthia*. MBio 3:00240-11. https://doi.org/10.1128/mBio.00240-11

Ros VID, Fleming VM, Feil EJ, Breeuwer JAJ (2009) How diverse is the genus *Wolbachia*? Multiple-gene sequencing reveals a putatively new *Wolbachia* supergroup recovered from spider mites (Acari: Tetranychidae). Appl Environ Microbiol 75:1036–1043. https://doi.org/10.1128/AEM.01109-08

Roubaud B (1919) Les particularites de la nutrition et de la vie symbiotique chez les mouches tsetse. Ann Inst Pasteur 33:489–537

Rousset F, Bouchon D, Pintureau B, Juchault P, Solignac M (1992) *Wolbachia* endosymbionts responsible for various alterations of sexuality in arthropods. Proc Biol Sci 250:91–98. https://doi.org/10.1098/rspb.1992.0135

Rowley SM, Raven RJ, McGraw EA (2004) *Wolbachia pipientis* in Australian spiders. Curr Microbiol 49:208–214. https://doi.org/10.1007/s00284-004-4346-z

Runyen-Janecky LJ, Brown AN, Ott B, Tujuba HG, Rio RV (2010) Regulation of high-affinity iron acquisition homologues in the tsetse fly symbiont *Sodalis glossinidius*. J Bacteriol 192:3780–3787. https://doi.org/10.1128/JB.00161-10

Schneider DI, Garschall KI, Parker AG, Abd-Alla AMM, Miller WJ (2013) Global *Wolbachia* prevalence, titer fluctuations and their potential of causing cytoplasmic incompatibilities in tsetse flies and hybrids of *Glossina morsitans* subgroup species. J Invertebr Pathol 112. Suppl: S104–S115. https://doi.org/10.1016/j.jip.2012.03.024

Schneider DI, Parker AG, Abd-Alla AM, Miller WJ (2018) High-sensitivity detection of cryptic Wolbachia in the African tsetse fly (*Glossina spp.*). BMC Microbiol 18:140. https://doi.org/10.1186/s12866-018-1291-8

Schneider DI et al (2019) Spatio-temporal distribution of *Spiroplasma* infections in the tsetse fly (*Glossina fuscipes fuscipes*) in northern Uganda. PLoS Negl Trop Dis 13:e0007340. https://doi.org/10.1371/journal.pntd.0007340

Schroder D et al (1996) Intracellular endosymbiotic bacteria of *Camponotus* species (carpenter ants): systematics, evolution and ultrastructural characterization. Mol Microbiol 21:479–489. https://doi.org/10.1111/j.1365-2958.1996.tb02557.x

Serbus LR, Sullivan W (2007) A cellular basis for *Wolbachia* recruitment to the host germline. PLoS Pathog 3:e190. https://doi.org/10.1371/journal.ppat.0030190

Serbus LR, Casper-Lindley C, Landmann F, Sullivan W (2008) The genetics and cell biology of *Wolbachia*-host interactions. Annu Rev Genet 42:683–707. https://doi.org/10.1146/annurev.genet.41.110306.130354

Serbus LR, Ferreccio A, Zhukova M, McMorris CL, Kiseleva E, Sullivan W (2011) A feedback loop between *Wolbachia* and the *Drosophila* gurken mRNP complex influences *Wolbachia* titer. J Cell Sci 124:4299–4308. https://doi.org/10.1242/jcs.092510

Simo G et al (2019) Molecular identification of *Wolbachia* and *Sodalis glossinidius* in the midgut of *Glossina fuscipes quanzensis* from the Democratic Republic of Congo. Parasite 26:5. https://doi.org/10.1051/parasite/2019005

Sinkins SP, Gould F (2006) Gene drive systems for insect disease vectors. Nat Rev Genet 7:427–435. https://doi.org/10.1038/nrg1870

Sinkins S, Curtis C, O'Neill SL (1997) The potential application of inherited symbiont systems to pest control. In: Influential passengers: inherited microorganisms and arthropod reproduction. Oxford University Press, Oxford, pp 155–175

Smith CL, Weiss BL, Aksoy S, Runyen-Janecky LJ (2013) Characterization of the achromobactin iron acquisition operon in *Sodalis glossinidius*. Appl Environ Microbiol 79:2872–2881. https://doi.org/10.1128/AEM.03959-12

Snyder AK, Rio RV (2015) "*Wigglesworthia morsitans*" folate (Vitamin B9) biosynthesis contributes to tsetse host fitness. Appl Environ Microbiol 81:5375–5386. https://doi.org/10.1128/AEM.00553-15

Snyder AK, Deberry JW, Runyen-Janecky L, Rio RV (2010) Nutrient provisioning facilitates homeostasis between tsetse fly (Diptera: Glossinidae) symbionts. Proc Biol Sci 277:2389–2397. https://doi.org/10.1098/rspb.2010.0364

Snyder AK, McLain C, Rio RV (2012) The tsetse fly obligate mutualist *Wigglesworthia morsitans* alters gene expression and population density via exogenous nutrient provisioning. Appl Environ Microbiol 78:7792–7797. https://doi.org/10.1128/AEM.02052-12

Soumana IH, Simo G, Njiokou F, Tchicaya B, Abd-Alla AM, Cuny G, Geiger A (2013) The bacterial flora of tsetse fly midgut and its effect on trypanosome transmission. J Invertebr Pathol 112. Suppl:S89–S93. https://doi.org/10.1016/j.jip.2012.03.029

Starr DJ, Cline TW (2002) A host parasite interaction rescues *Drosophila* oogenesis defects. Nature 418:76–79. https://doi.org/10.1038/nature00843

Stouthamer R, Breeuwer JA, Hurst GD (1999) *Wolbachia pipientis*: microbial manipulator of arthropod reproduction. Annu Rev Microbiol 53:71–102. https://doi.org/10.1146/annurev.micro.53.1.71

Strunov A, Kiseleva E (2016) *Drosophila melanogaster* brain invasion: pathogenic *Wolbachia* in central nervous system of the fly. Insect Sci 23:253–264. https://doi.org/10.1111/1744-7917.12187

Stuhlmann F (1907) Beitrage zur Kenntnis der Tsetsefliege (*G. fusca und G. tachinoides*). Arb Gesundh Amte (Berlin) xxxiii:489–536

Sugimoto TN, Kayukawa T, Shinoda T, Ishikawa Y, Tsuchida T (2015) Misdirection of dosage compensation underlies bidirectional sex-specific death in *Wolbachia*-infected *Ostrinia scapulalis*. Insect Biochem Mol Biol 66:72–76. https://doi.org/10.1016/j.ibmb.2015.10.001

Symula RE et al (2011) Influence of host phylogeographic patterns and incomplete lineage sorting on within-species genetic variability in *Wigglesworthia* species, obligate symbionts of tsetse flies. Appl Environ Microbiol 77:8400–8408. https://doi.org/10.1128/AEM.05688-11

Symula RE et al (2013) *Wolbachia* association with the tsetse fly, *Glossina fuscipes fuscipes*, reveals high levels of genetic diversity and complex evolutionary dynamics. BMC Evol Biol 13:31. https://doi.org/10.1186/1471-2148-13-31

Teixeira L, Ferreira A, Ashburner M (2008) The bacterial symbiont *Wolbachia* induces resistance to RNA viral infections in *Drosophila melanogaster*. PLoS Biol 6:e2. https://doi.org/10.1371/journal.pbio.1000002

Tobe SS (1978) Reproductive physiology of *Glossina*. Annu Rev Entomol 23:283–307. https://doi.org/10.1146/annurev.en.23.010178.001435

Toh H, Weiss BL, Perkin SA, Yamashita A, Oshima K, Hattori M, Aksoy S (2006) Massive genome erosion and functional adaptations provide insights into the symbiotic lifestyle of *Sodalis glossinidius* in the tsetse host. Genome Res 16:149–156. https://doi.org/10.1101/gr.4106106

Tram U, Sullivan W (2002) Role of delayed nuclear envelope breakdown and mitosis in *Wolbachia*-induced cytoplasmic incompatibility. Science 296:1124–1126. https://doi.org/10.1126/science.1070536

Tram U, Ferree PM, Sullivan W (2003) Identification of *Wolbachia*-host interacting factors through cytological analysis. Microbes Infect 5:999–1011. https://doi.org/10.1016/s1286-4579(03)00192-8

Tsagmo Ngoune JM et al (2019) The composition and abundance of bacterial communities residing in the gut of *Glossina palpalis palpalis* captured in two sites of southern Cameroon. Parasit Vectors 12:151. https://doi.org/10.1186/s13071-019-3402-2

Turelli M et al (2018) Rapid global spread of *w*Ri-like *Wolbachia* across multiple *Drosophila*. Curr Biol 28 e968:963–971. https://doi.org/10.1016/j.cub.2018.02.015

Unterman BM, Baumann P, McLean DL (1989) Pea aphid symbiont relationships established by analysis of 16S rRNAs. J Bacteriol 171:2970–2974. https://doi.org/10.1128/jb.171.6.2970-2974.1989

Vanderplank FL (1948) Experiments in crossbreeding tsetse-flies, *Glossina* species. Ann Trop Med Parasitol 42:131–152. https://doi.org/10.1080/00034983.1948.11685357

Vreysen MJ, Seck MT, Sall B, Bouyer J (2013) Tsetse flies: their biology and control using area-wide integrated pest management approaches. J Invertebr Pathol 112. Suppl:S15–S25. https://doi.org/10.1016/j.jip.2012.07.026

Walker T et al (2011) The *w*Mel *Wolbachia* strain blocks dengue and invades caged *Aedes aegypti* populations. Nature 476:450–453. https://doi.org/10.1038/nature10355

Wamwiri FN et al (2013) *Wolbachia, Sodalis* and trypanosome co-infections in natural populations of *Glossina austeni* and *Glossina pallidipes*. Parasit Vectors 6:232. https://doi.org/10.1186/1756-3305-6-232

Wang J, Aksoy S (2012) PGRP-LB is a maternally transmitted immune milk protein that influences symbiosis and parasitism in tsetse's offspring. Proc Natl Acad Sci U S A 109:10552–10557. https://doi.org/10.1073/pnas.1116431109

Wang W, Wen B, Gasparich GE, Zhu N, Rong L, Chen J, Xu Z (2004) A spiroplasma associated with tremor disease in the Chinese mitten crab (*Eriocheir sinensis*). Microbiology 150:3035–3040

Wang J, Wu Y, Yang G, Aksoy S (2009) Interactions between mutualist *Wigglesworthia* and tsetse peptidoglycan recognition protein (PGRP-LB) influence trypanosome transmission. Proc Natl Acad Sci U S A 106:12133–12138. https://doi.org/10.1073/pnas.0901226106

Wang Y, Carolan JC, Hao F, Nicholson JK, Wilkinson TL, Douglas AE (2010) Integrated metabonomic-proteomic analysis of an insect-bacterial symbiotic system. J Proteome Res 9:1257–1267. https://doi.org/10.1021/pr9007392

Weiss BL, Mouchotte R, Rio RV, Wu YN, Wu Z, Heddi A, Aksoy S (2006) Inter-specific transfer of bacterial endosymbionts between tsetse species: infection establishment and effect on host fitness. Appl Environ Microbiol 72:7013–7021

Weiss BL, Wu Y, Schwank JJ, Tolwinski NS, Aksoy S (2008) An insect symbiosis is influenced by bacterium-specific polymorphisms in outer-membrane protein A. Proc Natl Acad Sci U S A 105:15088–15093. https://doi.org/10.1073/pnas.0805666105

Weiss BL, Wang J, Aksoy S (2011) Tsetse immune system maturation requires the presence of obligate symbionts in larvae. PLoS Biol 9:e1000619. https://doi.org/10.1371/journal.pbio.1000619

Welburn SC, Maudlin I, Ellis DS (1987) In vitro cultivation of rickettsia-like-organisms from *Glossina* spp. Ann Trop Med Parasit 81:331–335

Werren JH (1997) *Wolbachia* run amok. Proc Natl Acad Sci U S A 94:11154–11155. https://doi.org/10.1073/pnas.94.21.11154

Werren JH, Zhang W, Guo LR (1995) Evolution and phylogeny of *Wolbachia*: reproductive parasites of arthropods. Proc Biol Sci 261:55–63. https://doi.org/10.1098/rspb.1995.0117

Wigglesworth VB (1929) Digestion in the tsetse-fly: a study of structure and function. Parasitology 21:288–321

Woolfit M, Iturbe-Ormaetxe I, McGraw EA, O'Neill SL (2009) An ancient horizontal gene transfer between mosquito and the endosymbiotic bacterium *Wolbachia pipientis*. Mol Biol Evol 26:367–374. https://doi.org/10.1093/molbev/msn253

Xi Z, Khoo CCH, Dobson SL (2005) *Wolbachia* establishment and invasion in an *Aedes aegypti* laboratory population. Science 310:326–328. https://doi.org/10.1126/science.1117607

Yen JH, Barr AR (1973) The etiological agent of cytoplasmic incompatibility in *Culex pipiens*. J Invertebr Pathol 22:242–250. https://doi.org/10.1016/0022-2011(73)90141-9

Yuan L-L et al (2015) Quantitative proteomic analyses of molecular mechanisms associated with Cytoplasmic Incompatibility in *Drosophila melanogaster* induced by *Wolbachia*. J Proteome Res 14:3835–3847. https://doi.org/10.1021/acs.jproteome.5b00191

Yun Y, Lei C, Peng Y, Liu F, Chen J, Chen L (2011) *Wolbachia* strains typing in different geographic population spider, *Hylyphantes graminicola* (Linyphiidae). Curr Microbiol 62:139–145. https://doi.org/10.1007/s00284-010-9686-2

Zabalou S, Apostolaki A, Livadaras I, Franz G, Robinson AS, Savakis C, Bourtzis K (2009) Incompatible insect technique: incompatible males from a *Ceratitis capitata* genetic sexing strain. Entomol Exp Appl 132:232–240. https://doi.org/10.1111/j.1570-7458.2009.00886.x

Zaidman-Remy A et al (2006) The *Drosophila* amidase PGRP-LB modulates the immune response to bacterial infection. Immunity 24:463–473. https://doi.org/10.1016/j.immuni.2006.02.012

Zhang D, Lees RS, Xi Z, Gilles JRL, Bourtzis K (2015a) Combining the Sterile Insect Technique with *Wolbachia*-based approaches: II-A safer approach to *Aedes albopictus* population suppression programmes, designed to minimize the consequences of inadvertent female release. PLoS One 10:e0135194. https://doi.org/10.1371/journal.pone.0135194

Zhang D, Zheng X, Xi Z, Bourtzis K, Gilles JRL (2015b) Combining the Sterile Insect Technique with the Incompatible Insect Technique: I-Impact of *Wolbachia* infection on the fitness of triple- and double-infected strains of *Aedes albopictus*. PLoS One 10:e0121126. https://doi.org/10.1371/journal.pone.0121126

Zhang D, Lees RS, Xi Z, Bourtzis K, Gilles JRL (2016) Combining the Sterile Insect Technique with the Incompatible Insect Technique: III-Robust mating competitiveness of irradiated triple *Wolbachia*-infected *Aedes albopictus* males under semi-field conditions. PLoS One 11: e0151864. https://doi.org/10.1371/journal.pone.0151864

Zheng Y, Wang J-L, Liu C, Wang C-P, Walker T, Wang Y-F (2011) Differentially expressed profiles in the larval testes of *Wolbachia* infected and uninfected *Drosophila*. BMC Genomics 12:595. https://doi.org/10.1186/1471-2164-12-595

Zhou JJ (2010) Odorant-binding proteins in insects. Vitam Horm 83:241–272. https://doi.org/10.1016/S0083-6729(10)83010-9

Zhou W, Rousset F, O'Neil S (1998) Phylogeny and PCR-based classification of *Wolbachia* strains using wsp gene sequences. Proc Biol Sci 265:509–515. https://doi.org/10.1098/rspb.1998.0324

Zientz E, Dandekar T, Gross R (2004) Metabolic interdependence of obligate intracellular bacteria and their insect hosts. Microbiol Mol Biol Rev 68:745–770. https://doi.org/10.1128/MMBR.68.4.745-770.2004

Part V
Symbiosis, Adaptive and Immune Responses, and Therapeutic Interventions

Chapter 20
Our Microbiome: On the Challenges, Promises, and Hype

Sara Federici, Jotham Suez, and Eran Elinav

Abstract The microbiome field is increasingly raising interest among scientists, clinicians, biopharmaceutical entities, and the general public. Technological advances from the past two decades have enabled the rapid expansion of our ability to characterize the human microbiome in depth, highlighting its previously under-appreciated role in contributing to multifactorial diseases including those with unknown etiology. Consequently, there is growing evidence that the microbiome could be utilized in medical diagnosis and patient stratification. Moreover, multiple gut microbes and their metabolic products may be bioactive, thereby serving as future potential microbiome-targeting or -associated therapeutics. Such therapies could include new generation probiotics, prebiotics, fecal microbiota transplantations, postbiotics, and dietary modulators. However, microbiome research has also been associated with significant limitations, technical and conceptual challenges, and, at times, "over-hyped" expectations that microbiome research will produce quick solutions to chronic and mechanistically complex human disorders. Herein, we summarize these challenges and also discuss some of the realistic promises associated with microbiome research and its applicability into clinical application.

Keywords Intestinal microbiome · Nutrition · Fecal microbiome transplant · Probiotics

Sara Federici and Jotham Suez contributed equally with all other contributors.

S. Federici · J. Suez · E. Elinav (✉)
Immunology Department, Weizmann Institute of Science, Rehovot, Israel
e-mail: eran.elinav@weizmann.ac.il

M. Kloc (ed.), *Symbiosis: Cellular, Molecular, Medical and Evolutionary Aspects*, Results and Problems in Cell Differentiation 69,
https://doi.org/10.1007/978-3-030-51849-3_20

20.1 Introduction

The last decade was marked by an extraordinary number of reports discussing how microbiome composition associates with human health. Fewer studies have demonstrated its causal role in the pathogenesis of several conditions. The microbiome plasticity, contrasting with that of the human genome, renders it an attractive target for the development of therapeutics. However, microbiome research also suffers from a descriptive level of evidence, lack of causality, molecular-level understanding of mechanisms, and empiric evidence, leading to premature claims of microbiome-mediated treatments. Thus, there is a sharp contrast between public expectations and perception of the microbiome field to actual applications already available. This "hyper-hype" situation enables the bloom of unregulated and unsupervised microbiome-targeting therapeutics. In this perspective, we will discuss limitations, challenges, and potential solutions supporting the utilization of the microbiome in several clinical contexts. Transforming the microbiome field toward a molecular-level mechanistic understanding of its role in physiological and pathophysiological processes may lead to the development of robust medical exploitation of the ecosystem toward better diagnosis, prophylaxis, and treatment of a myriad of "multifactorial" disorders.

20.2 Promises in Microbiome Research

Beneficial modulation of the microbiome for therapeutic purposes is currently a major focus of translational research in the field. In this section, we will discuss recent advances, the level of evidence for each application, and challenges to be addressed before widespread implementation (Fig. 20.1). A hallmark of many of these approaches is microbiome heterogeneity in the human population and its related challenges and advantages.

20.2.1 Dietary and "Prebiotic" Microbiome Interventions

While most evidence point to the stability of the gut microbiome configuration in healthy adults (Mehta et al. 2018), diet is among the strongest microbiome modulators, with robust effects observed even following short exposure to an intervention (Sonnenburg et al. 2016). The relative ease of altering one's diet and reports on beneficial health outcomes in the host following diet-induced microbiome alterations (Anhê et al. 2017) render it an attractive therapeutic approach. Of the various microbiome-modulating nutrients, dietary fibers emerge as key players. Individuals consuming a fiber-rich diet harbor a higher abundance of bacteria producing short-chain fatty acids (SCFA), which lead to improved metabolic health parameters (Zhao

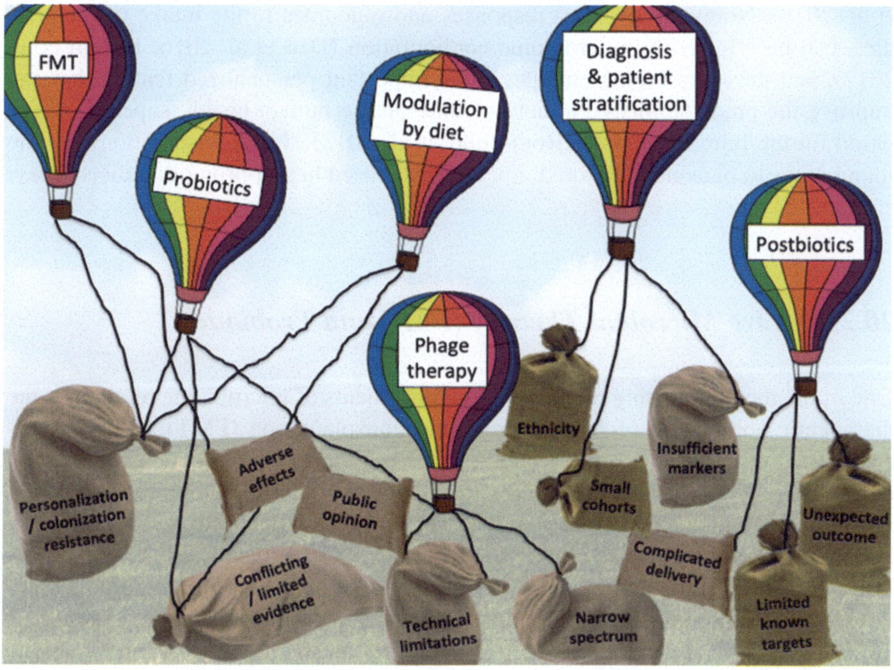

Fig. 20.1 Promises (balloons) and challenges (sacks) associated with microbiome research. The higher the balloon, the closest the applicability in the clinic

et al. 2018) and a beneficial outcome in IBD (Schroeder et al. 2018). Fibers could also improve gut barrier, either by restoring physical host–microbiome separation at the mucosal surface (Zou et al. 2018) or by correcting mucus layer defects (Schroeder et al. 2018), and protect against infections (Desai et al. 2016; Hryckowian et al. 2018). Nonetheless, dietary fiber may have a detrimental effect on hepatocellular carcinoma (Singh et al. 2018b). On the other side of the spectrum, significant reduction of carbohydrates to produce a ketogenic diet (KD) may be beneficial for the treatment of refractory epilepsy. This effect may be mediated by the microbiome, as KD promotes the bloom of *Akkermansia muciniphila* and *Parabacteroides spp.*, which were linked to systemic reductions in gamma-glutamylated amino acids, elevated hippocampal gamma-aminobutyric acid/glutamate levels, and consequently seizure protection (Olson et al. 2018). In addition to the nutrient balance of the diet, restricting the quantity and timing of feeding may beneficially affect the host through the microbiome. Individuals with obesity undergoing either short- (Dao et al. 2016) or long-term (Ruiz et al. 2017) caloric restriction diets experience a bloom of potentially beneficial taxa, and short-term restriction was also associated with improved insulin sensitivity. Microbiome transfer from mice undergoing experimental autoimmune encephalomyelitis and intermittent fasting ameliorated clinical score and spinal cord pathology; however, intermittent fasting did not improve clinical outcomes in individuals with multiple sclerosis (Cignarella et al. 2018). Noteworthily, host responses and outcomes to the intake of identical diets can be affected by microbiome configuration (Dao et al. 2016; Korem et al. 2017), and the microbiome emerges as an important personalized feature that can improve the predictability of a diet outcome on the human health, superior to that based on the human genome (Rothschild et al. 2018). These observations lay the foundations to personally tailored, microbiome-based health-promoting diets (Zeevi et al. 2015).

20.2.2 Live Microbial Therapy (FMT and Probiotics)

One of the most promising translational achievements of microbiome research is the therapeutic application of fecal microbiome transplantation (FMT). Following its established efficacy in treating recurrent *Clostridium difficile* infections (van Nood et al. 2013), FMT has demonstrated efficacy against other antibiotic-resistant pathogens, including extended-spectrum beta-lactamase (ESBL) producers (Singh et al. 2018a) and vancomycin-resistant *Enterococcus* (VRE) (Caballero et al. 2017). Following antibiotics, autologous FMT rapidly restores mucosal microbiome composition and function in both the upper and lower gastrointestinal tract (Suez et al. 2018). FMT is also gaining attention as the means for correcting dysbiotic microbiome and treating other noninfectious conditions. In individuals with metabolic syndrome, insulin sensitivity is improved by FMT from lean donors, though the effect is abated after 18 weeks (Kootte et al. 2017). Allogeneic FMT also improves symptoms in the majority of patients with IBS (Mizuno et al. 2017; Johnsen et al. 2018), ulcerative colitis (Fuentes et al. 2017; Jacob et al. 2017), hepatic encephalopathy (Bajaj et al. 2017), as well as GI and behavioral symptoms in children with autistic spectrum disorders (Kang et al. 2017). Nonetheless, the aforementioned effects are transient and mostly observed only in some of the transplanted individuals. The majority of clinical trials with FMT correlate between the extent to which the recipient microbiome shifted toward the donor configuration and improvement of clinical parameters (Fuentes et al. 2017; Mizuno et al. 2017; Zuo et al. 2018), which in turn may be related to person-specific colonization resistance (Li et al. 2016), attributed in some works to pre-FMT microbiome configuration (Kootte et al. 2017). In order to circumvent this microbiome-conferred resistance, several trials have applied pre-FMT antibiotics and/or bowel lavage (Bajaj et al. 2017; Kang et al. 2017), although the contribution of these practices to the outcome is yet unclear. As a result, these treatments are currently not available to the public, and patients turn to homemade self-treatments, which expose them to potentially serious adverse effects.

A common additional live microbial therapy approach is the consumption of a limited consortium of so-called probiotic microorganisms, mostly from the *Lactobacillus* and *Bifidobacterium* genera. Despite decades of research, studies on health claims or probiotics are often contested due to conflicting reports, for example, recent publicly funded large-scale studies showing no beneficial effect in the context of acute gastroenteritis (Freedman et al. 2018; Schnadower et al. 2018). Heterogeneity in therapeutic effects of probiotics may stem from variable capacity of probiotics to colonize the gut, either transiently during supplementation (Zmora et al. 2018) or in a persistent manner following cessation (Maldonado-Gómez et al. 2016; Zhang et al. 2016), as both were only observed in a subset of individuals (Zmora et al. 2018). Importantly, while fecal shedding does not reflect mucosal colonization, the fecal microbiome can be used to predict permissiveness or resistance to colonization (Zmora et al. 2018). Lack of colonization may limit the ability of probiotics to affect the gut microbiome (Zhang et al. 2016; Zmora et al. 2018), and both colonization and an effect on the microbiome may be required to produce a physiological effect in the context of experimental colitis (Suwal et al. 2018) or depression (Abildgaard et al. 2018). Colonization resistance to probiotics may be alleviated following antibiotics treatment, a common scenario in which probiotics are consumed as the means for the prevention of antibiotic-associated diarrhea and reconstitution of the pre-antibiotics configuration. Interestingly, rather than facilitating post-antibiotics microbiome reconstitution, probiotics may in fact delay the restoration of bacterial diversity in mice (Grazul et al. 2016; Suez et al. 2018) and in humans (Kabbani et al. 2017; Suez et al. 2018), which may explain some of the recent associations made between probiotics administered in the context of antibiotics and increased risk of infections (Spinler et al. 2016; Carvour et al. 2018; Oliveira and Widmer 2018).

To conclude, live microbial therapy is currently limited in efficacy. In parallel to addressing safety-related issues, through a better understanding of the interactions between the resident microbiome and supplemented microorganisms (either as probiotics or FMT) we can potentially tailor therapies that will bypass colonization resistance and successfully colonize the GI tract of the individual. An additional focus of research should be on the development of "new-generation probiotics," consisting of strains of gut-residing microbes that have shown benefits in pre-clinical models are being explored and tested in humans (O'Toole et al. 2017).

20.2.3 "Postbiotic" Approach

A more refined approach (termed "postbiotics") focuses on the administration of microbiome-derived bioactive molecules, which has the advantage of bypassing colonization resistance to the bacteria that express them. In addition, natural production of microbial metabolites often relies on the co-existence of a dietary nutrient (e.g., prebiotic fiber) and the presence of a metabolizing commensal, but the guts of individuals not harboring the commensal will not produce the metabolite. Administering the postbiotic product itself circumvents this personalization-related limitation. Noteworthy recent examples are as follows: flavonoid supplementation protected from diet-induced obesity (Thaiss et al. 2016); Muramyl dipeptide of Gram-positive bacteria reduced adipocyte inflammation and insulin tolerance in mice (Cavallari et al. 2017); and a membrane protein from *Akkermansia muciniphila* improved metabolism in obese and diabetic mice (Plovier et al. 2017). Bioactive molecules can also target the microbiome, as demonstrated by inhibition of trimethylamine production by the administration of a choline analog, potentially reducing atherosclerosis risk (Roberts et al. 2018b).

Major challenges to this approach are understanding the response of the microbiome and the host to the postbiotic metabolite, which may disrupt natural regulatory circuits of its levels or activity, potentially leading to resistance or loss of natural production. The pharmacokinetics of the metabolite should be dissected or improved for it to reach the target site in active concentrations. As with other drugs, the metabolite should be stable and available for mass production.

20.2.4 Microbiome Engineering

Multiple approaches fall under this broad definition, including the targeted elimination of pathogens, pathobionts, or commensals, e.g., using bacteriophages, or the introduction of strains with a novel engineered trait. Few recent in vivo examples of the latter include strains engineered to increase the immune response to tumors (Zheng et al. 2017), or as biosensors to detect markers of inflammation in the gut (Riglar et al. 2017). Coadministering a nutrient that the strains have been engineered to exclusively utilize in the gut may assist in circumventing colonization resistance to the newly introduced strains (Shepherd et al. 2018).

Utilizing bacteriophages to eliminate pathogens has several advantages over antibiotics: reduced risk of promoting the spread of antibiotics resistance; specificity to a bacterial epitope, thus not disrupting the microbial community or the host; phage infection is self-limiting; and finally, the ease of isolating phages from the environment results in lower costs. Efficacy of phage therapy against multiple pathogens has so far been demonstrated in vivo, with few anecdotal case reports in humans and clinical trials performed thus far (Furfaro et al. 2018). This approach may be further broadened to eliminate pathobionts and commensals. Nonetheless, efficient phage therapy will require overcoming bacterial anti-phage resistance (Asija and Teschke 2018), which may benefit from better understanding or recently described phage-cooperation mechanisms (Erez et al. 2017; Borges et al. 2018; Landsberger et al. 2018).

20.3 Microbiome in Patient Stratification

Diverging from the generalized "one size fits all" approach, precision medicine strives to utilize individual-specific traits, measurements, and preferences in order to achieve improved efficacy and minimize side effects of treatment and prophylaxis modalities. Inter-individual variations in the presence, absence, and quantity of commensal microorganisms offer a formidable additional array of markers that can be used for patient stratification or improved prophylaxis, with several notable advances made in recent years.

20.3.1 Microbiome as a Diagnostic Tool

The microbiome can harbor markers useful for early diagnosis and disease-risk prediction, superior to other, more invasive diagnostic tools. For example, children with a high risk for developing type 1 diabetes mellitus (T1DM) exhibit dysbiosis, decreased alpha diversity, and distinct microbiome-associated fecal and serum metabolites even before the overt manifestations of the disease (Vatanen et al. 2018). Levels of specific gut bacteria, including *Fusobacterium nucleatum*, could accurately distinguish between colorectal cancer patients and controls (Yu et al. 2017). In patients with nonalcoholic fatty liver disease, the microbiome can noninvasively classify advanced fibrosis or milder presentations (Loomba et al. 2017). In addition, microbial signatures can distinguish between patients with cirrhosis and early hepatocellular carcinoma (Ren et al. 2018b). In pediatric ulcerative colitis patients, microbiome markers were associated with remission, refractory disease, and severity (Schirmer et al. 2018). Translating these works into practice will require further validations in multiple cohorts, as well as identifying key taxonomic markers or metabolites from the gut microbiome.

20.3.2 A Drug for Each Bug?

Nonantibiotic drugs with a human target, especially proton-pump inhibitors and antipsychotics, can interact with the gut microbiome, potentially resulting in modulated activity or toxicity (Spanogiannopoulos et al. 2016; Maier et al. 2018). Prominent examples include the anti-diabetic drug Metformin, which was recently demonstrated to exert its beneficial effect by modulating the microbiome (Wu et al. 2017). Another example was recently described in the context of anti-Programmed cell death protein 1 (PD-1) checkpoint blockade immunotherapies, used as cancer therapies, but effective only in a subset of patients. Stratifying patients into "responders" and "nonresponders," specific microbiome signatures were found between these groups, with a causative role in mediating the effect of anti-PD1 therapy (Gopalakrishnan et al. 2018; Matson et al. 2018; Routy et al. 2018). Understanding drug–microbiome interactions could enable us to better choose between existing therapies and identify microorganisms or metabolites that may be used as novel adjuvants to improve drug efficacies.

20.4 Limitations and Challenges in Microbiome Research

In addition to challenges specific to each translational aspect of microbiome research noted above, there are further limitations to consider when addressing basic science questions in the young and still-developing field of microbiome research (Fig. 20.1).

20.4.1 Effect of Ethnicity and Geography

The majority of trials studying the role of the microbiome in human health have thus far focused on individuals from industrialized societies. However, differences in diets between individuals and populations play a major role in distinguishing between their respective microbiomes, as it was observed studying microbiome from hunter-gatherers (Smits et al. 2017), and U.S. immigrants (Vangay et al. 2018). The contribution of diet was challenged by the notion that microbiome composition of vegans and carnivores in an urban environment in USA is similar (Wu et al. 2016) although this lack of distinction was hypothesized to stem from broad dietary regimen descriptions (e.g., vegan) not being sufficiently descriptive of the diet contents. When the amount of consumed plant material is taken into consideration, the effect of diet is observed (McDonald et al. 2018). Indeed, disentangling ethnicity, diet, lifestyle, and genetics is not a trivial task, especially if small groups residing in distinct regions are characterized. A study of more than 2000 adults from six ethnicities living in Amsterdam identified an effect of ethnicity on the microbiome configuration, which was also partly explained by diet or lifestyle alone (Deschasaux et al. 2018). Confounding effects of ethnicity or geographical location on microbiome configuration may be an important limitation when developing diagnostics based on microbial markers. A study encompassing 7000 individuals from 14 districts in China found that geographical signature on the microbiome surpassed that of conditions such as type-2 diabetes, metabolic syndrome, and fatty liver. Consequently, machine-learning algorithms for the prediction of disease status performed poorly when applied to a population geographically distinct than the one used for training the predictor (He et al. 2018). It is, therefore, crucial to increase the diversity of sampled cohort, not only to improve patient stratification but also to potentially recognize human ancestral health-promoting commensals that may have been lost due to industrialization (Bello et al. 2018).

20.4.2 Neglected Omes: Nonbacterial Microbiomes

Improved sequencing technologies now enable better characterization of nonbacterial members of the microbial community. The virome may affect human health and serve as a biomarker of disease. Successful treatment of CDI by sterile fecal filtrate suggests a protective role of the virome against *C. difficile* (Ott et al. 2017), and in another pilot study, FMT performed in CDI resulted in a successful outcome only in the case of high richness of the recipients' virome (Zuo et al. 2018). Enteric viruses may elicit protective immunity during gut inflammation and ameliorate colitis (Yang et al. 2016). Characterization of the gut mycome is still challenging, due to great variability in the outcome with different extraction methods, as well as poor annotation of the current fungal databases (Vesty et al. 2017). The mycome has been receiving attention as a potential marker for IBD, but with inconsistent results (Hoarau et al. 2016; Liguori et al. 2016; Sokol et al. 2017). Importantly, the mycome may exert its effects on the host through interaction with the bacterial domain (Hoarau et al. 2016). There is a great need in expanding our understanding of the nonbacterial microbiome and unlock its therapeutic potential.

20.4.3 Extraintestinal Microbiomes

Commensal bacteria may be found in any environment-associated niche of the host and were even suggested to be present in the placenta (Collado et al. 2016; Parnell et al. 2017) and the brain (Roberts et al. 2018a) although the former was recently refuted (Leiby et al. 2018). An analysis of six distinct body sites of healthy humans demonstrated temporal stability (Lloyd-Price et al. 2017), though pathology-associated shifts may occur. *Fusobacterium nucleatum*, a pathobiont of the oral microbiome, was shown to inhibit human T cell response in CRC (Nosho 2016); several studies suggested *Fusobacterium* as a good diagnostic marker for CRC, either quantifying the bacterium itself (Wong et al. 2017; Guo et al. 2018) or serum antibodies against it (Wang et al. 2016). DOCK8 deficiency causes in humans recurrent skin infections; recent metagenomic analyses of the skin virome of these patients revealed an increase in papillomavirus sequences, pointing toward the importance of biosurveillance over viral microorganisms in genetically susceptible individuals (Tirosh et al. 2018). Recently, an association was found among disease exacerbation, Th17 response, and highly transcriptionally active *Streptococcus* and *Pseudomonas* in COPD patients (Ren et al. 2018a).

20.4.4 Technical Limitations

While designing and interpreting microbiome-related trials, the following should be considered. First, the descriptive nature of the majority of microbiome studies does not enable to distinguish between incidences in which the microbiome has a causative role in a phenotype and passenger effects. Causality may be demonstrated through recapitulating a phenotype by transferring the microbial community in question to a naive animal, or the microbe-produced metabolites. Quantifying relative abundances of bacteria can be misleading, as an increase in the relative abundance of a taxon could reflect the decrease in other commensals rather than an absolute increase in the abundance of a specific bacterium in question, and differences in microbial load among samples can produce bias in the relative quantification (Gloor et al. 2017). Genuine bacterial quantification in a sample can be achieved through qPCR, statistical algorithms (Rothschild et al. 2018), or combining DNA sequencing with flow cytometry for enumeration (Vandeputte et al. 2017). Differences in the sequencing and computational analysis pipeline can lead to different results. Analyses of 16s rDNA is a well-tested, cost-effective technology that enables to obtain a taxonomic resolution of the microbiome composition. However, it bears some important limitations such as providing bacterial identity solely at the genus level, and being prone to biases due to over-amplification or diverse affinity of the primers for different species. New algorithms utilizing error profiles, such as DADA2 and Deblur, now enable higher resolution analyses (Callahan et al. 2016; Amir et al. 2017). In addition, taxonomic assignment is highly dependent on reference databases, which are incomplete. Shotgun metagenomics analysis provides considerably more information, including functional insights and strain-level resolution; however, it is also prone to bias, mostly due to the impact of the DNA extraction method (Costea et al. 2017).

20.5 Conclusions and Prospects

The aforementioned challenges may seem to be discouraging, yet they may serve as a guide that distinguishes between microbiome-related discoveries that are already or will potentially be ripe for clinical application in the near future, to basic science questions that still lack fundamental elements before they can be applied. Of the aforementioned, microbiome targeting through FMT or probiotics is already practiced although both methods are associated with inconsistent reports of efficacy for multiple conditions. There is a great need for additional clinical trials with FMT, as well as nonbiased, publicly funded trials regarding probiotics, yet both methods are likely to benefit from identifying factors mediating colonization resistance and how to circumvent it. Integrating the microbiome to precision medicine can assist in improving diagnosis, prophylaxis, and prognosis, but thus far is unrealistic on a broad scale due to the cost of sequencing an individual's microbiome and the complexity of the analysis. In addition, it is crucial to identify markers that are applicable across distinct populations. Basic science questions remain to be addressed before other therapeutic approaches, including postbiotics supplement, microbiome engineering, and phage therapy, are proven to be safe and efficacious. Importantly, maintaining an effect of microbiome-based therapies may be affected by the host genetics, diet, or lifestyle (Kootte et al. 2017; Smits et al. 2018). Thus, the microbiome may serve as the first step for disease amelioration, but long-term maintenance requires further adaptations from the patient's side.

Acknowledgments The authors thank the Elinav lab for fruitful discussions and apologize to those authors whose works could not be cited due to space limitations. S.F. was supported by a Sergio Lombroso Foundation fellowship. J.S. is the recipient of the Strauss Institute research fellowship. E.E. is supported by Y. and R. Ungar; the Abisch Frenkel Foundation for the Promotion of Life Sciences; the Gurwin Family Fund for Scientific Research; the Leona M. and Harry B. Helmsley Charitable Trust; the Crown Endowment Fund for Immunological Research; the Else Kroener Fresenius Foundation; the estate of J. Gitlitz; the estate of L. Hershkovich; the Benoziyo Endowment Fund for the Advancement of Science; the Adelis Foundation; J. L. and V. Schwartz; A. and G. Markovitz; A. and C. Adelson; the French National Center for Scientific Research (CNRS); D. L. Schwarz; the V. R. Schwartz Research Fellow Chair; L. Steinberg; J. N. Halpern; A. Edelheit, and by grants funded by the European Research Council; a Marie Curie Integration grant; the German–Israeli Foundation for Scientific Research and Development; the Israel Science Foundation; the Minerva Foundation; the Rising Tide Foundation; the Helmholtz Foundation; and the European Foundation for the Study of Diabetes. E.E. is a senior fellow, Canadian Institute of Advanced Research (CIFAR), and an international scholar, The Bill & Melinda Gates Foundation & Howard Hughes Medical Institute (HHMI).

Conflict of Interest Eran Elinav is a paid consultant to DayTwo and BiomX.

References

Abildgaard A, Kern T, Pedersen O, Hansen T, Wegener G, Lund S (2018) The antidepressant-like effect of probiotics and their faecal abundance may be modulated by the cohabiting gut microbiota in rats. Eur Neuropsychopharmacol 29(1):98–110. https://doi.org/10.1016/j.euroneuro.2018.10.011

Amir A, McDonald D, Navas-Molina JA, Kopylova E, Morton JT, Zech Xu Z, Kightley EP, Thompson LR, Hyde ER, Gonzalez A, Knight R (2017) Deblur rapidly resolves single-nucleotide community sequence patterns. mSystems 2:e00191-16. https://doi.org/10.1128/mSystems.00191-16

Anhê FF, Nachbar RT, Varin TV, Vilela V, Dudonné S, Pilon G, Fournier M, Lecours M-A, Desjardins Y, Roy D, Levy E, Marette A (2017) A polyphenol-rich cranberry extract reverses insulin resistance and hepatic steatosis independently of body weight loss. Mol Metab 6:1563–1573. https://doi.org/10.1016/j.molmet.2017.10.003

Asija K, Teschke CM (2018) Lessons from bacteriophages part 2: a saga of scientific breakthroughs and prospects for their use in human health. PLoS Pathog 14:e1006970. https://doi.org/10.1371/journal.ppat.1006970

Bajaj JS, Kassam Z, Fagan A, Gavis EA, Liu E, Cox IJ, Kheradman R, Heuman D, Wang J, Gurry T, Williams R, Sikaroodi M, Fuchs M, Alm E, John B, Thacker LR, Riva A, Smith M, Taylor-Robinson SD, Gillevet PM (2017) Fecal microbiota transplant from a rational stool donor improves hepatic encephalopathy: a randomized clinical trial. Hepatology 66:1727–1738. https://doi.org/10.1002/hep.29306

Bello MGD, Knight R, Gilbert JA, Blaser MJ (2018) Preserving microbial diversity. Science 362:33–34. https://doi.org/10.1126/science.aau8816

Borges AL, Zhang JY, Rollins MF, Osuna BA, Wiedenheft B, Bondy-Denomy J (2018) Bacteriophage Cooperation Suppresses CRISPR-Cas3 and Cas9 Immunity. Cell 174:917–925. https://doi.org/10.1016/j.cell.2018.06.013

Caballero S, Kim S, Carter RA, Leiner IM, Sušac B, Miller L, Kim GJ, Ling L, Pamer EG (2017) Cooperating commensals restore colonization resistance to vancomycin-resistant enterococcus faecium. Cell Host Microbe 21:592–602.e4. https://doi.org/10.1016/j.chom.2017.04.002

Callahan BJ, McMurdie PJ, Rosen MJ, Han AW, Johnson AJA, Holmes SP (2016) DADA2: high-resolution sample inference from Illumina amplicon data. Nat Methods 13:581–583. https://doi.org/10.1038/nmeth.3869

Carvour ML, Wilder SL, Ryan KL, Walraven C, Qeadan F, Brett M, Page K (2018) Predictors of Clostridium difficile infection and predictive impact of probiotic use in a diverse hospital-wide cohort. Am J Infect Control 47:2–8. https://doi.org/10.1016/j.ajic.2018.07.014

Cavallari JF, Fullerton MD, Duggan BM, Foley KP, Denou E, Smith BK, Desjardins EM, Henriksbo BD, Kim KJ, Tuinema BR, Stearns JC, Prescott D, Rosenstiel P, Coombes BK, Steinberg GR, Schertzer JD (2017) Muramyl dipeptide-based postbiotics mitigate obesity-induced insulin resistance via IRF4. Cell Metab 25:1063–1074.e3. https://doi.org/10.1016/j.cmet.2017.03.021

Cignarella F, Cantoni C, Ghezzi L, Salter A, Dorsett Y, Chen L, Phillips D, Weinstock GM, Fontana L, Cross AH, Zhou Y, Piccio L (2018) Intermittent fasting confers protection in CNS autoimmunity by altering the gut microbiota. Cell Metab 27:1222–1235.e6. https://doi.org/10.1016/j.cmet.2018.05.006

Collado MC, Rautava S, Aakko J, Isolauri E, Salminen S (2016) Human gut colonisation may be initiated in utero by distinct microbial communities in the placenta and amniotic fluid. Sci Rep 6:23129

Costea PI, Zeller G, Sunagawa S, Pelletier E, Alberti A, Levenez F, Tramontano M, Driessen M, Hercog R, Jung F-E, Kultima JR, Hayward MR, Coelho LP, Allen-Vercoe E, Bertrand L, Blaut M, Brown JRM, Carton T, Cools-Portier S, Daigneault M, Derrien M, Druesne A, de Vos WM, Finlay BB, Flint HJ, Guarner F, Hattori M, Heilig H, Luna RA, van Hylckama VJ, Junick J, Klymiuk I, Langella P, Le Chatelier E, Mai V, Manichanh C, Martin JC, Mery C, Morita H, O'Toole PW, Orvain C, Patil KR, Penders J, Persson S, Pons N, Popova M, Salonen A, Saulnier D, Scott KP, Singh B, Slezak K, Veiga P, Versalovic J, Zhao L, Zoetendal EG, Ehrlich SD, Dore J, Bork P (2017) Towards standards for human fecal sample processing in metagenomic studies. Nat Biotechnol 35:1069

Dao MC, Everard A, Aron-Wisnewsky J, Sokolovska N, Prifti E, Verger EO, Kayser BD, Levenez F, Chilloux J, Hoyles L, Dumas M-E, Rizkalla SW, Doré J, Cani PD, Clément K, MICRO-Obes Consortium, Dumas M-E, Rizkalla SW, Doré J, Cani PD, Clément K (2016) Akkermansia muciniphila and improved metabolic health during a dietary intervention in obesity: relationship with gut microbiome richness and ecology. Gut 65:426–436. https://doi.org/10.1136/gutjnl-2014-308778

Desai MS, Seekatz AM, Koropatkin NM, Kamada N, Hickey CA, Wolter M, Pudlo NA, Kitamoto S, Terrapon N, Muller A, Young VB, Henrissat B, Wilmes P, Stappenbeck TS, Núñez G, Martens EC (2016) A dietary fiber-deprived gut microbiota degrades the colonic mucus barrier and enhances pathogen susceptibility. Cell 167:1339–1353.e21. https://doi.org/10.1016/j.cell.2016.10.043

Deschasaux M, Bouter KE, Prodan A, Levin E, Groen AK, Herrema H, Tremaroli V, Bakker GJ, Attaye I, Pinto-Sietsma S-J, van Raalte DH, Snijder MB, Nicolaou M, Peters R, Zwinderman AH, Bäckhed F, Nieuwdorp M (2018) Depicting the composition of gut microbiota in a population with varied ethnic origins but shared geography. Nat Med 24:1526–1531. https://doi.org/10.1038/s41591-018-0160-1

Erez Z, Steinberger-Levy I, Shamir M, Doron S, Stokar-Avihail A, Peleg Y, Melamed S, Leavitt A, Savidor A, Albeck S, Amitai G, Sorek R (2017) Communication between viruses guides lysis-lysogeny decisions. Nature 541:488–493. https://doi.org/10.1038/nature21049

Freedman SB, Williamson-Urquhart S, Farion KJ, Gouin S, Willan AR, Poonai N, Hurley K, Sherman PM, Finkelstein Y, Lee BE, Pang X-L, Chui L, Schnadower D, Xie J, Gorelick M, Schuh S, PERC PROGUT Trial Group (2018) Multicenter trial of a combination probiotic for children with gastroenteritis. N Engl J Med 379:2015–2026. https://doi.org/10.1056/NEJMoa1802597

Fuentes S, Rossen NG, van der Spek MJ, Hartman JH, Huuskonen L, Korpela K, Salojärvi J, Aalvink S, de Vos WM, D'Haens GR, Zoetendal EG, Ponsioen CY (2017) Microbial shifts and signatures of long-term remission in ulcerative colitis after faecal microbiota transplantation. ISME J 11:1877–1889. https://doi.org/10.1038/ismej.2017.44

Furfaro LL, Payne MS, Chang BJ (2018) Bacteriophage therapy: clinical trials and regulatory hurdles. Front Cell Infect Microbiol 8:376. https://doi.org/10.3389/fcimb.2018.00376

Gloor GB, Macklaim JM, Pawlowsky-Glahn V, Egozcue JJ (2017) Microbiome datasets are compositional: and this is not optional. Front Microbiol 8:2224. https://doi.org/10.3389/fmicb.2017.02224

Gopalakrishnan V, Spencer CN, Nezi L, Reuben A, Andrews MC, Karpinets TV, Prieto PA, Vicente D, Hoffman K, Wei SC, Cogdill AP, Zhao L, Hudgens CW, Hutchinson DS, Manzo T, Petaccia de Macedo M, Cotechini T, Kumar T, Chen WS, Reddy SM, Szczepaniak Sloane R, Galloway-Pena J, Jiang H, Chen PL, Shpall EJ, Rezvani K, Alousi AM, Chemaly RF, Shelburne S, Vence LM, Okhuysen PC, Jensen VB, Swennes AG, McAllister F, Marcelo Riquelme Sanchez E, Zhang Y, Le Chatelier E, Zitvogel L, Pons N, Austin-Breneman JL, Haydu LE, Burton EM, Gardner JM, Sirmans E, Hu J, Lazar AJ, Tsujikawa T, Diab A, Tawbi H, Glitza IC, Hwu WJ, Patel SP, Woodman SE, Amaria RN, Davies MA, Gershenwald JE, Hwu P, Lee JE, Zhang J, Coussens LM, Cooper ZA, Futreal PA, Daniel CR, Ajami NJ, Petrosino JF, Tetzlaff MT, Sharma P, Allison JP, Jenq RR, Wargo JA (2018) Gut microbiome modulates response to anti-PD-1 immunotherapy in melanoma patients. Science 359:97–103. https://doi.org/10.1126/science.aan4236

Grazul H, Kanda LL, Gondek D (2016) Impact of probiotic supplements on microbiome diversity following antibiotic treatment of mice. Gut Microbes 7:101–114. https://doi.org/10.1080/19490976.2016.1138197

Guo S, Li L, Xu B, Li M, Zeng Q, Xiao H, Xue Y, Wu Y, Wang Y, Liu W, Zhang G (2018) A simple and novel fecal biomarker for colorectal cancer: ratio of fusobacterium nucleatum to probiotics populations, based on their antagonistic effect. Clin Chem 64:1327–1337. https://doi.org/10.1373/clinchem.2018.289728

He Y, Wu W, Zheng H-M, Li P, McDonald D, Sheng H-F, Chen M-X, Chen Z-H, Ji G-Y, Zheng Z-D-X, Mujagond P, Chen X-J, Rong Z-H, Chen P, Lyu L-Y, Wang X, Wu C-B, Yu N, Xu Y-J, Yin J, Raes J, Knight R, Ma W-J, Zhou H-W (2018) Regional variation limits applications of healthy gut microbiome reference ranges and disease models. Nat Med 24:1532–1535. https://doi.org/10.1038/s41591-018-0164-x

Hoarau G, Mukherjee PK, Gower-Rousseau C, Hager C, Chandra J, Retuerto MA, Neut C, Vermeire S, Clemente J, Colombel JF, Fujioka H, Poulain D, Sendid B, Ghannoum MA (2016) Bacteriome and mycobiome interactions underscore microbial dysbiosis in familial Crohn's disease. MBio 7:e01250-16. https://doi.org/10.1128/mBio.01250-16

Hryckowian AJ, Van Treuren W, Smits SA, Davis NM, Gardner JO, Bouley DM, Sonnenburg JL (2018) Microbiota-accessible carbohydrates suppress Clostridium difficile infection in a murine model. Nat Microbiol 3:662–669. https://doi.org/10.1038/s41564-018-0150-6

Jacob V, Crawford C, Cohen-Mekelburg S, Viladomiu M, Putzel GG, Schneider Y, Chabouni F, O'Neil S, Bosworth B, Woo V, Ajami NJ, Petrosino JF, Gerardin Y, Kassam Z, Smith M, Iliev ID, Sonnenberg GF, Artis D, Scherl E, Longman RS (2017) Single delivery of high-diversity fecal microbiota preparation by colonoscopy is safe and effective in increasing microbial diversity in active ulcerative colitis. Inflamm Bowel Dis 23:903–911. https://doi.org/10.1097/MIB.0000000000001132

Johnsen PH, Hilpüsch F, Cavanagh JP, Leikanger IS, Kolstad C, Valle PC, Goll R (2018) Faecal microbiota transplantation versus placebo for moderate-to-severe irritable bowel syndrome: a double-blind, randomised, placebo-controlled, parallel-group, single-centre trial. lancet. Gastroenterol Hepatol 3:17–24. https://doi.org/10.1016/S2468-1253(17)30338-2

Kabbani TA, Pallav K, Dowd SE, Villafuerte-Galvez J, Vanga RR, Castillo NE, Hansen J, Dennis M, Leffler DA, Kelly CP (2017) Prospective randomized controlled study on the effects of Saccharomyces boulardii CNCM I-745 and amoxicillin-clavulanate or the combination on the gut microbiota of healthy volunteers. Gut Microbes 8:17–32. https://doi.org/10.1080/19490976.2016.1267890

Kang D-W, Adams JB, Gregory AC, Borody T, Chittick L, Fasano A, Khoruts A, Geis E, Maldonado J, McDonough-Means S, Pollard EL, Roux S, Sadowsky MJ, Lipson KS, Sullivan MB, Caporaso JG, Krajmalnik-Brown R (2017) Microbiota Transfer Therapy alters gut ecosystem and improves gastrointestinal and autism symptoms: an open-label study. Microbiome 5:10. https://doi.org/10.1186/s40168-016-0225-7

Kootte RS, Levin E, Salojärvi J, Smits LP, Hartstra AV, Udayappan SD, Hermes G, Bouter KE, Koopen AM, Holst JJ, Knop FK, Blaak EE, Zhao J, Smidt H, Harms AC, Hankemeijer T, Bergman JJGHM, Romijn HA, Schaap FG, Olde Damink SWM, Ackermans MT, Dallinga-Thie GM, Zoetendal E, de Vos WM, Serlie MJ, Stroes ESG, Groen AK, Nieuwdorp M (2017) Improvement of insulin sensitivity after lean donor feces in metabolic syndrome is driven by baseline intestinal microbiota composition. Cell Metab 26:611–619.e6. https://doi.org/10.1016/j.cmet.2017.09.008

Korem T, Zeevi D, Zmora N, Weissbrod O, Bar N, Lotan-Pompan M, Avnit-Sagi T, Kosower N, Malka G, Rein M, Suez J, Goldberg BZ, Weinberger A, Levy AA, Elinav E, Segal E (2017) Bread affects clinical parameters and induces gut microbiome-associated personal glycemic responses. Cell Metab 25:1243–1253. https://doi.org/10.1016/j.cmet.2017.05.002

Landsberger M, Gandon S, Meaden S, Rollie C, Chevallereau A, Chabas H, Buckling A, Westra ER, van Houte S (2018) Anti-CRISPR phages cooperate to overcome CRISPR-cas immunity. Cell 174:908–916. https://doi.org/10.1016/j.cell.2018.05.058

Leiby JS, McCormick K, Sherrill-Mix S, Clarke EL, Kessler LR, Taylor LJ, Hofstaedter CE, Roche AM, Mattei LM, Bittinger K, Elovitz MA, Leite R, Parry S, Bushman FD (2018) Lack of detection of a human placenta microbiome in samples from preterm and term deliveries. Microbiome 6:196. https://doi.org/10.1186/s40168-018-0575-4

Li SS, Zhu A, Benes V, Costea PI, Hercog R, Hildebrand F, Huerta-Cepas J, Nieuwdorp M, Salojärvi J, Voigt AY, Zeller G, Sunagawa S, de Vos WM, Bork P (2016) Durable coexistence of donor and recipient strains after fecal microbiota transplantation. Science 352:586–589. https://doi.org/10.1126/science.aad8852

Liguori G, Lamas B, Richard ML, Brandi G, da Costa G, Hoffmann TW, Di Simone MP, Calabrese C, Poggioli G, Langella P, Campieri M, Sokol H (2016) Fungal dysbiosis in mucosa-associated microbiota of Crohn's disease patients. J Crohn's Colitis 10:296–305. https://doi.org/10.1093/ecco-jcc/jjv209

Lloyd-Price J, Mahurkar A, Rahnavard G, Crabtree J, Orvis J, Hall AB, Brady A, Creasy HH, McCracken C, Giglio MG, McDonald D, Franzosa EA, Knight R, White O, Huttenhower C (2017) Strains, functions and dynamics in the expanded human microbiome project. Nature 550:61

Loomba R, Seguritan V, Li W, Long T, Klitgord N, Bhatt A, Dulai PS, Caussy C, Bettencourt R, Highlander SK, Jones MB, Sirlin CB, Schnabl B, Brinkac L, Schork N, Chen C-H, Brenner DA, Biggs W, Yooseph S, Venter JC, Nelson KE (2017) Gut microbiome-based metagenomic signature for non-invasive detection of advanced fibrosis in human nonalcoholic fatty liver disease. Cell Metab 25:1054–1062.e5. https://doi.org/10.1016/j.cmet.2017.04.001

Maier L, Pruteanu M, Kuhn M, Zeller G, Telzerow A, Anderson EE, Brochado AR, Fernandez KC, Dose H, Mori H, Patil KR, Bork P, Typas A (2018) Extensive impact of non-antibiotic drugs on human gut bacteria. Nature 555:623–628. https://doi.org/10.1038/nature25979

Maldonado-Gómez MX, Martínez I, Bottacini F, O'Callaghan A, Ventura M, van Sinderen D, Hillmann B, Vangay P, Knights D, Hutkins RW, Walter J (2016) Stable engraftment of bifidobacterium longum AH1206 in the human gut depends on individualized features of the resident microbiome. Cell Host Microbe 20:515–526. https://doi.org/10.1016/j.chom.2016.09.001

Matson V, Fessler J, Bao R, Chongsuwat T, Zha Y, Alegre M-L, Luke JJ, Gajewski TF (2018) The commensal microbiome is associated with anti-PD-1 efficacy in metastatic melanoma patients. Science 359:104–108. https://doi.org/10.1126/science.aao3290

McDonald D, Hyde E, Debelius JW, Morton JT, Gonzalez A, Ackermann G, Aksenov AA, Behsaz B, Brennan C, Chen Y, DeRight GL, Dorrestein PC, Dunn RR, Fahimipour AK,

Gaffney J, Gilbert JA, Gogul G, Green JL, Hugenholtz P, Humphrey G, Huttenhower C, Jackson MA, Janssen S, Jeste DV, Jiang L, Kelley ST, Knights D, Kosciolek T, Ladau J, Leach J, Marotz C, Meleshko D, Melnik AV, Metcalf JL, Mohimani H, Montassier E, Navas-Molina J, Nguyen TT, Peddada S, Pevzner P, Pollard KS, Rahnavard G, Robbins-Pianka A, Sangwan N, Shorenstein J, Smarr L, Song SJ, Spector T, Swafford AD, Thackray VG, Thompson LR, Tripathi A, Vázquez-Baeza Y, Vrbanac A, Wischmeyer P, Wolfe E, Zhu Q, Knight R (2018) American gut: an open platform for citizen science microbiome research. mSystems 3:e00031-18

Mehta RS, Abu-Ali GS, Drew DA, Lloyd-Price J, Subramanian A, Lochhead P, Joshi AD, Ivey KL, Khalili H, Brown GT, DuLong C, Song M, Nguyen LH, Mallick H, Rimm EB, Izard J, Huttenhower C, Chan AT (2018) Stability of the human faecal microbiome in a cohort of adult men. Nat Microbiol 3:347–355. https://doi.org/10.1038/s41564-017-0096-0

Mizuno S, Masaoka T, Naganuma M, Kishimoto T, Kitazawa M, Kurokawa S, Nakashima M, Takeshita K, Suda W, Mimura M, Hattori M, Kanai T (2017) Bifidobacterium-rich fecal donor may be a positive predictor for successful fecal microbiota transplantation in patients with irritable bowel syndrome. Digestion 96:29–38. https://doi.org/10.1159/000471919

Nosho K (2016) Association of Fusobacterium nucleatum with immunity and molecular alterations in colorectal cancer. World J Gastroenterol 22:557. https://doi.org/10.3748/wjg.v22.i2.557

O'Toole PW, Marchesi JR, Hill C (2017) Next-generation probiotics: the spectrum from probiotics to live biotherapeutics. Nat Microbiol 2:17057

Oliveira BCM, Widmer G (2018) Probiotic product enhances susceptibility of mice to cryptosporidiosis. Appl Environ Microbiol 84(21):e01408-18. https://doi.org/10.1128/AEM.01408-18

Olson CA, Vuong HE, Yano JM, Liang QY, Nusbaum DJ, Hsiao EY (2018) The gut microbiota mediates the anti-seizure effects of the ketogenic diet. Cell 174:497. https://doi.org/10.1016/j.cell.2018.06.051

Ott SJ, Waetzig GH, Rehman A, Moltzau-Anderson J, Bharti R, Grasis JA, Cassidy L, Tholey A, Fickenscher H, Seegert D, Rosenstiel P, Schreiber S (2017) Efficacy of sterile fecal filtrate transfer for treating patients with clostridium difficile infection. Gastroenterology 152:799–811.e7. https://doi.org/10.1053/j.gastro.2016.11.010

Parnell LA, Briggs CM, Cao B, Delannoy-Bruno O, Schrieffer AE, Mysorekar IU (2017) Microbial communities in placentas from term normal pregnancy exhibit spatially variable profiles. Sci Rep 7:11200. https://doi.org/10.1038/s41598-017-11514-4

Plovier H, Everard A, Druart C, Depommier C, Van Hul M, Geurts L, Chilloux J, Ottman N, Duparc T, Lichtenstein L, Myridakis A, Delzenne NM, Klievink J, Bhattacharjee A, van der Ark KCH, Aalvink S, Martinez LO, Dumas M-E, Maiter D, Loumaye A, Hermans MP, Thissen J-P, Belzer C, de Vos WM, Cani PD (2017) A purified membrane protein from Akkermansia muciniphila or the pasteurized bacterium improves metabolism in obese and diabetic mice. Nat Med 23:107–113. https://doi.org/10.1038/nm.4236

Ren L, Zhang R, Rao J, Xiao Y, Zhang Z, Yang B, Cao D, Zhong H, Ning P, Shang Y, Li M, Gao Z, Wang J (2018a) Transcriptionally active lung microbiome and its association with bacterial biomass and host inflammatory status. mSystems 3:e00199-18

Ren Z, Li A, Jiang J, Zhou L, Yu Z, Lu H, Xie H, Chen X, Shao L, Zhang R, Xu S, Zhang H, Cui G, Chen X, Sun R, Wen H, Lerut JP, Kan Q, Li L, Zheng S (2018b) Gut microbiome analysis as a tool towards targeted non-invasive biomarkers for early hepatocellular carcinoma. Gut. https://doi.org/10.1136/gutjnl-2017-315084

Riglar DT, Giessen TW, Baym M, Kerns SJ, Niederhuber MJ, Bronson RT, Kotula JW, Gerber GK, Way JC, Silver PA (2017) Engineered bacteria can function in the mammalian gut long-term as live diagnostics of inflammation. Nat Biotechnol 35:653–658. https://doi.org/10.1038/nbt.3879

Roberts C, Farmer CB, Walter CK (2018a) The human brain microbiome; there are bacteria in our brains! Birmingham, AL, Psychiatry Behav Neurobio, Univ Alabama

Roberts AB, Gu X, Buffa JA, Hurd AG, Wang Z, Zhu W, Gupta N, Skye SM, Cody DB, Levison BS, Barrington WT, Russell MW, Reed JM, Duzan A, Lang JM, Fu X, Li L, Myers AJ, Rachakonda S, DiDonato JA, Brown JM, Gogonea V, Lusis AJ, Garcia-Garcia JC, Hazen SL

(2018b) Development of a gut microbe-targeted nonlethal therapeutic to inhibit thrombosis potential. Nat Med 24:1407–1417. https://doi.org/10.1038/s41591-018-0128-1

Rothschild D, Weissbrod O, Barkan E, Kurilshikov A, Korem T, Zeevi D, Costea PI, Godneva A, Kalka IN, Bar N, Shilo S, Lador D, Vila AV, Zmora N, Pevsner-Fischer M, Israeli D, Kosower N, Malka G, Wolf BC, Avnit-Sagi T, Lotan-Pompan M, Weinberger A, Halpern Z, Carmi S, Fu J, Wijmenga C, Zhernakova A, Elinav E, Segal E (2018) Environment dominates over host genetics in shaping human gut microbiota. Nature 555:210–215. https://doi.org/10.1038/nature25973

Routy B, Le Chatelier E, Derosa L, Duong CPM, Alou MT, Daillère R, Fluckiger A, Messaoudene M, Rauber C, Roberti MP, Fidelle M, Flament C, Poirier-Colame V, Opolon P, Klein C, Iribarren K, Mondragón L, Jacquelot N, Qu B, Ferrere G, Clémenson C, Mezquita L, Masip JR, Naltet C, Brosseau S, Kaderbhai C, Richard C, Rizvi H, Levenez F, Galleron N, Quinquis B, Pons N, Ryffel B, Minard-Colin V, Gonin P, Soria J-C, Deutsch E, Loriot Y, Ghiringhelli F, Zalcman G, Goldwasser F, Escudier B, Hellmann MD, Eggermont A, Raoult D, Albiges L, Kroemer G, Zitvogel L (2018) Gut microbiome influences efficacy of PD-1-based immunotherapy against epithelial tumors. Science 359:91–97. https://doi.org/10.1126/science.aan3706

Ruiz A, Cerdó T, Jáuregui R, Pieper DH, Marcos A, Clemente A, García F, Margolles A, Ferrer M, Campoy C, Suárez A (2017) One-year calorie restriction impacts gut microbial composition but not its metabolic performance in obese adolescents. Environ Microbiol 19:1536–1551. https://doi.org/10.1111/1462-2920.13713

Schirmer M, Denson L, Vlamakis H, Franzosa EA, Thomas S, Gotman NM, Rufo P, Baker SS, Sauer C, Markowitz J, Pfefferkorn M, Oliva-Hemker M, Rosh J, Otley A, Boyle B, Mack D, Baldassano R, Keljo D, LeLeiko N, Heyman M, Griffiths A, Patel AS, Noe J, Kugathasan S, Walters T, Huttenhower C, Hyams J, Xavier RJ (2018) Compositional and temporal changes in the gut microbiome of pediatric ulcerative colitis patients are linked to disease course. Cell Host Microbe 24:600–610. https://doi.org/10.1016/j.chom.2018.09.009

Schnadower D, Tarr PI, Casper TC, Gorelick MH, Dean JM, O'Connell KJ, Mahajan P, Levine AC, Bhatt SR, Roskind CG, Powell EC, Rogers AJ, Vance C, Sapien RE, Olsen CS, Metheney M, Dickey VP, Hall-Moore C, Freedman SB (2018) Lactobacillus rhamnosus GG versus placebo for acute gastroenteritis in children. N Engl J Med 379:2002–2014. https://doi.org/10.1056/NEJMoa1802598

Schroeder BO, Birchenough GMH, Ståhlman M, Arike L, Johansson MEV, Hansson GC, Bäckhed F (2018) Bifidobacteria or fiber protects against diet-induced microbiota-mediated colonic mucus deterioration. Cell Host Microbe 23:27–40.e7. https://doi.org/10.1016/j.chom.2017.11.004

Shepherd ES, DeLoache WC, Pruss KM, Whitaker WR, Sonnenburg JL (2018) An exclusive metabolic niche enables strain engraftment in the gut microbiota. Nature 557:434–438. https://doi.org/10.1038/s41586-018-0092-4

Singh R, de Groot PF, Geerlings SE, Hodiamont CJ, Belzer C, Ten Berge IJM, de Vos WM, Bemelman FJ, Nieuwdorp M (2018a) Fecal microbiota transplantation against intestinal colonization by extended spectrum beta-lactamase producing Enterobacteriaceae: a proof of principle study. BMC Res Notes 11:190. https://doi.org/10.1186/s13104-018-3293-x

Singh V, Yeoh BS, Chassaing B, Xiao X, Saha P, Aguilera Olvera R, Lapek JD, Zhang L, Wang W-B, Hao S, Flythe MD, Gonzalez DJ, Cani PD, Conejo-Garcia JR, Xiong N, Kennett MJ, Joe B, Patterson AD, Gewirtz AT, Vijay-Kumar M (2018b) Dysregulated microbial fermentation of soluble fiber induces cholestatic liver cancer. Cell 175:679–694.e22. https://doi.org/10.1016/j.cell.2018.09.004

Smits SA, Leach J, Sonnenburg ED, Gonzalez CG, Lichtman JS, Reid G, Knight R, Manjurano A, Changalucha J, Elias JE, Dominguez-Bello MG, Sonnenburg JL (2017) Seasonal cycling in the gut microbiome of the Hadza hunter-gatherers of Tanzania. Science 357:802–806. https://doi.org/10.1126/science.aan4834

Smits LP, Kootte RS, Levin E, Prodan A, Fuentes S, Zoetendal EG, Wang Z, Levison BS, Cleophas MCP, Kemper EM, Dallinga-Thie GM, Groen AK, Joosten LAB, Netea MG, Stroes ESG, de Vos WM, Hazen SL, Nieuwdorp M (2018) Effect of vegan fecal microbiota transplantation on carnitine- and choline-derived trimethylamine-N-oxide production and vascular inflammation in patients with metabolic syndrome. J Am Heart Assoc 7:e008342. https://doi.org/10.1161/JAHA.117.008342

Sokol H, Leducq V, Aschard H, Pham H-P, Jegou S, Landman C, Cohen D, Liguori G, Bourrier A, Nion-Larmurier I, Cosnes J, Seksik P, Langella P, Skurnik D, Richard ML, Beaugerie L (2017) Fungal microbiota dysbiosis in IBD. Gut 66:1039–1048. https://doi.org/10.1136/gutjnl-2015-310746

Sonnenburg ED, Smits SA, Tikhonov M, Higginbottom SK, Wingreen NS, Sonnenburg JL (2016) Diet-induced extinctions in the gut microbiota compound over generations. Nature 529:212–215. https://doi.org/10.1038/nature16504

Spanogiannopoulos P, Bess EN, Carmody RN, Turnbaugh PJ (2016) The microbial pharmacists within us: a metagenomic view of xenobiotic metabolism. Nat Rev Microbiol 14:273–287. https://doi.org/10.1038/nrmicro.2016.17

Spinler JK, Brown A, Ross CL, Boonma P, Conner ME, Savidge TC (2016) Administration of probiotic kefir to mice with clostridium difficile infection exacerbates disease. Anaerobe 40:54–57. https://doi.org/10.1016/j.anaerobe.2016.05.008

Suez J, Zmora N, Zilberman-Schapira G, Mor U, Dori-Bachash M, Bashiardes S, Zur M, Regev-Lehavi D, Ben-Zeev Brik R, Federici S, Horn M, Cohen Y, Moor AE, Zeevi D, Korem T, Kotler E, Harmelin A, Itzkovitz S, Maharshak N, Shibolet O, Pevsner-Fischer M, Shapiro H, Sharon I, Halpern Z, Segal E, Elinav E (2018) Post-antibiotic gut mucosal microbiome reconstitution is impaired by probiotics and improved by autologous FMT. Cell 174:1406–1423.e16. https://doi.org/10.1016/j.cell.2018.08.047

Suwal S, Wu Q, Liu W, Liu Q, Sun H, Liang M, Gao J, Zhang B, Kou Y, Liu Z, Wei Y, Wang Y, Zheng K (2018) The probiotic effectiveness in preventing experimental colitis is correlated with host gut microbiota. Front Microbiol 9:2675. https://doi.org/10.3389/fmicb.2018.02675

Thaiss CA, Itav S, Rothschild D, Meijer M, Levy M, Moresi C, Dohnalová L, Braverman S, Rozin S, Malitsky S, Dori-Bachash M, Kuperman Y, Biton I, Gertler A, Harmelin A, Shapiro H, Halpern Z, Aharoni A, Segal E, Elinav E (2016) Persistent microbiome alterations modulate the rate of post-dieting weight regain. Nature 540:544–551. https://doi.org/10.1038/nature20796

Tirosh O, Conlan S, Deming C, Lee-Lin S-Q, Huang X, Barnabas BB, Bouffard GG, Brooks SY, Marfani H, Dekhtyar L, Guan X, Han J, Ho S, Legaspi R, Maduro QL, Masiello CA, McDowell JC, Montemayor C, Mullikin JC, Park M, Riebow NL, Schandler K, Scharer C, Schmidt B, Sison C, Stantripop S, Thomas JW, Thomas PJ, Vemulapalli M, Young AC, Su HC, Freeman AF, Segre JA, Kong HH, Program NCS (2018) Expanded skin virome in DOCK8-deficient patients. Nat Med. https://doi.org/10.1038/s41591-018-0211-7

van Nood E, Vrieze A, Nieuwdorp M, Fuentes S, Zoetendal EG, de Vos WM, Visser CE, Kuijper EJ, Bartelsman JFWM, Tijssen JGP, Speelman P, Dijkgraaf MGW, Keller JJ (2013) Duodenal infusion of donor feces for recurrent clostridium difficile. N Engl J Med 368:407–415. https://doi.org/10.1056/NEJMoa1205037

Vandeputte D, Kathagen G, D'hoe K, Vieira-Silva S, Valles-Colomer M, Sabino J, Wang J, Tito RY, De Commer L, Darzi Y, Vermeire S, Falony G, Raes J (2017) Quantitative microbiome profiling links gut community variation to microbial load. Nature 551:507–511. https://doi.org/10.1038/nature24460

Vangay P, Johnson AJ, Ward TL, Al-Ghalith GA, Shields-Cutler RR, Hillmann BM, Lucas SK, Beura LK, Thompson EA, Till LM, Batres R, Paw B, Pergament SL, Saenyakul P, Xiong M, Kim AD, Kim G, Masopust D, Martens EC, Angkurawaranon C, McGready R, Kashyap PC, Culhane-Pera KA, Knights D (2018) US immigration westernizes the human gut microbiome. Cell 175:962–972.e10. https://doi.org/10.1016/j.cell.2018.10.029

Vatanen T, Franzosa EA, Schwager R, Tripathi S, Arthur TD, Vehik K, Lernmark Å, Hagopian WA, Rewers MJ, She J-X, Toppari J, Ziegler A-G, Akolkar B, Krischer JP, Stewart CJ, Ajami

NJ, Petrosino JF, Gevers D, Lähdesmäki H, Vlamakis H, Huttenhower C, Xavier RJ (2018) The human gut microbiome in early-onset type 1 diabetes from the TEDDY study. Nature 562:589–594. https://doi.org/10.1038/s41586-018-0620-2

Vesty A, Biswas K, Taylor MW, Gear K, Douglas RG (2017) Evaluating the impact of DNA extraction method on the representation of human oral bacterial and fungal communities. PLoS One 12:e0169877. https://doi.org/10.1371/journal.pone.0169877

Wang H-F, Li L-F, Guo S-H, Zeng Q-Y, Ning F, Liu W-L, Zhang G (2016) Evaluation of antibody level against Fusobacterium nucleatum in the serological diagnosis of colorectal cancer. Sci Rep 6:33440

Wong SH, Kwong TNY, Chow T-C, Luk AKC, Dai RZW, Nakatsu G, Lam TYT, Zhang L, Wu JCY, Chan FKL, Ng SSM, Wong MCS, Ng SC, Wu WKK, Yu J, Sung JJY (2017) Quantitation of faecal Fusobacterium improves faecal immunochemical test in detecting advanced colorectal neoplasia. Gut 66:1441–1448. https://doi.org/10.1136/gutjnl-2016-312766

Wu GD, Compher C, Chen EZ, Smith SA, Shah RD, Bittinger K, Chehoud C, Albenberg LG, Nessel L, Gilroy E, Star J, Weljie AM, Flint HJ, Metz DC, Bennett MJ, Li H, Bushman FD, Lewis JD (2016) Comparative metabolomics in vegans and omnivores reveal constraints on diet-dependent gut microbiota metabolite production. Gut 65:63–72. https://doi.org/10.1136/gutjnl-2014-308209

Wu H, Esteve E, Tremaroli V, Khan MT, Caesar R, Mannerås-Holm L, Ståhlman M, Olsson LM, Serino M, Planas-Fèlix M, Xifra G, Mercader JM, Torrents D, Burcelin R, Ricart W, Perkins R, Fernàndez-Real JM, Bäckhed F (2017) Metformin alters the gut microbiome of individuals with treatment-naive type 2 diabetes, contributing to the therapeutic effects of the drug. Nat Med 23:850–858. https://doi.org/10.1038/nm.4345

Yang J-Y, Kim M-S, Kim E, Cheon JH, Lee Y-S, Kim Y, Lee S-H, Seo S-U, Shin S-H, Choi SS, Kim B, Chang S-Y, Ko H-J, Bae J-W, Kweon M-N (2016) Enteric viruses ameliorate gut inflammation via toll-like receptor 3 and toll-like receptor 7-mediated interferon-β production. Immunity 44:889–900. https://doi.org/10.1016/j.immuni.2016.03.009

Yu J, Feng Q, Wong SH, Zhang D, Liang QY, Qin Y, Tang L, Zhao H, Stenvang J, Li Y, Wang X, Xu X, Chen N, Wu WKK, Al-Aama J, Nielsen HJ, Kiilerich P, Jensen BAH, Yau TO, Lan Z, Jia H, Li J, Xiao L, Lam TYT, Ng SC, Cheng AS-L, Wong VW-S, Chan FKL, Xu X, Yang H, Madsen L, Datz C, Tilg H, Wang J, Brünner N, Kristiansen K, Arumugam M, Sung JJ-Y, Wang J (2017) Metagenomic analysis of faecal microbiome as a tool towards targeted non-invasive biomarkers for colorectal cancer. Gut 66:70–78. https://doi.org/10.1136/gutjnl-2015-309800

Zeevi D, Korem T, Zmora N, Israeli D, Rothschild D, Weinberger A, Ben-Yacov O, Lador D, Avnit-Sagi T, Lotan-Pompan M, Suez J, Mahdi JA, Matot E, Malka G, Kosower N, Rein M, Zilberman-Schapira G, Dohnalova L, Pevsner-Fischer M, Bikovsky R, Halpern Z, Elinav E, Segal E (2015) Personalized nutrition by prediction of glycemic responses. Cell 163:1079–1094. https://doi.org/10.1016/j.cell.2015.11.001

Zhang C, Derrien M, Levenez F, Brazeilles R, Ballal SA, Kim J, Degivry M-C, Quéré G, Garault P, van Hylckama Vlieg JET, Garrett WS, Doré J, Veiga P (2016) Ecological robustness of the gut microbiota in response to ingestion of transient food-borne microbes. ISME J 10:2235–2245. https://doi.org/10.1038/ismej.2016.13

Zhao L, Zhang F, Ding X, Wu G, Lam YY, Wang X, Fu H, Xue X, Lu C, Ma J, Yu L, Xu C, Ren Z, Xu Y, Xu S, Shen H, Zhu X, Shi Y, Shen Q, Dong W, Liu R, Ling Y, Zeng Y, Wang X, Zhang Q, Wang J, Wang L, Wu Y, Zeng B, Wei H, Zhang M, Peng Y, Zhang C (2018) Gut bacteria selectively promoted by dietary fibers alleviate type 2 diabetes. Science 359:1151–1156. https://doi.org/10.1126/science.aao5774

Zheng JH, Nguyen VH, Jiang S-N, Park S-H, Tan W, Hong SH, Shin MG, Chung I-J, Hong Y, Bom H-S, Choy HE, Lee SE, Rhee JH, Min J-J (2017) Two-step enhanced cancer immunotherapy with engineered Salmonella typhimurium secreting heterologous flagellin. Sci Transl Med 9:9537. https://doi.org/10.1126/scitranslmed.aak9537

Zmora N, Zilberman-Schapira G, Suez J, Mor U, Dori-Bachash M, Bashiardes S, Kotler E, Zur M, Regev-Lehavi D, Brik RB-Z, Federici S, Cohen Y, Linevsky R, Rothschild D, Moor AE,

Ben-Moshe S, Harmelin A, Itzkovitz S, Maharshak N, Shibolet O, Shapiro H, Pevsner-Fischer-M, Sharon I, Halpern Z, Segal E, Elinav E (2018) Personalized gut mucosal colonization resistance to empiric probiotics is associated with unique host and microbiome features. Cell 174:1388–1405.e21. https://doi.org/10.1016/j.cell.2018.08.041

Zou J, Chassaing B, Singh V, Pellizzon M, Ricci M, Fythe MD, Kumar MV, Gewirtz AT (2018) Fiber-mediated nourishment of gut microbiota protects against diet-induced obesity by restoring IL-22-mediated colonic health. Cell Host Microbe 23:41–53.e4. https://doi.org/10.1016/j.chom.2017.11.003

Zuo T, Wong SH, Lam K, Lui R, Cheung K, Tang W, Ching JYL, Chan PKS, Chan MCW, Wu JCY, Chan FKL, Yu J, Sung JJY, Ng SC (2018) Bacteriophage transfer during faecal microbiota transplantation in Clostridium difficile infection is associated with treatment outcome. Gut 67:634–643. https://doi.org/10.1136/gutjnl-2017-313952

Chapter 21
Endosymbiont-Mediated Adaptive Responses to Stress in Holobionts

Siao Ye and Evan Siemann

Abstract Endosymbiosis is found in all types of ecosystems and it can be sensitive to environmental changes due to the intimate interaction between the endosymbiont and the host. Indeed, global climate change disturbs the local ambient environment and threatens endosymbiotic species, and in some cases leads to local ecosystem collapse. Recent studies have revealed that the endosymbiont can affect holobiont (endosymbiont and host together) stress tolerance as much as the host does, and manipulation of the microbial partners in holobionts may mitigate the impacts of the environmental stress. Here, we first show how the endosymbiont presence affects holobiont stress tolerance by discussing three well-studied endosymbiotic systems, which include plant-fungi, aquatic organism-algae, and insect-bacteria systems. We then review how holobionts are able to alter their stress tolerance via associated endosymbionts by changing their endosymbiont composition, by adaptation of their endosymbionts, or by acclimation of their endosymbionts. Finally, we discuss how different transmission modes (vertical or horizontal transmission) might affect the adaptability of holobionts. We propose that the endosymbiont is a good target for modifying holobiont stress tolerance, which makes it critical to more fully investigate the role of endosymbionts in the adaptive responses of holobionts to stress.

21.1 Background

Endosymbiosis is a ubiquitous phenomenon found in nature, which refers to two distantly related species that exhibit close physical contact in the form of one species living inside the other (Wernegreen 2012a). While a comprehensive definition of endosymbiosis involves interactions that range from parasitic to mutualistic, the latter receives increasing attention in recent years because beneficial symbionts have been found to have significant ecological and evolutionary impacts on their hosts

S. Ye (✉) · E. Siemann
Department of Biosciences, Rice University, Houston, TX, USA
e-mail: siemann@rice.edu

M. Kloc (ed.), *Symbiosis: Cellular, Molecular, Medical and Evolutionary Aspects*, Results and Problems in Cell Differentiation 69,
https://doi.org/10.1007/978-3-030-51849-3_21

(Brucker and Bordenstein 2012). Due to the strong interdependency between the host and the symbiont, researchers have proposed that the two parties can be viewed as one entity called the holobiont and that they jointly determine the holobiont phenotypes (Zilber-Rosenberg and Rosenberg 2008).

Since the holobiont's existence relies on the well-being of both the host and the symbiont, the stress response of one party could negatively affect the other. Thus, species that are highly endosymbiosis-dependent could be especially vulnerable to climate change or other types of environmental change because they may experience mutual downfall (Coyte et al. 2015). Such vulnerability of holobionts to stress has been observed in various endosymbiotic species. For example, corals (marine invertebrates in the class Anthozoa) that host endosymbiotic algae (*Symbiodinium dinoflagellates*) have experienced massive bleaching (loss of algae) in past decades because of the abnormal environmental conditions, such as high temperature or high salinity, damage the algae (Brown 1997; Baird and Marshall 2002). Similarly, short-term exposure to heat can deplete insects of their endosymbiotic bacteria (that provide essential amino acids to sap-feeding insect hosts (Feng et al. 2019) and reduce their fecundity (Dunbar et al. 2007; Wernegreen 2012b). Whether holobionts can rapidly shift their stress tolerance is receiving increasing attention, and the answers may shed light on holobiont adaptation and future holobiont management.

According to the hologenome theory, holobiont stress tolerance could be rapidly altered through adaptive responses of the symbiont (Zilber-Rosenberg and Rosenberg 2008). Compared to the host, the endosymbiont is more likely to shift holobiont stress tolerance in the short term because of their larger population and shorter generation times. Thus, more mutations may occur in symbionts than in hosts for a given period, and adaptive evolution is more likely to take place in symbionts and to occur more quickly (Lynch et al. 1991; Desai and Fisher 2007). In addition, the holobiont may be able to respond to stress without any changes in the host, but rather through changes in their endosymbiont composition (e.g., species or strains), which could be the key to holobiont persistence in changing climates if host adaptation is slow. From a holobiont perspective, as long as the endosymbiont affects the holobiont phenotype, any genetic or non-genetic changes in the endosymbiont could influence holobiont fitness and have ecological or evolutionary consequences. In this review, we will discuss symbionts' impacts on holobiont stress tolerances, and demonstrate symbiont-mediated adaptive responses in holobionts.

21.2 Endosymbionts' Contributions to Holobiont Stress Tolerance

While the nutrient supplement function of the endosymbiont has long been well-recognized, the topic of endosymbiont-mediated stress tolerance is gaining more attention since researchers discovered that endosymbionts can play a significant role in host development or physiology (Montgomery and McFall-Ngai 1994; Dale and

Moran 2006). The holobiont could exhibit distinctive stress tolerance from the host or the endosymbiont alone, providing a distinct, emergent holobiont phenotype that is subjected to selection. For example, lichens exhibit significant higher stress tolerance compared to that of the alga and the fungus forming the holobionts (Lawrey 2009), which could probably be explained by the complex morphology formed by the two partners, and their complicated biochemical interactions (Kranner et al. 2008). Association with endosymbionts seems to alter the ecological niches of hosts, which further facilitates their diversification into diverse environments (Bennett and Moran 2015).

However, having endosymbionts does not always enhance holobiont stress tolerance. Whether the presence of endosymbionts has a positive or negative impact on the holobiont's stress tolerance could depend on the species as well as the stress type. In other words, even the same type of endosymbiont could have opposite impacts on the holobiont fitness with regard to different kinds of stress. Here, we would like to introduce three of the most well-studied endosymbiotic systems to show how endosymbionts could affect holobiont stress tolerance.

21.2.1 Plant-Fungi System

Almost all plants in nature are associated with fungi that can be categorized into mycorrhizal fungi or endophytic fungi (Rodriguez and Redman 2008). Mycorrhizal fungi usually reside in plant roots and rhizosphere, while the endophytic fungi may be found throughout the entire plant (Singh et al. 2011). Although mycorrhizal fungi are well-known for their nutrient supplementary function, both mycorrhizal and endophytic fungi confer abiotic stress tolerance to plants, which enhance their performance under drought, heat, or high salinity (Read 1999; Redman et al. 2002; Waller et al. 2005; Rodriguez et al. 2008). Surprisingly, such positive impacts could be transgenerational even when the offspring is fungus-free, apparently through epigenetic modifications of the host (Hubbard et al. 2014; Kumari et al. 2018). In some plants, endosymbionts are necessary for the plants to grow in its native habitat. For example, *Dichanthelium lanuginosum* grasses growing in the geothermal regions are colonized by fungal endophyte *Curvularia protuberata*, without which they will not be able to survive in high temperature (Redman et al. 2002). Similarly, the coastal dune grass *Leymus mollis* hosts the fungal endophyte *Fusarium culmorum*, which confers salt tolerance to the host (Rodriguez et al. 2008). Endophytes confer such tolerance by regulating plant osmotic pressure and stomata, stimulating the generation of antioxidative enzymes, or supplementing anti-stress chemicals (Singh et al. 2011; Lata et al. 2018; Nanda et al. 2019). Mycorrhizal fungi are also able to alleviate abiotic stress such as drought or high salinity by restructuring the soil and improving water/nutrients uptake (Ruiz-Lozano and Azcon 1995; Miller and Jastrow 2000; Begum et al. 2019). For example, arbuscular mycorrhizal fungi (AMF)-infected *Cucumis sativus* cucumber plants contain higher concentrations of inorganic nutrients and up-regulated activity of antioxidant enzymes than AMF-free controls under salt stress, which contributes to their higher biomass in such environments (Hashem et al. 2018). Compared to endophytic or mycorrhizal fungus-free plants, plants hosting these types of fungi usually exhibit improved tolerance of stressful environmental conditions, such as salinity, heat, drought, or heavy metals. (Nanda et al. 2019; Begum et al. 2019).

21.2.2 Aquatic Organism-Algae System

Invertebrate animals or protists in oligotrophic aquatic ecosystems often form symbioses with algae that provide fixed forms of carbon in exchange for physical protection and nutrients (Muscatine et al. 1967; Grube et al. 2017a). These organisms have wide distributions from freshwater to marine ecosystems, and some of the most well-studied organisms include Porifera (sponges), Cnidaria (corals, sea anemone, and hydra), Ciliophora (parameciums and other ciliated protists) and Foraminifera (amoeboid protists) (Venn et al. 2008; Nowack and Melkonian 2010; Pita et al. 2018). The interactions between algae and these organisms have received great attention in the past few decades because some of the algae-bearing animals, corals, for example, are keystone species in aquatic ecosystems and are subjected to anthropogenic disturbances (Brown 1997; Loya et al. 2001; Mieog et al. 2009). Symbiosis breakdown in these organisms may cause ecological cascades because they are important producers or degraders that support the growth of other creatures and promote biodiversity (Lipps and Valentine 1970; Wichterman and Wichterman 1986; Done 1992; Hallock et al. 2006).

There are two main types of algae in photosynthetic symbiosis: *Chlorella* species (green algae) and *Symbiodinium* species, which are mainly found in freshwater and marine water, respectively (Venn et al. 2008). Under normal conditions, the algae fix carbon and release oxygen (Muscatine et al. 1967; Cernichiari et al. 1969); however, the photosynthetic system could be disrupted under heat stress or strong irradiance, which produce harmful reactive oxygen species (ROS) that can lead to symbiosis breakdown (Murata et al. 2007; Grube et al. 2017b). Studies in corals and sea anemone have revealed that stressed algae in the animal hosts accumulate ROS and induce cell apoptosis (Dunn et al. 2007; Weis 2008). The hosts then digest or expel algae as a protective mechanism, which is known as bleaching (Buddemeier et al. 2004; Smith et al. 2005; Downs et al. 2009; Császár et al. 2010). In contrast, host cells without algae showed limited stress response under elevated temperatures (Strychar and Sammarco 2009; Sammarco and Strychar 2009). Coral larvae that are free of algae lack transcriptional responses to oxidative stress during heat stress (Rodriguez-Lanetty et al. 2009). In addition, Yakovleva et al. (2009) showed that algal-free coral larvae had higher survival compared to algal-bearing larvae in heat-stress experiments. Green hydra (*Hydra viridissima*) also exhibited a similar pattern with the presence of algae having non-positive impacts on hydra thermal tolerance, although contrasting results were found in *Paramecium* species with positive effects of algae presence (Iwatsuki et al. 1998; Salsbery and Delong 2018; Ye et al. 2019b).

The high sensitivity of photosynthetic systems to heat and light stress indicates that the photosynthetic endosymbionts could negatively impact the host under certain abiotic stresses, and make the holobiont more susceptible to environmental perturbations (Murata et al. 2007; Mathur et al. 2014). On the other hand, studies on *Paramecium bursaria* and sea anemone show that endosymbiotic algae can protect the holobiont from UV radiation by both physical and biochemical mechanisms and hosts that had their endosymbionts removed received more damage (Hörtnagl and Sommaruga 2007; Summerer et al. 2009; Shinzato et al. 2011). Thus, whether the presence of photosynthetic endosymbionts increases or decreases, the host stress tolerance depends on the stress type.

21.2.3 Insect-Bacteria System

Insects are a highly diverse and abundant group of animals and many of them associate with endosymbionts (Bahrndorff et al. 2016; Stork 2018). Usually, these endosymbionts provide nutrients such as vitamins or amino acids to hosts, but they are also involved in host development and immunity (Bahrndorff et al. 2016). For example, almost all aphids host *Buchnera aphidicola* bacteria, which supplement essential compounds lacking in aphids' diet (Douglas 2009). In weevils, a symbiotic bacteria *Nardonella* species is required for the development of hard cuticles (Anbutsu et al. 2017). Being small poikilotherms, insects are susceptible to temperature changes (Wernegreen 2012b), and temperature fluctuations could easily affect endosymbionts that reside in them (Kikuchi et al. 2016). Impacts on these endosymbionts could potentially alter whole insect communities and ecosystems, as the effects cascade through the food webs (McLean et al. 2016).

The endosymbiotic bacteria in insects are generally categorized into two groups: primary endosymbionts and secondary endosymbionts (Su et al. 2013). Primary endosymbionts are obligate microbes that are transmitted vertically, from mothers to offspring; secondary endosymbionts are facultative microbes that are not necessary for insect survival and can be acquired horizontally (Baumann 2005; Feldhaar 2011). Primary endosymbionts usually have coevolved with hosts for millions of years and are integrated into their hosts' life cycles (Baumann 2005). The long-term coevolution with hosts drives the endosymbiont to lose genes that are redundant with those of the host, so the bacteria might lack necessary genes to maintain functional stability on their own (Bennett and Moran 2015). The specialized organs (bacteriocytes) hosting these endosymbionts are also susceptible to stress (Shan et al. 2017). Thus, these key endosymbionts could be easily lost during environmental stress and negatively impact insect hosts (Moran 2016). For example, high temperature has shown similar effects as antibiotics on an obligate bacterial symbiont in the stinkbug *Nezara viridula*, suppression of which reduces the host growth and body size (Kikuchi et al. 2016). Aphids that were exposed to 25–30 °C also experienced declines in their obligate endosymbiotic *Buchnera* bacteria (Montllor et al. 2002). Moreover, such declines in endosymbionts could have further negative impacts on offspring because the transmission rate of endosymbionts is also largely reduced (Feldhaar 2011). It is given that global climate change could dramatically reshape the current climate, such as increasing the occurrence of extreme temperatures and insects that are highly dependent on primary endosymbionts would face great challenges and may even become extinct (Moran 2016). In contrast, species that are not associated with primary endosymbionts may be more likely to expand their ranges and potentially become invasive (Moran 2016). While the primary endosymbionts may fail during environmental stress, secondary endosymbionts may rescue the hosts from the loss of their obligate endosymbionts (Koga et al. 2003; Wernegreen 2012b). For instance, aphids that were eliminated of their primary *Buchnera* endosymbionts due to heat stress experienced a decrease in their fecundity, but inoculation of secondary endosymbionts partially restored their fertility (Montllor et al. 2002). The secondary endosymbionts could also confer novel traits to hosts and improve their stress tolerance (Russell and Moran 2006; Heyworth and Ferrari 2015). Since the primary endosymbionts and secondary endosymbionts differ in their evolutionary history and intimacy with hosts, their impacts on host stress tolerance can be quite distinctive (Baumann 2005; Wernegreen 2012b; Su et al. 2013).

21.3 Adaptive Responses of Holobionts to Stress Via Endosymbionts

Due to the intimacy between the endosymbiont and the host, the holobiont phenotype can be largely influenced by the former in addition to the host (Zilber-Rosenberg and Rosenberg 2008). The presence of endosymbionts could confer novel traits to hosts and enable the holobionts to expand into new niches, but could also increase their sensitivity to stress (Stoecker et al. 2009; Wernegreen 2012a; Moran 2016). Since large-scale endosymbiosis breakdowns have already occurred with severe ecological consequences, one key question is whether holobionts are able to alter their stress tolerance via symbiont changes to alleviate the stress (Császár et al. 2010; Singh et al. 2013; Hurst 2017; Simon et al. 2019). Here, we will discuss some common strategies of holobionts responding to stress.

21.3.1 Impacts of Existing Symbiont Composition Alteration

Hosts could associate with multiple potential endosymbiont genotypes or species, and one-to-one obligate endosymbiosis is rare in nature (Douglas 1998; Fabina et al. 2012). The diverse symbionts serve as a pool from which holobionts could choose optimal partners because these symbionts exhibit distinctive phenotypes in vitro (i.e., when grown outside a host) and thus, the tolerance they can confer in vivo (i.e., to their hosts) (Baskett et al. 2009; Wietheger 2012). According to the hologenome theory, holobionts can achieve rapid shifts in stress tolerance by taking advantage of the existing symbiont variation in three ways: (1) symbiont amplification (change in relative abundances of diverse associated symbionts); (2) novel symbiont acquisition (uptake of novel symbionts horizontally, i.e. from the environment or other holobionts); (3) horizontal gene transfer (transfer of genes among symbionts or between symbionts and hosts) (Zilber-Rosenberg and Rosenberg 2008).

21.3.1.1 Endosymbiont Amplification

Holobionts hosting diverse endosymbionts can adjust to environmental changes through the amplification of those that maximize holobiont fitness (Zilber-Rosenberg and Rosenberg 2008). Endosymbiont diversity could be higher than expected because low-density microbes are hard to detect (Rosenberg and Zilber-Rosenberg 2018). For example, critical endosymbiotic bacteria that are part of diverse assemblages may be rare in cicadas and at risk of failing to be passed on to offspring by their mothers, so these types of bacteria may be passed on in relatively much higher numbers than bacteria groups that have low diversity (and high per strain abundance) in the host (Campbell et al. 2018). In corals, *Symbiodinium* diversity is high not only in the community but can also be high within an individual (Quigley et al. 2014; Boulotte et al. 2016; LaJeunesse et al. 2018). As coral thermal tolerance is largely determined by the endosymbionts, amplification of certain types of endosymbionts may greatly alter the holobiont stress tolerance (Cunning and Baker 2013; Zhao et al. 2013; Rosenberg and Zilber-Rosenberg 2018). Researchers have revealed that *Symbiodinium* species vary in their stress tolerances (Goyen et al. 2017). Heat sensitive *Symbiodinium* produces more ROS when subjected to heat stress, and their growth rates in vitro are also suppressed (Wietheger 2012; Karim et al. 2015; Goyen et al. 2017). Corals associated with these sensitive *Symbiodinium* are more likely to bleach under stress, but they can become more resistant by shifting to stress-tolerant endosymbionts (Berkelmans and van Oppen 2006; Jones et al. 2008). Jones et al. (2008) found that the relative abundance of stress-tolerant endosymbionts increased at both community and individual levels after stress occurred even though they were previously dominated by stress-sensitive endosymbionts. In addition, corals that become dominated by stress-tolerant endosymbionts after stress can better withstand future stress (Cunning et al. 2015). The improved stress tolerance is not only a result of the increase in tolerant endosymbionts but also due to the elimination of intolerant endosymbionts (Kinzie et al. 2001; Buddemeier et al. 2004). Similarly, insects usually harbor heat-sensitive endosymbionts under normal conditions, which are crucial to maintaining physiological function (Baumann 2005; Moran 2016). While these primary endosymbionts are eliminated by heat stress, some tolerant secondary endosymbionts could survive and rescue the host from malfunction (Montllor et al. 2002; Wernegreen 2012b; Shan et al. 2017). Amplification of rarer but more stress-tolerant endosymbionts appears to be a common method for holobionts to achieve higher stress tolerance and persist in climate change but the dominance of stress-sensitive endosymbionts in benign conditions suggests a trade-off between stress tolerance and other potentially beneficial traits (Baskett et al. 2009).

21.3.1.2 Novel Endosymbiont Acquisition

Novel endosymbiont acquisition is another mechanism for holobionts to alter their stress tolerance (Zilber-Rosenberg and Rosenberg 2008; Sudakaran et al. 2017). Many endosymbionts have a free-living stage in their life cycles, during which they are released into the environment and can be picked up by hosts (Bright and Bulgheresi 2010; Drown et al. 2013). For example, endosymbiotic fungi are rich in soil, and settled plants can form associations with nearby fungi (Rodriguez and Redman 2008). Corals are able to conduct both vertical transmission and horizontal transmission, through the latter they are able to pick up novel endosymbionts (Byler et al. 2013; Suzuki et al. 2013). Such acquisition of novel endosymbionts appears to be common and may be key to species radiation (Richardson 2001; Sudakaran et al. 2017). Chong and Moran (2018) discovered *Geopemphigus* aphids lost their obligate *Buchenera* endosymbionts over evolutionary time but established endosymbiosis with a new type of bacteria. Since endosymbionts vary in their response to stress, tolerances conferred to hosts by different endosymbionts could also vary. Ye et al. (2019a) found that when the same aposymbiotic green-hydra strain was inoculated with algae isolated from different hydra populations, the hydra varied widely in their thermal tolerances. This indicates hydra may be able to alter their thermal tolerance by switching which algae they harbor. Marine ecologists have proposed that such endosymbiont switching could rescue corals from extinction under climate change because some *Symbiodinium* species are more resistant to stress (Berkelmans and van Oppen 2006; Gilbert et al. 2010). Plants are also able to acquire stress tolerance by associating with specific endosymbiotic fungi (Rodriguez and Redman 2008). *Curvularia* isolates (a type of fungal endophyte) from the geothermal regions can confer thermal tolerance to fungal-free plants by establishing endosymbiosis with hosts, which also suggests that stress tolerance could be acquired horizontally (Redman et al. 2002). In aphids, the artificial inoculation of two facultative endosymbionts increased their resistance to heat stress, in contrast, those that only hosted obligate endosymbionts had significantly lower reproduction success (Montllor et al. 2002). These data indicate that association with novel endosymbionts may alleviate stress in holobionts.

21.3.1.3 Horizontal Gene Transfer Among Endosymbionts

Horizontal gene transfer is an important force that shapes the endosymbiont phenotype, and thus, the holobiont phenotype (Gogarten and Townsend 2005; Singh et al. 2013). Such gene exchanges are common in both eukaryotes and prokaryotes, which indicates all kinds of endosymbionts, from bacteria to algae and fungi, are able to acquire novel traits horizontally (Keeling and Palmer 2008; Boto 2014). Research in plant-associated endosymbionts revealed the signs of intra-clade horizontal gene transfer, which affected plants' secondary metabolism (Pinto-Carbó et al. 2016). In an endosymbiotic *Chlorella* species from *Paramecium*, researchers detected gene fragments that might have originated from a virus (Rautian et al. 2009). Since viruses can be inserted into multiple kinds of species and package additional cellular genes, the endosymbiont could acquire novel genes indirectly (Lambowitz and Belfort 1993; Keese 2008). For example, three phages were found to contain functional-photosynthesis genes probably from different cyanobacteria, and cyanobacteria form endosymbioses with a variety of species, including corals, algae, and fungi (Lindell et al. 2004; Kranner et al. 2008; Kvennefors and Roff 2009; Grube et al. 2017a). In addition, the horizontal gene transfer could happen between endosymbionts and their hosts. Moran and Jarvik (2010) discovered aphids may acquire their carotenoid biosynthetic genes from fungi, and Chapman et al. (2010) detected that the hydra genome contains several bacteria-originated genes. The horizontal gene transfer could contribute to endosymbiont stress tolerance and impact holobiont persistence under stress (Webster and Reusch 2017).

21.3.2 Impacts of Symbiont Adaptation and Acclimation

In addition to utilizing existing genetic variation in endosymbionts, holobionts could respond to environmental stress through adaptation and acclimation responses in their microbial partners (Webster and Reusch 2017). Adaptation and acclimation are two major mechanisms that alleviate organisms' stress in changing environments, which could take place in both the endosymbiont and the host (Zilber-Rosenberg and Rosenberg 2008; Fitt et al. 2009; Bellantuono et al. 2012; Palumbi et al. 2014). However, microbes have higher plasticity and adaptability compared to host organisms (Terzaghi and O'Hara 1990), which suggests holobiont stress tolerance can be altered simply through symbiont acclimation or adaptation, which is especially critical for species hosting low-diversities of endosymbionts (Drown et al. 2013; Morrow et al. 2017).

21.3.2.1 Adaptation

Adaptation involves genetic changes due to natural selection on the population level, which results in a shift in the mean phenotypes. As microbes typically have very short generation times, their adaptation could be relatively fast compared to hosts (Zilber-Rosenberg and Rosenberg 2008; Allemand and Furla 2018). For instance, the doubling time of *Symbiodium* species is typically between days to months, but up to years for their coral hosts (Berkelmans and van Oppen 2006; Allemand and Furla 2018; Kumari et al. 2018). In addition, the endosymbiont population size far exceeds that of the host because one individual holobiont could contain thousands of endosymbionts (Bossert and Dunn 1986; Fitt et al. 2000; Mira and Moran 2002). This suggests that mutations are more likely to occur in the endosymbionts and provide an opportunity for adaptation.

Studies on endosymbiotic species have confirmed intra-species adaptation in endosymbionts. Rodriguez et al. (2008) found that the same fungal species may or may not confer stress tolerance based on habitats from which they are collected. Fungi that were isolated from geothermal habitats or costal habitats conferred thermal tolerance or drought/salinity tolerance to host plants, respectively. However, the same species collected from non-stressful habitats did not improve the plant stress tolerance when they were introduced into hosts. Local adaptation of endosymbionts was also observed in corals (Howells et al. 2012). Howells et al. (2012) compared the stress response of a general *Symbiodinium* type from a warmer reef and a control reef. They found even after multiple-asexual generations cultivated in the lab, the endosymbionts from the warmer region exhibited lower stress response, such as higher survival and quantum yield, to elevated temperature than endosymbionts from the control. The endosymbiotic bacteria *Buchnera* in aphids are also found to control aphid thermal tolerance (Dunbar et al. 2007). A single nucleotide mutation in *Buchnera* determines the thermal tolerance of aphids, and both high-temperature adapted, and low-temperature adapted strains can be found in the wild population. This suggests aphids hosting different *Buchnera* strains might diverge in their ecological niches and adapt to different environments. Attempts have also been made to test the adaptability of endosymbionts. Chakravarti et al. (2017) conducted experimental evolution on *Symbiodinium* species, and they found that the endosymbionts were able to improve their stress tolerance over two years. They raised algae in in vitro at an elevated temperature of 31 °C for about 80 generations and observed improvement in the growth rate and decrease in ROS production when subjected to stress. However, they did not detect significant differences in thermal tolerance between corals inoculated with the selected algae and corals with wild-type algae. Since the host may fail to adapt fast enough to keep up with environmental change, endosymbiont adaptation may allow the holobiont more time to adapt.

21.3.2.2 Acclimation

Acclimation is a reversible physiological process of organisms to deal with stress, which involves no genetic changes but adjustments in gene expression (Collier et al. 2019). The word "acclimation" is sometimes used interchangeably with the word "acclimatization", but the latter usually refers to a more comprehensive concept, and in holobiont studies, it may include a broader range of microbial responses, such as endosymbiont amplification or switching (Webster and Reusch 2017; Collier et al. 2019). Here, we use the term "acclimation" to describe such phenotypic plasticity without genetic changes.

Acclimation has been observed in holobionts by investigating their responses under stress over prolonged periods. For example, corals exposed to high temperature are able to adjust their metabolisms in both the host and the endosymbiont, which reflects non-genetic adaptive responses to stress in both parties (Gibbin et al. 2018; Morikawa and Palumbi 2019). In another acclimation experiment, while the coral calcification rate was strongly suppressed under short-term high CO_2 exposure, it was able to recover over long-term exposure (Form and Riebesell 2012). In addition, the holobiont can acclimate not only to increase in stress intensity but also in stress variability (Mayfield et al. 2012), and the acclimatory ability differs among populations (Kenkel and Matz 2016). In fact, acclimation can effectively assist holobionts to persist in stress because preconditioning to sub-lethal stress improves the holobiont's tolerance towards subsequent acute stress (Bellantuono et al. 2012; Hawkins and Warner 2017).

Recent studies have shown the acclimatory responses of endosymbionts. Takahashi et al. (2013) inspected how long-term elevated temperature affected photosynthesis systems of *Symbiodinium* species in vitro under heat stress, and they found acclimation to high temperature could improve the thermal stability of the endosymbiont. Transcriptome analysis of endosymbionts in corals and clams reveals gene expressions related to photosynthesis, membrane lipid synthesis, and ROS scavenging are mostly altered (Mayfield et al. 2014; Levin et al. 2016; Gierz et al. 2017; Alves Monteiro et al. 2019). In contrast, Palumbi et al. (2014) analyzed the transcriptome profiles of the host and the endosymbiont in corals reciprocally transplanted between high variation pools and low variation pools. They found the acclimation was mainly attributed to the host instead of to the endosymbiont. *Symbiodinium* species seem to be more responsive to higher stress because their transcriptome changes increased with the concentration of a toxic pollutant (Gust et al. 2014).

However, studies on how acclimation in the endosymbiont affects holobiont stress tolerance are still limited because most studies measured responses of the holobiont where the symbiont and the host are usually treated as an intact entity. Due to the intimacy between the symbiont and the host, it is usually difficult to isolate the net impacts of endosymbiont acclimation. Recently, Bui and Franken discovered that the endosymbiotic fungi can acclimate to heavy-metal stress (Zn) in culture, and confer Zn tolerance to host plants when introduced into hosts (Bui and Franken 2018). Using green hydra as a model system, Ye et al. (2019b) found that although the presence of endosymbionts reduced green hydra thermal tolerance before acclimation, it did not affect the acclimation rate or maximum thermal tolerance after acclimation. They also constructed green hydra that had only the endosymbionts (algae) or hosts (hydra) acclimated to high temperature and showed that acclimation in either party improved green hydra thermal tolerance and that these positive effects could last for multiple asexual generations. Together, these studies provide evidence that acclimation in endosymbionts alone could improve stress tolerance in holobionts, and we suggest that more attention to be given to this relatively poorly studied topic.

21.4 Heritability of Symbiont-Mediated Traits

Whether the holobiont can be considered as an integrated unit experiencing natural selection is still under debate, because it addresses a core question: what is a biological individual? (Douglas and Werren 2016; Theis et al. 2016; Suárez 2018). A key question that should be answered is how heritable the microbes associated with hosts are, which is critical for the reproduction of an adapted holobiont. Some researchers have claimed that host-associated microbes could be shaped simply by opportunity and environmental filters, and the transmission fidelity is low (Moran and Sloan 2015; Douglas and Werren 2016). In this review, we focus specifically on endosymbiont–host interactions in which the endosymbiont is a necessary component of the holobiont phenotype.

There are two endosymbiont transmission modes in holobionts: vertical and horizontal transmission. In vertical transmission, offspring inherit symbionts directly from their parents (Ferdy and Godelle 2005). An extreme case could be the mitochondria, which originated from bacteria but can now be found in most eukaryotes (Birky 1983; Gray et al. 2001; Ferdy and Godelle 2005). Green hydra transmit their algae to offspring during both sexual and asexual reproduction, and offspring are born with algae from parents (Campbell 1990; Hamada et al. 2018). Aphids are able to selectively transmit their obligate *Buchnera* endosymbionts to offspring through complicated cellular regulations while preventing facultative endosymbionts from transmission (Koga et al. 2012). Vertically transmitted endosymbionts can be considered as heritable extended genomes with high fidelity, and traits conferred by this type of symbiont are also heritable (Hurst 2017). In contrast, holobionts conducting horizontal transmission may not receive symbionts from their parents but may take up random ambient symbionts (Bright and Bulgheresi 2010). The disassociation between the host and the symbiont could result in the loss of acquired beneficial symbionts, and random symbiont colonization of the host would disqualify the holobiont being a selection unit (Douglas 1998). Surprisingly, recent studies reveal that inheritance is not completely random in horizontally transmitted endosymbionts. A study on corals found both vertically and horizontally transmitted symbionts are partially heritable for which the host genes might be responsible (Quigley et al. 2017, 2019). They discovered corals inherited shuffled (changes in relative abundances) endosymbionts years after bleaching happened, which indicates holobionts are able to maintain the stability between endosymbionts and hosts. Schweitzer et al. (2008) also discovered non-random assembly of below-ground microbes associated with plants, suggesting an unknown mechanism of microbe recruitment. In a stinkbug, offspring is able to pick up the beneficial symbiont from the environment every generation, which demonstrates a high fidelity between the microbe and the host can be kept in horizontal transmission (Kikuchi et al. 2007). All of these suggest potential inheritance of horizontally transmitted symbionts, even if it is not as perfect as that of vertically transmitted ones.

Perhaps, more important questions are: do both transmission modes allow holobionts to adapt to the changing environment, and how do they impact the adaptability? Roughgarden (2018) used a simulation model to show that holobionts using either transmission mode are able to evolve, regardless of whether there is a disassociation between the host and the symbiont. Yet, it does not tell how holobionts may react to increasing environmental stress with different transmission modes. On one hand, the vertical transmission allows the direct inheritance of adaptive endosymbionts in offspring, but it imposes a strong bottleneck effect on the population of endosymbionts the offspring could get (Mira and Moran 2002; Drown et al. 2013). Endosymbiont diversity could be depleted over time, and the microbes may become maladaptive once the ambient condition changes (Dusi et al. 2014). For example, psyllid insects that conduct vertical transmission have low diversity in their bacterial endosymbionts, and in some cases become clonal within the host (Morrow et al. 2017). Loh et al. (2001) also showed higher endosymbiont genetic heterogeneity among populations that exist in vertically transmitted corals than in horizontally transmitted corals. On the other hand, horizontal transmission enables the holobiont to associate with a great variety of endosymbionts, so they may have access to more adaptive endosymbionts (Bright and Bulgheresi 2010; Byler et al. 2013). However, these holobionts might lose adaptive endosymbionts between generations by chance, which reduces their fitness over time (Douglas 1998). Research in human culture transmission could be enlightening in which similar concepts such as vertical transmission and oblique transmission of cultures or knowledge are studied (Fogarty and Feldman 2015). Some of the major findings are vertical culture transmission is favored under stable conditions, while the horizontal transmission is favored under fluctuating conditions (McElreath and Strimling 2008; Xue and Leibler 2016; Ram et al. 2018, 2019). Nevertheless, there could be fundamental differences between culture transmission and endosymbiont transmission, and research in a holobiont context is needed.

21.5 Summary

Endosymbiotic species make significant contributions to our ecosystems, and disruption in endosymbiosis due to their prevalence in nature would be devastating to humans (Wernegreen 2012a). Climate change has already threatened some of the most important ecosystems on the Earth, which are built on endosymbiosis (Hoegh-Guldberg et al. 2007; Veron et al. 2009). Understanding how stress tolerance of endosymbiotic species is controlled will shed light on conservation methods and ways to mitigate damage. Since the endosymbiont plays a critical role in holobiont stress tolerance and is more easily manipulated, targeting the microbes seems to be a promising approach (Mueller and Sachs 2015). Here, we have summarized how endosymbionts can affect holobiont stress tolerance and demonstrated that we can alter holobiont stress tolerance via manipulating the existing associated endosymbionts, as well as inducing adaptive responses in them. For example, we can alter the microbial composition of holobionts, subject microbes to selection or acclimation, or engineer them to improve holobiont fitness under stress (Levin et al. 2017;

Chakravarti and van Oppen 2018; Ye et al. 2019a, b). We believe that there is much potential in endosymbiotic microbes of which we can take advantage to improve our practices in agriculture, forestry, and conservation, especially as environmental conditions are rapidly changing.

References

Allemand D, Furla P (2018) How does an animal behave like a plant? Physiological and molecular adaptations of zooxanthellae and their hosts to symbiosis. C R Biol 341:276–280. https://doi.org/10.1016/J.CRVI.2018.03.007

Alves Monteiro HJ, Brahmi C, Mayfield AB et al (2019) Molecular mechanisms of acclimation to long-term elevated temperature exposure in marine symbioses. Glob Chang Biol 00:1–14. https://doi.org/10.1111/gcb.14907

Anbutsu H, Moriyama M, Nikoh N et al (2017) Small genome symbiont underlies cuticle hardness in beetles. Proc Natl Acad Sci U S A 114:E8382–E8391. https://doi.org/10.1073/pnas.1712857114

Bahrndorff S, Alemu T, Alemneh T, Lund Nielsen J (2016) The microbiome of animals: implications for conservation biology. Int J Genomics 2016:1–7. https://doi.org/10.1155/2016/5304028

Baird AH, Marshall PA (2002) Mortality, growth and reproduction in scleractinian corals following bleaching on the great barrier reef. Mar Ecol Prog Ser 237:133–141. https://doi.org/10.3354/meps237133

Baskett ML, Gaines SD, Nisbet RM (2009) Symbiont diversity may help coral reefs survive moderate climate change. Ecol Appl 19:3–17

Baumann P (2005) Biology of bacteriocyte-associated endosymbionts of plant sap-sucking insects. Annu Rev Microbiol 59:155–189. https://doi.org/10.1146/annurev.micro.59.030804.121041

Begum N, Qin C, Ahanger MA et al (2019) Role of arbuscular mycorrhizal fungi in plant growth regulation: implications in abiotic stress tolerance. Front Plant Sci 10:1068

Bellantuono AJ, Hoegh-Guldberg O, Rodriguez-Lanetty M (2012) Resistance to thermal stress in corals without changes in symbiont composition. Proceedings Biol Sci 279:1100–1107. https://doi.org/10.1098/rspb.2011.1780

Bennett GM, Moran NA (2015) Heritable symbiosis: the advantages and perils of an evolutionary rabbit hole. Proc Natl Acad Sci U S A 112:10169–10176. https://doi.org/10.1073/pnas.1421388112

Berkelmans R, van Oppen MJ (2006) The role of zooxanthellae in the thermal tolerance of corals: a 'nugget of hope' for coral reefs in an era of climate change. Proc R Soc B Biol Sci 273:2305–2312. https://doi.org/10.1098/rspb.2006.3567

Birky CW (1983) Relaxed cellular controls and organelle heredity. Science 222:468–476

Bossert P, Dunn KW (1986) Regulation of intracellular algae by various strains of the symbiotic *Hydra viridissima*. J Cell Sci 85:187–195

Boto L (2014) Horizontal gene transfer in the acquisition of novel traits by metazoans. Proc R Soc B Biol Sci 281(1777):20132450

Boulotte NM, Dalton SJ, Carroll AG et al (2016) Exploring the *Symbiodinium* rare biosphere provides evidence for symbiont switching in reef-building corals. ISME J 10:2693–2701. https://doi.org/10.1038/ismej.2016.54

Bright M, Bulgheresi S (2010) A complex journey: transmission of microbial symbionts. Nat Rev Microbiol 8:218–230. https://doi.org/10.1038/nrmicro2262

Brown BE (1997) Coral bleaching: causes and consequences. Coral Reefs 16:S129–S138. https://doi.org/10.1007/s003380050249

Brucker RM, Bordenstein SR (2012) Speciation by symbiosis. Trends Ecol Evol 27:443–451. https://doi.org/10.1016/J.TREE.2012.03.011

Buddemeier RW, Baker AC, Fautin DG, Jacobs JR (2004) The adaptive hypothesis of bleaching. In: Rosenberg PE, Loya PY (eds) Coral health and disease. Springer, Berlin Heidelberg, pp 427–444

Bui VC, Franken P (2018) Acclimatization of *Rhizophagus irregularis* enhances Zn tolerance of the fungus and the mycorrhizal plant partner. Front Microbiol 9:3156. https://doi.org/10.3389/fmicb.2018.03156

Byler KA, Carmi-Veal M, Fine M, Goulet TL (2013) Multiple symbiont acquisition strategies as an adaptive mechanism in the coral *Stylophora pistillata*. PLoS One 8:e59596. https://doi.org/10.1371/journal.pone.0059596

Campbell RD (1990) Transmission of symbiotic algae through sexual reproduction in hydra: movement of algae into the oocyte. Tissue Cell 22:137–147. https://doi.org/10.1016/0040-8166(90)90017-4

Campbell MA, Łukasik P, Meyer MC et al (2018) Changes in endosymbiont complexity drive host-level compensatory adaptations in cicadas. MBio 9:e02104–e02118. https://doi.org/10.1128/mBio.02104-18

Cernichiari E, Muscatine L, Smith DC (1969) Maltose excretion by the symbiotic algae of *Hydra viridis*. Proc R Soc B Biol Sci 173:557–576. https://doi.org/10.1098/rspb.1969.0077

Chakravarti LJ, van Oppen MJH (2018) Experimental evolution in coral photosymbionts as a tool to increase thermal tolerance. Front Mar Sci 5(227). https://doi.org/10.3389/fmars.2018.00227

Chakravarti LJ, Beltran VH, van Oppen MJH (2017) Rapid thermal adaptation in photosymbionts of reef-building corals. Glob Chang Biol 23:4675–4688. https://doi.org/10.1111/gcb.13702

Chapman JA, Kirkness EF, Simakov O et al (2010) The dynamic genome of *Hydra*. Nature 464:592–596. https://doi.org/10.1038/nature08830

Chong RA, Moran NA (2018) Evolutionary loss and replacement of *Buchnera*, the obligate endosymbiont of aphids. ISME J 12:898–908. https://doi.org/10.1038/s41396-017-0024-6

Collier RJ, Baumgard LH, Zimbelman RB, Xiao Y (2019) Heat stress: physiology of acclimation and adaptation. Anim Front 9:12–19. https://doi.org/10.1093/af/vfy031

Coyte KZ, Schluter J, Foster KR (2015) The ecology of the microbiome: networks, competition, and stability. Science 350:663–666. https://doi.org/10.1126/science.aad2602

Császár NBM, Ralph PJ, Frankham R et al (2010) Estimating the potential for adaptation of corals to climate warming. PLoS One 5:e9751. https://doi.org/10.1371/journal.pone.0009751

Cunning R, Baker AC (2013) Excess algal symbionts increase the susceptibility of reef corals to bleaching. Nat Clim Chang 3:259–262. https://doi.org/10.1038/nclimate1711

Cunning R, Silverstein RN, Baker AC (2015) Investigating the causes and consequences of symbiont shuffling in a multi-partner reef coral symbiosis under environmental change. Proc R Soc B Biol Sci 282:20141725. https://doi.org/10.1098/rspb.2014.1725

Dale C, Moran NA (2006) Molecular interactions between bacterial symbionts and their hosts. Cell 126(3):453–465. https://doi.org/10.1016/j.cell.2006.07.014

Desai MM, Fisher DS (2007) Beneficial mutation selection balance and the effect of linkage on positive selection. Genetics 176:1759–1798. https://doi.org/10.1534/genetics.106.067678

Done TJ (1992) Phase shifts in coral reef communities and their ecological significance. Hydrobiologia 247:121–132. https://doi.org/10.1007/BF00008211

Douglas AE (1998) Host benefit and the evolution of specialization in symbiosis. Heredity (Edinb) 81:599–603

Douglas AE (2009) The microbial dimension in insect nutritional ecology. Funct Ecol 23:38–47. https://doi.org/10.1111/j.1365-2435.2008.01442.x

Douglas AE, Werren JH (2016) Holes in the hologenome: why host-microbe symbioses are not holobionts. MBio 7:e02099. https://doi.org/10.1128/mBio.02099-15

Downs CA, Kramarsky-Winter E, Martinez J et al (2009) Symbiophagy as a cellular mechanism for coral bleaching. Autophagy 5:211–216. https://doi.org/10.4161/auto.5.2.7405

Drown DM, Zee PC, Brandvain Y, Wade MJ (2013) Evolution of transmission mode in obligate symbionts. Evol Ecol Res 15:43–59

Dunbar HE, Wilson ACC, Ferguson NR, Moran NA (2007) Aphid thermal tolerance is governed by a point mutation in bacterial symbionts. PLoS Biol 5:e96. https://doi.org/10.1371/journal.pbio.0050096

Dunn SR, Schnitzler CE, Weis VM (2007) Apoptosis and autophagy as mechanisms of dinoflagellate symbiont release during cnidarian bleaching: every which way you lose. Proceedings Biol Sci 274:3079–3085. https://doi.org/10.1098/rspb.2007.0711

Dusi E, Krenek S, Schrallhammer M et al (2014) Vertically transmitted symbiont reduces host fitness along temperature gradient. J Evol Biol 27:796–800. https://doi.org/10.1111/jeb.12336

Fabina NS, Putnam HM, Franklin EC et al (2012) Transmission mode predicts specificity and interaction patterns in coral-*Symbiodinium* networks. PLoS One 7:e44970. https://doi.org/10.1371/journal.pone.0044970

Feldhaar H (2011) Bacterial symbionts as mediators of ecologically important traits of insect hosts. Ecol Entomol 36:533–543. https://doi.org/10.1111/j.1365-2311.2011.01318.x

Feng H, Edwards N, Anderson CMH et al (2019) Trading amino acids at the aphid–*Buchnera* symbiotic interface. Proc Natl Acad Sci U S A 116:16003–16011. https://doi.org/10.1073/pnas.1906223116

Ferdy J-B, Godelle B (2005) Diversification of transmission modes and the evolution of mutualism. Am Nat 166:613–627. https://doi.org/10.1086/491799

Fitt WK, McFarland FK, Warner ME, Chilcoat GC (2000) Seasonal patterns of tissue biomass and densities of symbiotic dinoflagellates in reef corals and relation to coral bleaching. Limnol Oceanogr 45:677–685. https://doi.org/10.4319/lo.2000.45.3.0677

Fitt WK, Gates RD, Hoegh-Guldberg O et al (2009) Response of two species of indo-Pacific corals, *Porites cylindrica* and *Stylophora pistillata*, to short-term thermal stress: the host does matter in determining the tolerance of corals to bleaching. J Exp Mar Bio Ecol 373:102–110. https://doi.org/10.1016/j.jembe.2009.03.011

Fogarty L, Feldman MW (2015) Cultural evolution: theory and models. In: International encyclopedia of the social and behavioral sciences, 2nd edn. Elsevier, Amsterdam, pp 401–408

Form AU, Riebesell U (2012) Acclimation to ocean acidification during long-term <scp> CO_2 exposure in the cold-water coral *Lophelia pertusa*. Glob Chang Biol 18:843–853. https://doi.org/10.1111/j.1365-2486.2011.02583.x

Gibbin EM, Krueger T, Putnam HM et al (2018) Short-term thermal acclimation modifies the metabolic condition of the coral holobiont. Front Mar Sci 5(10). https://doi.org/10.3389/fmars.2018.00010

Gierz SL, Forêt S, Leggat W (2017) Transcriptomic analysis of thermally stressed *Symbiodinium* reveals differential expression of stress and metabolism genes. Front Plant Sci 8(271). https://doi.org/10.3389/fpls.2017.00271

Gilbert SF, McDonald E, Boyle N et al (2010) Symbiosis as a source of selectable epigenetic variation: taking the heat for the big guy. Philos Trans R Soc Lond B Biol Sci 365:671–678. https://doi.org/10.1098/rstb.2009.0245

Gogarten JP, Townsend JP (2005) Horizontal gene transfer, genome innovation and evolution. Nat Rev Microbiol 3:679–687

Goyen S, Pernice M, Szabó M et al (2017) A molecular physiology basis for functional diversity of hydrogen peroxide production amongst *Symbiodinium* spp. (*Dinophyceae*). Mar Biol 164:46. https://doi.org/10.1007/s00227-017-3073-5

Gray MW, Burger G, Franz Lang B (2001) The origin and early evolution of mitochondria. Genome Biol 2:reviews1018.1. https://doi.org/10.1186/gb-2001-2-6-reviews1018

Grube M, Seckbach J, Muggia L (2017a) Algal and cyanobacteria symbioses. World Scientific, Europe

Grube M, Seckbach J, Muggia L et al (2017b) Trade-offs of symbiotic relationships between aquatic hosts and algae in a changing world. In: Algal and cyanobacteria symbioses. World Scientific, Europe, pp 241–276

Gust KA, Najar FZ, Habib T et al (2014) Coral-zooxanthellae meta-transcriptomics reveals integrated response to pollutant stress. BMC Genomics 15:591. https://doi.org/10.1186/1471-2164-15-591

Hallock P, Williams DE, Fisher EM, Toler SK (2006) Bleaching in foraminifera with algal symbionts: implications for reef monitoring and risk assessment. Anuario Instit Geociênc 29:108–128

Hamada M, Schröder K, Bathia J et al (2018) Metabolic co-dependence drives the evolutionarily ancient *Hydra–Chlorella* symbiosis. elife 7:e35122. https://doi.org/10.7554/eLife.35122

Hashem A, Alqarawi AA, Radhakrishnan R et al (2018) Arbuscular mycorrhizal fungi regulate the oxidative system, hormones and ionic equilibrium to trigger salt stress tolerance in *Cucumis sativus* L. Saudi J Biol Sci 25:1102–1114. https://doi.org/10.1016/j.sjbs.2018.03.009

Hawkins TD, Warner ME (2017) Warm preconditioning protects against acute heat-induced respiratory dysfunction and delays bleaching in a symbiotic sea anemone. J Exp Biol 220:969–983. https://doi.org/10.1242/jeb.150391

Heyworth ER, Ferrari J (2015) A facultative endosymbiont in aphids can provide diverse ecological benefits. J Evol Biol 28:1753–1760. https://doi.org/10.1111/jeb.12705

Hoegh-Guldberg O, Mumby PJ, Hooten AJ et al (2007) Coral reefs under rapid climate change and ocean acidification. Science 318:1737–1742

Hörtnagl PH, Sommaruga R (2007) Photo-oxidative stress in symbiotic and aposymbiotic strains of the ciliate *Paramecium bursaria*. Photochem Photobiol Sci Off J Eur Photochem Assoc Eur Soc Photobiol 6:842–847. https://doi.org/10.1039/b703119j

Howells EJ, Beltran VH, Larsen NW et al (2012) Coral thermal tolerance shaped by local adaptation of photosymbionts. Nat Clim Chang 2:116–120. https://doi.org/10.1038/nclimate1330

Hubbard M, Germida JJ, Vujanovic V (2014) Fungal endophytes enhance wheat heat and drought tolerance in terms of grain yield and second-generation seed viability. J Appl Microbiol 116:109–122. https://doi.org/10.1111/jam.12311

Hurst GDD (2017) Extended genomes: symbiosis and evolution. Interface Focus 7:20170001

Iwatsuki K, Nishidoi M, Suehiro K (1998) Symbiotic *Chlorella* enhances the thermal tolerance in *Paramecium bursaria*. Comp Biochem Physiol Part A Mol Integr Physiol 121:405–409. https://doi.org/10.1016/S1095-6433(98)10151-4

Jones AM, Berkelmans R, van OMJH et al (2008) A community change in the algal endosymbionts of a scleractinian coral following a natural bleaching event: field evidence of acclimatization. Proc R Soc London B Biol Sci 275:1359–1365. https://doi.org/10.1098/rspb.2008.0069

Karim W, Nakaema S, Hidaka M et al (2015) Temperature effects on the growth rates and photosynthetic activities of *Symbiodinium* cells. J Mar Sci Eng 3:368–381. https://doi.org/10.3390/jmse3020368

Keeling PJ, Palmer JD (2008) Horizontal gene transfer in eukaryotic evolution. Nat Rev Genet 9:605–618

Keese P (2008) Risks from GMOs due to horizontal gene transfer. Environ Biosaf Res 7:123–149

Kenkel CD, Matz MV (2016) Gene expression plasticity as a mechanism of coral adaptation to a variable environment. Nat Ecol Evol 1:0014. https://doi.org/10.1038/s41559-016-0014

Kikuchi Y, Hosokawa T, Fukatsu T (2007) Insect-microbe mutualism without vertical transmission: a stinkbug acquires a beneficial gut symbiont from the environment every generation. Appl Environ Microbiol 73:4308–4316. https://doi.org/10.1128/AEM.00067-07

Kikuchi Y, Tada A, Musolin DL et al (2016) Collapse of insect gut symbiosis under simulated climate change. MBio 7:e01578-16. https://doi.org/10.1128/mBio.01578-16

Kinzie RA, Takayama M, Santos SR, Coffroth MA (2001) The adaptive bleaching hypothesis: experimental tests of critical assumptions. Biol Bull 200:51–58. https://doi.org/10.2307/1543084

Koga R, Tsuchida T, Fukatsu T (2003) Changing partners in an obligate symbiosis: a facultative endosymbiont can compensate for loss of the essential endosymbiont *Buchnera* in an aphid. Proc Biol Sci 270(1533):2543–2550. https://doi.org/10.1098/rspb.2003.2537

Koga R, Meng X-Y, Tsuchida T, Fukatsu T (2012) Cellular mechanism for selective vertical transmission of an obligate insect symbiont at the bacteriocyte-embryo interface. Proc Natl Acad Sci U S A 109:E1230–E1237. https://doi.org/10.1073/pnas.1119212109

Kranner I, Beckett R, Hochman A, Nash TH III (2008) Desiccation-tolerance in lichens: a review. Bryologist 111:576–593

Kumari V, Germida J, Vujanovic V (2018) Legume endosymbionts: drought stress tolerance in second-generation chickpea (*Cicer arietinum*) seeds. J Agron Crop Sci 204:529–540. https://doi.org/10.1111/jac.12283

Kvennefors ECE, Roff G (2009) Evidence of cyanobacteria-like endosymbionts in *Acroporid* corals from the great barrier reef. Coral Reefs 28:547

LaJeunesse TC, Parkinson JE, Gabrielson PW et al (2018) Systematic revision of *Symbiodiniaceae* highlights the antiquity and diversity of coral endosymbionts. Curr Biol 28:2570–2580.e6. https://doi.org/10.1016/J.CUB.2018.07.008

Lambowitz AM, Belfort M (1993) Introns as mobile genetic elements. Annu Rev Biochem 62:587–622. https://doi.org/10.1146/annurev.bi.62.070193.003103

Lata R, Chowdhury S, Gond SK, White JF (2018) Induction of abiotic stress tolerance in plants by endophytic microbes. Lett Appl Microbiol 66:268–276

Lawrey J (2009) Chemical defense in lichen symbioses. In: White JF, Torres MS (eds) Defensive mutualism in microbial Symbiosis. CRC Press, Boca Raton, FL, pp 167–181

Levin RA, Beltran VH, Hill R et al (2016) Sex, scavengers, and chaperones: transcriptome secrets of divergent *Symbiodinium* thermal tolerances. Mol Biol Evol 33:2201–2215. https://doi.org/10.1093/molbev/msw119

Levin RA, Voolstra CR, Agrawal S et al (2017) Engineering strategies to decode and enhance the genomes of coral symbionts. Front Microbiol 8:1200. https://doi.org/10.3389/fmicb.2017.01220

Lindell D, Sullivan MB, Johnson ZI et al (2004) Transfer of photosynthesis genes to and from Prochlorococcus viruses. Proc Natl Acad Sci U S A 101:11013–11018. https://doi.org/10.1073/pnas.0401526101

Lipps JH, Valentine JW (1970) The role of foraminifera in the trophic structure of marine communities. Lethaia 3:279–286. https://doi.org/10.1111/j.1502-3931.1970.tb01271.x

Loh WKW, Loi T, Carter D, Hoegh-Guldberg O (2001) Genetic variability of the symbiotic dinoflagellates from the wide ranging coral species *Seriatopora hystrix* and *Acropora longicyathus* in the indo-West Pacific. Mar Ecol Prog Ser 222:97–107. https://doi.org/10.3354/meps222097

Loya Y, Sakai K, Yamazato K et al (2001) Coral bleaching: the winners and the losers. Ecol Lett 4:122–131. https://doi.org/10.1046/j.1461-0248.2001.00203.x

Lynch M, Gabriel W, Wood AM (1991) Adaptive and demographic responses of plankton populations to environmental change. Limnol Oceanogr 36:1301–1312. https://doi.org/10.4319/lo.1991.36.7.1301

Mathur S, Agrawal D, Jajoo A (2014) Photosynthesis: response to high temperature stress. J Photochem Photobiol B Biol 137:116–126. https://doi.org/10.1016/j.jphotobiol.2014.01.010

Mayfield AB, Chan P-H, Putnam HM et al (2012) The effects of a variable temperature regime on the physiology of the reef-building coral *Seriatopora hystrix*: results from a laboratory-based reciprocal transplant. J Exp Biol 215:4183–4195. https://doi.org/10.1242/jeb.071688

Mayfield AB, Bin WY, Chen CS et al (2014) Compartment-specific transcriptomics in a reef-building coral exposed to elevated temperatures. Mol Ecol 23:5816–5830. https://doi.org/10.1111/mec.12982

McElreath R, Strimling P (2008) When natural selection favors imitation of parents. Curr Anthropol 49:307–316. https://doi.org/10.1086/524364

McLean AHC, Parker BJ, Hrček J et al (2016) Insect symbionts in food webs. Philos Trans R Soc B Biol Sci 371:20150325

Mieog JC, Olsen JL, Berkelmans R et al (2009) The roles and interactions of symbiont, host and environment in defining coral fitness. PLoS One 4:e6364. https://doi.org/10.1371/journal.pone.0006364

Miller RM, Jastrow JD (2000) Mycorrhizal fungi influence soil structure. In: Arbuscular mycorrhizas: physiology and function. Springer, Berlin, pp 3–18

Mira A, Moran NA (2002) Estimating population size and transmission bottlenecks in maternally transmitted endosymbiotic bacteria. Microb Ecol 44:137–143. https://doi.org/10.1007/s00248-002-0012-9

Montgomery MK, McFall-Ngai M (1994) Bacterial symbionts induce host organ morphogenesis during early postembryonic development of the squid *Euprymna scolopes*. Development 120:1719–1729

Montllor CB, Maxmen A, Purcell AH (2002) Facultative bacterial endosymbionts benefit pea aphids *Acyrthosiphon pisum* under heat stress. Ecol Entomol 27:189–195. https://doi.org/10.1046/j.1365-2311.2002.00393.x

Moran NA (2016) When obligate partners melt down. MBio 7:e01904-16. https://doi.org/10.1128/MBIO.01904-16

Moran NA, Jarvik T (2010) Lateral transfer of genes from fungi underlies carotenoid production in aphids. Science 328:624–627. https://doi.org/10.1126/science.1187113

Moran NA, Sloan DB (2015) The hologenome concept: helpful or hollow? PLoS Biol 13: e1002311. https://doi.org/10.1371/journal.pbio.1002311

Morikawa MK, Palumbi SR (2019) Using naturally occurring climate resilient corals to construct bleaching-resistant nurseries. Proc Natl Acad Sci U S A 116:10586–10591. https://doi.org/10.1073/pnas.1721415116

Morrow JL, Hall AAG, Riegler M (2017) Symbionts in waiting: the dynamics of incipient endosymbiont complementation and replacement in minimal bacterial communities of psyllids. Microbiome 5:58. https://doi.org/10.1186/s40168-017-0276-4

Mueller UG, Sachs JL (2015) Engineering microbiomes to improve plant and animal health. Trends Microbiol 23:606–617. https://doi.org/10.1016/j.tim.2015.07.009

Murata N, Takahashi S, Nishiyama Y, Allakhverdiev SI (2007) Photoinhibition of photosystem II under environmental stress. Biochim Biophys Acta Bioenerg 1767:414–421. https://doi.org/10.1016/j.bbabio.2006.11.019

Muscatine L, Karakashian SJ, Karakashian MW (1967) Soluble extracellular products of algae symbiotic with a ciliate, a sponge and a mutant hydra. Comp Biochem Physiol 20:1–6. https://doi.org/10.1016/0010-406x(67)90720-7

Nanda S, Mohanty B, Joshi RK (2019) Endophyte-mediated host stress tolerance as a means for crop improvement. Springer, Berlin, pp 677–701

Nowack ECM, Melkonian M (2010) Endosymbiotic associations within protists. Philos Trans R Soc Lond Ser B Biol Sci 365:699–712. https://doi.org/10.1098/rstb.2009.0188

Palumbi SR, Barshis DJ, Traylor-Knowles N, Bay RA (2014) Mechanisms of reef coral resistance to future climate change. Science 344:895–898. https://doi.org/10.1126/science.1251336

Pinto-Carbó M, Sieber S, Dessein S et al (2016) Evidence of horizontal gene transfer between obligate leaf nodule symbionts. ISME J 10:2092–2105. https://doi.org/10.1038/ismej.2016.27

Pita L, Rix L, Slaby BM et al (2018) The sponge holobiont in a changing ocean: from microbes to ecosystems. Microbiome 6:46. https://doi.org/10.1186/s40168-018-0428-1

Quigley KM, Davies SW, Kenkel CD et al (2014) Deep-sequencing method for quantifying background abundances of Symbiodinium types: exploring the rare *Symbiodinium* biosphere in reef-building corals. PLoS One 9:e94297. https://doi.org/10.1371/journal.pone.0094297

Quigley KM, Willis BL, Bay LK (2017) Heritability of the *Symbiodinium* community in vertically- and horizontally-transmitting broadcast spawning corals. Sci Rep 7:8219. https://doi.org/10.1038/s41598-017-08179-4

Quigley KM, Willis BL, Kenkel CD (2019) Transgenerational inheritance of shuffled symbiont communities in the coral *Montipora digitata*. Sci Rep 9:13328. https://doi.org/10.1038/s41598-019-50045-y

Ram Y, Liberman U, Feldman MW (2018) Evolution of vertical and oblique transmission under fluctuating selection. Proc Natl Acad Sci U S A 115:E1174–E1183. https://doi.org/10.1073/pnas.1719171115

Ram Y, Liberman U, Feldman MW (2019) Vertical and oblique cultural transmission fluctuating in time and in space. Theor Popul Biol 125:11–19. https://doi.org/10.1016/J.TPB.2018.11.001

Rautian M, Vorobyev K, Andronov E et al (2009) An atypical *Chlorella* symbiont from *Paramecium bursaria*. Protistology 6:39–44

Read DJ (1999) Mycorrhiza – the state of the art. In: Mycorrhiza. Springer, Berlin, pp 3–34

Redman RS, Sheehan KB, Stout RG et al (2002) Thermotolerance generated by plant / fungal symbiosis. Science 298:1581. https://doi.org/10.1126/science.1078055

Richardson SL (2001) Endosymbiont change as a key innovation in the adaptive radiation of *Soritida* (foraminifera). Paleobiology 27:262–289. https://doi.org/10.1666/0094-8373(2001)027<0262:ECAAKI>2.0.CO;2

Rodriguez R, Redman R (2008) More than 400 million years of evolution and some plants still can't make it on their own: plant stress tolerance via fungal symbiosis. J Exp Bot 59:1109–1114

Rodriguez RJ, Henson J, Van Volkenburgh E et al (2008) Stress tolerance in plants via habitat-adapted symbiosis. ISME J 2:404–416. https://doi.org/10.1038/ismej.2007.106

Rodriguez-Lanetty M, Harii S, Hoegh-Guldberg O (2009) Early molecular responses of coral larvae to hyperthermal stress. Mol Ecol 18:5101–5114. https://doi.org/10.1111/j.1365-294X.2009.04419.x

Rosenberg E, Zilber-Rosenberg I (2018) The hologenome concept of evolution after 10 years. Microbiome 6:78

Roughgarden J (2018) Holobiont evolution: model with lineal vs collective hologenome inheritance. bioRxiv:465310. https://doi.org/10.1101/465310

Ruiz-Lozano JM, Azcon R (1995) Hyphal contribution to water uptake in mycorrhizal plants as affected by the fungal species and water status. Physiol Plant 95:472–478. https://doi.org/10.1111/j.1399-3054.1995.tb00865.x

Russell JA, Moran NA (2006) Costs and benefits of symbiont infection in aphids: variation among symbionts and across temperatures. Proc R Soc B Biol Sci 273:603–610. https://doi.org/10.1098/rspb.2005.3348

Salsbery ME, Delong JP (2018) The benefit of algae endosymbionts in *Paramecium bursaria* is temperature dependent. Evol Ecol Res 19:669–678

Sammarco PW, Strychar KB (2009) Effects of climate change/global warming on coral reefs: adaptation/exaptation in corals, evolution in zooxanthellae, and biogeographic shifts. Environ Bioindic 4:9–45. https://doi.org/10.1080/15555270902905377

Schweitzer JA, Bailey JK, Fischer DG et al (2008) Plant-soil-microorganism interactions: heritable relationship between plant genotype and associated soil microorganisms. Ecology 89:773–781. https://doi.org/10.1890/07-0337.1

Shan HW, Deng WH, Luan JB et al (2017) Thermal sensitivity of bacteriocytes constrains the persistence of intracellular bacteria in whitefly symbiosis under heat stress. Environ Microbiol Rep 9:706–716. https://doi.org/10.1111/1758-2229.12554

Shinzato C, Shoguchi E, Kawashima T et al (2011) Using the *Acropora digitifera* genome to understand coral responses to environmental change. Nature 476:320–323. https://doi.org/10.1038/nature10249

Simon JC, Marchesi JR, Mougel C, Selosse MA (2019) Host-microbiota interactions: from holobiont theory to analysis. Microbiome 7:5. https://doi.org/10.1186/s40168-019-0619-4

Singh LP, Gill SS, Tuteja N (2011) Unraveling the role of fungal symbionts in plant abiotic stress tolerance. Plant Signal Behav 6:175–191

Singh Y, Ahmad J, Musarrat J et al (2013) Emerging importance of holobionts in evolution and in probiotics. Gut Pathog 5:12

Smith DJ, Suggett DJ, Baker NR (2005) Is photoinhibition of zooxanthellae photosynthesis the primary cause of thermal bleaching in corals? Glob Chang Biol 11:1–11

Stoecker DK, Johnson MD, De Vargas C, Not F (2009) Acquired phototrophy in aquatic protists. Aquat Microb Ecol 57:279–310. https://doi.org/10.3354/ame01340

Stork NE (2018) How many species of insects and other terrestrial arthropods are there on earth? Annu Rev Entomol 63:31–45. https://doi.org/10.1146/annurev-ento-020117-043348

Strychar KB, Sammarco PW (2009) Exaptation in corals to high seawater temperatures: low concentrations of apoptotic and necrotic cells in host coral tissue under bleaching conditions. J Exp Mar Bio Ecol 369:31–42. https://doi.org/10.1016/J.JEMBE.2008.10.021

Su Q, Zhou X, Zhang Y (2013) Symbiont-mediated functions in insect hosts. Commun Integr Biol 6:e23804. https://doi.org/10.4161/cib.23804

Suárez J (2018) The importance of symbiosis in philosophy of biology: an analysis of the current debate on biological individuality and its historical roots. Symbiosis 76:77–96

Sudakaran S, Kost C, Kaltenpoth M (2017) Symbiont acquisition and replacement as a source of ecological innovation. Trends Microbiol 25:375–390

Summerer M, Sonntag B, Hörtnagl P, Sommaruga R (2009) Symbiotic ciliates receive protection against UV damage from their algae: a test with *Paramecium bursaria* and *Chlorella*. Protist 160:233–243. https://doi.org/10.1016/j.protis.2008.11.005

Suzuki G, Yamashita H, Kai S et al (2013) Early uptake of specific symbionts enhances the post-settlement survival of *Acropora* corals. Mar Ecol Prog Ser 494:149–158. https://doi.org/10.3354/meps10548

Takahashi S, Yoshioka-Nishimura M, Nanba D, Badger MR (2013) Thermal acclimation of the symbiotic alga *Symbiodinium* spp. alleviates photobleaching under heat stress. Plant Physiol 161:477–485. https://doi.org/10.1104/pp.112.207480

Terzaghi E, O'Hara M (1990) Microbial plasticity the relevance to microbial ecology. In: Advances in microbial ecology. Springer, Boston, MA, pp 431–460

Theis KR, Dheilly NM, Klassen JL et al (2016) Getting the hologenome concept right: an eco-evolutionary framework for hosts and their microbiomes. mSystems 1:e00028–e00016. https://doi.org/10.1128/mSystems00028-16

Venn AA, Loram JE, Douglas AE (2008) Photosynthetic symbioses in animals. J Exp Bot 59:1069–1080

Veron JEN, Hoegh-Guldberg O, Lenton TM et al (2009) The coral reef crisis: the critical importance of <350 ppm CO_2. Mar Pollut Bull 58:1428–1436. https://doi.org/10.1016/j.marpolbul.2009.09.009

Waller F, Achatz B, Baltruschat H et al (2005) The endophytic fungus *Piriformospora* indica reprograms barley to salt-stress tolerance, disease resistance, and higher yield. Proc Natl Acad Sci U S A 102:13386–13391. https://doi.org/10.1073/pnas.0504423102

Webster NS, Reusch TBH (2017) Microbial contributions to the persistence of coral reefs. ISME J 11:2167–2174

Weis VM (2008) Cellular mechanisms of *Cnidarian* bleaching: stress causes the collapse of symbiosis. J Exp Biol 211:3059–3066

Wernegreen JJ (2012a) Endosymbiosis. Curr Biol 22:R555–R561. https://doi.org/10.1016/j.cub.2012.06.010

Wernegreen JJ (2012b) Mutualism meltdown in insects: Bacteria constrain thermal adaptation. Curr Opin Microbiol 15:255–262

Wichterman R, Wichterman R (1986) Organisms living in and upon *Paramecium*: the endosymbionts. In: The biology of Paramecium. Springer US, Berlin, pp 375–420

Wietheger A (2012) Coral bleaching resistance vs susceptibility : the role of antioxidant activity in symbiotic dinoflagellates. Thesis, University of Victoria Wellington

Xue B, Leibler S (2016) Evolutionary learning of adaptation to varying environments through a transgenerational feedback. Proc Natl Acad Sci U S A 113:11266–11271. https://doi.org/10.1073/pnas.1608756113

Yakovleva IIM, Baird AAH, Yamamoto HHH et al (2009) Algal symbionts increase oxidative damage and death in coral larvae at high temperatures. Mar Ecol Prog Ser 378:105–112

Ye S, Badhiwala KN, Robinson JT et al (2019a) Thermal plasticity of a freshwater cnidarian holobiont: detection of trans-generational effects in asexually reproducing hosts and symbionts. ISME J 13:2058–2067. https://doi.org/10.1038/s41396-019-0413-0

Ye S, Bhattacharjee M, Siemann E (2019b) Thermal tolerance in green hydra: identifying the roles of algal endosymbionts and hosts in a freshwater holobiont under stress. Microb Ecol 77:537–545. https://doi.org/10.1007/s00248-018-01315-1

Zhao DX, Zhang XF, Chen DS et al (2013) *Wolbachia*-host interactions: host mating patterns affect *Wolbachia* density dynamics. PLoS One 8:e66373. https://doi.org/10.1371/journal.pone.0066373

Zilber-Rosenberg I, Rosenberg E (2008) Role of microorganisms in the evolution of animals and plants: the hologenome theory of evolution. FEMS Microbiol Rev 32:723–735. https://doi.org/10.1111/j.1574-6976.2008.00123.x

Chapter 22
Microbial Metabolites as Molecular Mediators of Host-Microbe Symbiosis in Colorectal Cancer

J. M. Keane, S. A. Joyce, C. G. M. Gahan, N. P. Hyland, and A. Houston

Abstract The symbiosis between the gut microbiota and the host has been identified as an integral part of normal human physiology and physiological development. Research in germ-free or gnotobiotic animals has demonstrated the importance of this symbiosis in immune, vascular, hepatic, respiratory and metabolic systems. Disruption of the microbiota can also contribute to disease, and the microbiota has been implicated in numerous intestinal and extra-intestinal pathologies including colorectal cancer. Interactions between host and microbiota can occur either directly or indirectly, via microbial-derived metabolites. In this chapter, we focus on two major products of microbial metabolism, short-chain fatty acids and bile acids, and

---

N. P. Hyland and A. Houston contributed equally with all other contributors.

J. M. Keane
APC Microbiome Ireland, University College Cork, Cork, Ireland

School of Microbiology, University College Cork, Cork, Ireland

Department of Medicine, University College Cork, Cork, Ireland

S. A. Joyce
APC Microbiome Ireland, University College Cork, Cork, Ireland

School of Biochemistry and Cell Biology, University College Cork, Cork, Ireland

C. G. M. Gahan
APC Microbiome Ireland, University College Cork, Cork, Ireland

School of Microbiology, University College Cork, Cork, Ireland

School of Pharmacy, University College Cork, Cork, Ireland

N. P. Hyland (✉)
APC Microbiome Ireland, University College Cork, Cork, Ireland

Department of Physiology, University College Cork, Cork, Ireland
e-mail: n.hyland@ucc.ie

A. Houston
APC Microbiome Ireland, University College Cork, Cork, Ireland

Department of Medicine, University College Cork, Cork, Ireland

M. Kloc (ed.), *Symbiosis: Cellular, Molecular, Medical and Evolutionary Aspects*,
Results and Problems in Cell Differentiation 69,
https://doi.org/10.1007/978-3-030-51849-3_22

their role in colorectal cancer. Short-chain fatty acids are the products of microbial fermentation of complex carbohydrates and confer protection against cancer risk, while bile acids are compounds which are endogenous to the host, but undergo microbial modification in the large intestine leading to alterations in their bioactivity. Lastly, we discuss the ability of microbial modulation to mediate cancer risk and the potential to harness this ability as a prophylactic or therapeutic treatment in colorectal cancer.

Keywords Butyrate · Bile · Gut microbiota · Colon · Tumorigenesis

22.1 The Gut Microbiota

The human microbiota is a community of bacteria, archaea, protists, fungi and viruses that live in and on the human body (Group NHW et al. 2009). The term gut "microbiome" is sometimes used synonymously with the gut "microbiota" but can also refer to the full collection of genes present in the microbiota of a community. The cells of our microbiota are estimated to outnumber our nucleated human cells by a ratio of approximately 13:1, about 70% of which occupy our gastrointestinal (GI) tract (Sekirov et al. 2010). A symbiotic relationship exists between the microbiota and host, and this relationship plays a vital role in host immune modulation, metabolism, inhibition of pathogens and structural development (Quigley 2013; Patel and Lin 2010). Members of the microbiota may be classified by the nature of their symbiotic relationship with the host, ranging from harmful pathogens to beneficial probiotics. These probiotic bacteria are characterised as "live microorganisms that, when administered in adequate amounts, confer a health benefit on the host", while prebiotics are "selectively fermented ingredients that result in specific changes in the composition and/or activity of the gastrointestinal microbiota, thus conferring benefit(s) upon host health" (Davani-Davari et al. 2019; Marteau et al. 2001; Food and Agriculture Organization of the United Nations, World Health Organization 2002).

The gut microbiota comprises over 5000 bacterial species and 3 million genes in a typical individual, with possibly over 35,000 species in the collective human microbiome (Pasolli et al. 2019; Frank et al. 2007). It is dominated by the phyla Firmicutes and Bacteroidetes, featuring smaller proportions of Proteobacteria, Verrucomicrobia, Actinobacteria, Fusobacteria and Cyanobacteria (Quigley 2013; Eckburg et al. 2005). This consistency of phyla, combined with significant inter-individual variation within the phyla, suggests a selective pressure to maintain the higher taxonomic structure with a functional redundancy at lower levels (Eckburg et al. 2005; Gill et al. 2006). The upper GI tract contains relatively few microbial inhabitants. The stomach and duodenum contain approximately 10^2 organisms per gram of contents. This rises to 10^4–10^7 in the jejunum, finally reaching ~10^9 colony-forming units (CFUs)/mL in the terminal ileum and

~10^{12} CFU/mL of primarily anaerobic bacteria in the colon (Quigley 2013; O'Hara and Shanahan 2006). The composition also changes along the length of the GI tract, with *Bacillus* and Actinobacteria enriched in the small intestine, while Bacteroidetes and *Lachnospiraceae* are enriched in the large intestine (Frank et al. 2007).

The intestinal tract is generally considered sterile at birth, with colonisation beginning immediately through contact with the mother and environmental bacteria. Recent research, however, has suggested colonisation of the placenta by *Streptococcus agalactiae* in approximately 5% of pregnancies. However, the possibility remains that this is a result of sample contamination (Mandar and Mikelsaar 1996; de Goffau et al. 2019). The newborn microbiota is reflective of the mode of delivery, with babies delivered by Caesarean section having a microbiota characterised by fewer *Bifidobacterium* species compared to vaginal births (Huurre et al. 2008). The shift towards an adult microbial composition begins during weaning before the microbiota stabilises at approximately 1–2.5 years of age (Voreades et al. 2014). The microbiota then remains largely stable in the absence of disruptions such as long-term dietary changes or migration (Maskarinec and Noh 2004; Turnbaugh et al. 2008). Further changes to the microbiota are observed later in life, such as a reduction in diversity and in the number of symbiotic species, and an increase in enteric bacteria, which may be associated with the age-related physiological decline (Nagpal et al. 2018; O'Toole and Claesson 2010; Claesson et al. 2012).

22.1.1 Host-Microbe Symbiosis and Physiological Development

The ancient association and co-evolution between host and microbe have led to the deep integration of the microbiota into normal physiological processes and development. This is illustrated by germ-free (GF) animals, which, in the absence of normal gut microbiota, display several developmental abnormalities including an immature immune system (Smith et al. 2007). Potential mechanisms by which the neonatal microbiota mediate the development of the immune system differ between bacterial species, and likely involve the interacting influences of many different taxa. GF mice have a suppressed *T* Helper Type 1 (T_h1) cell response that can be restored by monocolonisation with *Listeria monocytogenes*, which stimulates interleukin (IL)-12 production in macrophages. Likewise, the reduction in T_h17 cells observed in these animals can be normalised by colonisation by segmented filamentous bacteria (SFB), leading to the release of serum amyloid A from intestinal epithelial cells (Tibbs et al. 2019). Colonisation with SFB also upregulated the production of immunoglobulin A, which is crucial for a tolerance of commensal microbiota by the mucosal immune system (Klaasen et al. 1993; Mathias et al. 2014).

The host immune response also modulates the composition of the gut microbiota, and the ability of the mucosal immune system to differentiate between commensal and pathogenic bacteria is a topic of ongoing research (Yap and Mariño 2018). Members of the gut microbiota interact with the host directly by signalling through pathogen recognition receptors, such as Toll-like Receptors (Gold et al. 2004). The gut microbiota also produces a wide array of bioactive bacteria-derived metabolites, both from compounds endogenous to the host, e.g. bile acids, or exogenous compounds such as those found in the diet or environment, which allow them to interact indirectly with the host. These metabolites can also play an important role in host health and disease, including colorectal cancer (CRC) (discussed in Sect. 22.3).

22.2 Host-Microbiota Interactions in Colorectal Cancer

There is precedence for the involvement of bacteria in GI cancer. *Helicobacter pylori*, for example, is the strongest known risk factor for gastric cancer (Wroblewski et al. 2010). Given the close apposition between the gut microbiome and colonic epithelium, in particular, research efforts have focussed on the role of the microbiota in colon cancer (Table 22.1) (Hope et al. 2005). The proposed mechanisms by which the microbiota may impact CRC include effects on the immune system and proto-oncogenic pathways such as proliferation and apoptosis, while microbial metabolites can have pro- and anti-tumorigenic associations (Macarthur et al. 2004). The strongest links between the microbiota and potentially cancer-promoting inflammation involve pathogenic species such as *Fusobacterium nucleatum* or enterotoxigenic *Bacteroides fragilis*, both of which have been positively correlated with CRC (Wu et al. 2019; Haghi et al. 2019). The role of the microbiota in proliferation is evident in GF mice which display smaller intestinal crypts with a lower mitotic index (Nowacki 1993), while the microbiota can mediate apoptosis via a number of mechanisms including the production of butyrate (Sect. 22.3.1). Moreover, tumour formation is reduced in GF animals (Yang et al. 2017), with faecal microbial transfer from CRC patients to GF mice increasing tumorigenesis in these animals (Ellmerich et al. 2000; Horie et al. 1999; Scanlan et al. 2008). This capacity to regulate both intestinal proliferation and apoptosis highlights the importance of this delicate symbiotic relationship, which could contribute to cell cycle disruption if dysregulated.

Substantial evidence exists in animal models for the role of gut bacteria in promoting CRC. These studies primarily utilise mouse models either genetically predisposed to CRC such as the APC^{MIN} mouse or use genotoxic compounds such as azoxymethane (AOM), or its precursor dimethylhydrazine (DMH), to chemically induce CRC. AOM can also be combined with dextran sodium sulphate (DSS) to model colitis-associated CRC. In this AOM/DSS model, manipulation of the microbiota with antibiotics reduced tumorigenesis, but had conflicting effects in APC^{MIN} mice (Zackular et al. 2013; Kaur et al. 2018). Antibiotic treatment was

Table 22.1 Bacteria associated with human CRC (modified from Jahani-Sherafat et al. (2018))

Sample	Bacteria	References
Faecal	*Fusobacterium nucleatum*	Suehiro et al. (2017), Wong et al. (2017), Liang et al. (2017)
Tumour	*F. nucleatum*	Castellarin et al. (2012), Li et al. (2016), Mima et al. (2016)
Faecal	*F. nucleatum, Clostridium difficile*	Fukugaiti et al. (2015)
Tumour	*F. nucleatum, Bacteroides fragilis*	Wei et al. (2016)
Faecal, mucosal	*F. nucleatum, Enterobacteriaceae*	Mira-Pascual et al. (2015)
CRC	*F. nucleatum, Pan-fusobacterium*	Tahara et al. (2014)
Tumour	*Fusobacterium*	Marchesi et al. (2011), Kostic et al. (2012)
Tumour	*Fusobacterium, Providencia*	Burns et al. (2015)
Tumour	*Fusobacterium,* Firmicutes	Gao et al. (2015)
Faecal	*Fusobacterium, Atopobium/Porphyromonas*	Ahn et al. (2013)
Faecal	*Fusobacterium, Porphyromonas*	Sinha et al. (2016)
CRC	*Fusobacterium,* Enterotoxigenic *B. fragilis* (ETBF)	Viljoen et al. (2015)
Tumour	*Fusobacterium, Roseburia*	Geng et al. (2013)
Tumour	*Fusobacterium, Enterococcus faecalis, ETBF*	Zhou et al. (2016)
CRC	*Fusobacterium, Leptotrichia, Campylobacter*	Warren et al. (2013)
Faecal	*Bacteroides/Prevotella*	Sobhani et al. (2011)
Faecal	*E. faecalis*	Balamurugan et al. (2008)
Faecal, mucosal	Bacteroidetes, *Prevotella*	Flemer et al. (2017)
Tumour, mucosal	*Escherichia coli*	Swidsinski et al. (1998)
Meta-analysis	*Helicobacter pylori*	Zumkeller et al. (2006)
Bloodstream	*Streptococcus gallolyticus, Clostridium perfringens, Clostridium septicum, Peptostreptococcus*	Kwong et al. (2018)
Bloodstream	*Streptococcus bovis*	Gold et al. (2004)
Oral	*Haemophilus, Parvimonas, Prevotella, Alloprevotella, Lachnoanaerobaculum, Neisseria, Streptococcus*	Flemer et al. (2018)

protective in APC^{MIN} mice when compound mutations in DNA repair or interleukin receptor genes were present (Kaur et al. 2018; Xiao et al. 2010; Belcheva et al. 2014). Furthermore, Onoue et al. observed decreased numbers of aberrant crypt foci (ACF) in DMH-treated GF rats compared to conventional rats (Onoue et al. 1997). Conversely, the administration of bacteria associated with cancer risk, for example, *Streptococcus bovis* or *F. nucleatum*, to susceptible animals was shown to increase proliferation, inflammation and tumorigenesis (Yang et al. 2017; Ellmerich et al. 2000). Tumour multiplicity was also increased in gnotobiotic (GB) rats colonised by enterococci compared to GB rats without enterococci, with the tumour numbers in the former group significantly decreased by the inclusion of probiotic strain *Bifidobacterium breve* (Onoue et al. 1997). A similar result was achieved by Horie et al. concerning adenomas, with the lowest incidence of adenoma development observed in rats mono-associated with probiotic *Lactobacillus acidophilus* (Horie et al. 1999).

Human studies present associative evidence for the role of the microbiota in CRC. The microbiota is altered in the colon of CRC patients and in the tumour tissue compared to healthy controls, with adenomatous polyps representing an intermediate step between the two states (Scanlan et al. 2008). The colonic mucosa is the symbiotic interface between host and microbiota, and studies have shown colonisation of this interface by adherent and invasive *Escherichia coli* in carcinoma patients (Swidsinski et al. 1998; Martin et al. 2004). Moreover, CRC patients had increased carcinogenic microbial metabolites in their faeces compared to healthy individuals despite both groups having similar diets, with the difference ascribed to their different levels of enzymatically active anaerobic bacteria (Kanazawa et al. 1996). Similarly, *Lactobacillus* species have been shown to reduce faecal and urinary mutagenicity induced by fried meat consumption and to reduce faecal β-glucuronidase, β-glucosidase, nitroreductase and glycocholic acid hydrolase activity (Hayatsu and Hayatsu 1993; Ling et al. 1994; Spanhaak et al. 1998). The gut microbiota can also modulate the production of mucus in the intestinal lumen, which in itself can play an important role in CRC by regulating the interaction of the gut bacteria and luminal contents with the colonic epithelium (Velcich et al. 2002).

The composition of the microbiota has also been investigated as a potential predictive biomarker for human CRC. Two meta-analyses of human faecal shotgun sequencing studies identified microbial taxonomic signatures with sensitivity to, and specificity for, CRC, which was comparable to common non-invasive clinical screening tests (Wirbel et al. 2019; Thomas et al. 2019). Models based on the functional gene content of the faecal microbiome were also generated, and enrichment of the bile acid-inducible operon, which is involved in microbial bile acid metabolism, was demonstrated at both the genomic and transcriptomic levels (Wirbel et al. 2019; Thomas et al. 2019). Additionally, bacterial species associated with the oral cavity are frequently enriched in the gut microbiota of CRC patients and a model combining data from oral and faecal microbiota was highly predictive of CRC (Flemer et al. 2018).

22.3 Microbial Metabolites as Mediators of Host-Microbe Symbiosis in Colorectal Cancer

A key interaction between the host and the microbiota occurs through the production of microbial-derived metabolites (Louis et al. 2014). Here, we focus on two major products of microbial metabolism, short-chain fatty acids and bile acids, and their role in CRC.

22.3.1 Short-Chain Fatty Acids

Commensal bacteria contribute to host-microbial homeostasis and resistance to CRC via the production of short-chain fatty acids (SCFAs). SCFAs are fatty acids with less than six carbon atoms and are primarily the product of fermentation of dietary fibre by anaerobic bacteria in the proximal colon (Topping and Clifton 2001). The three most common SCFAs are acetate, propionate and butyrate, with butyrate shown to play a predominant role in CRC (Hinnebusch et al. 2002). The majority of butyrate is produced by bacteria in *Clostridium* clusters XIVa and IV, particularly *Roseburia/Eubacterium rectale*-related bacteria in cluster XIVa and *Faecalibacterium prausnitzii* relatives in cluster IV (Louis et al. 2010). In a screen of butyryl-CoA:acetate CoA-transferase sequences from human faecal samples, 88% of sequences belonged to *E. rectale*, *Roseburia faecis*, *Eubacterium hallii* and an unnamed species, with the remainder coming from uncultured strains (Louis et al. 2010).

Butyrate is the primary energy source for normal colonic epithelial cells and has been associated with positive health effects, including in CRC (Wu et al. 2018; Donohoe et al. 2011). Concentrations of SCFAs are highest in the caecum and proximal colon, where the incidence of tumours is low (Macfarlane et al. 1992). The lowest intracolonic levels of SCFAs are found in the distal colon and rectum, the site of the majority of human CRC. Butyrate was also reduced in a rat model of CRC, where it correlated negatively with tumour mass (McIntyre et al. 1993). Moreover, protein feeding increased tumour number in AOM-treated rats which was ameliorated by resistant starch, which is a substrate for microbial butyrate production (Le Leu et al. 2007). Mechanisms by which butyrate protect against CRC are presented in Table 22.2.

Whilst predominantly protective against the development of CRC, butyrate can have pro-tumorigenic effects following CRC onset. One such mechanism involves its ability to act as a histone deacetylase (HDAC) inhibitor in vivo where it epigenetically promotes cell proliferation (Bultman 2017). The contrasting effects of butyrate in normal epithelial cells versus CRC cells can be explained by the metabolic fate of intracellular butyrate. The ability to use butyrate as an energy source is lost in malignant colonocytes (Onoue et al. 1997). Instead, these cells perform glycolysis in what is termed the Warburg effect. This causes the accumulation of intracellular butyrate which generates concentrations sufficient to allow butyrate to act as an HDAC inhibitor. This effect in CRC cells is amplified by glucose-induced metabolism of butyrate by ATP citrate lyase to acetyl-CoA, which acts as a histone acetyltransferase in cells exhibiting the Warburg effect (Berwick et al. 2002; Donohoe et al. 2012).

Table 22.2 The role of butyrate in CRC

Butyrate effect	References
Stimulates proliferation in the base of normal colonic crypts	Kripke et al. (1989)
Inhibits deoxycholic acid (DCA)-induced proliferation at the crypt surface	Velazquez et al. (1997)
Induces apoptosis in cancer cell lines at physiological concentrations	Hague et al. (1995)
Induces apoptosis in a p53-independent manner	Hague et al. (1993)
Reduces apoptosis in colonic ACF of AOM-treated rats	Caderni et al. (1998)
Absence induces BAX-mediated apoptosis in guinea pig colon	Hass et al. (1997)
Induces differentiation in HT29 cells, possibly by reducing levels of c-myc	Augeron and Laboisse (1984), Taylor et al. (1992)
Inhibits growth at G1 stage	Heerdt et al. (1997), Gamet et al. (1992), Barnard and Warwick (1993), Siavoshian et al. (1997)
Promotes differentiation in cancer cells but suppresses it in normal cells	Gibson et al. (1992)
Improves barrier function by upregulating Claudin-1	Wang et al. (2012)
Anti-inflammatory and may protect against IBD	Wachtershauser and Stein (2000)
Lowers intestinal pH which protects against DCA-induced epithelial damage	Campbell et al. (1997), Rafter et al. (1986)
Reduces damage induced by H_2O_2 in normal colonocytes	Abrahamse et al. (1999)

22.3.2 *Bile acids*

Bile acids are endogenous steroid molecules that are conjugated to a glycine or taurine amino acid residue to form bile salts and stored in the gallbladder for postprandial release into the duodenum to aid lipid digestion. They are derived from cholesterol and are the major route of cholesterol elimination from the body. The major human bile acids are cholic acid (CA) and chenodeoxycholic acid (CDCA), while in mice the majority of CDCA is converted into muricholic acid (MCA) (Li and Chiang 2012). Although most bile salts are reabsorbed in the distal ileum, around 5% escape to the large intestine where they can be modified by intestinal bacteria (Dawson et al. 2003). These bile acids undergo deconjugation of the amino acid residue by *bile salt hydrolase* to form free bile acids, followed by 7α-dehydroxylation to form cytotoxic secondary bile acids, as well as a number of other minor modifications (Hill 1990). 7α-dehydroxylation of the major human bile acids CA and CDCA forms deoxycholic acid (DCA) and lithocholic acid (LCA), respectively. These modifications can alter the biochemistry and bioactivity of bile acids, as well as their receptor specificities, which affect their role in CRC. The synthesis and microbial metabolism of bile acids are presented in Fig. 22.1.

Secondary bile acids are hydrophobic, cytotoxic molecules and evidence suggests they play a role in CRC. For example, numerous epidemiological studies have highlighted higher faecal bile acid content in populations with increased CRC rates (Jensen et al. 1982; Cheah 1990; Crowther et al. 1976). Moreover, DCA is higher in patients with colorectal adenomas and was proposed as a carcinogen as early as 1940 based on its induction of tumours in mice (Bayerdörffer et al. 1995; Cook et al. 1940). Bile acids were initially classified as tumour promoters rather than tumour initiators, as studies primarily demonstrated their action when co-administered with chemical carcinogens such as AOM (Magnuson et al. 1993; Reddy et al. 1976a). However, the role of bile acids as aetiologic agents of cancer in their own right is now emerging (Bernstein et al. 2005a). For example, a diet high in fat and low in fibre is a known risk factor for colon cancer (Reddy et al. 1975). This diet was also associated with increased secondary bile acids, as well as increased glucuronidase deconjugation (Reddy et al. 1975). Also of note, GF rats are generally resistant to chemical carcinogen-induced CRC (Sumi and Miyakawa 1979). However, GF rats treated with the chemical carcinogen methylnitronitrosoguanidine (MNNG) and DCA displayed colonic adenocarcinomas, suggesting microbial production of DCA could play a role in tumorigenesis and may explain, in part, the resistance to CRC observed in GF animals (Reddy et al. 1976b).

Bile acids can increase cancer risk by several mechanisms. DCA and CDCA were shown to upregulate pro-inflammatory cyclooxygenase-2 and its downstream inflammatory product prostaglandin E_2 in a protein kinase C-dependent manner, whilst activating c-Jun and AP-1 (Glinghammar et al. 2002; Zhang et al. 1998). Bile acids also generate reactive oxygen and nitrogen species via a detergent effect on cell membranes and activation of inducible nitric oxide synthase (Bernstein et al. 2005b). Additionally, bile acids may induce apoptosis in the short term but select for apoptosis-resistant cells in the longer term (Bernstein et al. 2009). This ability appears to be related to their hydrophobicity, with the most powerful effect displayed by the most hydrophobic bile acids (Powell et al. 2001). Indeed, normal cells adjacent to tumour tissue in colon cancer patients were shown to display resistance to bile salt- and bile acid-induced apoptosis, and this is mediated by an upregulation of the anti-apoptotic protein B-cell lymphoma-extra large (Bernstein et al. 2009; Badvie et al. 2006).

Bile acids can also induce chromosomal abnormalities such as aneuploidy and micronucleus formation (Jenkins et al. 2007; Assinder and Upshall 1982). In yeast, DCA, LCA, CDCA and CA each induced mitotic chromosome aneuploidy, while tauro- or glyco-conjugated DCA did not (Ferguson and Parry 1984). Oxidative stress is a well-established source of chromosomal instability and this is a plausible mechanism of bile acid-induced DNA damage and increased CRC risk (Limoli and Giedzinski 2003; Hunt et al. 1998). LCA was also shown to inhibit the repair activity of DNA polymerase β which could exacerbate the consequences of bile acid-induced DNA damage (Ogawa et al. 1998). Finally, a proteomic study of CRC cell

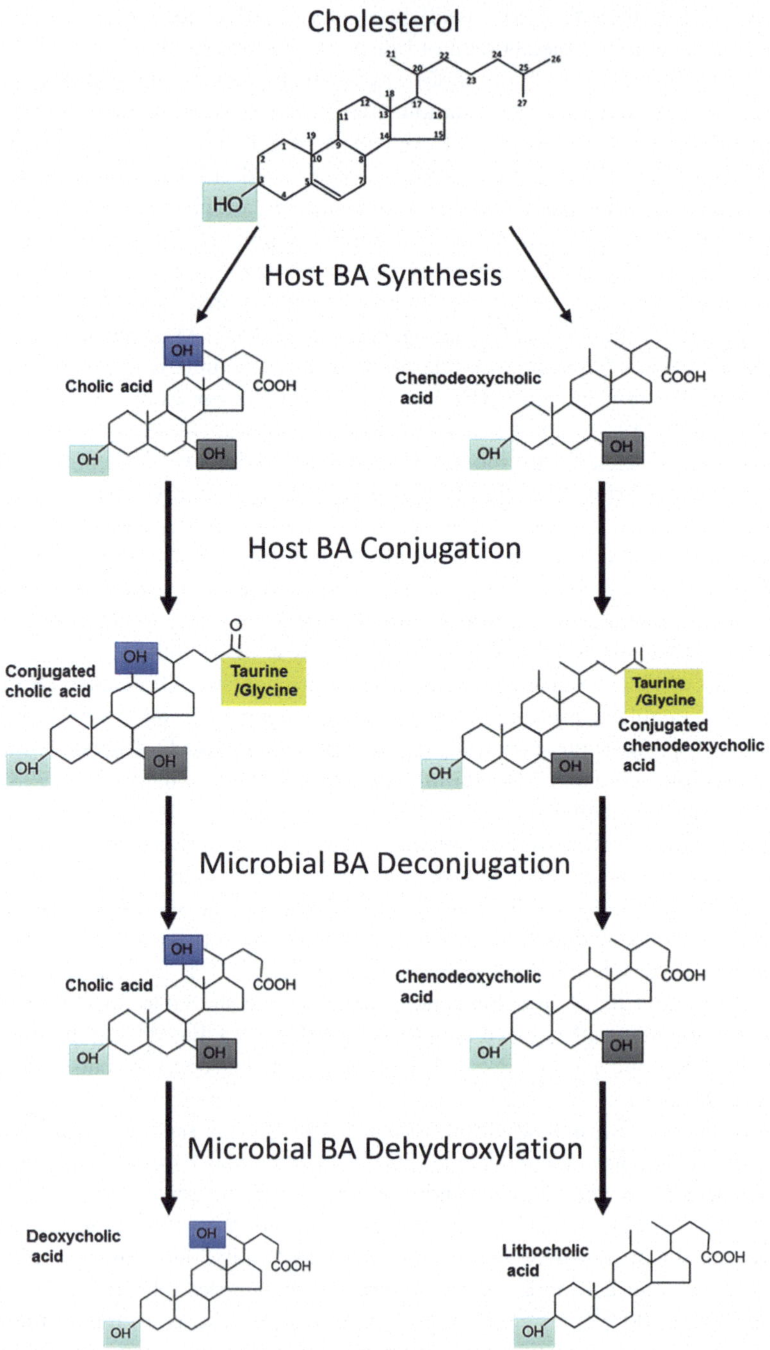

Fig. 22.1 Synthesis and microbial metabolism of bile acids. Cholic acid (CA) and chenodeoxycholic acid (CDCA) are the primary human bile acids (BAs) which are formed from cholesterol in the liver by a multistep pathway. After conjugation and release into the duodenum,

lines induced with DCA identified alterations in ten proteins involved in DNA repair and cell cycle checkpoints (Bernstein et al. 2004).

Bile acids have also been associated with cancer through Farnesoid X Receptor (FXR) signalling (Degirolamo et al. 2011). Bile acid homeostasis is regulated by FXR, which is a nuclear receptor expressed by liver hepatocytes and small intestine enterocytes (Makishima et al. 1999). *FXR* expression is downregulated in human colorectal tumours and colon cancer cell lines (De Gottardi et al. 2004), while $Fxr^{-/-}$ mice are predisposed to multiple cancers, including that of CRC (Maran et al. 2009; Kim et al. 2007). Moreover, administration of tauro-conjugated βMCA, which is an FXR antagonist bile acid, increased stem cell proliferation by activating Wnt signalling, impaired intestinal integrity, accelerated tumour growth, induced dysplastic morphology and chromosome instability, and increased the serum levels of pro-inflammatory cytokines in APC^{MIN} mice (Fu et al. 2019). FXR agonists, in turn, promoted apoptosis, downregulated intestinal stem cell genes and inhibited Wnt signalling (Modica et al. 2008). FXR agonists also delayed tumour progression, reduced tumour multiplicity, proliferation and serum cytokines, and improved intestinal morphology, differentiation, barrier function and bile acid homeostasis (Fu et al. 2019). Microbial modification of bile acids plays a role in their interaction with FXR, as FXR displays greater affinity for conjugated bile acids, with reducing affinity for CDCA $>$ DCA $=$ LCA $>$ CA (Ding et al. 2015). As a result, bacterial modification of bile acids can influence their specificity for FXR and hence their influence on cancer risk. FXR has also been demonstrated to modulate the microbiota as FXR antagonism increased the proportion of Bacteroidetes compared to Firmicutes (Bervoets et al. 2013; Jiang et al. 2015). FXR can also suppress expression of pro-inflammatory cytokines (Stojancevic et al. 2012), to the extent that a synthetic FXR ligand protected mice from DSS-induced colitis (Vavassori et al. 2009).

22.4 Pre- and Pro-Biotics as Modulators of Host-Microbe Symbiosis: Implications for Colorectal Cancer

Clinical trials have provided evidence for the beneficial role of pre- and pro-biotics in CRC (Table 22.3). One such trial using a combination of pre- and pro-biotics comprising inulin, *Lactobacillus* and *Bifidobacterium* administered to individuals at high risk of CRC development showed that the combination treatment resulted in a decrease in colonic epithelial proliferation, decreased abundance of *Clostridium*

Fig. 22.1 (continued) 95% are reabsorbed at the terminal ileum but 5% escape to the large intestine to be modified by the microbiota. This includes deconjugation and 7α-dehydroxylation to form deoxycholic acid and lithocholic acid from CA and CDCA, respectively. Figure modified from Long et al. (2017)

Table 22.3 Proposed protective mechanisms by which prebiotics and probiotics affect CRC development (modified from Wollowski et al. (2001))

Probiotic/Prebiotic	Protective mechanism	References
Lactobacillus casei, omniflora or yoghurt	Mutations in the Ames test decreased	Pool-Zobel et al. (1993a), Bodana and Rao (1990)
Strains of *Lactobacillus* and *Bifidobacterium*, cellular components and metabolites of lactic acid bacteria	DNA damage in colon cells decreased (antigenotoxicity)	Pool-Zobel et al. (1993b), Pool-Zobel et al. (1996)
Fermented milk with *L. acidophilus*, *Bifidobacterium bifidum*, *Streptococcus lactis* and *Streptococcus cremoris*; lactulose	Procarcinogenic enzyme activity decreased: β-glucuronidase, nitroreductase, azoreductase and detoxifying enzyme activity increased; glutathione-S-transferase	Goldin and Gorbach (1984), Goldin et al. (1992), Benno and Mitsuoka (1992), Bouhnik et al. (1996)
L. acidophilus, *S. cremoris*, cell wall of lactic acid bacteria	Binding of mutagens	Orrhage et al. (1994), Zhang and Ohta (1991), Morotomi and Mutai (1986)
Milk fermented with *L. acidophilus*	Excretion of mutagens decreased	Lidbeck et al. (1992)
Milk fermented with *L. acidophilus* and *Bifidobacterium*	Immune stimulation increased	Link-Amster et al. (1994)
Fermentation of prebiotics	SCFAs increased, pH decreased, probiotics increased	Segal et al. (1995), Baghurst et al. (1996)
Butyrate	Decreased proliferation, increased apoptosis, epigenetic changes (Sect. 22.3.1)	Hague et al. (1995), Hass et al. (1997), Marchetti et al. (1997)

perfringens and reduced ability of faecal water to induce necrosis in colon cells in vitro (Rafter et al. 2007). Epithelial barrier function, which is deficient in CRC, was also improved (Soler et al. 1999).

This beneficial effect of pre- and pro-biotics has been replicated in several studies (Roessler et al. 2012; Gianotti et al. 2010; Xia et al. 2010; Roller et al. 2007). Moreover, a prebiotic mixture decreased chemotherapy-associated side effects including diarrhoea and enterocolitis in CRC patients (Mego et al. 2015). Furthermore, the administration of probiotics can have potential cancer-preventative effects. For example, a mixture of *Lactobacillus* and *Propionibacterium* administered to healthy subjects reduced faecal levels of the bacterial enzyme β-glucuronidase, which is implicated in the activation of carcinogens in the colon (Ishikawa et al. 2003).

In animal studies, *Bifidobacterium longum* has been shown to ameliorate AOM/DMH-induced colon carcinogenesis, an effect that is enhanced by co-administration with the prebiotics inulin and lactulose (Challa et al. 1997; Rowland et al. 1998a). A similar effect was seen with *Lactobacillus* species, although this effect was absent when probiotic administration was delayed until 9 weeks into DMH-administration, suggesting *Lactobacillus* was only protective in the early stages of tumorigenesis (Goldin et al. 1996; Goldin and Gorbach 1980).

The ability of probiotics to affect early-stage cancer development could be due to their function as anti-mutagenic agents. For instance, *Lactobacillus casei* gavage attenuated DNA damage induced by MNNG in rat colonic and gastric mucosa, while in another study, a selection of lactic acid bacteria (LAB) inhibited the genotoxic effects of MNNG and DMH in the rat colon (Pool-Zobel et al. 1993b, 1996). Heat treatment eliminated the protective effect of the bacteria in both studies, suggesting that viable bacteria are required for this effect, although the peptidoglycan fraction and whole freeze-dried *L. acidophilus* were also anti-genotoxic. Arimochi et al. also demonstrated a reduction in ACF in AOM-treated rats after the administration of *L. acidophilus* and *C. perfringens* (Arimochi et al. 1997). In particular, *L. acidophilus* improved DNA repair by DNA methyltransferase. Other potential mechanisms include the ability of LAB to bind dietary mutagens which limits their ability to interact with the colonic epithelium (Orrhage et al. 1994; Morotomi and Mutal 1986). For example, toxic compounds are detoxified by glucuronidation in the liver, but bacterial β-glucuronidase activity may hydrolyse these molecules and liberate carcinogens. The activity of this enzyme was shown to be reduced in AOM- and DMH-treated rats following gavage with the probiotic *B. longum*. This effect was enhanced by co-administration with the prebiotic inulin, possibly as a result of acidification of the intestinal environment and displacement of bacteria expressing β-glucuronidase (Kulkarni and Reddy 1994; Abdelali et al. 1995; Rowland et al. 1998b).

Probiotic and commensal bacteria, including species that are indigenous to the normal human microbiota, can also provide health benefits by competing with more harmful organisms and preventing them from becoming established in the GI tract (Collado et al. 2007). LAB have been shown to inhibit the growth of coliforms in the GI tract and return *E. coli*-infected rats to a normal microbiota composition while reducing β-glucuronidase activity (Sreekumar and Hosono 2000). Probiotics can also produce antimicrobial compounds that inhibit enteric pathogens (Spinler et al. 2008; O'Shea et al. 2012).

Chronic inflammation has been shown to promote CRC and this can be ameliorated by probiotic bacteria (Drago 2019). This can be mediated by the production of anti-inflammatory metabolites such as butyrate (Sect. 22.3.1). Some probiotic bacteria have also been shown to suppress the production of inflammatory factors by host immune cells, with *Lactobacillus reuteri* being shown to suppress the production of tumour necrosis factor-alpha (TNFα) and monocyte chemoattractant protein 1 production by lipopolysaccharide-activated monocytes and macrophages (Lin et al. 2008). A similar anti-inflammatory effect was also observed in rat pups (Liu et al. 2010). As well as inhibiting pro-tumorigenic inflammation, probiotics may also induce the targeted production of immune-activating cytokines to suppress tumorigenesis. For instance, the *L. casei* strain Shirota, when administered into the intrapleural cavity of tumour-bearing mice, induced the production of interferon gamma, IL-1β and TNFα, which in turn inhibited tumour growth and increased survival (Matsuzaki 1998).

22.5 Conclusions

In summary, the gut microbiota is an integral part of normal human physiology. This microbial reservoir of genes and metabolic functions is larger and more dynamic than the human genome, and from this grows a complex symbiosis between microbiota and host. Disruption of this relationship can have widespread negative effects on human health. This chapter has presented evidence of both protective and harmful influences of gut bacteria and their metabolites in CRC, with a particular focus on SCFAs and bile acids. Manipulation of this symbiosis with pre- and pro-biotics has the potential to have considerable health benefits as we begin to better understand the crosstalk between the gut microbiota and the host in the maintenance of a healthy symbiotic relationship.

Acknowledgements The authors and their work is supported in part by a research grant from Science Foundation Ireland (SFI) under Grant Number SFI/12/RC/2273.

References

Abdelali H, Cassand P, Soussotte V, Daubeze M, Bouley C, Narbonne JF (1995) Effect of dairy products on initiation of precursor lesions of colon cancer in rats. Nutr Cancer 24(2):121–132

Abrahamse SL, Pool-Zobel BL, Rechkemmer G (1999) Potential of short chain fatty acids to modulate the induction of DNA damage and changes in the intracellular calcium concentration by oxidative stress in isolated rat distal colon cells. Carcinogenesis 20(4):629–634

Ahn J, Sinha R, Pei Z, Dominianni C, Wu J, Shi J et al (2013) Human gut microbiome and risk for colorectal cancer. J Natl Cancer Inst 105(24):1907–1911

Arimochi H, Kinouchi T, Kataoka K, Kuwahara T, Ohnishi Y (1997) Effect of intestinal bacteria on formation of azoxymethane-induced aberrant crypt foci in the rat colon. Biochem Biophys Res Commun 238(3):753–757

Assinder SJ, Upshall A (1982) Mitotic aneuploidy induced by sodium deoxycholate in Aspergillus nidulans. Mutat Res 93(1):101–108

Augeron C, Laboisse CL (1984) Emergence of permanently differentiated cell clones in a human colonic cancer cell line in culture after treatment with sodium butyrate. Cancer Res 44 (9):3961–3969

Badvie S, Hanna-Morris A, Andreyev HJ, Cohen P, Saini S, Allen-Mersh TG (2006) A "field change" of inhibited apoptosis occurs in colorectal mucosa adjacent to colorectal adenocarcinoma. J Clin Pathol 59(9):942–946

Baghurst PA, Baghurst KI, Record SJ (1996) Dietary fibre, non-starch polysaccharides and resistant starch: a review. Food Australia 48:S3–S35

Balamurugan R, Rajendiran E, George S, Samuel GV, Ramakrishna BS (2008) Real-time polymerase chain reaction quantification of specific butyrate-producing bacteria, Desulfovibrio and Enterococcus faecalis in the feces of patients with colorectal cancer. J Gastroenterol Hepatol 23 (8 Pt 1):1298–1303

Barnard JA, Warwick G (1993) Butyrate rapidly induces growth inhibition and differentiation in HT-29 cells. Cell Growth Differ 4(6):495–501. The molecular biology journal of the American association for cancer research

Bayerdörffer E, Mannes GA, Ochsenkühn T, Dirschedl P, Wiebecke B, Paumgartner G (1995) Unconjugated secondary bile acids in the serum of patients with colorectal adenomas. Gut 36 (2):268–273

Belcheva A, Irrazabal T, Robertson Susan J, Streutker C, Maughan H, Rubino S et al (2014) Gut microbial metabolism drives transformation of Msh2-deficient colon epithelial cells. Cell 158 (2):288–299

Benno Y, Mitsuoka T (1992) Impact of Bifidobacterium longum on human fecal microflora. Microbiol Immunol 36(7):683–694

Bernstein H, Payne CM, Kunke K, Crowley-Weber CL, Waltmire CN, Dvorakova K et al (2004) A proteomic study of resistance to deoxycholate-induced apoptosis. Carcinogenesis 25 (5):681–692

Bernstein H, Bernstein C, Payne CM, Dvorakova K, Garewal H (2005a) Bile acids as carcinogens in human gastrointestinal cancers. Mutat Res 589(1):47–65

Bernstein H, Bernstein C, Payne CM, Dvorakova K, Garewal H (2005b) Bile acids as carcinogens in human gastrointestinal cancers. Mutat Res/Rev Mutat Res 589(1):47–65

Bernstein H, Bernstein C, Payne CM, Dvorak K (2009) Bile acids as endogenous etiologic agents in gastrointestinal cancer. World J Gastroenterol 15(27):3329–3340

Bervoets L, Van Hoorenbeeck K, Kortleven I, Van Noten C, Hens N, Vael C et al (2013) Differences in gut microbiota composition between obese and lean children: a cross-sectional study. Gut pathogens 5(1):10

Berwick DC, Hers I, Heesom KJ, Moule SK, Tavare JM (2002) The identification of ATP-citrate lyase as a protein kinase B (Akt) substrate in primary adipocytes. J Biol Chem 277 (37):33895–33900

Bodana AR, Rao DR (1990) Antimutagenic activity of milk fermented by Streptococcus thermophilus and Lactobacillus bulgaricus. J Dairy Sci 73(12):3379–3384

Bouhnik Y, Flourie B, Andrieux C, Bisetti N, Briet F, Rambaud JC (1996) Effects of Bifidobacterium sp fermented milk ingested with or without inulin on colonic bifidobacteria and enzymatic activities in healthy humans. Eur J Clin Nutr 50(4):269–273

Bultman SJ (2017) Interplay between diet, gut microbiota, epigenetic events, and colorectal cancer. Mol Nutr Food Res 61(1):1500902

Burns MB, Lynch J, Starr TK, Knights D, Blekhman R (2015) Virulence genes are a signature of the microbiome in the colorectal tumor microenvironment. Genome Med 7(1):55

Caderni G, Luceri C, Lancioni L, Tessitore L, Dolara P (1998) Slow-release pellets of sodium butyrate increase apoptosis in the colon of rats treated with azoxymethane, without affecting aberrant crypt foci and colonic proliferation. Nutr Cancer 30(3):175–181

Campbell JM, Fahey GC Jr, Wolf BW (1997) Selected indigestible oligosaccharides affect large bowel mass, cecal and fecal short-chain fatty acids, pH and microflora in rats. J Nutr 127 (1):130–136

Castellarin M, Warren RL, Freeman JD, Dreolini L, Krzywinski M, Strauss J et al (2012) Fusobacterium nucleatum infection is prevalent in human colorectal carcinoma. Genome Res 22(2):299–306

Challa A, Rao DR, Chawan CB, Shackelford L (1997) Bifidobacterium longum and lactulose suppress azoxymethane-induced colonic aberrant crypt foci in rats. Carcinogenesis 18 (3):517–521

Cheah PY (1990) Hypotheses for the etiology of colorectal cancer – an overview. Nutr Cancer 14 (1):5–13

Claesson MJ, Jeffery IB, Conde S, Power SE, O'Connor EM, Cusack S et al (2012) Gut microbiota composition correlates with diet and health in the elderly. Nature 488:178

Collado MC, Meriluoto J, Salminen S (2007) Role of commercial probiotic strains against human pathogen adhesion to intestinal mucus. Lett Appl Microbiol 45(4):454–460

Cook JW, Kennaway EL, Kennaway NM (1940) Production of tumours in mice by deoxycholic acid. Nature 145(3677):627

Crowther JS, Drasar BS, Hill MJ, Maclennan R, Magnin D, Peach S et al (1976) Faecal steroids and bacteria and large bowel cancer in Hong Kong by socio-economic groups. Br J Cancer 34 (2):191–198
Davani-Davari D, Negahdaripour M, Karimzadeh I, Seifan M, Mohkam M, Masoumi SJ et al (2019) Prebiotics: definition, types, sources, mechanisms, and clinical applications. Foods (Basel, Switzerland) 8(3):92
Dawson PA, Haywood J, Craddock AL, Wilson M, Tietjen M, Kluckman K et al (2003) Targeted Deletion of the Ileal Bile Acid Transporter Eliminates Enterohepatic Cycling of Bile Acids in Mice. J Biol Chem 278(36):33920–33927
de Goffau MC, Lager S, Sovio U, Gaccioli F, Cook E, Peacock SJ et al (2019) Human placenta has no microbiome but can contain potential pathogens. Nature 572(7769):329–334
De Gottardi A, Touri F, Maurer CA, Perez A, Maurhofer O, Ventre G et al (2004) The bile acid nuclear receptor FXR and the bile acid binding protein IBABP are differently expressed in colon cancer. Dig Dis Sci 49(6):982–989
Degirolamo C, Modica S, Palasciano G, Moschetta A (2011) Bile acids and colon cancer: solving the puzzle with nuclear receptors. Trends Mol Med 17(10):564–572
Ding L, Yang L, Wang Z, Huang W (2015) Bile acid nuclear receptor FXR and digestive system diseases. Acta Pharmaceutica Sinica B 5(2):135–144
Donohoe DR, Garge N, Zhang X, Sun W, O'Connell TM, Bunger MK et al (2011) The microbiome and butyrate regulate energy metabolism and autophagy in the mammalian colon. Cell Metab 13 (5):517–526
Donohoe DR, Collins LB, Wali A, Bigler R, Sun W, Bultman SJ (2012) The Warburg effect dictates the mechanism of butyrate-mediated histone acetylation and cell proliferation. Mol Cell 48(4):612–626
Drago L (2019) Probiotics and colon cancer. Microorganisms:7(3)
Eckburg PB, Bik EM, Bernstein CN, Purdom E, Dethlefsen L, Sargent M et al (2005) Diversity of the human intestinal microbial flora. Science (New York, NY) 308(5728):1635–1638
Ellmerich S, Scholler M, Duranton B, Gosse F, Galluser M, Klein JP et al (2000) Promotion of intestinal carcinogenesis by Streptococcus bovis. Carcinogenesis 21(4):753–756
Ferguson LR, Parry JM (1984) Mitotic aneuploidy as a possible mechanism for tumour promoting activity in bile acids. Carcinogenesis 5(4):447–451
Flemer B, Lynch DB, Brown JMR, Jeffery IB, Ryan FJ, Claesson MJ et al (2017) Tumour-associated and non-tumour-associated microbiota in colorectal cancer. Gut 66(4):633–643
Flemer B, Warren RD, Barrett MP, Cisek K, Das A, Jeffery IB et al (2018) The oral microbiota in colorectal cancer is distinctive and predictive. Gut 67(8):1454–1463
Food and Agriculture Organization of the United Nations, World Health Organization (2002) Guidelines for the evaluation of probiotics in food. Food and Agriculture Organization of the United Nations, World Health Organization, Rome
Frank DN, St Amand AL, Feldman RA, Boedeker EC, Harpaz N, Pace NR (2007) Molecular-phylogenetic characterization of microbial community imbalances in human inflammatory bowel diseases. Proc Natl Acad Sci U S A 104(34):13780–13785
Fu T, Coulter S, Yoshihara E, Oh TG, Fang S, Cayabyab F et al (2019) FXR regulates intestinal cancer stem cell proliferation. Cell 176(5):1098–112.e18
Fukugaiti MH, Ignacio A, Fernandes MR, Ribeiro Júnior U, Nakano V, Avila-Campos MJ (2015) High occurrence of Fusobacterium nucleatum and Clostridium difficile in the intestinal microbiota of colorectal carcinoma patients. Braz J Microbiol 46(4):1135–1140
Gamet L, Daviaud D, Denis-Pouxviel C, Remesy C, Murat JC (1992) Effects of short-chain fatty acids on growth and differentiation of the human colon-cancer cell line HT29. Int J Cancer 52 (2):286–289
Gao Z, Guo B, Gao R, Zhu Q, Qin H (2015) Microbiota disbiosis is associated with colorectal cancer. Front Microbiol 6:20
Geng J, Fan H, Tang X, Zhai H, Zhang Z (2013) Diversified pattern of the human colorectal cancer microbiome. Gut pathogens 5(1):2

Gianotti L, Morelli L, Galbiati F, Rocchetti S, Coppola S, Beneduce A et al (2010) A randomized double-blind trial on perioperative administration of probiotics in colorectal cancer patients. World J Gastroenterol 16(2):167–175

Gibson PR, Moeller I, Kagelari O, Folino M, Young GP (1992) Contrasting effects of butyrate on the expression of phenotypic markers of differentiation in neoplastic and non-neoplastic colonic epithelial cells in vitro. J Gastroenterol Hepatol 7(2):165–172

Gill SR, Pop M, Deboy RT, Eckburg PB, Turnbaugh PJ, Samuel BS et al (2006) Metagenomic analysis of the human distal gut microbiome. Science (New York, NY) 312(5778):1355–1359

Glinghammar B, Inoue H, Rafter JJ (2002) Deoxycholic acid causes DNA damage in colonic cells with subsequent induction of caspases, COX-2 promoter activity and the transcription factors NF-kB and AP-1. Carcinogenesis 23(5):839–845

Gold JS, Bayar S, Salem RR (2004) Association of Streptococcus bovis bacteremia with colonic neoplasia and extracolonic malignancy. Arch Surg (Chicago, Ill: 1960) 139(7):760–765

Goldin BR, Gorbach SL (1980) Effect of Lactobacillus acidophilus dietary supplements on 1,2-dimethylhydrazine dihydrochloride-induced intestinal cancer in rats. J Natl Cancer Inst 64 (2):263–265

Goldin BR, Gorbach SL (1984) The effect of milk and lactobacillus feeding on human intestinal bacterial enzyme activity. Am J Clin Nutr 39(5):756–761

Goldin BR, Gorbach SL, Saxelin M, Barakat S, Gualtieri L, Salminen S (1992) Survival of Lactobacillus species (strain GG) in human gastrointestinal tract. Dig Dis Sci 37(1):121–128

Goldin BR, Gualtieri LJ, Moore RP (1996) The effect of Lactobacillus GG on the initiation and promotion of DMH-induced intestinal tumors in the rat. Nutr Cancer 25(2):197–204

Group NHW, Peterson J, Garges S, Giovanni M, McInnes P, Wang L et al (2009) The NIH Human Microbiome Project. Genome Res 19(12):2317–2323

Haghi F, Goli E, Mirzaei B, Zeighami H (2019) The association between fecal enterotoxigenic B. fragilis with colorectal cancer. BMC Cancer 19(1):879

Hague A, Manning AM, Hanlon KA, Huschtscha LI, Hart D, Paraskeva C (1993) Sodium butyrate induces apoptosis in human colonic tumour cell lines in a p53-independent pathway: implications for the possible role of dietary fibre in the prevention of large-bowel cancer. Int J Cancer 55 (3):498–505

Hague A, Elder DJ, Hicks DJ, Paraskeva C (1995) Apoptosis in colorectal tumour cells: induction by the short chain fatty acids butyrate, propionate and acetate and by the bile salt deoxycholate. Int J Cancer 60(3):400–406

Hass R, Busche R, Luciano L, Reale E, Engelhardt WV (1997) Lack of butyrate is associated with induction of Bax and subsequent apoptosis in the proximal colon of guinea pig. Gastroenterology 112(3):875–881

Hayatsu H, Hayatsu T (1993) Suppressing effect of Lactobacillus casei administration on the urinary mutagenicity arising from ingestion of fried ground beef in the human. Cancer Lett 73 (2):173–179

Heerdt BG, Houston MA, Augenlicht LH (1997) Short-chain fatty acid-initiated cell cycle arrest and apoptosis of colonic epithelial cells is linked to mitochondrial function. Cell Growth Differ 8(5):523–532. The molecular biology journal of the American association for cancer research

Hill MJ (1990) Bile flow and colon cancer. Mutat Res/Rev Genet Toxicol 238(3):313–320

Hinnebusch BF, Meng S, Wu JT, Archer SY, Hodin RA (2002) The effects of short-chain fatty acids on human colon cancer cell phenotype are associated with histone hyperacetylation. J Nutr 132(5):1012–1017

Hope ME, Hold GL, Kain R, El-Omar EM (2005) Sporadic colorectal cancer – role of the commensal microbiota. FEMS Microbiol Lett 244(1):1–7

Horie H, Kanazawa K, Okada M, Narushima S, Itoh K, Terada A (1999) Effects of intestinal bacteria on the development of colonic neoplasm: an experimental study. Eur J Cancer Prev 8 (3):237–245. The official journal of the European Cancer Prevention Organisation (ECP)

Hunt CR, Sim JE, Sullivan SJ, Featherstone T, Golden W, Von Kapp-Herr C et al (1998) Genomic instability and catalase gene amplification induced by chronic exposure to oxidative stress. Cancer Res 58(17):3986–3992

Huurre A, Kalliomaki M, Rautava S, Rinne M, Salminen S, Isolauri E (2008) Mode of delivery – effects on gut microbiota and humoral immunity. Neonatology 93(4):236–240

Ishikawa H, Akedo I, Umesaki Y, Tanaka R, Imaoka A, Otani T (2003) Randomized controlled trial of the effect of bifidobacteria-fermented milk on ulcerative colitis. J Am Coll Nutr 22(1):56–63

Jahani-Sherafat S, Alebouyeh M, Moghim S, Ahmadi Amoli H, Ghasemian-Safaei H (2018) Role of gut microbiota in the pathogenesis of colorectal cancer; a review article. Gastroenterol Hepatol Bed Bench 11(2):101–109

Jenkins GJ, D'Souza FR, Suzen SH, Eltahir ZS, James SA, Parry JM et al (2007) Deoxycholic acid at neutral and acid pH, is genotoxic to oesophageal cells through the induction of ROS: the potential role of anti-oxidants in Barrett's oesophagus. Carcinogenesis 28(1):136–142

Jensen OM, MacLennan R, Wahrendorf J (1982) Diet, bowel function, fecal characteristics, and large bowel cancer in Denmark and Finland. Nutr Cancer 4(1):5–19

Jiang C, Xie C, Lv Y, Li J, Krausz KW, Shi J et al (2015) Intestine-selective farnesoid X receptor inhibition improves obesity-related metabolic dysfunction. Nat Commun 6:10166

Kanazawa K, Konishi F, Mitsuoka T, Terada A, Itoh K, Narushima S et al (1996) Factors influencing the development of sigmoid colon cancer. Bacteriologic and biochemical studies. Cancer 77(8 Suppl):1701–1706

Kaur K, Saxena A, Debnath I, O'Brien JL, Ajami NJ, Auchtung TA et al (2018) Antibiotic-mediated bacteriome depletion in Apc(Min/+) mice is associated with reduction in mucus-producing goblet cells and increased colorectal cancer progression. Cancer Med 7 (5):2003–2012

Kim I, Morimura K, Shah Y, Yang Q, Ward JM, Gonzalez FJ (2007) Spontaneous hepatocarcinogenesis in farnesoid X receptor-null mice. Carcinogenesis 28(5):940–946

Klaasen HL, Van der Heijden PJ, Stok W, Poelma FG, Koopman JP, Van den Brink ME et al (1993) Apathogenic, intestinal, segmented, filamentous bacteria stimulate the mucosal immune system of mice. Infect Immun 61(1):303–306

Kostic AD, Gevers D, Pedamallu CS, Michaud M, Duke F, Earl AM et al (2012) Genomic analysis identifies association of Fusobacterium with colorectal carcinoma. Genome Res 22(2):292–298

Kripke SA, Fox AD, Berman JM, Settle RG, Rombeau JL (1989) Stimulation of intestinal mucosal growth with intracolonic infusion of short-chain fatty acids. JPEN J Parenter Enter Nutr 13 (2):109–116

Kulkarni N, Reddy BS (1994) Inhibitory Effect of Bifidobacterium Iongum Cultures on the Azoxymethane-Induced Aberrant Crypt Foci Formation and Fecal Bacterial β-Glucuronidase. Proc Soc Exp Biol Med 207(3):278–283

Kwong TNY, Wang X, Nakatsu G, Chow TC, Tipoe T, Dai RZW et al (2018) Association between bacteremia from specific microbes and subsequent diagnosis of colorectal cancer. Gastroenterology 155(2):383-90.e8

Le Leu RK, Brown IL, Hu Y, Morita T, Esterman A, Young GP (2007) Effect of dietary resistant starch and protein on colonic fermentation and intestinal tumourigenesis in rats. Carcinogenesis 28(2):240–245

Li T, Chiang JYL (2012) Bile Acid Signaling in Liver Metabolism and Diseases. Journal of Lipids 2012:9

Li Y-Y, Ge Q-X, Cao J, Zhou Y-J, Du Y-L, Shen B et al (2016) Association of Fusobacterium nucleatum infection with colorectal cancer in Chinese patients. World J Gastroenterol 22 (11):3227–3233

Liang Q, Chiu J, Chen Y, Huang Y, Higashimori A, Fang J et al (2017) Fecal bacteria act as novel biomarkers for noninvasive diagnosis of colorectal cancer. Clin Cancer Res 23(8):2061–2070. An official journal of the American association for cancer research

Lidbeck A, Övervik E, Rafter J, Nord CE, Gustafsson JÅ (1992) Effect of lactobacillus acidophilus supplements on mutagen excretion in faeces and urine in humans. Microb Ecol Health Dis 5 (1):59–67

Limoli CL, Giedzinski E (2003) Induction of chromosomal instability by chronic oxidative stress. Neoplasia (New York, NY) 5(4):339–346

Lin YP, Thibodeaux CH, Pena JA, Ferry GD, Versalovic J (2008) Probiotic Lactobacillus reuteri suppress proinflammatory cytokines via c-Jun. Inflamm Bowel Dis 14(8):1068–1083

Ling WH, Korpela R, Mykkänen H, Salminen S, Hänninen O (1994) Lactobacillus strain GG supplementation decreases colonic hydrolytic and reductive enzyme activities in healthy female adults. J Nutr 124(1):18–23

Link-Amster H, Rochat F, Saudan KY, Mignot O, Aeschlimann JM (1994) Modulation of a specific humoral immune response and changes in intestinal flora mediated through fermented milk intake. FEMS Immunol Med Microbiol 10(1):55–63

Liu Y, Fatheree NY, Mangalat N, Rhoads JM (2010) Human-derived probiotic Lactobacillus reuteri strains differentially reduce intestinal inflammation. Am J Physiol Gastrointest Liver Physiol 299(5):G1087–G1096

Long SL, Gahan CGM, Joyce SA (2017) Interactions between gut bacteria and bile in health and disease. Mol Aspects Med 56:54–65

Louis P, Young P, Holtrop G, Flint HJ (2010) Diversity of human colonic butyrate-producing bacteria revealed by analysis of the butyryl-CoA:acetate CoA-transferase gene. Environ Microbiol 12(2):304–314

Louis P, Hold GL, Flint HJ (2014) The gut microbiota, bacterial metabolites and colorectal cancer. Nat Rev Microbiol 12:661

Macarthur M, Hold GL, El-Omar EM (2004) Inflammation and Cancer II. Role of chronic inflammation and cytokine gene polymorphisms in the pathogenesis of gastrointestinal malignancy. Am J Physiol Gastrointest Liver Physiol 286(4):G515–G520

Macfarlane GT, Gibson GR, Cummings JH (1992) Comparison of fermentation reactions in different regions of the human colon. J Appl Bacteriol 72(1):57–64

Magnuson BA, Carr I, Bird RP (1993) Ability of aberrant crypt foci characteristics to predict colonic t\imor incidence in rats fed cholic acid. Cancer Res 53(19):4499–4504

Makishima M, Okamoto AY, Repa JJ, Tu H, Learned RM, Luk A et al (1999) Identification of a nuclear receptor for bile acids. Science (New York, NY) 284(5418):1362–1365

Mandar R, Mikelsaar M (1996) Transmission of mother's microflora to the newborn at birth. Biol Neonate 69(1):30–35

Maran RRM, Thomas A, Roth M, Sheng Z, Esterly N, Pinson D et al (2009) Farnesoid X receptor deficiency in mice leads to increased intestinal epithelial cell proliferation and tumor development. J Pharmacol Exp Ther 328(2):469–477

Marchesi JR, Dutilh BE, Hall N, Peters WHM, Roelofs R, Boleij A et al (2011) Towards the human colorectal cancer microbiome. PLoS One 6(5):e20447-e

Marchetti C, Migliorati G, Moraca R, Riccardi C, Nicoletti I, Fabiani R et al (1997) Deoxycholic acid and SCFA-induced apoptosis in the human tumor cell-line HT-29 and possible mechanisms. Cancer Lett 114(1-2):97–99

Marteau PR, Md V, Cellier CJ, Schrezenmeir J (2001) Protection from gastrointestinal diseases with the use of probiotics. Am J Clin Nutr 73(2):430s-6s

Martin HM, Campbell BJ, Hart CA, Mpofu C, Nayar M, Singh R et al (2004) Enhanced Escherichia coli adherence and invasion in Crohn's disease and colon cancer. Gastroenterology 127 (1):80–93

Maskarinec G, Noh JJ (2004) The effect of migration on cancer incidence among Japanese in Hawaii. Ethn Dis 14(3):431–439

Mathias A, Pais B, Favre L, Benyacoub J, Corthesy B (2014) Role of secretory IgA in the mucosal sensing of commensal bacteria. Gut microbes 5(6):688–695

Matsuzaki T (1998) Immunomodulation by treatment with Lactobacillus casei strain Shirota. Int J Food Microbiol 41(2):133–140

McIntyre A, Gibson PR, Young GP (1993) Butyrate production from dietary fibre and protection against large bowel cancer in a rat model. Gut 34(3):386–391

Mego M, Chovanec J, Vochyanova-Andrezalova I, Konkolovsky P, Mikulova M, Reckova M et al (2015) Prevention of irinotecan induced diarrhea by probiotics: A randomized double blind, placebo controlled pilot study. Complement Ther Med 23(3):356–362

Mima K, Cao Y, Chan AT, Qian ZR, Nowak JA, Masugi Y et al (2016) Fusobacterium nucleatum in colorectal carcinoma tissue according to tumor location. Clin Transl Gastroenterol 7(11): e200-e

Mira-Pascual L, Cabrera-Rubio R, Ocon S, Costales P, Parra A, Suarez A et al (2015) Microbial mucosal colonic shifts associated with the development of colorectal cancer reveal the presence of different bacterial and archaeal biomarkers. J Gastroenterol 50(2):167–179

Modica S, Murzilli S, Salvatore L, Schmidt DR, Moschetta A (2008) Nuclear Bile Acid Receptor FXR Protects against Intestinal Tumorigenesis. Cancer Res 68(23):9589–9594

Morotomi M, Mutai M (1986) In vitro binding of potent mutagenic pyrolysates to intestinal bacteria. J Natl Cancer Inst 77(1):195–201

Morotomi M, Mutal M (1986) In Vitro Binding of Potent Mutagenic Pyrolyzates to Intestinal Bacteria. JNCI: J Nat Cancer Inst 77(1):195–201

Nagpal R, Mainali R, Ahmadi S, Wang S, Singh R, Kavanagh K et al (2018) Gut microbiome and aging: Physiological and mechanistic insights. Nutr Healthy Aging 4(4):267–285

Nowacki MR (1993) Cell proliferation in colonic crypts of germ-free and conventional mice--preliminary report. Folia Histochem Cytobiol 31(2):77–81

Ogawa A, Murate T, Suzuki M, Nimura Y, Yoshida S (1998) Lithocholic acid, a putative tumor promoter, inhibits mammalian DNA polymerase β. Jpn J Cancer Res 89(11):1154–1159

O'Hara AM, Shanahan F (2006) The gut flora as a forgotten organ. EMBO Rep 7(7):688–693

Onoue M, Kado S, Sakaitani Y, Uchida K, Morotomi M (1997) Specific species of intestinal bacteria influence the induction of aberrant crypt foci by 1,2-dimethylhydrazine in rats. Cancer Lett 113(1-2):179–186

Orrhage K, Sillerstrom E, Gustafsson JA, Nord CE, Rafter J (1994) Binding of mutagenic heterocyclic amines by intestinal and lactic acid bacteria. Mutat Res 311(2):239–248

O'Shea EF, Cotter PD, Stanton C, Ross RP, Hill C (2012) Production of bioactive substances by intestinal bacteria as a basis for explaining probiotic mechanisms: bacteriocins and conjugated linoleic acid. Int J Food Microbiol 152(3):189–205

O'Toole PW, Claesson MJ (2010) Gut microbiota: Changes throughout the lifespan from infancy to elderly. Int Dairy J 20(4):281–291

Pasolli E, Asnicar F, Manara S, Zolfo M, Karcher N, Armanini F et al (2019) Extensive unexplored human microbiome diversity revealed by over 150,000 genomes from metagenomes spanning age, geography, and lifestyle. Cell 176(3):649-62.e20

Patel RM, Lin PW (2010) Developmental biology of gut-probiotic interaction. Gut microbes 1 (3):186–195

Pool-Zobel BL, Munzner R, Holzapfel WH (1993a) Antigenotoxic properties of lactic acid bacteria in the S. typhimurium mutagenicity assay. Nutr Cancer 20(3):261–270

Pool-Zobel BL, Bertram B, Knoll M, Lambertz R, Neudecker C, Schillinger U et al (1993b) Antigenotoxic properties of lactic acid bacteria in vivo in the gastrointestinal tract of rats. Nutr Cancer 20(3):271–281

Pool-Zobel BL, Neudecker C, Domizlaff I, Ji S, Schillinger U, Rumney C et al (1996) Lactobacillus- and bifidobacterium-mediated antigenotoxicity in the colon of rats. Nutr Cancer 26 (3):365–380

Powell AA, LaRue JM, Batta AK, Martinez JD (2001) Bile acid hydrophobicity is correlated with induction of apoptosis and/or growth arrest in HCT116 cells. Biochem J 356(Pt 2):481–486

Quigley EMM (2013) Gut bacteria in health and disease. Gastroenterol Hepatol (N Y) 9 (9):560–569

Rafter JJ, Eng VW, Furrer R, Medline A, Bruce WR (1986) Effects of calcium and pH on the mucosal damage produced by deoxycholic acid in the rat colon. Gut 27(11):1320–1329

Rafter J, Bennett M, Caderni G, Clune Y, Hughes R, Karlsson PC et al (2007) Dietary synbiotics reduce cancer risk factors in polypectomized and colon cancer patients. Am J Clin Nutr 85 (2):488–496

Reddy BS, Weisburger JH, Wynder EL (1975) Effects of high risk and low risk diets for colon carcinogenesis on fecal microflora and steroids in man. J Nutr 105(7):878–884

Reddy BS, Narasawa T, Weisburger JH, Wynder EL (1976a) Promoting effect of sodium deoxycholate on colon adenocarcinomas in germfree Rats2. JNCI: J Natl Cancer Inst 56 (2):441–442

Reddy BS, Narasawa T, Weisburger JH, Wynder EL (1976b) Promoting effect of sodium deoxycholate on colon adenocarcinomas in germfree rats. J Natl Cancer Inst 56(2):441–442

Roessler A, Forssten SD, Glei M, Ouwehand AC, Jahreis G (2012) The effect of probiotics on faecal microbiota and genotoxic activity of faecal water in patients with atopic dermatitis: a randomized, placebo-controlled study. Clin Nutr (Edinburgh, Scotland) 31(1):22-9

Roller M, Clune Y, Collins K, Rechkemmer G, Watzl B (2007) Consumption of prebiotic inulin enriched with oligofructose in combination with the probiotics Lactobacillus rhamnosus and Bifidobacterium lactis has minor effects on selected immune parameters in polypectomised and colon cancer patients. Br J Nutr 97(4):676–684

Rowland IR, Rumney CJ, Coutts JT, Lievense LC (1998a) Effect of Bifidobacterium longum and inulin on gut bacterial metabolism and carcinogen-induced aberrant crypt foci in rats. Carcinogenesis 19(2):281–285

Rowland IR, Rumney CJ, Coutts JT, Lievense LC (1998b) Effect of Bifidobacterium longum and inulin on gut bacterial metabolism and carcinogen-induced aberrant crypt foci in rats. Carcinogenesis 19(2):281–285

Scanlan PD, Shanahan F, Clune Y, Collins JK, O'Sullivan GC, O'Riordan M et al (2008) Culture-independent analysis of the gut microbiota in colorectal cancer and polyposis. Environ Microbiol 10(3):789–798

Segal I, Hassan H, Walker AR, Becker P, Braganza J (1995) Fecal short chain fatty acids in South African urban Africans and whites. Dis Colon Rectum 38(7):732–734

Sekirov I, Russell SL, Antunes LCM, Finlay BB (2010) Gut Microbiota in Health and Disease. Physiol Rev 90(3):859–904

Siavoshian S, Blottiere HM, Cherbut C, Galmiche JP (1997) Butyrate stimulates cyclin D and p21 and inhibits cyclin-dependent kinase 2 expression in HT-29 colonic epithelial cells. Biochem Biophys Res Commun 232(1):169–172

Sinha R, Ahn J, Sampson JN, Shi J, Yu G, Xiong X et al (2016) Fecal microbiota, fecal metabolome, and colorectal cancer interrelations. PLoS One 11(3):e0152126-e

Smith K, McCoy KD, Macpherson AJ (2007) Use of axenic animals in studying the adaptation of mammals to their commensal intestinal microbiota. Semin Immunol 19(2):59–69

Sobhani I, Tap J, Roudot-Thoraval F, Roperch JP, Letulle S, Langella P et al (2011) Microbial dysbiosis in colorectal cancer (CRC) patients. PLoS One 6(1):e16393-e

Soler AP, Miller RD, Laughlin KV, Carp NZ, Klurfeld DM, Mullin JM (1999) Increased tight junctional permeability is associated with the development of colon cancer. Carcinogenesis 20 (8):1425–1431

Spanhaak S, Havenaar R, Schaafsma G (1998) The effect of consumption of milk fermented by Lactobacillus casei strain Shirota on the intestinal microflora and immune parameters in humans. Eur J Clin Nutr 52(12):899–907

Spinler JK, Taweechotipatr M, Rognerud CL, Ou CN, Tumwasorn S, Versalovic J (2008) Human-derived probiotic Lactobacillus reuteri demonstrate antimicrobial activities targeting diverse enteric bacterial pathogens. Anaerobe 14(3):166–171

Sreekumar O, Hosono A (2000) Immediate effect of Lactobacillus acidophilus on the intestinal flora and fecal enzymes of rats and the in vitro inhibition of Escherichia coli in coculture. J Dairy Sci 83(5):931–939

Stojancevic M, Stankov K, Mikov M (2012) The impact of farnesoid X receptor activation on intestinal permeability in inflammatory bowel disease. Can J Gastroenterol 26(9):631–637

Suehiro Y, Sakai K, Nishioka M, Hashimoto S, Takami T, Higaki S et al (2017) Highly sensitive stool DNA testing of Fusobacterium nucleatum as a marker for detection of colorectal tumours in a Japanese population. Ann Clin Biochem 54(1):86–91

Sumi Y, Miyakawa M (1979) Gastrointestinal carcinogenesis in germ-free rats given N-methyl-N′-nitro-N-nitrosoguanidine in drinking water. Cancer Res 39(7 Pt 1):2733–2736

Swidsinski A, Khilkin M, Kerjaschki D, Schreiber S, Ortner M, Weber J et al (1998) Association between intraepithelial Escherichia coli and colorectal cancer. Gastroenterology 115 (2):281–286

Tahara T, Yamamoto E, Suzuki H, Maruyama R, Chung W, Garriga J et al (2014) Fusobacterium in colonic flora and molecular features of colorectal carcinoma. Cancer Res 74(5):1311–1318

Taylor CW, Kim YS, Childress-Fields KE, Yeoman LC (1992) Sensitivity of nuclear c-myc levels and induction to differentiation-inducing agents in human colon tumor cell lines. Cancer Lett 62 (2):95–105

Thomas AM, Manghi P, Asnicar F, Pasolli E, Armanini F, Zolfo M et al (2019) Metagenomic analysis of colorectal cancer datasets identifies cross-cohort microbial diagnostic signatures and a link with choline degradation. Nat Med 25(4):667–678

Tibbs TN, Lopez LR, Arthur JC (2019) The influence of the microbiota on immune development, chronic inflammation, and cancer in the context of aging. Microbial cell (Graz, Austria) 6 (8):324–334

Topping DL, Clifton PM (2001) Short-chain fatty acids and human colonic function: roles of resistant starch and nonstarch polysaccharides. Physiol Rev 81(3):1031–1064

Turnbaugh PJ, Backhed F, Fulton L, Gordon JI (2008) Diet-induced obesity is linked to marked but reversible alterations in the mouse distal gut microbiome. Cell Host Microbe 3(4):213–223

Vavassori P, Mencarelli A, Renga B, Distrutti E, Fiorucci S (2009) The bile acid receptor FXR is a modulator of intestinal innate immunity. J Immunol 183(10):6251–6261

Velazquez OC, Seto RW, Bain AM, Fisher J, Rombeau JL (1997) Deoxycholate inhibits in vivo butyrate-mediated BrDU labeling of the colonic crypt. J Surg Res 69(2):344–348

Velcich A, Yang W, Heyer J, Fragale A, Nicholas C, Viani S et al (2002) Colorectal cancer in mice genetically deficient in the mucin Muc2. Science (New York, NY) 295(5560):1726–1729

Viljoen KS, Dakshinamurthy A, Goldberg P, Blackburn JM (2015) Quantitative profiling of colorectal cancer-associated bacteria reveals associations between fusobacterium spp., enterotoxigenic Bacteroides fragilis (ETBF) and clinicopathological features of colorectal cancer. PLoS One 10(3):e0119462

Voreades N, Kozil A, Weir TL (2014) Diet and the development of the human intestinal microbiome. Front Microbiol 5:494

Wachtershauser A, Stein J (2000) Rationale for the luminal provision of butyrate in intestinal diseases. Eur J Nutr 39(4):164–171

Wang HB, Wang PY, Wang X, Wan YL, Liu YC (2012) Butyrate enhances intestinal epithelial barrier function via up-regulation of tight junction protein Claudin-1 transcription. Dig Dis Sci 57(12):3126–3135

Warren RL, Freeman DJ, Pleasance S, Watson P, Moore RA, Cochrane K et al (2013) Co-occurrence of anaerobic bacteria in colorectal carcinomas. Microbiome 1(1):16

Wei Z, Cao S, Liu S, Yao Z, Sun T, Li Y et al (2016) Could gut microbiota serve as prognostic biomarker associated with colorectal cancer patients' survival? A pilot study on relevant mechanism. Oncotarget 7(29):46158–46172

Wirbel J, Pyl PT, Kartal E, Zych K, Kashani A, Milanese A et al (2019) Meta-analysis of fecal metagenomes reveals global microbial signatures that are specific for colorectal cancer. Nat Med 25(4):679–689

Wollowski I, Rechkemmer G, Pool-Zobel BL (2001) Protective role of probiotics and prebiotics in colon cancer. Am J Clin Nutr 73(2 Suppl):451s-5s

Wong SH, Kwong TNY, Chow T-C, Luk AKC, Dai RZW, Nakatsu G et al (2017) Quantitation of faecal Fusobacterium improves faecal immunochemical test in detecting advanced colorectal neoplasia. Gut 66(8):1441–1448

Wroblewski LE, Peek RM Jr, Wilson KT (2010) Helicobacter pylori and gastric cancer: factors that modulate disease risk. Clin Microbiol Rev 23(4):713–739

Wu X, Wu Y, He L, Wu L, Wang X, Liu Z (2018) Effects of the intestinal microbial metabolite butyrate on the development of colorectal cancer. J Cancer 9(14):2510–2517

Wu J, Li Q, Fu X (2019) Fusobacterium nucleatum Contributes to the Carcinogenesis of Colorectal Cancer by Inducing Inflammation and Suppressing Host Immunity. Translational oncology 12 (6):846–851

Xia Y, Yang Z, Chen HQ, Qin HL (2010) [Effect of bowel preparation with probiotics on intestinal barrier after surgery for colorectal cancer]. Zhonghua wei chang wai ke za zhi =. Chin J Gastrointes Surg 13(7):528–531

Xiao H, Yin W, Khan MA, Gulen MF, Zhou H, Sham HP et al (2010) Loss of single immunoglobulin interlukin-1 receptor-related molecule leads to enhanced colonic polyposis in Apc(min) mice. Gastroenterology 139(2):574–585

Yang Y, Weng W, Peng J, Hong L, Yang L, Toiyama Y et al (2017) Fusobacterium nucleatum increases proliferation of colorectal cancer cells and tumor development in mice by activating toll-like receptor 4 signaling to nuclear factor-κB, and Up-regulating expression of MicroRNA-21. Gastroenterology 152(4):851-66.e24

Yap YA, Mariño E (2018) An Insight Into the Intestinal Web of Mucosal Immunity, Microbiota, and Diet in Inflammation. Front Immunol 9:2617

Zackular JP, Baxter NT, Iverson KD, Sadler WD, Petrosino JF, Chen GY et al (2013) The gut microbiome modulates colon tumorigenesis. MBio 4(6):e00692–e00613

Zhang XB, Ohta Y (1991) In vitro binding of mutagenic pyrolyzates to lactic acid bacterial cells in human gastric juice. J Dairy Sci 74(3):752–757

Zhang F, Subbaramaiah K, Altorki N, Dannenberg AJ (1998) Dihydroxy Bile Acids Activate the Transcription of Cyclooxygenase-2. J Biol Chem 273(4):2424–2428

Zhou Y, He H, Xu H, Li Y, Li Z, Du Y et al (2016) Association of oncogenic bacteria with colorectal cancer in South China. Oncotarget 7(49):80794–80802

Zumkeller N, Brenner H, Zwahlen M, Rothenbacher D (2006) Helicobacter pylori infection and colorectal cancer risk: a meta-analysis. Helicobacter 11(2):75–80

Chapter 23
The Macrophages and Intestinal Symbiosis

Malgorzata Kloc, Ahmed Uosef, Mahmoud Elshawwaf, Ahmed Adel Abbas Abdelshafy, Kamal Mamdoh Kamal Elsaid, Jacek Z. Kubiak, and Rafik Mark Ghobrial

Abstract The human intestinal tract is inhabited by trillions of microorganisms and houses the largest pool of macrophages in the human body. Being a part of the innate immune system, the macrophages, the professional phagocytes, vigorously respond to the microbial and dietary antigens present in the intestine. Because such a robust immune response poses the danger to the survival of the non-harmful and beneficial gut microbiota, the macrophages developed mechanisms of recognition and hyposensitivity toward the non-harmful/beneficial inhabitants of the gut. We will discuss the evolution and identity of some of these mechanisms in the following chapter.

M. Kloc (✉)
The Houston Methodist Research Institute, Houston, TX, USA

Department of Surgery, Houston Methodist Hospital, Houston, TX, USA

Department of Genetics, The University of Texas, M.D. Anderson Cancer Center, Houston, TX, USA
e-mail: mkloc@houstonmethodist.org

A. Uosef · M. Elshawwaf · R. M. Ghobrial
The Houston Methodist Research Institute, Houston, TX, USA

Department of Surgery, Houston Methodist Hospital, Houston, TX, USA

A. A. A. Abdelshafy · K. M. K. Elsaid
The Houston Methodist Research Institute, Houston, TX, USA

Department of Surgery, Houston Methodist Hospital, Houston, TX, USA

Department of General Surgery, Faculty of Medicine, Ain-Shams University, Cairo, Egypt

J. Z. Kubiak
Department of Regenerative Medicine and Cell Biology, Military Institute of Hygiene and Epidemiology (WIHE), Warszawa, Poland

Faculty of Medicine, Cell Cycle Group, Institute of Genetics and Development of Rennes, (IGDR) UnivRennes, CNRS, UMR 6290, Rennes, France

M. Kloc (ed.), *Symbiosis: Cellular, Molecular, Medical and Evolutionary Aspects*, Results and Problems in Cell Differentiation 69,
https://doi.org/10.1007/978-3-030-51849-3_23

23.1 Introduction

Because of its enormous length (~ 5–10 m) and surface (~ 180–300 square meters), the human gastrointestinal tract (GI) represents the largest interface between the host and the microbial and dietary antigens in the human body (Helander and Fändriks 2014; Hounnou et al. 2002; Rooks and Garrett 2016; Sears 2005). This deluge of antigens requires an extremely robust immunological response both in the immune cell number and their activity. The immune cells, which represent the first-line responders to such an enormous immunologic challenge, are the gut macrophages (Hine and Loke 2019). The quantity of the immune cells present in the gut wall is so great that from the histological perspective, the gut wall looks like a strongly inflamed tissue—the state described as the "physiological inflammation" (Fiocchi 2003, 2008; Medzhitov 2008). Indeed, the human gastrointestinal tract is the largest pool of macrophages in the body (Bain and Mowat 2014; Bain and Schridde 2018). The macrophages, being the voracious "professional" phagocytes (Stuart and Ezekowitz 2005), engulf the microorganisms, dead cells, toxins, and foreign debris, and send the activation/inflammatory signals to other types of the immune cells. However, such an avid immunologic response against pathogens poses also a danger to the symbiotic microorganisms inhabiting the GI. The gastrointestinal tract of humans is the ecosystem of over a hundred trillion microorganisms collectively called "the gut microbiota" (Thursby and Juge 2017). This represents over 10 times more microbial cells than the number of human cells in the body, and over 100 times microbial genomic content than the human genome.

Thus, the question arises how the macrophages distinguish the harmful from the non-harmful or beneficial organisms and how they modify, accordingly, their response. In this chapter, we will discuss the origin and functions of gut macrophages, the mechanisms modifying their response to the harmful versus symbiotic microorganisms, and the hypotheses on the evolutionary origin of the GI macrophages and such responses.

23.2 How Macrophages Recognize Microorganisms?

All microorganisms have on their surface various molecules specific for a given microorganism species, which articulate the microorganism "signature" or the Pathogen-Associated Molecular Pattern (PAMP; Herwald and Egesten 2016). The examples of PAMPs are bacterial lipopolysaccharides (LPSs, present on the bacterial membrane), bacterial endotoxins, lipoteichoic acid (LTA, a major component of the cell wall of gram-positive bacteria), flagellin (a component of bacterial flagellum), peptidoglycan (murein, a polymer of amino acids and sugars in the bacterial cell wall), and various forms of viruses' nucleic acids. The macrophages have a whole spectrum of pattern recognition receptors (PRRs), which recognize PAMPs (Takeuchi and Akira 2010). The binding of the PAMPs to their complementary PRRs switches on the production and release of the cytokines, which signal to the immune cells the presence of the pathogens and infected cells, which need to be engulfed (phagocytosed) and destroyed. In humans, there are over 40 different types of cytokines including various interleukins and interferons. Interestingly, the macrophages not only recognize PAMPs but are also able, using their membrane ruffles, to identify the size and shape of the microorganisms (Doshi and Mitragotri 2010).

The macrophages and other immune cells also produce an array of soluble proteins called the complement system because they "complement" the antibody response (Fujita 2002; Markiewski and Lambris 2007; Sarma and Ward 2011). The complement proteins bind the microorganisms especially those which are already bound to the antibodies. The microorganism coated by the complement proteins, the process called the "opsonization", is marked as a phagocytotic target. In addition, some of the complement proteins make pores in the microbe membrane causing the leakage of the microbe's content and death (Markiewski and Lambris 2007; Sarma and Ward 2011).

23.3 Phylogenetic/Evolutionary Origin of GI Macrophages and Development of Hyposensitivity toward Symbionts

In 1882, a Russian zoologist Ilya Ilitch Metchnikov (Elie Metschnikoff in French) introduced the concept of phagocytosis and phagocytes (he received the Nobel Prize for this discovery in 1908) and postulated that the vertebrate macrophages evolved from the invertebrate enteric phagocytes. The Metchnikov's observations and hypothesis were subsequently developed into the concept of a common origin of immunity and digestion (Broderick 2015; Hartenstein and Martinez 2019; Hoffmann et al. 1999). The ancestral connection between the immunity and digestion is exemplified by a unicellular amoeba in which the initial stages of food ingestion and an engulfment of the pathogens are indistinguishable, and by the fact that many enzymes involved in the digestion of food are also involved in the immune response (Broderick 2015; Gaudet et al. 2016; Hartenstein and Martinez 2019; Hoffmann et al. 1999). The examples of such dual-purpose molecules are the toll-interleukin-1 receptor (TIR) domain proteins, such as TirA in the soil-dwelling ameba *Dictyostelium*, and SARM in *C. elegans* (Broderick 2015; Chen et al. 2007; Shivers et al. 2009). Another example is the β1,3-glucanase digestive enzymes, which also, as the pattern recognition receptors (βGRPs), activate the immune response in the invertebrates (Broderick 2015; Rosengaus et al. 2014; Sun et al. 2011). In this context, one can imagine that some microorganisms, which were resistant to the digestion and produced compounds beneficial to the host were, during evolution, recognized by the macrophages as a benign or beneficial. This led to the development of hyposensitivity (macrophage "anergy") toward them, leading to the immunologic tolerance, and a permanent cohabitation of the microorganisms with the macrophages and the host (Chang et al. 2014; Fava and Danese 2011; Nyholm and Graf 2012; Scott et al. 2018; Smith et al. 2011; Traylor-Knowles et al. 2019, Wang et al. 2019; Wells et al. 2010; Wynn et al. 2013). The phenomenon of an emergency utilization of the gut microbiota as a nutritional reserve in the starving animals (Broderick 2015; Conway et al. 1986) also argues for the evolutionary advantage for the development and retention of a large pool of gut microbiota.

Although the theory of the common origin of immunity and digestion has been criticized by some researchers (van Niekerk and Engelbrecht 2015) for being overly simplistic, it gives a stimulating perspective on the phylogenetic origin of macrophages and the innate immunity.

23.4 Distribution of Macrophages in the Gastrointestinal Tract

The wall of the human gastrointestinal tract (GI) is built of four layers (Fig. 23.1; Hine and Loke 2019). The first, most internal (facing the gut lumen) layer is *the mucosa*. The mucosa consists of three sublayers: the epithelium sublayer, which lines the GI lumen, is covered by the bactericidal mucous, and forms a main protective barrier against the microorganisms; the lamina propria (LP) sublayer, which underlines the epithelium and is built of connective tissue, blood capillaries (which absorb digestion-derived nutrients), and mucous glands; and the muscularis mucosa sublayer, which is built of the smooth muscle cells providing local movement of the mucosa and facilitating transport of the food particles. The second layer of the GI wall is *the submucosa*, which is built of the connective tissue and contains blood and lymphatic vessels, nerves, and mucous glands. The third layer of the GI wall is *the muscularis propria (muscular layer),* which is built of two (circular and longitudinal) layers of smooth muscle cells providing the peristaltic movement of the GI. The fourth (the most internal) layer of the GI, *the adventitia (serosa)*, is covered by the visceral peritoneum and contains connective tissue, nerves, and blood and lymphatic vessels (Fig. 23.1). The macrophages are most abundant in the LP of the mucosa (Hine and Loke 2019). In human and rodents, there is an ascending oral-anal gradient of macrophage quantity; the LP of the small intestine contains fewer macrophages than the LP of the colon. The immediate proximity of the LP macrophages to the epithelial layer allows them to phagocyte the microorganisms or compounds that cross the epithelial barrier. In addition, the macrophages (and also the dendritic cells) can sample the content of the gut lumen through the formation of the trans-epithelial lamellipodia/dendrites (TEDs). Because the macrophages and dendritic cells express the tight junction proteins (Blank et al. 2011; Rescigno et al.

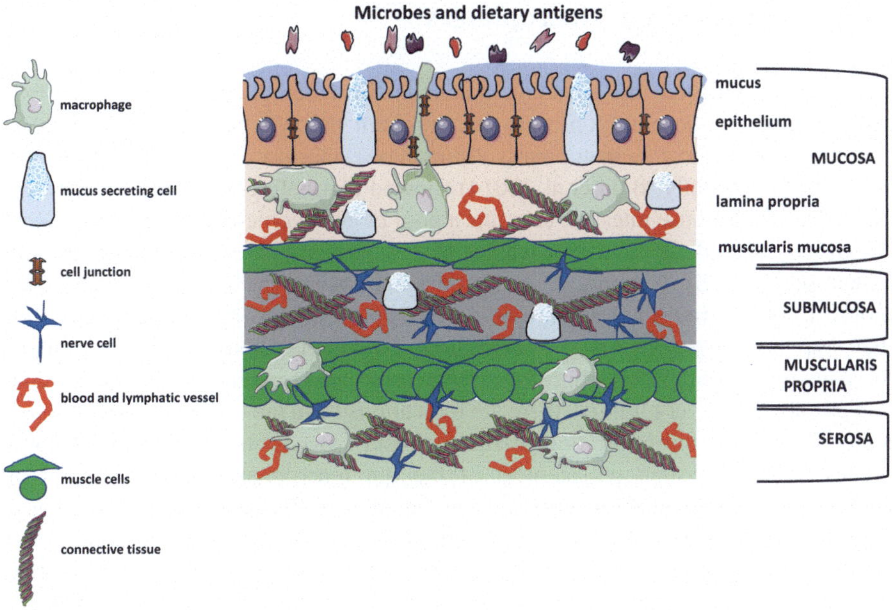

Fig. 23.1 Schematic representation of the mammalian intestine wall. The epithelium of the mucosa consists of tightly connected by the intercellular junctions epithelial cells interspaced with the mucus-producing goblet cells. The epithelium forms a physical barrier against the invading pathogens and the mucus has the bactericidal properties. The macrophages are most abundant in the lamina propria of the mucosa, the muscularis propria, and the serosa. The lamina propria macrophages form the dendritic extensions (TEDs), which allow them to directly sample the content of the gut lumen. Because macrophages express cell junction proteins, it is possible that TEDs might form the temporary junctions with the epithelial cells, which prevent disruption of the epithelial barrier. Beside the phagocytotic and immune activities, the macrophages also interact with the nerve cells and influence the peristaltic movement of the gut (For details, see the text; Bain and Mowat 2014; Hine and Loke 2019)

2001), they are probably able to form the temporary junctions between TEDs and epithelial cells, which allow them to reach the gut lumen without destroying the epithelial barrier (Bain and Schridde 2018; Blank et al. 2011; Chieppa et al. 2006; Gross et al. 2015; Niess et al. 2005; Rescigno et al. 2001). The macrophages are also located within the mucosa-associated lymphoid tissue (MALT), and in the muscularis propria and serosa, where they communicate with the muscle and nerve cells responsible for the gut peristalsis (Bain and Mowat 2014; Bain and Schridde 2018; Grainger et al. 2017; Gross et al. 2015; Hume et al. 1984; Mikkelsen and Rumessen 1992; Muller et al. 2014; Nagashima et al. 1996).

23.5 Sources of GI Macrophages

It is well established that in adult mammals the macrophages derive from two distinct sources: the embryonic progenitors from the yolk sac and/or fetal liver, and the conventional hematopoiesis (blood/bone marrow) monocytes (McGrath et al. 2015). Recently, Bain et al. (2014) showed that before and immediately after birth, the LP of mouse intestine contains a pool of embryonic-derived macrophages, which are, around the time of weaning at 2–3 weeks after birth, replaced by the hematopoietic-derived monocytes/macrophages. During the first few weeks after mouse birth, the intestinal embryonic-derived macrophages proliferate in situ, but they stop proliferating around 3 weeks after birth, which correlates with the influx of the hematopoietic-derived monocytes (Bain et al. 2014). Although the human data on this subject are very limited, there are indications from the transplantation studies, which monitored the infiltration of the recipient's macrophages into the donor's organs, that a constant replacement of the gut macrophages from the pool of hematopoietic progenitors occurs also in humans (references listed in Bain and Schridde 2018). It is still unclear why certain tissues and organs such as brain and skin retain a large pool of the embryonic-derived macrophages while the intestine does not. The most reasonable theory is that the constant exposure to a massive quantity of food- and microorganism-derived antigens causes the "physiological"/ "controlled" inflammation of the intestinal tract, which drives constant influx of the monocytes but keeps the inflammatory response in check (a "low grade" inflammation) to prevent an excessive inflammatory damage to the intestine (Fiocchi 2003, 2008; Medzhitov 2008). Evidently, the influx of the monocytes and thus macrophage turnover is regulated by the gut microbiota; the sterile (germ-free) mice have fewer macrophages, and the antibiotic treatments perturbs monocyte/macrophage turnover in the gut wall (Bain et al. 2014; Bain and Mowat 2014; Schmidt et al. 2019; Scott et al. 2018).

Despite numerous studies confirming a continuous replenishment of the gut macrophages by the incoming monocytes, the most recent studies showed that the intestine wall contains also a pool of the self-maintaining, long-living macrophages (Shaw et al. 2018; de Schepper et al. 2018). These self-maintaining and self-proliferating macrophages are localized not in the LP but in the deeper layers of the gut wall, i.e. in the submucosa and the muscularis in the proximity of the blood vessels, and are necessary for the survival of the enteric neurons and a proper functioning of the submucosal vasculature, secretory glands, and peristalsis (de Schepper et al. 2018; Sieweke and Allen 2013).

23.6 Hyporesponsiveness/Anergy of GI Macrophages Toward the Commensals and Symbionts

The hyporesponsiveness (anergy) of the intestinal macrophages toward the non-harmful (commensal and symbiotic) microorganisms develops progressively. The monocytes which enter the GI wall are fully immunologically competent—they express a whole spectrum of immune signaling/regulatory molecules. These monocytes progressively switch off some immune signaling/regulatory molecules, switch on certain anti-inflammatory molecules, and mature into the hyporesponsive/anergic macrophages with a limited repertoire of immune responses (Fig. 23.2; Bain and Mowat 2014). In mice, the development of macrophage hyporesponsiveness correlates with the increased synthesis of interleukin 10 (IL-10), a potent anti-inflammatory cytokine (Iyer and Cheng 2012). This demonstrates that in the absence of an infection or injury threat, the fully competent monocytes, which continuously infiltrate intestine wall as the surveillance agents, mature into the anti-inflammatory/anergic phenotype macrophages.

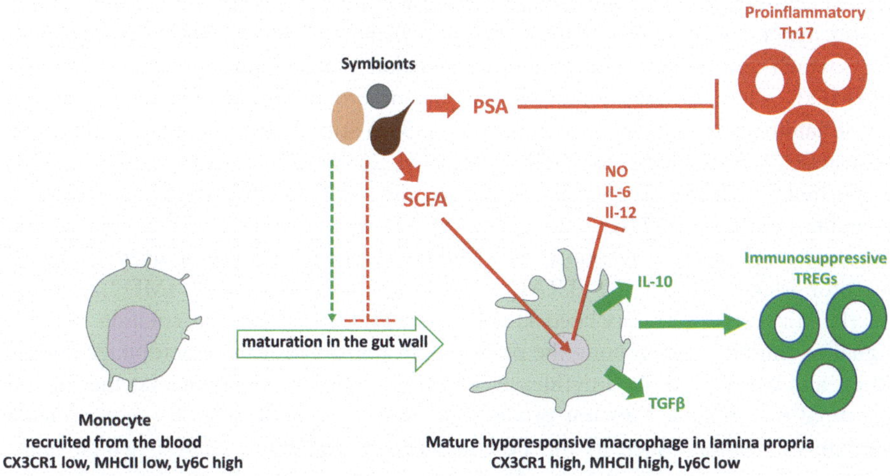

Fig. 23.2 Diagram of some of the effects of symbionts on the macrophage immune responses. The monocytes recruited from the blood to the gut wall are fully immune responsive. They have been identified based on the level of the markers they express: a low level of fractalkine receptor CX3CR1, and major histocompatibility complex MHCII, and a high level of lymphocyte antigen 6 complex, locus C1, Ly6c. Under the influence of microbiota and the intestine wall niche, these monocytes progressively mature into the hyporesponsive (anergic) mature macrophages, which express a high level of CX3CR1 and MHCII and a low level of Ly6c. The mature macrophages express a high level of anti-inflammatory IL-10 and TGF β and activate the immunosuppressive regulatory T cells, Tregs. The symbionts inhabiting the gut produce a variety of metabolites such as the short-chain fatty acids (SCFA) and polysaccharide A (PSA). One of the SCFA is the n-butyrate, which decreases, through the inhibition of histone deacetylases, expression of the proinflammatory molecules such as a nitric oxide (NO), IL-6, and IL-12. The PSA suppresses the pro-inflammatory Th17 helper cells producing interleukin 17 (IL-17) (Chang et al. 2014; Wang et al. 2019)

The immunomodulation/silencing of the intestinal monocytes/macrophages and other types of immune cells is partially the effect of various metabolites produced by the symbionts inhabiting the gut. Some of such molecules are the short-chain fatty acids (SCFA; Fig. 23.2). One of the SCFA is the n-butyrate, a bacterial metabolite produced in large quantities by the commensals inhabiting the colon (Chang et al. 2014). The n-butyrate, decreases, through the inhibition of histone deacetylases, which change chromatin conformation, synthesis of the proinflammatory molecules such as nitric oxide, IL-6, and IL-12, which are produced in the macrophages in response to the lipopolysaccharides (LPS) present in the outer membrane of Gram-negative bacteria (Chang et al. 2014). Another example is the polysaccharide A (PSA; Fig. 23.2) produced by a colon commensal, a Gram-negative, rod-shaped bacterium *Bacteroides fragilis*. During bacterial colonization of the colon, the PSA suppresses the pro-inflammatory Th17 helper cells producing interleukin 17 (IL-17) (Wang et al. 2019). Other factors inducing macrophage anergy are the growth factors found in the gut niche where these macrophages reside. For example, the transforming growth factor-beta (TGF β), produced by the monocytes/macrophages, other immune cells, and the intestinal stromal cells of the gut wall, and abundantly present in the lamina propria, decreases the production of cytokines but does not affect macrophage phagocytic or bactericidal activities (Smythies et al. 2015). TGF β causes hyporesponsiveness of the monocytes to the Toll-like receptor (TLR) signaling, which recognizes the pathogen-associated molecular pattern molecules (PAMPs), such as LPS, present on the microorganisms and activates the pro-inflammatory cytokines, chemokines, and co-stimulatory molecules (Bain and Mowat 2014; Smythies et al. 2010; Wang et al. 2019). The downregulation of TLR downstream effector molecules such as the TIR-domain-containing adapter-inducing interferon-β (TRIF), the cluster of differentiation 14 (CD14), the myeloid differentiation primary response 88 (MyD88) protein, the tumor necrosis factor receptor-associated factor 6 (TRAF6), the lymphocyte antigen 96 (MD2), and the interleukin-1 receptor-associated kinase 1 (IRAK1), and the upregulation of TLR signaling inhibitors are among the pathways responsible for the macrophage anergy (Bain and Mowat 2014; Smythies et al. 2010). It is also known that the human intestinal macrophages do not produce or produce very low levels of the innate immunity receptors, such as the Fc-alfa receptor 1 (FcαRI, CD89), which binds immunoglobulin A antibodies; the Fc-gamma receptor 1 (FcγRI, CD64), and the Fc receptors FcγRIIIa (CD16a), which bind IgG IgG-type antibodies; the CD32, which downregulates antibody production in the presence of IgG; the complement receptors 3 (CR3, CD11b, CD18) and 4 (CR4, CD11c, CD18), which detect PAMPs (including LPS) on the microbes without mediation by the antibodies; and the LPS co-receptor, the cluster of differentiation 14 (CD14).

Another mechanism preventing a robust immune response of macrophages toward the gut microbiota is lack of activation of TREM-1—the receptor expressed on myeloid cells (granulocytes and monocytes), which enhances inflammatory responses toward bacteria and fungi (Bouchon et al. 2000; Colonna and Facchetti 2003; Wang et al. 2019). The absence of TREM-1 prevents upregulation of co-stimulatory cluster of differentiation molecules CD40, CD80, CD86, and proinflammatory factors such as the tumor necrosis factor TNF, IL-1β, and IL-6 (Bouchon et al. 2000; Wang et al. 2019). This allows the intestinal macrophages to efficiently scavenge and phagocyte microbes without triggering inflammation and intestinal damage (Smythies et al. 2010, 2015; Wang et al. 2019). In this context, it is easy to imagine how a disruption of macrophage anergy may affect gut microbiota, and vice versa, how a disruption of microbiota homeostasis may affect the macrophages and lead to the autoimmune diseases and the pathological inflammation of the intestine and severe inflammatory bowel diseases such as Crohn's disease (CD) and ulcerative colitis (UC).

In summary, although we currently know many molecular factors and pathways involved in the development of macrophage hyporesponsiveness/anergy toward the gut microbiota, we are still very far from comprehending the entire picture of immense and extremely sophisticated interactions between the macrophages, intestinal wall niches, and the gut inhabiting microorganisms. A better understanding of these processes not only will satisfy our scientific curiosity but also help to prevent and/or cure the autoimmune and inflammatory diseases.

Acknowledgments We are grateful for the support from The William Stamps Farish Fund and Polish Ministry of National Defense project "Kościuszko" no: 5508/2017/DA to JZ Kubiak. We acknowledge that some of the images used to make figures were from the Servier Medical ART: SMART, smart.servier.com.

Conflict of Interests None of the authors have conflict of interests.

References

Bain CC, Mowat AM (2014) Macrophages in intestinal homeostasis and inflammation. Immunol Rev 260:102–117. https://doi.org/10.1111/imr.12192

Bain CC, Schridde A (2018) Origin, differentiation, and function of intestinal macrophages. Front Immunol 9:2733. https://doi.org/10.3389/fimmu.2018.02733

Bain CC, Bravo-Blas A, Scott CL, Gomez Perdiguero E, Geissmann F, Henri S et al (2014) Constant replenishment from circulating monocytes maintains the macrophage pool in the intestine of adult mice. Nat Immunol 15:929–937. https://doi.org/10.1038/ni.2967

Blank F, Wehrli M, Lehmann A, Baum O, Gehr P, von Garnier C, Rothen-Rutishauser BM (2011) Macrophages and dendritic cells express tight junction proteins and exchange particles in an in vitro model of the human airway wall. Immunobiology 216:86–95. https://doi.org/10.1016/j.imbio.2010.02.006

Bouchon A, Dietrich J, Colonna M (2000) Cutting edge: inflammatory responses can be triggered by TREM-1, a novel receptor expressed on neutrophils and monocytes. J Immunol 164:4991–4995. https://doi.org/10.4049/jimmunol.164.10.4991

Broderick NA (2015) A common origin for immunity and digestion. Front Immunol 6(72). https://doi.org/10.3389/fimmu.2015.00072. eCollection 2015

Chang PV, Hao L, Offermanns S, Medzhitov R (2014) The microbial metabolite butyrate regulates intestinal macrophage function via histone deacetylase inhibition. Proc Natl Acad Sci U S A 111:2247–2252. https://doi.org/10.1073/pnas.1322269111

Chen G, Zhuchenko O, Kuspa A (2007) Immune-like phagocyte activity in the social amoeba. Science 317:678–681. https://doi.org/10.1126/science.1143991

Chieppa M, Rescigno M, Huang AY, Germain RN (2006) Dynamic imaging of dendritic cell extension into the small bowel lumen in response to epithelial cell TLR engagement. J Exp Med 203:2841–2852. https://doi.org/10.1084/jem.20061884

Colonna M, Facchetti F (2003) TREM-1 (triggering receptor expressed on myeloid cells): a new player in acute inflammatory responses. J Infect Dis 187(Suppl 2):S397–S401. https://doi.org/10.1086/374754

Conway PL, Maki J, Mitchell R, Kjelleberg S (1986) Starvation of marine flounder, squid and laboratory mice and its effect on the intestinal microbiota. FEMS Microbiol Ecol 2:187–195. https://doi.org/10.1111/j.1574-6968.1986.tb01728.x

De Schepper S, Verheijden S, Aguilera-Lizarraga J, Viola MF, Boesmans W, Stakenborg N et al (2018) Self-maintaining gut macrophages are essential for intestinal homeostasis. Cell 175:400–15.e13. https://doi.org/10.1016/j.cell.2018.07.048

Doshi N, Mitragotri S (2010 Apr 6) Macrophages recognize size and shape of their targets. PLoS One 5(4):e10051. https://doi.org/10.1371/journal.pone.0010051

Fava F, Danese S (2011) Intestinal microbiota in inflammatory bowel disease: friend of foe? World J Gastroenterol 17:557–566. https://doi.org/10.3748/wjg.v17.i5.557

Fiocchi C (2003) The normal intestinal mucosa: a state of 'controlled inflammation'. In: Targan SR, Shanahan F (eds) Inflammatory bowel disease: from bench to bedside, 2nd edn. Kluwer Academic, Dordrecht, pp 101–120. https://doi.org/10.1007/0-387-25808-6_5

Fiocchi C (2008) What is "physiological" intestinal inflammation and how does it differ from "pathological" inflammation? Inflamm Bowel Dis 14(Suppl. 2):S77–S78. https://doi.org/10.1002/ibd.20618

Fujita T (2002) Evolution of the lectin-complement pathway and its role in innate immunity. Nat Rev Immunol 2:346–353. https://doi.org/10.1038/nri800. PMID:12033740

Gaudet RG, Bradfield CJ, MacMicking JD (2016) Evolution of cell-autonomous effector mechanisms in macrophages versus non-immune cells. Microbiol Spectr 4(6). https://doi.org/10.1128/microbiolspec.MCHD-0050-2016

Grainger JR, Konkel JE, Zangerle-Murray T, Shaw TN (2017) Macrophages in gastrointestinal homeostasis and inflammation. Pflugers Arch 469:527–539. https://doi.org/10.1007/s00424-017-1958-2

Gross M, Salame TM, Jung S (2015) Guardians of the gut – murine intestinal macrophages and dendritic cells. Front Immunol 6:254. https://doi.org/10.3389/fimmu.2015.00254. eCollection 2015

Hartenstein V, Martinez P (2019) Phagocytosis in cellular defense and nutrition: a food-centered approach to the evolution of macrophages. Cell Tissue Res 377:527–547. https://doi.org/10.1007/s00441-019-03096-6

Helander HF, Fändriks L (2014) Surface area of the digestive tract – revisited. Scand J Gastroenterol 49:681–689. https://doi.org/10.3109/00365521.2014.898326

Herwald H, Egesten A (2016) On PAMPs and DAMPs. J Innate Immun 8:427–428. https://doi.org/10.1159/000448437

Hine AM, Loke P (2019) Intestinal macrophages in resolving inflammation. J Immunol 203:593–599. https://doi.org/10.4049/jimmunol.1900345

Hoffmann JA, Kafatos FC, Janeway CA, Ezekowitz RA (1999) Phylogenetic perspectives in innate immunity. Science 284:1313–1318

Hounnou G, Destrieux C, Desmé J, Bertrand P, Velut S (2002) Anatomical study of the length of the human intestine. Surg Radiol Anat 24:290–294. https://doi.org/10.1007/s00276-002-0057-y. ISSN 0930-1038

Hume DA, Perry VH, Gordon S (1984) The mononuclear phagocyte system of the mouse defined by immunohistochemical localisation of antigen F4/80: macrophages associated with epithelia. Anat Rec 210:503–512

Iyer SS, Cheng G (2012) Role of interleukin 10 transcriptional regulation in inflammation and autoimmune disease. Crit Rev Immunol 32:23–63. PMID: 22428854

Markiewski MM, Lambris JD (2007) The role of complement in inflammatory diseases from behind the scenes into the spotlight. Am J Pathol 171:715–727. https://doi.org/10.2353/ajpath.2007.070166

McGrath KE, Frame JM, Palis J (2015) Early hematopoiesis and macrophage development. Semin Immunol 27:379–387. https://doi.org/10.1016/j.smim.2016.03.013

Medzhitov R (2008) Origin and physiological roles of inflammation. Nature 454:428–435. https://doi.org/10.1038/nature07201

Mikkelsen HB, Rumessen JJ (1992) Characterization of macrophage-like cells in the external layers of human small and large intestine. Cell Tissue Res 270:273–279

Muller PA, Koscso B, Rajani GM, Stevanovic K, Berres ML, Hashimoto D, Mortha A, Leboeuf M, Li XM, Mucida D, Stanley ER, Dahan S, Margolis KG, Gershon MD, Merad M, Bogunovic M (2014) Crosstalk between muscularis macrophages and enteric neurons regulates gastrointestinal motility. Cell 158:300–313. https://doi.org/10.1016/j.cell.2014.04.050

Nagashima R, Maeda K, Imai Y, Takahashi T (1996) Lamina propria macrophages in the human gastrointestinal mucosa: their distribution, immunohistological phenotype, and function. J Histochem Cytochem 44:721–731

Niess JH, Brand S, Gu X, Landsman L, Jung S, McCormick BA et al (2005) CX3CR1-mediated dendritic cell access to the intestinal lumen and bacterial clearance. Science 307:254–258. https://doi.org/10.1126/science.1102901

Nyholm SV, Graf J (2012) Knowing your friends: invertebrate innate immunity fosters beneficial bacterial symbioses. Nat Rev Microbiol 10:815–827. https://doi.org/10.1038/nrmicro2894

Rescigno M, Urbano M, Valzasina B, Francolini M, Rotta G, Bonasio R, Granucci F, Kraehenbuhl JP, Ricciardi-Castagnoli P (2001) Dendritic cells express tight junction proteins and penetrate gut epithelial monolayers to sample bacteria. Nat Immunol 2:361–367. https://doi.org/10.1038/86373

Rooks MG, Garrett WS (2016) Gut microbiota, metabolites and host immunity. Nat Rev Immunol 16:341–352. https://doi.org/10.1038/nri.2016.42

Rosengaus RB, Schultheis KF, Yalonetskaya A, Bulmer MS, DuComb WS, Benson RW, Thottam JP, Godoy-Carter V (2014) Symbiont-derived β-1,3-glucanases in a social insect: mutualism beyond nutrition. Front Microbiol 5:607. https://doi.org/10.3389/fmicb.2014.00607. eCollection 2014

Sarma JV, Ward PA (2011) The complement system. Cell Tissue Res 343:227–235. https://doi.org/10.1007/s00441-010-1034-0. Epub 2010 Sep 14

Schmidt F, Dahlke K, Batra A, Keye J, Wu H, Friedrich M, Glauben R, Ring C, Loh G, Schaubeck M, Hackl H, Trajanoski Z, Schumann M, Kühl AA, Blaut M, Siegmund B (2019) Microbial colonization in adulthood shapes the intestinal macrophage compartment. J Crohns Colitis 13:1173–1185. https://doi.org/10.1093/ecco-jcc/jjz036

Scott NA, Andrusaite A, Andersen P, Lawson M, Alcon-Giner C, Leclaire C, Caim S, Le Gall G et al (2018) Antibiotics induce sustained dysregulation of intestinal T cell immunity by perturbing macrophage homeostasis. Sci Transl Med 10(464):pii:eaao4755. https://doi.org/10.1126/scitranslmed.aao4755

Sears CL (2005) A dynamic partnership: celebrating our gut flora. Anaerobe 11:247–251. https://doi.org/10.1016/j.anaerobe.2005.05.001

Shaw TN, Houston SA, Wemyss K, Bridgeman HM, Barbera TA, Zangerle-Murray T et al (2018) Tissue-resident macrophages in the intestine are long lived and defined by Tim-4 and CD4 expression. J Exp Med 215:1507–1518. https://doi.org/10.1084/jem.20180019

Shivers RP, Kooistra T, Chu SW, Pagano DJ, Kim DH (2009) Tissue specific activities of an immune signaling module regulate physiological responses to pathogenic and nutritional bacteria in C. elegans. Cell Host Microbe 6:321–330. https://doi.org/10.1016/j.chom.2009.09.001

Sieweke MH, Allen JE (2013) Beyond stem cells: self-renewal of differentiated macrophages. Science 342:1242974. https://doi.org/10.1126/science.1242974

Smith PD, Smythies LE, Shen R, Greenwell-Wild T, Gliozzi M, Wahl SM (2011) Intestinal macrophages and response to microbial encroachment. Mucosal Immunol 4:31–42. https://doi.org/10.1038/mi.2010.66

Smythies LE, Shen R, Bimczok D, Novak L, Clements RH, Eckhoff DE, Bouchard P, George MD, Hu WK, Dandekar S, Smith PD (2010) Inflammation anergy in human intestinal macrophages is due to Smad-induced IkappaBalpha expression and NF-kappaB inactivation. J Biol Chem 285:19593–19604. https://doi.org/10.1074/jbc.M109.069955

Smythies LE, Sellers M, Clements RH, Mosteller-Barnum M, Meng G, Benjamin WH, Orenstein JM, Smith PD (2015) Human intestinal macrophages display profound inflammatory anergy despite avid phagocytic and bacteriocidal activity. J Clin Invest 115:66–75

Stuart LM, Ezekowitz RA (2005) Phagocytosis: elegant complexity. Immunity 22:539–550. https://doi.org/10.1016/j.immuni.2005.05.002

Sun Z, Wu W, Zhang G (2011) Structure and expression of β-1,3-glucan recognition proteins from the ghost moth, Thitarodes pui (Hepialidae), and their response to Beauveria bassiana infection. J Insect Physiol 57:1660–1669. https://doi.org/10.1016/j.jinsphys.2011.08.019

Takeuchi O, Akira S (2010) Pattern recognition receptors and inflammation. Cell 140:805–820. https://doi.org/10.1016/j.cell.2010.01.022

Thursby E, Juge N (2017) Introduction to the human gut microbiota. Biochem J 474:1823–1836. https://doi.org/10.1042/BCJ20160510

Traylor-Knowles N, Vandepas LE, Browne WE (2019) Still enigmatic: innate immunity in the ctenophore *Mnemiopsis leidyi*. Integr Comp Biol 59(4):811–818. https://doi.org/10.1093/icb/icz116

van Niekerk G, Engelbrecht AM (2015) Commentary on: "a common origin for immunity and digestion". Front Microbiol 6:531. https://doi.org/10.3389/fmicb.2015.00531. eCollection 2015

Wang S, Ye Q, Zeng X, Qiao S (2019) Functions of macrophages in the maintenance of intestinal homeostasis. J Immunol Res 2019:1512969. https://doi.org/10.1155/2019/1512969

Wells JM, Loonen LM, Karczewski JM (2010) The role of innate signaling in the homeostasis of tolerance and immunity in the intestine. Int J Med Microbiol 300:41–48. https://doi.org/10.1016/j.ijmm.2009.08.008

Wynn TA, Chawla A, Pollard JW (2013) Macrophage biology in development, homeostasis and disease. Nature 496:445–455. https://doi.org/10.1038/nature12034

Correction to: The Photosynthetic Adventure of *Paulinella* Spp

Przemysław Gagat, Katarzyna Sidorczuk, Filip Pietluch, and Paweł Mackiewicz

Correction to: Chapter 13 in: M. Kloc (ed.), *Symbiosis: Cellular, Molecular, Medical and Evolutionary Aspects*, https://doi.org/10.1007/978-3-030-51849-3_13

The funder information was missing from the acknowledgement section and the funding agency has asked the author to change this. The acknowledgement should read as:

Acknowledgments We are grateful to the reviewers for their helpful comments. We also would like to thank Dr. Malgorzata Kloc for inviting us to write this chapter. This work was supported by the National Science Centre grant UMO-2017/26/D/NZ8/00444 to P. Gagat and UMO-2018/31/N/NZ2/01338 to K. Sidorczuk.

The original chapter has been corrected.

The updated online version of this chapter can be found at https://doi.org/10.1007/978-3-030-51849-3_13

M. Kloc (ed.), *Symbiosis: Cellular, Molecular, Medical and Evolutionary Aspects*, Results and Problems in Cell Differentiation 69,
https://doi.org/10.1007/978-3-030-51849-3_24

The manufacturer's authorised representative in the EU is Springer Nature Customer Service Centre GmbH, Europaplatz 3, 69115 Heidelberg, Germany. If you have any concerns regarding our products, please contact ProductSafety@springernature.com

Printed and bound by CPI Group (UK) Ltd, Croydon, CR0 4YY
30/09/2026
02225695-0020